Lecture Notes in Computer Science 2139

Edited by G. Goos, J. Hartmanis, and J. van Leeuwen

Springer
Berlin
Heidelberg
New York
Barcelona
Hong Kong
London
Milan
Paris
Singapore
Tokyo

Joe Kilian (Ed.)

Advances in Cryptology – CRYPTO 2001

21st Annual International Cryptology Conference, Santa Barbara, California, USA, August 19-23, 2001
Proceedings

Volume Editor

Joe Kilian
Yianilos Labs.
707 State Rd., Rt. 206, Suite 212
Princeton, NJ 08540, USA
E-mail: joe@pnylab.com

Cataloging-in-Publication Data applied for

Die Deutsche Bibliothek - CIP-Einheitsaufnahme

Advances in cryptology : proceedings / CRYPTO 2001, 21st Annual International Cryptology Conference, Santa Barbara, California, USA, August 19 - 23, 2001. Joe Kilian (ed.). [IACR]. - Berlin ; Heidelberg ; New York ; Barcelona ; Hong Kong ; London ; Milan ; Paris ; Singapore ; Tokyo : Springer, 2001
(Lecture notes in computer science ; Vol. 2139)
ISBN 3-540-42456-3

CR Subject Classification (1998): E.3, G.2.1, F.2.1-2, D.4.6, K.6.5, C.2, J.1

ISSN 0302-9743
ISBN 3-540-42456-3 Springer-Verlag Berlin Heidelberg New York

Springer-Verlag Berlin Heidelberg New York
a member of BertelsmannSpringer Science+Business Media GmbH

http://www.springer.de

Printed in Germany

Typesetting: Camera-ready by author, data conversion by PTP-Berlin, Stefan Sossna
Printed on acid-free paper SPIN 10840151 06/3142 5 4 3 2 1 0

Preface

Crypto 2001, the 21st Annual Crypto conference, was sponsored by the International Association for Cryptologic Research (IACR) in cooperation with the IEEE Computer Society Technical Committee on Security and Privacy and the Computer Science Department of the University of California at Santa Barbara.

The conference received 156 submissions, of which the program committee selected 34 for presentation; one was later withdrawn. These proceedings contain the revised versions of the 33 submissions that were presented at the conference. These revisions have not been checked for correctness, and the authors bear full responsibility for the contents of their papers.

The conference program included two invited lectures. Mark Sherwin spoke on, "Quantum information processing in semiconductors: an experimentalist's view." Daniel Weitzner spoke on, "Privacy, Authentication & Identity: A recent history of cryptographic struggles for freedom." The conference program also included its perennial "rump session," chaired by Stuart Haber, featuring short, informal talks on late–breaking research news.

As I try to account for the hours of my life that flew off to oblivion, I realize that most of my time was spent cajoling talented innocents into spending even more time on my behalf. I have accumulated more debts than I can ever hope to repay. As mere statements of thanks are certainly insufficient, consider the rest of this preface my version of Chapter 11.

I would like to first thank the many researchers from all over the world who submitted their work to this conference. Without them, Crypto is just a pile of shrimp and chocolate covered strawberries.

I thank David Balenson, the general chair, for shielding me from innumerable logistical headaches, and showing great generosity in supporting my efforts.

Selecting from so many submissions is a daunting task. My deepest thanks go to the members of the program committee, for their knowledge, wisdom, and near-masochistic work ethic. We in turn have relied heavily on the expertise of the many outside reviewers who assisted us in our deliberations. My thanks to all those listed on the following pages, and my thanks and apologies to any I have missed.

I thank Rebecca Wright for hosting the program committee meeting in New York City, AT&T for providing the space, and Sandy Barbu for helping out with the local arrangements. Thanks also go to Ran Canetti, my favorite native culinary guide, wherever I go, for organizing the post–deliberations dinner.

I thank the people who, by their past and continuing work, have greatly streamlined the submission and review process. All but one of the submissions were handled using Chanathip Namprempre's web-based submission software. Reviews were administered using software written by Wim Moreau and Joris Claessens, developed under the guidance of Bart Preneel. These software packages have made the process idiot proof, and nearly theorist-proof. My thanks also go to Sam Rebelsky for writing the email-based predecessor of the submission software. He and the other members of the SIGACT Electronic Publications Board have for many years made program committee chairs' lives much more bearable.

I am grateful to Mihir Bellare, last year's program chair, and Kevin McCurley and Josh Benaloh, my main contacts with the IACR board, for patiently trying to teach me my job.

Even if I can't really account for what I, personally, was doing, the hours did go somewhere. I thank my boss, Peter Yianilos, for being so supportive of my efforts, and so absurdly forgiving of the time it has taken away from my work. Last, and more importantly, I'd like to thank my family, Dina, Gersh, and Pearl, for their support, understanding, and love.

June 2001 Joe Kilian

CRYPTO 2001

August 19–23, 2001, Santa Barbara, California, USA

Sponsored by the
International Association for Cryptologic Research (IACR)

in cooperation with
IEEE Computer Society Technical Committee on Security and Privacy,
Computer Science Department, University of California, Santa Barbara

General Chair
David Balenson, NAI Labs, Network Associates, Inc., USA

Program Chair
Joe Kilian, Yianilos Labs, USA

Program Committee

Bill Aiello .... AT&T Research, USA
Don Beaver .... CertCo, USA
Josh Benaloh .... Microsoft Research, USA
Antoon Bosselaers .... Katholieke Universiteit Leuven, Belgium
Jan Camenisch .... IBM Zurich, Switzerland
Ran Canetti .... IBM T. J. Watson, USA
Claude Crépeau .... McGill University, Canada
Alfredo De Santis .... Università di Salerno, Italy
Marc Girault .... France Telecom, France
Stuart Haber .... InterTrust STAR Lab, USA
Tatsuaki Okamoto .... NTT Labs, Japan
Jacques Patarin .... BULL, France
Erez Petrank .... Technion, Israel
Omer Reingold .... AT&T Research, USA
Kazue Sako .... NEC C&C Media Research Lab, Japan
Tomas Sander .... InterTrust STAR Lab, USA
Doug Stinson .... University of Waterloo, Canada
Yacov Yacobi .... Microsoft Research, USA

Advisory Members

Mihir Bellare (Crypto 2000 program chair) .... UCSD, USA
Moti Yung (Crypto 2002 program chair) .... CertCo, USA

External Reviewers

Masayuki Abe
Mehdi-Laurent Akkar
Seigo Arita
Mihir Bellare
Juergen Bierbrauer
Eli Biham
Daniel Bleichenbacher
Carlo Blundo
Dan Boneh
Eric Brier
Daniel Brown
Christian Cachin
Anne Canteaut
William Chambers
Joris Claessens
Henry Cohn
Don Coppersmith
Jean-Sébastien Coron
Ronald Cramer
Paolo D'Arco
Annalisa De Bonis
Patrick Dehornoy
Giovanni Di Crescenzo
Markus Dichtl
Marc Fischlin
Roger Fischlin
Gerhard Frey
Eiichiro Fujisaki
Jun Furukawa
Clemente Galdi
Rosario Gennaro
Oded Goldreich
Dieter Gollmann
Roberto Gorrieri
Louis Goubin
Gaëtan Haché
Shai Halevi
Helena Handschuh
Nick Howgrave-Graham
Yuval Ishai
Stanislaw Jarecki
Thomas Johansson
Jakob Jonsson
Marc Joye
Ari Juels
Charanjit Jutla
Jonathan Katz
Darko Kirovski
Andrew Klapper
Francis Klay
Tetsutaro Kobayashi
Hugo Krawczyk
Eyal Kushilevitz
Kristin Lauter
Arjen Lenstra
Yehuda Lindell
Anna Lysyanskaya
David M'Raihi
Spyros Magliveras
Dahlia Malkhi
Tal Malkin
Barbara Masucci
Minoru Matsui
Ueli Maurer
Rita Mayer-Sommer
Alfred Menezes
Niodrag Mihaljevic
Kazuhiko Minematsu
Jean-François Misarsky
Peter Montgomery
Guglielmo Morgari
Shiho Moriai
David Naccache
Moni Naor
Satoshi Obana
Rafail Ostrovsky
Christof Paar
Béatrice Peirani
Pino Persiano
Benny Pinkas
David Pointcheval
Bart Preneel
Jean-Jacques Quisquater
Tal Rabin
Raj Rajagopalan
Vincent Rijmen
Matt Robshaw
Phillip Rogaway
Pankaj Rohatgi
Alon Rosen
Amit Sahai
Taiichi Saitoh
Louis Salvail
Palash Sarkar
Claus Schnorr
Peter Shor
Victor Shoup
Alice Silverberg
Dan Simon
Dawn Song
Markus Stadler
Jacques Stern
Koutarou Suzuki
Paul Syverson
Jacques Traoré
Luca Trevisan
Shigenori Uchiyama
Serge Vaudenay
Ramarathnam Venkatesan
Frederik Vercauteren
Eric Verheul
Ivan Visconti
David Wagner
Michael Waidner
Yongge Wang
Ruizhong Wei
Gideon Yuval

Table of Contents

Foundations

Traitor Tracing

Multi-party Computation

Two-Party Computation

Elliptic Curves

Applications of Groups and Codes

Broadcast and Secret Sharing

Soundness and Zero-Knowledge

On the (Im)possibility of Obfuscating Programs
(Extended Abstract)

Boaz Barak[1], Oded Goldreich[1], Rusell Impagliazzo[2], Steven Rudich[3], Amit Sahai[4], Salil Vadhan[5], and Ke Yang[3]

[1] Department of Computer Science, Weizmann Institute of Science, Rehovot, ISRAEL. {boaz,oded}@wisdom.weizmann.ac.il
[2] Department of Computer Science and Engineering, University of California, San Diego, La Jolla, CA 92093-0114. russell@cs.ucsd.edu
[3] Computer Science Department, Carnegie Mellon University, 5000 Forbes Ave. Pittsburgh, PA 15213. {rudich,yangke}@cs.cmu.edu
[4] Department of Computer Science, Princeton University, 35 Olden St. Princeton, NJ 08540. sahai@cs.princeton.edu
[5] Division of Engineering and Applied Sciences, Harvard University, 33 Oxford Street, Cambridge, MA 02138. salil@eecs.harvard.edu

Abstract. Informally, an *obfuscator* $\mathcal{O}$ is an (efficient, probabilistic) "compiler" that takes as input a program (or circuit) P and produces a new program $\mathcal{O}(P)$ that has the same functionality as P yet is "unintelligible" in some sense. Obfuscators, if they exist, would have a wide variety of cryptographic and complexity-theoretic applications, ranging from software protection to homomorphic encryption to complexity-theoretic analogues of Rice's theorem. Most of these applications are based on an interpretation of the "unintelligibility" condition in obfuscation as meaning that $\mathcal{O}(P)$ is a "virtual black box," in the sense that anything one can efficiently compute given $\mathcal{O}(P)$, one could also efficiently compute given oracle access to P.

In this work, we initiate a theoretical investigation of obfuscation. Our main result is that, even under very weak formalizations of the above intuition, obfuscation is impossible. We prove this by constructing a family of functions $\mathcal{F}$ that are *inherently unobfuscatable* in the following sense: there is a property $\pi : \mathcal{F} \rightarrow \{0,1\}$ such that (a) given *any program* that computes a function $f \in \mathcal{F}$, the value $\pi(f)$ can be efficiently computed, yet (b) given *oracle access* to a (randomly selected) function $f \in \mathcal{F}$, no efficient algorithm can compute $\pi(f)$ much better than random guessing.

We extend our impossibility result in a number of ways, including even obfuscators that (a) are not necessarily computable in polynomial time, (b) only *approximately* preserve the functionality, and (c) only need to work for very restricted models of computation ($\mathbf{TC}_0$). We also rule out several potential applications of obfuscators, by constructing "unobfuscatable" signature schemes, encryption schemes, and pseudorandom function families.

J. Kilian (Ed.): CRYPTO 2001, LNCS 2139, pp. 1–18, 2001.

1 Introduction

The past few decades of cryptography research has had amazing success in putting most of the classical cryptographic problems — encryption, authentication, protocols — on complexity-theoretic foundations. However, there still remain several important problems in cryptography about which theory has had little or nothing to say. One such problem is that of *program obfuscation.* Roughly speaking, the goal of (program) obfuscation is to make a program "unintelligible" while preserving its functionality. Ideally, an obfuscated program should be a "virtual black box," in the sense that anything one can compute from it one could also compute from the input-output behavior of the program.

The hope that some form of obfuscation is possible arises from the fact that analyzing programs expressed in rich enough formalisms is hard. Indeed, any programmer knows that total unintelligibility is the natural state of computer programs (and one must work hard in order to keep a program from deteriorating into this state). Theoretically, results such as Rice's Theorem and the hardness of the HALTING PROBLEM and SATISFIABILITY all seem to imply that the only useful thing that one can do with a program or circuit is to run it (on inputs of one's choice). However, this informal statement is, of course, an over-generalization, and the existence of obfuscators requires its own investigation.

To be a bit more clear (though still informal), an *obfuscator* $\mathcal{O}$ is an (efficient, probabilistic) "compiler" that takes as input a program (or circuit) P and produces a new program $\mathcal{O}(P)$ satisfying the following two conditions:

- (functionality) $\mathcal{O}(P)$ computes the same function as P.
- ("virtual black box" property) "Anything that can be efficiently computed from $\mathcal{O}(P)$ can be efficiently computed given oracle access to P."

While there are heuristic approaches to obfuscation in practice (cf., Figure 1 and [CT00]), there has been little theoretical work on this problem. This is unfortunate, since obfuscation, if it were possible, would have a wide variety of cryptographic and complexity-theoretic applications.

```
#include<stdio.h> #include<string.h> main(){char*O,l[999]=
"'‘acgo\177~|xp .-\0R^8)NJ6%K4O+A2M(*0ID57$3G1FBL";while(O=
fgets(l+45,954,stdin)){*l=O[strlen(O)[O-1]=0,strspn(O,l+11)];
while(*O)switch((*l&&isalnum(*O))-!*l){case-1:{char*I=(O+=
strspn(O,l+12)+1)-2,O=34;while(*I&3&&(O=(O-16<<1)+*I---'-')<80);
putchar(O&93?*I&8||!( I=memchr( l , O , 44 ) ) ?'?':I-l+47:32);
break;case 1: ;}*l=(*O&31)[l-15+(*O>61)*32];while(putchar(45+*l%2),
(*l=*l+32>>1)>35);case 0:putchar((++O,32));}putchar(10);}}
```

Fig. 1. The winning entry of the 1998 *International Obfuscated C Code Contest*, an ASCII/Morse code translator by Frans van Dorsselaer [vD98] (adapted for this paper).

In this work, we initiate a theoretical investigation of obfuscation. We examine various formalizations of the notion, in an attempt to understand what we can and cannot hope to achieve. Our main result is a negative one, showing that obfuscation (as it is typically understood) is *impossible*. Before describing this result and others in more detail, we outline some of the potential applications of obfuscators, both for motivation and to clarify the notion.

1.1 Some Applications of Obfuscators

Software Protection. The most direct applications of obfuscators are for various forms of software protection. By definition, obfuscating a program protects it against reverse engineering. For example, if one party, Alice, discovers a more efficient algorithm for factoring integers, she may wish to sell another party, Bob, a program for apparently weaker tasks (such as breaking the RSA cryptosystem) that use the factoring algorithm as a subroutine without actually giving Bob a factoring algorithm. Alice could hope to achieve this by obfuscating the program she gives to Bob.

Intuitively, obfuscators would also be useful in *watermarking* software (cf., [CT00,NSS99]). A software vendor could modify a program's behavior in a way that uniquely identifies the person to whom it is sold, and then obfuscate the program to guarantee that this "watermark" is difficult to remove.

Homomorphic Encryption. A long-standing open problem is whether *homomorphic* encryption schemes exist (cf., [RAD78,FM91,DDN00,BL96,SYY99]). That is, we seek a secure public-key cryptosystem for which, given encryptions of two bits (and the public key), one can compute an encryption of any binary Boolean operation of those bits. Obfuscators would allow one to convert any public-key cryptosystem into a homomorphic one: use the secret key to construct an algorithm that performs the required computations (by decrypting, applying the Boolean operation, and encrypting the result), and publish an obfuscation of this algorithm together with the public key.[1]

Removing Random Oracles. The *Random Oracle Model* [BR93] is an idealized cryptographic setting in which all parties have access to a truly random function. It is (heuristically) hoped that protocols designed in this model will remain secure when implemented using an efficient, publicly computable cryptographic hash function in place of the random function. While it is known that this is not true in general [CGH98], it is unknown whether there exist efficiently computable functions with strong enough properties to be securely used in place of the random function in various *specific* protocols (e.g., in Fiat-Shamir type schemes [FS87]). One might hope to obtain such functions by obfuscating a

[1] There is a subtlety here, caused by the fact that encryption algorithms must be *probabilistic* to be semantically secure in the usual sense [GM84]. However, both the "functionality" and "virtual black box" properties of obfuscators become more complex for probabilistic algorithms, so in this work, we restrict our attention to obfuscating deterministic algorithms. This restriction only makes our main (impossibility) result stronger.

family of pseudorandom functions [GGM86], whose input-output behavior is by definition indistinguishable from that of a truly random function.

Transforming Private-Key Encryption into Public-Key Encryption. Obfuscation can also be used to create new public-key encryption schemes by obfuscating a private-key encryption scheme. Given a secret key K of a private-key encryption scheme, one can publish an obfuscation of the encryption algorithm Enc_K.[2] This allows everyone to encrypt, yet only one possessing the secret key K should be able to decrypt.

1.2 Our Results

The Basic Impossibility Result. Most of the above applications rely on the intuition that an obfuscated program is a "virtual black box." That is, anything one can efficiently compute from the obfuscated program, one should be able to efficiently compute given just oracle access to the program.

Our main result shows that it is impossible to achieve this notion of obfuscation. We prove this by constructing (from any one-way function) a family $\mathcal{F}$ of functions which is *inherently unobfuscatable* in the sense that there is some property $\pi : \mathcal{F} \to \{0,1\}$ such that:

- Given *any* program (circuit) that computes a function $f \in \mathcal{F}$, the value $\pi(f)$ can be efficiently computed;
- Yet, given oracle access to a (randomly selected) function $f \in \mathcal{F}$, no efficient algorithm can compute $\pi(f)$ much better than by random guessing.

Thus, there is no way of obfuscating the programs that compute these functions, even if (a) the obfuscation is meant to hide only one bit of information about the function (namely $\pi(f)$), and (b) the obfuscator itself has unbounded computation time.

We believe that the existence of such functions shows that the "virtual black box" paradigm for obfuscators is inherently flawed. Any hope for positive results about obfuscator-like objects must abandon this viewpoint, or at least be reconciled with the existence of functions as above.

Approximate Obfuscators. The basic impossibility result as described above applies to obfuscators $\mathcal{O}$ for which we require that the obfuscated program $\mathcal{O}(P)$ computes exactly the same function as the original program P. However, for some applications it may suffice that, for every input x, $\mathcal{O}(P)$ and P agree on x with high probability (over the coin tosses of $\mathcal{O}$). Using some additional ideas, our impossibility result extends to such *approximate obfuscators*.

[2] This application involves the same subtlety pointed out in Footnote 1. Thus, our results regarding the (un)obfuscatability of private-key encryption schemes (described later) refer to a relaxed notion of security in which multiple encryptions of the same message are not allowed (which is consistent with a deterministic encryption algorithm).

Impossibility of Applications. To give further evidence that our impossibility result is not an artifact of definitional choices, but rather that there is something inherently flawed in the "virtual black box" idea, we also demonstrate that several of the applications of obfuscators are also impossible. We do this by constructing *inherently unobfuscatable* signature schemes, encryption schemes, and pseudorandom functions. These are objects satisfying the standard definitions of security (except for the subtlety noted in Footnote 2), but for which one can efficiently compute the secret key K from *any program* that signs (or encrypts or evaluates the pseudorandom function, resp.) relative to K. (Hence handing out "obfuscated forms" of these keyed-algorithms is highly insecure.)

In particular, we complement Canetti et. al.'s critique of the Random Oracle Methodology [CGH98]. They show that there exist (contrived) protocols that are secure in the idealized Random Oracle Model (of [BR93]), but are *insecure* when the random oracle is replaced with *any* (efficiently computable) function. Our results imply that for even for *natural* protocols that are secure in the random oracle model (e.g., Fiat-Shamir type schemes [FS87]), there exist (contrived) pseudorandom functions, such that these protocols are insecure when the random oracle is replaced with *any program* that computes the contrived function.

Obfuscating restricted complexity classes. Even though obfuscation of general programs/circuits is impossible, one may hope that it is possible to obfuscate more restricted classes of computations. However, using the pseudorandom functions of [NR97] in our construction, we can show that the impossibility result holds even when the input program P is a constant-depth threshold circuit (i.e., is in $\mathbf{TC_0}$), under widely believed complexity assumptions (e.g., the hardness of factoring).

Obfuscating Sampling Algorithms. Another way in which the notion of obfuscators can be weakened is by changing the functionality requirement. Until now, we have considered programs in terms of the functions they compute, but sometimes one is interested in other kinds of behavior. For example, one sometimes considers *sampling algorithms*, i.e. probabilistic programs that take no input (other than, say, a length parameter) and produce an output according to some desired distribution. We consider two natural definitions of obfuscators for sampling algorithms, and prove that the stronger definition is impossible to meet. We also observe that the weaker definition implies the nontriviality of statistical zero knowledge.

Software Watermarking. As mentioned earlier, there appears to be some connection between the problems of software watermarking and code obfuscation. In the full version of the paper [BGI$^+$01], we consider a couple of formalizations of the watermarking problem and explore their relationship to our results on obfuscation.

1.3 Discussion

Our work rules out the standard, "virtual black box" notion of obfuscators as impossible, along with several of its applications. However, it does not mean that there is no method of making programs "unintelligible" in some meaningful and precise sense. Such a method could still prove useful for software protection.

Thus, we consider it to be both important and interesting to understand whether there are alternative senses (or models) in which some form of obfuscation is possible. Towards this end, in the full version of the paper we suggest two weaker definitions of obfuscators that avoid the "virtual black box" paradigm (and hence are not ruled out by our impossibility proof). These definitions could be the subject of future investigations, but we hope that other alternatives will also be proposed and examined.

As is usually the case with impossibility results and lower bounds, we show that obfuscators (in the "virtual black box" sense) do not exist by supplying a somewhat contrived counterexample of a function ensemble that cannot be obfuscated. It is interesting whether obfuscation is possible for a restricted class of algorithms, which nonetheless contains some "useful" algorithms. If we try to restrict the algorithms by their computational complexity, then there's not much hope for obfuscation. Indeed, as mentioned above, we show that (under widely believed complexity assumptions) our counterexample can be placed in $\mathbf{TC_0}$. In general, the complexity of our counterexample is essentially the same as the complexity of pseudorandom functions, and so a complexity class which does not contain our example will also not contain many cryptographically useful algorithms.

1.4 Additional Related Work

There are a number of heuristic approaches to obfuscation and software watermarking in the literature, as described in the survey of Collberg and Thomborson [CT00]. A theoretical study of software protection was previously conducted by Goldreich and Ostrovsky [GO96], who considered *hardware-based* solutions.

Hada [Had00] gave some definitions for code obfuscators which are stronger than the definitions we consider in this paper, and showed some implications of the existence of such obfuscators. (Our result rules out also the existence of obfuscators according to the definitions of [Had00].)

Canetti, Goldreich and Halevi [CGH98] showed another setting in cryptography where getting a function's description is provably more powerful than black-box access. As mentioned above, they have shown that there exist protocols that are secure when executed with black-box access to a random function, but insecure when instead the parties are given a description of any hash function.

1.5 Organization of the Paper

In Section 2, we give some basic definitions along with (very weak) definitions of obfuscators. In Section 3, we prove the impossibility of obfuscators by constructing an inherently unobfuscatable function ensemble. Other extensions and results are deferred to the full version of the paper [BGI+01].

2 Definitions

2.1 Preliminaries

TM is shorthand for Turing machine. *PPT* is shorthand for probabilistic polynomial-time Turing machine. For algorithms A and M and a string x, we denote by $A^M(x)$ the output of A when executed on input x and oracle access to M. If A is a probabilistic Turing machine then by $A(x;r)$ we refer to the result of running A on input x and random tape r. By $A(x)$ we refer to the distribution induced by choosing r uniformly and running $A(x;r)$. If D is a distribution then by $x \stackrel{R}{\leftarrow} D$ we mean that x is a random variable distributed according to D. If S is a set then by $x \stackrel{R}{\leftarrow} S$ we mean that x is a random variable that is distributed uniformly over the elements of S. Supp(D) denotes the *support* of distribution D, i.e. the set of points that have nonzero probability under D. A function $\mu : \mathbb{N} \to \mathbb{N}$ is called *negligible* if it grows slower than the inverse of any polynomial. That is, for any positive polynomial $p(\cdot)$ there exists $N \in \mathbb{N}$ such that $\mu(n) < 1/p(n)$ for any $n > N$. We'll sometimes use neg$(\cdot)$ to denote an unspecified negligible function. We will identify Turing machines and circuits with their canonical representations as strings in $\{0,1\}^*$.

2.2 Obfuscators

In this section, we aim to formalize the notion of obfuscators based on the "virtual black box" property as described in the introduction. Recall that this property requires that "anything that an adversary can compute from an obfuscation $\mathcal{O}(P)$ of a program P, it could also compute given just oracle access to P." We shall define what it means for the adversary to successfully compute something in this setting, and there are several choices for this (in decreasing order of generality):

- (computational indistinguishability) The most general choice is not to restrict the nature of what the adversary is trying to compute, and merely require that it is possible, given just oracle access to P, to produce an output distribution that is computationally indistinguishable from what the adversary computes when given $\mathcal{O}(P)$.
- (satisfying a relation) An alternative is to consider the adversary as trying to produce an output that satisfies an arbitrary (possibly polynomial-time) relation with the original program P, and require that it is possible, given just oracle access to P, to succeed with roughly the same probability as the adversary does when given $\mathcal{O}(P)$.
- (computing a function) A weaker requirement is to restrict the previous requirement to relations which are functions; that is, the adversary is trying to compute some function of the original program.
- (computing a predicate) The weakest is to restrict the previous requirement to $\{0,1\}$-valued functions; that is, the adversary is trying to decide some property of the original program.

Since we will be proving impossibility results, our results are strongest when we adopt the weakest requirement (i.e., the last one). This yields two definitions for obfuscators, one for programs defined by Turing machines and one for programs defined by circuits.

Definition 2.1 (TM obfuscator). *A probabilistic algorithm $\mathcal{O}$ is a* TM obfuscator *if the following three conditions hold:*

- *(functionality) For every TM M, the string $\mathcal{O}(M)$ describes a TM that computes the same function as M.*
- *(polynomial slowdown) The description length and running time of $\mathcal{O}(M)$ are at most polynomially larger than that of M. That is, there is a polynomial p such that for every TM M, $|\mathcal{O}(M)| \leq p(|M|)$, and if M halts in t steps on some input x, then $\mathcal{O}(M)$ halts within $p(t)$ steps on x.*
- *("virtual black box" property) For any PPT A, there is a PPT S and a negligible function α such that for all TMs M*

$$\left|\Pr\left[A(\mathcal{O}(M))=1\right]-\Pr\left[S^{M}(1^{|M|})=1\right]\right| \leq \alpha(|M|).$$

We say that $\mathcal{O}$ is efficient *if it runs in polynomial time.*

Definition 2.2 (circuit obfuscator). *A probabilistic algorithm $\mathcal{O}$ is a* (circuit) obfuscator *if the following three conditions hold:*

- *(functionality) For every circuit C, the string $\mathcal{O}(C)$ describes a circuit that computes the same function as C.*
- *(polynomial slowdown) There is a polynomial p such that for every circuit C, $|\mathcal{O}(C)| \leq p(|C|)$.*
- *("virtual black box" property) For any PPT A, there is a PPT S and a negligible function α such that for all circuits C*

$$\left|\Pr\left[A(\mathcal{O}(C))=1\right]-\Pr\left[S^{C}(1^{|C|})=1\right]\right| \leq \alpha(|C|).$$

We say that $\mathcal{O}$ is efficient *if it runs in polynomial time.*

We call the first two requirements (functionality and polynomial slowdown) the *syntactic requirements* of obfuscation, as they do not address the issue of security at all.

There are a couple of other natural formulations of the "virtual black box" property. The first, which more closely follows the informal discussion above, asks that for every predicate π, the probability that $A(\mathcal{O}(C)) = \pi(C)$ is at most the probability that $S^C(1^{|C|}) = \pi(C)$ plus a negligible term. It is easy to see that this requirement is equivalent to the ones above. Another formulation refers to the distinguishability between obfuscations of two TMs/circuits: ask that for every C_1 and C_2, $|\Pr[A(\mathcal{O}(C_1)) = 1] - \Pr[A(\mathcal{O}(C_2))]|$ is approximately equal to $|\Pr[S^{C_1}(1^{|C_1|}, 1^{|C_2|}) = 1] - \Pr[S^{C_2}(1^{|C_1|}, 1^{|C_2|})]|$. This definition appears to be slightly weaker than the ones above, but our impossibility proof also rules it out.

Note that in both definitions, we have chosen to simplify the definition by using the size of the TM/circuit to be obfuscated as a security parameter. One can always increase this length by padding to obtain higher security.

The main difference between the circuit and TM obfuscators is that a circuit computes a function with finite domain (all the inputs of a particular length) while a TM computes a function with infinite domain. Note that if we had not restricted the size of the obfuscated circuit $\mathcal{O}(C)$, then the (exponential size) list of all the values of the circuit would be a valid obfuscation (provided we allow S running time $\text{poly}(|\mathcal{O}(C)|)$ rather than $\text{poly}(|C|)$). For Turing machines, it is not clear how to construct such an obfuscation, even if we are allowed an exponential slowdown. Hence obfuscating TMs is intuitively harder. Indeed, it is relatively easy to prove:

Proposition 2.3. *If a TM obfuscator exists, then a circuit obfuscator exists.*

Thus, when we prove our impossibility result for circuit obfuscators, the impossibility of TM obfuscators will follow. However, considering TM obfuscators will be useful as motivation for the proof.

We note that, from the perspective of applications, Definitions 2.1 and 2.2 are already too weak to have the wide applicability discussed in the introduction. The point is that they are nevertheless *impossible* to satisfy (as we will prove).

3 The Main Impossibility Result

To state our main result we introduce the notion of inherently unobfuscatable function ensemble.

Definition 3.1. *An* inherently unobfuscatable function ensemble *is an ensemble $\{\mathcal{H}_k\}_{k\in\mathbb{N}}$ of distributions $\mathcal{H}_k$ on finite functions (from, say, $\{0,1\}^{l_{\text{in}}(k)}$ to $\{0,1\}^{l_{\text{out}}(k)}$) such that:*

- *(efficiently computable) Every function $f \xleftarrow{R} \mathcal{H}_k$ is computable by a circuit of size* $\text{poly}(k)$*. (Moreover, a distribution on circuits consistent with $\mathcal{H}_k$ can be sampled uniformly in time* $\text{poly}(k)$*.)*
- *(unobfuscatability) There exists a function $\pi : \bigcup_{k\in\mathbb{N}} \text{Supp}(\mathcal{H}_k) \to \{0,1\}$ such that*
  1. *$\pi(f)$ is hard to compute with black-box access to f: For any PPT S*

$$\Pr_{f \xleftarrow{R} \mathcal{H}_k} [S^f(1^k) = \pi(f)] \leq \frac{1}{2} + \text{neg}(k)$$

  2. *$\pi(f)$ is easy to compute with access to any circuit that computes f: There exists a PPT A such that for any $f \in \bigcup_{k\in\mathbb{N}} \text{Supp}(\mathcal{H}_k)$ and for any circuit C that computes f*

$$A(C) = \pi(f)$$

We prove in Theorem 3.9 that, assuming one-way functions exist, there exists an inherently unobfuscatable function ensemble. This implies that, under the same assumption, there is no obfuscator that satisfies Definition 2.2 (actually we prove the latter fact directly in Theorem 3.6). Since the existence of an *efficient* obfuscator implies the existence of one-way functions (Lemma 3.7), we conclude that efficient obfuscators do not exist (unconditionally).

However, the existence of inherently unobfuscatable function ensemble has even stronger implications. As mentioned in the introduction, these functions can not be obfuscated even if we allow the following relaxations to the obfuscator:

1. As mentioned above, the obfuscator does not have to run in polynomial time — it can be any random process.
2. The obfuscator has only to work for functions in $\mathrm{Supp}(\mathcal{H}_k)$ and only for a non-negligible fraction of these functions under the distributions $\mathcal{H}_k$.
3. The obfuscator has only to hide an *a priori* fixed property π from an *a priori* fixed adversary A.

Structure of the Proof of the Main Impossibility Result. We shall prove our result by first defining obfuscators that are secure also when applied to several (e.g., two) algorithms and proving that they do not exist. Then we shall modify the construction in this proof to prove that TM obfuscators in the sense of Definition 2.1 do not exist. After that, using an additional construction (which requires one-way functions), we will prove that a circuit obfuscator as defined in Definition 2.2 does not exist if one-way functions exist. We will then observe that our proof actually yields an unobfuscatable function ensemble (Theorem 3.9).

3.1 Obfuscating Two TMs/Circuits

Obfuscators as defined in the previous section provide a "virtual black box" property when a single program is obfuscated, but the definitions do not say anything about what happens when the adversary can inspect more than one obfuscated program. In this section, we will consider extensions of those definitions to obfuscating two programs, and prove that they are impossible to meet. The proofs will provide useful motivation for the impossibility of the original one-program definitions.

Definition 3.2 (2-TM obfuscator). *A* 2-TM obfuscator *is defined in the same way as a TM obfuscator, except that the "virtual black box" property is strengthened as follows:*

– *("virtual black box" property) For any PPT A, there is a PPT S and a negligible function α such that for all TMs M, N*

$$\left|\Pr\left[A(\mathcal{O}(M),\mathcal{O}(N))=1\right]-\Pr\left[S^{M,N}(1^{|M|+|N|})=1\right]\right|\leq\alpha(\min\{|M|,|N|\})$$

2-circuit obfuscators are defined by modifying the definition of circuit obfuscators in an analogous fashion.

Proposition 3.3. *Neither 2-TM nor 2-circuit obfuscators exist.*

Proof. We begin by showing that 2-TM obfuscators do not exist. Suppose, for sake of contradiction, that there exists a 2-TM obfuscator $\mathcal{O}$. The essence of this proof, and in fact of all the impossibility proofs in this paper, is that there is a fundamental difference between getting black-box access to a function and getting a program that computes it, no matter how obfuscated: A program is a succinct description of the function, on which one can perform computations (or run other programs). Of course, if the function is (exactly) learnable via oracle queries (i.e., one can acquire a program that computes the function by querying it at a few locations), then this difference disappears. Hence, to get our counterexample, we will use a function that cannot be exactly learned with oracle queries. A very simple example of such an unlearnable function follows. For strings $\alpha, \beta \in \{0,1\}^k$, define the Turing machine

$$C_{\alpha,\beta}(x) \stackrel{\text{def}}{=} \begin{cases} \beta & x = \alpha \\ 0^k & \text{otherwise} \end{cases}$$

We assume that on input x, $C_{\alpha,\beta}$ runs in $10 \cdot |x|$ steps (the constant 10 is arbitrary). Now we will define a TM $D_{\alpha,\beta}$ that, given the code of a TM C, can distinguish between the case that C computes the same function as $C_{\alpha,\beta}$ from the case that C computes the same function as $C_{\alpha',\beta'}$ for any $(\alpha', \beta') \neq (\alpha, \beta)$.

$$C_{\alpha,\beta}(x) \stackrel{\text{def}}{=} \begin{cases} 1 & C(\alpha) = \beta \\ 0 & \text{otherwise} \end{cases}$$

(Actually, this function is uncomputable. However, as we shall see below, we can use a modified version of $D_{\alpha,\beta}$ that only considers the execution of $C(\alpha)$ for $\text{poly}(k)$ steps, and outputs 0 if C does not halt within that many steps, for some fixed polynomial $\text{poly}(\cdot)$. We will ignore this issue for now, and elaborate on it later.) Note that $C_{\alpha,\beta}$ and $D_{\alpha,\beta}$ have description size $\Theta(k)$.

Consider an adversary A, which, given two (obfuscated) TMs as input, simply runs the second TM on the first one. That is, $A(C, D) = D(C)$. (Actually, like we modified $D_{\alpha,\beta}$ above, we also will modify A to only run D on C for $\text{poly}(|C|, |D|)$ steps, and output 0 if D does not halt in that time.) Thus, for any $\alpha, \beta \in \{0,1\}^k$,

$$\Pr\left[A(\mathcal{O}(C_{\alpha,\beta}), \mathcal{O}(D_{\alpha,\beta})) = 1\right] = 1 \tag{1}$$

Observe that any $\text{poly}(k)$-time algorithm S which has oracle access to $C_{\alpha,\beta}$ and $D_{\alpha,\beta}$ has only exponentially small probability (for a random α and β) of querying either oracle at a point where its value is nonzero. Hence, if we let Z_k be a Turing machine that always outputs 0^k, then for every PPT S,

$$\left|\Pr\left[S^{C_{\alpha,\beta}, D_{\alpha,\beta}}(1^k) = 1\right] - \Pr\left[S^{Z_k, D_{\alpha,\beta}}(1^k) = 1\right]\right| \leq 2^{-\Omega(k)}, \tag{2}$$

where the probabilities are taken over α and β selected uniformly in $\{0,1\}^k$ and the coin tosses of S. On the other hand, by the definition of A we have:

$$\Pr\left[A(\mathcal{O}(Z_k), \mathcal{O}(D_{\alpha,\beta})) = 1\right] = 0 \tag{3}$$

The combination of Equations (1), (2), and (3) contradict the fact that $\mathcal{O}$ is a 2-TM obfuscator.

In the above proof, we ignored the fact that we had to truncate the running times of A and $D_{\alpha,\beta}$. When doing so, we must make sure that Equations (1) and (3) still hold. Equation (1) involves executing (a) $A(\mathcal{O}(D_{\alpha,\beta}), \mathcal{O}(C_{\alpha,\beta}))$, which in turn amounts to executing (b) $\mathcal{O}(D_{\alpha,\beta})(\mathcal{O}(C_{\alpha,\beta}))$. By definition (b) has the same functionality as $D_{\alpha,\beta}(\mathcal{O}(C_{\alpha,\beta}))$, which in turn involves executing (c) $\mathcal{O}(C_{\alpha,\beta})(\alpha)$. Yet the functionality requirement of the obfuscator definition assures us that (c) has the same functionality as $C_{\alpha,\beta}(\alpha)$. By the polynomial slowdown property of obfuscators, execution (c) only takes $\text{poly}(10 \cdot k) = \text{poly}(k)$ steps, which means that $D_{\alpha,\beta}(\mathcal{O}(C_{\alpha,\beta}))$ need only run for $\text{poly}(k)$ steps. Thus, again applying the polynomial slowdown property, execution (b) takes $\text{poly}(k)$ steps, which finally implies that A need only run for $\text{poly}(k)$ steps. The same reasoning holds for Equation (3), using Z_k instead of $C_{\alpha,\beta}$.[3] Note that all the polynomials involved are *fixed* once we fix the polynomial $p(\cdot)$ of the polynomial slowdown property.

The proof for the 2-circuit case is very similar to the 2-TM case, with a related, but slightly different subtlety. Suppose, for sake of contradiction, that $\mathcal{O}$ is a 2-circuit obfuscator. For $k \in \mathbb{N}$ and $\alpha, \beta \in \{0,1\}^k$, define Z_k, $C_{\alpha,\beta}$ and $D_{\alpha,\beta}$ in the same way as above but as circuits rather than TMs, and define an adversary A by $A(C, D) = D(C)$. (Note that the issues of A and $D_{\alpha,\beta}$'s running times go away in this setting, since circuits can always be evaluated in time polynomial in their size.) The new subtlety here is that the definition of A as $A(C, D) = D(C)$ only makes sense when the input length of D is larger than the size of C (note that one can always pad C to a larger size). Thus, for the analogues of Equations (1) and (3) to hold, the input length of $D_{\alpha,\beta}$ must be larger than the sizes of the *obfuscations* of $C_{\alpha,\beta}$ and Z_k. However, by the polynomial slowdown property of obfuscators, it suffices to let $D_{\alpha,\beta}$ have input length $\text{poly}(k)$ and the proof works as before.

■

3.2 Obfuscating One TM/Circuit

Our approach to extending the two-program obfuscation impossibility results to the one-program definitions is to combine the two programs constructed above into one. This will work in a quite straightforward manner for TM obfuscators, but will require new ideas for circuit obfuscators.

Combining functions and programs. For functions, TMs, or circuits $f_0, f_1 : X \to Y$, define their *combination* $f_0 \# f_1 : \{0,1\} \times X \to Y$ by $(f_0 \# f_1)(b, x) \stackrel{\text{def}}{=} f_b(x)$. Conversely, if we are given a TM (resp., circuit) $C : \{0,1\} \times X \to Y$, we can efficiently decompose C into $C_0 \# C_1$ by setting $C_b(x) \stackrel{\text{def}}{=} C(b, x)$; note that C_0 and C_1 have size and running time essentially the same as that of C. Observe that having oracle access to a combined function $f_0 \# f_1$ is equivalent to having oracle access to f_0 and f_1 individually.

[3] Another, even more minor subtlety that we ignored is that, strictly speaking, A only has running time polynomial in the description of the *obfuscations* of $C_{\alpha,\beta}$, $D_{\alpha,\beta}$, and Z_k, which could conceivably be shorter than the original TM descriptions. But a counting argument shows that for all but an exponentially small fraction of pairs $(\alpha, \beta) \in \{0,1\}^k \times \{0,1\}^k$, $\mathcal{O}(C_{\alpha,\beta})$ and $\mathcal{O}(D_{\alpha,\beta})$ must have description size $\Omega(k)$.

Theorem 3.4. *TM obfuscators do not exist.*

Proof Sketch: Suppose, for sake of contradiction, that there exists a TM obfuscator $\mathcal{O}$. For $\alpha, \beta \in \{0,1\}^k$, let $C_{\alpha,\beta}$, $D_{\alpha,\beta}$, and Z_k be the TMs defined in the proof of Proposition 3.3. Combining these, we get the TMs $F_{\alpha,\beta} = C_{\alpha,\beta} \# D_{\alpha,\beta}$ and $G_{\alpha,\beta} = Z_k \# C_{\alpha,\beta}$.

We consider an adversary A analogous to the one in the proof of Proposition 3.3, augmented to first decompose the program it is fed. That is, on input a TM F, algorithm A first decomposes F into $F_0 \# F_1$ and then outputs $F_1(F_0)$. (As in the proof of Proposition 3.3, A actually should be modified to run in time poly($|F|$).) Let S be the PPT simulator for A guaranteed by Definition 2.1. Just as in the proof of Proposition 3.3, we have:

$$\Pr\left[A(\mathcal{O}(F_{\alpha,\beta})) = 1\right] = 1 \text{ and } \Pr\left[A(\mathcal{O}(G_{\alpha,\beta})) = 1\right] = 0$$
$$\left|\Pr\left[S^{F_{\alpha,\beta}}(1^k) = 1\right] - \Pr\left[S^{G_{\alpha,\beta}}(1^k) = 1\right]\right| \leq 2^{-\Omega(k)},$$

where the probabilities are taken over uniformly selected $\alpha, \beta \in \{0,1\}^k$, and the coin tosses of A, S, and $\mathcal{O}$. This contradicts Definition 2.1. □

There is a difficulty in trying to carry out the above argument in the circuit setting. (This difficulty is related to (but more serious than) the same subtlety regarding the circuit setting discussed earlier.) In the above proof, the adversary A, on input $\mathcal{O}(F_{\alpha,\beta})$, attempts to evaluate $F_1(F_0)$, where $F_0 \# F_1 = \mathcal{O}(F_{\alpha,\beta}) = \mathcal{O}(C_{\alpha,\beta} \# D_{\alpha,\beta})$. In order for this to make sense in the circuit setting, the size of the circuit F_0 must be at most the input length of F_1 (which is the same as the input length of $D_{\alpha,\beta}$). But, since the output $F_0 \# F_1$ of the obfuscator can be polynomially larger than its input $C_{\alpha,\beta} \# D_{\alpha,\beta}$, we have no such guarantee. Furthermore, note that if we compute F_0, F_1 in the way we described above (i.e., $F_b(x) \stackrel{\text{def}}{=} \mathcal{O}(F_{\alpha,\beta})(b,x)$) then we'll have $|F_0| = |F_1|$ and so F_0 will necessarily be larger than F_1's input length.

To get around this, we modify $D_{\alpha,\beta}$ in a way that will allow A, when given $D_{\alpha,\beta}$ and a circuit C, to test whether $C(\alpha) = \beta$ even when C is larger than the input length of $D_{\alpha,\beta}$. Of course, oracle access to $D_{\alpha,\beta}$ should not reveal α and β, because we do not want the simulator S to be able to test whether $C(\alpha) = \beta$ given just oracle access to C and $D_{\alpha,\beta}$. We will construct such functions $D_{\alpha,\beta}$ based on pseudorandom functions [GGM86].

Lemma 3.5. *If one-way functions exist, then for every $k \in \mathbb{N}$ and $\alpha, \beta \in \{0,1\}^k$, there is a distribution $\mathcal{D}_{\alpha,\beta}$ on circuits such that:*

1. *Every $D \in \mathrm{Supp}(\mathcal{D}_{\alpha,\beta})$ is a circuit of size* poly(k).
2. *There is a polynomial-time algorithm A such that for any circuit C, and any $D \in \mathrm{Supp}(\mathcal{D}_{\alpha,\beta})$, $A^D(C, 1^k) = 1$ iff $C(\alpha) = \beta$.*
3. *For any PPT S, $\Pr\left[S^D(1^k) = \alpha\right] = \mathrm{neg}(k)$, where the probability is taken over $\alpha, \beta \xleftarrow{R} \{0,1\}^k$, $D \xleftarrow{R} \mathcal{D}_{\alpha,\beta}$, and the coin tosses of S.*

Proof. Basically, the construction implements a private-key "homomorphic encryption" scheme. More precisely, the functions in $\mathcal{D}_{\alpha,\beta}$ will consist of three parts. The first part gives out an encryption of the bits of α (under some private-key encryption scheme). The second part provides the ability to perform binary Boolean operations on encrypted bits, and the third part tests whether a sequence of encryptions consists of encryptions of the bits of β. These operations will enable one to efficiently test whether a given circuit C satisfies $C(\alpha) = \beta$, while keeping α and β hidden when only oracle access to C and $D_{\alpha,\beta}$ is provided.

We begin with any one-bit (probabilistic) private-key encryption scheme (Enc, Dec) that satisfies indistinguishability under *chosen plaintext* and *non-adaptive chosen ciphertext* attacks. Informally, this means that an encryption of 0 should be indistinguishable from an encryption of 1 even for adversaries that have access to encryption and decryption oracles prior to receiving the challenge ciphertext, and access to just an encryption oracle after receiving the challenge ciphertext. (See [KY00] for formal definitions.) We note that such encryptions schemes exist if one-way functions exist; indeed, the "standard" encryption scheme $\mathrm{Enc}_K(b) = (r, f_K(r) \oplus b)$, where $r \stackrel{\mathrm{R}}{\leftarrow} \{0,1\}^{|K|}$ and f_K is a pseudorandom function, has this property.

Now we consider a "homomorphic encryption" algorithm Hom, which takes as input a private-key K and two ciphertexts c and d (w.r.t. this key K), and a binary boolean operation $\odot$ (specified by its 2×2 truth table). We define

$$\mathrm{Hom}_K(c, d, \odot) \stackrel{\mathrm{def}}{=} \mathrm{Enc}_K(\mathrm{Dec}_K(c) \odot \mathrm{Dec}_K(d)).$$

It can be shown that such an encryption scheme retains its security even if the adversary is given access to a Hom oracle. This is formalized in the following claim:

Claim. For every PPT A,

$$\left|\Pr\left[A^{\mathrm{Hom}_K,\mathrm{Enc}_K}(\mathrm{Enc}_K(0)) = 1\right] - \Pr\left[A^{\mathrm{Hom}_K,\mathrm{Enc}_K}(\mathrm{Enc}_K(1)) = 1\right]\right| \leq \mathrm{neg}(k).$$

Proof of claim: Suppose there were a PPT A violating the claim. First, we argue that we can replace the responses to all of A'S Hom_K-oracle queries with encryptions of 0 with only a negligible effect on A's distinguishing gap. This follows from indistinguishability under chosen plaintext and ciphertext attacks and a hybrid argument: Consider hybrids where the first i oracle queries are answered according to Hom_K and the rest with encryptions of 0. Any advantage in distinguishing two adjacent hybrids must be due to distinguishing an encryption of 1 from an encryption of 0. The resulting distinguisher can be implemented using oracle access to encryption and decryption oracles prior to receiving the challenge ciphertext (and an encryption oracle afterwards).

Once we have replaced the Hom_K-oracle responses with encryptions of 0, we have an adversary that can distinguish an encryption of 0 from an encryption of 1 when given access to just an encryption oracle. This contradicts indistinguishability under chosen plaintext attack. □

Now we return to the construction of our circuit family $\mathcal{D}_{\alpha,\beta}$. For a key K, let $E_{K,\alpha}$ be an algorithm which, on input i outputs $\mathrm{Enc}_K(\alpha_i)$, where α_i is the

i'th bit of α. Let $B_{K,\beta}$ be an algorithm which when fed a k-tuple of ciphertexts $(c_1, \ldots, c_k)$ outputs 1 if for all i, $\mathrm{Dec}_K(c_i) = \beta_i$, where $\beta_1, \ldots, \beta_k$ are the bits of β. A random circuit from $\mathcal{D}_{\alpha,\beta}$ will essentially be the algorithm

$$D_{K,\alpha,\beta} \stackrel{\text{def}}{=} E_{K,\alpha} \# \mathrm{Hom}_K \# B_{K,\beta}$$

(for a uniformly selected key K). One minor complication is that $D_{K,\alpha,\beta}$ is actually a *probabilistic* algorithm, since $E_{K,\alpha}$ and Hom_K employ probabilistic encryption, whereas the lemma requires deterministic functions. This can be solved in the usual way, by using pseudorandom functions. Let $q = q(k)$ be the input length of $D_{K,\alpha,\beta}$ and $m = m(k)$ the maximum number of random bits used by $D_{K,\alpha,\beta}$ on any input. We can select a pseudorandom function $f_{K'} : \{0,1\}^q \to \{0,1\}^m$, and let $D'_{K,\alpha,\beta,K'}$ be the (determinstic) algorithm, which on input $x \in \{0,1\}^q$ evaluates $D_{K,\alpha,\beta}(x)$ using randomness $f_{K'}(x)$.

Define the distribution $\mathcal{D}_{\alpha,\beta}$ to be $D'_{K,\alpha,\beta,K'}$, over uniformly selected keys K and K'. We argue that this distribution has the properties stated in the lemma. By construction, each $D'_{K,\alpha,\beta,K'}$ is computable by circuit of size $\mathrm{poly}(k)$, so Property 1 is satisfied.

For Property 2, consider an algorithm A that on input C and oracle access to $D'_{K,\alpha,\beta,K'}$ (which, as usual, we can view as access to (deterministic versions of) the three separate oracles $E_{K,\alpha}$, Hom_K, and $B_{K,\alpha}$), proceeds as follows: First, with k oracle queries to the $E_{K,\alpha}$ oracle, A obtains encryptions of each of the bits of α. Then, A uses the Hom_K oracle to do a gate-by-gate emulation of the computation of $C(\alpha)$, in which A obtains encryptions of the values at each gate of C. In particular, A obtains encryptions of the values at each output gate of C (on input α). It then feeds these output encryptions to $D_{K,\beta}$, and outputs the response to this oracle query. By construction, A outputs 1 iff $C(\alpha) = \beta$.

Finally, we verify Property 3. Let S be any PPT algorithm. We must show that S has only a negligible probability of outputting α when given oracle access to $D'_{K,\alpha,\beta,K'}$ (over the choice of K, α, β, K', and the coin tosses of S). By the pseudorandomness of $f_{K'}$, we can replace oracle access to the function $D'_{K,\alpha,\beta,K'}$ with oracle access to the probabilistic algorithm $D_{K,\alpha,\beta}$ with only a negligible effect on S's success probability. Oracle access to $D_{K,\alpha,\beta}$ is equivalent to oracle access to $E_{K,\alpha}$, Hom_K, and $B_{K,\beta}$. Since β is independent of α and K, the probability that S queries $B_{K,\beta}$ at a point where its value is nonzero (i.e., at a sequence of encryptions of the bits of β) is exponentially small, so we can remove S's queries to $B_{K,\beta}$ with only a negligible effect on the success probability. Oracle access to $E_{K,\alpha}$ is equivalent to giving S polynomially many encryptions of each of the bits of α. Thus, we must argue that S cannot compute α with nonnegligible probability from these encryptions and oracle access to Hom_K. This follows from the fact that the encryption scheme remains secure in the presence of a Hom_K oracle (Claim 3.2) and a hybrid argument. ■

Theorem 3.6. *If one-way functions exist, then circuit obfuscators do not exist.*

Proof. Suppose, for sake of contradiction, that there exists a circuit obfuscator $\mathcal{O}$. For $k \in \mathbb{N}$ and $\alpha, \beta \in \{0,1\}^k$, let Z_k and $C_{\alpha,\beta}$ be the circuits defined in the proof of Proposition 3.3, and let $\mathcal{D}_{\alpha,\beta}$ be the distribution on circuits given by Lemma 3.5. Fer each $k \in \mathbb{N}$, consider the following two distributions on circuits of size poly(k):

$\mathcal{F}_k$: Choose α and β uniformly in $\{0,1\}^k$, $D \stackrel{R}{\leftarrow} \mathcal{D}_{\alpha,\beta}$. Output $C_{\alpha,\beta}\#D$.
$\mathcal{G}_k$: Choose α and β uniformly in $\{0,1\}^k$, $D \stackrel{R}{\leftarrow} \mathcal{D}_{\alpha,\beta}$. Output $Z_k\#D$.

Let A be the PPT algorithm guaranteed by Property 2 in Lemma 3.5, and consider a PPT A' which, on input a circuit F, decomposes $F = F_0\#F_1$ and evaluates $A^{F_1}(F_0, 1^k)$, where k is the input length of F_0. Thus, when fed a circuit from $\mathcal{O}(\mathcal{F}_k)$ (resp., $\mathcal{O}(\mathcal{G}_k)$), A' is evaluating $A^D(C, 1^k)$ where D computes the same function as some circuit from $\mathcal{D}_{\alpha,\beta}$ and C computes the same function as $C_{\alpha,\beta}$ (resp., Z_k). Therefore, by Property 2 in Lemma 3.5, we have:

$$\Pr[A'(\mathcal{O}(\mathcal{F}_k)) = 1] = 1 \qquad \Pr[A'(\mathcal{O}(\mathcal{G}_k)) = 1] = 0.$$

We now argue that for any PPT algorithm S

$$\left|\Pr\left[S^{\mathcal{F}_k}(1^k) = 1\right] - \Pr\left[S^{\mathcal{G}_k}(1^k) = 1\right]\right| \leq 2^{-\Omega(k)},$$

which will contradict the definition of circuit obfuscators. Having oracle access to a circuit from $\mathcal{F}_k$ (respectively, $\mathcal{G}_k$) is equivalent to having oracle access to $C_{\alpha,\beta}$ (resp., Z_k) and $D \stackrel{R}{\leftarrow} \mathcal{D}_{\alpha,\beta}$, where α, β are selected uniformly in $\{0,1\}^k$. Property 3 of Lemma 3.5 implies that the probability that S queries the first oracle at α is negligible, and hence S cannot distinguish that oracle being $C_{\alpha,\beta}$ from it being Z_k. ∎

We can remove the assumption that one-way functions exist for *efficient* circuit obfuscators via the following (easy) lemma (proven in the full version of the paper).

Lemma 3.7. *If efficient obfuscators exist, then one-way functions exist.*

Corollary 3.8. *Efficient circuit obfuscators do not exist (unconditionally).*

As stated above, our impossibility proof can be cast in terms of "inherently unbfuscatable functions":

Theorem 3.9 (inherently unobfuscatable functions). *If one-way functions exist, then there exists an inherently unobfuscatable function ensemble.*

Proof. Let $\mathcal{F}_k$ and $\mathcal{G}_k$ be the distributions on functions in the proof of Theorem 3.6,and let $\mathcal{H}_k$ be the distribution that, with probability 1/2 outputs a sample of $\mathcal{F}_k$ and with probability 1/2 outputs a sample of $\mathcal{G}_k$. We claim that $\{\mathcal{H}_k\}_{k\in\mathbb{N}}$ is an inherently unobfuscatable function ensemble.

The fact that $\{\mathcal{H}_k\}_{k\in\mathbb{N}}$ is efficiently computable is obvious. We define $\pi(f)$ to be 1 if $f \in \bigcup_k \mathrm{Supp}(\mathcal{F}_k)$ and 0 otherwise (note that $(\bigcup_k \mathrm{Supp}(\mathcal{F}_k)) \cap (\bigcup_k \mathrm{Supp}(\mathcal{G}_k)) = \emptyset$ and so $\pi(f) = 0$ for any $f \in \bigcup_k \mathrm{Supp}(\mathcal{G}_k)$). The algorithm A' given in the proof of Theorem 3.6 shows that $\pi(f)$ can be computed in polynomial time from any circuit computing $f \in \mathrm{Supp}(\mathcal{H}_k)$. Because oracle access to $\mathcal{F}_k$ cannot be distinguished from oracle access to $\mathcal{G}_k$ (as shown in the proof of Theorem 3.6), it follows that $\pi(f)$ cannot be computed from an oracle for $f \stackrel{R}{\leftarrow} \mathcal{H}_k$ with probability noticeably greater than 1/2. ∎

Acknowledgments. We are grateful to Luca Trevisan for collaboration at an early stage of this research. We also thank Dan Boneh, Ran Canetti, and Yacov Yacobi for helpful discussions and comments.

This work was partially supported by the following funds: Oded Goldreich was supported by the Minerva Foundation, Germany; Salil Vadhan (at the time at MIT) was supported by a DOD/NDSEG Graduate Fellowship and an NSF Mathematical Sciences Postdoctoral Research Fellowship.

References

[BGI+01] Boaz Barak, Oded Goldreich, Russell Impagliazzo, Steven Rudich, Amit Sahai, Salil Vadhan, and Ke Yang. On the (im)possibility of obfuscating programs. Technical report, Electronic Colloquium on Computational Complexity, 2001. `http://www.eccc.uni-trier.de/eccc`.

[BR93] Mihir Bellare and Phillip Rogaway. Random oracles are practical: A paradigm for designing efficient protocols. In *Proceedings of the First Annual Conference on Computer and Communications Security.* ACM, November 1993.

[BL96] Dan Boneh and Richard Lipton. Algorithms for black-box fields and their applications to cryptography. In M. Wiener, editor, *Advances in Cryptology—CRYPTO '96*, volume 1109 of *Lecture Notes in Computer Science*, pages 283–297. Springer-Verlag, August 1996.

[CGH98] Ran Canetti, Oded Goldreich, and Shai Halevi. The random oracle methodology, revisited. In *Proceedings of the Thirtieth Annual ACM Symposium on Theory of Computing*, pages 209–218, Dallas, 23–26 May 1998.

[CT00] Christian Collberg and Clark Thomborson. Watermarking, tamper-proofing, and obfuscation – tools for software protection. Technical Report TR00-03, The Department of Computer Science, University of Arizona, February 2000.

[DDN00] Danny Dolev, Cynthia Dwork, and Moni Naor. Nonmalleable cryptography. *SIAM Journal on Computing*, 30(2):391–437 (electronic), 2000.

[FM91] Joan Feigenbaum and Michael Merritt, editors. *Distributed computing and cryptography*, Providence, RI, 1991. American Mathematical Society.

[FS87] Amos Fiat and Adi Shamir. How to prove yourself: practical solutions to identification and signature problems. In *Advances in cryptology—CRYPTO '86 (Santa Barbara, Calif., 1986)*, pages 186–194. Springer, Berlin, 1987.

[GGM86] Oded Goldreich, Shafi Goldwasser, and Silvio Micali. How to construct random functions. *Journal of the Association for Computing Machinery*, 33(4):792–807, 1986.

[GO96] Oded Goldreich and Rafail Ostrovsky. Software protection and simulation on oblivious RAMs. *Journal of the ACM*, 43(3):431–473, 1996.

[GM84] Shafi Goldwasser and Silvio Micali. Probabilistic encryption. *Journal of Computer and System Sciences*, 28(2):270–299, April 1984.

[Had00] Satoshi Hada. Zero-knowledge and code obfuscation. In T. Okamoto, editor, *Advances in Cryptology - ASIACRYPT ' 2000*, Lecture Notes in Computer Science, pages 443–457, Kyoto, Japan, 2000. International Association for Cryptologic Research, Springer-Verlag, Berlin Germany.

[KY00] Jonathan Katz and Moti Yung. Complete characterization of security notions for private-key encryption. In *Proceedings of the 32nd Annual ACM Symposium on Theory of Computing*, pages 245–254, Portland, OR, May 2000. ACM.

[NSS99] David Naccache, Adi Shamir, and Julien P. Stern. How to copyright a function? In H. Imai and Y. Zheng, editors, *Public Key Cryptography—PKC '99*, volume 1560 of *Lecture Notes in Computer Science*, pages 188–196. Springer-Verlag, March 1999.

[NR97] Moni Naor and Omer Reingold. Number-theoretic constructions of efficient pseudo-random functions. In *38th Annual Symposium on Foundations of Computer Science*, pages 458–467, Miami Beach, Florida, 20–22 October 1997. IEEE.

[RAD78] Ronald L. Rivest, Len Adleman, and Michael L. Dertouzos. On data banks and privacy homomorphisms. In *Foundations of secure computation (Workshop, Georgia Inst. Tech., Atlanta, Ga., 1977)*, pages 169–179. Academic, New York, 1978.

[SYY99] Thomas Sander, Adam Young, and Moti Yung. Non-interactive cryptocomputing for NC^1. In *40th Annual Symposium on Foundations of Computer Science*, pages 554–566, New York, NY, 17–19 October 1999. IEEE.

[vD98] Frans van Dorsselaer. Obsolescent feature. Winning entry for the *1998 International Obfuscated C Code Contest*, 1998. `http://www.ioccc.org/`.

Universally Composable Commitments
(Extended Abstract)

Ran Canetti[1] and Marc Fischlin[2*]

[1] IBM T.J. Watson Research Center
canetti@watson.ibm.com
[2] Goethe-University of Frankfurt
marc@mi.informatik.uni-frankfurt.de

Abstract. We propose a new security measure for commitment protocols, called Universally Composable (UC) Commitment. The measure guarantees that commitment protocols behave like an "ideal commitment service," even when concurrently composed with an arbitrary set of protocols. This is a strong guarantee: it implies that security is maintained even when an unbounded number of copies of the scheme are running concurrently, it implies non-malleability (not only with respect to other copies of the same protocol but even with respect to other protocols), it provides resilience to selective decommitment, and more.
Unfortunately, two-party UC commitment protocols do not exist in the plain model. However, we construct two-party UC commitment protocols, based on general complexity assumptions, in the *common reference string model* where all parties have access to a common string taken from a predetermined distribution. The protocols are non-interactive, in the sense that both the commitment and the opening phases consist of a single message from the committer to the receiver.

Keywords: Commitment schemes, concurrent composition, non-malleability, security analysis of protocols.

1 Introduction

Commitment is one of the most basic and useful cryptographic primitives. It is an essential building block in Zero-Knowledge protocols (e.g., [GMW91,BCC88, D89]), in general function evaluation protocols (e.g., [GMW87,GHY88,G98]), in contract-signing and electronic commerce, and more. Indeed, commitment protocols have been studied extensively in the past two decades (e.g., [N91,DDN00, NOVY92,B99,DIO98,FF00,DKOS01]).

The basic idea behind the notion of commitment is attractively simple: A *committer* provides a *receiver* with the digital equivalent of a "sealed envelope" containing a value x. From this point on, the committer cannot change the value inside the envelope, and, as long as the committer does not assist the receiver

* part of this work done while visiting IBM T.J. Watson Research Center.

J. Kilian (Ed.): CRYPTO 2001, LNCS 2139, pp. 19–40, 2001.

in opening the envelope, the receiver learns nothing about x. When both parties cooperate, the value x is retrieved in full.

Formalizing this intuitive idea is, however, non-trivial. Traditionally, two quite distinct basic flavors of commitment are formalized: *unconditionally binding* and *unconditionally secret* commitment protocols (see, e.g., [G95]). These basic definitions are indeed sufficient for some applications (see there). But they treat commitment as a "stand alone" task and do not in general guarantee security when a commitment protocol is used as a building-block within other protocols, or when multiple copies of a commitment protocol are carried out together. A good first example for the limitations of the basic definitions is the *selective decommitment* problem [DNRS99], that demonstrates our inability to prove some very minimal composition properties of the basic definitions.

Indeed, the basic definitions turned out to be inadequate in some scenarios, and stronger variants that allow to securely "compose" commitment protocols —both with the calling protocol and with other invocations of the commitment protocol— were proposed and successfully used in some specific contexts. One such family of variants make sure that knowledge of certain trapdoor information allows "opening" commitments in more than a single way. These include *chameleon commitments* [BCC88], *trapdoor commitments* [FS90] and *equivocable commitments* [B99]. Another strong variant is *non-malleable commitments* [DDN00], where it is guaranteed that a dishonest party that receives an unopened commitment to some value x will be unable to commit to a value that depends on x in any way, except for generating another commitment to x. (A more relaxed variant of the [DDN00] notion of non-malleability is *non-malleability with respect to opening* [DIO98,FF00,DKOS01].)

These stronger measures of security for commitment protocols are indeed very useful. However they only solve specific problems and stop short of guaranteeing that commitment protocols maintain the expected behavior in general cryptographic contexts, or in other words when composed with arbitrary protocols. To exemplify this point, notice for instance that, although [DDN00] remark on more general notions of non-malleability, the standard notion of non-malleability considers only other copies of the same protocol. There is no guarantee that a malicious receiver is unable to "maul" a given commitment by using a totally different commitment protocol. And it is indeed easy to come up with two commitment protocols $\mathcal{C}$ and $\mathcal{C}'$ such that both are non-malleable with respect to themselves, but an adversary that plays a receiver in $\mathcal{C}$ can generate a $\mathcal{C}'$-commitment to a related value.

This work proposes a measure of security for commitment protocols that guarantees the "envelope-like" intuitive properties of commitment even when the commitment protocol is *concurrently composed* with an arbitrary set of protocols. In particular, protocols that satisfy this measure (called universally composable (UC) commitment protocols) remain secure even when an unbounded number of copies of the protocol are executed concurrently in an adversarially controlled way; they are resilient to selective decommitment attacks; they are non-malleable both with respect to other copies of the same protocol and with respect to arbitrary commitment protocols. In general, a UC commitment protocol successfully emulates an "ideal commitment service" for any application protocol (be it a Zero-Knowledge protocol, a general function evaluation protocol, an e-commerce application, or any combination of the above).

This measure of security for commitment protocols is very strong indeed. It is perhaps not surprising then that UC commitment protocols which involve only the committer and the receiver do not exist in the standard "plain model" of computation where no set-up assumptions are provided. (We formally prove this fact.) However, in the *common reference string* (CRS) model things look better. (The CRS model is a generalization of the *common random string* model. Here all parties have access to a common string that was chosen according to some predefined distribution. Other equivalent terms include the *reference string* model [D00] and the *public parameter* model [FF00].) In this model we construct UC commitment protocols based on standard complexity assumptions. A first construction, based on any family of trapdoor permutations, requires the length of the reference string to be linear in the number of invocations of the protocol throughout the lifetime of the system. A second protocol, based on any claw-free pair of trapdoor permutations, uses a short reference string for an unbounded number of invocations. The protocols are non-interactive, in the sense that both the commitment and the decommitment phases consist of a single message from the committer to the receiver. We also note that UC commitment protocols can be constructed in the plain model, if the committer and receiver are assisted by third parties (or, "servers") that participate in the protocol without having local inputs and outputs, under the assumption that a majority of the servers remain uncorrupted.

1.1 On the New Measure

Providing meaningful security guarantees under composition with arbitrary protocols requires using an appropriate framework for representing and arguing about such protocols. Our treatment is based in a recently proposed such general framework [C00a]. This framework builds on known definitions for function evaluation and general tasks [GL90,MR91,B91,PW94,C00,DM00,PW01], and allows defining the security properties of practically any cryptographic task. Most importantly, in this framework security of protocols is maintained under general *concurrent* composition with an unbounded number of copies of arbitrary protocols. We briefly summarize the relevant properties of this framework. See more details in Section 2.1 and in [C00a].

As in prior general definitions, the security requirements of a given task (i.e., the functionality expected from a protocol that carries out the task) are captured via a set of instructions for a "trusted party" that obtains the inputs of the participants and provides them with the desired outputs. However, as opposed to the standard case of secure function evaluation, here the trusted party (which is also called the ideal functionality) runs an arbitrary algorithm and in particular may interact with the parties in several iterations, while maintaining state in between. Informally, a protocol securely carries out a given task if running the protocol amounts to "emulating" an ideal process where the parties hand their inputs to the appropriate ideal functionality and obtain their outputs from it, without any other interaction.

In order to allow proving the concurrent composition theorem, the notion of emulation in [c00a] is considerably stronger than previous ones. Traditionally, the model of computation includes the parties running the protocol and an adversary, $\mathcal{A}$, and "emulating an ideal process" means that for any adversary $\mathcal{A}$ there should exist an "ideal process adversary" (or, simulator) $\mathcal{S}$ that results in similar distribution on the outputs for the parties. Here an additional adversarial entity, called the environment $\mathcal{Z}$, is introduced. The environment generates the inputs to all parties, reads all outputs, and in addition interacts with the adversary in an arbitrary way throughout the computation. A protocol is said to securely realize a given ideal functionality $\mathcal{F}$ if for any adversary $\mathcal{A}$ there exists an "ideal-process adversary" $\mathcal{S}$, such that *no environment* $\mathcal{Z}$ can tell whether it is interacting with $\mathcal{A}$ and parties running the protocol, or with $\mathcal{S}$ and parties that interact with $\mathcal{F}$ in the ideal process. (In a sense, here $\mathcal{Z}$ serves as an "interactive distinguisher" between a run of the protocol and the ideal process with access to $\mathcal{F}$. See [c00a] for more motivating discussion on the role of the environment.) Note that the definition requires the "ideal-process adversary" (or, simulator) $\mathcal{S}$ to interact with $\mathcal{Z}$ throughout the computation. Furthermore, $\mathcal{Z}$ cannot be "rewound".

The following *universal composition* theorem is proven in [c00a]. Consider a protocol π that operates in a *hybrid* model of computation where parties can communicate as usual, and in addition have ideal access to (an unbounded number of copies of) some ideal functionality $\mathcal{F}$. Let ρ be a protocol that securely realizes $\mathcal{F}$ as sketched above, and let π^ρ be the "composed protocol". That is, π^ρ is identical to π with the exception that each interaction with some copy of $\mathcal{F}$ is replaced with a call to (or an invocation of) an appropriate instance of ρ. Similarly, ρ-outputs are treated as values provided by the appropriate copy of $\mathcal{F}$. Then, π and π^ρ have essentially the same input/output behavior. In particular, if π securely realizes some ideal functionality $\mathcal{G}$ given ideal access to $\mathcal{F}$ then π^ρ securely realizes $\mathcal{G}$ from scratch.

To apply this general framework to the case of commitment protocols, we formulate an ideal functionality $\mathcal{F}_{com}$ that captures the expected behavior of commitment. Universally Composable (UC) commitment protocols are defined to be those that securely realize $\mathcal{F}_{com}$. Our formulation of $\mathcal{F}_{com}$ is a straightforward transcription of the "envelope paradigm": $\mathcal{F}_{com}$ first waits to receive a request from some party C to commit to value x for party R. (C and R are identities of two parties in a multiparty network). When receiving such a request, $\mathcal{F}_{com}$ records the value x and notifies R that C has committed to some value for him. When C later sends a request to open the commitment, $\mathcal{F}_{com}$ sends the recorded value x to R, and halts. (Some other variants of $\mathcal{F}_{com}$ are discussed within.) The general composition theorem now implies that running (multiple copies of) a UC commitment protocol π is essentially equivalent to interacting with the same number of copies of $\mathcal{F}_{com}$, regardless of what the calling protocol does. In particular, the calling protocol may run other commitment protocols and may use the committed values in any way. As mentioned above, this implies a strong version of non-malleability, security under concurrent composition, resilience to selective decommitment, and more.

The definition of security and composition theorem carry naturally to the CRS model as well. However, this model hides a caveat: The composition operation requires that each copy of the UC commitment protocol will have its own copy of the CRS. Thus, a protocol that securely realizes $\mathcal{F}_{com}$ as described above is highly wasteful of the reference string. In order to capture protocols where multiple commitments may use the same reference string we formulate a natural extension of $\mathcal{F}_{com}$ that handles multiple commitment requests. Let $\mathcal{F}_{mcom}$ denote this extension.

We remark that UC commitment protocols need not, by definition, be neither unconditionally secret nor unconditionally binding. Indeed, one of the constructions presented here has neither property.

1.2 On the Constructions

At a closer look, the requirements from a UC commitment protocol boil down to the following two requirements from the ideal-process adversary (simulator) $\mathcal{S}$. *(a)*. When the committer is corrupted (i.e., controlled by the adversary), $\mathcal{S}$ must be able to "extract" the committed value from the commitment. (That is, $\mathcal{S}$ has to come up with a value x such that the committer will almost never be able to successfully decommit to any $x' \neq x$.) This is so since in the ideal process $\mathcal{S}$ has to explicitly provide $\mathcal{F}_{com}$ with a committed value. *(b)*. When the committer is uncorrupted, $\mathcal{S}$ has to be able to generate a kosher-looking "simulated commitment" c that can be "opened" to any value (which will become known only later). This is so since $\mathcal{S}$ has to provide adversary $\mathcal{A}$ and environment $\mathcal{Z}$ with the simulated commitment c before the value committed to is known. All this needs to be done *without rewinding the environment $\mathcal{Z}$*. (Note that non-malleability is not explicitly required in this description. It is, however, implied by the above requirements.)

From the above description it may look plausible that no simulator $\mathcal{S}$ exists that meets the above requirements in the plain model. Indeed, we formalize and prove this statement for the case of protocols that involve only a committer and a receiver. (In the case where the committer and the receiver are assisted by third parties, a majority of which is guaranteed to remain uncorrupted, standard techniques for multiparty computation are sufficient for constructing UC commitment protocols. See [c00a] for more details.)

In the CRS model the simulator is "saved" by the ability to choose the reference string and plant trapdoors in it. Here we present two UC commitment protocols. The first one (that securely realizes functionality $\mathcal{F}_{com}$) is based on the equivocable commitment protocols of [DIO98], while allowing the simulator to have trapdoor information that enables it to extract the values committed to by corrupted parties. However, the equivocability property holds only with respect to a single usage of the CRS. Thus this protocol fails to securely realize the multiple commitment functionality $\mathcal{F}_{mcom}$.

In the second protocol (that securely realizes $\mathcal{F}_{mcom}$), the reference string contains a description of a claw-free pair of trapdoor permutations and a public encryption key of an encryption scheme that is secure against adaptive chosen ciphertext attacks (CCA) (as in, say, [DDN00,RS91,BDPR98,CS98]). Commitments are generated via standard use of a claw-free pair, combined with encrypting potential decommitments. The idea to use CCA-secure encryption in this context is taken from [L00,DKOS01].

Both protocols implement commitment to a single bit. Commitment to arbitrary strings is achieved by composing together several instances of the basic protocol. Finding more efficient UC commitment protocols for string commitment is an interesting open problem.

Applicability of the notion. In addition to being an interesting goal in their own right, UC commitment protocols can potentially be very useful in constructing more complex protocols with strong security and composability properties. To demonstrate the applicability of the new notion, we show how UC commitment protocols can be used in a simple way to construct strong Zero-Knowledge protocols *without any additional cryptographic assumptions.*

Related work. Pfitzmann et. al. [PW94,PW01] present another definitional framework that allows capturing the security requirements of general reactive tasks, and prove a concurrent composition theorem with respect to their framework. Potentially, our work could be cast in their framework as well; however, the composition theorem provided there is considerably weaker than the one in [C00a].

Organization. Section 2 shortly reviews the general framework of [C00a] and presents the ideal commitment functionalities $\mathcal{F}_{com}$ and $\mathcal{F}_{mcom}$. Section 3 presents and proves security of the protocols that securely realize $\mathcal{F}_{com}$ and $\mathcal{F}_{mcom}$. Section 4 demonstrates that functionalities $\mathcal{F}_{com}$ and $\mathcal{F}_{mcom}$ cannot be realized in the plain model by a two-party protocol. Section 5 presents the application to constructing Zero-Knowledge protocols. For lack of space most proofs are omitted. They appear in [CF01].

2 Definitions

Section 2.1 shortly summarizes the relevant parts of the general framework of [C00a], including the definition of security and the composition theorem. Section 2.2 defines the ideal commitment functionalities, $\mathcal{F}_{com}$ and $\mathcal{F}_{mcom}$.

2.1 The General Framework

Protocol syntax. Following [GMRa89,G95], protocols are represented as a set of interactive Turing machines (ITMs). Specifically, the input and output tapes model inputs and outputs that are received from and given to other programs running on the same machine, and the communication tapes model messages sent to and received from the network. Adversarial entities are also modeled as ITMs; we concentrate on a non-uniform complexity model where the adversaries have an arbitrary additional input, or an "advice".

The adversarial model. [c00a] discusses several models of computation. We concentrate on one main model, aimed at representing current realistic communication networks (such as the Internet). Specifically, the network is *asynchronous* without guaranteed delivery of messages. The communication is public (i.e., all messages can be seen by the adversary) but ideally authenticated (i.e., messages cannot be modified by the adversary). In addition, parties have unique *identities.*[1] The adversary is adaptive in corrupting parties, and is active (or, *Byzantine*) in its control over corrupted parties. Finally, the adversary and environment are restricted to probabilistic polynomial time (or, "feasible") computation.

Protocol execution in the real-life model. We sketch the "mechanics" of executing a given protocol π (run by parties $P_1, ..., P_n$) with some adversary $\mathcal{A}$ and an environment machine $\mathcal{Z}$ with input z. All parties have a security parameter $k \in \mathbf{N}$ and are polynomial in k. The execution consists of a sequence of *activations,* where in each activation a single participant (either $\mathcal{Z}$, $\mathcal{A}$, or some P_i) is activated. The activated participant reads information from its input and incoming communication tapes, executes its code, and possibly writes information on its outgoing communication tapes and output tapes. In addition, the environment can write information on the input tapes of the parties, and read their output tapes. The adversary can read messages off the outgoing message tapes of the parties and *deliver* them by copying them to the incoming message tapes of the recipient parties. The adversary can also corrupt parties, with the usual consequences that it learns the internal information known to the corrupted party and that from now on it controls that party.

The environment is activated first; once activated, it may choose to activate either one of the parties (with some input value) or to activate the adversary. Whenever the adversary delivers a message to some party P, this party is activated next. Once P's activation is complete, the environment is activated. Throughout, the environment and the adversary may exchange information freely using their input and output tapes. The output of the protocol execution is the output of $\mathcal{Z}$. (Without loss of generality $\mathcal{Z}$ outputs a single bit.)

Let $\text{REAL}_{\pi,\mathcal{A},\mathcal{Z}}(k, z, \boldsymbol{r})$ denote the output of environment $\mathcal{Z}$ when interacting with adversary $\mathcal{A}$ and parties running protocol π on security parameter k, input z and random input $\boldsymbol{r} = r_{\mathcal{Z}}, r_{\mathcal{A}}, r_1 \ldots r_n$ as described above (z and $r_{\mathcal{Z}}$ for $\mathcal{Z}$, $r_{\mathcal{A}}$ for $\mathcal{A}$; r_i for party P_i). Let $\text{REAL}_{\pi,\mathcal{A},\mathcal{Z}}(k, z)$ denote the random variable

[1] Indeed, the communication in realistic networks is typically *unauthenticated,* in the sense that messages may be adversarially modified en-route. In addition, there is no guarantee that identities will be unique. Nonetheless, since authentication and the guarantee of unique identities can be added independently of the rest of the protocol, we allow ourselves to assume ideally authenticated channels and unique identities. See [c00a] for further discussion.

describing $\text{REAL}_{\pi,\mathcal{A},\mathcal{Z}}(k, z, \boldsymbol{r})$ when $\boldsymbol{r}$ is uniformly chosen. Let $\text{REAL}_{\pi,\mathcal{A},\mathcal{Z}}$ denote the ensemble $\{\text{REAL}_{\pi,\mathcal{A},\mathcal{Z}}(k, z)\}_{k\in\mathbf{N},z\in\{0,1\}^*}$.

The ideal process. Security of protocols is defined via comparing the protocol execution in the real-life model to an *ideal process* for carrying out the task at hand. A key ingredient in the ideal process is the ideal functionality that captures the desired functionality, or the specification, of that task. The ideal functionality is modeled as another ITM that interacts with the environment and the adversary via a process described below. More specifically, the ideal process involves an ideal functionality $\mathcal{F}$, an ideal process adversary $\mathcal{S}$, an environment $\mathcal{Z}$ on input z and a set of dummy parties $\tilde{P}_1, ..., \tilde{P}_n$. The dummy parties are fixed and simple ITMS: Whenever a dummy party is activated with input x, it forwards x to $\mathcal{F}$, say by copying x to its outgoing communication tape; whenever it is activated with incoming message from $\mathcal{F}$ it copies this message to its output. $\mathcal{F}$ receives information from the (dummy) parties by reading it off their outgoing communication tapes. It hands information back to the parties by sending this information to them. The ideal-process adversary $\mathcal{S}$ proceeds as in the real-life model, except that it has no access to the contents of the messages sent between $\mathcal{F}$ and the parties. In particular, $\mathcal{S}$ is responsible for delivering messages, and it can corrupt dummy parties, learn the information they know, and control their future activities.

The order of events in the ideal process is as follows. As in the real-life model, the environment is activated first. As there, parties are activated when they receive new information (here this information comes either from the environment or from $\mathcal{F}$). In addition, whenever a dummy party P sends information to $\mathcal{F}$, then $\mathcal{F}$ is activated. Once $\mathcal{F}$ completes its activation, P is activated again. Also, $\mathcal{F}$ may exchange messages directly with the adversary. It is stressed that in the ideal process there is no communication among the parties. The only "communication" is in fact idealized transfer of information between the parties and the ideal functionality. The output of the ideal process is the (one bit) output of $\mathcal{Z}$.

Let $\text{IDEAL}_{\mathcal{F},\mathcal{S},\mathcal{Z}}(k, z, \boldsymbol{r})$ denote the output of environment $\mathcal{Z}$ after interacting in the ideal process with adversary $\mathcal{S}$ and ideal functionality $\mathcal{F}$, on security parameter k, input z, and random input $\boldsymbol{r} = r_{\mathcal{Z}}, r_{\mathcal{S}}, r_{\mathcal{F}}$ as described above (z and $r_{\mathcal{Z}}$ for $\mathcal{Z}$, $r_{\mathcal{S}}$ for $\mathcal{S}$; $r_{\mathcal{F}}$ for $\mathcal{F}$). Let $\text{IDEAL}_{\mathcal{F},\mathcal{S},\mathcal{Z}}(k, z)$ denote the random variable describing $\text{IDEAL}_{\mathcal{F},\mathcal{S},\mathcal{Z}}(k, z, \boldsymbol{r})$ when $\boldsymbol{r}$ is uniformly chosen. Let $\text{IDEAL}_{\mathcal{F},\mathcal{S},\mathcal{Z}}$ denote the ensemble $\{\text{IDEAL}_{\mathcal{F},\mathcal{S},\mathcal{Z}}(k, z)\}_{k\in\mathbf{N},z\in\{0,1\}^*}$.

Securely realizing an ideal functionality. We say that a protocol ρ securely realizes an ideal functionality $\mathcal{F}$ if for any real-life adversary $\mathcal{A}$ there exists an ideal-process adversary $\mathcal{S}$ such that no environment $\mathcal{Z}$, on any input, can tell with non-negligible probability whether it is interacting with $\mathcal{A}$ and parties running ρ in the real-life process, or it is interaction with $\mathcal{A}$ and $\mathcal{F}$ in the ideal process. This means that, from the point of view of the environment, running protocol ρ is 'just as good' as interacting with an ideal process for $\mathcal{F}$. (In a way, $\mathcal{Z}$ serves as an "interactive distinguisher" between the two processes. Here it is important that $\mathcal{Z}$ can provide the process in question with *adaptively chosen* inputs throughout the computation.)

Definition 1. *Let* $\mathcal{X} = \{X(k,a)\}_{k\in\mathbf{N},a\in\{0,1\}^*}$ *and* $\mathcal{Y} = \{Y(k,a)\}_{k\in\mathbf{N},a\in\{0,1\}^*}$ *be two distribution ensembles over* $\{0,1\}$*. We say that* $\mathcal{X}$ *and* $\mathcal{Y}$ *are* indistinguishable *(written* $\mathcal{X} \overset{c}{\approx} \mathcal{Y}$*) if for any* $c \in \mathbf{N}$ *there exists* $k_0 \in \mathbf{N}$ *such that* $|\Pr(X(k,a) = 1) - \Pr(Y(k,a) = 1)| < k^{-c}$ *for all* $k > k_0$ *and all* a*.*

Definition 2 ([c00a]). *Let* $n \in \mathbf{N}$*. Let* $\mathcal{F}$ *be an ideal functionality and let* π *be an* n*-party protocol. We say that* π securely realizes $\mathcal{F}$ *if for any adversary* $\mathcal{A}$ *there exists an ideal-process adversary* $\mathcal{S}$ *such that for any environment* $\mathcal{Z}$ *we have* $\text{IDEAL}_{\mathcal{F},\mathcal{S},\mathcal{Z}} \overset{c}{\approx} \text{REAL}_{\pi,\mathcal{A},\mathcal{Z}}$.

The common reference string (CRS) model. In this model it is assumed that all the participants have access to a common string that is drawn from some specified distribution. (This string is chosen ahead of time and is made available before any interaction starts.) It is stressed that the security of the protocol depends on the fact that the reference string is generated using a pre-specified randomized procedure, and no "trapdoor information" related to the string exists in the system. This in turn implies full trust in the entity that generates the reference string. More precisely, the CRS model is formalized as follows.

- The real-life model of computation is modified so that all participants have access to a common string that is chosen in advance according to some distribution (specified by the protocol run by the parties) and is written in a special location on the input tape of each party.
- The ideal process is modified as follows. In a preliminary step, the ideal-model adversary chooses a string in some arbitrary way and writes this string on the input tape of the environment machine. After this initial step the computation proceeds as before. It is stressed that the ideal functionality has no access to the reference string.

Justification of the CRS model. Allowing the ideal-process adversary (i.e., the simulator) to choose the reference string is justified by the fact that the behavior of the ideal functionality does not depend on the reference string. This means that the security guarantees provided by the ideal process hold regardless of how the reference string is chosen and whether trapdoor information regarding this string is known.

On the composition theorem: The hybrid model. In order to state the composition theorem, and in particular in order to formalize the notion of a real-life protocol with access to an ideal functionality, the hybrid model of computation with access to an ideal functionality $\mathcal{F}$ (or, in short, the $\mathcal{F}$-hybrid model) is formulated. This model is identical to the real-life model, with the following exceptions. In addition to sending messages to each other, the parties may send messages to and receive messages from an unbounded number of copies of $\mathcal{F}$. Each copy of $\mathcal{F}$ is identified via a unique session identifier (SID); all messages addressed to this copy and all message sent by this copy carry the corresponding SID. (The SIDs are chosen by the protocol run by the parties.)

The communication between the parties and each one of the copies of $\mathcal{F}$ mimics the ideal process. That is, once a party sends a message to some copy of $\mathcal{F}$, that copy is immediately activated and reads that message off the party's tape. Furthermore, although the adversary in the hybrid model is responsible for delivering the messages from the copies of $\mathcal{F}$ to the parties, it does not have access to the contents of these messages.

Replacing a call to $\mathcal{F}$ with a protocol invocation. Let π be a protocol in the $\mathcal{F}$-hybrid model, and let ρ be a protocol that securely realizes $\mathcal{F}$ (with respect to some class of adversaries). The composed protocol π^ρ is constructed by modifying the code of each ITM in π so that the first message sent to each copy of $\mathcal{F}$ is replaced with an invocation of a new copy of π with fresh random input, and with the contents of that message as input. Each subsequent message to that copy of $\mathcal{F}$ is replaced with an activation of the corresponding copy of ρ, with the contents of that message given to ρ as new input. Each output value generated by a copy of ρ is treated as a message received from the corresponding copy of $\mathcal{F}$.

Theorem statement. In its general form, the composition theorem basically says that if ρ securely realizes $\mathcal{F}$ then an execution of the composed protocol π^ρ "emulates" an execution of protocol π in the $\mathcal{F}$-hybrid model. That is, for any real-life adversary $\mathcal{A}$ there exists an adversary $\mathcal{H}$ in the $\mathcal{F}$-hybrid model such that no environment machine $\mathcal{Z}$ can tell with non-negligible probability whether it is interacting with $\mathcal{A}$ and π^ρ in the real-life model or it is interacting with $\mathcal{H}$ and π in the $\mathcal{F}$-hybrid model..

A more specific corollary of the general theorem states that if π securely realizes some functionality $\mathcal{G}$ in the $\mathcal{F}$-hybrid model, and ρ securely realizes $\mathcal{F}$ in the real-life model, then π^ρ securely realizes $\mathcal{G}$ in the real-life model. (Here one has to define what it means to securely realize functionality $\mathcal{G}$ in the $\mathcal{F}$-hybrid model. This is done in the natural way.)

Theorem 1 ([c00a]). *Let $\mathcal{F}, \mathcal{G}$ be ideal functionalities. Let π be an n-party protocol that realizes $\mathcal{G}$ in the $\mathcal{F}$-hybrid model and let ρ be an n-party protocol that securely realizes $\mathcal{F}$ Then protocol π^ρ securely realizes $\mathcal{G}$.*

Protocol composition in the CRS model. Some words of clarification are in order with respect to the composition theorem in the CRS model. Specifically, it is stressed that each copy of protocol ρ within the composed protocol π^ρ should have its own copy of the reference string, or equivalently uses a separate portion of a long string. (If this is not the case then the theorem no longer holds in general.) As seen below, the behavior of protocols where several copies of the protocol use the same instance of the reference string can be captured using ideal functionalities that represent multiple copies of the protocol within a single copy of the functionality.

2.2 The Commitment Functionalities

We propose ideal functionalities that represent the intuitive "envelope-like" properties of commitment, as sketched in the introduction. Two functionalities are presented: functionality $\mathcal{F}_{com}$ that handles a single commitment-decommitment process, and functionality $\mathcal{F}_{mcom}$ that handles multiple such processes.. (Indeed, in the plain model functionality $\mathcal{F}_{mcom}$ would be redundant, since one can use the composition theorem to obtain protocols that securely realize $\mathcal{F}_{mcom}$ from any protocol that securely realizes $\mathcal{F}_{com}$. However, in the CRS model realizing $\mathcal{F}_{mcom}$ is considerably more challenging than realizing $\mathcal{F}_{com}$.) Some further discussion on the functionalities and possible variants appears in [CF01].

Both functionalities are presented as *bit* commitments. Commitments to strings can be obtained in a natural way using the composition theorem. It is also possible, in principle, to generalize $\mathcal{F}_{com}$ and $\mathcal{F}_{mcom}$ to allow commitment to strings. Such extensions may be realized by string-commitment protocols that are more efficient than straightforward composition of bit commitment protocols. Finding such protocols is an interesting open problem.

Functionality $\mathcal{F}_{com}$

$\mathcal{F}_{com}$ proceeds as follows, running with parties $P_1, ..., P_n$ and an adversary $\mathcal{S}$.

1. Upon receiving a value $(\texttt{Commit}, sid, P_i, P_j, b)$ from P_i, where $b \in \{0,1\}$, record the value b and send the message $(\texttt{Receipt}, sid, P_i, P_j)$ to P_j and $\mathcal{S}$. Ignore any subsequent Commit messages.
2. Upon receiving a value $(\texttt{Open}, sid, P_i, P_j)$ from P_i, proceed as follows: If some value b was previously recoded, then send the message $(\texttt{Open}, sid, P_i, P_j, b)$ to P_j and $\mathcal{S}$ and halt. Otherwise halt.

Fig. 1. The Ideal Commitment functionality for a single commitment

Functionality $\mathcal{F}_{com}$, described in Figure 1, proceeds as follows. The commitment phase is modeled by having $\mathcal{F}_{com}$ receive a value $(\texttt{Commit}, sid, P_i, P_j, b)$, from some party P_i (the committer). Here *sid* is a Session ID used to distinguish among various copies of $\mathcal{F}_{com}$, P_j is the identity of another party (the receiver), and $b \in \{0,1\}$ is the value committed to. In response, $\mathcal{F}_{com}$ lets the receiver P_j *and the adversary* $\mathcal{S}$ know that P_i has committed to some value, and that this value is associated with session ID *sid*. This is done by sending the message $(\texttt{Receipt}, sid, P_i, P_j)$ to P_j and $\mathcal{S}$. The opening phase is initiated by the committer sending a value $(\texttt{Open}, sid, P_i, P_j)$ to $\mathcal{F}_{com}$. In response, $\mathcal{F}_{com}$ hands the value $(\texttt{Open}, sid, P_i, P_j, b)$ to P_j and $\mathcal{S}$.

Functionality $\mathcal{F}_{mcom}$, presented in Figure 2, essentially mimics the operation of $\mathcal{F}_{com}$ for an unbounded number of times. In addition to the session ID *sid*, functionality $\mathcal{F}_{mcom}$ uses an additional identifier, a Commitment ID *cid*, that is used to distinguish among the different commitments that take place within a single run of $\mathcal{F}_{mcom}$. The record for a committed value now includes the Commitment ID, plus the identities of the committer and receiver. To avoid ambiguities, no two commitments with the same committer and verifier are allowed to have the same Commitment ID. It is stressed that the various Commit and Open requests may be interleaved in an arbitrary way. Also, note that $\mathcal{F}_{mcom}$ allows a committer to open a commitment several times (to the same receiver).

Functionality $\mathcal{F}_{mcom}$

$\mathcal{F}_{mcom}$ proceeds as follows, running with parties $P_1, ..., P_n$ and an adversary $\mathcal{S}$.

1. Upon receiving a value $(\texttt{Commit}, sid, cid, P_i, P_j, b)$ from P_i, where $b \in \{0,1\}$, record the tuple (cid, P_i, P_j, b) and send the message $(\texttt{Receipt}, sid, cid, P_i, P_j)$ to P_j and $\mathcal{S}$. Ignore subsequent $(\texttt{Commit}, sid, cid, P_i, P_j, ...)$ values.
2. Upon receiving a value $(\texttt{Open}, sid, cid, P_i, P_j)$ from P_i, proceed as follows: If the tuple (cid, P_i, P_j, b) is recorded then send the message $(\texttt{Open}, sid, cid, P_i, P_j, b)$ to P_j and $\mathcal{S}$. Otherwise, do nothing.

Fig. 2. The Ideal Commitment functionality for multiple commitments

Definition 3. *A protocol is a* universally composable (UC) commitment *protocol if it securely realizes functionality $\mathcal{F}_{com}$. If the protocol securely realizes $\mathcal{F}_{mcom}$ then it is called a* reusable-CRS UC *commitment protocol.*

Remark: On duplicating commitments. Notice that functionalities $\mathcal{F}_{com}$ and $\mathcal{F}_{mcom}$ disallow "copying commitments". That is, assume that party A commits to some value x for party B, and that the commitment protocol in use allows B to commit to the same value x for some party C, before A decommitted to x. Once A decommits to x for B, B will decommit to x for C. Then this protocol does not securely realize $\mathcal{F}_{com}$ or $\mathcal{F}_{mcom}$. This requirement may seem hard to enforce at first, since B can always play "man in the middle" (i.e., forward A's messages to C and C's messages to A.) We enforce it using the unique identities of the parties. (Recall that unique identities are assumed to be provided via an underlying lower-level protocol that also guarantees authenticated communication.)

3 Universally Composable Commitment Schemes

We present two constructions of UC commitment protocols in the common reference string model. The protocol presented in Section 3.1 securely realizes functionality $\mathcal{F}_{com}$, i.e. each part of the public string can only be used for a single commitment. It is based on any trapdoor permutation. The protocol presented in Section 3.2 securely realizes $\mathcal{F}_{mcom}$, i.e. it reuses the public string for multiple commitments. This protocol requires potentially stronger assumptions (either existence of claw-free pairs of trapdoor permutations or alternatively the hardness of discrete log).

3.1 One-Time Common Reference String

The construction in this section works in the common random string model where each part of the commitment can be used only once for a commitment. It is based on the equivocable bit commitment scheme of Di Crescenzo et al. [DIO98], which in turn is a clever modification of Naor's commitment scheme [N91].

Let G be a pseudorandom generator stretching n-bit inputs to $4n$-bit outputs. For security parameter n the receiver in [N91] sends a random $4n$-bit string σ to the sender, who picks a random $r \in \{0,1\}^n$, computes $G(r)$ and returns $G(r)$ or $G(r) \oplus \sigma$ to commit to 0 and 1, respectively. To decommit, the sender transmits b and r. By the pseudorandomness of G the receiver cannot distinguish both cases, and with probability 2^{-2n} over the choice of σ it is impossible to find openings r_0 and r_1 such that $G(r_0) = G(r_1) \oplus \sigma$.

In [DIO98] an equivocable version of Naor's scheme has been proposed. Suppose that σ is not chosen by the receiver, but rather is part of the common random string. Then, if instead we set $\sigma = G(r_0) \oplus G(r_1)$ for random r_0, r_1, and let the sender give $G(r_0)$ to the receiver, it is later easy to open this commitment as 0 with r_0 as well as 1 with r_1 (because $G(r_0) \oplus \sigma = G(r_1)$). On the other hand, choosing σ in that way in indistinguishable from a truly random choice.

We describe a UC bit commitment protocol $\mathsf{UCC}_{\mathsf{OneTime}}$ (for universally composable commitment scheme in the one-time-usable common reference string model). The idea is to use the [DIO98] scheme with a special pseudorandom generator, namely, the Blum-Micali-Yao generator based on any trapdoor permutation [Y82,BM84]. Let KGen denote an efficient algorithm that on input 1^n generates a random public key pk and the trapdoor td. The key pk describes a trapdoor permutation f_{pk} over $\{0,1\}^n$. Let $B(\cdot)$ be a hard core predicate for f_{pk}. Define a pseudorandom generator expanding n bits to $4n$ bits with public description pk by

$$G_{pk}(r) = \left(f_{pk}^{(3n)}(r), B\big(f_{pk}^{(3n-1)}(r)\big), \ldots, B\big(f_{pk}(r)\big), B(r)\right)$$

where $f_{pk}^{(i)}(r)$ is the i-th fold application of f_{pk} to r. An important feature of this generator is that given the trapdoor td to pk it is easy to recognize images $y \in \{0,1\}^{4n}$ under G_{pk}.

The public random string in our scheme consists of a random $4n$-bit string σ, together with two public keys pk_0, pk_1 describing trapdoor pseudorandom generators G_{pk_0} and G_{pk_1}; both generators stretch n-bit inputs to $4n$-bit output. The public keys pk_0, pk_1 are generated by two independent executions of the key generation algorithm KGen on input 1^n. Denote the corresponding trapdoors by td_0 and td_1, respectively.

In order to commit to a bit $b \in \{0,1\}$, the sender picks a random string $r \in \{0,1\}^n$, computes $G_{pk_b}(r)$, and sets $y = G_{pk_b}(r)$ if $b = 0$, or $y = G_{pk_b}(r) \oplus \sigma$ for $b = 1$. The sender passes y to the receiver. In the decommitment step the sender gives (b, r) to the receiver, who verifies that $y{=}G_{pk_b}(r)$ for $b = 0$ or that $y = G_{pk_b}(r) \oplus \sigma$ for $b = 1$. See also Figure 3.

Commitment scheme $\mathsf{UCC}_{\mathsf{OneTime}}$

public string:

σ — random string in $\{0,1\}^{4n}$
pk_0, pk_1 — keys for generators $G_{pk_0}, G_{pk_1} : \{0,1\}^n \to \{0,1\}^{4n}$

commitment for $b \in \{0,1\}$ *with SID sid:*

compute $G_{pk_b}(r)$ for random $r \in \{0,1\}^n$
set $y = G_{pk_b}(r)$ for $b = 0$, or $y = G_{pk_b}(r) \oplus \sigma$ for $b = 1$
send $(\mathtt{Com}, sid, y)$ to receiver

decommitment for y*:*

send b, r to receiver
receiver checks $y \stackrel{?}{=} G_{pk_b}(r)$ for $b = 0$, or $y \stackrel{?}{=} G_{pk_b}(r) \oplus \sigma$ for $b = 1$

Fig. 3. Commitment Scheme in the One-Time-Usable Common Reference String Model

Clearly, the scheme is computationally hiding and statistically binding. An important observation is that our scheme inherits the equivocability property of [DIO98]. In a simulation we replace σ by $G_{pk_0}(r_0) \oplus G_{pk_1}(r_1)$ and therefore, if we impersonate the sender and transmit $y = G_{pk}(r_0)$ to a receiver, then we can later open this value with 0 by sending r_0 and with 1 via r_1.

Moreover, if we are given a string y^*, e.g., produced by the adversary, and we know the trapdoor td_0 to pk_0, then it is easy to check if y^* is an image under G_{pk_0} and therefore represents a 0-commitment. Unless y^* belongs to G_{pk_0} and, simultaneously, $y^* \oplus \sigma$ belongs to G_{pk_1}, the encapsulated bit is unique and we can extract the correct value with td_0. (We stress, however, that this property will not be directly used in the proof. This is so since there the CRS has a different distribution, so a more sophisticated argument is needed.)

To summarize, our commitment scheme supports equivocability and extraction. The proof of the following theorem appears in [CF01].

Theorem 2. *Protocol* $\mathsf{UCC}_{\mathsf{OneTime}}$ *securely realizes functionality* $\mathcal{F}_{com}$ *in the* CRS *model.*

3.2 Reusable Common Reference String

The drawback of the construction in the previous section is that a fresh part of the random string must be reserved for each committed bit. In this section, we overcome this disadvantage under a potentially stronger assumption, namely the existence of claw-free trapdoor permutation pairs. We concentrate on a solution that only works for erasing parties in general, i.e., security is based on the parties' ability to irrevocably erase certain data as soon as they are supposed to. At the end of this section we sketch a solution that does not require data erasures. This solution is based on the Decisional Diffie-Hellman assumption.

Basically, a claw-free trapdoor permutation pair is a pair of trapdoor permutations with a common range such that it is hard to find two elements that are preimages of the same element under the two permutations. More formally, a key generation $\mathsf{KGen}_{\mathsf{claw}}$ outputs a random public key pk_{claw} and a trapdoor td_{claw}. The public key defines permutations $f_{0,pk_{\mathsf{claw}}}, f_{1,pk_{\mathsf{claw}}} : \{0,1\}^n \to \{0,1\}^n$, whereas the secret key describes the inverse functions $f^{-1}_{0,pk_{\mathsf{claw}}}, f^{-1}_{1,pk_{\mathsf{claw}}}$. It should be infeasible to find a claw x_0, x_1 with $f_{0,pk_{\mathsf{claw}}}(x_0) = f_{1,pk_{\mathsf{claw}}}(x_1)$ given only pk_{claw}. For ease of notation we usually omit the keys and write $f_0, f_1, f_0^{-1}, f_1^{-1}$ instead. Claw-free trapdoor permutation pairs exist for example under the assumption that factoring is hard [GMRi88]. For a more formal definition see [G95].

We also utilize an encryption scheme $\mathcal{E} = (\mathsf{KGen}, \mathsf{Enc}, \mathsf{Dec})$ secure against adaptive-chosen ciphertext attacks, i.e., in the notation of [BDPR98] the encryption system should be IND-CCA2. On input 1^n the key generation algorithm KGen returns a public key $pk_{\mathcal{E}}$ and a secret key $sk_{\mathcal{E}}$. An encryption of a message m is given by $c \leftarrow \mathsf{Enc}_{pk_{\mathcal{E}}}(m)$, and the decryption of a ciphertext c is $\mathsf{Dec}_{sk_{\mathcal{E}}}(c)$. It should always hold that $\mathsf{Dec}_{sk_{\mathcal{E}}}(c) = m$ for $c \leftarrow \mathsf{Enc}_{pk_{\mathcal{E}}}(m)$, i.e., the system supports errorless decryption. Again, we abbreviate $\mathsf{Enc}_{pk_{\mathcal{E}}}(\cdot)$ by $\mathsf{Enc}(\cdot)$ and $\mathsf{Dec}_{sk_{\mathcal{E}}}(\cdot)$ by $\mathsf{Dec}(\cdot)$. IND-CCA2 encryption schemes exist for example under the assumption that trapdoor permutations exist [DDN00]. A more efficient solution is based on the decisional Diffie-Hellman assumption [CS98]. Both schemes have errorless decryption.

The commitment scheme $\mathsf{UCC}_{\mathsf{ReUse}}$ (for universally composable commitment with reusable reference string) is displayed in Figure 4. The (reusable) public string contains random public keys pk_{claw} and $pk_{\mathcal{E}}$. For a commitment to a bit b the sender P_i applies the trapdoor permutation f_b to a random $x \in \{0,1\}^n$, computes $c_b \leftarrow \mathsf{Enc}_{pk_{\mathcal{E}}}(x, P_i)$ and $c_{1-b} \leftarrow \mathsf{Enc}_{pk_{\mathcal{E}}}(0^n, P_i)$, and sends the tuple (y, c_0, c_1) with $y = f_b(x)$ to the receiver. The sender is also instructed to erase the randomness for the encryption of $(0^n, P_i)$ before the commitment message is sent. This ciphertext is called a dummy ciphertext.

To open the commitment, the committer P_i sends b, x and the randomness used for encrypting (x, P_i). The receiver P_j verifies that $y = f_b(x)$, that the encryption randomness is consistent with c_b, and that cid was never used before in a commitment of P_i to P_j.

We remark that including the sender's identity in the encrypted strings plays an important role in the analysis. Essentially, this precaution prevents a corrupted committer from "copying" a commitment generated by an uncorrupted party.

The fact that the dummy ciphertext is never opened buys us equivocability. Say that the ideal-model simulator knows the trapdoor of the claw-free permutation pair. Then it can compute the pre-images x_0, x_1 of some y under both functions f_0, f_1 and send y as well as encryptions of (x_0, P_i) and (x_1, P_i). To open it as 0 hand $0, x_0$ and the randomness for ciphertext (x_0, P_i) to the receiver and claim to have erased the randomness for the other encryption. For a 1-decommitment send $1, x_1$, the randomness for the encryption of (x_1, P_i) and

Commitment scheme $\mathsf{UCC}_{\mathsf{ReUse}}$

public string:

pk_{claw} — public key for claw-free trapdoor permutation pair f_0, f_1
$pk_{\mathcal{E}}$ — public key for encryption algorithm Enc

commitment by party P_i to party P_j to $b \in \{0,1\}$ with identifier sid, cid:

compute $y = f_b(x)$ for random $x \in \{0,1\}^n$;
compute $c_b \leftarrow \mathsf{Enc}(x, P_i)$ with randomness r_b;
compute $c_{1-b} \leftarrow \mathsf{Enc}(0^n, P_i)$ with randomness r_{1-b};
erase r_{1-b};
send $(\mathtt{Com}, sid, cid, (y, c_0, c_1))$, and record (sid, cid, b, x, r_b).
Upon receiving $(\mathtt{Com}, sid, cid, (y, c_0, c_1))$ from P_i,
P_j outputs $(\mathtt{Receipt}, sid, cid, P_i, P_j))$

decommitment for $(P_i, P_j, sid, cid, b, x, r_b)$:

Send $(\mathtt{Dec}, sid, cid, b, x, r_b)$ to P_j.
Upon receiving $(\mathtt{Dec}, sid, cid, b, x, r_b)$, P_j verifies that $y \stackrel{?}{=} f_b(x)$,
that c_b is encryption of (x, P_i) under randomness r_b
where P_i is the committer's identity
and that *cid* has not been used with this committer before.

Fig. 4. Commitment Scheme with Reusable Reference String

deny to know the randomness for the other ciphertext. If the encryption scheme is secure then it is intractable to distinguish dummy and such fake encryptions. Hence, this procedure is indistinguishable from the actual steps of the honest parties.

Analogously to the extraction procedure for the commitment scheme in the previous section, here an ideal-process adversary can also deduce the bit from an adversarial commitment (y^*, c_0^*, c_1^*) if it knows the secret key of the encryption scheme. Specifically, decrypt c_0^* to obtain (x_0^*, P_i^*); if x_0^* maps to y^* under f_0 then let the guess be 0, else predict 1. This decision is only wrong if the adversary has found a claw, which happens only with negligible probability. The proof of the following theorem appears in [CF01].

Theorem 3. *Protocol $\mathsf{UCC}_{\mathsf{ReUse}}$ securely realizes functionality $\mathcal{F}_{mcom}$ in the* CRS *model.*

A solution for non-erasing parties. The security of the above scheme depends on the ability and good-will of parties to securely erase sensitive data (specifically, to erase the randomness used to generate the dummy ciphertext). A careful look shows that it is possible to avoid the need to erase: It is sufficient to be able to generate a ciphertext without knowing the plaintext. Indeed, it would be enough to enable the parties to obliviously generate a string that is indistinguishable from a ciphertext. Then the honest parties can use this mechanism to produce the dummy ciphertext, while the simulator is still able to place the fake encryption into the commitment. For example, the Cramer-Shoup system in subgroup G of $\mathcal{Z}_p^*$ has this property under the decisional Diffie-Hellman assumption: To generate a dummy ciphertext simply generate four random elements in G.

Relaxing the need for claw-free pairs. The above scheme was presented and proven using any claw-free pair of trapdoor permutations. However, it is easy to see that the claw-free pair can be substituted by chameleon commitments a la [BCC88], thus basing the security of the scheme on the hardness of the discrete logarithm or factoring. Further relaxing the underlying hardness assumptions is an interesting task.

4 Impossibility of UC Commitments in the Plain Model

This section demonstrates that in the plain model there cannot exist universally composable commitment protocols that do not involve third parties in the interaction and allow for successful completion when both the sender and the receiver are honest. This impossibility result holds even under the more liberal requirement that for any real-life adversary and any environment there should be an ideal-model adversary (i.e., under a relaxed definition where the ideal-model simulator may depend on the environment).

We remark that universally composable commitment protocols exist in the plain model if the protocol makes use of third parties, as long as a majority of the parties remain uncorrupted. This follows from a general result in [c00a], where it is shown that practically any functionality can be realized in this setting.

Say that a protocol π between n parties $P_1, \ldots, P_n$ is *bilateral* if all except two parties stay idle and do not transmit messages. A bilateral commitment protocol π is called *terminating* if, with non-negligible probability, the receiver P_j accepts a commitment of the honest sender P_i and outputs $(\texttt{Receipt}, sid, P_i, P_j)$, and moreover if the receiver, upon getting a valid decommitment for a message m and sid from the honest sender, outputs $(\texttt{Open}, sid, P_i, P_j, m)$ with non-negligible probability.

Theorem 4. *There exists no bilateral, terminating protocol π that securely realizes functionality $\mathcal{F}_{\text{com}}$ in the plain model. This holds even if the ideal-model adversary $\mathcal{S}$ is allowed to depend on the environment $\mathcal{Z}$.*

Proof. The idea of the proof is as follows. Consider a protocol execution between an adversarially controlled committer P_i and an honest receiver P_j, and assume that the adversary merely sends messages that are generated by the environment. The environment secretly picks a random bit b at the beginning and generates the messages for P_i by running the protocol of the honest committer for b and P_j's answers. In order to simulate this behavior, the ideal-model adversary $\mathcal{S}$ must be able to provide the ideal functionality with a value for the committed bit. For this purpose, the simulator has to "extract" the committed bit from the messages generated by the environment, without the ability to rewind the environment. However, as will be seen below, if the commitment scheme allows the simulator to successfully extract the committed bit, then the commitment is not secure in the first place (in the sense that a corrupted receiver can obtain the value of the committed bit from interacting with an honest committer).

More precisely, let the bilateral protocol π take place between the sender P_i and the receiver P_j. Consider the following environment $\mathcal{Z}$ and real-life adversary $\mathcal{A}$. At the outset of the execution the adversary $\mathcal{A}$ corrupts the committer P_i. Then, in the sequel, $\mathcal{A}$ has the corrupted committer send every message it receives from $\mathcal{Z}$, and reports any reply received by P_j to $\mathcal{Z}$. The environment $\mathcal{Z}$ secretly picks a random bit b and follows the program of the honest sender to commit to b, as specified by π. Once the the honest receiver has acknowledged the receipt of a commitment, $\mathcal{Z}$ lets $\mathcal{A}$ decommit to b by following protocol π. Once the receiver outputs $(\texttt{Open}, sid, P_i, P_j, b')$, $\mathcal{Z}$ outputs 1 if $b = b'$ and outputs 0 otherwise.

Formally, suppose that there is an ideal-model adversary $\mathcal{S}$ such that $\mathrm{REAL}_{\pi,\mathcal{A},\mathcal{Z}} \approx \mathrm{IDEAL}_{\mathcal{F}_{com},\mathcal{S},\mathcal{Z}}$. Then we construct a new environment $\mathcal{Z}'$ and a new real-life adversary $\mathcal{A}'$ for which there is no appropriate ideal-model adversary for π. This time, $\mathcal{A}'$ corrupts the receiver P_j at the beginning. During the execution $\mathcal{A}'$ obtains messages form the honest committer P_i and feeds these messages into a virtual copy of $\mathcal{S}$. The answers of $\mathcal{S}$, made on behalf of an honest receiver, are forwarded to P_i in the name of the corrupted party P_j. At some point, $\mathcal{S}$ creates a submission $(\texttt{Commit}, sid, P_i, P_j, b')$ to $\mathcal{F}_{com}$; the adversary $\mathcal{A}'$ outputs b' and halts. If $\mathcal{S}$ halts without creating such a submission then $\mathcal{A}'$ outputs a random bit and halts.

The environment $\mathcal{Z}'$ instructs the honest party P_i to commit to a randomly chosen secret bit b. (No decommitment is ever carried out.) Conclusively, $\mathcal{Z}'$ outputs 1 iff the adversary's output b' satisfies $b = b'$.

By the termination property, we obtain from the virtual simulator $\mathcal{S}$ a bit b' with non-negligible probability. This bit is a good approximation of the actual bit b, since $\mathcal{S}$ simulates the real protocol π except with negligible error. Hence, the guess of $\mathcal{A}'$ for b is correct with 1/2 plus a non-negligible probability. But for a putative ideal-model adversary $\mathcal{S}'$ predicting this bit b with more than non-negligible probability over 1/2 is impossible, since the view of $\mathcal{S}'$ in the ideal process is statistically independent from the bit b. (Recall that the commitment to b is never opened).

5 Application to Zero-Knowledge

In order to exemplify the power of UC commitments we show how they can be used to construct simple Zero-Knowledge (ZK) protocols with strong security properties. Specifically, we formulate an ideal functionality, $\mathcal{F}_{zk}$, that captures the notion of Zero-Knowledge in a very strong sense. (In fact, $\mathcal{F}_{zk}$ implies concurrent and non-malleable Zero-Knowledge proofs of knowledge.) We then show that in the $\mathcal{F}_{com}$-hybrid model (i.e., in a model with ideal access to $\mathcal{F}_{com}$) there is a 3-round protocol that securely realizes $\mathcal{F}_{zk}$ with respect to any NP relation. Using the composition theorem of [c00a], we can replace $\mathcal{F}_{com}$ with any UC commitment protocol. (This of course requires using the CRS model, unless we involve third parties in the interaction. Also, using functionality $\mathcal{F}_{mcom}$ instead of $\mathcal{F}_{com}$ is possible and results in a more efficient use of the common string.)

Functionality $\mathcal{F}_{zk}$, described in Figure 5, is parameterized by a binary relation $R(x, w)$. It first waits to receive a message $(\texttt{verifier}, id, P_i, P_j, x)$ from some party P_i, interpreted as saying that P_i wants P_j to prove to P_i that it knows a value w such that $R(x, w)$ holds. Next, $\mathcal{F}_{zk}$ waits for P_j to explicitly provide a value w, and notifies P_i whether $R(x, w)$ holds. (Notice that the adversary is notified whenever either the prover or the verifier starts an interaction. It is also notified whether the verifier accepts. This represents the fact that ZK is not traditionally meant to hide this information.)

Functionality $\mathcal{F}_{zk}$

$\mathcal{F}_{zk}$ proceeds as follows, running with parties $P_1, ..., P_n$ and an adversary $\mathcal{S}$. The functionality is parameterized by a binary relation R.

1. Wait to receive a value $(\texttt{verifier}, id, P_i, P_j, x)$ from some party P_i. Once such a value is received, send $(\texttt{verifier}, id, P_i, P_j, x)$ to $\mathcal{S}$, and ignore all subsequent $(\texttt{verifier}\ldots)$ values.
2. Upon receipt of a value $(\texttt{prover}, id, P_j, P_i, x', w)$ from P_j, let $v = 1$ if $x = x'$ and $R(x, w)$ holds, and $v = 0$ otherwise. Send (id, v) to P_i and $\mathcal{S}$, and halt.

Fig. 5. The Zero-Knowledge functionality, $\mathcal{F}_{zk}$

We demonstrate a protocol for securely realizing $\mathcal{F}_{zk}^R$ for any NP relation R. The protocol is a known one: It consists of n parallel repetitions of the 3-round protocol of Blum for graph Hamiltonicity, where the provers commitments are replaced by invocations of $\mathcal{F}_{com}$. The protocol (in the $\mathcal{F}_{com}$-hybrid model) is presented in Figure 6.

We remark that in the $\mathcal{F}_{com}$-hybrid model the protocol securely realizes $\mathcal{F}_{zk}$ *without any computational assumptions,* and even if the adversary and the environment are computationally unbounded. (Of course, in order to securely realize $\mathcal{F}_{com}$ the adversary and environment must be computationally bounded.) Also, in the $\mathcal{F}_{com}$-hybrid model there is no need in a common reference string. That is, the CRS model is needed only for realizing $\mathcal{F}_{com}$.

Let $\mathcal{F}_{zk}^H$ denote functionality $\mathcal{F}_{zk}$ parameterized by the Hamiltonicity relation H. (I.e., $H(G, h) = 1$ iff h is a Hamiltonian cycle in graph G.) The following theorem is proven in [CF01].

Theorem 5. *Protocol* HC *securely realizes $\mathcal{F}_{zk}^H$ in the $\mathcal{F}_{com}$-hybrid model.*

Acknowledgements. We thank Yehuda Lindell for suggesting to use non-malleable encryptions for achieving non-malleability of commitments in the common reference string model. This idea underlies our scheme that allows to reuse the common string for multiple commitments. (The same idea was independently suggested in [DKOS01].)

Protocol Hamilton-Cycle (HC)

1. Given input (`Prover`, id, P, V, G, h), where G is a graph over nodes $1, ..., n$, the prover P proceeds as follows. If h is not a Hamiltonian cycle in G, then P sends a message `reject` to V. Otherwise, P proceeds as follows for $k = 1, ..., n$:
   a) Choose a random permutation π_k over $[n]$.
   b) Using $\mathcal{F}_{com}$, commit to the edges of the permuted graph. That is, for each $(i, j) \in [n]^2$ send (`Commit`, $(i, j, k), P, V, e$) to $\mathcal{F}_{com}$, where $e = 1$ if there is an edge between $\pi_k(i)$ and $\pi_k(j)$ in G, and $e = 0$ otherwise.
   c) Using $\mathcal{F}_{com}$, commit to the permutation π_k. That is, for $l = 1, ..., L$ send (`Commit`, $(l, k), P, V, p_l$) to $\mathcal{F}_{com}$ where $p_1, ..., p_L$ is a representation of π_k in some agreed format.
2. Given input (`Verifier`, id, V, P, G), the verifier V waits to receive either `reject` from P, or (`Receipt`, $(i, j, k), P, V$) and (`Receipt`, $(l, k), P, V$) from $\mathcal{F}_{com}$, for $i, j, k = 1, ..., n$ and $l = 1, ..., L$. If `reject` is received, then V output 0 and halts. Otherwise, once all the (`Receipt`,...) messages are received V randomly chooses n bits $c_1, ..., c_n$ and sends to P.
3. Upon receiving $c_1, ..., c_n$ from V, P proceeds as follows for $k = 1, ..., n$:
   a) If $c_k = 0$ then send (`Open`, $(i, j, k), P, V$) and (`Open`, $(l, k), P, V$) to $\mathcal{F}_{com}$ for all $i, j = 1, ..., n$ and $l = 1, ..., L$.
   b) If $c_k = 1$ then send (`Open`, $(i, j, k), P, V$) to $\mathcal{F}_{com}$ for all $i, j = 1, ..., n$ such that the edge $\pi_k(i), \pi_k(j)$ is in the cycle h.
4. Upon receiving the appropriate (`Open`,...) messages from $\mathcal{F}_{com}$, the verifier V verifies that for all k such that $c_k = 0$ the opened edges agree with the input graph G and the opened permutation π_k, and for all k such that $c_k = 1$ the opened edges are all 1 and form a cycle. If verification succeeds then output 1, otherwise output 0.

Fig. 6. The protocol for proving Hamiltonicity in the $\mathcal{F}_{com}$-hybrid model

References

[B91] D. Beaver, "Secure Multi-party Protocols and Zero-Knowledge Proof Systems Tolerating a Faulty Minority", J. Cryptology, Springer-Verlag, (1991) 4: 75-122.

[B99] D. Beaver, "Adaptive Zero-Knowledge and Computational Equivocation", *28th Symposium on Theory of Computing (STOC),* ACM, 1996.

[BBM00] M. Bellare, A. Boldyreva and S. Micali, "Public-Key Encryption in a Multi-user Setting: Security Proofs and Improvements," *Eurocrypt 2000, pp. 259–274, Springer LNCS 1807,* 2000.

[BDJR97] M Bellare, A. Desai, E. Jokipii and P. Rogaway, "A concrete security treatment of symmetric encryption: Analysis of the DES modes of operations," 38*th Annual Symp. on Foundations of Computer Science (FOCS),* IEEE, 1997.

[BDPR98] M. Bellare, A. Desai, D. Pointcheval and P. Rogaway, "Relations among notions of security for public-key encryption schemes", *CRYPTO '98,* 1998, pp. 26-40.

[BM84] M.Blum, S.Micali: How to Generate Cryptographically Strong Sequences of Pseudorandom Bits, *SIAM Journal on Computation, Vol. 13, pp. 850–864*, 1984.

[BCC88] G. Brassard, D. Chaum and C. Crépeau. Minimum Disclosure Proofs of Knowledge. *JCSS*, Vol. 37, No. 2, pages 156–189, 1988.

[C00] R. Canetti, "Security and composition of multi-party cryptographic protocols", *Journal of Cryptology,* Vol. 13, No. 1, winter 2000.

[C00a] R. Canetti, "A unified framework for analyzing security of Protocols", manuscript, 2000. Available at http://eprint.iacr.org/2000/067.

[CF01] R. Canetti and M. Fischlin, "Universally Composable Commitments". Available at http://eprint.iacr.org/2001.

[CS98] R. Cramer and V. Shoup, "A paractical public-key cryptosystem provably secure against adaptive chosen ciphertext attack", *CRYPTO '98,* 1998.

[D89] I. Damgard, On the existence of bit commitment schemes and zero-knowledge proofs, Advances in Cryptology - Crypto '89, pp. 17–29, 1989.

[D00] I. Damgard. Efficient Concurrent Zero-Knowledge in the Auxiliary String Model. *Eurocrypt 00*, LNCS, 2000.

[DIO98] G. Di Crescenzo, Y. Ishai and R. Ostrovsky, Non-interactive and non-malleable commitment, *30th STOC*, 1998, pp. 141-150.

[DKOS01] G. Di Crecenzo, J. Katz, R. Ostrovsky and A. Smith. Efficient and Perfectly-Hiding Non-Interactive, Non-Malleable Commitment. *Eurocrypt '01,* 2001.

[DM00] Y. Dodis and S. Micali, "Secure Computation", *CRYPTO '00,* 2000.

[DDN00] D. Dolev, C. Dwork and M. Naor, Non-malleable cryptography, *SIAM.. J. Computing,* Vol. 30, No. 2, 2000, pp. 391-437. Preliminary version in *23rd Symposium on Theory of Computing (STOC),* ACM, 1991.

[DNRS99] C. Dwork, M. Naor, O. Reingold, and L. Stockmeyer. Magic functions. In *40th Annual Symposium on Foundations of Computer Science*, pages 523–534. IEEE, 1999.

[FS90] U. Feige and A. Shamir. Witness Indistinguishability and Witness Hiding Protocols. In *22nd STOC*, pages 416–426, 1990.

[FF00] M. Fischlin and R. Fischlin, "Efficient non-malleable commitment schemes", *CRYPTO '00, LNCS 1880*, 2000, pp. 413-428.

[GHY88] Z. Galil, S. Haber and M. Yung, Cryptographic computation: Secure faut-tolerant protocols and the public-key model, *CRYPTO '87, LNCS 293*, Springer-Verlag, 1988, pp. 135-155.

[G95] O. Goldreich, *"Foundations of Cryptography (Fragments of a book)"*, Weizmann Inst. of Science, 1995. (Avaliable at http://philby.ucsd.edu)

[G98] O. Goldreich. *"Secure Multi-Party Computation"*, 1998. (Avaliable at http://philby.ucsd.edu)

[GMW91] O. Goldreich, S. Micali and A. Wigderson, "Proofs that yield nothing but their validity or All Languages in NP Have Zero-Knowledge Proof Systems", *Journal of the ACM,* Vol 38, No. 1, ACM, 1991, pp. 691–729. Preliminary version in *27th Symp. on Foundations of Computer Science (FOCS),* IEEE, 1986, pp. 174-187.

[GMW87] O. Goldreich, S. Micali and A. Wigderson, "How to Play any Mental Game", *19th Symposium on Theory of Computing (STOC),* ACM, 1987, pp. 218-229.

[GL90] S. Goldwasser, and L. Levin, "Fair Computation of General Functions in Presence of Immoral Majority", *CRYPTO '90, LNCS 537,* Springer-Verlag, 1990.

[GMRa89] S. Goldwasser, S. Micali and C. Rackoff, "The Knowledge Complexity of Interactive Proof Systems", *SIAM Journal on Comput.*, Vol. 18, No. 1, 1989, pp. 186-208.

[GMRi88] S.Goldwasser, S.Micali, R.Rivest: A Digital Signature Scheme Secure Against Adaptive Chosen-Message Attacks, *SIAM Journal on Computing, Vol. 17, No. 2, pp. 281–308*, 1988.

[L00] Y. Lindell, private communication, 2000.

[MR91] S. Micali and P. Rogaway, "Secure Computation", unpublished manuscript, 1992. Preliminary version in *CRYPTO '91, LNCS 576,* Springer-Verlag, 1991.

[N91] M.Naor: Bit Commitment Using Pseudo-Randomness, *Journal of Cryptology, vol. 4, pp. 151–158*, 1991.

[NOVY92] M. Naor, R. Ostrovsky, R. Venkatesan, and M. Yung, Perfect zero-knowledge arguments for NP can be based on general complexity assumptions, Advances in Cryptology - Crypto '92, pp. 196–214, 1992.

[PW94] B. Pfitzmann and M. Waidner, "A general framework for formal notions of secure systems", Hildesheimer Informatik-Berichte 11/94, Universität Hildesheim, 1994. Available at http://www.semper.org/sirene/lit.

[PW01] B. Pfitzmann and M. Waidner, "A model for asynchronous reactive systems and its application to secure message transmission", IEEE Symposium on Security and Privacy, 2001. See also IBM Research Report RZ 3304 (#93350), IBM Research, Zurich, December 2000.

[RS91] C. Rackoff and D. Simon, "Non-interactive zero-knowledge proof of knowledge and chosen ciphertext attack", *CRYPTO '91*, 1991.

[Y82] A. Yao, Theory and applications of trapdoor functions, In *Proc. 23rd Annual Symp. on Foundations of Computer Science (FOCS),* pages 80–91. IEEE, 1982.

Revocation and Tracing Schemes for Stateless Receivers*

Dalit Naor[1], Moni Naor[2]**, and Jeff Lotspiech[1]

[1] IBM Almaden Research Center
650 Harry Road, San-Jose, CA. 95120
{lots, dalit}@almaden.ibm.com
[2] Department of Computer Science and Applied Math
Weizmann Institute, Rehovot Israel.
naor@wisdom.weizmann.ac.il

Abstract. We deal with the problem of a center sending a message to a group of users such that some subset of the users is considered revoked and should not be able to obtain the content of the message. We concentrate on the *stateless receiver* case, where the users do not (necessarily) update their state from session to session. We present a framework called the *Subset-Cover* framework, which abstracts a variety of revocation schemes including some previously known ones. We provide sufficient conditions that guarantees the security of a revocation algorithm in this class.

We describe two explicit Subset-Cover revocation algorithms; these algorithms are very flexible and work for any number of revoked users. The schemes require storage at the receiver of $\log N$ and $\frac{1}{2}\log^2 N$ keys respectively (N is the total number of users), and in order to revoke r users the required message lengths are of $r \log N$ and $2r$ keys respectively. We also provide a general *traitor tracing* mechanism that can be integrated with any Subset-Cover revocation scheme that satisfies a "bifurcation property". This mechanism does not need an a priori bound on the number of traitors and does not expand the message length by much compared to the revocation of the same set of traitors.

The main improvements of these methods over previously suggested methods, when adopted to the stateless scenario, are: (1) reducing the message length to $O(r)$ *regardless* of the coalition size while maintaining a single decryption at the user's end (2) provide a seamless integration between the revocation and tracing so that the tracing mechanisms does not require any change to the revocation algorithm.

Keywords: Broadcast Encryption, Revocation scheme, Tracing scheme, Copyright Protection.

---

* A full version of the paper is available at the IACR Crypto Archive http://eprint.iacr.org/ and at http://www.wisdom.weizmann.ac.il/~naor/

** Work done while the author was visiting IBM Almaden Research Center and Stanford University. Partially supported by DARPA contract F30602-99-1-0530.

J. Kilian (Ed.): CRYPTO 2001, LNCS 2139, pp. 41–62, 2001.

1 Introduction

The problem of a Center transmitting data to a large group of receivers so that only a predefined subset is able to decrypt the data is at the heart of a growing number of applications. Among them are pay-TV applications, multicast communication, secure distribution of copyright-protected material (e.g. music) and audio streaming. The area of Broadcast Encryption deals with methods to efficiently broadcast information to a dynamically changing group of users who are allowed to receive the data. It is often convenient to think of it as a Revocation Scheme, which addresses the case where some subset of the users are excluded from receiving the information. In such scenarios it is also desirable to have a Tracing Mechanism, which enables the efficient tracing of leakage, specifically, the source of keys used by illegal devices, such as pirate decoders or clones.

One special case is when the receivers are stateless. In such a scenario, a (legitimate) receiver is not capable of recording the past history of transmissions and change its state accordingly. Instead, its operation must be based on the current transmission and its initial configuration. Stateless receivers are important for the case where the receiver is a device that is not constantly on-line, such as a media player (e.g. a CD or DVD player where the "transmission" is the current disc), a satellite receiver (GPS) and perhaps in multicast applications. The stateless scenario is particularly relevant to the application of Copyright Protection.

This paper introduces very efficient revocation schemes which are especially suitable for stateless receivers. Our approach is quite general. We define a framework of such algorithms, called Subset-Cover algorithms, and provide a sufficient condition for an algorithm in this family to be secure. We suggest two particular constructions of schemes in this family; the performance of the second method is substantially better than any previously known algorithm for this problem (see Section 1.1). We also provide a general property ('bifurcation') of revocation algorithms in our framework that allows efficient tracing methods, *without* modifying the underlying revocation scheme.

Notation: Let N be the total number of users in the system let r be the size of the revoked set $\mathcal{R}$.

1.1 Related Work

Broadcast Encryption. The area of Broadcast Encryption was first formally studied (and coined) by Fiat and Naor in [12] and has received much attention since then. To the best of our knowledge the scenario of stateless receivers has not been considered explicitly in the past in a scientific paper. In principle any scheme that works for the connected mode, where receivers can remember past communication, may be converted to a scheme for stateless receivers (such a conversion may require to include with any transmission the entire 'history' of revocation events). Hence, when discussing previously proposed schemes we will consider their performance as adapted to the stateless receiver scenario.

A parameter that was often considered is t, the upper bound on the size of the coalition an adversary can assemble. The algorithms in this paper do not require such a bound and we can think of $t = r$; on the other hand some previously proposed schemes depend on t but are independent of r. The Broadcast Encryption method of [12] allows the removal of any number of users as long as at most t of them collude; the message length is $O(t \log^2 t)$, a user must store a number of keys that is logarithmic in t and is required to perform $\tilde{O}(r/t)$ decryptions.

The logical-tree-hierarchy (LKH) scheme, suggested independently by Wallner et al. [29] and Wong et al. [30], is designed for the connected mode for multicast applications. If used in the stateless scenario it requires to transmit $2r \log N$, store $\log N$ keys at each user and perform $r \log N$ encryptions (these bounds are somewhat improved in [5,6,20]). The key assignment of this scheme and the key assignment of our first method are similar (see Sect. 3.1 for comparison).

Luby and Staddon [19] considered the information theoretic setting and devised bounds for any revocation algorithms under this setting. Their "Or Protocol" fits our Subset-Cover framework; our second algorithm (the Subset Difference method) which is *not* information theoretic, beats their lower bound (Theorem 12 in [19]). In Garay et al. [16] keys of compromised decoders are no longer used and the scheme is adapted so as to maintain security for the good users. The method of Kumar et. al. [18] enables one-time revocation of up to r users with message lengths of $O(r \log N)$ and $O(r^2)$. CPRM [10] is one of the methods that explicitly considers the stateless scenario.

Tracing Mechanisms. Tracing systems, introduced by Chor et al. [8] and later refined to the Threshold Traitor model [23], [9], distribute decryption keys to the users so as to allow the detection of at least one 'identity' of a key that is used in a pirate box which was constructed using keys of at most t users. *Black-box tracing* assumes that only the outcome of the decoding box can be examined. The construction of [23] guarantees tracing with high probability; it required $O(t \log N)$ keys at each user, a single decryption operation and message length is $4t$. The public key tracing scheme of Boneh and Franklin [3] provides a number-theoretic deterministic method for tracing. Note that in all of the above methods t is an *a-priori* bound. Another notion, the one of *Content Tracing*, attempts to detect illegal users who redistribute the content *after* it is decoded (see [4,13,2,26]).

Integration of tracing and revocation. Broadcast encryption can be combined with tracing schemes to yield trace-and-revoke schemes[1], a powerful approach to prevent illegal leakage of keys (others include the *legal* approach [25] and the *self enforcement* approach [11]). While Gafni et al. [15] and Stinson and Wei [28] consider combinatorial constructions, the schemes in Naor and Pinkas [24] are computational constructions and hence more general. The previously best known

[1] However it is *not* the case that every system which enables revocation and enables tracing is a trace-and-revoke scheme.

trace-and-revoke algorithm of [24] can tolerate a coalition of at most t users. It requires to store $O(t)$ keys at each user and to perform $O(r)$ decryptions; the message length is r keys, however these keys are elements in a group where the Decisional Diffie-Hellman problem is difficult, and hence these keys are longer than symmetric ones. The tracing model of [24] is not a "pure" black-box model. (Anzai et al. [1] employs a similar method for revocation, but without tracing capabilities.)

1.2 Summary of Results

In this paper we define a generic framework encapsulating several previously proposed revocation methods (e.g. the "Or Protocol" of [19]), called Subset-Cover algorithms. These algorithms are based on the principle of covering all non-revoked users by disjoint subsets from a predefined collection, together with a method for assigning (long-lived) keys to subsets in the collection. We define the security of a revocation scheme and provide a sufficient condition (key-indistinguishability) for a revocation algorithm in the Subset-Cover Framework to be secure. An important consequence of this framework is the separation between long-lived keys and short-term keys. The framework can be easily extended to the public-key scenario.

We provide two new instantiations of revocation schemes in the Subset-Cover Framework, with a different performance tradeoff (summarized in Table 3[2]). Both instantiations are tree-based, namely the subsets are derived from a virtual tree structure imposed on all devices in the system[3]. The first requires a message length of $r \log N$ and storage of $\log N$ keys at the receiver and constitutes a moderate improvement over previously proposed schemes; the second exhibits a substantial improvement: it requires a message length of $2r-1$ (in the worst case, or $1.38r$ in the average case) and storage of $\frac{1}{2}\log^2 N$ keys at the receiver. This improvement is (provably) due to the fact that the key assignment is computational and not information theoretic (for the information theoretic case there exists a lower bound which exhibits its limits, see [21]). Furthermore, these algorithms are r-flexible, namely they do not assume an upper bound of the number of revoked receivers.

Thirdly, we present a tracing mechanism that works in tandem with a Subset-Cover revocation scheme. We identify the *bifurcation property* for a Subset-Cover scheme. Our two constructions of revocation schemes posses this property. We show that every scheme that satisfies the bifurcation property can be combined with the tracing mechanism to yield a trace-and-revoke scheme. The integration

[2] Note that the comparison in the processing time between the two methods treats an application of a pseudo-random generator and a lookup operation as having the same cost, even though they might be quite different. More explicitly, the processing of both methods consists of $O(\log\log N)$ lookups; in addition, the Subset Difference method requires at most $\log N$ applications of a pseudo-random generator.

[3] An alternative view is to map the receivers to points on a line and the subsets as segments.

of the two mechanisms is seamless in the sense that no change is required for any one of them. Moreover, no a-priori bound on the number of traitors is needed for our tracing scheme. In order to trace t illegal users, the first revocation method requires a message length of $t \log N$, and the second revocation method requires a message length of $5t$.

Main Contributions: the main improvements that our methods achieve over previously suggested methods, when adopted to the stateless scenario, are:

- Reducing the message length to linear in r regardless of the coalition size, while maintaining a single decryption at the user's end. This applies also to the case where public keys are used, *without* a substantial length increase.
- The seamless integration between revocation and tracing: the tracing mechanism does not require any change of the revocation algorithm and no a priori bound on the number of traitors, even when all traitors cooperate among themselves.
- The rigorous treatment of the security of such schemes, identifying the effect of parameter choice on the overall security of the scheme.

Method	Message Length	Storage@Receiver	Processing time	decryptions
Complete Subtree	$r \log \frac{N}{r}$	$\log N$	$O(\log \log N)$	1
Subset Difference	$2r - 1$	$\frac{1}{2} \log^2 N$	$O(\log N)$	1

Fig. 1. Performance tradeoff for the Complete Subtree method and the Subset Difference method

Organization of the paper. Section 2 describes the framework for Subset-Cover algorithms and a sketch of the main theorem characterizing the security of a revocation algorithm in this family (the security is described in details in the full version of the paper). Section 3 describes two specific implementations of such algorithms. Section 3.3 gives an overview of few implementation issues, public-key methods and hierarchical revocation, as well as applications to copy protection and secure multicast. Section 4 provides a traitors-tracing algorithm that works for every revocation algorithm in the Subset-Cover framework and an improvement specifically suited for the Subset-Difference revocation algorithm.

2 The Subset-Cover Revocation Framework

2.1 Preliminaries - Problem Definition

Let $\mathcal{N}$ be the set of all users, $|\mathcal{N}| = N$, and $\mathcal{R} \subset \mathcal{N}$ be a group of $|\mathcal{R}| = r$ users whose decryption privileges should be revoked. The goal of a revocation algorithm is to allow a center to transmit a message M to all users such that any user $u \in \mathcal{N} \setminus \mathcal{R}$ can decrypt the message correctly, while even a coalition consisting of all members of $\mathcal{R}$ cannot decrypt it. The definition of the latter is provided in Sect. 2.3.

A system consists of three parts: (1) An initiation scheme, which is a method for assigning the receivers secret information that will allow them to decrypt. (2) The broadcast algorithm - given a message M and the set $\mathcal{R}$ of users that should be revoked outputs a ciphertext message M' that is broadcast to all receivers. (3) A decryption algorithm - a (non-revoked) user that receives ciphertext M' using its secret information should produce the original message M. Since the receivers are stateless, the output of the decryption should be based on the current message and the secret information only.

2.2 The Framework

We present a framework for algorithms which we call *Subset-Cover*. In this framework an algorithm defines a collection of subsets $S_1, \ldots, S_w$, $S_j \subseteq \mathcal{N}$. Each subset S_j is assigned (perhaps implicitly) a long-lived key L_j; each member u of S_j should be able to deduce L_j from its secret information. Given a revoked set $\mathcal{R}$, the remaining users $\mathcal{N} \setminus \mathcal{R}$ are partitioned into disjoint subsets $S_{i_1}, \ldots, S_{i_m}$ so that $\mathcal{N} \setminus \mathcal{R} = \bigcup_{j=1}^{m} S_{i_j}$ and a session key K is encrypted m times with $L_{i_1}, \ldots, L_{i_m}$.

Specifically, an algorithm in the framework uses two encryption schemes:

- A method $F_K : \{0,1\}^* \mapsto \{0,1\}^*$ to encrypt the message itself. The key K used will be chosen fresh for each message M - a session key - as a random bit string. F_K should be a fast method and should not expand the plaintext. The simplest implementation is to Xor the message M with a stream cipher generated by K.
- A method to deliver the session key to the receivers, for which we will employ an encryption scheme. The keys L here are long-lived. The simplest implementation is to make $E_L : \{0,1\}^\ell \mapsto \{0,1\}^\ell$ a block cipher.

A discussion of the security requirements of these primitives is given in Sect. 2.3. Suggestions for the implementation of F_K and E_L are outlined in Sect. 3.3 and given in [21]. The algorithm consists of three components:

Scheme Initiation: Every receiver u is assigned private information I_u. For all $1 \leq i \leq w$ such that $u \in S_i$, I_u allows u to deduce the key L_i corresponding to the set S_i. Note that the keys L_i can be chosen either (i) uniformly at random and independently from each other (which we call the *information-theoretic* case) or (ii) as a function of other (secret) information (which we call the *computational* case), and thus may not be independent of each other.

The **Broadcast algorithm** at the Center: The center chooses a session encryption key K. Given a set $\mathcal{R}$ of revoked receivers, it finds a partition $S_{i_1}, \ldots, S_{i_m}$ covering all users in $\mathcal{N} \setminus \mathcal{R}$. Let $L_{i_1}, \ldots, L_{i_m}$ be the keys associated with the above subsets. The center encrypts K with keys $L_{i_1}, \ldots, L_{i_m}$ and sends the ciphertext

$$\langle [i_1, i_2, \ldots, i_m, E_{L_{i_1}}(K), E_{L_{i_2}}(K), \ldots, E_{L_{i_m}}(K)], F_K(M) \rangle$$

The portion in square brackets preceding $F_K(M)$ is called the *header* and $F_K(M)$ is called the *body*.

The **Decryption step** at the receiver u, upon receiving a broadcast message $\langle[i_1, i_2, \ldots, i_m, C_1, C_2, \ldots, C_m], M'\rangle$: the receiver finds i_j such that $u \in S_{i_j}$ (in case $u \in \mathcal{R}$ the result is **null**). It then extracts the corresponding key L_{i_j} from I_u, computes $D_{L_{i_j}}(C_j))$ to obtain K and computes $D_K(M')$ to obtain and output M.

A particular implementation of such scheme is specified by (1) the collection of subsets $S_1, \ldots, S_w$ (2) the key assignment to each subset in the collection (3) a method to cover the non-revoked receivers $\mathcal{N} \setminus \mathcal{R}$ by disjoint subsets from this collection, and (4) A method that allows each user u to find its cover S and compute its key L_S from I_u. The algorithm is evaluated based upon three parameters:

1. Message Length - the length of the header that is attached to $F_K(M)$, which is proportional to m, the number of sets in the partition covering $\mathcal{N} \setminus \mathcal{R}$.
2. Storage size at the receiver - how much private information (typically, keys) does a receiver need to store. For instance, I_u could simply consists of all the keys S_i such that $u \in S_i$, or if the key assignment is more sophisticated it should allow the computation of all such keys.
3. Message processing time at receiver. We often distinguish between decryption and other types of operations.

It is important to characterize the dependence of the above three parameters in both N and r. Specifically, we say that a revocation scheme is *flexible with respect to* r if the storage at the receiver is not a function of r. Note that the efficiency of setting up the scheme and computing the partition (given $\mathcal{R}$) is not taken into account in the algorithm's analysis. However, for all schemes presented in this paper the computational requirements of the sender are rather modest: finding the partition takes time linear in $|\mathcal{R}| \log N$ and the encryption is proportional to the number of subsets in the partition. In this framework we demonstrate the substantial gain that can be achieved by using a computational key-assignment scheme as opposed to an information-theoretic one [4].

2.3 Security of the Framework

The definition of the Subset-Cover framework allows a rigorous treatment of the security of any algorithm in this family. Unfortunately, due to lack of space, this discussion must be omitted and is included in the full version of the paper [21]. A summary of this analysis follows.

Our contribution is twofold. We first define the notion of revocation-scheme security, namely specify the adversary's power in this scenario and what is considered a successful break. This roughly corresponds to an adversary that may pool the secret information of several users and may have some influence on the

[4] Note that since the assumptions on the security of the encryption primitives are computational, a computational key-assignment method is a natural.

choice of messages encrypted in this scheme (chosen plaintext). Also it may create bogus messages and see how legitimate users (that will not be revoked) react. Finally, to say that the adversary has broken the scheme means that when the users who have provided it their secret information are all revoked (otherwise it is not possible to protect the plaintext) the adversary can still learn something about the encrypted message. Here we define "learn" as distinguishing its encryption from random (this is equivalent to semantic security).

Second, we state the security assumptions on the primitives used in the scheme (these include the encryptions primitives E_L and F_K and the key assignment method in the subset-cover algorithm.) We identify a critical property that is required from the key-assignment method: a subset-cover algorithm satisfies the "key-indistinguishability" property if for every subset S_i its key L_i is *indistinguishable from a random key* given all the information of all users that are *not* in S_i. Note that any scheme in which the keys to all subsets are chosen independently (trivially) satisfies this property. To obtain our security theorem, we require two different sets of properties from E_L and F_K, since F_K uses short lived keys whereas E_L uses long-lived ones. Specifically, E_L is required to be semantically secure against chosen ciphertext attacks in the pre-processing mode, and F_K to be chosen-plaintext, one-message semantically secure (see [21] for details). We then proceed to show that if the subset-cover algorithm satisfies the key-indistinguishability property and if E_L and F_K satisfy their security requirements, then the revocation scheme is secure under the above definition.

Theorem 1. *Let $\mathcal{A}$ be a Subset-Cover revocation algorithm where (i) the key assignment satisfies the key-indistinguishability property (ii) E_L is semantically secure against chosen ciphertext attacks in the pre-processing mode, and (iii) F_K is chosen-plaintext, one-message semantically secure. Then $\mathcal{A}$ satisfies the notion of revocation scheme security defined above.*

3 Two Subset-Cover Revocation Algorithms

We describe two schemes in the Subset-Cover framework with a different performance tradeoff, as summarized in table 3[5]. Each is defined over a different collection of subsets. Both schemes are r-flexible, namely they work with any number of revocations. In the first scheme, the key assignment is information-theoretic whereas in the other scheme the key assignment is computational. While the first method is relatively simple, the second method is more involved, and exhibits a *substantial improvement over previous methods.*

In both schemes the subsets and the partitions are obtained by imagining the receivers as the leaves in a rooted full binary tree with N leaves (assume that N is a power of 2). Such a tree contains $2N-1$ nodes (leaves plus internal nodes) and for any $1 \leq i \leq 2N-1$ we assume that v_i is a node in the tree. We denote

[5] Recently a method exhibiting various tradeoffs between the measures (bandwidth, storage and processing time) was proposed [22]. In particular it is possible to reduce the device storage down to $\log^2 n/\log D$ by increasing processing time to $D \log n$.

by $ST(\mathcal{R})$ the (directed) Steiner Tree induced by the set $\mathcal{R}$ of vertices and the root, i.e. the minimal subtree of the full binary tree that connects all the leaves in $\mathcal{R}$ ($ST(\mathcal{R})$ is unique). The systems differ in the collections of subsets they consider.

3.1 The Complete Subtree Method

The collection of subsets $S_1, \ldots, S_w$ in our first scheme corresponds to all complete subtrees in the full binary tree with N leaves. For any node v_i in the full binary tree (either an internal node or a leaf, $2N-1$ altogether) let the subset S_i be the collection of receivers u that correspond to the leaves of the subtree rooted at node v_i. The key assignment method simply assigns an independent and random key L_i to every node v_i in the complete tree. Provide every receiver u with the $\log N + 1$ keys associated with the nodes along the path from the root to leaf u.

For a given set $\mathcal{R}$ of revoked receivers, let $S_{i_1}, \ldots, S_{i_m}$ be all the subtrees of the original tree whose roots are adjacent to nodes of outdegree 1 in $ST(\mathcal{R})$, but they are not in $ST(\mathcal{R})$. It follows immediately that this collection covers all nodes in $\mathcal{N} \setminus \mathcal{R}$ and only them. The cover size is at most $r \log(N/r)$. This is also the average number of subsets in the cover.

At decryption, given a message $\langle [i_1, \ldots, i_m, E_{L_{i_1}}(K), \ldots, E_{L_{i_m}}(K)], F_K(M)] \rangle$ a receiver u needs to find whether any of its ancestors is among $i_1, i_2, \ldots i_m$; note that there can be only one such ancestor, so u may belong to at most one subset. This lookup can be facilitated efficiently by using hash-table lookups with *perfect hash functions*.

The key assignment in this method is information theoretic, that is keys are assigned randomly and independently. Hence the "key-indistinguishability" property of this method follows from the fact that no $u \in \mathcal{R}$ is contained in any of the subsets $i_1, i_2, \ldots i_m$.

Theorem 2. *The Complete Subtree Revocation method requires (i) message length of at most* $r \log \frac{N}{r}$ *keys (ii) to store* $\log N$ *keys at a receiver and (iii)* $O(\log \log N)$ *operations plus a* single *decryption operation to decrypt a message. Moreover, the method is secure in the sense of the definition outlined in 2.3.*

Comparison to the Logical Key Hierarchy (LKH) approach: Readers familiar with the LKH method of [29,30] may find it instructive to compare it to the Complete Subtree Scheme. The main similarity lies in the key assignment - an independent label is assigned to each node in the binary tree. However, these labels are used quite differently - in the multicast re-keying LKH scheme some of these labels change at every revocation. In the Complete Subtree method labels are *static*; what changes is a single session key.

Consider an extension of the LKH scheme which we call the *clumped re-keying method*: here, r revocations are performed at a time. For a batch of r revocations, no label is changed more than once, i.e. only the "latest" value is transmitted and used. In this variant the number of encryptions is roughly the same as in the Complete Subtree method, but it requires $\log N$ decryptions at the user, (as opposed to a single decryption in our framework). An additional advantage of the Complete Subtree method is the separation of the labels and the session key which has a consequence on the message length; see discussion about Prefix-Truncation in [21].

3.2 The Subset Difference Method

The main disadvantage of the Complete Subtree method is that $\mathcal{N} \setminus \mathcal{R}$ may be partitioned into a number of subsets that is too large. The goal is now to reduce the partition size. We show an improved method that partitions the non-revoked receivers into at most $2r - 1$ subsets (or $1.25r$ on average), thus getting rid of a $\log N$ factor and effectively reducing the message length accordingly. In return, the number of keys stored by each receiver increases by a factor of $\frac{1}{2} \cdot \log N$. The key characteristic of the Subset-Difference method, which essentially leads to the reduction in message length, is that in this method any user belongs to *substantially* more subsets than in the first method ($O(N)$ instead of $\log N$). The challenge is then to devise an efficient procedure to succinctly encode this large set of keys at the user, which is achieved by using a computational key assignment.

The subset description. As in the previous method, the receivers are viewed as leaves in a complete binary tree. The collection of subsets $S_1, \ldots, S_w$ defined by this algorithm corresponds to subsets of the form "a group of receivers G_1 minus another group G_2", where $G_2 \subset G_1$. The two groups G_1, G_2 correspond to leaves in two full binary subtrees. Therefore a valid subset S is represented by two nodes in the tree (v_i, v_j) such that v_i is an ancestor of v_j. We denote such subset as $S_{i,j}$. A leaf u is in $S_{i,j}$ iff it is in the subtree rooted at v_i but *not* in the subtree rooted at v_j, or in other words $u \in S_{i,j}$ iff v_i is an ancestor of u but v_j is not. Figure 2 depicts $S_{i,j}$. Note that all subsets from the Complete Subtree Method are also subsets of the Subset Difference Method; specifically, a subtree appears here as the difference between its parent and its sibling. The only exception is the full tree itself, and we will add a special subset for that. We postpone the description of the key assignment till later; for the time being assume that each subset $S_{i,j}$ has an associated key $L_{i,j}$.

The Cover. For a set $\mathcal{R}$ of revoked receivers, the following Cover algorithm finds a collection of disjoint subsets $S_{i_1,j_1}, S_{i_2,j_2} \ldots, S_{i_m,j_m}$ which partitions $\mathcal{N} \setminus \mathcal{R}$. The method builds the subsets collection iteratively, maintaining a tree T which is a subtree of $ST(\mathcal{R})$ with the property that any $u \in \mathcal{N} \setminus \mathcal{R}$ that is below a leaf of T has been covered. We start by making T be equal to $ST(\mathcal{R})$ and then iteratively remove nodes from T (while adding subsets to the collection) until T consists of just a single node:

1. Find two leaves v_i and v_j in T such that the least-common-ancestor v of v_i and v_j does not contain any other leaf of T in its subtree. Let v_l and v_k be

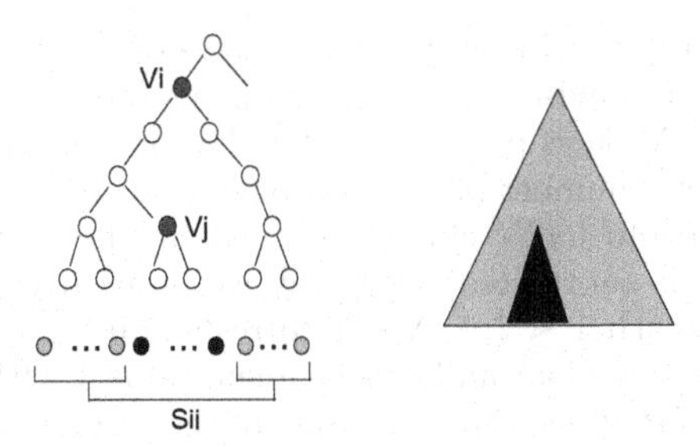

Fig. 2. The Subset Difference method: subset $S_{i,j}$ contains all marked leaves (non-black).

the two children of v such that v_i a descendant of v_l and v_j a descendant of v_k. (If there is only one leaf left, make $v_i = v_j$ to the leaf, v to be the root of T and $v_l = v_k = v$.)
2. If $v_l \not\equiv v_i$ then add the subset $S_{l,i}$ to the collection; likewise, if $v_k \not\equiv v_j$ add the subset $S_{k,j}$ to the collection.
3. Remove from T all the descendants of v and make it a leaf.

An alternative description of the cover algorithm is as follows: consider maximal chains of nodes with outdegree 1 in $ST(\mathcal{R})$. More precisely, each such chain is of the form $[v_{i_1}, v_{i_2}, \ldots v_{i_\ell}]$ where (i) all of $v_{i_1}, v_{i_2}, \ldots v_{i_{\ell-1}}$ have outdegree 1 in $ST(\mathcal{R})$ (ii) v_{i_ℓ} is either a leaf or a node with outdegree 2 and (iii) the parent of v_{i_1} is either a node of outdegree 2 or the root. For each such chain where $\ell \geq 2$ add a subsets S_{i_1,i_ℓ} to the cover. Note that all nodes of outdegree 1 in $ST(\mathcal{R})$ are members of precisely one such chain.

We state, without a proof, that a cover can contain at most $2r - 1$ subsets for any set of r revocations. Moreover, if the set of revoked leaves is *random*, then average-case analysis bounds the cover size by $1.38r$, whereas simulation experiments tighten the bound to $1.25r$.

The next lemma is concerned with covering more general sets than those obtained by removing users. Rather it assumes that we are removing a collection of subsets from the Subset Difference collection. It is applied later in Section 4.2.

Lemma 1. *Let $\mathcal{S} = S_{i_1}, S_{i_2}, \ldots S_{i_m}$ be a collection of m* **disjoint** *subsets from the underlying collection defined by the Subset Difference method, and $\mathcal{U} = \cup_{j=1}^{m} S_{i_j}$. Then the leaves in $\mathcal{N} \setminus \mathcal{U}$ can be covered by at most $3m - 1$ subsets from the underlying Subset Difference collection.*

Key assignment to the subsets. We now define what information each receiver must store. If we try and repeat the information-theoretic approach of the previous scheme where each receiver needs to store *explicitly* the keys of all the subsets it belongs to, the storage requirements would expand tremendously: consider a receiver u; for each complete subtree T_k it belongs to, u must store a number of keys proportional to the number of nodes in the subtree T_k that are *not* on the path from the root of T_k to u. There are $\log N$ such trees, one for each height $1 \leq k \leq \log N$, yielding a total of $\sum_{k=1}^{\log N}(2^k - k)$ which is $O(N)$ keys. We therefore devise a key assignment method that requires a receiver to store only $O(\log N)$ keys per subtree, for the total of $O(\log^2 N)$ keys.

While the total number of subsets to which a user u belongs is $O(N)$, these can be grouped into $\log N$ clusters defined by the first subset i (from which another subsets is subtracted). The way we proceed with the keys assignment is to choose for each $1 \leq i \leq N - 1$ corresponding to an internal node in the full binary tree a random and independent value LABEL_i. This value should *induce* the keys for all legitimate subsets of the form $S_{i,j}$. The idea is to employ the method used by Goldreich, Goldwasser and Micali [17] to construct pseudo-random functions, which was also used by Fiat and Naor [12] for purposes similar to ours.

Let G be a (cryptographic) pseudo-random sequence generator (see definition below) that *triples* the input, i.e. whose output length is *three times* the length of the input; let $G_L(S)$ denote the left third of the output of G on seed S, $G_R(S)$ the right third and $G_M(S)$ the middle third. We say that $G : \{0,1\}^n \mapsto \{0,1\}^{3n}$ is a pseudo-random sequence generator if no polynomial-time adversary can distinguish the output of G on a randomly chosen seed from a truly random string of similar length. Let ε_4 denote the bound on the distinguishing probability.

Consider now the subtree T_i (rooted at v_i). We will use the following top-down labeling process: the root is assigned a label LABEL_i. Given that a parent was labeled S, its two children are labeled $G_L(S)$ and $G_R(S)$ respectively. Let $\text{LABEL}_{i,j}$ be the label of node v_j derived in the subtree T_i from LABEL_i. Following such a labeling, the key $L_{i,j}$ assigned to set $S_{i,j}$ is G_M of $\text{LABEL}_{i,j}$. Note that each label induces three parts: G_L - the label for the left child, G_R - the label for the right child, and G_M the key at the node. The process of generating labels and keys for a particular subtree is depicted in Fig. 3. For such a labeling process, given the label of a node it is possible to compute the labels (and keys) of all its descendants. On the other hand, without receiving the label of an ancestor of a node, its label is pseudo-random and for a node j, given the labels of all its descendants (but not including itself) the key $L_{i,j}$ is pseudo-random ($\text{LABEL}_{i,j}$, the label of v_j, is not pseudo-random given this information simply because one can check for consistency of the labels). It is important to note that given LABEL_i, computing $L_{i,j}$ requires at most $\log N$ invocations of G.

We now describe the information I_u that each receiver u gets in order to derive the key assignment described above. For each subtree T_i such that u is a leaf of T_i the receiver u should be able to compute $L_{i,j}$ iff j is *not* an ancestor of u. Consider the path from v_i to u and let $v_{i_1}, v_{i_2}, \ldots v_{i_k}$ be the nodes just "hanging off" the path, i.e. they are adjacent to the path but not ancestors of u (see Fig. 3). Each j in T_i that is not an ancestor of u is a descendant of one of these nodes. Therefore if u receives the labels of $v_{i_1}, v_{i_2}, \ldots v_{i_k}$ as part of I_u, then invoking G at most $\log N$ times suffices to compute $L_{i,j}$ for any j that is not an ancestor of u.

As for the total number of keys (in fact, labels) stored by receiver u, each tree T_i of depth k that contains u contributes $k - 1$ keys (plus one key for the case where there are no revocations), so the total is $1 + \sum_{k=1}^{\log N+1} k - 1 = 1 + \frac{(\log N+1)\log N}{2} = \frac{1}{2}\log^2 N + \frac{1}{2}\log N + 1$.

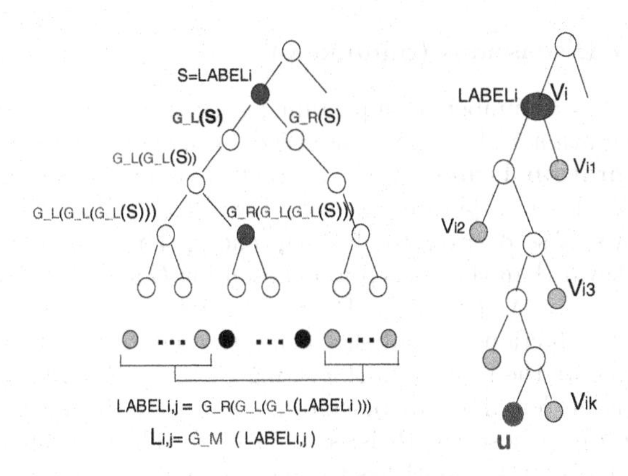

Fig. 3. Key assignment in the Subset Difference method. *Left:* generation of $\text{LABEL}_{i,j}$ and the key $L_{i,j}$. *Right:* leaf u receives the labels of $v_{i_1}, \ldots v_{i_k}$ that are induced by the label LABEL_i of v_i.

At decryption time, a receiver u first finds the subset $S_{i,j}$ such that $u \in S_{i,j}$, and computes the key corresponding to $L_{i,j}$. Using the techniques described in the complete subtree method for table lookup structure, this subset can be found in $O(\log \log N)$. The evaluation of the subset key takes now at most $\log N$ applications of a pseudo-random generator. After that, a single decryption is needed.

Security. In order to prove security we have to show that the key indistinguishability condition (outlined in Sect. 2.3) holds for this method, namely that each key is indistinguishable from a random key for all users not in the corresponding subset.

Observe first that for any $u \in \mathcal{N}$, u never receives keys that correspond to subtrees to which it does not belong. Let S_i denote the set of leaves in the subtree T_i rooted at v_i. For any set $S_{i,j}$ the key $L_{i,j}$ is (information theoretically) independent of all I_u for $u \notin S_i$. Therefore we have to consider only the combined secret information of all $u \in S_j$. This is specified by at most $\log N$ labels - those hanging on the path from v_i to v_j plus the two children of v_j - which are sufficient to derive all other labels in the combined secret information. Note that these labels are $\log N$ strings that were generated independently by G, namely it is never the case that one string is derived from another. Hence, a hybrid argument implies that the probability of distinguishing $L_{i,j}$ from random can be at most $\varepsilon_4/\log N$, where ε_4 is the bound on distinguishing outputs of G from random strings.

Theorem 3. *The Subset Difference method requires (i) message length of at most $2r-1$ keys (ii) to store $\frac{1}{2}\log^2 N + \frac{1}{2}\log N + 1$ keys at a receiver and (iii) $O(\log N)$ operations plus a single decryption operation to decrypt a message. Moreover, the method is secure in the sense of definition outlined in 2.3.*

3.3 Further Discussions (Summary)

In [21] we discuss a number of important issues related to the above schemes, their implementation and applications. Below is a short summary of the topics.

Implementation Issues: A key characteristic of the Subset-Cover framework is that it clearly separates the long-term keys from the short, one time, key. This allows, if so desired, to chose an encryption F that might be weaker (uses shorter keys) than the encryption chosen for E and to reduce the message length appropriately. We provide a "Prefix-Truncation" specification for E to implement such a reduction without sacrificing the security of the long-lived keys. Let $\text{Prefix}_i S$ denote the first i bits of a string S. choose $\mathcal{U}$ to be a random string whose length is the length of the block of E_L and let K be a relatively short key for the cipher F_K (whose length is, say, 56 bits). Then, $[\text{Prefix}_{|K|} E_L(\mathcal{U})] \oplus K$ provides an encryption that satisfies the requirements of E, as described in Sect. 2.3. The Prefix-Truncated header is therefore:

$$\langle [\, i_1, \ldots, i_m, \mathcal{U}, [\text{Prefix}_{|K|} E_{L_{i_1}}(\mathcal{U})] \oplus K, \ldots, [\text{Prefix}_{|K|} E_{L_{i_m}}(\mathcal{U})] \oplus K \,], F_K(M) \rangle$$

Note that the length of the header is reduced to about $m \times |K|$ bits long (say $56m$) instead of $m \times |L|$.

Hierarchical Revocation: We point out that the schemes are well suited to efficiently support hierarchical revocations of large groups of clustered-users; this is useful, for instance, to revoke all devices of a certain manufacturer.

Public Key methods: A revocation scheme that is used in a public key mode is appropriate in scenarios where the the party that generated the ciphertext is not necessarily trustworthy. This calls for implementing E with a public-key cryptosystem; however, a number of difficulties arise such as the public-key generation process, the size of the public key file and the header reduction. As we show, using a Diffie-Hellman like scheme solves most of these problems (except the public key file size).

An interesting point is that prefix truncation is still applicable and we get that the length of public-key encryption is hardly longer than the private-key case. This can be done as follows: Let G be a group with a generator g, $g^{y_{i_j}}$ be the public key of subset S_{i_j} and y_{i_j} the secret key. Choose h as a pairwise-independent function $h : G \mapsto \{0,1\}^{|K|}$, thus elements which are uniformly distributed over G are mapped to uniformly distributed strings of the desired length. The encryption E is done by picking a new element x from G, publicizing g^x, and encrypting K as $E_{L_{i_j}}(K) = h(g^{xy_{i_j}}) \oplus K$. That is, the header now becomes

$$\langle [\, i_1, i_2, \ldots, i_m, g^x, h,\ h(g^{xy_{i_1}}) \oplus K, \ldots, h(g^{xy_{i_m}}) \oplus K \,], F_K(M) \rangle$$

In terms of the broadcast length such system hardly increases the number of bits in the header as compared with a shared-key system - the only difference is g^x and the description of h. Therefore this difference is fixed and does not grow with the number of revocations. Note however that the scheme as defined above is not immune to chosen-ciphertext attacks, but only to chosen plaintext ones.

Coming up with public-key schemes where prefix-truncation is possible that are immune to chosen ciphertext attacks of either kind is an interesting challenge[6].

Copy Protection and CPRM. Copy protection is a natural application for trace-and-revoke schemes, and the stateless scenario is especially appropriate when content is distributed on pre-recorded media. CPRM/CPPM (Content Protection for Recordable Media and Pre-Recorded Media) is a technology developed and licensed by the "4C" group - IBM, Intel, MEI (Panasonic) and Toshiba [10]. It defines a method for protecting content on physical media such as recordable DVD, DVD Audio, Secure Digital Memory Card and Secure CompactFlash. A licensing Entity (the Center) provides a unique set of secret device keys to be included in each device at manufacturing time. The licensing Entity also provides a Media Key Block (MKB) to be placed on each compliant media (for example, on the DVD). The MKB is essentially the header of the ciphertext which encrypts the session key. It is assumed that this header resides on a write-once area on the media, e.g. a Pre-embossed lead-in area on the recordable DVD. When the compliant media is placed in a player/recorder device, it computes the session key from the header (MKB) using its secret keys; the content is then encrypted/decrypted using this session key.

The algorithm employed by CPRM is essentially a Subset-Cover scheme. Consider a table with A rows and C columns. Every device (receiver) is viewed as a collection of C entries from the table, exactly one from each column, that is $u = [u_1, \ldots, u_C]$ where $u_i \in \{0, 1, \ldots, A-1\}$. The collection of subsets $S_1, \ldots, S_w$ defined by this algorithm correspond to subsets of receivers that share the same entry at a given column, namely $S_{r,i}$ contains all receivers $u = [u_1, \ldots, u_C]$ such that $u_i = r$. For every $0 \leq i \leq A-1$ and $1 \leq j \leq C$ the scheme associates a key denoted by $L_{i,j}$. The private information I_u that is provided to a device $u = [u_1, \ldots, u_C]$ consists of C keys $L_{u_1,1}, L_{u_2,2}, \ldots, L_{u_C,C}$.

For a given set $\mathcal{R}$ of revoked devices, the method partitions $\mathcal{N} \setminus \mathcal{R}$ as follows: $S_{i,j}$ is in the cover iff $S_{i,j} \bigcap \mathcal{R} = \emptyset$. While this partition guarantees that a revoked device is never covered, there is a low probability that a non-revoked device $u \notin \mathcal{R}$ will not be covered as well and therefore become non-functional[7].

The CPRM method is a Subset-Cover method with two exceptions: (1) the subsets in a cover are not necessarily disjoint and (2) the cover is not always perfect as a non-revoked device may be uncovered. Note that the CPRM method is not *r-flexible*: the probability that a non-revoked device is uncovered grows with r, hence in order to keep it small enough the number of revocations must be bounded by A.

For the sake of comparing the performance of CPRM with the two methods suggested in this paper, assume that $C = \log N$ and $A = r$. Then, the message is composed of $r \log N$ encryptions, the storage at the receiver consists of $\log N$ keys and the computation at the receiver requires a single decryption. These

[6] Both the scheme of Cramer and Shoup [7] and the random oracle based scheme [14] require some specific information for each recipient; a possible approach with random oracles is to follow the lines of [27].

[7] This is similar to the scenario considered in [16]

bounds are similar to the Complete Subtree method; however, unlike CPRM, the Complete Subtree method is r-flexible and achieves perfect coverage. The advantage of the Subset Difference Method is much more substantial: in addition to the above, the message consists of $1.25r$ encryptions on average, or of at most $2r - 1$ encryptions, rather than $r \log N$.

For example, in DVD Audio, the amount of storage that is dedicated for its MKB (the header) is 3 MB. This constrains the maximum allowed message length. Under a certain choice of parameters, such as the total number of manufactured devices and the number of distinct manufacturers, with the current CPRM algorithm the system can revoke up to about 10,000 devices. In contrast, for the same set of parameters and the same 3MB constraint, a Subset-Difference algorithm achieves up to 250,000 (!) revocations, a factor of 25 improvement over the currently used method. This major improvement is partly due to fact that hierarchical revocation can be done very effectively, a property that the current CPRM algorithm does not have.

Applications to Multicast. The difference between key management for the scenario considered in this paper and for the Logical Key Hierarchy for multicast is that in the latter the users (i.e. receivers) may update their keys [30,29]. This update is referred to as a re-keying event and it requires all users to be connected during this event and change their internal state (keys) accordingly. However, even in the multicast scenario it is not reasonable to assume that all the users receive all the messages and perform the required update. Therefore some mechanism that allows individual update must be in place. Taking the stateless approach gets rid of the need for such a mechanism: simply add a header to each message denoting who are the legitimate recipients by revoking those who should not receive it. In case the number of revocations is not too large this may yield a more manageable solution. This is especially relevant when there is a single source for the sending messages or when public-keys are used.
Backward secrecy: Note that revocation in itself lacks backward secrecy in the following sense: a constantly listening user that has been revoked from the system records all future transmission (which it can't decrypt anymore) and keeps all ciphertexts. At a later point it gains a valid *new* key (by re-registering) which allows decryption of all past communication. Hence, a newly acquired user-key can be used to decrypt all past session keys and ciphertexts. The way that [30,29] propose to achieve backward secrecy is to perform re-keying when new users are added to the group (such a re-keying may be reduced to only one way chaining, known as LKH+), thus making such operations non-trivial. We point out that in the subset-cover framework and especially in the two methods we proposed it may be easier: At any given point of the system include in the set of revoked receivers all identities that have not been assigned yet. As a result, a newly assigned user-key cannot help in decrypting an earlier ciphertext. Note that this is feasible since we assume that new users are assigned keys in a consecutive order of the leaves in the tree, so unassigned keys are consecutive leaves in the complete tree and can be covered by at most $\log N$ sets (of either type, the Complete-Subtree method or the Subtree-Difference method). Hence, the unassigned leaves can be treated with the hierarchical revocation technique, resulting in *adding* at most $\log N$ revocations to the message.

4 Tracing Traitors

It is highly desirable that a revocation mechanism could work in tandem with a tracing mechanism to yield a *trace and revoke* scheme. We show a tracing method that works for many schemes in the subset-cover framework. The method is quite efficient. The goal of a tracing algorithm is to find the identities of those that contributed their keys to an illicit decryption box[8] and revoke them; short of identifying them we should render the box useless by finding a "pattern" that does not allow decryption using the box, but still allows broadcasting to the legitimate users. Note that this is a slight relaxation of the requirement of a tracing mechanism, say in [23] (which requires an identification of the traitor's identity) and in particular it lacks *self enforcement* [11]. However as a mechanism that works in conjunction with the revocation scheme it is a powerful tool to combat piracy.

The model. Suppose that we have found an illegal decryption-box (decoder, or clone) which contains the keys associated with at most t receivers $u_1, \ldots, u_t$ known as the "traitors".

We are interested in "black-box" tracing, i.e. one that does not take the decoder apart but by providing it with an encrypted message and observing its output (the decrypted message) tries to figure out who leaked the keys. A pirate decoder is of interest if it correctly decodes with probability p which is at least some threshold q, say $q > 0.5$. We assume that the box has a "reset button", i.e. that its internal state may be retrieved to some initial configuration. In particular this excludes a "locking" strategy on the part of the decoder which says that in case it detects that it is under test, it should refuse to decode further. Clearly software-based systems can be simulated and therefore have the reset property.

The result of a tracing algorithm is either a subset consisting of traitors or a partition into subsets that renders the box useless i.e. given an encryption with the given partition it decrypts with probability smaller than the threshold q while all good users can still decrypt.

In particular, a "subsets based" tracing algorithm devises a sequence of queries which, given a black-box that decodes with probability above the threshold q, produces the results mentioned above. It is based on constructing useful sets of revoked devices $\mathcal{R}$ which will ultimately allow the detection of the receiver's identity or the configuration that makes the decoder useless. A tracing algorithm is evaluated based on (i) the level of performance downgrade it imposes on the revocation scheme (ii) number of queries needed.

4.1 The Tracing Algorithm

Subset tracing: An important procedure in our tracing mechanism is one that given a partition $\mathcal{S} = S_{i_1}, S_{i_2}, \ldots S_{i_m}$ and an illegal box outputs one of two possible outputs: either (1) that the box cannot decrypt with probability greater than the threshold when the encryption is done with partition $\mathcal{S}$ or (ii) Finds

[8] Our algorithm also works for more than one box.

a subset S_{i_j} such that S_{i_j} contains a traitor. Such a procedure is called subset tracing and is described below.

Bifurcation property: Given a subset-tracing procedure, we describe a tracing strategy that works for many Subset-Cover revocation schemes. The property that the revocation algorithm should satisfy is that for any subset $S_i, 1 \leq i \leq w$, it is possible to partition S_i into two (or constant) roughly equal sets, i.e. that there exists $1 \leq i_1, i_2 \leq w$ such that $S_i = S_{i_1} \cup S_{i_2}$ and $|S_{i_1}|$ is roughly the same as $|S_{i_2}|$. For a Subset Cover scheme, let the *bifurcation value* be the relative size of the largest subset in such a split.

Both the Complete Subtree and the Subtree Difference methods satisfy this requirement: in the case of the Complete Subtree Method each subset, which is a complete subtree, can be split into exactly two equal parts, corresponding to the left and right subtrees. Therefore the bifurcation value is 1/2. As for the Subtree Difference Method, Each subset $S_{i,j}$ can be split into two subsets each containing between one third and two thirds of the elements. Here, again, this is done using the left and right subtrees of node i. See Fig. 4. The only exception is when i is a parent of j, in which case the subset is the complete subtree rooted at the other child; such subsets can be perfectly split. The worst case of $(1/3, 2/3)$ occurs when i is the grandparent of j. Therefore the bifurcation value is $2/3$.

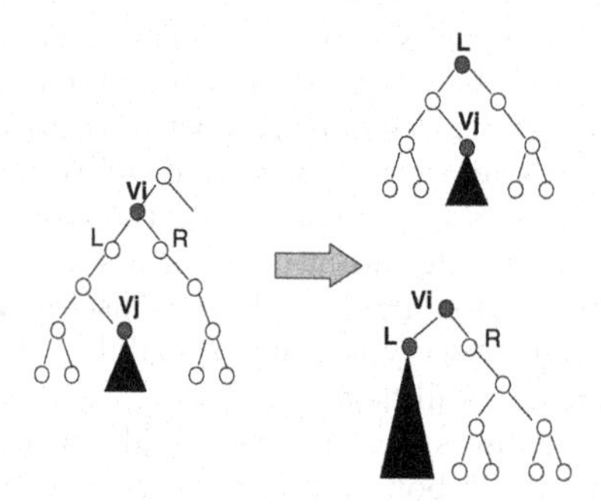

Fig. 4. Bifurcating a Subset Difference set $S_{i,j}$, depicted in the left. The black triangle indicates the excluded subtree. L and R are the left and the right children of v_i. The resulting sets $S_{L,j}$ and $S_{i,L}$ are depicted to the right.

The Tracing Algorithm: We now describe the general tracing algorithm, assuming that we have a good subset tracing procedure. The algorithm maintains a partition $S_{i_1}, S_{i_2}, \ldots S_{i_m}$. At each phase one of the subsets is partitioned, and the goal is to partition a subset only if it contains a traitor.

Each phase initially applies the subset-tracing procedure with the current partition $\mathcal{S} = S_{i_1}, S_{i_2}, \ldots S_{i_m}$. If the procedure outputs that the box cannot decrypt with $\mathcal{S}$ then we are done, in the sense that we have found a way to disable the box without hurting any legitimate user. Otherwise, let S_{i_j} be the set output by the procedure, namely S_{i_j} contains the a traitor.

If S_{i_j} contains only one possible candidate - it must be a traitor and we permanently revoke this user; this doesn't hurt a legitimate user. Otherwise we split S_{i_j} into two roughly equal subsets and continue with the new partitioning. The existence of such a split is assured by the bifurcation property.

Analysis: Since a partition can occur only in a subset that has a traitor and contains more than one element, it follows that the number of iterations can be at most $t \log_a N$, where a is the inverse of the bifurcation value (a more refined expression is $t(\log_a N - log_2 t)$, the number of edges in a binary tree with t leaves and depth $\log_a N$.)

The Subset Tracing Procedure. The Subset Tracing procedure first tests whether the box decodes a message with the partition $\mathcal{S} = S_{i_1}, S_{i_2}, \ldots S_{i_m}$ with sufficient probability greater than the threshold, say > 0.5. If not, then it concludes (and outputs) that the box cannot decrypt with $\mathcal{S}$. Otherwise, it needs to find a subset S_{i_j} that contains a traitor.

Let p_j be the probability that the box decodes the ciphertext

$$\langle [i_1, i_2, \ldots, i_m, E_{L_{i_1}}(R_K), \ldots, E_{L_{i_j}}(R_K), E_{L_{i_{j+1}}}(K), \ldots, E_{L_{i_m}}(K)], F_K(M) \rangle$$

where R_K is a random string of the same length as the key K. That is, p_j is the probability of decoding when the first j subsets are noisy and the remaining subsets encrypt the correct key. Note that $p_0 = p$ and $p_m = 0$, hence there must be some $0 < j \leq m$ for which $|p_{j-1} - p_j| \geq \frac{p}{m}$. It can be shown that if p_{j-1} is different from p_j by more than ε, where ε is an upper bound on the sum of the probabilities of breaking the encryption scheme E and key assignment method, then the set S_{i_j} must contain a traitor. It also provides a binary-search-like method that efficiently finds a pair of values p_j, p_{j-1} among $p_0, \ldots, p_m$ satisfying $|p_{j-1} - p_j| \geq \frac{p}{m}$.

4.2 Improving the Tracing Algorithm

The basic traitors tracing algorithm described above requires $t \log(N/t)$ iterations. Furthermore, since at each iteration the number of subsets in the partition increases by one, tracing t traitors may result with up to $t \log(N/t)$ subsets and hence in messages of length $t \log(N/t)$. This bound holds for any Subset-Cover method satisfying the *Bifurcation property*, and both the Complete Subtree and the Subset Difference methods satisfy this property. What is the bound on the number of traitors that the algorithm can trace?

Recall that the message length required by the Complete Subtree method is $r \log(N/r)$ for r revocations, hence the tracing algorithm can trace up to r traitors if it uses the Complete Subtree method. However, since the message length of the Subset Difference method is at most $2r - 1$, only $\frac{2r-1}{\log N/r}$ traitors can be traced if Subset Difference is used. We now describe an improvement on the basic tracing algorithm that reduces the number of subsets in the partition to $5t-1$ for the Subset Difference method (although the number of iterations remains $t\log(N/t)$). With this improvement the algorithm can trace up to $r/5$ traitors.

Note that among the $t\log N/t$ subsets generated by the basic tracing algorithm, only t actually contain a traitor. The idea is to repeatedly merge those subsets which are not known to contain a traitor.[9] Specifically, we maintain at each iteration a *frontier* of at most $2t$ subsets plus $3t-1$ additional subsets. In the following iteration a subset that contains a traitor is further partitioned; as a result, a new *frontier* is defined and the remaining subsets are re-grouped.

Frontier subsets: Let $S_{i_1}, S_{i_2}, \ldots S_{i_m}$ be the partition at the current iteration. A pair of subsets $(S_{i_{j1}}, S_{i_{j2}})$ is said to be in the frontier if $S_{i_{j1}}$ and $S_{i_{j2}}$ resulted from a split-up of a single subset at an earlier iteration. Also neither ($S_{i_{j1}}$ nor $S_{i_{j2}}$) was singled out by the subset tracing procedure so far. This definition implies that the frontier is composed of k disjoint pairs of *buddy subsets.* Since buddy-subsets are disjoint and since each pair originated from a single subset that contained a traitor (and therefore has been split) $k \le t$.

We can now describe the improved tracing algorithm which proceeds in iterations. Every iteration starts with a partition $\mathcal{S} = S_{i_1}, S_{i_2}, \ldots S_{i_m}$. Denote by $F \subset S$ the frontier of S. An iteration consists of the following steps, by the end of which a new partition $\mathcal{S}'$ and a new frontier F' is defined.

- As before, use the Subset Tracing procedure to find a subset S_{i_j} that contains a traitor. If the tracing procedure outputs that the box can not decrypt with $\mathcal{S}$ then we are done. Otherwise, split S_{i_j} into $S_{i_{j1}}$ and $S_{i_{j2}}$.
- $F' = F \cup S_{i_{j1}} \cup S_{i_{j2}}$ ($S_{i_{j1}}$ and $S_{i_{j2}}$ are now in the frontier). Furthermore, if S_{i_j} was in the frontier F and S_{i_k} was its buddy-subset in F then $F' = F' \setminus S_{i_k}$ (remove S_{i_k} from the frontier).
- Compute a cover $\mathcal{C}$ for all receivers that are not covered by F'. Define the new partition $\mathcal{S}'$ as the union of $\mathcal{C}$ and F'.

To see that the process described above converges, observe that at each iteration the number of new *small* frontier sets always increases by at least one. More precisely, at the end of each iteration construct a vector of length N describing how many sets of size i, $1 \le i \le N$, constitute the frontier. It is easy to see that these vectors are lexicographically increasing. The process must stop when or before all sets in the frontier are singletons.

By definition, the number of subsets in a frontier can be at most $2t$. Furthermore, they are paired into at most t disjoint buddy subsets. As for non-frontier subsets ($\mathcal{C}$), Lemma 1 shows that covering the remaining elements can be done by at most $|F| \le 3t-1$ subsets (note that we apply the lemma so as to cover all elements that are not covered by the buddy subsets, and there are at most t of them). Hence the partition at each iteration is composed of at most $5t-1$ subsets.

[9] This idea is similar to the second scheme of [13], Sect. 3.3. However, in [13] the merge is straightforward as their model allows any subset. In our model only members from the Subset Difference are allowed, hence a merge which produces subsets of this particular type is non-trivial.

Acknowledgements. We thank Omer Horvitz for many comments regarding the paper and the implementation of the system. We thank Ravi Kumar, Nelly Fazio, Florian Pestoni and Victor Shoup for useful comments.

References

1. J. Anzai, N. Matsuzaki and T. Matsumoto, A Quick Group Key Distribution Sceheme with "Entity Revocation", Advances in Cryptology - Asiacrypt '99, LNCS 1716, Springer, 1999, pp. 333–347.
2. O. Berkman, M. Parnas and J. Sgall, Efficient Dynamic Traitor Tracing, Proc. of the 11th ACM-SIAM Symp. on Discrete Algorithms (SODA), pp. 586–595, 2000.
3. D. Boneh and M. Franklin, An efficient public key traitor tracing scheme, Advances in Cryptology - Crypto '99, LNCS 1666, Springer, 1999, pp. 338–353.
4. D. Boneh, and J. Shaw, *Collusion Secure Fingerprinting for Digital Data*, IEEE Transactions on Information Theory, Vol 44, No. 5, pp. 1897–1905, 1998.
5. R. Canetti, J. Garay, G. Itkis, D. Micciancio, M. Naor and B. Pinkas, Multicast Security: A Taxonomy and Some Efficient Constructions, Proc. of INFOCOM '99, Vol. 2, pp. 708–716, New York, NY, March 1999.
6. R. Canetti, T. Malkin, K. Nissim, Efficient Communication-Storage Tradeoffs for Multicast Encryption, Advances in Cryptology - EUROCRYPT '99, LNCS 1592, Springer, 1999, pp. 459–474.
7. R. Cramer and V. Shoup, A Practical Public Key Cryptosystem Provably Secure Against Adaptive Chosen Ciphertext Attack. Advances in Cryptology - CRYPTO 1999, Lecture Notes in Computer Science 1462, Springer, pp. 13–25.
8. B. Chor, A. Fiat and M. Naor, Tracing traitors, Advances in Cryptology - CRYPTO '94, LNCS 839, Springer, pp. 257–270, 1994.
9. B. Chor, A. Fiat, M. Naor and B. Pinkas, Tracing traitors, IEEE Transactions on Information Theory, Vol. 46, No. 3, May 2000.
10. Content Protection for Recordable Media. Available: http://www.4centity.com/4centity/tech/cprm
11. C. Dwork, J. Lotspiech and M. Naor, Digital Signets: Self-Enforcing Protection of Digital Information, 28th Symp. on the Theory of Computing, 1996, pp. 489– 498.
12. A. Fiat and M. Naor, *Broadcast Encryption*, Advances in Cryptology - CRYPTO '93, LNCS 773, Springer, 1994, pp. 480—491.
13. A. Fiat and T. Tassa, *Dynamic Traitor Tracing* Advances in Cryptology - CRYPTO '99, LNCS 1666, 1999, pp. 354–371.
14. E. Fujisaki and T. Okamoto, *Secure Integration of Asymmetric and Symmetric Encryption Schemes*, Advances in Cryptology - CRYPTO 1999, LNCS 1666, 1999, pp. 537–554.
15. E. Gafni, J. Staddon and Y. L. Yin, *Efficient Methods for Integrating Traceability and Broadcast Encryption*, Advances in Cryptology - CRYPTO'99, LNCS 1666, Springer, 1999, pp. 372–387.
16. J.A. Garay, J. Staddon and A. Wool, Long-Lived Broadcast Encryption. Advances in Cryptology - CRYPTO'2000, LNCS 1880, pp. 333–352, 2000.
17. O. Goldreich, S. Goldwasser and S. Micali, How to Construct Random Functions. JACM 33(4): 792–807 (1986)
18. R. Kumar, R. Rajagopalan and A. Sahai, Coding Constructions for blacklisting problems without Copmutational Assumptions. Advances in Cryptology - CRYPTO '99, LNCS 1666, 1999, pp. 609–623.

19. M. Luby and J. Staddon, Combinatorial Bounds for Broadcast Encryption. Advances in Cryptology - EUROCRYPT '98, LNCS vol 1403, 1998, pp. 512–526.
20. D. McGrew, A. T. Sherman, *Key Establishment in Large Dynamic Groups Using One-Way Function Trees*, submitted to IEEE Transactions on Software Engineering (May 20, 1998).
21. D. Naor, M. Naor, J. Lotspiech, *Revocation and Tracing Schemes for Stateless Receivers*, full version available at the IACR Crypto Archive http://eprint.iacr.org/.
22. M. Naor, *Tradeoffs in Subset-Cover Revocation Schemes*, manuscript, 2001.
23. M. Naor and B. Pinkas, Threshold traitor tracing, Advances in Cryptology - Crypto '98, LNCS 1462, pp. 502–517.
24. M. Naor and B. Pinkas, Efficient Trace and Revoke Schemes Financial Cryptography '2000, LNCS , Springer.
25. B. Pfitzmann, Trials of Traced Traitors, Information Hiding Workshop, First International Workshop, Cambridge, UK, LNCS 1174, Springer, 1996, pp. 49–64.
26. R. Safavi-Naini and Y. Wang, Sequential Traitor Tracing, Advances in Cryptology - CRYPTO 2000, LNCS 1880, pp. 316–332, 2000.
27. V. Shoup and R. Gennaro, Securing threshold cryptosystems against chosen ciphertext attack, Advances in Cryptology - EUROCRYPT '98, LNCS 1403, 1998, pp. 1–16.
28. D.R. Stinson and R. Wei, Key Preassigned Traceability Schemes for Broadcast Encryption, Proc. Fifth Annual Workshop on Selected Areas in Cryptography, LNCS 1556 (1999), pp. 144–156.
29. D.M. Wallner, E.J. Harder and R.C. Agee, *Key Management for Multicast: Issues and Architectures*, Internet Request for Comments 2627, June, 1999. Available: `ftp.ietf.org/rfc/rfc2627.txt`
30. C. K. Wong, M. Gouda and S. Lam, Secure Group Communications Using Key Graphs, Proc. ACM SIGCOMM'98, pp. 68–79.

Self Protecting Pirates and Black-Box Traitor Tracing

Aggelos Kiayias[1] and Moti Yung[2]

[1] Graduate Center, CUNY, NY USA,
akiayias@gc.cuny.edu
[2] CertCo, NY USA
moti@cs.columbia.edu

Abstract. We present a new generic black-box traitor tracing model in which the pirate-decoder employs a self-protection technique. This mechanism is simple, easy to implement in any (software or hardware) device and is a natural way by which a pirate (an adversary) which is black-box accessible, may try to evade detection. We present a necessary combinatorial condition for black-box traitor tracing of self-protecting devices. We constructively prove that any system that fails this condition, is incapable of tracing pirate-decoders that contain keys based on a superlogarithmic number of traitor keys. We then combine the above condition with specific properties of concrete systems. We show that the Boneh-Franklin (BF) scheme as well as the Kurosawa-Desmedt scheme have no black-box tracing capability in the self-protecting model when the number of traitors is superlogarithmic, unless the ciphertext size is as large as in a trivial system, namely linear in the number of users. This partially settles in the negative the open problem of Boneh and Franklin regarding the general black-box traceability of the BF scheme: at least for the case of superlogarithmic traitors. Our negative result does not apply to the Chor-Fiat-Naor (CFN) scheme (which, in fact, allows tracing in our self-protecting model); this separates CFN black-box traceability from that of BF. We also investigate a weaker form of black-box tracing called single-query "black-box confirmation." We show that, when suspicion is modeled as a confidence weight (which biases the uniform distribution of traitors), such single-query confirmation is essentially not possible against a self-protecting pirate-decoder that contains keys based on a superlogarithmic number of traitor keys.

1 Introduction

The problem of Traitor Tracing can be understood best in the context of Pay-TV. In such a system there are n subscribers, each one possessing a decryption box (decoder). The authority scrambles digital data and broadcasts it to all subscribers, who use their decryption boxes to descramble the data. It is possible for some of the users to collude and produce a pirate decoder: a device not registered with the authority that can decrypt the scrambled digital content. The goal of Traitor Tracing is to provide a method so that the authority, given

J. Kilian (Ed.): CRYPTO 2001, LNCS 2139, pp. 63–79, 2001.

a pirate decoder, is able to recover the identity of some of the legitimate users that participated in the construction of the decoder (traitors). In such a system piracy would be reduced due to the fear of exposure.

A standard assumption is that each user's decoder is "open" (to the user) so that the decryption key is recoverable. A set of users can combine their keys in order to construct a pirate decoder. It is immediately clear that each user should have a distinct private key, otherwise distinguishing traitors from non-traitors would be impossible. Given the contents of a pirate decoder the authority should be able to recover one of the traitors' keys. A scheme that allows this, is called a Traitor Tracing Scheme (TTS). A standard measure of the efficiency of a TTS is the size of the ciphertexts. Constructing a TTS with linear (in the number of users) ciphertexts is trivial; as a result the focus is on how to achieve traitor tracing when the ciphertext size is sublinear in the number of users. An additional requirement for TTSs is black-box traitor tracing, namely, a system where tracing is done using only black-box access to the pirate decoder (namely, only an input/ output access is allowed). To keep tracing cheap, it is extremely desirable that the tracing algorithm is black-box.

Previous Work

Let us first review the work of the various notions of traitor tracing. Traitor Tracing was introduced in [CFN94,CFNP00], with the presentation of a generic TTS. Explicit constructions based on combinatorial designs were given in [SW98b]. A useful variation of the [CFN94] scheme was presented in [NP98]. Public key Traitor Tracing Schemes based on ElGamal encryption were presented in [KD98, BF99]. In most settings (here also) it is assumed that the tracing authority is trusted (i.e. the authority does not need to obtain a *proof* that a certain user is a traitor); the case where the authority is not trusted was considered in [Pfi96, PS96,PW97]. An online approach to tracing, targeting pirate re-broadcasting (called dynamic traitor tracing) was presented in [FT99]. A method of discouraging users from sharing their decryption keys with other parties, called self-enforcement, was introduced in [DLN96]. A traitor tracing scheme along the lines of [KD98,BF99] combining self-enforcement and revocation capabilities was presented in [NP00]. Further combinatorial constructions of traitor tracing schemes in combination with revocation methods were discussed in [GSY99].

Previous work on black-box traitor tracing is as follows: a black-box traitor tracing scheme successful against any resettable[1] pirate decoder was presented in [CFN94,CFNP00]. In [BF99], a black-box traitor tracing scheme was presented against a restricted model called "single-key pirates": the pirate-decoder uses a single key for decryption *without* any other side computation (note that this single key could have been a combination of many traitors' keys). In the same paper, a weaker form of black-box traitor tracing was presented: "black-box confirmation." In this setting the tracer has a set of suspects and it wants to confirm that the traitors that constructed the pirate decoder are indeed included in the set of suspects. The work in [BF99] presented a single-query black-box confir-

[1] A pirate decoder is called resettable if the tracer has a means of resetting the device to its initial state for each trial.

mation method: using a single query to the pirate decoder the tracer solves the problem; multiple queries may be used to increase confidence. Black-box confirmation can be used for general tracing by trying all possible subsets. However the resulting traitor tracing algorithm needs exponential time (unless the number of traitors is a constant). In [Pfi96], a piracy prevention behavior was noted, dealing with the possibility of pirate decoders shutting down whenever an invalid ciphertext (used for tracing, perhaps) is detected. In [BF01] a combination of black-box confirmation and tracing appeared: extending the methods of [BF99] it was shown how one can trace within the suspect set (which is assumed to include all traitors) and recover one of the traitors. In addition, a new mode of black box tracing was considered in [BF01] called minimal access black box tracing: for any query to the pirate decoder, the tracer does not obtain the plaintext but merely whether the pirate-decoder can decrypt the ciphertext and "play" it or not (e.g. the case of a pirate cable-box incorporating a TV-set).

Our Results

THE MODEL: Our perspective on black-box traitor tracing is as follows: under normal operation all users decrypt the same message; we say that in this case all users are *colored* in the same way. As we will see, in order to trace a pirate decoder in a black-box manner we have to disrupt this uniformity: color the users using more than one color. A ciphertext that induces such a coloring over the user population, will be called an "invalid" ciphertext. Tracing algorithms will have to probe with invalid ciphertexts (we assume our tracing methods to be aware of this fact). We consider a simple self-protection mechanism that can be used by any pirate decoder in order to detect tracing: *before decrypting, the pirate decoder computes the projection of the induced coloring onto the set of traitor keys (for some systems the stored keys can actually be combinations of traitor keys). If the traitor keys are colored by two colors or more, then the decoder knows that it is probed by the tracer, and can take actions to protect itself.* Computing the projection of the coloring onto the traitor keys is typically not a time-consuming operation and can be implemented within any software or hardware pirate decoder: prior to giving output the pirate decoder decrypts the given input with all available traitor keys (or combinations thereof) that are stored in its code. Since the decoder is black-box accessible, the presence of the keys internally, does not reduce its evasion power.

NECESSARY COMBINATORIAL CONDITION AND NEGATIVE RESULTS: By adding the above simple self-protecting mechanism to the capabilities of pirate decoders together with an appropriate reaction mechanism we present a condition that has to be satisfied by any TTS in order to be able to black-box trace a pirate decoder that contains $\omega(\log n)$ traitor keys. Namely, the condition that *most users should be colored in the same way.* If this is not the case, we present a strategy that can be followed by a pirate decoder of any type (involving the previously stated self-protection mechanism) that defeats any black-box tracing method with high probability, assuming randomly chosen traitors.

NECESSARY CONDITION AND NEGATIVE RESULTS FOR CONFIRMATION: The assumption above which underlies our negative result is that the choice of keys available to the pirate is randomly distributed over the keys of the user population, i.e. the tracer has no a-priori idea about the identities of the traitors. In the context of black-box confirmation the situation is different because it is assumed that the tracer has a set of suspects, that are traitors with higher probability compared to a user chosen at random. We formalize this setting (differently from [BF01]) by assigning a "confidence level" function to the set of suspects that measures the amplification of the probability that a user is a traitor given that he belongs to the suspect set. Using this formalization we show that single-query black-box confirmation fails against any pirate-decoder, provided that the decoder contains a superlogarithmic number of traitor keys, and the confidence level of the tracer is below a certain (explicitely defined) threshold. We note that the confidence level exhibits a trade-off with the size of the suspect set, i.e. for small suspect sets, the confidence of the tracer should be very high in order to be successful in black-box confirmation. An immediate corollary of our result is that single-query black-box confirmation can be successful against decoders including a superlogarithmic number of traitor keys only in the case that the confidence level of the tracer is so high that *the probability that a user is a suspect given that it is a traitor is arbitrarily close to 1.* Note that in this case, confirmation becomes quite localized (the tracer knows already that the suspect set contains all traitors with very high probability; this type of confirmation is covered in [BF01]).

APPLYING THE RESULTS TO CONCRETE SYSTEMS: We continue by combining our negative results with specific properties of concrete schemes which we analyze. First, we consider the Boneh-Franklin scheme [BF99] which possesses many attractive properties (based on public key, small ciphertext size, deterministic tracing). We show that the scheme is incapable of black-box traitor tracing when there are $\omega(\log n)$ traitors in the self-protecting model, unless the scheme becomes trivial (i.e. with ciphertexts of size linear in the number of users). This partially (for the $\omega(\log n)$ traitor case) settles in the negative the open problem from [BF99] who asked whether [BF99] traceability can be extended to the general black-box traitor tracing model of [CFN94,CFNP00] (i.e. black-box tracing of *any* resettable pirate decoder). Note that this is not an inconsistency with the black-box traitor tracing methods of [BF99], since they apply tracing against pirate decoders of an explicit construction or against a constant number of traitors. Similar negative results hold for the scheme of [KD98]. We note that our negative results do not apply to the black-box tracing methods of [CFN94,CFNP00] since their scheme is proved to work against any resettable pirate-decoder by (obviously not coincidentally) using colorings that satisfy the condition we show to be necessary (most users are colored in the same way). Thus, our work can be seen as retrofitting a design criterion for the early work of [CFN94] and it provides a separation with respect to black-box traceability between [CFN94, CFNP00] and [BF99,KD98]. Additionally, we show that black-box confirmation fails for both [BF99,KD98] against a superlogarithmic number of traitors unless the confidence level of the tracer is extremely high. Note again that this is not an inconsistency with the black-box confirmation result of [BF99] which allows the differently modeled tracer's confidence to be quite large.

Organization. To state negative results, careful modeling is required. We define Multicast Encryption Schemes and non-black-box traitor tracing in section 2, whereas in section 3 we formalize the concepts of black-box tracing and coloring, and we provide the groundwork for the rest of the paper. In section 4 we prove the necessary condition for black-box traitor tracing (section 4.1), and we identify families of TTSs that are incapable for black-box tracing (section 4.2). Black-box confirmation is discussed in section 4.3. The negative results regarding the black-box traceability of the [BF99] and [KD98] schemes in the "self-protecting" pirate-decoder model, are proven in section 5.1 and section 5.2 respectively.

2 Multicast Encryption Schemes

Any traitor tracing scheme is based on a Multicast Encryption Scheme (MES) – a cryptographic primitive we formalize in this section. Let $\mathcal{U} := \{1, \ldots, n\}$ be the set of users. Let $\{\mathcal{G}_w\}_{w \in \mathbb{N}}$ be some a family of sets of elements of length w (e.g. $\mathcal{G}_w = \{0,1\}^w$). For a certain w, we fix the following sets: the message space $\mathcal{M} \subseteq \mathcal{G}_w$; the ciphertext-space $\mathcal{C} \subseteq \mathcal{G}_w^v$; the user key-space $\mathcal{D} \subseteq \mathcal{G}_w^u$; v, u express the dimension of ciphertext space and user key space respectively over the message space. Without loss of generality we will assume that $u \leq v$ i.e. a user key does not have to be "longer" than a ciphertext (this is justified by all concrete MESs in the literature). Note that in a concrete MES $\mathcal{M}, \mathcal{C}, \mathcal{D}$ may be of slightly different structure e.g. in the [BF99]-scheme $\mathcal{M} \subseteq G_q$, $\mathcal{C} \subseteq G_q^v$ but $\mathcal{D} \subseteq Z_q^{v-1}$ (see section 5.1), but these differences are of minor importance here. A function $\sigma(n)$ will be called *negligible* if $\sigma(n) < n^{-c}$ for all c, for sufficiently large n. For brevity we make the assumption that 1^w is polynomially related to n. A Multicast Encryption Scheme (MES) is a triple (G, E, D) of probabilistic polynomial time algorithms with the following properties:

- Key Generation. On input 1^w and n, G produces a pair (e, K) with $K \subseteq \mathcal{D}$, $|K| = n$.
- Encryption. $c \leftarrow E(1^w, m, e)$; $m \in \mathcal{M}$, $e : (e, K) \leftarrow G(1^w, n)$, $(c \in \mathcal{C})$.
- Decryption. For any $m \in \mathcal{M}, (e, K) \leftarrow G(1^w, n)$, if $c \leftarrow E(1^w, m, e)$, then the probabilities $\mathbf{Prob}[m' \neq m : m' \leftarrow D(1^w, d, c)]$ and $\mathbf{Prob}[m' \neq m'' : m' \leftarrow D(1^w, c, d), m' \leftarrow D(1^w, c, d')]$ are negligible, for any keys $d, d' \in K$. The first probability states that incorrect decryption event is negligible whereas the second probability states that all user keys decrypt the same word but with negligible error.

Note that the above scheme can be either public or secret key. It is easy to adapt the standard notions of semantic security or chosen-ciphertext security for MESs.

Let $\mathcal{F}$ be the set of functions of $(\mathbb{N} \to \mathbb{N})$ s.t. $f \in \mathcal{F}$ if and only if f is non-decreasing and constructible (i.e., there is an algorithm M s.t. on input n,

M outputs the string $0^{f(n)}$). Moreover, for any $f, g \in \mathcal{F}$ it holds that either (a) $\exists n_0 \; \forall n \geq n_0 \; (f(n) = g(n))$ (b) $\exists n_0 \; \forall n \geq n_0 \; (f(n) > g(n))$ (c) $\exists n_0 \; \forall n \geq n_0 \; (f(n) < g(n))$ (i.e. it is possible to define a total order over $\mathcal{F}$). Since we are interested only in functions less than n, we assume that $\forall f \in \mathcal{F}$ it holds that $\forall n(f(n) \leq n)$. To facilitate traitor tracing, some additional security requirements have to be imposed.

Non-Triviality of Decryption. For any probabilistic polynomial time algorithm A the following probability is negligible for almost all messages m: $\mathbf{Prob}[m = m' : m' \leftarrow A(1^w, c); c \leftarrow E(1^w, m, e)]$. This property ensures that there are no "shortcuts" in the decryption process. Namely, decryption without access to a key amounts to reversing a one-way function, thus for effective decryption one needs some or a combination of the designated user keys.

Key-User correspondence. It should be guaranteed that each user does not divulge its own key; more generally that a user is responsible when its key is being used for decryption. This should apply to collusions of users as well. More specifically, given $t \in \mathcal{F}$, there should be no probabilistic polynomial-time algorithm working with non-negligible success probability that given the keys of a set of subscribers $d_{i_1}, \ldots, d_{i_k}$ with $k \leq t(n)$, and all other public information, and is able to compute one additional private key d_j with $j \notin \{i_1, \ldots, i_k\}$.

Non-Ambiguity of Collusions. The user keys are drawn from a key-space $\mathcal{D}_e$ defined for each encryption key e; i.e. $\mathcal{D}_e \subseteq \mathcal{G}_w^u$ contains all d that can be used to invert e. Obviously $\mathcal{D}_e \supseteq K$, if $(e, K) \leftarrow G(1^w, n)$. Then, the following holds: Given $t \in \mathcal{F}$; let A, B be probabilistic polynomial algorithms. Given T_1, T_2 two disjoint subsets of K, of cardinality less or equal to $t(n)$. Let I_1, I_2 be all private and public information available to T_1, T_2 correspondingly. Then the following probability is negligible $\mathbf{Prob}[d = d' \wedge (d \in \mathcal{D}_e) : d \leftarrow A(T_1, I_1, 1^w), d' \leftarrow B(T_2, I_2, 1^w)]$.

Non-ambiguity of collusions requires that two disjoint sets of users cannot generate the same decryption key. It is an essential property of any traitor-tracing scheme, since if it fails it is immediately possible to generate instances where tracing is impossible due to ambiguity.

Definition 1. Traitor Tracing Scheme (non-black-box). *Given $t, f, v \in \mathcal{F}$, a MES satisfying non-triviality of decryption, key-user correspondence for $t(n)$, non-ambiguity of collusions for $t(n)$ and, in addition, has $wv(n)$ ciphertext size, is called a $\langle t(n), f(n), v(n)\rangle$-Traitor Tracing Scheme (TTS) if there exists a probabilistic polynomial time algorithm B (tracing algorithm) s.t. for any set $T \subseteq K$, $(e, K) \leftarrow G(1^w, n)$, with $|T| \leq t(n)$ and any probabilistic polynomial time algorithm A that given T and all public information outputs $d \in \mathcal{D}_e$, it holds that:* $\mathbf{Prob}[\tau \in T : \tau \leftarrow B(d, K, 1^w), d \leftarrow A(T, 1^w)] \geq 1/f(n)$.

Because of key-user correspondence, the recovery of τ is equivalent to exposing a traitor. Note that in the non-black-box setting it is assumed that the decoder is "open" and because of the non-triviality of decryption a decryption key is available to the tracer. Black-Box Traitor Tracing Schemes where the tracing algorithm does not have access to keys (but only black box access to devices) are discussed in the next section.

3 Black-Box Traitor Tracing: Preliminaries

3.1 Colorings

Consider an MES with given $w, (e, K)$. A coloring of the user population is a partition $\cup_i C_i$ of $\mathcal{U}$. Let $s \in \mathcal{G}_w^v$ (an element from the extended ciphertext space) induces a coloring over $\mathcal{U}$ as follows: Define a relation over K: $d \equiv d'$ iff $D(1^w, d, s) = D(1^w, d', s)$. Note that if D is deterministic then this is an equivalence relation. The coloring can be defined as the set of all the equivalence classes of $\equiv$. If D is probabilistic (with negligible error) we define $\equiv$ as $d \equiv d'$ iff $\mathbf{Prob}[D(1^w, d, s) \neq D(1^w, d', s)]$ is negligible.

If $c \leftarrow E(1^w, m, e)$ for some $m \in \mathcal{M}$ (i.e., c is a "real or valid ciphertext") then it holds that for all $d, d' \in K$, $D(1^w, d, c) = D(1^w, d', c)$ (with high probability if D is probabilistic), therefore there is only one equivalence class induced by c, i.e. all users are colored by the same color (we call such a coloring *trivial*). Let $\mathcal{X}_1$ be the subset of $\mathcal{G}_w^v$ s.t. $\forall s \in \mathcal{X}_1$, s induces a trivial coloring (with negligible error). Obviously the valid ciphertexts constitute a subset of $\mathcal{X}_1$.

We say that an MES can induce a coloring $\cup_i C_i$ if there is an algorithm that produces a string s s.t. the string s induces the coloring $\cup_i C_i$ over the user population. Note that a decryption algorithm of some sort may not necessarily return one of the "color labels" i.e. the elements of the set $\{D(1^w, d, s) \mid d \in K\}$ (this can happen if the decryption algorithm operates with some "compound" decryption key – that has been derived from combining more than one of the users' keys).

3.2 Black-Box Traitor Tracing Schemes

The black-box tracing algorithm $\mathcal{R}$ and the pirate decoder algorithm $\mathcal{B}$ are probabilistic polynomial-time Turing machines with communication and output tapes. $\mathcal{B}$ incorporates a correct decoding algorithm: i.e. given a valid ciphertext it decrypts it, by running the decryption algorithm D with some key d that inverts e (note that d is not necessarily one of the user keys, but it is an element of $\mathcal{D}_e$ by the non-triviality of decryption property; also note that d may change from one decryption to the next). In the terminology of the previous section this means that if all traitor keys are colored in the same way the pirate decoder is bound to decrypt properly. If $\mathcal{B}$, on the other hand, finds that something is wrong with the encryption it may take measures to protect itself, e.g. it may return a random word. The set of user keys that are employed in the construction of $\mathcal{B}$ is denoted by $\mathcal{T}$ (due to key-user correspondence the set $\mathcal{T}$ can be also defined to be the set of traitor users). The tracing algorithm $\mathcal{R}$ is allowed oracle access to $\mathcal{B}$, namely, $\mathcal{R}$ can adaptively generate input strings s (queries) for $\mathcal{B}$ and $\mathcal{B}$, in response, will return a value (which is a correct decryption if s is a valid ciphertext).

From now on we will use the following notation: $\cup_{i=1}^{k(n)} C_i^n$ denotes a coloring induced over the user population by some s of $\mathcal{G}_w^v$; $c_i(n)$ will denote the cardinality of C_i^n. Note that for any n, it holds that $k(n), c_i(n) \in \{1, \ldots, n\}$; with this in mind we will use standard asymptotic notation to express the relation of these functions to n, e.g. $k(n) = \Theta(n)$ means that the number of colors is linear in n etc. We make the assumption that the functions $k(n), c_i(n)$ that are related to colorings produced by $\mathcal{R}$ are always in $\mathcal{F}$. Note that occasionally we may suppress "(n)" and write k instead of $k(n)$ etc.

Definition 2. *For $t, f \in \mathcal{F}$, we say that a polynomial-time (in n) probabilistic algorithm $\mathcal{R}$ is a $\langle t(n), f(n) \rangle$-tracer if for any set of traitors $\mathcal{T} \subseteq \mathcal{U}$ s.t. $|\mathcal{T}| \leq t(n)$ and for any polynomial-time pirate-decoder algorithm $\mathcal{B}$ that was created using the keys of $\mathcal{T}$, $\mathcal{R}^{\mathcal{B}}$ given all user keys, outputs a user with non-negligible probability in n, who is in the traitor set with probability at least $1/f(n)$.*

In this paper we consider tracers R which are non-ambiguous, i.e., when they probe the decoder they know that their queries are valid ciphertexts or invalid ones.

We will refer to the function f as the *uncertainty* of the tracer. Obviously obtaining a tracer with $\Theta(n)$ uncertainty for any MES is very simple: merely output any user at random achieves that. The other extreme is a tracer with uncertainty $\Theta(1)$ (ideally uncertainty=1), that no matter how large is the user population it returns a traitor with constant probability of success.

Remark 3. Consider the tracing approach of accusing any user at random. As stated above this has linear uncertainty and is obviously not useful in any setting. Suppose now that we have a lower bound on the number of traitors $\omega(t'(n))$; the uncertainty of this tracing approach becomes $n/t'(n)$ which can be sublinear if $t'(n)$ is not a constant. Nevertheless because we would like to rule it out as a way of tracing we say that the uncertainty is still linear — and therefore not acceptable (but it is linear in $n' = n/t'(n)$ instead of n); abusing the notation we may continue to write that the uncertainty in this case is $\Theta(n)$).

Definition 4. *For some $t, f, v \in \mathcal{F}$, a $\langle t(n), f(n), v(n) \rangle$-Black-Box Traitor Tracing Scheme (BBTTS), is an MES that (1) satisfies key-user correspondence and non-ambiguity of collusions for $t(n)$, (2) satisfies non-triviality of decryption, (3) it has $v(n)w$ ciphertext size, and (4) there is an $\langle t(n), f(n) \rangle$-tracer so that all colorings used by the tracer can be induced by the MES.*

We say that an MES is *incapable of Black-Box Tracing collusions of size* $t(n)$ if any polynomial-time tracer $\mathcal{R}$ has linear uncertainty (i.e., it is a $\langle t(n), \Theta(n) \rangle$-tracer).

The proof technique for establishing the fact that a BBTTS is incapable of black-box traitor tracing is the following: for any tracer $\mathcal{R}$ that can be defined in the BBTTS there is another algorithm $\mathcal{R}'$ that operates *without* oracle access to $\mathcal{B}$ so that the outputs of $\mathcal{R}$ and $\mathcal{R}'$ are essentially identical (i.e. they can be different in at most a negligible fraction of all inputs). More specifically the oracle $\mathcal{B}$ can be simulated without knowing any information pertaining to $\mathcal{B}$. In such a case we will state that the tracer essentially operates without interacting with the decoder and as a result it will be immediate that it has linear uncertainty (similar to the fact that any algorithm trying to guess a result of a coin flip without interacting with any agents which know the result of the coin cannot have probability greater than 1/2). A preliminary result on tracing follows; we show that strings that induce the trivial coloring over the user population are useless for tracing:

Proposition 5. *Queries which are elements of $\mathcal{X}_1$ do not help in reducing the uncertainty of a tracer.*

Proof. If the $\mathcal{R}$ algorithm uses an element of $\mathcal{X}_1$ for querying the pirate decoder then, the pirate decoder decrypts normally. This answer can be simulated by any decryption box. In particular, since the tracer is non-ambiguous it knows that it can generate the answer itself using any of the user keys (since it knows all user keys). □

We will assume that the number of traitors in any pirate decoder is sublinear in n, and as it is customary, we will give to the tracer the advantage of knowing a (sublinear) upper bound on the number of traitors. Additionally we would like to point out that our negative results on traitor tracing are not based on history-recording capabilities of the pirate decoder (i.e. $\mathcal{B}$ as an oracle does not have access to the previously asked queries). As a result the tracer is allowed to reset the decoder in its initial state after each query. In addition, our results apply even when the tracer has access to the randomness used by the pirate decoder.

4 Necessary Conditions for Black-Box Traitor Tracing

4.1 Combinatorial Condition

In this section we establish the fact that if the number of traitor keys is superlogarithmic in the user population size, it is not possible to trace without the decoder noticing it, unless queries of a specific type are used. We denote by $\cup_{i=1}^{k(n)} C_i^n \downarrow \mathcal{T}$ the projection of a coloring onto the traitor keys. Any pirate decoder can easily compute $\cup C_i^n \downarrow \mathcal{T}$; this is done by merely applying the decryption algorithm with each traitor key onto the given element s. Since this is a straightforward process we assume that any pirate decoder implements it. Obviously, if $\cup C_i^n \downarrow \mathcal{T}$ contains more than one color then the decoder "understands" it is being traced. In some systems, rather than projecting on individual traitor keys, one can project on combinations thereof (and thus reduce storage and computation requirements).

Theorem 6. *Suppose that a pirate decoder containing $t(n) = \omega(\log n)$ traitor keys, randomly distributed over all user keys, is given a query $s \in \mathcal{G}_w^v$ that induces a non-trivial coloring $\cup_{i=1}^{k(n)} C_i^n$ over the user population. Suppose further, that the coloring has the property $\neg(\exists i\ c_i(n) = n - o(n))$. Then, the probability that the pirate decoder does not detect it is being queried by the tracer is negligible in n.*

Proof. (recall that $|C_i^n| = c_i(n)$ for $i = 1, \ldots, k$; $c_1(n) + \ldots + c_k(n) = n$) Since $t(n)$ and $c_i(n)$ for $i = 1, \ldots, k$ are elements of $\mathcal{F}$, without loss of generality we assume that $c_i(n) \geq t(n)$ for all $i = 1, \ldots, \ell$ with $\ell \leq k$, for sufficiently large n. Obviously if $\ell = 0$ the decoder detects it is being traced.

Recall that we occasionally write t instead of $t(n)$ and c_i instead of $c_i(n)$. The total number of ways the pirate keys may be distributed over the user population are $\binom{n}{t}$. Similarly, the number of ways in which the decoder cannot detect that it is being traced is $\sum_{i=1}^{\ell} \binom{c_i}{t}$. The probability that the decoder cannot detect that it is being traced is $P := \frac{\sum_{i=1}^{\ell} \binom{c_i}{t}}{\binom{n}{t}} = \frac{(c_1)_t + \ldots + (c_\ell)_t}{(n)_t}$, where $(m)_v := m!/(m-v)!$. For sufficiently large n there will be a $m \in \{1, \ldots, \ell\}$ s.t. $c_m(n) \geq c_i(n)$ for all $i = 1, \ldots, k$.

The probability P is then: $\frac{(c_1)_t + \ldots + (c_\ell)_t}{(n)_t} \leq \frac{\ell (c_m)_t}{(n)_t} \leq \frac{n (c_m)_t}{(n)_t}$. Therefore we only need to show that $(c_m(n))_t/(n)_t$ is negligible in n. We consider two sub-cases: (i) There exists a real number $\alpha > 1$ such that $n \geq \alpha c_m(n)$ for sufficiently large n. Then, $(c_m)_t/(n)_t \leq (c_m)_t/(\alpha c_m)_t$. It holds that $\frac{c_m - i}{\alpha c_m - i} \leq \frac{1}{\alpha}$ for any $i = 0, \ldots, t-1$, (recall that $c_m \geq t$). Then $(c_m)_t/(n)_t \leq 1/\alpha^t$ which obviously is negligible since $\alpha > 1$ and $t = \omega(\log n)$: in details, $1/\alpha^t < 1/n^d$ for any constant d and sufficiently large n; equivalently $n^d < \alpha^t$ or $\alpha^{d \log_\alpha n} < \alpha^t$ or $t := t(n) > d \log_\alpha n$, which is true since $t(n) = \omega(\log n)$.
(ii) There is no $\alpha > 1$ with $n \geq \alpha c_m(n)$. Since $c_m(n) \leq n$ though, there has to be a function $f(n) \in \mathcal{F}$ s.t. $c_m(n) = n - f(n)$. If $f(n) = \Theta(n)$ there is a $0 < \beta \leq 1$ s.t. $f(n) \geq \beta n$. The case $\beta = 1$ is not possible since we deal with elements which induce coloring and $c_m = 0$ is impossible. In the case $\beta < 1$ we have that $n - f(n) \leq n - \beta n$ or equivalently $n \geq 1/(\beta - 1) \cdot c_m(n)$ therefore we are in case (i) since $1/(\beta - 1) > 1$ (i.e. $\alpha := 1/(\beta - 1)$). Finally if $f(n) = o(n)$ we fall into the case excluded by the theorem. □

The Theorem asserts that a decoder detects that it is being queried unless *most users are colored in the same way.* Namely, the negation of the Theorem's condition $\neg(\exists i \; c_i(n) = n - o(n))$ is that there is an i s.t. almost all users are colored in the same way ($c_i(n) = n - o(n)$). By "almost all" we mean that $c_i(n)/n \to 1$ when $n \to \infty$.

4.2 Negative Results

In this section we discuss how a pirate decoder can take advantage of Theorem 6 in order to protect itself. Specifically we show that there is a deterministic self-protecting strategy for any pirate decoder: *when the pirate decoder detects tracing it returns "0" (a predetermined output).* This strategy is successful for decoders containing enough traitor keys. The next Theorem asserts that any BBTTS whose underlying MES can only produce ciphertexts that are either valid or do not color most users in the same way (as discussed in the previous section) has $\Theta(n)$ uncertainty for any pirate decoder that incorporates $t(n) = \omega(\log n)$ traitor keys.

Theorem 7. *Given an $\langle t(n), f(n), v(n)\rangle$-BBTTS s.t. the underlying MES can only induce colorings $\cup_{i=1}^{k(n)} C_i^n$ with the property $(k(n) = 1) \vee \neg(\exists i\ c_i(n) = n - o(n))$ then it holds that if $t(n) = \omega(\log n)$ then $f(n) = \Theta(n)$.*

Proof. Assume that the decoder employs $t(n)$ traitor keys. The algorithm followed by the decoder is the following: before decrypting, it computes $\cup C_i^n \downarrow \mathcal{T}$. If all traitor keys are colored in the same way, it decrypts using any key. If there is more than one color the decoder returns "0".

The coloring conditions on the MES assures that an invalid ciphertext will be detected by the pirate decoder based on Theorem 6. Consequently the decoder on an invalid ciphertext will return "0" with overwhelming probability. On the other hand, any element in $\mathcal{X}_1$ will be properly decrypted. Since the tracer is non-ambiguous, the oracle can be simulated with overwhelming probability. So the tracer essentially operates without interacting with the decoder. By remark 3 the uncertainty of the scheme is $\Theta(n)$. □

The pirate decoder strategy used in the proof above can be defeated by a tracer that is able to produce colorings s.t. $n - o(n)$ users are colored in the same way. This is achieved in the MES of [CFNP00], and a black-box traitor tracing method which uses such colorings is presented there.

4.3 Negative Results for Black-Box Confirmation

Black-Box Confirmation is an alternative form of revealing some information about the keys hidden in the pirate decoder. Suppose that the tracer has some information that traitors are included in a set of suspects $\mathcal{S}$ and wants to confirm this. The fact that the tracer has some information about the traitor keys means that they are not randomly distributed over all users' keys and therefore Theorem 6 is not applicable (in fact, biasing the distribution of a potential adversary is, at times, a way to model suspicion). Under such modeling, we can show a strong negative result for single-query black-box confirmation, i.e. when a single query is sent to the pirate-decoder that induces the same color on the suspects and different color(s) on other users. If the pirate decoder returns the color label associated to the suspect set then the suspicion is confirmed (note that this is exactly the black-box confirmation method used in [BF99]).

The change of the distribution of the traitor keys can be modeled as follows: the probability $\mathbf{Prob}[i \in \mathcal{T} | i \in \mathcal{S}] = \alpha(n)\mathbf{Prob}[i \in \mathcal{T}]$ where $\alpha(n) > 1$ for sufficiently large n; note that when the tracer has no information it holds that $\alpha(n) = 1$. Let us fix t the size of the traitor set. We will denote the distribution of t-sets of potential traitor keys by $\mathcal{D}_{\mathcal{S},\alpha}$, and refer to $\alpha(n)$ as the advantage of the tracer. For example, for $t = 1$ the probability of all $\mathcal{T}$ inside $\mathcal{S}$ is α/n, whereas the probability of all other $\mathcal{T}$'s is $\frac{n-\alpha s}{n(n-s)}$. As usual, we allow the tracer to know an upper bound on the number of traitors' keys and therefore $|\mathcal{S}| \geq |\mathcal{T}|$.

Lemma 8. *Let $\mathcal{S}$ be a set of users such that $s(n) := |\mathcal{S}|$ and an $\alpha(n) \in \mathcal{F}$ such that $s(n)\alpha(n) \leq cn$ for some $c \in (0, 1)$. Suppose that a pirate decoder employing*

$t(n) = \omega(\log n)$ traitor keys, distributed according to $\mathcal{D}_{\mathcal{S},\alpha}$, is given a query that induces the following coloring over the user population: the users in $\mathcal{S}$ are colored in the same way and the remaining users in different color(s). Then, the probability that the traitor set is included in the suspect set is negligible in n.

Proof. For simplicity we write s, α instead of $s(n), \alpha(n)$ respectively. We show that the probability $\mathbf{Prob}[\mathcal{T} \subseteq \mathcal{S}]$, when $\mathcal{T}$ is distributed according to $\mathcal{D}_{\mathcal{S},\alpha}$, is negligible.

It is easy to see that $\mathbf{Prob}[\mathcal{T}] = \alpha^t/\binom{n}{t}$ (when $\mathcal{T}$ is distributed according to $\mathcal{D}_{\mathcal{S},\alpha}$ and $\mathcal{T} \subseteq \mathcal{S}$), and as a result $\mathbf{Prob}[\mathcal{T} \subseteq \mathcal{S}] = \alpha^t\binom{s}{t}/\binom{n}{t}$. The fact that $s\alpha \leq cn$ implies $\frac{\alpha(s-i)}{n-i} \leq c$ for any $i > 0$; as a result it holds that $\alpha^t\binom{s}{t}/\binom{n}{t} < c^t$. Since $0 < c < 1$ and $t = \omega(\log n)$ the probability is negligible. □

Theorem 9. *Single-Query Black-Box Confirmation with a suspect set $\mathcal{S}$ and confidence $\alpha(n)$ is not possible against any pirate-decoder which contains $t(n) = \omega(\log n)$ traitor keys, provided that $|\mathcal{S}|\alpha(n) \leq cn$ for some constant $c \in (0,1)$.*

Proof. Suppose that the pirate decoder returns "0" when it detects an invalid ciphertext. Then, by lemma 8 with overwhelming probability not all the traitors are in the suspect set, thus the pirate decoder will return the color label of the suspect set with negligible probability in n. As a result single-query black-box confirmation will fail. □

Note the trade-off between the size of $\mathcal{S}$ and the advantage $\alpha(n)$. How large should be the advantage of the tracer so that single-query black-box confirmation is possible? it should hold that $\alpha(n)|\mathcal{S}| = n - o(n)$. In this case it holds that $\mathbf{Prob}[i \in \mathcal{S} | i \in \mathcal{T}] = \mathbf{Prob}[(i \in \mathcal{S}) \wedge (i \in \mathcal{T})]/\mathbf{Prob}[i \in \mathcal{T}] = \alpha(n)\mathbf{Prob}[(i \in \mathcal{S}) \wedge (i \in \mathcal{T})]/\mathbf{Prob}[i \in \mathcal{T} | i \in \mathcal{S}] = \alpha(n)\mathbf{Prob}[i \in \mathcal{S}] \rightarrow 1$, when $n \rightarrow \infty$ (under the condition that $\alpha(n)|\mathcal{S}| = n - o(n)$). This, together with the above Theorem imply:

Corollary 10. *Single-query Black-box confirmation is impossible against any pirate decoder that includes $t(n) = \omega(\log n)$ traitor keys, unless the probability that a user is a suspect given that it is a traitor is arbitrarily close to 1.*

Some remarks should be placed herein: (1) $\mathbf{Prob}[i \in \mathcal{S} | i \in \mathcal{T}]$ is arbitrarily close to 1, means that the confidence level of the tracer is so high that it "forces" $\mathcal{T}$ to be a subset of $\mathcal{S}$ (for more discussion on confirmation in this case and the relation to the black-box confirmation results of [BF01] see subsection 5.1). (2) We do not rule-out black-box confirmation with smaller confidence levels in different models or by multiple-queries that do not directly color the suspect set in a single color and the remaining users differently.

5 From Necessary Conditions to Concrete Systems

In this section, we apply our generic necessary condition results to concrete systems. We actually analyze specific properties of the schemes of [BF99,KD98];

these properties in combination with the generic results reveal inherent black-box tracing limitations of these schemes in the self-protecting model. This demonstrates that these schemes are, in fact, sensitive to the self-protection property of our model and the number of traitors. This shows the power of the self-protecting pirate model, since in more restricted pirate models (restricting the power of the pirate decoder or the number of traitors) tracing was shown possible, whereas we get negative results for the more general model defined here. We note that below we will assume that self-protection involves decryption with traitor keys. However, achieving self-protection using a linear combination of traitor keys is possible as well; in which case the actual traitor keys are not necessarily stored and the storage as well as the computation of the pirate can be reduced.

Our results can be seen as a separation of the schemes of [BF99,KD98] and the scheme of [CFN94,CFNP00] with respect to black-box traceability. In the latter scheme our self-protection method fails to evade tracing, since the ciphertext messages induce colorings which fall into the exception case of Theorem 6 and the tracing method, in fact, employs such ciphertexts.

5.1 The [BF99]-Scheme

Description. We present the basic idea of the Boneh and Franklin scheme [BF99]. All base operations are done in a multiplicative group G_q in which finding discrete logs is presumed hard, whereas exponent operations are done in Z_q. Vectors (denoted in bold face) are in Z_q^v and $\boldsymbol{a} \cdot \boldsymbol{b}$ denotes the inner product of $\boldsymbol{a}$ and $\boldsymbol{b}$. Given a set $\Gamma := \{\boldsymbol{\gamma_1}, \ldots, \boldsymbol{\gamma_n}\}$ where $\boldsymbol{\gamma_i}$ is a vector of length v, and given random $\boldsymbol{r} := \langle r_1, \ldots, r_v \rangle$ and $c \in Z_q$, we select $\boldsymbol{d_i} = \theta_i \boldsymbol{\gamma_i}$, $i = 1, \ldots, n$ such that $\forall i \; \boldsymbol{r} \cdot \boldsymbol{d_i} = c$, where n is the number of users (i.e. we select $\theta_i := c/(\boldsymbol{r} \cdot \boldsymbol{\gamma_i})$). The vector $\boldsymbol{\gamma_i}$ is selected as the i-th row of an $(n \times v)$-matrix B where the columns of B form a base for the null space of A, where A is an $(n - v) \times n$ matrix where the i-th row of A is the vector $\langle 1^i, 2^i, \ldots, n^i \rangle$, $i = 0, \ldots, n - v - 1$.

The public key is $\langle y, h_1, \ldots, h_v \rangle$, where $h_j = g^{r_j}$ and $y = g^c$, where g is a generator of G_q. Note that all vectors $\boldsymbol{d_i}$ are representations of y w.r.t the base $h_1, \ldots, h_v$. Vector $\boldsymbol{d_i}$ is the secret key of user i. Encryption is done as follows: given a message $M \in G_q$, a random $a \in Z_q$ is selected and the ciphertext is $\langle My^a, h_1^a, \ldots, h_v^a \rangle$. Given a ciphertext, decryption is done by applying $\boldsymbol{d_i}$ to the "tail" of the ciphertext: $h_1^a, \ldots, h_v^a$ pointwise, in order to obtain y^a by multiplication of the resulting points, and then M is recoverable by division (cf. ElGamal encryption). In [BF99] a tracing algorithm is presented showing that the scheme described above is a $\langle t(n), 1, 2t(n) \rangle$-TTS. It is also shown that their scheme is black-box against pirate decoders of specific implementations ("single-key pirate", "arbitrary pirates"). We next investigate further black-box capabilities of the [BF99]-scheme.

Negative Results. Suppose that we want to induce a coloring $\cup_{i=1}^{k(n)} C_i^n$ in the [BF99] scheme. Given a (possibly invalid) ciphertext $\langle C, g^{r_1 x_1}, \ldots, g^{r_v x_v} \rangle$, user i decrypts as follows: $C/g^{r_1 x_1 (\boldsymbol{d_i})_1 + \ldots + r_v x_v (\boldsymbol{d_i})_v}$. Thus, we can color user i by the color label $C/g^{\theta_i c_i}$ (the value of the decryption by the user) provided that we find the $x_1, \ldots, x_v$ such that $r_1 x_1 (\boldsymbol{d_i})_1 + \ldots + r_v x_v (\boldsymbol{d_i})_v = \theta_i c_i$. This can be done by finding a $\boldsymbol{z} := \langle z_1, \ldots, z_v \rangle$ s.t. $\boldsymbol{\gamma_i} \cdot \boldsymbol{z} = c_i$ for all $i = 1, \ldots, n$. Given such a $\boldsymbol{z}$ we can compute the appropriate x-values to use in the ciphertext as follows: $x_j = z_j (r_j)^{-1}$ for $j = 1, \ldots, v$. Note that for valid ciphertexts it holds that $\boldsymbol{z} = a\boldsymbol{r}$ for some $a \in Z_q$ (and as a result $x_1 = \ldots = x_v = a$).

Next we present a property of the Boneh-Franklin scheme, showing that an invalid ciphertext (namely, a ciphertext which induces more than one color), cannot color too many users by the same color.

Theorem 11. *In the [BF99]-MES, given a (possibly invalid) ciphertext that induces a coloring over the user population so that v users are labelled by the same color then all users are labelled by the same color.*

Proof. Suppose that the ciphertext $\langle C, g^{r_1 x_1}, \ldots, g^{r_v x_v} \rangle$ colors user i by label $C/g^{\theta_i c_i}$ to user i, and that v users are colored by the same label. Let $c'_i := \boldsymbol{r} \cdot \boldsymbol{\gamma_i}$, for $i = 1, \ldots, n$. Without loss of generality assume that users $1, \ldots, v$ are colored by the same label. Then it holds that $\theta_1 c_1 = \ldots = \theta_v c_v$ or equivalently $c_1/c'_1 = \ldots = c_v/c'_v$. Let $a := c_1/c'_1$. Then we have that $c_1 = ac'_1, \ldots, c_v = ac'_v$.

Define $\boldsymbol{z} = \langle z_1, \ldots, z_v \rangle$ s.t. $z_j = r_j x_j$ for $j = 1, \ldots, v$. It follows that $\boldsymbol{\gamma_i} \cdot \boldsymbol{z} = c_i$, for $i = 1, \ldots, n$ (we call this system of equations system 1). Because it holds that $\boldsymbol{\gamma_i} \cdot (a\boldsymbol{r}) = ac'_i$ for $i = 1, \ldots, v$ (and this will hold for any v users) it follows that $\boldsymbol{z} = a\boldsymbol{r}$ provided (which we show next) that $\boldsymbol{\gamma_1}, \ldots, \boldsymbol{\gamma_v}$ are linearly independent (since in this case system 1 is of full rank, and as a result it has a unique solution). Since $\boldsymbol{z} = a\boldsymbol{r}$ it follows that $x_1 = \ldots = x_v = a$, i.e. the ciphertext $\langle C, g^{r_1 x_1}, \ldots, g^{r_v x_v} \rangle$ is valid.

To complete the proof we have to show that any v vectors of $\Gamma = \{\boldsymbol{\gamma_1}, \ldots, \boldsymbol{\gamma_n}\}$ are linearly independent. Suppose, for the sake of contradiction, that $\boldsymbol{\gamma_1}, \ldots, \boldsymbol{\gamma_v}$ are linearly dependent. Recall that $\boldsymbol{\gamma_i}$ is the i-th row of a $(n \times v)-$matrix B where the columns of B constitute a base of the null space of the $(n-v) \times n$-matrix A. Let us construct another base as follows: the null space of A contains all n-vectors $\boldsymbol{x} := \langle x_1, \ldots, x_n \rangle$ such that $A\boldsymbol{x}^T = 0$. Choose $x_1, \ldots, x_v$ as free variables and solve the system $A\boldsymbol{x}^T = 0$ (the system is solvable since if we exclude any v columns of A the matrix becomes the transpose of a Vandermonde matrix of size $n - v$; due to this fact the choice of the "first" v $\boldsymbol{\gamma}$ vectors is without loss of generality). Solving the system like this will generate a base B' for the null space of A so that the first v rows of B' contain the identity matrix of size v. But then it is easy to see that there are vectors in the span of B' that do not belong in the span of B, a contradiction. As a result $\boldsymbol{\gamma_1}, \ldots, \boldsymbol{\gamma_v}$ should be linearly independent. The same argument holds for any other v vectors of Γ. □

By theorem 7 we know that almost all users $(n - o(n))$ should be colored in the same way in order for the pirate-decoder to be unable to detect tracing. However, by the previous Theorem it holds that at most $v-1$ users can be colored in the same way (otherwise the coloring becomes trivial which means that the ciphertext does not constitute a query which helps in tracing by Proposition 5). As a result it should hold that $v = n - o(n)$; note that in this case $v/n \to 1$ if $n \to \infty$. As a result we obtain the following corollary:

Corollary 12. *Let $\langle t(n), f(n), v(n)\rangle$-BBTTS be a scheme based on the [BF99]-MES. If $t(n) = \omega(\log n)$ then it holds that either $f(n) = \Theta(n)$ or that $v(n) = n - o(n)$.*

Essentially this means that the [BF99]-scheme is incapable of black-box tracing superlogarithmic self-protecting traitor collusions unless the ciphertext size is linear in the number of users.

Regarding single-query black-box confirmation (introduced in [BF99]) we showed that when suspicion is modeled as biasing the uniform distribution, where suspects are distinguished by increasing the probabilistic confidence in them being traitors, then as a result of section 4.3 it holds that:

Corollary 13. *In the [BF99]-scheme, Single-query Black-box confirmation is impossible against a pirate decoder which includes $t(n) = \omega(\log n)$ traitor keys, unless the probability that a user is a suspect given that it is a traitor, is arbitrarily close to 1.*

Note: in [BF01], a more sophisticated combination of black-box confirmation with traitor tracing is presented. The scheme is a single-query black-box confirmation in principle, but multiple queries that induce different colorings *within* the suspect set are employed, until a traitor is pinned down. Our negative results for black-box confirmation (in the self-protecting model variant) apply to this setting as well. The arguments in [BF01] are plausible in the "arbitrary pirates" model (including self-protecting one). For the method to work, however, they assume "compactness" (called confirmation requirement), namely that it is *given* that all traitors are within the suspect list. Our results point out that without this compactness, relying solely on likelihood (modeled as probability), successful confirmation is unlikely unless there is a very high confidence level (which will enforce the "compactness condition" almost always). Our results do not dispute black-box confirmation under compactness but rather point to the fact that obtaining (namely, biasing a uniform distribution to get) a "tight" suspect set $\mathcal{S}$ which satisfies compactness at the same time can be hard.

5.2 The [KD98]-Scheme

Description. The scheme of Kurosawa and Desmedt is defined as follows: a random secret polynomial $f(x) = a_0 + a_1 x + \ldots a_v x^v$ is chosen and the values $g^{a_0}, \ldots, g^{a_v}$ are publicized (the public key of the system). User i is given $f(i)$ as its secret key. A message s is encrypted as follows: $\langle g^r, sg^{ra_0}, g^{ra_1}, \ldots, g^{ra_v}\rangle$, were r is chosen at random. User i decrypts as follows: $sg^{ra_0}g^{ra_1 i}\ldots g^{ra_v i^v}/g^{rf(i)} = s$. It is more convenient to think of the secret key of user i as $\langle f(i), \boldsymbol{i}\rangle$ where $\boldsymbol{i} := \langle 1, i, i^2, \ldots, i^v\rangle$. In [KD98] it was proven that their scheme satisfies key-user correspondence for collusions of up to v users provided the discrete-log problem is hard. However non-ambiguity of collusions was overlooked, something pointed out in [SW98a] and in [BF99].

The problem arises from the fact that the set of possible keys used also includes linear combinations of user keys: $\langle\sum_{m=1}^{t}\alpha_m f(i_m), \sum_{m=1}^{t}\alpha_m \boldsymbol{i_m}\rangle$ where

$\alpha_m \in Z_q$ with $\sum_{m=1}^{t} \alpha_m = 1$ and $i_1, \ldots, i_t \in \{1, \ldots, n\}$. This tuple can also be used for decryption since: given $\langle g^r, sg^{ra_0}, g^{ra_1}, \ldots, g^{ra_v} \rangle$, one may compute

$$sg^{ra_0} g^{ra_1 \sum_{m=1}^{t} \alpha_m (\boldsymbol{i}_m)_1} \ldots g^{ra_v \sum_{m=1}^{t} \alpha_m (\boldsymbol{i}_m)_v} / g^{r \sum_{m=1}^{t} \alpha_m f(i_m)} = s$$

To achieve non-ambiguity of collusions we would like to show that given any two subsets of users $i_1, \ldots, i_t$ and $j_1, \ldots, j_t$ it should hold that $\{\sum_{m=1}^{t} \alpha_m \boldsymbol{i}_m \mid \alpha_1, \ldots, \alpha_m\} \cap \{\sum_{m=1}^{t} \alpha_m \boldsymbol{j}_m \mid \alpha_1, \ldots, \alpha_m\} = \emptyset$. Something that can be true only if $v \geq 2t$ i.e. v should be twice the size of the biggest traitor collusion allowed. In the light of this, it is not known if it is possible to trace traitors in this scheme (even in the non-black-box setting). The only known approach is the brute-force "black-box confirmation" for all possible traitor subsets suggested in [BF99] that needs exponential time (unless the number of traitors is assumed to be a constant). Despite this shortcoming the [KD98]-scheme is a very elegant public-key MES that inspired further work as seen in the schemes of [BF99, NP00]. In the rest of the section we show that the [KD98]-scheme has similar black-box traitor tracing limitations as the [BF99]-scheme.

Negative Results. Suppose we want to induce the coloring $\cup_{i=1}^{k} C_i^n$ in the [KD98]-MES. Given a (possibly invalid) ciphertext $\langle g^r, sg^{x_0 a_0}, g^{x_1 a_1}, \ldots, g^{x_v a_v} \rangle$, user i applies $\langle f(i), \boldsymbol{i} \rangle$ to obtain $sg^{\sum_{j=0}^{v} x_j a_j (\boldsymbol{i})_j - rf(i)} = sg^{\sum_{j=0}^{v} (x_j - r) a_j (\boldsymbol{i})_j}$. So we can color each user by a color-label sg^{c_i}, if we find a $\boldsymbol{z}$ s.t. $\boldsymbol{z} \cdot \boldsymbol{i} = c_i$ for all $i = 1, \ldots, n$; given such a $\boldsymbol{z}$ we can compute the appropriate $x_0, \ldots, x_v$ values to use in the ciphertext as follows: $x_j = z_j (a_j)^{-1} + r$ for $j = 0, \ldots, v$. The set of all valid ciphertexts corresponds to the choice $\boldsymbol{z} = \boldsymbol{0}$ (and in this case it follows that $x_0 = \ldots = x_r$), nevertheless the choice of $\boldsymbol{z} = \langle a, 0, \ldots, 0 \rangle$ also colors all users in the same way although in this case the decryption yields sg^a (instead of s).

Next we present a property of the Kurosawa-Desmedt scheme, showing that an invalid ciphertext (which induces more than one color), cannot color too many users by the same color.

Theorem 14. *In the [KD98]-MES, given a (possibly invalid) ciphertext that induces a coloring over the user population so that $v+1$ users are labelled by the same color then all users are labelled by the same color.*

Proof. Suppose that the ciphertext $\langle g^r, sg^{x_0 a_0}, g^{x_1 a_1}, \ldots, g^{x_v a_v} \rangle$ induces a color-label sg^{c_i} on user i so that $v+1$ users are colored in the same way. Without loss of generality we assume that $c_1 = \ldots = c_{v+1}$. Define $z_j := (x_j - r) a_j$ for $j = 0, \ldots, v$. It follows that $\boldsymbol{i} \cdot \boldsymbol{z} = c_i$ for $i = 1, \ldots, n$. Seen as a linear system with $\boldsymbol{z}$ as the unknown vector the equations $\boldsymbol{i} \cdot \boldsymbol{z} = c_i$ for $i = 1, \ldots, n$ suggest that $\boldsymbol{z}$ corresponds to the coefficients of a polynomial $p(x) := z_0 + z_1 x + \ldots z_v x^v$ such that $p(i) = c_i$ for $i = 1, \ldots, n$. Because $p(1) = \ldots = p(v+1)$ and the degree of p is at most v it follows immediately that p has to be a constant polynomial, i.e. $\boldsymbol{z} = \langle a, 0, \ldots, 0 \rangle$ with $a = p(1) = \ldots = p(v+1)$. (Any $v+1$ equal value points on the polynomial will imply the above, which justifies the arbitrary choice of users). If follows immediately that user i receives the color label $sg^{c_i} = sg^{\boldsymbol{i} \cdot \boldsymbol{z}} = sg^a$ and as a result all users are labeled by the same color. □

With similar arguments as in section 5.1 we conclude:

Corollary 15. *Let $\langle t(n), f(n), v(n)\rangle$-BBTTS be a scheme based on the [KD98]-MES. If $t(n) = \omega(\log n)$ then it holds that either $f(n) = \Theta(n)$ or that $v(n) = n - o(n)$.*

Essentially this means that the [KD98]-scheme is incapable of black-box tracing superlogarithmic self-protecting traitor collusions unless the ciphertext size is linear in the number of users.

Corollary 16. *In the [KD98]-scheme, Single-query Black-box confirmation is impossible against a pirate decoder which includes $t(n) = \omega(\log n)$ traitor keys, unless the probability that a user is a suspect given that it is a traitor, is arbitrarily close to 1.*

References

[BF99] Dan Boneh and Matthew Franklin, *An Efficient Public Key Traitor Tracing Scheme*, CRYPTO 1999.

[BF01] Dan Boneh and Matthew Franklin, *An Efficient Public Key Traitor Tracing Scheme*, manuscript, full-version of [BF99], 2001.

[CFN94] Benny Chor, Amos Fiat, and Moni Naor, *Tracing Traitors*, CRYPTO 1994.

[CFNP00] Benny Chor, Amos Fiat, and Moni Naor, and Benny Pinkas, *Tracing Traitors*, IEEE Transactions on Information Theory, Vol. 46, no. 3, pp. 893-910, 2000. (journal version of [CFN94,NP98]).

[DLN96] Cynthia Dwork, Jeff Lotspiech and Moni Naor, *Digital Signets: Self-Enforcing Protection of Digital Content*, STOC 1996.

[FT99] Amos Fiat and T. Tassa, *Dynamic Traitor Tracing*, CRYPTO 1999.

[GSY99] Eli Gafni, Jessica Staddon and Yiqun Lisa Yin, *Efficient Methods for Integrating Traceability and Broadcast Encryption*, CRYPTO 1999.

[KD98] Kaoru Kurosawa and Yvo Desmedt, *Optimum Traitor Tracing and Asymmetric Schemes*, Eurocrypt 1998.

[NP98] Moni Naor and Benny Pinkas, *Threshold Traitor Tracing*, CRYPTO 1998.

[NP00] Moni Naor and Benny Pinkas, *Efficient Trace and Revoke Schemes* , In the Proceedings of Financial Crypto '2000, Anguilla, February 2000.

[Pfi96] Birgit Pfitzmann, *Trials of Traced Traitors*, Information Hiding Workshop, Spring LNCS 1174, pp. 49-63, 1996.

[PS96] Birgit Pfitzmann and Matthias Schunter, *Asymmetric Fingerprinting*, Eurocrypt 1996.

[PW97] Birgit Pfitzmann and M. Waidner, *Asymmetric fingerprinting for larger collusions*, in proc. ACM Conference on Computer and Communication Security, pp. 151–160, 1997.

[SW98a] Douglas Stinson and Ruizhong Wei, *Key preassigned traceability schemes for broadcast encryption*, In the Proceedings of SAC'98, Lecture Notes in Computer Science 1556, Springer Verlag, pp.144–156, 1998.

[SW98b] Douglas R. Stinson and R. Wei, *Combinatorial Properties and Constructions of Traceability Schemes and Frameproof Codes*, SIAM J. on Discrete Math, Vol. 11, no. 1, 1998.

Minimal Complete Primitives for Secure Multi-party Computation

Matthias Fitzi[1], Juan A. Garay[2], Ueli Maurer[1], and Rafail Ostrovsky[3]

[1] Dept. of Computer Science, Swiss Federal Institute of Technology (ETH), CH-8092 Zürich, Switzerland. {fitzi,maurer}@inf.ethz.ch.
[2] Bell Labs – Lucent Technologies, 600 Mountain Ave., Murray Hill, NJ 07974, USA. garay@research.bell-labs.com.
[3] Telcordia Technologies Inc., 445 South Street, Morristown, New Jersey 07960-6438, USA. rafail@research.telcordia.com

Abstract. The study of minimal cryptographic primitives needed to implement secure computation among two or more players is a fundamental question in cryptography. The issue of complete primitives for the case of two players has been thoroughly studied. However, in the multi-party setting, when there are $n > 2$ players and t of them are corrupted, the question of what are the simplest complete primitives remained open for $t \geq n/3$. We consider this question, and introduce complete primitives *of minimal cardinality* for secure multi-party computation. The cardinality issue (number of players accessing the primitive) is essential in settings where the primitives are implemented by some other means, and the simpler the primitive the easier it is to realize it. We show that our primitives are complete and of minimal cardinality possible.

1 Introduction

In this paper, with respect to the strongest, active adversary, we initiate the study of minimal complete primitives for multi-party computation from the point of view of the cardinality of the primitive — i.e., the number of players accessing it. A primitive is called *complete* if any computation can be carried out by the players having access (only) to the primitive and local computation. A primitive is called *minimal* if any primitive involving less players is not complete.

For n players, t of which might be corrupted, the question is well understood for $t < n/3$. In this paper we consider this question for $t \geq n/3$. We show that in fact there are three interesting "regions" for t: $t < n/3$, $n/3 \leq t < n/2$, and $n/2 \leq t < n$, and present, for each region, minimal complete primitives for t-resilient unconditional multi-party computation.

1.1 Prior and Related Work

Secure multi-party computation. Secure multi-party computation (MPC) has been actively studied since the statement of the problem by Yao in [Yao82].

J. Kilian (Ed.): CRYPTO 2001, LNCS 2139, pp. 80–100, 2001.

For the standard model with secure pairwise channels between the players, the first general solution of the problem was given by Goldreich, Micali, and Wigderson [GMW87] with respect to computational security. Ben-Or, Goldwasser, and Wigderson [BGW88] and Chaum, Crépeau, and Damgård [CCD88] constructed the first general protocols with unconditional security. Additionally, it was proven in [BGW88] that unconditionally secure MPC was possible if and only if less than half (one third) of the players are corrupted passively (actively).

For the model where, in addition to the pairwise secure channels, a global broadcast channel is available, Rabin and Ben-Or [RB89] constructed a protocol that tolerates (less than) one half of the players being actively corrupted. Their solution is not perfect, as it carries a small probability of error. However, it was later shown by Dolev, Dwork, Waarts and Yung [DDWY93] that this is unavoidable for the case $t \geq \lceil n/3 \rceil$ (and the assumed communication primitives), as there exist problems with no error-free solutions in this setting. Fitzi and Maurer [FM00] recently proved that, instead of global broadcast, broadcast among three players is sufficient in order to achieve unconditionally secure MPC for $t < n/2$.

Complete primitives. Another line of research deals with the completeness of primitives available to the players. Kilian [Kil88] proved that oblivious transfer (OT) [Rab81] is complete for two-party computation in the presence of an active adversary. A complete characterization of complete functions for two-party computation, for both active and passive adversaries, was given in [Kil00] based on [Kil91] and results by Beimel, Micali, and Malkin [BMM99]. These results are stated with respect to *asymmetric* multi-party computation in the sense that the result of the function is provided to one single (predefined) player.

A first generalization of completeness results to the more general n-party case was made by Kilian, Kushilevitz, Micali, and Ostrovsky [KMO94,KKMO99], who characterized all complete boolean functions for multi-party computation secure against a passive adversary that corrupts any number of players.

With the noted exception of Goldreich's treatment of reductions in [Gol00], previous work on complete primitives typically assumes that the cardinality of the primitive is the same as the number of players involved in the computation. In contrast, in this paper we are concerned with the minimal cardinality of complete primitives for multi-party computation.

1.2 Our Results

In this paper, for any primitive cardinality k, $2 \leq k \leq n$, we give upper and lower bounds on t such that there is a complete primitive g_k for multi-party computation secure against an active adversary corrupting that many players. With one exception, all these bounds are tight. In particular, for each resiliency "region" $t < \frac{n}{3}$, $t < \frac{n}{2}$, and $t < n$, we present minimal complete primitives for t-resilient unconditional multi-party computation. To our knowledge, this is the first time that the power of the cardinality of cryptographic primitives — and their minimality — is rigorously studied.

Table 1. *Complete primitives of cardinality k*

Primitive cardinality	Resiliency		Primitive	Number of instances
	Efficient reduction	Lower bound		
$k = 2$	$t < n/3$	$t < n/3$	SC_2	$2\binom{n}{2}$
$k = 3$	$t < n/2$	$t < n/2$	OC_3/CC_3	$3\binom{n}{3}$
$4 \leq k \leq n-1$	$t < n/2$	$t < n-2$	OC_3/CC_3	$3\binom{n}{3}$
$k = n$	$t < n$	$t < n$	UBB_n	1

SC_2: Secure Channel, OC_3: Oblivious Cast, CC_3: Converge Cast, UBB_n: Universal Black Box. OC_3, CC_3 and UBB_n are primitives introduced in this paper.

When $k = 2$, it is well known that *secure pairwise channels* (or, more generally, OT) are enough (complete) for $t < n/3$, as it follows from [BGW88, CCD88] and [Kil88]. We show that, for $n > 2$, no primitive of cardinality 2 can go above this resiliency bound, including primitives that are complete for 2-party computation.

The case $k = 3$ is of special interest. We introduce two primitives: *oblivious cast* [Bla96], a natural generalization of oblivious transfer to the three-party case, and *converge cast*, a primitive that is related to the anonymous channel of [Cha88], and show that they are complete for $t < n/2$. In light of the impossibility result for $k = 2$, these primitives are also minimal.

For the case $k = n$ we introduce a new primitive, which we call the *universal black box* (UBB), and show that it is complete for arbitrary resiliency ($t < n$). This primitive has interesting implications for computations involving a trusted third party (TTP), in that it enables *oblivious* TTPs, i.e., trusted parties that do not require any prior knowledge of the function to be computed by the players — even if a majority of the players are corrupted. The UBB is also minimal, since we also show that no primitive of cardinality $n - 1$ can be complete for $t < n$. These results are summarized in Table 1.

Multicast and "convergecast," with a single sender and a single recipient, respectively, constitute two natural communication models. We also show that no primitive that conforms to these types — even of full cardinality — can achieve more than $t < n/2$, and therefore be more powerful than our primitives of cardinality 3. In other words, with respect to these types, Table 1 "collapses" to two equivalence classes: $k = 2$ and $3 \leq k \leq n$.

All the primitives we present allow for *efficient* multi-party computation.

2 Model and Definitions

In this paper we focus on *secure function evaluation* (SFE) [Yao82] by a set P of n players, where each player p_i has an input value x_i and obtains an output value $f_i(x_1, x_2, \cdots x_n)$, for a (probabilistic) function f_i. We are interested in unconditional security against an active adversary who may corrupt up to t of the players; i.e., the adversary may make the corrupted players deviate from the protocol in an arbitrarily malicious way, and no assumptions are made about his computational power.

In contrast to the treatment of two-party computation (e.g. [Kil91,Kil00] and [BMM99]), where only one predefined player receives the final result of the computation, our model allows every player to receive his own (in general, different) result — which corresponds to the general notion of multi-party computation in [Yao82,CCD88,BGW88]. Similarly, our definition of a primitive, as given in the next paragraph, also allows every involved player to provide an input and get an output, as opposed to just one player. Nonetheless, our constructions apply to the former model as well since for each of our complete multiple-output primitives there is also a single-output primitive that is complete with respect to single-output SFE.

Primitives of arbitrary cardinality. Our communication model is based on ideal *primitives* that can be accessed by k players, $2 \leq k \leq n$, implementing the secure computation of some k-ary, possibly probabilistic function; k is called the *cardinality* of the primitive. Besides this primitive, no other means of communication is assumed among the players.

We view primitives as "black boxes" in the sense that all implementation details are hidden from the players. Depending on the function being implemented, of the k players accessing the primitive one or more may secretly enter an input to it, and one or more may secretly receive the value(s) of the function.

We use $g_k[i,j]$ to denote the primitive implementing k-ary function g, in which $i \leq k$ players provide an input, and where $j \leq k$ players receive the output of the function.[1] We call $[i,j]$ the *type* of the primitive. We will drop the type when clear from the context. We focus on the following types: $[1,1]$, $[1,k]$, $[k,1]$, and $[k,k]$.[2]

Note that a primitive of a given cardinality can always be simulated (when applicable) by the same primitive with a larger cardinality by cutting some of the "wires." More formally, the following domination relation exists: Let $(k',i',j') \supseteq (k,i,j)$ (meaning $k' \geq k$, $i' \geq i$ and $j' \geq j$); then for every primitive $g_k[i,j]$ there exists a primitive $g'_{k'}[i',j']$ that is as powerful as $g_k[i,j]$.

We assume that every subset $S \subset P$ of k players shares $k!$ instances of the primitive — one for each permutation of the players; thus, we assume $\binom{n}{k}k!$ instances of the primitive in total. However, we will show that there is always a (minimal complete) primitive such that, overall, polynomially-many instances (specifically, less than n^3) of the primitive are sufficient.

Security model. Several formal definitions of secure function evaluation exist (e.g., [Bea92,Can00,Gol00,GL90,MR92]). The process is assumed to be synchronous, a fact that simplifies the task of reasoning about security. In [Can00]

[1] A complete specification of the primitive should include additional aspects, such as which i (j) out of the k players provide an input (resp., receive an output), etc., but the simpler notation will be expressive enough for the primitives we will consider.

[2] In the case of $[1,1]$, we always ignore the "reflexive" case (same player providing input and receiving the output).

(and in a nutshell), the computation to be performed among the n players is specified with respect to an incorruptible trusted party τ who interacts securely with the players. For the special case of secure function evaluation where a function on the players' inputs is to be computed and revealed, such a process can be defined by the players first secretly handing their inputs to τ, τ computing the output corresponding to the (possibly probabilistic) function, and then handing it back to the players. Such a protocol among $P \cup \{\tau\}$ is called an *ideal process*.

Of course, the goal of multi-party computation is to perform the same task without the need for a trusted party; thus, a multi-party computation protocol for evaluating a function is called *secure* if it emulates the ideal evaluation process of the function, i.e., if for every strategy of the adversary in the real protocol there is a corresponding adversary strategy that, with similar cost, achieves the same effect in the ideal process. In particular, this means that whenever the ideal process satisfies some consistency or privacy property with respect to the players (e.g., privately computes some specific function on the players' inputs), then the secure protocol also satisfies them. This notion of security can then be refined further by distinguishing among the different types of similarities between the global outputs in both the ideal and real life computations. We are interested in unconditional security, which is obtained by requiring that these output distributions be indistinguishable, except for a negligible function of the security parameter, independently of the adversary's computational power.

The trusted party τ is assumed to be equivalent to a probabilistic Turing machine with a memory tape of fixed (limited) size. This implies that τ can perform any task a standard computer can but not more. On the other hand, τ is also equivalent to an arithmetic circuit (though of potentially large size) and hence can be modeled as a (stateless) circuit. Thus, the multi-party computation specification simply defines a sequence of circuit evaluations on the players' inputs. Note that this ideal computation model, and hence the set of problems computable with an SFE protocol, is as strong as the "standard" one (e.g., [BGW88,CCD88,RB89]).

Reducibility and completeness. A main theme in this paper is that of reductions "across" cardinalities. The notion of reduction generalizes to the case of an n-ary function (n-player protocol) invoking another k-ary function (primitive of cardinality k, resp.), with $k \leq n$, in a natural way [Gol00]:

Definition 1 (Reductions). *An n-player protocol* unconditionally reduces *f_n to g_k for a given $t < n$, if it computes f_n unconditionally t-securely just by black-box calls to g_k and local computation. In such a case we say that f_n* unconditionally reduces *(for short,* reduces*) to g_k for that t.*[3]

The notion of completeness also generalizes to the different cardinality setting in a natural way: if g_k is complete one can use g_k to perform secure n-party computation. More formally:

[3] Note that the definition of reduction also admits the opposite direction, i.e., from smaller cardinality to larger cardinality. Occasionally in our constructions we will also use this direction (for example, by implementing secure pairwise channels using a three-player primitive).

Definition 2 (Completeness). *We say a primitive g_k is* unconditionally complete *(for short,* complete*) for a given $t < n$, if every n-ary function unconditionally reduces to g_k (for the same t).*

Typically, the reduction step is applied more than once, by reducing a primitive already known to be complete to another, perhaps simpler primitive. For example, this is the case in the two-party case, where protocols are given that implement oblivious transfer using a different primitive (see, e.g., [Kil00]). This is also the approach we will follow in this paper, by showing how to implement, using our primitives, the "resources" that are known to be required for SFE.

Furthermore, all our reductions will be unconditionally secure in a way that the simulation can fail with some negligible probability, but, in the non-failure case, it *perfectly* provides the desired functionality; i.e., compared to an ideal implementation of the functionality, the reductions leak no additional information and provide perfect correctness. (Note that this allows for parallel composition.) Hence, by estimating the overall error probability of the complete reduction from the given SFE problem to the complete primitive as the probability that at least one single implementation of a reduction step fails, we actually get an upper bound on the probability that the whole protocol does not provide perfect security. Since our reductions keep this probability negligibly small, we achieve unconditional security according to the definition above.

Finally, we note that all our reductions are *efficient*, i.e., polynomial in n and a security parameter σ such that the overall error probability is smaller than $2^{-\sigma}$.

3 Primitives of Cardinality 2

It is well known that secure channels (SC_2) are sufficient for unconditional SFE [BGW88,CCD88] with $t < n/3$. That is, in our parlance:

Proposition 1. *For any n, there is a primitive of cardinality 2, the secure channel, that is complete for $t < n/3$.*

Since we are assuming that every permutation of the players share a primitive, the type of a secure channel is $[1,1]$; hence, for $t < n/3$, the complete primitive is of the weakest type. We now prove that, for $t \geq \lceil n/3 \rceil$, no primitive of cardinality 2 can be complete (if $n > 2$). This is done by showing that there is a problem, namely broadcast (aka Byzantine agreement) [PSL80], that cannot be solved in a model where players are connected by "g_2-channels" for any two-party primitive g_2. We first recall the definition of broadcast.

Definition 3. Broadcast *is a primitive among n players, one sender and $n-1$ recipients. The sender sends an input bit $b \in \{0,1\}$ and the recipients get an output (decision) value $v \in \{0,1\}$ such that the following conditions hold:*

Agreement*: All correct recipients decide on the same value $v \in \{0,1\}$.*

Validity*: If the sender is correct, then all correct recipients decide on the sender's input bit ($v = b$).*

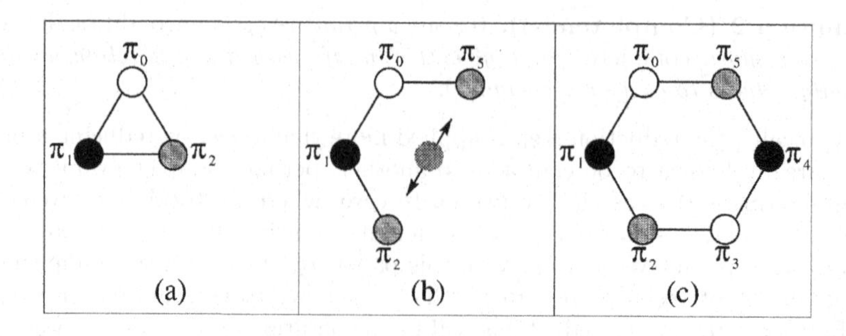

Fig. 1. Rearrangement of processors in proof of Lemma 1

We first consider the special case of $n = 3$ and $t \geq 1$, and then reduce the general case of $n \geq 3$ and $t \geq \lceil n/3 \rceil$ to this special case. The impossibility proof (for $n = 3$ and $t \geq 1$) is based on the impossibility proof in [FLM86], where it is shown that broadcast for $t \geq \lceil n/3 \rceil$ is not achievable in a model with pairwise authentic channels. In the new model, however, every pair of players can perform secure two-party computation. The idea in the proof is to assume that there exists an unconditionally secure broadcast protocol involving three players — interconnected by such a "g_2 channel", which then can be used to build a different system with contradictory behavior, hence proving that such a protocol cannot exist.

Lemma 1. *Let $n = 3$. For any two-player primitive connecting each pair of players, unconditionally secure broadcast is not possible if $t \geq 1$.*

Proof (sketch). Suppose, for the sake of contradiction, that there is a protocol that achieves broadcast for three players p_0, p_1, and p_2, with p_0 being the sender, even if one of the players is actively corrupted.

Let π_0, π_1, π_2 denote the players' corresponding processors with their local programs and, for each $i \in \{0, 1, 2\}$, let π_{i+3} be an identical copy of processor π_i; and let the (set of) given two-party primitive(s) between two processors π_i and π_j be called the *channel* between π_i and π_j. Instead of connecting the original processors as required for the broadcast setting, we build a network involving all six processors (i.e., the original ones together with their copies) by arranging them in a circle, i.e., each processor π_i ($i \in \{0, \ldots, 5\}$) is connected (exactly) by one channel with $\pi_{(i-1) \bmod 6}$ and one with $\pi_{(i+1) \bmod 6}$.

We now prove that for every pair of adjacent processors π_i and $\pi_{(i+1) \bmod 6}$ in the new system and without the presence of an adversary, their common view is indistinguishable from their view as two processors $\pi_{i \bmod 3}$ and $\pi_{(i+1) \bmod 3}$ in the original system with respect to an adversary that corrupts the remaining processor $\pi_{(i+2) \bmod 3}$ in an admissible way.[4] Refer to Figure 1. The original system is depicted in Figure 1-(a). Let the processors π_0 and π_1 be correct. An admissible

[4] I.e., for every pair of original processors, the rearrangement simultaneously simulates some particular adversary strategy by corrupting the third processor.

adversary strategy is to "split" π_2 and to make it behave independently with respect to π_0 and π_1 (Figure 1-(b)). Finally, by arranging the six processors in a circle as described above and shown in Figure 1-(c), this particular adversary strategy is simulated with respect to every pair π_i and $\pi_{(i+1) \bmod 6}$.

The new system involves two processors of the type corresponding to the sender, namely, π_0 and π_3, and these are the only processors that enter an input. Let now π_0 and π_3 be initialized with different inputs, i.e., let's assume that π_0 has input $v_0 \in \{0, 1\}$ and that π_3 has input $v_3 = 1 - v_0$.[5] We now show that there are at least two pairs of adjacent processors in the new system, i.e., one third among all six such pairs, for which the broadcast conditions are not satisfied despite being completely consistent with two correct processors in the original system.

First, suppose that agreement holds with respect to every pair on, wlog, the value v_0. Then the validity condition is violated with respect to both pairs involving processor π_3 since $v_3 \neq v_0$. On the other hand, suppose that the agreement condition is violated with respect to at least one pair. Then there must exist at least two such pairs because the processors are arranged in a circle.

Hence, on inputs $v_0 \in \{0, 1\}$ and $v_3 = 1 - v_0$, there must be some pair of adjacent processors $(\alpha, \beta) = (\pi_i, \pi_{(i+1) \bmod 6})$ that fails with a probability of at least $\frac{1}{3}$. Otherwise, strictly less than two pairs would fail per such invocation of the new system. The view of pair (α, β) is consistent with the view of the pair $(\alpha_0, \beta_0) = (\pi_i, \pi_{(i+1) \bmod 3})$ in the original system for one of the cases where the sender inputs either $v_0 = 0$ or $v_0 = 1$. Let Pr_0 be the probability that the sender selects input 0 in the original system. Then, in the original system, the adversary can force the pair (α_0, β_0) to be inconsistent with a probability of at least $\frac{1}{3} \min(\mathrm{Pr}_0, 1 - \mathrm{Pr}_0)$, which is non-negligible, since Pr_0 and $1 - \mathrm{Pr}_0$ are non-negligible by assumption. □

Theorem 1. *Let $n \geq 3$. For any primitive g_2, unconditionally secure broadcast is not possible if $t \geq \lceil n/3 \rceil$.*

Proof. Assume that there is an unconditionally secure broadcast protocol Π for $n \geq 3$ players and some $t \geq \lceil n/3 \rceil$ with arbitrarily small error probability $\varepsilon > 0$. Then we can let three players p_0, p_1, and p_2 each simulate up to $\lceil n/3 \rceil$ of the players in Π, with the sender in Π being simulated by p_0. Thus, this protocol among players $\{p_0, p_1, p_2\}$ achieves broadcast (with sender p_0) secure against one corrupted player because he simulates at most $\lceil n/3 \rceil$ of the players in Π, which is tolerated by assumption. Since this contradicts Lemma 1, the theorem follows. □

4 Primitives of Cardinality 3

Evidently, a primitive $g_3[1, 1]$ is equivalent to $g_2[1, 1]$ since in g_3 one of the players neither provides an input nor receives an output. Hence, in this section

[5] We assume that any input value from $\{0, 1\}$ will be selected by the sender with some non-negligible probability. Otherwise, the broadcast problem could be trivially solved for any $t < n$.

we consider primitives (of cardinality 3) of type different from [1, 1]. In fact, it turns out that either two inputs (and single output) or two outputs (and single input) is sufficient. For each type we introduce a primitive and show it to be complete for $t < n/2$. Moreover, we show that no primitive of cardinality 3 can be complete for $t \geq \lceil n/2 \rceil$.

It follows from [RB89,CDD+99] that pairwise secure channels and a global broadcast channel are sufficient for SFE secure against $t < n/2$ active corruptions. Hence, it is sufficient to show that the primitives introduced in this section imply both, unconditionally secure pairwise channels and global broadcast.

4.1 $g_3[1, *]$ Primitives: Oblivious Cast

Definition 4. Oblivious cast (OC_3) *is a primitive among three players: a sender* s *who sends a bit* $b \in \{0, 1\}$ *and two recipients* r_0 *and* r_1*, such that the following conditions are satisfied:*

(1) *The bit* b *is received by exactly one of the recipients,* r_0 *or* r_1*, each with probability* $\frac{1}{2}$.

(2) *While both recipients learn who got the bit, the other recipient gets no information about* b*. In case there are other players (apart from* s*,* r_0 *and* r_1*), they get no information about* b.

Implementing secure channels using oblivious cast. Secure pairwise channels can be achieved by the simulation of authentic channels and the implementation of a pairwise key-agreement protocol between every pair of players p_i and p_j. Players p_i and p_j can then use the key (e.g., a one-time pad) to encrypt the messages to be sent over the authentic channel.

Lemma 2. *Let* $n \geq 3$*. Then authentic channel reduces to oblivious cast for* $t < n/2$.

Proof (sketch). An authentic channel between players p_i and p_j can be achieved from oblivious cast among p_i, p_j, and some arbitrary third player $p_k \in P \setminus \{p_i, p_j\}$, by p_i (or p_j) oblivious-casting his bit (or whole message) σ times. Finally, p_j decides on the first bit he has received in those oblivious casts.

Since it is sufficient to achieve authentic channels only between pairs of correct players we can assume that the sender is correct. The invocation of this channel fails if p_j does not receive any of the bits being sent by oblivious cast, and this happens with a probability of at most $\Pr_{err}^{auth} = 2^{-\sigma}$. □

In order to generate a one-time pad (OTP) s_{ij} of one bit between two players p_i and p_j, we can let p_i generate some m random bits $b_1, \ldots, b_m$ and oblivious-cast them to p_j and some arbitrary third player p_k, where m is chosen such that, with overwhelming probability, p_j receives at least one of those random bits (every bit b_x is received by p_j with probability $\frac{1}{2}$). Finally, p_j uses his authentic channel to p_i (Lemma 2) to send to p_i the index $x \in \{1, \ldots, m\}$ of the first bit b_x that p_j received. Since p_k gets no information about the bit, bit b_x

can be used as an OTP-bit between p_i and p_j. In order to get an OTP of length $\ell > 1$ this process can be repeated ℓ times.[6]

In order to guarantee that the transmission of a bit through the secure channel thus obtained fails with an error probability of at most $\Pr_{err} = 2^{-\sigma}$, we can parameterize m and the security parameter for the invocations of the authentic channel, σ_{auth}, as follows:

- $\Pr_{err}^{oc} \leq 2^{-\sigma-1}$ — the probability that none of the m bits transmitted by oblivious cast is received by player p_j.
- $\Pr_{err}^{auth} \leq 2^{-\sigma-1}$ — the probability that at least one of the invocations of the authentic channel fails.

So we can choose $m = \sigma + 1$. The number of invocations of the authentic channel is $\ell = \lceil \log m \rceil + 1$ ($\lceil \log m \rceil$ for the transmission of index x plus one for the final transmission of the encrypted bit). Hence, σ_{auth} can be chosen as $\sigma_{auth} = \sigma + \lceil \log \ell \rceil + 1$.

Lemma 3. *Let $n \geq 3$. Then secure channel reduces to oblivious cast for $t < n/2$.*

Proof. From Lemma 2 and the discussion above it follows that the secure channel construction has an error probability of $\Pr_{err} \leq \Pr_{err}^{auth} + \Pr_{err}^{oc} \leq 2^{-\sigma}$. □

Implementing broadcast using oblivious cast. It is shown in [FM00] that a three-party primitive called *weak 2-cast*, defined below, yields global broadcast secure against $t < n/2$ active corruptions. Thus, it is sufficient to show that, using oblivious cast, an implementation of weak 2-cast in any set $S \subset P$, $|S| = 3$, and for any selection of a sender among those players, is possible. We first recall the definition of weak 2-cast from [FM00].

Definition 5. Weak 2-cast *is a primitive among three players: one sender and two recipients. The sender sends an input bit $b \in \{0, 1\}$ and both recipients get an output (decision) value $v \in \{0, 1, \bot\}$ such that the following conditions hold:*

(1) *If both recipients are correct and decide on different values, then one of them decides on $\bot$.*

(2) *If the sender is correct, then all correct recipients decide on his input bit.*

The idea behind the implementation of weak 2-cast using oblivious cast is to have the sender repeatedly oblivious-cast his bit a given number of times. Hence, a recipient who receives two different bits reliably detects that the sender is faulty and may safely decide on $\bot$. On the other hand, in order to make the two recipients decide on different bits, a corrupted sender must oblivious-cast 0's and 1's in such a way that each recipient gets one value, but not the other one. However, since the sender cannot influence which of the recipients gets a bit, he can enforce this situation only with exponentially small probability. We now describe the implementation in more detail.

[6] A more efficient way to generate an OTP of length ℓ is to choose a larger m and have p_j send to p_i the indices of the first ℓ bits he received. For simplicity we restrict ourselves to the less efficient but simpler method.

Protocol `Weak-2-Cast-Impl-1`$(s, \{r_0, r_1\}, \sigma)$:

1. Sender s oblivious-casts his bit $(\sigma + 1)$ times to the recipients.
2. Recipients r_i $(i \in \{0,1\})$ decide $v_i = \begin{cases} 0 & \text{if 0 received at least once, and no 1's;} \\ 1 & \text{if 1 received at least once, and no 0's;} \\ \perp & \text{otherwise.} \end{cases}$

Lemma 4. *Protocol* `Weak-2-Cast-Impl-1` *achieves weak 2-cast with an error probability of at most $2^{-\sigma}$, by only using oblivious cast and local computation.*

Proof. If the sender is correct, the protocol can only fail if one of the recipients does not receive any bit from the sender, because the sender always transmits the same bit. This happens with probability $\Pr_{err_1} = 2^{-\sigma}$.

If the sender is incorrect, the protocol may fail only if he manages to make one of the recipients receive all 0's and make the other one receive all 1's. In order to achieve this, after having transmitted the first bit, the sender must correctly guess in advance the recipient of every subsequent bit. This happens with probability $\Pr_{err_2} = 2^{-\sigma}$.

Hence, the error probability is $\Pr_{err} \leq \max(\Pr_{err_1}, \Pr_{err_2}) = 2^{-\sigma}$. □

Lemma 4 together with the reduction of broadcast to weak 2-cast in [FM00] (which does not require pairwise channels) immediately yield

Lemma 5. *Broadcast among $n \geq 3$ players reduces to oblivious cast for $t < n/2$.*

Lemmas 3 and 5 and the constructions of [RB89,CDD+99] yield

Theorem 2. *Let $n \geq 3$. Then there is a single-input two-output primitive of cardinality 3, oblivious cast, that is complete for $t < n/2$.*

4.2 $g_3[*, 1]$ Primitives: Converge Cast

We now show that a cardinality-3 primitive with two inputs and a single output — i.e., the converse of oblivious cast (in several ways) — is also complete for $t < n/2$. Specifically, we introduce *converge cast*, a primitive related to the "anonymous channel" of [Cha88], defined as follows:

Definition 6. Converge cast (CC_3) *is a primitive among three players: two senders s_0 and s_1 and one recipient r. The senders send a value x_i, $i \in \{0,1\}$, from a finite domain $\mathcal{D}$, $|\mathcal{D}| \geq 3$, such that the following conditions hold:*

(1) *The recipient r receives either x_0 or x_1, each with probability $\frac{1}{2}$.*

(2) *Neither sender learns the other sender's input value, and none of the players learns which of the senders was successful. In case there are other players (apart from s_0, s_1 and r), they get no information about the input values or the successful sender's identity.*

As in the previous section, we show how to implement secure channels and broadcast (weak 2-cast). We use "$p_i, p_j \xrightarrow{?} p_k : (x_i, x_j)$" to denote an invocation of converge cast with senders p_i and p_j sending values x_i and x_j, respectively, and recipient p_k. Furthermore, for two sequences s_a and s_b of elements in $\{0,1,2\}$ of same length, we use $\mathcal{H}(s_a, s_b)$ to denote the Hamming distance (difference) between the sequences.

Implementing secure channels using converge cast. We now present a protocol to implement a secure channel from p_0 to p_1 for the transmission of one bit x_0. The idea is as follows: first, p_1 and some other player, say, p_2, choose two random keys of an adequate length, one for 0 and for 1, and converge-cast them to p_0. p_0 stores the two received keys (note that each received key may contain elements from both senders), using the corresponding key as input to a converge cast with p_1 as the recipient to communicate the desired bit.

Protocol `Secure-Channel-Impl-2`(p_0, p_1, ℓ):

1. Player p_i, $i = 1, 2$, computes random keys $s_i^{(0)}$ and $s_i^{(1)}$ of length ℓ over $\{0, 1, 2\}$
2. $p_1, p_2 \overset{?}{\longrightarrow} p_0$: $(s_1^{(0)}, s_2^{(0)})$; $(s_1^{(1)}, s_2^{(1)})$ (element-wise) (p_0 receives $s_0^{(0)}; s_0^{(1)}$)
3. $p_0, p_2 \overset{?}{\longrightarrow} p_1$: $(s_0^{(x_0)}, *)$ (element-wise) (p_1 receives s_1')
4. p_1: **if** $\mathcal{H}(s_1', s_1^{(0)}) < \frac{7}{12}\ell$ **then** $y_1 = 0$, **else** $y_1 = 1$ **fi**

The proof of the following lemma follows from elementary probability, independently of p_2's strategy:

Lemma 6. *Consider protocol* `Secure-Channel-Impl-2`. *If p_0 and p_1 are correct, then for every k, $k \in \{1, \ldots, \ell\}$,*
(1) $s_1'[k] = s_1^{(x_0)}[k]$ *with probability* $\frac{1}{2}$;

(2) $s_1'[k] = s_1^{(1-x_0)}[k]$ *with probability* $\frac{1}{3}$.

Lemma 7. *Let $n \geq 3$. Then secure channels reduces to converge cast for $t < n/2$.*

Proof. Consider protocol `Secure-Channels-Impl-2`. First, it is easy to see that p_2 gets no information about bit x_0. We now show that the channel also provides authenticity. The only ways the protocol can fail is that either $x_0 = 0$ and $\mathcal{H}(s_1', s_1^{(0)}) \geq \frac{7}{12}\ell$ (probability $\Pr_0$), or that $x_0 = 1$ and $\mathcal{H}(s_1', s_1^{(0)}) < \frac{7}{12}\ell$ (probability $\Pr_1$). These probabilities can be estimated by Chernoff bounds:

- $\Pr_0$: By Lemma 6(1), $s_1'[k] = s_1^{(0)}[k]$ holds with probability $\frac{1}{2}$. Hence, $\Pr_0$ is the probability that out of ℓ trials with expected value $\frac{1}{2}$, at most $\frac{5}{12}l$ do match. We get $\Pr_0 \leq e^{-\frac{\ell}{4}(\frac{1}{6})^2} = e^{-\frac{\ell}{144}}$.
- $\Pr_1$: By Lemma 6(2), $s_1'[k] = s_1^{(0)}[k]$ holds with probability $\frac{1}{3}$. Hence, $\Pr_1$ is the probability that out of ℓ trials with expected value $\frac{1}{3}$, at least $1 - \frac{7}{12} = \frac{5}{12}\ell$ do match. We get $\Pr_1 \leq e^{-\frac{\ell}{9}(\frac{1}{4})^2} = e^{-\frac{\ell}{144}}$.

Thus, the overall error probability is $\Pr_{err}^{auth} \leq \max(\Pr_0, \Pr_1) = e^{-\frac{\ell}{144}}$. □

Implementing broadcast using converge cast. We now show how weak 2-cast of a bit x_0 from p_0 to p_1 and p_2 can be simulated using CC_3. Roughly, he protocol can be described as follows: First, p_1 and p_2 choose two random keys of an adequate length, one for 0 and for 1, and converge-cast them to p_0. p_0 stores the two received (mixed) keys. p_0 then sends his input bit to p_1 and p_2 using secure channels. Additionally, p_0 sends to p_1 the (received) key corresponding to his input bit. This key can then be used by p_1 to "prove" to p_2 which value he received from p_0. If things "look" consistent to p_2 (see protocol below), he adopts this value; otherwise, he outputs the value received directly from p_0. Let "$p_i \xrightarrow{!} p_j$" denote the secure channel from p_i to p_j (by means of protocol `Secure-Channels-Impl-2`).

Protocol `Weak-2-Cast-Impl-2`$(p_0, \{p_1, p_2\}, \ell)$:

1. Player p_i, $i = 1, 2$, computes random keys $s_i^{(0)}$ and $s_i^{(1)}$ of length 2ℓ over $\{0, 1, 2\}$
2. $p_1, p_2 \xrightarrow{?} p_0$: $(s_1^{(0)}, s_2^{(0)})$; $(s_1^{(1)}, s_2^{(1)})$ (element-wise) (p_0 receives $s_0^{(0)}; s_0^{(1)}$)
3. $p_0 \xrightarrow{!} p_i$ $(i = 1, 2)$: $x_0 \in \{0, 1\}$ (p_i receives $x_i \in \{0, 1\}$)
4. $p_0 \xrightarrow{!} p_1$: $s_0^{(x_0)}$ (p_1 receives s_1')
5. p_1: `if` $\mathcal{H}(s_1', s_1^{(x_1)}) < \frac{4}{5}\ell$ `then` $y_1 = x_1$, `else` $y_1 = \perp$ `fi`
6. $p_1 \xrightarrow{!} p_2$: y_1; s_1' (p_2 receives y_2; s_2')
7. p_2: `if` $(y_2 = \perp) \vee \left(\mathcal{H}(s_2', s_2^{(y_2)}) > \frac{5}{4}\ell\right)$ `then` $y_2 = x_2$ `fi`

Lemma 8. *Let $n \geq 3$. Then weak 2-cast reduces to converge cast for $t < n/2$.*

The proof appears in Section A.

As before, Lemmas 7 and 8 and the constructions of [RB89,CDD+99,FM00] yield

Theorem 3. *Let $n \geq 3$. Then there is a two-input single-output primitive of cardinality 3, converge cast, that is complete for $t < n/2$.*

We note that allowing the inputs of converge cast to be from a larger domain (than $\{0, 1, 2\}$) considerably improves the efficiency of our reductions.

4.3 Impossibility of $t \geq \lceil n/2 \rceil$ for g_3

We now show that no primitive of cardinality 3 can be complete with respect to half resiliency. We do so by generalizing the impossibility proof in [FM00] for broadcast with $t \geq \lceil n/2 \rceil$ using primitive 2-Cast, to arbitrary primitives of cardinality 3.

Theorem 4. *Let $n \geq 4$. For any primitive g_3, unconditionally secure broadcast is not possible if $t \geq \lceil n/2 \rceil$.*

Proof (sketch). Impossibility for $n = 4$ and $t = 2$: Suppose that there are four processors (with local programs) that achieve broadcast for $t = 2$. Again, we build a new system with the four processors and one copy of each, arranged in a circle. Analogously to Lemma 1, the 3-player primitives can be reconnected such that the view of any two adjacent processors is indistinguishable from their view in the original system (i.e., they can be reconnected in the same way as in the proof in [FM00]), and by assigning different inputs to the sender and its copy we get the same kind of contradiction.

Impossibility for $n > 4$ and $t \geq \lceil n/2 \rceil$: Suppose now that there are n processors and we want to achieve broadcast with $t \geq \lceil n/2 \rceil$. The processors are partitioned into four sets and each set is duplicated. Instead of reconnecting single processors, the connections between different sets are reconnected so that the common view of all the processors in two adjacent sets is indistinguishable from their view in the original system, and we get a contradiction along the lines of [FM00]. □

5 Primitives of Full Cardinality

In this section, we first show that even cardinality n does not help if the primitive is restricted, in the sense of having either a single input or a single output. Such a primitive is no more powerful than a primitive of cardinality 3 (Section 4). We then introduce a new primitive of type $[n, n]$, the *universal black box* (UBB_n), which allows for arbitrary resiliency ($t < n$). The UBB_n has an interesting application to computations involving a trusted third party: its functionality enables *oblivious* trusted third parties, that is, trusted parties which do not require any prior knowledge of the function to be computed by the players. Finally, we show that full cardinality is necessary to achieve arbitrary resiliency. We start with the impossibility results for restricted primitives.

5.1 $\mathbf{g_n[1,*]}$ and $\mathbf{g_n[*,1]}$ Primitives

Theorem 5. *There is no $g_n[1,*]$ primitive complete for $t \geq \lceil n/2 \rceil$.*

Proof (sketch). Assume that a particular primitive $g_n[1,*]$ is complete for $t \geq \lceil n/2 \rceil$, and consider two players, p and q, who want to compute the logical OR of their input bits. We can have both players each simulate up to $\lceil n/2 \rceil$ of the players involved in the complete primitive g_n (in such a way that every original player is simulated either by p or q) which allows them to securely compute the OR function. Since there is only one input to g_n (to be given either by p or by q), there must be a first invocation of the primitive that reveals some input information to the other player. This is a contradiction to [BGW88], where it is shown that no player may reveal any information about his input to the other player unless he knows that the other player's input is 0. □

Theorem 6. *There is no $g_n[*,1]$ primitive complete for $t \geq \lceil n/2 \rceil$.*

Proof (sketch). The proof is again by contradiction. Suppose that there is a primitive of type $[*,1]$ that is complete for $t \geq \lceil n/2 \rceil$. We can have two players p and q each simulate up to $\lceil n/2 \rceil$ of the players involved in this primitive which allows them to securely compute any function on their inputs. Thus, there is a two-player primitive with a single output that is complete with respect to any computation where both players learn the same result. This is a direct contradiction to the "one-sidedness" observation in [BMM99] that a protocol based on an asymmetric two-player primitive cannot guarantee that both players learn the result. □

5.2 $\mathbf{g_n[n,n]}$ Primitives: The Universal Black Box

We now introduce the *universal black box* (UBB_n), a complete primitive for $t < n$. At first, it might seem trivial to build a complete primitive for arbitrary t by just implementing the functionality of a trusted party. However, computations by trusted parties are generally based on the fact that the trusted party already knows the function to be computed. But since the primitive must be universally applicable, it cannot have any prior information about what is to be computed, i.e., what step of what computation is to be executed. Hence, the specification of the computation step to be performed by the black box must be entered by the players at every invocation of the black box. Although there seems to be no apparent solution to this problem since a dishonest majority might always overrule the honest players' specification, we now describe how the UBB_n effectively overcomes this problem.

For simplicity, we first assume that exactly one function is to be computed on the inputs of all players, and that exactly one player, p_0, is to learn the result of the computation. The more general cases (multiple functions, multiple/different outputs) can be obtained by simple extensions to this case.

The main idea behind the UBB_n is simple: It contains a universal circuit [Val76], and has two inputs per involved player,

- the *function input*, wherein the player specifies the function to be computed on all argument inputs, and
- the *argument input*, where the player inputs his argument to the function.

The UBB_n now computes the function specified by player p_0, but for every player that does not input the same function as p_0, it replaces his argument input by some fixed default value. Finally, the function is computed by evaluating the universal circuit on p_0's function and all argument inputs, and its output is sent to player p_0. Note that only one invocation of the UBB_n is required per computation.

Theorem 7. *The universal black box is a complete primitive for $t < n$.*

Proof (sketch). We show that privacy and correctness hold for arbitrary t.
Privacy: Trivially, no $p_i \neq p_0$ learns anything. On the other hand, p_0's output can give information about player p_i's argument input only if p_i entered the same function input as p_0 (which means that p_i had "agreed" on exactly this computation). Hence, p_0 would get the same information about p_i's argument as in an ideal process involving a trusted party. If p_0 is corrupted and inputs a wrong function input, no argument from a correct player will be used for this computation.
Correctness: The function to be computed is selected by p_0. Hence, if he's correct, the UBB_n does compute the desired function. Corrupted players that input a different function only achieve that their input be replaced by a default value — a strategy that is also (easily) achievable in an ideal process by selecting the default argument. □

Corollary 1. [Oblivious TTPs] *Computations involving a trusted third party do not require the trusted party to have any prior knowledge of the task to be completed by the players.*

The single-output version of a UBB_n can be generalized to a multi-output UBB_n by the following modification. The function input specifies n functions to be computed on the inputs — one function per player. The function f_i to be computed and output to player p_i is determined by player p_i himself, and for the computation of f_i the argument inputs of only those players are considered by the UBB_n who agree on the same n functions $f_1, \ldots, f_n$ to be computed with respect to the n players, i.e., whose function inputs match with the function input of p_i.

Finally, we show that full cardinality is necessary in order to achieve arbitrary resiliency (proof in Section A):

Theorem 8. *For $k < n$, there is no primitive g_k complete for $t < n$.*

Moreover, there is strong evidence that even a primitive of the most powerful category for cardinalities $k < n$, i.e., a $g_{n-1}[n-1, n-1]$, cannot be complete for $t \geq \lceil n/2 \rceil$, but we have no formal proof for it.

Conjecture 1. For $k < n$, there is no primitive g_k complete for $t \geq \lceil n/2 \rceil$.

6 Summary and Open Problems

Originally (Section 2), we assumed that one primitive instance was available for every permutation of every k-tuple of players, i.e., $\binom{n}{k}k!$ instances. In contrast, it follows from Proposition 1 and from the constructions for the proofs of Theorems 2, 3, and 7 that there is always a minimal complete primitive such that at most $3\binom{n}{3}$ instances of the primitive are required for the computation of any function.

Corollary 2. *For each cardinality k, $2 \leq k \leq n$, and each primitive type, there is a complete primitive such that at most $3\binom{n}{3}$ instances are sufficient for unconditional SFE.*

In this paper we have put forward the concept of minimal cardinality of primitives that are complete for SFE. Since this is a new line of research, several questions remain open.

We completely characterized the cases of types $[1,1]$, $[1,k]$, and $[k,1]$, for all cardinalities $k \leq n$. In particular, for $t < n/3$ there is a complete primitive $SC_2[1,1]$ and no g_2 can do any better; and, for $t < n/2$, there are complete primitives $OC_3[1,2]$ and $CC_3[2,1]$ and no g_k, $k \leq n$, can do any better. For the case of type $[k,k]$ it remains to prove Conjecture 1, that no $g_{n-1}[n-1, n-1]$ is complete for $t \geq \lceil n/2 \rceil$. This would partition the whole hierarchy into three equivalence classes of cardinalities $k = 2$ $(t < n/3)$, $2 < k < n$ $(t < n/2)$, and $k = n$ $(t < n)$.

It would also be interesting to analyze the completeness of primitives as a function of the size of the input and output domains. For example, if the primitive CC_3 were restricted to one single input bit per player and one single output bit, it would not be complete for $t < n/2$. Also the completeness of the UBB_n for $t < n$ relies on the fact that inputs of large size are allowed.

Acknowledgements. We thank the anonymous referees for their valuable comments. The work of Matthias Fitzi was partly done while visiting Bell Labs.

References

[Bea92] D. Beaver. Foundations of secure interactive computation. In *Advances in Cryptology – CRYPTO '91*, LNCS, pp. 377–391. Springer-Verlag, 1992.

[BGW88] M. Ben-Or, S. Goldwasser, and A. Wigderson. Completeness theorems for non-cryptographic fault-tolerant distributed computation. In *Proc. 20th ACM Symp. on the Theory of Computing*, pp. 1–10, 1988.

[Bla96] M. Blaze. Oblivious Key Escrow. In R. Anderson, editor, *Proc.First Info-hiding*, LNCS, pp. 335–343, Cambridge, U.K., 1996. Springer-Verlag.

[BMM99] A. Beimel, T. Malkin, and S. Micali. The all-or-nothing nature of two-party secure computation. In *Advances in Cryptology - CRYPTO '99, volume 1666 of LNCS*, pp. 80–97. Springer-Verlag, 1999.

[Can00] R. Canetti. Security and composition of multiparty cryptographic protocols. *Journal of Cryptology*, 13(1):143–202, 2000.

[CCD88] D. Chaum, C. Crépeau, and I. Damgård. Multiparty unconditionally secure protocols (extended abstract). In *Proc. 20th ACM Symp. on the Theory of Computing*, pp. 11–19, 1988.

[CDD+99] R. Cramer, I. Damgård, S. Dziembowski, M. Hirt, and T. Rabin. Efficient multiparty computations secure against an adaptive adversary. In *Advances in Cryptology — EUROCRYPT '99*, LNCS, 1999.

[Cha88] D. Chaum. The Dining Cryptographers Problem: Unconditional sender and recipient untraceability. *Journal of Cryptology*, 1(1):65–75, 1988.

[Chv79] V. Chvátal. The tail of the hypergeometric distribution. *Discrete Mathematics*, 25:285–287, 1979.

[DDWY93] D. Dolev, C. Dwork, O. Waarts, and M. Yung. Perfectly secure message transmission. *Journal of the ACM*, 40(1):17–47, Jan. 1993.

[FLM86] M. J. Fischer, N. A. Lynch, and M. Merritt. Easy impossibility proofs for distributed consensus problems. *Distributed Computing*, 1:26–39, 1986.

[FM00] M. Fitzi and U. Maurer. From partial consistency to global broadcast. In *32nd Annual Symp. on the Theory of Computing*, pp. 494–503, 2000.

[GL90] S. Goldwasser and L. Levin. Fair computation of general functions in presence of immoral majority. In *Advances in Cryptology — CRYPTO '90*, volume 537 of *LNCS*. Springer-Verlag, 1990. g

[GMW87] O. Goldreich, S. Micali, and A. Wigderson. How to play any mental game. In Proc. 19th *ACM Symp. on the Theory of Computing*, pp. 218–229, 1987.

[Gol00] O. Goldreich. Secure multi-party computation, working draft, version 1.2, Mar. 2000.

[Kil88] J. Kilian. Founding cryptography on oblivious transfer. In *Proc.20th Annual ACM Symp. on the Theory of Computing*, pp. 20–31, 2–4 May 1988.

[Kil91] J. Kilian. A general completeness theorem for two-party games. In *Proc.23rd Annual ACM Symposium on the Theory of Computing*, pp. 553–560, New Orleans, Louisiana, 6–8 May 1991.

[Kil00] J. Kilian. More general completeness theorems for secure two-party computation. In *Proc.32nd Annual ACM Symp. on the Theory of Computing*, pp. 316–324, Portland, Oregon, 21–23 May 2000.

[KKMO99] J. Kilian, E. Kushilevitz, S. Micali, and R. Ostrovsky. Reducibility and completeness in private computations. *SIAM Journal on Computing*, 29, 1999.

[KMO94] E. Kushilevitz, S. Micali, and R. Ostrovsky. Reducibility and completeness in multi-party private computations. In *Proc. 35th Annual IEEE Symp. on the Foundations of Computer Science*, pp. 478–491, Nov. 1994.

[MR92] S. Micali and P. Rogaway. Secure computation. In *Advances in Cryptology — CRYPTO '91*, volume 576 of LNCS, pp. 392–404. Springer-Verlag, 1992.

[PSL80] M. Pease, R. Shostak, and L. Lamport. Reaching agreement in the presence of faults. *Journal of the ACM*, 27(2):228–234, Apr. 1980.

[Rab81] M. O. Rabin. How to exchange secrets by oblivious transfer. Technical Report TR-81, Harvard Aiken Computation Laboratory, 1981. g

[RB89] T. Rabin and M. Ben-Or. Verifiable secret sharing and multiparty protocols with honest majority. In *Proc. 21st ACM Symp. on the Theory of Computing*, pp. 73–85, 1989.

[Val76] L. G. Valiant. Universal circuits. In *ACM Symposium on Theory of Computing (STOC '76)*, pp. 196–203, New York, May 1976. ACM Press.

[Yao82] A. C. Yao. Protocols for secure computations. In *Proc. 23rd IEEE Symp. on the Foundations of Computer Science*, pp. 160–164. IEEE, 1982.

A Proofs

We repeat the statements here for convenience.

Lemma 8. *Let $n \geq 3$. Then weak 2-cast reduces to converge cast for $t < n/2$.*

In the proof we will be using the following two lemmas. Their validity follows from elementary probability.

Lemma 9. *The probability that in protocol* `Weak-2-Cast-Impl-2` *p_0 receives at least $\frac{9}{8}\ell$ elements of $s_1^{(x_0)}$ and at most $\frac{7}{8}\ell$ elements of $s_2^{(x_0)}$, or vice versa, is* $\mathrm{Pr}_1 \leq 2e^{-\frac{\ell}{128}}$.

Lemma 10. *Given is a set of pairs $S = \{(x_1, y_1), \ldots, (x_m, y_m)\}$ with elements $x_i, y_i \in \{0, 1, 2\}$. If at least all elements x_i or all elements y_i are selected uniformly at random from $\{0, 1, 2\}$, then the following holds:*

(1) *If $|S| \geq \frac{9}{8}\ell$, then the probability that there are $\frac{13}{40}\ell$ or less indices i such that $x_i = y_i$ is $\mathrm{Pr}_2 \leq e^{-\frac{\ell}{300}}$.*

(2) *If $|S| \leq \frac{7}{8}\ell$, then the probability that there are $\frac{7}{20}\ell$ or more indices i such that $x_i = y_i$ is $\mathrm{Pr}_3 \leq e^{-\frac{\ell}{258}}$.*

Proof of Lemma 8. Consider protocol `Weak-2-Cast-Impl-2`. Let us neglect the error probabilities of the secure channel invocations (protocol `Secure-Channels-Impl-2`) until the end of the proof. Assume that p_0 receives at least $\frac{7}{8}\ell$ elements from each of the players' key during step 2 (by Lemma 9, this happens with probability at least $1 - \mathrm{Pr}_1$). Since the conditions for weak 2-cast are trivially satisfied if more than one player is corrupted, we can distinguish three cases.

• *All players correct or at most p_2 corrupted.* The only way the protocol can fail is if p_1 decides on $\perp$ ($\mathcal{H}(s_1', s_1^{(x_1)}) \geq \frac{4}{5}\ell$); i.e., that at most $\frac{6}{5}\ell$ elements of s_1' match with p_1's key. We assume that p_0 receives at least $k \geq \frac{7}{8}\ell$ elements of p_1's key during step 2, $k = \frac{7}{8}\ell$ in the worst case. Hence, of all other $2\ell - \frac{7}{8}\ell = \frac{9}{8}\ell$ elements (i.e., the elements of s_1' originating from p_2), at most $\frac{6}{5}\ell - \frac{7}{8}\ell = \frac{13}{40}\ell$ are identical to the element that p_1 chose for the same invocation of "$p_1, p_2 \xrightarrow{?} p_0$". By Lemmas 9 and 10, this happens with probability $\mathrm{Pr}_4 \leq \mathrm{Pr}_1 + \mathrm{Pr}_2 \leq 3e^{\frac{\ell}{300}}$.

• *p_1 corrupted.* In order to achieve that p_2 decides on a wrong output it must hold that $y_2 = 1 - x_0$ and $\mathcal{H}(s_2', s_2^{(y_2)}) \leq \frac{5}{4}\ell$ before step 7; i.e., that at least

$\frac{3}{4}\ell$ elements of s'_2 match p_2's key. But since p_1 does not learn anything about the elements in $s_2^{(y_2)}$, he must guess $\frac{3}{8}$ or more of those 2ℓ elements correctly — otherwise p_2 would decide on x_0. This probability can be estimated by Chernoff bounds (2ℓ trials with expected value $\frac{1}{3}$) and we get $\Pr_5 \leq e^{-\frac{\ell}{288}}$. (In fact, this holds independently of the assumption at the beginning of the proof.)

• p_0 *corrupted.* Since p_1 and p_2 are correct, the following equalities hold before step 7: $s'_1 = s'_2 \stackrel{\text{def}}{=} s'$ and $x_1 = y_1 = y_2 \stackrel{\text{def}}{=} y$. In order to achieve that p_1 and p_2 decide on different bits p_0 must select $x_2 = 1-y$ and achieve that $\mathcal{H}(s', s_1^{(y)}) < \frac{4}{5}\ell$ and $\mathcal{H}(s', s_2^{(y)}) > \frac{5}{4}\ell$; i.e., p_0 must prepare s' such that at least $\frac{6}{5}\ell$ elements still match p_1's key but at most $\frac{3}{4}\ell$ elements match p_2's key. Given that the CC_3 statistics are good (which happens with probability $1 - \Pr_1$; see above) $s_0^{(y)}$ contains at most $\frac{9}{8}\ell$ elements originating from player p_1 and at least $\frac{7}{8}\ell$ elements originating from player p_2. In the sequel we assume that these quantities exactly hold, which constitutes the best case for p_0 to succeed (maximal number of matches with p_1's key and minimal number for p_2's).

Suppose now that s' is selected such that $h = \mathcal{H}(s', s_0^{(y)}) > \frac{3}{4}\ell$. We show that then $\mathcal{H}(s', s_1^{(y)}) < \frac{4}{5}\ell$ cannot be achieved almost certainly. By Lemma 10, with probability $1 - \Pr_3$, there are at most $\frac{7}{20}\ell$ elements in s' that match $s_1^{(y)}$ at positions where the corresponding element of $s_0^{(y)}$ was actually received from p_2. Hence at least $x > (2 - \frac{4}{5})\ell - \frac{7}{20}\ell = \frac{17}{20}\ell$ elements of s' must match $s_1^{(y)}$ at positions where the corresponding element was actually received from p_1; i.e., of the $h > \frac{3}{4}\ell$ differences between s' and $s_0^{(y)}$, at most $y = \frac{9}{8}\ell - x < \frac{11}{40}\ell$ may be made up at positions where the corresponding element was received from p_1. The probability of this event can be estimated by Hoeffding bounds [Chv79] (Hypergeometric distribution; $N = 2\ell$ elements, $n = \frac{3}{4}\ell$ trials, $K = \frac{9}{8}\ell$ "good" elements, less than $k = \frac{11}{40}\ell$ hits), and we get $\Pr_a < e^{-\frac{\ell}{18}}$.

Suppose now that s' is selected such that $h = \mathcal{H}(s', s_0^{(y)}) \leq \frac{3}{4}\ell$. We show that then $\mathcal{H}(s', s_2^{(y)}) > \frac{5}{4}\ell$ cannot be achieved almost certainly. By Lemma 10, with probability $1 - \Pr_2$, there are at least $\frac{13}{40}\ell$ elements in s' that match $s_2^{(y)}$ at positions where the corresponding element of $s_0^{(y)}$ was actually received from p_1. Hence at most $x < (2 - \frac{5}{4})\ell - \frac{13}{40}\ell = \frac{17}{40}\ell$ elements of s' may match $s_2^{(y)}$ at positions where the corresponding element was actually received from p_2; in other words, of the $h \leq \frac{3}{4}\ell$ differences between s' and $s_0^{(y)}$, at least $y = \frac{7}{8}\ell - x > \frac{9}{20}\ell$ must be made up at positions where the corresponding element was received from p_2. The probability of this event can be estimated by the Hoeffding bound (hypergeometric distribution; $N = 2\ell$ elements, $n = \frac{3}{4}\ell$ trials, $K = \frac{7}{8}\ell$ "good" elements, more than $k = \frac{9}{20}\ell$ hits), giving $\Pr_a < e^{-\frac{\ell}{26}}$.

Hence, when p_0 is corrupted, we get an error probability of at most $\Pr_6 \leq \Pr_1 + \max(\Pr_a + \Pr_3, \Pr_b + \Pr_2) \leq 4e^{\frac{\ell}{300}}$.

Since the error probability of protocol `Secure-Channels-Impl-2` can be made negligibly small, it can be parameterized such that the overall probability that at least one invocation fails satisfies $\Pr_{SC} \leq e^{-\frac{\ell}{300}}$. Thus, the overall error probability of the weak 2-cast construction is at most

$$\Pr_{err} \leq \Pr_{SC} + \max(\Pr_4, \Pr_5, \Pr_6) \leq 5e^{-\frac{\ell}{300}} .$$

For security parameter σ, we let $\ell \geq 300(\sigma + \lceil \ln 5 \rceil)$, and hence $\Pr_{err} \leq e^{-\sigma}$. □

Theorem 8. *For $k < n$, there is no primitive g_k complete for $t < n$.*

Proof (sketch). Consider a UBB of cardinality $k = n-1$. Since it is complete for any computation among $n-1$ players, it can securely simulate the functionality of any k-player primitive with $k < n$; i.e., the existence of any k-player primitive ($k < n$) complete for $t < n$ would imply that the UBB_{n-1} is also complete. Hence it is sufficient to show that there is no complete $(n-1)$-player primitive for $t < n$.

Suppose, for the sake of contradiction, that there is an $(n-1)$-player black box BB_{n-1} such that broadcast among the n players $p_0, \ldots, p_{n-1}$ is reducible to BB_{n-1} for $t < n$ — and hence also for $t = n-2$, i.e., in the presence of exactly two correct players.[7] Let $\pi_0, \ldots, \pi_{n-1}$ denote the players' processors with their local programs and, for each $i \in \{0, \ldots, n-1\}$ let π_{i+n} be an identical copy of processor π_i. For every processor π_k, $k \in \{0, \ldots, 2n-1\}$, let the number ($k \bmod n$) be called the *type* of p_k. Similarly to the proof of Lemma 1, we now build a new system involving all $2n$ processors but, instead of reconnecting them with pairwise channels, the instances of BB_{n-1} have to be reconnected in such a way that, again, for each pair of adjacent processors, π_i and $\pi_{(i+1) \bmod 2n}$, their view in the new system is indistinguishable from their view in the original system, for some particular strategy of an adversary corrupting all the remaining $n-2$ processors.

In order to guarantee that the view of every processor pair π_i and $\pi_{(i+1) \bmod 2n}$ is consistent with their view in the original system, the following two conditions must be satisfied:

1. For every processor π_i, $i \in \{0, \ldots, 2n-1\}$, and for every selection of $n-2$ processors of types different from ($i \bmod n$), π_i shares exactly one BB_{n-1} with these processors (as it does in the original system).
2. If processor π_i, $i \in \{0, \ldots, 2n-1\}$, shares an instance of BB_{n-1} with a processor of type $\pi_{(i \pm 1) \bmod n}$ (i.e., an adjacent type in the original system), then it shares it with the concrete processor $\pi_{(i \pm 1) \bmod 2n}$ (i.e., its adjacent processor of this type in the new system).

This can be achieved by applying the following rule for every processor π_i, $i \in \{0, \ldots, 2n-1\}$. For each δ, $\delta = 1, \ldots, n-1$, there is a BB_{n-1} that originally connects $\pi_{i \bmod n}$ with all other processors but $\pi_{(i+\delta) \bmod n}$. For every such δ, exactly one BB_{n-1} now connects π_i with processors $\pi_{(i+1) \bmod 2n}, \ldots, \pi_{(i+\delta-1) \bmod 2n}$ and $\pi_{(i-1) \bmod 2n}, \ldots, \pi_{(i-n+\delta+1) \bmod 2n}$. This principle is depicted in Figure 2 for the special case of $n = 4$ and BB_3.

[7] Since, by definition, broadcast is trivial if strictly less than two players are correct, this is the non-trivial case that involves the least number of correct players.

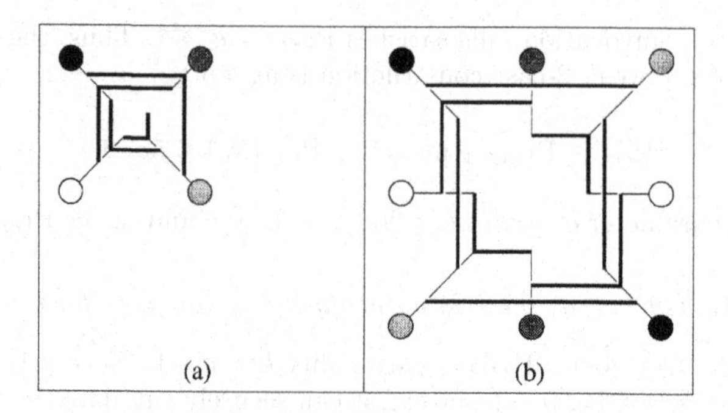

Fig. 2. Reconnection of processors in the proof of Theorem 8: special case $n = 4$.

Now the proof proceeds as the proof of Lemma 1 by assigning different input values to both sender processors, and concluding that the broadcast conditions are not satisfied with respect to at least 2 of the $2n$ pairs of adjacent processors. This contradicts the assumption that broadcast is possible with an arbitrarily small error probability. □

Robustness for Free in Unconditional Multi-party Computation

Martin Hirt and Ueli Maurer

ETH Zurich
Department of Computer Science
{hirt,maurer}@inf.ethz.ch

Abstract. We present a very efficient multi-party computation protocol unconditionally secure against an active adversary. The security is maximal, i.e., active corruption of up to $t < n/3$ of the n players is tolerated. The communication complexity for securely evaluating a circuit with m multiplication gates over a finite field is $\mathcal{O}(mn^2)$ field elements, including the communication required for simulating broadcast, but excluding some overhead costs (independent of m) for sharing the inputs and reconstructing the outputs. This corresponds to the complexity of the best known protocols for the passive model, where the corrupted players are guaranteed not to deviate from the protocol. The complexity of our protocol may well be optimal. The constant overhead factor for robustness is small and the protocol is practical.

1 Introduction

1.1 Secure Multi-party Computation

Secure multi-party computation (MPC), as introduced by Yao [Yao82], allows a set of n players to compute an arbitrary agreed function of their private inputs, even if an adversary may corrupt up to t arbitrary players. Almost any distributed cryptographic protocol can be seen as a multi-party computation, and can be realized with a general MPC protocol. Multi-party computation protocols are an important building block for reducing the required trust and building secure distributed systems. While currently special-purpose protocols (e.g., for collective signing) are considered practical, this paper suggests also that general-purpose protocols may well be practical for realistic applications.

Two different notions of corruption are usually considered. A passive (or curious) adversary may only read the information stored by the corrupted players, without controlling the player's behavior. Hence only privacy of the inputs is an issue to consider, but not the correctness of the result. In contrast, an active adversary can take full control of the corrupted players. Assuring not only the privacy of the inputs, but also the correctness of the outputs (robustness) appears to entail a substantial overhead. For instance, all known protocols make (usually heavy) use of a broadcast sub-protocol for which the optimal known communication complexity is $\mathcal{O}(n^2)$.

J. Kilian (Ed.): CRYPTO 2001, LNCS 2139, pp. 101–118, 2001.

We briefly review the classical results on secure MPC. Goldreich, Micali, and Wigderson [GMW87] presented a protocol, based on cryptographic intractability assumptions, which allows n players to securely compute an arbitrary function even if an active adversary corrupts any $t < n/2$ of the players. In the secure-channels model, where bilateral secure channels between every pair of players are assumed, Ben-Or, Goldwasser, and Wigderson [BGW88] and independently Chaum, Crépeau, and Damgård [CCD88] proved that unconditional security is possible if at most $t < n/3$ of the players are corrupted. In a model where additionally physical broadcast channels are available, unconditional security is achievable if at most $t < n/2$ players are corrupted [RB89,Bea91b,CDD+99].

1.2 Previous Work on Efficiency

In the past, both the round complexity and the communication complexity of secure multi-party protocol were subject to many investigations: Protocols with low round complexity [BB89,BFKR90,FKN94,IK00] suffer either from an unacceptably high communication complexity (even quadratic in the number of multiplication gates), or tolerate only a very small number of cheaters.

First steps towards better communication complexity were taken by Franklin and Yung [FY92] and Gennaro, Rabin, and Rabin [GRR98], where first a private but non-resilient computation is performed (for the whole protocol in [FY92], and for a segment of the protocol in [GRR98]), and only in case of faults the computation is repeated with a slow but resilient protocol. Although this approach can improve the best-case complexity of the protocol (when no adversary is present), it cannot speed up the protocol in the presence of a malicious adversary: a single corrupted player can persistently enforce the robust but slow execution, annihilating any efficiency gain.

Recently, Hirt, Maurer, and Przydatek [HMP00] proposed a new protocol for perfectly secure multi-party computation with considerably better communication complexity than previous protocols: A set of n players can compute any function (over a finite field $\mathbb{F}$) which is specified as a circuit with m multiplication gates (and any number of linear gates) by communicating $\mathcal{O}(mn^3)$ field elements, contrasting the previously best complexity of $\mathcal{O}(mn^6)$. Subsequently, the same complexity was achieved by Cramer, Damgård, and Nielsen [CDN01] in the cryptographic model (where more cheaters can be tolerated).

1.3 Contributions

The main open question in this line of research was whether security against active cheaters can be achieved with the same communication complexity as security against passive cheaters, namely with $\mathcal{O}(mn^2)$. For sufficiently large circuits, we answer this question in the affirmative: The only (and unavoidable) price for robustness is a reduction in the number of tolerable cheaters ($t < n/3$ instead of $t < n/2$). The computation complexity of the new protocol is on the order of the communication complexity and hence not relevant. The achieved communication complexity of $\mathcal{O}(mn^2)$ may well be optimal as even in the passive model, it appears very difficult to avoid that for each multiplication gate, every player sends a value to every other player.

The new protocol uses Beaver's circuit-randomization technique [Bea91a] and the player-elimination framework from [HMP00].

2 Model

We consider the well-known secure-channels model as introduced in [BGW88, CCD88]: The set $\mathcal{P} = \{P_1, \ldots, P_n\}$ of n players is connected by bilateral synchronous reliable secure channels. Broadcast channels are not assumed to be available. The goal of the protocol is to compute an agreed function, specified as an arithmetic circuit over a finite field $\mathbb{F}$ with $|\mathbb{F}| > n$. The number of multiplication gates in the circuit is denoted by m. To each player P_i a unique public value $\alpha_i \in \mathbb{F} \setminus \{0\}$ is assigned. The computation of the function is secure with respect to a computationally unbounded active adversary that is allowed to corrupt up to t of the players, where t is a given threshold with $t < n/3$. Once a player is corrupted, the adversary can read all his information and can make the player misbehave arbitrarily. We consider both static and adaptive adversaries, and distinguish both cases in the analysis whenever necessary. The security of our protocol is unconditional with an arbitrarily small probability of error. More precisely, there is an event that occurs with negligible probability, and as long as this event does not occur, the security of the protocol is perfect.

3 Protocol Overview

The protocol proceeds in two phases: In a preparation phase, which could actually be performed as a pre-computation independent of the circuit (except an upper bound on the number m of multiplication gates must be known), m random triples $\left(a^{(i)}, b^{(i)}, c^{(i)}\right)$ (for $i = 1, \ldots, m$) with $c^{(i)} = a^{(i)}b^{(i)}$ are t-shared among the players. In the computation phase, the circuit is evaluated gate by gate, where for each multiplication gate one shared triple from the preparation phase is used [Bea91a].

In the preparation phase, the triples are generated in a very efficient but non-robust manner (essentially with techniques from the passive protocol of [BGW88]). The generation is divided into blocks, and after each block, the consistency of all triples in this block is verified in a single verification procedure. If a block contains inconsistent triples, this is detected with overwhelming probability, and a set of two players that accuse each other of cheating is identified and eliminated from the protocol execution. The triples from the erroneous block are of course not used. At the end of the preparation phase, we have m triples that are t-shared among the set $\mathcal{P}' \subseteq \mathcal{P}$ of remaining players, and it will be guaranteed that the number t' of corrupted players in $\mathcal{P}'$ is smaller than $(|\mathcal{P}'| - t)/2$, which is sufficient for evaluating the circuit.

In the computation stage, for every multiplication one random triple is required. Two linear combinations of the values in this triple must be reconstructed. Therefore, it is important that all triples are shared *with the same degree* t (for privacy), and that $2t' < |\mathcal{P}'| - t$ (for reconstructibility).

The fault-localization procedure of the preparation phase is rather involved because it identifies a set of *two* players, one of whom is corrupted, whereas finding such a set of *three* players would be easy. However, eliminating a set of three players would violate the condition $2t' < n' - t$, and the t-shared triples would be useless.

As the underlying secret-sharing scheme we use the scheme of Shamir [Sha79], like in most threshold protocols: In order to t-share a value s, the dealer selects a random polynomial f of degree *at most* t with $f(0) = s$, and hands the share $s_i = f(\alpha_i)$ to player P_i for $i = 1, \ldots, n$. Selecting a random polynomial of degree at most t means to select t random coefficients $a_1, \ldots, a_t \in \mathbb{F}$ and to set $f(x) = s + a_1 x + \ldots + a_t x^t$. We say that a value s is t-shared among the players if there exists a polynomial $f(x)$ of degree at most t such that $f(0) = s$ and the share s_i held by player P_i satisfies $s_i = f(\alpha_i)$ for $i = 1, \ldots, n$. Such a t-shared value can be efficiently reconstructed by a set $\mathcal{P}' \subseteq \mathcal{P}$ of players, as long as less than $(|\mathcal{P}'| - t)/2$ of them misbehave (e.g., see [BW86]).

4 Preparation Phase

4.1 Overview

The goal of this phase is to generate m t-shared random triples $\left(a^{(i)}, b^{(i)}, c^{(i)}\right)$ with $c^{(i)} = a^{(i)} b^{(i)}$ in such a way that the adversary obtains no information about $a^{(i)}$, $b^{(i)}$, and $c^{(i)}$ (except that $c^{(i)}$ is the product of $a^{(i)}$ and $b^{(i)}$). The generation of these triples makes extensive use of the player-elimination framework of [HMP00]:

The triples are generated in blocks of $\ell = \lceil m/n \rceil$ triples. The triples of a block are generated (in parallel) in a non-robust manner; only at the end of the block, consistency is checked jointly for all triples of the block in a single verification procedure (*fault detection*). In case of an inconsistency, a set $\mathcal{D} \subseteq \mathcal{P}$ of *two* players, at least one of whom is corrupted, is identified (*fault localization*) and excluded from further computations (*player elimination*). The triples of the failed block are discarded. Player elimination ensures that at most t blocks fail, and hence in total at most $(n + t)$ blocks must be processed.

More precisely, the consistency verification takes place in *two* steps. In the first step (fault detection I), the degree of all involved sharings is verified. In other words, the players jointly verify that all sharings produced for generating the triples are of appropriate degree. The second verification step (fault detection II) is performed only if the first verification step is successful. Here, the players jointly verify that for every triple $\left(a^{(i)}, b^{(i)}, c^{(i)}\right)$, every player shared the correct values such that $c^{(i)} = a^{(i)} b^{(i)}$. If a fault is detected (in either fault-detection step), then all triples in the actual block are discarded. Furthermore, a set $\mathcal{D} \subseteq \mathcal{P}$ of two players, one of whom is corrupted, is found (fault localization I, resp. fault localization II) and eliminated from further computations. Note that in the fault-localization procedure, the privacy of the triples is not maintained. The triples contain completely random values unrelated to all values of the actual computation.

Both verification steps use n "blinding triples", and the privacy of these triples is annihilated in the verification procedure. Therefore, in each block, $\ell + 2n$ triples are generated. The first verification step verifies the degree of all sharings of the first $\ell + n$ triples, using (and destroying) the remaining n triples for blinding. The second verification step verifies the first ℓ triples, using the remaining n triples for blinding. Note that the second verification step requires that the sharings of all $\ell + n$ involved triples are verified to be correct.

During the generation of the blocks, players can be eliminated. At a given step, we denote the current set of players by $\mathcal{P}'$, the current number of players by $n' = |\mathcal{P}'|$, and the maximum number of cheaters in $\mathcal{P}'$ by t'. Without loss of generality, we assume that $\mathcal{P}' = \{P_1, \ldots, P_{n'}\}$. During the computation, the inequality $2t' < n' - t$ will hold as an invariant. In the beginning, $\mathcal{P}' = \mathcal{P}$, $n' = n$, and $t' = t$, and trivially $2t' < n' - t$ is satisfied. In player elimination, n' will be decreased by 2, and t' by 1. Clearly, this preserves the invariant.

0. Set $\mathcal{P}' = \mathcal{P}$, $n' = n$, and $t' = t$.
1. Repeat until n blocks (i.e., $n\ell \geq m$ triples) succeeded:
   - 1.1 Generate $\ell + 2n'$ triples (in parallel) in a non-robust manner (Sect. 4.2).
   - 1.2 Verify the consistency of all sharings involved in the first $\ell + n'$ triples (fault detection I, Sect. 4.3). If a fault is detected, identify a set $\mathcal{D} \subseteq \mathcal{P}'$ of two players such that at least one player in $\mathcal{D}$ is a cheater, and set $\mathcal{P}'$ to $\mathcal{P}' \setminus D$, n' to $n' - 2$ and t' to $t' - 1$ (fault localization I).
   - 1.3 If no fault was detected in Step 1.2, then verify that in the first ℓ triples, every player shared the correct values (fault detection II, Sect. 4.4). If a fault is detected, identify a set $\mathcal{D} \subseteq \mathcal{P}'$ of two players, at least one of whom is corrupted, and set $\mathcal{P}'$ to $\mathcal{P}' \setminus D$, n' to $n' - 2$ and t' to $t' - 1$ (fault localization II).
   - 1.4 If both verification steps were successful, then the generation of the block was successful, and the first ℓ triples can be used. If either verification procedure failed, then all triples of the actual block are discarded.

4.2 Generate One t-Shared Triple (a, b, c)

The purpose of this protocol is to generate one t-shared triple (a, b, c), where $c = ab$. The generation of this triple is non-robust: verification will take place only at the end of the block. In particular, in order to share a value, the dealer simply computes the shares and sends them to the players; the consistency verification of the sent shares is delayed.

The generation of the triple is straight-forward: First, the players jointly generate t'-sharings of two random values a and b. This is achieved by having every player share two random values, one for a and one for b, which are then summed up. Then, a t'-sharing of $c = ab$ is computed along the lines of [BGW88, GRR98] (passive model): Every player computes the product of his share of a and his share of b. These product shares define a $2t'$-sharing of c, and c can be computed by Lagrange interpolation. This interpolation is a linear function on the product shares. Hence, a t'-sharing of c can be computed as a linear combination of t'-sharings of the product shares. Finally, the degrees of the sharings of a, b, and c must be increased from t' to t. In order to do so, the players jointly generate three random sharings of 0, each with degree t, and add one of them to the t'-sharings of a, b, and c, respectively. These random t-sharings of 0 are generated by first selecting a random $t-1$-sharing of a random value, and then multiplying this polynomial by the monomial x.

Note that the protocol for computing a sharing of $c = ab$ relies on the fact that the degree of the sharings of a and b is less than one third of the number of actual players, and it would not work if a and b would be shared with degree t for $3t \geq n'$. On the other hand, it is important that finally the sharings of *all* blocks have the same degree (otherwise the multiplication protocol of Section 5 would leak information about the factors), and t' can decrease from block to block. Therefore, first the triple is generated with degree t', and then this degree is increased to t.

Protocol "Generate"

We give the exact protocol for generating one t-shared triple (a, b, c):

1. The players jointly generate t'-sharings of random values a and b:
   1.1 Every player $P_i \in \mathcal{P}'$ selects two random degree-t' polynomials $\widetilde{f}_i(x)$ and $\widetilde{g}_i(x)$, and hands the shares $\widetilde{a}_{ij} = \widetilde{f}_i(\alpha_j)$ and $\widetilde{b}_{ij} = \widetilde{g}_i(\alpha_j)$ to player P_j for $j = 1, \ldots, n'$.
   1.2 The polynomial for sharing a is $\widetilde{f}(x) = \sum_{i=1}^{n'} \widetilde{f}_i(x)$ (thus $a = \widetilde{f}(0)$), and the polynomial for sharing b is $\widetilde{g}(x) = \sum_{i=1}^{n'} \widetilde{g}_i(x)$ (thus $b = \widetilde{g}(0)$), and every player $P_j \in \mathcal{P}'$ computes his shares of a and b as
$$\widetilde{a}_j = \sum_{i=1}^{n'} \widetilde{a}_{ij}, \quad \text{and} \quad \widetilde{b}_j = \sum_{i=1}^{n'} \widetilde{b}_{ij}.$$
2. The players jointly compute a t'-sharing of $c = ab$:
   2.1 Every player $P_i \in \mathcal{P}'$ computes his product share $\widetilde{e}_i = \widetilde{a}_i \widetilde{b}_i$, and shares it among the players with the random degree-t' polynomial $\widetilde{h}_i(x)$ (with $\widetilde{h}_i(0) = \widetilde{e}_i$), i.e., sends the share $\widetilde{e}_{ij} = \widetilde{h}_i(\alpha_j)$ to player P_j for $j = 1, \ldots, n'$.
   2.2 Every player P_j computes his share $\widetilde{c}_j$ of c as
$$\widetilde{c}_j = \sum_{i=1}^{n'} w_i \widetilde{e}_{ij}, \quad \text{where } w_i = \prod_{\substack{j=1 \\ j \neq i}}^{n'} \frac{\alpha_j}{\alpha_j - \alpha_i}.$$

3. The players jointly increase the degree of the sharings of a, b, and c to t (this step is performed only if $t' < t$):
   3.1 Every player $P_i \in \mathcal{P}'$ selects three polynomials $\bar{f}_i(x)$, $\bar{g}_i(x)$, $\bar{h}_i(x)$ of degree $t-1$ at random, and sends the shares $\bar{a}_{ij} = \bar{f}_i(\alpha_j)$, $\bar{b}_{ij} = \bar{g}_i(\alpha_j)$, and $\bar{c}_{ij} = \bar{h}_i(\alpha_j)$ to player P_j for $j = 1, \ldots, n'$.
   3.2 Every player $P_j \in \mathcal{P}'$ computes his t-shares a_j, b_j, and c_j of a, b, and c, respectively, as follows:

$$a_j = \widetilde{a}_j + \alpha_j \sum_{i=1}^{n'} \bar{a}_{ij}, \quad b_j = \widetilde{b}_j + \alpha_j \sum_{i=1}^{n'} \bar{b}_{ij}, \quad c_j = \widetilde{c}_j + \alpha_j \sum_{i=1}^{n'} \bar{c}_{ij}.$$

Analysis

At the end of the block, two verifications will take place: First, it will be verified that the degree of all sharings is as required (t', respectively $t-1$, Section 4.3). Second, it will be verified that in Step 2.1, every player P_i indeed shares his correct product share $\widetilde{e}_i = \widetilde{a}_i \widetilde{b}_i$ (Section 4.4). In the sequel, we analyze the security of the above protocol under the assumption that these two conditions are satisfied.

After Step 1, obviously the assumption that the degree of all sharings is as required immediately implies that the resulting shares $\widetilde{a}_1, \ldots, \widetilde{a}_{n'}$ (respectively $\widetilde{b}_1, \ldots, \widetilde{b}_{n'}$) lie on a polynomial of degree t', and hence define a valid sharing. Furthermore, if at least one player in $P_i \in \mathcal{P}'$ honestly selected *random* polynomials $\widetilde{f}_i(x)$ and $\widetilde{g}_i(x)$, then a and b are random and unknown to the adversary.

In Step 2, we need the observation that c can be computed by Lagrange interpolation [GRR98]:

$$c = \sum_{i=1}^{n'} w_i \widetilde{e}_i, \quad \text{where } w_i = \prod_{\substack{j=1 \\ j \neq i}}^{n'} \frac{\alpha_j}{\alpha_j - \alpha_i}.$$

Assuming that every player P_i really shares his correct product share $\widetilde{e}_i$ with a polynomial $\widetilde{h}_i(x)$ of degree t', it follows immediately that the polynomial $\widetilde{h}(x) = \sum_{i=1}^{n'} w_i \widetilde{h}_i(x)$ is also of degree t', and furthermore

$$\widetilde{h}(0) = \sum_{i=1}^{n'} w_i \widetilde{h}_i(0) = \sum_{i=1}^{n'} w_i \widetilde{e}_i = c.$$

The privacy is guaranteed because the adversary does not obtain information about more than t' shares of any polynomial $\widetilde{h}_i(x)$ (for any $i = 1, \ldots, n'$).

Step 3 is only performed if $t' < t$. Assuming that the polynomials $\bar{f}_i(x)$, $\bar{g}_i(x)$, and $\bar{h}_i(x)$ of every player $P_i \in \mathcal{P}'$ have degree at most $t-1$, it immediately follows that all the polynomials defined as

$$\bar{f}(x) = \sum_{i=1}^{n'} \bar{f}_i(x), \quad \bar{g}(x) = \sum_{i=1}^{n'} \bar{g}_i(x), \quad \bar{h}(x) = \sum_{i=1}^{n'} \bar{h}_i(x)$$

also all have degree at most $t-1$. Hence, the polynomials $x\bar{f}(x)$, $x\bar{g}(x)$, and $x\bar{h}(x)$ have degree at most t, and they all share the secret 0. Thus, the sums $\tilde{f}(x)+x\bar{f}(x)$, $\tilde{g}(x)+x\bar{g}(x)$, and $\tilde{h}(x)+x\bar{h}(x)$ are of degree t and share a, b, and c, respectively. The privacy of the protocol is obvious for $t' \leq t-1$.

We briefly analyze the communication complexity of the above protocol: Every sharing requires n field elements to be sent, and in total there are $6n$ sharings, which results in a total of $6n^2$ field elements to be communicated per triple.

4.3 Verification of the Degrees of All Sharings in a Block

The goal of this fault-detection protocol is to verify the degree of the sharings of $\ell+n'$ triples in a single step, using (and destroying) another n' triples.

The basic idea of this protocol is to verify the degree of a *random linear combination* of the polynomials. More precisely, every player distributes a random challenge vector of length $\ell+n'$ with elements in $\mathbb{F}$, and the corresponding linear combinations of each involved polynomial is reconstructed towards the challenging player, who then checks that the resulting polynomial is of appropriate degree. In order to preserve the privacy of the involved polynomials, for each verifier one additional blinding polynomial of appropriate degree is added. If a verifier detects a fault (i.e., one of the linearly combined polynomials has too high degree), then the triples of the actual block are discarded, and in a fault-localization protocol, a set $\mathcal{D} \subseteq \mathcal{P}'$ of two players, at least one of whom is corrupted, is found and eliminated.

Protocol "Fault-Detection I"

The following steps for verifying the degree of all sharings in one block are performed in parallel, once for every verifier $P_v \in \mathcal{P}'$:

1. The verifier P_v selects a random vector $[r_1,\ldots,r_{\ell+n'}]$ with elements in $\mathbb{F}$ and sends it to each player $P_j \in \mathcal{P}'$.
2. Every player P_j computes and sends to P_v the following corresponding linear combinations (plus the share of the blinding polynomial) for every $i=1,\ldots,n'$:

$$\tilde{a}_{ij}^{(\Sigma)} = \sum_{k=1}^{\ell+n'} r_k \tilde{a}_{ij}^{(k)} + \tilde{a}_{ij}^{(\ell+n'+v)} \qquad \bar{a}_{ij}^{(\Sigma)} = \sum_{k=1}^{\ell+n'} r_k \bar{a}_{ij}^{(k)} + \bar{a}_{ij}^{(\ell+n'+v)}$$

$$\tilde{b}_{ij}^{(\Sigma)} = \sum_{k=1}^{\ell+n'} r_k \tilde{b}_{ij}^{(k)} + \tilde{b}_{ij}^{(\ell+n'+v)} \qquad \bar{b}_{ij}^{(\Sigma)} = \sum_{k=1}^{\ell+n'} r_k \bar{b}_{ij}^{(k)} + \bar{b}_{ij}^{(\ell+n'+v)}$$

$$\tilde{c}_{ij}^{(\Sigma)} = \sum_{k=1}^{\ell+n'} r_k \tilde{c}_{ij}^{(k)} + \tilde{c}_{ij}^{(\ell+n'+v)} \qquad \bar{c}_{ij}^{(\Sigma)} = \sum_{k=1}^{\ell+n'} r_k \bar{c}_{ij}^{(k)} + \bar{c}_{ij}^{(\ell+n'+v)}$$

3. P_v verifies whether for each $i=1,\ldots,n'$, the shares $\tilde{a}_{i1}^{(\Sigma)},\ldots,\tilde{a}_{in'}^{(\Sigma)}$ lie on a polynomial of degree at most t'. The same verification is performed for the

shares $\widetilde{b}_{i1}^{(\Sigma)}, \ldots, \widetilde{b}_{in'}^{(\Sigma)}$ and for the shares $\widetilde{c}_{i1}^{(\Sigma)}, \ldots, \widetilde{c}_{in'}^{(\Sigma)}$, for $i = 1, \ldots, n'$. Furthermore, P_v verifies whether for each $i = 1, \ldots, n'$, the shares $\bar{a}_{i1}^{(\Sigma)}, \ldots, \bar{a}_{in'}^{(\Sigma)}$ lie on a polynomial of degree at most $t-1$. The same verification is performed for the shares $\bar{b}_{i1}^{(\Sigma)}, \ldots, \bar{b}_{in'}^{(\Sigma)}$ and for the shares $\bar{c}_{i1}^{(\Sigma)}, \ldots, \bar{c}_{in'}^{(\Sigma)}$ for $i = 1, \ldots, n'$.

4. Finally, P_v broadcasts (using an appropriate sub-protocol) one bit according to whether all the $6n'$ verified polynomials have degree at most t', respectively $t-1$ (confirmation), or at least one polynomial has too high degree (complaint).

Protocol "Fault-Localization I"

This protocol is performed if and only if at least one verifier has broadcasts a complaint in Step 4 of the above fault-detection protocol. We denote with P_v the verifier who has reported a fault. If there are several such verifiers, the one with the smallest index v is selected.

5. The verifier P_v selects one of the polynomials of too high degree and broadcasts the location of the fault, consisting of the index i and the "name" of the sharing ($\widetilde{a}$, $\widetilde{b}$, $\widetilde{c}$, $\bar{a}$, $\bar{b}$, or $\bar{c}$). Without loss of generality, we assume that the fault was observed in the sharing $\widetilde{a}_{i1}^{(\Sigma)}, \ldots, \widetilde{a}_{in'}^{(\Sigma)}$.
6. The owner P_i of this sharing (i.e., the player who acted as dealer for this sharing) sends to the verifier P_v the correct linearly combined polynomial $\widetilde{f}_i^{(\Sigma)}(x) = \sum_{k=1}^{\ell+n'} r_k \widetilde{f}_i^{(k)}(x) + \widetilde{f}_i^{(\ell+n'+v)}(x)$.
7. P_v finds the (smallest) index j such that $\widetilde{a}_{ij}^{(\Sigma)}$ (received from P_j in Step 2) does not lie on the polynomial $\widetilde{f}_i^{(\Sigma)}(x)$ (received from the owner P_i in Step 6), and broadcasts j among the players in $\mathcal{P}'$.
8. Both P_i and P_j send the list $\widetilde{a}_{ij}^{(1)}, \ldots, \widetilde{a}_{ij}^{(\ell+n')}, \widetilde{a}_{ij}^{(\ell+n'+v)}$ to P_v.
9. P_v verifies that the linear combination $[r_1, \ldots, r_{\ell+n'}]$ applied to the values received from P_i is equal to $\widetilde{f}_i^{(\Sigma)}(\alpha_j)$. Otherwise, P_v broadcasts the index i, and the set of players to be eliminated is $\mathcal{D} = \{P_i, P_v\}$. Analogously, P_v verifies the values received from P_j to be consistent with $\widetilde{a}_{ij}^{(\Sigma)}$ received in Step 2, and in case of failure broadcasts the index j, and $\mathcal{D} = \{P_j, P_v\}$.
10. P_v finds the (smallest) index k such that the values $\widetilde{a}_{ij}^{(k)}$ received from P_i and P_j differ, and broadcasts k and both values $\widetilde{a}_{ij}^{(k)}$ from P_i and $\widetilde{a}_{ij}^{(k)}$ from P_j.
11. Both P_i and P_j broadcast their value of $\widetilde{a}_{ij}^{(k)}$.
12. If the values broadcast by P_i and P_j differ, then the localized set is $\mathcal{D} = \{P_i, P_j\}$. If the value broadcast by P_i differs from the value that P_v broadcast (and claimed to be the value received from P_i), then $\mathcal{D} = \{P_i, P_v\}$. Else, $\mathcal{D} = \{P_j, P_v\}$.

Analysis

It follows from simple algebra that if all players are honest, then the above fault-detection protocol will always pass. On the other hand, if at least one of the involved sharings (in any of the $\ell+n'$ triples) has too high degree, then every honest verifier will detect this fault with probability at least $1-1/|\mathbb{F}|$.

The correctness of the fault-localization protocol can be verified by inspection. There is no privacy issue; the generated triples are discarded.

The fault-detection protocol requires $n(n(\ell+n)+6n^2)= n^2\ell+7n^3$ field elements to be sent and n bits to be broadcast. For fault localization, up to $n+2(\ell+n+1)=2\ell+3n+2$ field elements must be sent and $2\log n+\log 6+\log(\ell+n+1)+4\log|\mathbb{F}|$ bits must be broadcast.

4.4 Verification That All Players Share the Correct Product Shares

It remains to verify that in each triple $k=1,\dots,\ell$, every player P_i shared the correct product share $\widetilde{e}_i^{(k)}=\widetilde{a}_i^{(k)}\widetilde{b}_i^{(k)}$ (Step 2.1 of protocol Generate). Since it is already verified that the sharings of all factor shares are of degree t', it is sufficient to verify that the shares $\widetilde{e}_1^{(k)},\dots,\widetilde{e}_{n'}^{(k)}$ lie on a polynomial of degree at most $2t'$. Note that the at least $n'-t'>2t'$ shares of the honest players uniquely define this polynomial. The key idea of this verification protocol is the same as in the previous verification protocol: Every verifier P_v distributes a random challenge vector, and the corresponding linear combination of the polynomials (plus one blinding polynomial) is opened towards P_v. If a fault is detected, then a set $\mathcal{D}$ of two players (one of whom is corrupted) can be found with the fault-localization protocol.

Protocol "Fault-Detection II"
The following steps are performed for each verifier $P_v\in\mathcal{P}'$ in parallel.

1. The verifier P_v selects a random vector $[r_1,\dots,r_\ell]$ with elements in $\mathbb{F}$ and sends it to each player $P_j\in\mathcal{P}'$.
2. Every player P_j computes and sends to P_v the following linear combinations (with blinding) for every $i=1,\dots,n'$:

$$\widetilde{e}_{ij}^{(\Sigma)}=\sum_{k=1}^{\ell} r_k\widetilde{e}_{ij}^{(k)} \;+\; \widetilde{e}_{ij}^{(\ell+v)}.$$

3. P_v verifies whether for each $i=1,\dots,n'$ the shares $\widetilde{e}_{i1}^{(\Sigma)},\dots,\widetilde{e}_{in'}^{(\Sigma)}$ lie on a polynomial of degree at most t', and if so, whether the secrets $\widetilde{e}_1^{\Sigma},\dots,\widetilde{e}_{n'}^{\Sigma}$ of the above sharings (computed by interpolating the corresponding share-shares) lie on a polynomial of degree at most $2t'$. P_v broadcasts one bit according to whether all polynomials have appropriate degree (confirmation), or at least one polynomial has too high degree (complaint).

Protocol "Fault-Localization II"
We denote with P_v the verifier who has reported a fault in Step 3 of the above fault-detection protocol. If there are several such verifiers, the one with the smallest index v is selected.

4. If in Step 3, the degree of one of the second-level sharings $\widetilde{e}_{i1}^{(\Sigma)}, \ldots, \widetilde{e}_{in'}^{(\Sigma)}$ was too high, then P_v applies error-correction to find the smallest index j such that $\widetilde{e}_{ij}^{(\Sigma)}$ must be corrected. Since all sharings have been verified to have correct degree, P_v can conclude that P_j has sent the wrong value $\widetilde{e}_{ij}^{(\Sigma)}$. P_v broadcasts the index j, and the set of players to be eliminated is $\mathcal{D} = \{P_j, P_v\}$ (and the following steps need not be performed).
5. Every player P_i sends to P_v all his factor shares $\widetilde{a}_i^{(1)}, \ldots, \widetilde{a}_i^{(\ell)}, \widetilde{a}_i^{(\ell+v)}$ and $\widetilde{b}_i^{(1)}, \ldots, \widetilde{b}_i^{(\ell)}, \widetilde{b}_i^{(\ell+v)}$.
6. P_v verifies for every $k = 1, \ldots, \ell, \ell+v$ whether the shares $\widetilde{a}_1^{(k)}, \ldots, \widetilde{a}_{n'}^{(k)}$ lie on a polynomial of degree t'. If not, then P_v applies error-correction and finds and broadcasts the (smallest) index j such that $\widetilde{a}_j^{(k)}$ must be corrected. The set of players to be eliminated is $\mathcal{D} = \{P_j, P_v\}$. The same verification is performed for the shares $\widetilde{b}_1^{(k)}, \ldots, \widetilde{b}_{n'}^{(k)}$ for $k = 1, \ldots, \ell, \ell + v$.
7. P_v verifies for every $i = 1, \ldots, n'$ whether the value $\widetilde{e}_i^{\Sigma}$ computed in Step 4 is correct, i.e., whether

$$\widetilde{e}_i^{\Sigma} \stackrel{?}{=} \sum_{k=1}^{\ell} r_k \widetilde{a}_i^{(k)} \widetilde{b}_i^{(k)} \quad + \quad \widetilde{a}_i^{(\ell+v)} \widetilde{b}_i^{(\ell+v)}.$$

This test will fail for at least one i, and P_v broadcasts this index i. The players in $\mathcal{D} = \{P_i, P_v\}$ are eliminated.

Analysis

The above fault-detection protocol always passes when all players are honest. If the degree of at least one of the involved sharings is higher than $2t'$, then every honest verifier will detect this fault with probability at least $1 - 1/|\mathbb{F}|$. The correctness of the fault-localization protocol follows by inspection.

The fault-detection protocol requires $n(n\ell + n^2) = n^2\ell + n^3$ elements to be sent, and n bits to be broadcast. The fault-localization protocol requires $2n(\ell+1)$ field elements to be sent and $\log n$ bits to be broadcast.

4.5 Error Probabilities and Repetitive Verifications

We first calculate the probability that a static adversary can introduce a bad triple into a block, without being detected. So assume that in a block, at least one triple is bad. This is detected by every honest player with probability $1 - 1/|\mathbb{F}|$. Hence, the probability that no honest player detects (and reports) the inconsistency is at most $|\mathbb{F}|^{-(n'-t')}$. Once the bad block is detected, one corrupted player is eliminated. Hence the adversary can try t times to make pass a bad block, and the probability that (at least) one of these trials is not detected (and the protocol is disrupted) is at most $\sum_{i=0}^{t-1} |\mathbb{F}|^{-(n-t-i)} \leq (1/|\mathbb{F}|)^{n-2t}$.

If the adversary is adaptive, he can decide whether or not to corrupt the verifier *after* the challenge vector is known. Hence, a bad block passes the verification step if at least $n' - t'$ of the challenge vectors cannot discover the fault, and this happens with probability at most $\sum_{i=0}^{t'}\binom{n'}{i}(1-1/|\mathbb{F}|)^i(1/|\mathbb{F}|)^{n'-i} \leq (3/|\mathbb{F}|)^{n'-t'}$. Again, the adversary can try t times to make pass a bad block, which results in an overall error probability of $\sum_{i=0}^{t-1}(3/|\mathbb{F}|)^{n-t-i} \leq (3/|\mathbb{F}|)^{n-2t}$.

If the above error probabilities are too high, they can further be decreased by simply repeating the fault-detection protocols (with new and independent blinding triples). By repeating the protocol k times, the error probability is lowered to $(1/|\mathbb{F}|)^{k(n-2t)}$ (static case), respectively $(3/|\mathbb{F}|)^{k(n-2t)}$ (adaptive case).

5 Computation Phase

The evaluation of the circuit is along the lines of the protocol of [Bea91a]. Slight modifications are needed because the degree t of the sharings and the upper bound t' on the number of cheaters need not be equal. Furthermore, special focus is given to the fact that in our protocol, also eliminated players must be able to give input to and receive output from the computation.

From the preparation phase, we have m random triples $\left(a^{(i)}, b^{(i)}, c^{(i)}\right)$ with $c^{(i)} = a^{(i)}b^{(i)}$, where the sharings are of degree t among the set $\mathcal{P}'$ of players. The number of corrupted players in $\mathcal{P}'$ is at most t' with $2t' < n' - t$, where $n' = |\mathcal{P}'|$. This is sufficient for efficient computation of the circuit.

5.1 Input Sharing

First, every player who has input secret-shares it (with degree t) among the set $\mathcal{P}'$ of players. We use the verifiable secret-sharing protocol of [BGW88] (with perfect security), with a slight modification to support $t \neq t'$. The dealer is denoted by P, and the secret to be shared by s. We do not assume that $P \in \mathcal{P}'$ (neither $P \in \mathcal{P}$).

1. The dealer P selects at random a polynomial $f(x, y)$ of degree t in both variables, with $p(0, 0) = s$, and sends the polynomials $f_i(x) = f(\alpha_i, x)$ and $g_i(x) = p(x, \alpha_i)$ to player P_i for $i = 1, \ldots, n'$.
2. Every player $P_i \in \mathcal{P}'$ sends to P_j for $j = i + 1, \ldots, n'$ the values $f_i(\alpha_j)$ and $g_i(\alpha_j)$.
3. Every player P_j broadcasts one bit according to whether all received values are consistent with the polynomials $f_j(x)$ and $g_j(x)$ (confirmation) or not (complaint).
4. If no player has broadcast a complaint, then the secret-sharing is finished, and the share of player P_j is $f_j(0)$. Otherwise, every player P_j who has complaint broadcasts a bit vector of length n', where a 1-bit in position i means that one of the values received from P_i was not consistent with $f_j(x)$ or $g_j(x)$. The dealer P must answer all complaints by broadcasting the correct values $f(\alpha_i, \alpha_j)$ and $f(\alpha_j, \alpha_i)$.
5. Every player P_i checks whether the values broadcast by the dealer in Step 4 are consistent with his polynomials $f_i(x)$ and $g_i(x)$, and broadcasts either a confirmation or an accusation. The dealer P answers every accusation by

broadcasting both polynomials $f_i(x)$ and $g_i(x)$ of the accusing player P_i, and P_i replaces his polynomials by the broadcast ones.

6. Every player P_i checks whether the polynomials broadcast by the dealer in Step 5 are consistent with his polynomials $f_i(x)$ and $g_i(x)$, and broadcasts either a confirmation or an accusation.
7. If in Steps 5 and 6, there are in total at most t' accusations, then every player P_i takes $f_i(0)$ as his share of s. Otherwise, clearly the dealer is faulty, and the players take a default sharing (e.g., the constant sharing of 0).

It is clear that an honest player never accuses an honest dealer. On the other hand, if there are at most t' accusations, then the polynomials of at least $n' - 2t' > t$ honest players are consistent, and these polynomials uniquely define the polynomial $f(x, y)$ with degree t. Hence, the polynomials of all honest players are consistent, and their shares $f_1(0), \ldots, f_{n'}(0)$ lie on a polynomial of degree t.

This protocol communicates $3n^2$ field elements, and it broadcasts n bits (in the best case), respectively $n^2 + 3n + 2t^2 \log |\mathbb{F}|$ bits (in the worst case).

5.2 Evaluation of the Circuit

The circuit is evaluated gate by gate. Linear gates can be evaluated without any communication due to the linearity of the used sharing. Multiplication gates are evaluated according to [Bea91a]: Assume that the factors x and y are t-shared among the players. Furthermore, a t-shared triple (a, b, c) with $c = ab$ is used. The product xy can be written as follows:

$$xy = \big((x-a)+a\big)\big((y-b)+b\big) = \big((x-a)(y-b)\big) + (x-a)b + (y-b)a + c.$$

The players in $\mathcal{P}'$ reconstruct the differences $d_x = x - a$ and $d_y = y - b$. This reconstruction is possible because $2t' < n' - t$ (e.g., see [BW86]). Note that reconstructing these values does not give any information about x or y, because a and b are random. Then, the following equation holds:

$$xy = d_x d_y + d_x b + d_y a + c.$$

This equation is linear in a, b, and c, and we can compute linear combinations on shared values without communication. This means that the players can compute the above linear combination on their respective shares of x and y and they receive a t-sharing of the product xy. More details can be found in [Bea91a].

This multiplication protocol requires two secret-reconstructions per multiplication gate. Secret-reconstruction requires every player in $\mathcal{P}'$ to send his share to every other player (who then applies error-correction to the received shares and interpolates the secret). The communication costs per multiplication gate are hence $2n^2$. Broadcast is not needed.

5.3 Output Reconstruction

Any player P can receive output (not only players in $\mathcal{P}'$ or in $\mathcal{P}$). In order to reconstruct a shared value x towards player P, every player in $\mathcal{P}'$ sends his share of x to P, who then applies error-correction and interpolation to compute the output x. In the error-correction procedure, up to $(n' - t - 1)/2 \geq t'$ errors can be corrected (e.g., see [BW86]).

Reconstructing one value requires n field elements of communication, and no broadcast.

5.4 Probabilistic Functions

The presented protocol is for deterministic functions only. In order to capture probabilistic functions, one can generate one (or several) blocks with single values $a^{(i)}$ only (with simplified verification), and use these values as shared randomness.

Alternatively, but somewhat wastefully, one just picks the value $a^{(i)}$ from a shared triple $(a^{(i)}, b^{(i)}, c^{(i)})$, and discards the rest of the triple. Then, m denotes the number of multiplication gates plus the number of "randomness gates".

5.5 On-Going Computations

In an on-going computation, inputs and outputs can be given and received at any time during the computation, not only at the beginning and at the end. Furthermore, it might even not be specified beforehand which function will be computed. An example of an on-going computation is the simulation of a fair stock market.

In contrast to the protocol of [HMP00], the proposed protocol can easily be extended to capture the scenario of on-going computations. First, the players generate ℓ triples (a, b, c) with $c = ab$, and perform the computation until all triples are exhausted. Then, a new block of ℓ triples is generated, and so on.

6 Complexity Analysis

A detailed complexity analysis is given in Appendix A. Here we summarize the most important results: Let n denote the number of players, $\mathbb{F}$ the field over which the function (circuit) is defined, m the number of multiplication gates in the circuit, C_d the depth of the circuit, n_I the number of inputs and n_O the number of outputs of the function. Evaluating this circuit securely with respect to an active adversary corrupting any $t < n/3$ of the players is possible with communicating $14mn^2 + \mathcal{O}(n_I n^4 + n_O n + n^4)$ field elements. The number of communication rounds is $C_d + \mathcal{O}(n^2)$. All complexities include the costs for simulating broadcast. If the field $\mathbb{F}$ is too small (and the resulting error probability is too high), then fault-detection protocols are repeated, and the overall communication complexity increases accordingly.

This complexity should be compared with the complexity of the most efficient protocols. In the secure-channels model, the most efficient protocol for unconditionally secure multi-party protocols [HMP00] requires $\mathcal{O}(mn^3)$ field elements in $\mathcal{O}(C_d + n^2)$ rounds (where both hidden constants are slightly higher than ours).

For completeness, we also compare the complexity of our protocol with the complexity of the most efficient protocol for the cryptographic model [CDN01]. This protocol requires a communication complexity of $\mathcal{O}(mn^3)$ field elements in $\mathcal{O}(C_d n)$ rounds. The high round complexity results from the fact that the protocol invokes a broadcast sub-protocol for each multiplication gate. The most efficient broadcast protocols require $\mathcal{O}(n)$ rounds. Constant-round broadcast protocols are known [FM88], but they have higher communication complexities and would results in a communication complexity of $\mathcal{O}(mn^5)$ field elements.

Finally, we compare the protocol with the most efficient known protocol for passive security, namely [BGW88] with the simplification of [GRR98]. This protocol communicates $mn^2 + \mathcal{O}(n_I n + n_O n)$ field elements. Hence, for large enough circuits, robustness can be achieved with a communication overhead factor of about 14.

7 Conclusions and Open Problems

We have presented a protocol for secure multi-party computation unconditionally secure against an active adversary which is (up to a small constant factor) as efficient as protocols with passive security. The protocol provides some (arbitrarily small) probability of error. Note that due to the player-elimination technique, this error-probability does not grow with the length of the protocol (like in all previous MPC protocols with error probability), but only in the upper bound t of the number of corrupted players.

It remains open whether quadratic complexity can also be achieved in other models. In the unconditional model with perfect security, the most efficient protocol requires communication of $\mathcal{O}(n^3)$ field elements per multiplication gate [HMP00]. In the unconditional model with broadcast (with small error probability), the most efficient protocol requires $\mathcal{O}(n^4)$ field elements to be broadcast per multiplication gate [CDD+99,Feh00]. In the cryptographic model (where up to $t < n/2$ of the players may be corrupted), the most efficient protocol requires communication of $\mathcal{O}(n^3)$ field elements (and $O(n)$ rounds!) per multiplication gate [CDN01]. A very recent result for Boolean circuits achieves essentially the same communication complexity per multiplication, but in a constant number of rounds for the whole circuit [DN01].

Also, it would be interesting to combine the techniques of this paper with the techniques of papers with protocols that require a constant number of rounds only (but have a high communication complexity), to achieve a multi-party protocol which has both low communication complexity and very low round complexity.

Furthermore, the presented protocol is for the synchronous model. Some real-world networks appear to be more appropriately modeled by the asynchronous model, and the protocol must be adapted for this setting. It seems that this can be done along the lines of [BCG93,Can95,SR00].

Finally, it would be interesting to have a proof that quadratic complexity is optimal for passive security. This would immediately imply that the protocol of this paper is optimally efficient (up to a constant factor).

Acknowledgments. We would like to thank Serge Fehr and Matthias Fitzi for many fruitful discussions, and the anonymous referees for their helpful comments.

References

[BB89] J. Bar-Ilan and D. Beaver. Non-cryptographic fault-tolerant computing in a constant number of rounds of interaction. In *Proc. 8th ACM Symposium on Principles of Distributed Computing (PODC)*, pp. 201–210, Aug. 1989.

[BCG93] M. Ben-Or, R. Canetti, and O. Goldreich. Asynchronous secure computation. In *Proc. 25th ACM Symposium on the Theory of Computing (STOC)*, pp. 52–61, 1993.

[Bea91a] D. Beaver. Efficient multiparty protocols using circuit randomization. In *Advances in Cryptology — CRYPTO '91*, volume 576 of *Lecture Notes in Computer Science*, pp. 420–432, 1991.

[Bea91b] D. Beaver. Secure multiparty protocols and zero-knowledge proof systems tolerating a faulty minority. *Journal of Cryptology*, pp. 75–122, 1991.

[BW86] E. R. Berlekamp and L. Welch. Error correction of algebraic block codes. US Patent Number 4,633,470, 1986.

[BFKR90] D. Beaver, J. Feigenbaum, J. Kilian, and P. Rogaway. Security with low communication overhead. In *Advances in Cryptology — CRYPTO '90*, volume 537 of *Lecture Notes in Computer Science*. Springer-Verlag, 1990.

[BGP89] P. Berman, J. A. Garay, and K. J. Perry. Towards optimal distributed consensus (extended abstract). In *Proc. 21st ACM Symposium on the Theory of Computing (STOC)*, pp. 410–415, 1989.

[BGW88] M. Ben-Or, S. Goldwasser, and A. Wigderson. Completeness theorems for non-cryptographic fault-tolerant distributed computation. In *Proc. 20th ACM Symposium on the Theory of Computing (STOC)*, pp. 1–10, 1988.

[Can95] R. Canetti. *Studies in Secure Multiparty Computation and Applications*. PhD thesis, Weizmann Institute of Science, Rehovot 76100, Israel, June 1995.

[CCD88] D. Chaum, C. Crépeau, and I. Damgård. Multiparty unconditionally secure protocols (extended abstract). In *Proc. 20th ACM Symposium on the Theory of Computing (STOC)*, pp. 11–19, 1988.

[CDD+99] R. Cramer, I. Damgård, S. Dziembowski, M. Hirt, and T. Rabin. Efficient multiparty computations secure against an adaptive adversary. In *Advances in Cryptology — EUROCRYPT '99*, volume 1592 of *Lecture Notes in Computer Science*, pp. 311–326, 1999.

[CDN01] R. Cramer, I. Damgård, and J. B. Nielsen. Multiparty computation from threshold homomorphic encryption. In *Advances in Cryptology — EUROCRYPT '01*, volume 2045 of *Lecture Notes in Computer Science*, pp. 280–300, 2001.

[DN01] I. Damgård and J. B. Nielsen. An efficient pseudo-random generator with applications to public-key encryption and constant-round multiparty computation. Manuscript, May 2001.

[Feh00] Serge Fehr. Personal communications, 2000.

[FKN94] U. Feige, J. Kilian, and M. Naor. A minimal model for secure computation. In *Proc. 26th ACM Symposium on the Theory of Computing (STOC)*, pp. 554–563, 1994.

[FM88] P. Feldman and S. Micali. Optimal algorithms for Byzantine agreement. In *Proc. 20th ACM Symposium on the Theory of Computing (STOC)*, pp. 148–161, 1988.

[FY92] M. K. Franklin and M. Yung. Communication complexity of secure computation. In *Proc. 24th ACM Symposium on the Theory of Computing (STOC)*, pp. 699–710, 1992.

[GMW87] O. Goldreich, S. Micali, and A. Wigderson. How to play any mental game — a completeness theorem for protocols with honest majority. In *Proc. 19th ACM Symposium on the Theory of Computing (STOC)*, pp. 218–229, 1987.

[GRR98] R. Gennaro, M. O. Rabin, and T. Rabin. Simplified VSS and fast-track multiparty computations with applications to threshold cryptography. In *Proc. 17th ACM Symposium on Principles of Distributed Computing (PODC)*, pp. 101–111, 1998.

[HMP00] M. Hirt, U. Maurer, and B. Przydatek. Efficient secure multi-party computation. In T. Okamoto, editor, *Advances in Cryptology — ASIACRYPT '00*, volume 1976 of *Lecture Notes in Computer Science*, pp. 143–161. Springer-Verlag, Dec. 2000.

[IK00] Y. Ishai and E. Kushilevitz. Randomizing polynomials: A new representation with applications to round-efficient secure computation. In *Proc. 41st IEEE Symposium on the Foundations of Computer Science (FOCS)*, Oct. 2000.

[RB89] T. Rabin and M. Ben-Or. Verifiable secret sharing and multiparty protocols with honest majority. In *Proc. 21st ACM Symposium on the Theory of Computing (STOC)*, pp. 73–85, 1989.

[Sha79] A. Shamir. How to share a secret. *Communications of the ACM*, 22:612–613, 1979.

[SR00] K. Srinathan and C. P. Rangan. Efficient asynchronous secure multiparty distributed computation. In *Indocrypt 2000*, volume 1977 of *Lecture Notes in Computer Science*, Dec. 2000.

[Yao82] A. C. Yao. Protocols for secure computations. In *Proc. 23rd IEEE Symposium on the Foundations of Computer Science (FOCS)*, pp. 160–164. IEEE, 1982.

A Detailed Complexity Analysis

We summarize the complexities of all involved sub-protocols. For each sub-protocol, we indicate both the message complexity (MC, in communicated field elements) and the broadcast complexity (BC, in bits) of the protocol involved once, and specify how often the protocol is called at least (when no adversary is present) and at most (when the corrupted players misbehave in the most effective way). The complexity of the verifiable secret-sharing protocol of [BGW88], which is used for giving input, depends on whether or not some of the players misbehave. We list both complexities.

In the table, n denotes the number of players, t the upper bound on the number of actively corrupted players, m the total number of multiplication gates, ℓ the number of multiplication gates per block, n_I the number of inputs to the function, and n_O the number of outputs of the function.

The indicated complexities are upper bounds: In particular, when a player has to send a message to all players, we count this as n messages (instead of $n-1$).

What	MC (field elements)	BC (bits)	#Calls (min...max)	
Generate triples	$6n^2$	—	$n(\ell+2n)\ldots(n+t)(\ell+2n)$	(1)
Fault detection I	ℓn^2+7n^3	n	$n\ldots n+t$	(2)
Fault localization I	$2\ell+3n+2$	$2\log n+4\log\lvert\mathbb{F}\rvert+\log(\ell+n+1)+\log 6$	$0\ldots t$	(3)
Fault detection II	ℓn^2+n^3	n	$n\ldots n+t$	(4)
Fault localization II	$2\ell n+2n$	$\log n$	$0\ldots t$	(5)
Give input (best)	$3n^2$	n	n_I	(6)
Give input (worst)	$3n^2$	$n^2+3n+2t^2\log\lvert\mathbb{F}\rvert$	n_I	(7)
Multiply	$2n^2$	—	m	(8)
Get output	n	—	n_O	(9)

We add up the above complexities for $\ell\leq m/n+1$, $n\geq 4$, and $t\leq n/3$. In order to simplify the expressions, some of the terms are slightly rounded up. Furthermore, for the sake of simplicity, we assume that the field $\mathbb{F}$ is large such that the resulting failure probability of the fault-detection protocols is small enough and there is no need to repeat the protocol.

In the best case (when no cheating occurs), $10mn^2+22n^4+3n_In^2+n_On$ field elements are communicated and $2n^2+n_In$ bits are broadcast. Applying the broadcast protocol of [BGP89] (which communicates $9n^2$ bits for broadcasting one bit), this results in a total complexity of less than $10mn^2\log\lvert\mathbb{F}\rvert+22n^4(\log\lvert\mathbb{F}\rvert+1)+n_In^2(3\log\lvert\mathbb{F}\rvert+9n)+n_On\log\lvert\mathbb{F}\rvert$ bits.

In the worst case, the protocol communicates $13mn^2+30n^4+3n_In^2+n_On$ field elements and broadcasts $3n^2+2n\log\lvert\mathbb{F}\rvert+\frac{n}{3}\log m+n_In^2\log\lvert\mathbb{F}\rvert$ bits. Simulating broadcast with [BGP89], this gives less than $14mn^2\log\lvert\mathbb{F}\rvert+35n^4(\log\lvert\mathbb{F}\rvert+1)+9n_In^4\log\lvert\mathbb{F}\rvert+n_On\log\lvert\mathbb{F}\rvert$ bits. This is about $14mn^2+\mathcal{O}(n_In^4+n_On+n^4)$ field elements.

Secure Distributed Linear Algebra in a Constant Number of Rounds

Ronald Cramer and Ivan Damgård

Comp. Sc. Dept., Aarhus University & BRICS (Basic Research in Computer Science, center of the Danish National Research Foundation), cramer@daimi.aau.dk
ivan@daimi.aau.dk

Abstract. Consider a network of processors among which elements in a finite field K can be verifiably shared in a constant number of rounds. Assume furthermore constant-round protocols are available for generating random shared values, for secure multiplication and for addition of shared values. These requirements can be met by known techniques in all standard models of communication.
In this model we construct protocols allowing the network to securely solve standard computational problems in linear algebra. In particular, we show how the network can securely, efficiently and in constant-round compute determinant, characteristic polynomial, rank, and the solution space of linear systems of equations. Constant round solutions follow for all problems which can be solved by direct application of such linear algebraic methods, such as deciding whether a graph contains a perfect match.
If the basic protocols (for shared random values, addition and multiplication) we start from are unconditionally secure, then so are our protocols. Our results offer solutions that are significantly more efficient than previous techniques for secure linear algebra, they work for arbitrary fields and therefore extend the class of functions previously known to be computable in constant round and with unconditional security. In particular, we obtain an unconditionally secure protocol for computing a function f in constant round, where the protocol has complexity polynomial in the span program size of f over an arbitrary finite field.

1 Introduction

In this paper we consider the problem of secure multiparty computation (MPC), where n players, each holding a secret input, want to compute an agreed function of the inputs, in such a way that the correct result is computed, and no additional information about the inputs is released. This should hold, even in presence of an *adversary* who can *corrupt* some of the players, this means he can see all their internal data and can (if he is *active*) even make them behave as he likes.

A main distinction between different kinds of MPC protocols concerns the model for communication: In the *cryptographic* model (first studied in [25,14]), the adversary may see all messages sent, and security can then only be guaranteed under a computational assumption. In the *information-theoretic* model (first

J. Kilian (Ed.): CRYPTO 2001, LNCS 2139, pp. 119–136, 2001.

studied in [5,7]), one assumes a private channel between every pair of players, and security can then be guaranteed unconditionally.

Two measures of complexity are important for MPC protocols, namely the communication complexity (total number of bits sent) and the round complexity (where a round is a phase where each player is allowed to send one message to each other player).

In this paper, we focus on the round complexity of MPC protocols, in particular on building constant-round protocols. Kilian [19] showed that Boolean formulas can be securely and efficiently evaluated in constant rounds in the two-party case, with secure computations based on Oblivious Transfer. Under a complexity assumption, it was shown in [2] by Beaver, Micali and Rogaway that any function that can be computed in polynomial time can also be securely computed in a constant number of rounds (and polynomial communication). The result works under minimal complexity assumptions, but leads in practice to very inefficient protocols. Thus, for computationally secure MPC in constant round, the question is not which functions can be securely computed, but rather how efficiently it can be done.

The situation is different for *unconditionally* secure MPC: in this model, it is not known which functions can be securely computed in constant-round. However, Bar-Ilan and Beaver [1] showed that it can be done for any arithmetic formula.

Later results by Feige, Kilian and Naor [12] and Ishai and Kushilevitz [16, 17] and Beaver [3] extend this to functions in NL and some related counting classes. More precisely, their protocols are polynomial in the modular branching program size of the function computed. Their methods also apply to the more general arithmetic branching program model of Beimel and Gal[4].

2 Our Work

In this paper, we start from the assumption that we are given an efficient, constant round method to share securely between the players values in a finite field K and to reveal them. For an active adversary, this would be a verifiable secret sharing (VSS). In the following, we write $[a]$ for a sharing of a, i.e. $[a]$ denotes the collection of all information related to a held by the players. When M is a matrix over K, $[M]$ will denote a sharing of each of the coordinates of M. Whenever we say "let $[x]$ be a sharing" we mean that either some processor has distributed shares of his private input x, or that $[x]$ is the result of previous secure computations on certain private inputs of the processors. An expression such as "$[f(x)]$ is securely computed from $[x]$" means that the processors in the network perform secure computations on a sharing of x, as a result of which they obtain a (random) sharing of $f(x)$.

We show how to design efficient constant-round protocols for a number of standard linear algebra problems:

- Given a shared matrix $[A]$ over an arbitrary finite field K, we show how to compute securely a sharing $[\det(A)]$ of the determinant of A. More generally, $[\mathbf{f}]$ is computed where $\mathbf{f}$ denotes the vector containing the coefficients of the characteristic polynomial of A.

- Given a shared (not necessarily square) matrix $[A]$ over a finite field K, we show how to securely compute the rank of A, concretely we can compute $[\mathbf{r}]$ where $\mathbf{r}$ is a unary encoding of the rank of A, padded with zeroes.
- Given also a shared vector $[\mathbf{y}]$ we show how to securely compute $[b]$ where b is a bit that indicates whether the system of equations $A\mathbf{x} = \mathbf{y}$ is solvable. Finally, we show how to solve the system by securely computing $[\mathbf{x}], [B]$, where $\mathbf{x}$ is a solution and $[B]$ generates A's kernel.

Our protocols work for arbitrary fields and do not use any cryptographic assumptions, so if the basic sharing method we start from is unconditionally secure, then so are the protocols we construct.

It is easy to see that our results allow handling all functions computable in constant round and with unconditional security using the most general previous methods [16,17]: for instance, our protocol for subspace membership immediately implies a constant-round protocol for computing a function f, of complexity polynomial in the span program[18] size of f. By the results from [4] span programs are always at least as powerful as the modular and arithmetic branching programs to which the methods from [16,17] apply. For fields with fixed characteristic, all three models are equivalent in power. However, since this is not known to hold for arbitrary fields, our results extend the class of function known to be computable in constant round and with unconditional security.

What is equally important, however, is that the standard linear algebra problems we can handle are problems that occur naturally in practice. For instance, deciding if a determinant is non-zero allows to decide if a bipartite graph contains a perfect match. Moreover, privacy is a natural requirement in matching type problems that occur in practice.

We therefore believe it is of interest to be able to do linear algebra securely and efficiently. Our work leads to a protocols with better efficiency compared to solutions based on combinations of known techniques. Please refer to Section 6.5 for more details in the case of determinant and characteristic polynomial.

We note that our results apply to the cryptographic model as well as the information theoretic, the only difference being the implementation of the underlying sharing and multiplication protocols. And because we attack the problems directly rather than going through reductions (to, e.g., Boolean circuits for the problems) we get much more efficient solutions than what one gets from, e.g, [2].

3 Some Basic Protocols

For convenience in describing our main protocols, we assume that secure constant round protocols are available for the following tasks:

- Computing (from scratch) a sharing $[r]$ where $r \in K$ is random and unknown to the adversary.
- Computing from sharings $[a], [b]$ a sharing of $[a + b]$.
- Computing from sharings $[a], [b]$ a sharing of $[ab]$.

Also, these protocols must remain secure when composed in parallel.

The first two requirements are always met if the sharing method used is *linear* over K, in the sense that from $[a], [b]$ and a known constant c, we can *non-interactively compute* new sharings $[a + b]$ and $[ac]$, and more generally, arbitrary linear functionals. For standard examples of VSS, this just translates to locally adding shares of a to corresponding shares of b, and to multiplying the shares of a by c.

In fact, all three requirements can be met by known techniques in all standard models of communication. We give here a few examples of existing efficient, constant round, linear MPC protocols: The classical unconditionally secure MPC protocols of Ben-Or, Goldwasser and Wigderson [5] and Chaum, Crépeau and Damgaard [7] are examples in the secure channels model satisfying all our requirements, tolerating an active, adaptive threshold adversary that corrupts less than a third of the processors.

MPC protocols secure against general adversaries [15] are given by Cramer, Damgaard and Maurer [9]. Their protocols make no restriction on the field size, as opposed to [7,5] where this must be larger than the size of the network. [1] For the broadcast model of Rabin and Ben-Or [23], one can take the protocols of [10], tolerating an actively (and adaptively) corrupted minority at the expense of negligible errors and the assumption that a secure broadcast primitive is given. [2] An example in the cryptographic model is given by the protocols of Gennaro, Rabin and Rabin [13]. Here the size of the field is necessarily large. For the binary field an example given in [8]. This protocol, which is based on homomorphic threshold encryption, is quite efficient and tolerates an actively corrupted minority.

Note that parallel composition is not secure in general for all the models of communication mentioned here, unless extra properties are required. Nevertheless, the example protocols considered above are in fact secure under parallel composition.

A final basic protocol (called Π_1 in the following) that we will need is:

- Compute from a sharing $[a]$ a sharing $[h(a)]$ where h is the function on K defined by $h(a) = 0$ if $a = 0$ and $h(a) = 1$ if $a \neq 0$.

Later we show a constant-round realization of this protocol based only on the three requirements above. This realization is efficient if the characteristic of the field is polynomially bounded.

For *arbitrary fields*, we can do the following instead: assume first that $K = GF(q)$ for a (large) prime q. Represent an element $a \in K$ in the natural way as a bit string $a_0, ..., a_k$. Choose a new field $F = GF(p)$ where p is a small prime, all that is required is that p is larger than the number of players, in particular, p does not depend on the size of the input to the desired computation. Define $[a] = [a_0]_F, ..., [a_k]_F$, i.e., using any of the standard methods described above we share each bit in the representation of a over the field F.

We can now use the well-known fact that for a given, fixed q, there exist NC^1 Boolean circuits for elementary operations in $GF(q)$ (and even for unbounded

[1] In the full version of [9] it is pointed out that their VSS is actually constant round.
[2] One extra level of sub-sharing must be built in (which is no problem) to ensure constant rounds for their multiplication protocol.

fan-in addition). This, together with the result from [1] and the fact that Boolean operations can be simulated in a natural way by arithmetic in F immediately implies existence of constant round protocols for this sharing method meeting the three requirements above. Moreover, computing the function h is now trivial since we only have to compute the OR of all bits of the input value.

Finally, we show in Section 4 how the basic protocols for a field K can be lifted to any extension field of K.

Since most of the known MPC protocols are linear, we explain our protocols under this assumption, since it leads to more efficient and easier to explain protocols. At the end of Section 4, we argue that our results also hold in the more general model described earlier.

4 Known Techniques Used

Let a secure linear MPC protocol for elementary arithmetic (i.e., multiplication and addition) over a finite field K be given, that is efficient and constant round. Write $q = |K|$. We frequently use the following constant-round techniques from Bar-Ilan and Beaver [1].

Joint Secret Randomness is a protocol to generate a sharing $[\rho]$ where $\rho \in K$ is random and secret. This is just by letting all players in the network share a random element, and taking the sum as the result. This extends in a natural way to random vectors and matrices. *Secure Matrix Multiplication* is a protocol that starts from sharings $[A], [B]$ of matrices A and B, and generates a sharing $[AB]$ of their product. This protocol works in the obvious way. We denote this secure computation by $[AB] = [A] \cdot [B]$. By our assumptions on the basic MPC, it follows that if any of these matrices, say A, is publicly known, secure multiplication can be performed non-interactively. and we write $[AB] = A \cdot [B]$ instead.

Jointly Random Secret Invertible Elements and Matrices is a protocol that generates a sharing of a secret, random nonzero field element or an invertible matrix. The protocol securely generates two random elements (matrices), securely multiplies them, and reveals the result. If this is non-zero (invertible), one of the secret elements (matrices) is taken as the desired output of the protocol. The probability that a random matrix $A \in K^{n,n}$ is invertible is than $1/4$,[3] and is at least $1 - n/q$. In particular if n is negligible compared to q, almost all $A \in K^{n,n}$ are invertible. This is easy to verify (see also the counting arguments in Section 6.1).

Secure Inversion of Field Elements and Matrices is a protocol that starts from a sharing of an invertible field element or matrix, and results in a sharing of its inverse. We denote this secure computation by $[x^{-1}] = [x]^{-1}$, and $[A^{-1}] = [A]^{-1}$ respectively. This protocol first generates $[\rho]$ with $\rho \in K$ random and non-zero, securely computes $[\sigma] = [\rho] \cdot [x]$, and finally reveals σ. The result $[x^{-1}]$ is then non-interactively computed as $\sigma^{-1} \cdot [\rho]$. The same approach applies to the case of an invertible matrix.

[3] For instance, by simple counting and induction it follows that this probability is at least $1/4 + (1/2)^{n+1}$. Also, there are better estimates known from the literature.

Securely Solving Regular Systems is a protocol that starts from sharings of an invertible matrix and a vector, and generates a sharing of the unique pre-image of that vector under the given invertible matrix. This protocol follows immediately from the above protocols.

Secure Unbounded Fan-In Multiplication is a protocol that produces a sharing of the product of an unbounded number n of shared field elements $[x_i]$. First consider the case where all elements $[x_1], \ldots, [x_m]$ are invertible. The network generates sharings $[\rho_1], \ldots, [\rho_m]$ of independently random non-zero values, and subsequently sharings of their multiplicative inverses. Next, they compute $[\sigma_1] = [x_1] \cdot [\rho_1]$, and, for $i = 2 \ldots m$, $[\sigma_i] = [\rho_{i-1}^{-1}] \cdot [x_i] \cdot [\rho_i]$. Finally, they publicly reconstruct σ_i for $i = 1 \ldots m$, compute the product of the σ_i's, and multiply the result into the sharing of ρ_m^{-1} to get a sharing of the product of the x_i's. See [3] for a more efficient solution. Using a result by Ben-Or and Cleve, the general case (i.e., x_i's may be equal to 0) is reduced to the previous case. The resulting protocol comes down to unbounded secure multiplication of certain invertible 3×3-matrices. The overhead is essentially a multiplicative factor n. See Section 6.4 for an alternative approach.

Note also that the MPC protocol over K is easily "lifted" to an extension field L of K, as we show below. If the original is efficient and constant round, then so is the lifted protocol.

L may be viewed as a K-vectorspace, and let $b_0, \ldots, b_{d-1}$ be a fixed K-basis for L, where d is the degree of L over K. More precisely, let α be a root of an irreducible polynomial $f(X) \in K[X]$ of degree d. We set $b_i = \alpha^i$ for $i = 0 \ldots d-1$. Elements of L are represented by coordinate vectors with respect to this chosen basis. In particular, the vectors that are everywhere zero except possibly in the first coordinate correspond to the elements of K.

If $[x_0]_K, \ldots, [x_{d-1}]_K$ are sharings, with the x_i's in K, it is interpreted as $[x]_L$, where $x = \sum_{i=0}^{d-1} x_i \cdot b_i \in L$. Let $[x]_L = ([x_0]_K, \ldots, [x_{d-1}]_K)$, $[y]_L = ([y_0]_K, \ldots, [y_{d-1}]_K)$ be sharings of $x, y \in L$. Securely computing $[x+y]_L = [x]_L + [y]_L$ amounts to computing the sum of the vectors, which is be done by local operations. So we have the correspondence $[x+y]_L \leftrightarrow ([x_0+y_0]_K, \ldots, [x_{d-1}+y_{d-1}]_K)$.

Now consider multiplication. For $i = 0, \ldots, d-1$ let B_i be the matrix whose j-th column is the vector representation of the element $e_i \cdot e_j \in L$, and let $B = B_0 || \ldots || B_{d-1}$ (concatenation from left to right). Let $\mathbf{x} \otimes \mathbf{y} \in K^{n^2}$ be the (column-) vector whose j-th "block" is $x_j \cdot y_0, \ldots, x_j \cdot y_{d-1}$. Then: $x \cdot y = \sum_{i,j=0}^{d-1} x_i \cdot y_j \cdot e_i \cdot e_j = \sum_{i=0}^{d-1} \lambda_i \cdot e_i$, where $B(\mathbf{x} \otimes \mathbf{y}) = (\lambda_0, \ldots, \lambda_{d-1})^T \in K^d$, and so we have the correspondence: $[x \cdot y]_L \leftrightarrow [B(\mathbf{x} \otimes \mathbf{y})]_K$.

Over K, it is straightforward for the network to first securely compute $[\mathbf{x} \otimes \mathbf{y}] = [\mathbf{x}] \otimes [\mathbf{y}]$ efficiently in constant rounds. Since B is public, secure computation of $[B(\mathbf{x} \otimes \mathbf{y})]_K$ is then by local operations only. Hence, secure multiplication in the extension field can be carried out efficiently in constant rounds. Note that securely multiplying in a known constant is a special case, which is handled completely by local operations.

Finally, we note that if the linearity assumption on the MPC protocol is dropped, and instead we work with the model also described in Section 2, where more generally secure constant round protocols for generation of shared random element, addition and multiplication are assumed, the above sub-protocols still work. It is sufficient to argue that unbounded addition can be securely and efficiently performed in constant rounds.

Although this does not directly follow from the model, as is the case with linearity, it can be done in similar style as unbounded multiplication of non-zero field elements using the ideas of Bar-Ilan and Beaver. This is actually where the assumption on secure generation of a shared random value comes into play. It is easy to verify that all protocols to follow also work in this more general model.

5 Overview and Conventions

Throughout we assume efficient, constant round, secure linear MPC protocols over a finite field K. In the analysis we assume that the required properties are perfectly satisfied.

The linear algebraic problems of our interest are determinant, characteristic polynomial, rank, sub-space membership, random sampling and general linear systems. We first explain secure solutions with negligible error probabilities.

We will assume that K ("the field of interest") is "large enough", i.e., n ("dimension" of the linear algebraic problems) is negligible compared to $q = |K|$. Without loss of generality we may use the lifting technique to achieve this.

In all cases, solutions of the original problems defined over K can be recovered from the solutions of the lifted problem.

In Section 9 we argue how to obtain zero-error modifications of our protocols.

6 Secure MPC of Determinant

Let $[A]$ be a sharing, where $A \in K^{n,n}$. The goal of the network is to securely compute a sharing $[\det(A)]$, where $\det(A)$ is the determinant of A, efficiently in constant rounds.

Secure computation via the standard definition of determinant is inefficient, and a secure version of Gaussian elimination for instance, seems inherently sequential. After we give our efficient and constant round solution, we discuss some less efficient alternatives based on combinations of known techniques.

6.1 The Case of Invertible Matrices

We start by solving the problem under the assumption that the shared matrix A is promised to be invertible and that there exists an efficient constant round protocol Π_0 allowing the network to securely generate a pair $([R], [\det(R)])$ where $R \in K^{n,n}$ is an (almost) random invertible matrix and $\det(R)$ is its determinant. Note that we do *not* require that the network can securely compute the determinant of a random invertible matrix; we merely require that Π_0 securely constructs a sharing of a random invertible matrix *together* with its determinant.

In the following, let $\mathrm{GL}_n(K) \subset K^{n,n}$ denote the group of invertible matrices. Let $[A]$ be a sharing, with $A \in \mathrm{GL}_n(K)$.

1. Under the assumption that protocol Π_0 is given, the network securely generates $([R],[d])$, where $R \in \mathrm{GL}_n(K)$ is random and $d = \det(R)$.
2. By the method of Bar-Ilan and Beaver for secure inversion, the network securely computes $[d^{-1}] = [d]^{-1}$.
3. The network securely computes $[S] = [R] \cdot [A]$, and *reveals* S.
4. All compute $e = \det(S)$, and by local operations they securely compute $[\det(A)] = e \cdot [d^{-1}]$.

Note that (S, e) gives no information on A. Also note that the protocol is not private if A is not invertible, since $e = 0$ exactly when that is the case. A realization for Protocol Π_0 is shown below.

Realization of Protocol Π_0. We show an efficient, constant round protocol for securely generating pairs $([R],[d])$, where $R \in \mathrm{GL}_n(K)$ is random and $d = \det(R)$. It achieves perfect correctness. The distribution of (R, d) has a negligible bias.

Our solution is based on the idea of securely multiplying random matrices of a special form, and requires that n is negligible compared to q. The protocol goes as follows.

1. The network securely generates the pair of shared vectors $[\mathbf{x}_L], [\mathbf{x}_U]$, where $\mathbf{x_L}, \mathbf{x_U} \in K^n$ both consist of random non-zero entries, and securely computes $[d] = (\prod_{i=1}^n [x_L(i)]) \cdot (\prod_{i=1}^n [x_U(i)])$.
   This is done using the methods of Bar-Ilan and Beaver for secure unbounded fan-in multiplication of *non-zero* values.
2. The network securely generates $n^2 - n$ elements $[r_i]$, where the $r_i \in K$ are random.
   Next, the network defines $[L]$ such that $L \in K^{n,n}$ has $\mathbf{x}_L$ on its diagonal, while the elements below the diagonal are formed by the first $\frac{1}{2}(n^2 - n)$ of the $[r_i]$'s. The elements above the diagonal are set to 0.
   Similarly for the matrix $[U]$, but with $\mathbf{x}_U$ on its diagonal, and the remaining $[r_i]$'s placed above the diagonal. The elements below the diagonal are set to 0.
   Finally, the network securely computes $[R] = [L] \cdot [U]$, Note that $\det(R) = d \neq 0$. The result of the protocol is set to $([R],[d])$.

Correctness is clear. We now discuss privacy. Define $\mathcal{L}$, $\mathcal{U}$ as the sub-groups of $\mathrm{GL}_n(K)$ consisting of the invertible lower- and upper-triangular matrices, i.e., the matrices with non-zero diagonal elements, and zeroes above (resp. below) the diagonal. For $n > 1$ these groups are non-abelian. Let $\mathcal{D}$ denote the invertible diagonal matrices, i.e., the matrices with non-zero diagonal elements and zeroes elsewhere. We have $\mathcal{L} \cap \mathcal{U} = \mathcal{D}$, $|\mathcal{L}| = |\mathcal{U}| = q^{\frac{n^2-n}{2}}(q-1)^n$, $|\mathcal{D}| = (q-1)^n$.

Define the map $h : \mathcal{L} \times \mathcal{U} \longrightarrow \mathrm{GL}_n(K)$, $(L, U) \mapsto LU$, and write $\mathcal{R} = h(\mathcal{L} \times \mathcal{U})$ for the range of h, i.e., $\mathcal{R}$ consists of all invertible matrices that can be written as the product of a lower- and an upper-triangular matrix.

For each $R \in \mathcal{R}$, it holds that $|h^{-1}(R)| = |\mathcal{D}|$. Using the fact that $\mathcal{L}$, $\mathcal{U}$ and $\mathcal{D}$ are groups and that $\mathcal{L} \cap \mathcal{U} = \mathcal{D}$, this claim is easily proved as follows. Let

$R = LU$, and let $D \in \mathcal{D}$. Then $LD^{-1} \in \mathcal{L}$ and $DU \in \mathcal{U}$, and $R = (LD^{-1})(DU)$. This shows that R has at least $|\mathcal{D}|$ pre-images under h. On the other hand, if $R = L_0U_0 = L_1U_1$, then $L_1^{-1}L_0 = U_1U_0^{-1}$. Since $L_1^{-1}L_0 \in \mathcal{L}$ and $U_1U_0^{-1} \in \mathcal{U}$, both are equal to D for some $D \in \mathcal{D}$, and so we can write $L_1 = L_0D^{-1}$ and $U_1 = DU_0$.

As a consequence, $|\mathcal{R}| = \frac{|\mathcal{L}|\cdot|\mathcal{U}|}{|\mathcal{D}|}$. Thus we have $\frac{|\mathcal{R}|}{|K^{n,n}|} = (1-\frac{1}{q})^n$, and hence, $\frac{|\mathcal{R}|}{|\mathrm{GL}_n(K)|} > (1-\frac{1}{q})^n \geq 1 - \frac{n}{q}$.

These facts imply that if (L, U) is chosen uniformly at random from $\mathcal{L} \times \mathcal{U}$, then $R = LU$ is distributed uniformly in $\mathcal{R}$, which is almost all of the invertible matrices when n is negligible compared to q. [4]

We note that it is possible to devise an alternative for protocol Π_0, where each player in the network shares a random invertible matrix and a value he claims is the determinant. Invertibility is proved using Bar-Ilan and Beaver's techniques. Using cut-and-choose techniques it can be established that this value is indeed the determinant. The desired output is obtained by taking products. However, this method introduces correctness errors, and is less efficient compared to the above solution.

6.2 The General Case of Determinant

If $A \in K^{n,n}$ is no longer guaranteed to be invertible the situation becomes slightly more involved. Although the protocol would still compute the determinant correctly, security is not provided if the matrix is singular: by inspection of the previous protocol, the publicly available value e is equal to 0 exactly when A is singular. Moreover, any blinding technique in which a product of A with randomizing matrices is revealed, provides a lower-bound on A's rank. [5]

We now propose our solution for secure computation of determinant. Let $[A]$ be a sharing, where $A \in K^{n,n}$ is an arbitrary matrix. The purpose of the network is to securely compute a sharing $[\det(A)]$ efficiently in constant rounds.

Let $f_A(X) = \det(X \cdot I_n - A) \in K[X]$ denote the characteristic polynomial of A, where I_n denotes the $n \times n$ identity matrix. Then $f_A(0) = (-1)^n \cdot \det(A)$ and $\deg f = n$. By Lagrange Interpolation, for distinct $z_0, \ldots, z_n \in K$, there are $l_0, \ldots, l_n \in K$, only depending on the z_i's, such that $\det(A) = (-1)^n \cdot f_A(0) = (-1)^n \cdot \sum_{i=0}^n l_i \cdot f_A(z_i) = (-1)^n \cdot \sum_{i=0}^n l_i \cdot \det(z_i I_n - A)$.

Now, for $z \in K$, it holds that $zI_n - A \in \mathrm{GL}_n(K)$ if and only if $f_A(z) \neq 0$, i.e., z is not an eigenvalue of A.

Since A has at most n eigenvalues, the matrix $zI_n - A$ is invertible when z is randomly and independently chosen, except with probability at most $1/q$.

---

[4] We note that all invertible matrices can be brought into "LUP" form, where L and U are invertible matrices in lower-, resp. upper-triangular form, and P is a permutation matrix. However, choosing each of these at random, LUP does not have the uniform distribution on $\mathrm{GL}_n(K)$. Moreover, securely computing the sign of the permutation would pose a separate problem at this point.

[5] The rank of the product of matrices is at most equal to the smallest rank among them.

This enables a reduction of the problem of securely computing $[\det(A)]$ to that of secure computation of the determinant of a number of invertible matrices, which now we know how to do, and proceed as before.

1. In parallel, the network securely generates $[z_0], \ldots, [z_n]$, where the z_i are randomly distributed in K. They *reveal* the z_i's. Except with negligible probability, the z_i's are distinct (which can be checked of course) and all matrices $z_i I_n - A$ are invertible. For $i = 0, \ldots, n$, the network securely computes by local computations $[z_i I_n - A] = z_i I_n - [A]$
2. Using our protocol for securely computing the determinant of an invertible matrix, they securely compute in parallel $[\det(z_0 I_n - A)], \ldots, [\det(z_n I_n - A)]$.
3. Finally, the network securely computes $[\det(A)] = (-1)^n \cdot \sum_{i=0}^{n} l_i \cdot [\det(z_i I_n - A)]$, where the l_i's are the interpolation coefficients.

Note that if some z_i happens to be an eigenvalue of A, this becomes publicly known, since the sub-protocol for securely computing the determinant of an invertible matrix noticeably fails in case it is not invertible. On the other hand, it also means that if z_i is *not* an eigenvalue of A, this also becomes known, and the adversary can rule out all matrices A' of which z_i is an eigenvalue.

However, it is only with negligible probability that the adversary learns an eigenvalue. The actual probability depends on A, but this poses no privacy problems since it is negligible anyway.

Also, the adversary could predict with almost complete certainty in advance that z_i is not an eigenvalue. Hence we have almost perfect privacy, and perfect correctness.

6.3 Secure MPC of Characteristic Polynomial

Let $M \in K^{n+1,n+1}$ be the Vandermonde matrix whose i-th row is $(1, z_i, \ldots, z_i^n)$, and write $\mathbf{y}$ and $\mathbf{f}$ for the (column) vectors whose i-th coordinates are equal to y_i and to the coefficient of X^i in $f_A(X)$, respectively. Then $\mathbf{f} = M^{-1}\mathbf{y}$.

The protocol above not only securely computes the determinant $[\det(A)]$, but in fact the coefficient vector $\mathbf{f}$ of the characteristic polynomial, if we replace the last step by $[\mathbf{f}] = M^{-1} \cdot ([\det(z_0 I_n - A)], \ldots, [\det(z_n I_n - A)])^T$. Note that we might as well omit computation of the leading coefficient of $f_A(X)$ since it is equal to 1 anyway.

6.4 Alternative Protocol for Unbounded Multiplication

As an aside, we note that a similar reduction via interpolation yields an alternative protocol for unbounded multiplication. Namely, consider $[a_1], \ldots, [a_n]$ with the a_i's in K, and define $f(X) = \prod_{i=1}^{n} (X - a_i)$. By applying interpolation through random points on $f(X)$, we get a similar reduction to the much simpler case of unbounded multiplication of non-zero field elements. Zero-error can be obtained by a method described in Section 9.

6.5 Other Approaches

We discuss some interesting but less efficient alternatives based on combinations of known results, in particular from Parallel Computing.

First we consider a combination of techniques due to Mahajan and Vinay [21], Ishai and Kushilevitz [16,17], and Beimel and Gál [4].

For our purposes, an Arithmetic Program (AP) [4] is a weighted directed acyclic graph with two distinguished vertices s, t. Each edge is labelled by a variable that can take on a value in a finite field K. The function computed by an AP is defined by taking a path from s to t, multiplying the weights, and summing up over all such paths to finally obtain the function value. The computations take place in the finite field K. By elementary algebraic graph theory, the function value shows up as the (s,t)-entry in the matrix $(I-H)^{-1}$, where H is the adjacency matrix of the weighted graph. Ishai and Kushilevitz [16,17] nicely exploit this fact in their construction of representations of functions in terms of certain degree-3 randomized polynomials obtained from branching programs.

The result of [21] in particular says that there is an AP with roughly n^3 vertices for computing determinant. The weights are entries from the matrix of interest, where the correspondence does not depend on the actual matrix.

Therefore, determinant can in principle be securely computed using a single secure matrix inversion. Unfortunately, this matrix has dimension greater than n^3. Bar-Ilan and Beaver's inversion applied to the matrix $I-H$, essentially requires secure multiplication of two $n^3 \times n^3$ matrices. Methods for securely computing a sharing of just the (s,t)-entry of $(I-H)^{-1}$ rather than the whole matrix (via a classical identity relating inverse with determinants) seem to require secure computations of determinant in the first place.

Another approach can be based on Leverier's Lemma (see e.g. [20]), which retrieves the coefficients of the characteristic polynomial by inverting a certain lower-triangular matrix, where each entry below the diagonal is the trace of a power of the matrix of interest. This lemma is obtained by combining Newton's identities with the fact that these traces correspond to sums of powers of the characteristic roots.

If $K = p$, with p a prime greater that the dimension n of the matrix, it is possible to devise a secure protocol for characteristic polynomial whose complexity is dominated by securely computing all i-th powers of the matrix, for $i = 1 \ldots n$. These terms can be computed separately using techniques of Bar-Ilan and Beaver, or by using the observation that obtaining the n powers of an $n \times n$-matrix is no harder than inverting an $n^2 \times n^2$-matrix (see e.g. [20] for more details).

Note that our solution for large fields essentially just requires secure multiplication of two $n \times n$-matrices (due to Bar-Ilan and Beaver's matrix inversion) if the matrix is promised to be invertible, and n times that amount in the general case.

7 Secure MPC of Rank

The purpose of the network is to securely compute the *rank* of a matrix A efficiently in constant rounds. An important feature of our solution is that the network in fact securely computes a sharing $[\mathbf{r}]$, where $\mathbf{r} \in K^n$ encodes the rank of A in *unary*. This means that, when viewed as a column vector, all non-zero entries of $\mathbf{r}$ are all equal to 1 and occur in the bottom r positions. Rank encoded this way facilitates an easy way to securely compare the ranks of given matrices, as we show in an application to the subspace membership problem later on.

We note that Ishai and Kushilevitz [16,17] have proposed an elegant and efficient protocol for secure computation of rank. Their protocol produces a random shared matrix with the same rank as the shared input matrix. This particular way of encoding rank, however, seems to limit applicability in a scenario of ongoing secure multi-party computations.

In some special cases, such as when a square matrix A is in triangular form, its rank $\mathrm{r}(A)$ can be read off its characteristic polynomial $f_A(X)$, as $n-t$, where X^t is largest such that it divides $f_A(X)$, and n is its degree. This is not always the case.

Mulmuley [22] proved the following result. Let $S \in K^{m,m}$ be symmetric. Let Y be an indeterminate, and define the diagonal matrix $D = (d_{ii}) \in K[Y]^{m,m}$ with $d_{ii} = Y^{i-1}$. Let $f_{DS}(X) \in K[X, Y]$ denote the characteristic polynomial of $DS \in K[Y]^{m,m}$. Then $\mathrm{r}(S) = m - t$ where t is maximal such that X^t divides the characteristic polynomial $f_{DS}(X) \in K[X, Y]$ of DS. In other words, $f_{DS}(X) = X^{m-\mathrm{r}(S)} \cdot \sum_{i=0}^{\mathrm{r}(S)} f_i(Y)X^i$, where $f_0(Y) \neq 0$ and $f_{\mathrm{r}(S)}(Y) = (-1)^m$.

If S is not symmetric, it can be replaced by the symmetric matrix S^*, which has S^T in its lower-left corner and S in its upper-right corner, while the rest is set to 0. Both dimension and rank of S^* are twice that of S.

We exploit this result as follows. Let $[A]$ be a sharing with $A \in K^{n,n}$. [6].

The network first constructs a sharing $[A^*]$ of the symmetric matrix $A^* \in K^{2n,2n}$, which is done locally in a trivial manner. Next, they securely generate $[y_0]$ with y_0 random in K, and reveal it. Define $D_0 \in K^{2n,2n}$ as the matrix D from above, with the substitution $Y = y_0$.

If $f_0(y_0) \neq 0$, then $2 \cdot \mathrm{r}(A) = 2n - t$, with X^t largest such that it divides the characteristic polynomial $f_{D_0A^*}(X) \in K[X]$ of the matrix D_0A^*. Since the degree of $f_0(Y)$ is at most $n(2n-1)$ (as follows from simple inspection), $f_0(y_0) \neq 0$, except with probability $n(2n-1)/q$.

The next step for the network is to securely compute $f_{D_0A^*}(X)$. To this end, they publicly compute D_0 from y_0, and finally by local computations $[D_0A^*] = D_0[A^*]$. Using our Characteristic Polynomial Protocol they securely compute a sharing of the coefficient vector of the polynomial.

Viewing this as a column vector whose i-th entry is the coefficient of X^i in the polynomial, $i = 0 \ldots 2n-1$, and neglecting the coefficient of X^{2n}, it has its top t entries equal to zero, while the $t+1$-st is non-zero. By discarding "every

[6] Note that if A is not square, say $A \in K^{n,m}$, then we can easily extend A to a square matrix whose rank is the same, by appending all-zero rows or columns. This leads to an $s \times s$-matrix where $s = \max(n, m)$.

second" entry we obtain a vector $\mathbf{f} \in K^n$ whose top $n-r$ entries are zero, while its $n-r+1$-st entry is non-zero, where $r = \mathrm{r}(A)$.

DEFINITION 1 *Let* $\mathbf{r} \in K^n$ *be a column-vector, and let* $0 \leq r \leq n$ *be an integer. We say that* $\mathbf{r}$ *is an* almost-unary *encoding of* r *if its bottom* r *entries are non-zero, while it has zeroes elsewhere. If the non-zero entries are all equal to 1, we say that* $\mathbf{r}$ *is a* unary *encoding of* r.

If $H \in K^{n,n}$ is a random lower-triangular matrix, then $H\mathbf{f}$ has its top $n-r$ entries equal to 0, while its bottom r entries are randomly and independently distributed in K. Hence, except with probability at most $r/q \leq n/q$, $H\mathbf{f}$ is a random almost-unary encoding of A's rank r. The actual probability depends on the rank, but it is negligible anyway.

The network now simply securely generates a sharing $[H]$ of a random lower-triangular, reveals it, and non-interactively computes the almost-unary encoding of A's rank as $[H\mathbf{f}] = H[\mathbf{f}]$.

A unary encoding $[\mathbf{r}]$ of A's rank r can be securely computed from $[H\mathbf{f}]$ by applying the protocol Π_1 mentioned earlier. This protocol starts from a sharing $[x]$ with $x \in K$, and securely computes $[h(x)]$ in constant rounds, where $h(x) = 1$ if $x \neq 0$ and $h(x) = 0$ if $x = 0$. [7]

Applying protocol Π_1 in parallel to each of the entries of the almost-unary encoding $H\mathbf{f}$, we get the desired result. We show one realization of such a protocol below. A less efficient, but more general method was shown in Section 3.

7.1 Protocol Π_1 Based on Secure Exponentiation

We assume that the field K has "small" characteristic p, and that the MPC protocol over K run by the network can be viewed as a lifting from protocols over $\mathrm{GF}(p)$ to K.

Let $[x]$ with $x \in K$ be a sharing. Note that $h(x) = x^{q-1}$, where $q = |K|$.

The first idea that comes to mind is to securely perform repeated squaring. This requires $O(\log q)$ rounds of communication however. Applying the constant round protocol of Bar-Ilan and Beaver for unbounded fan-in secure multiplication to our problem is no option either, since in this case the communication overhead will be polynomial in q instead of $\log q$.

Another idea is to apply Bar-Ilan and Beaver's protocol for secure inversion. Namely, the network would securely compute $[y]$, where $y = x^{-1}$ if $x \neq 0$ and $y = 0$ if $x = 0$, and finally compute $[h(x)] = [x] \cdot [y]$. Unfortunately, the network would learn that $x = 0$ in the first step, as can be seen by inspection of the Bar-Ilan and Beaver method. Hence, the security requirements are contradicted. We note that the function h defined above is closely related to the Normalization Function defined in [1], which tells whether two elements are equal or not. They show how this function (and hence h as well) can be securely computed in constant rounds if the field K is small.

We need an alternative approach which works for exponentially large fields. Our solution comes at the expense of assuming small characteristic. Write d for

---

[7] As an aside, note that $h(x)$ is the rank of the 1×1-matrix x.

the degree of K over GF(p). So $q = p^d = |K|$. Let $1 \leq s \leq q-1$ be a given, public integer, and let $[x]$ with $x \in K$ be a sharing. This is how they can securely compute $[x]^s$, efficiently in constant rounds. Setting $s = q-1$, we get the desired protocol Π_1.

Taking p-th powers in K is a field automorphism of K that leaves GF(p) fixed. In particular this means that this map can be viewed as an automorphism of K as a GF(p)-vectorspace. Let $B \in \mathrm{GL}_d(\mathrm{GF}(p))$ denote the (public) matrix representing this map, with respect to the chosen basis. Then for $i \geq 1$, the matrix $B^i \in \mathrm{GL}_d(\mathrm{GF}(p))$ represents taking p^i-th powers.

For $i = 0 \ldots d-1$, write $z_i = x^{p^i}$. Let $s = \sum_{i=0}^{d-1} s_i p^i$ be the p-ary representation of s. Then we have $x^s = x^{\sum_{i=0}^{d-1} s_i p^i} = \prod_{i=0}^{d-1} \left(x^{p^i}\right)^{s_i} = \prod_{i=0}^{d-1} z_i^{s_i}$. If $(x_0, \ldots, x_{d-1}) \in \mathrm{GF}(p)^d$ is the vector representation of x, then the vector representation of z_i is $B^i(x_0, \ldots, x_{d-1})^T$. Since B^i is public and since the vector representation of x are available as sharings, the network can securely compute the vector representation of $[z_i]$ by local computations. Next, the network securely computes the s_i-th powers of the z_i, running Bar-Ilan's and Beaver's unbounded fan-in secure multiplication protocols in parallel. Each of these steps costs $O(p^2)$ secure multiplications, so the total number is $O(\log q \cdot p^2)$. But since p is "small" (for instance, constant or polynomial in $\log q$)) this is efficient. The protocol is finalized by securely multiplying the $d = O(\log q)$ results together using the same technique.

7.2 Application to Sub-space Membership Decisions

Using our Rank Protocol, the network can securely compute a shared decision bit $[b]$ from $[A]$ and $[\mathbf{y}]$, where $b = 1$ if the linear system $A\mathbf{x} = \mathbf{y}$ is solvable and $b = 0$ otherwise.

Defining $A_{\mathbf{y}}$ by concatenating $\mathbf{y}$ to A as the last column, we have $1 - b = \mathrm{r}(A_{\mathbf{y}}) - \mathrm{r}(A)$, where $\mathrm{r}(\cdot)$ denotes the rank of a matrix and $b = 1$ if the system is solvable, and $b = 0$ otherwise.

Suitably padding both matrices with zeroes, we make them both square of the same dimension, while their respective ranks are unchanged. Running the Rank Protocol in parallel, the network now securely computes unary encodings $[\mathbf{r}]$, $[\mathbf{r_y}]$ of the ranks of A and $A_{\mathbf{y}}$, respectively. It holds that $\mathrm{r}(A) = \mathrm{r}(A_{\mathbf{y}})$ exactly when $\mathbf{r} = \mathbf{r_y}$.

Next, the network securely computes $[\mathbf{u}] = [\mathbf{r}] - [\mathbf{r_y}]$ by local computations, securely generates $[\mathbf{v}]$ with $\mathbf{v}$ random in K^n, and finally securely computes $[v] = [\mathbf{u}] \cdot [\mathbf{v}^T]$. Except with negligible probability $1/q$, it holds that $v = 0$ if $b = 1$ and $v \neq 0$ if $b = 0$. The network securely computes $[b] = 1 - [h(v)]$, using protocol Π_1.

8 General Linear Systems

Let $[A]$ and $[\mathbf{y}]$ be sharings, where A is a square matrix, [8] say $A \in K^{n,n}$, and $\mathbf{y} \in K^n$. The purpose of the network is to securely compute $[b]$, $[\mathbf{x}]$ and $[B]$,

[8] As in the Rank Protocol, the assumption that A is a square matrix is not a limitation.

efficiently and in constant rounds, with the following properties. If the system is solvable, $b = 1$ and $\mathbf{x} \in K^n$ and $B \in K^{n,n}$ are such that $A\mathbf{x} = \mathbf{y}$ and the columns of B generate the null-space Ker A (optionally, the non-zero columns form a basis). If the system is not solvable, $b = 0$, and $[\mathbf{x}]$, $[B]$ are both all-zero.

Our solution is based on a Random Sampling Protocol which we describe first.

8.1 Secure Random Sampling

Let $\mathbf{y}$ be given to the network, and assume for the moment that the linear system $A\mathbf{x} = \mathbf{y}$ has a solution. The purpose of the network is to securely compute $[\mathbf{x}]$, where $\mathbf{x}$ is a random solution of $A\mathbf{x} = \mathbf{y}$, efficiently and in constant rounds. Note that in particular this implies a means for the network to securely sample random elements from Ker A by setting $\mathbf{y} = \mathbf{0}$.

Our approach is to reduce the problem to that of solving a regular system, since this can be handled by the methods of Bar-Ilan and Beaver. Using the Sub-Space Membership Protocol the network first securely computes the shared decision bit $[b]$ on whether the system has a solution at all. Applying that same protocol in an appropriate way, they are able to select a linearly independent generating subset of the columns of A, and to replace the other columns by random ones. With high probability θ, this new matrix T is invertible: if r is the rank of A, then $\theta = \frac{(q^n - q^r)\cdots(q^n - q^{n-1})}{q^n} \geq \left(1 - \frac{1}{q}\right)^{n-r} \geq 1 - \frac{n}{q}$, which differs from 1 only negligibly.

This means that, with high probability, the methods of Bar-Ilan and Beaver can be applied to the system $T\mathbf{x}_1 = \mathbf{y}$ in the unknown $\mathbf{x}_1$. More precisely, they are applied to the system $T\mathbf{x}_1 = \mathbf{y} - \mathbf{y}_0$, where $\mathbf{y}_0$ is a random linear combination over the columns of A that were replaced by columns of R in the construction of T. In other words, $\mathbf{y}_0 = A\mathbf{x}_0$ for $\mathbf{x}_0$ chosen randomly such that its i-th coordinate equals 0 if $c_i = 1$.

If T is indeed invertible and if $A\mathbf{x} = \mathbf{y}$ has a solution at all, then the coordinates of $\mathbf{x}_1$ corresponding to the "random columns" in T must be equal to 0. Then $\mathbf{x} = \mathbf{x}_0 + \mathbf{x}_1$ is a solution of $A\mathbf{x} = \mathbf{y}$, since $A\mathbf{x} = A\mathbf{x}_0 + A\mathbf{x}_1 = \mathbf{y}_0 + T\mathbf{x}_1 = \mathbf{y}_0 + (\mathbf{y} - \mathbf{y}_0) = \mathbf{y}$. It is also clearly random, since $\mathbf{x}_1$ is unique given $\mathbf{y}$ and random $\mathbf{x}_0$.

The result of the protocol is computed as $([b], [b] \cdot [\mathbf{x}])$, where b is the decision bit computed at the beginning.

Here are the details. Write $\mathbf{k}_1, \ldots, \mathbf{k}_n$ to denote the columns of A, and set $\mathbf{k}_0 = \mathbf{0}$. Define the vector $\mathbf{c} \in K^n$ by $c_i = 1$ if $\mathbf{k}_i$ is not a linear combination of $\mathbf{k}_0, \ldots, \mathbf{k}_{i-1}$, and $c_i = 0$ otherwise. Note that $\mathcal{B} = \{\mathbf{k}_i : c_i = 1\}$ is a basis for the space generated by the columns of A.

The shared vector $[\mathbf{c}]$ is securely computed by applying the Sub-Space Membership Protocol in parallel to the pairs $([A_{i-1}], [\mathbf{k}_i])$, where A_i is the matrix consisting of the columns $\mathbf{k}_0, \ldots, \mathbf{k}_{i-1}$, and "negating" the resulting shared decision bits.

Write $[C]$ for the shared diagonal matrix with $\mathbf{c}$ on its diagonal. We 'll use this matrix as a selector as follows. After generating a random shared matrix

$[R]$, the network replaces the columns in A that do not belong to the basis $\mathcal{B}$ by corresponding columns from R, by securely computing $[T] = [A] \cdot [C] + [R] \cdot ([I - C])$. As argued before, T is invertible with high probability.

Next, they securely generate a random shared vector $[\mathbf{x}_0]$ with zeroes at the coordinates i with $\mathbf{k}_i \in \mathcal{B}$, by generating $[\mathbf{x}'_0]$ randomly, and securely multiplying its i-th coordinate by $[1 - c_i]$, $i = 1, \ldots, n$. The shared vector $[\mathbf{y}_0]$ is now securely computed as $[\mathbf{y}_0] = [A] \cdot [\mathbf{x}_0]$.

Using Bar-Ilan and Beaver's method for securely solving a regular system, the network computes $[\mathbf{x}_1] = [T]^{-1} \cdot ([\mathbf{y}] - [\mathbf{y}_0])$, and finally $[\mathbf{x}] = [\mathbf{x}_0] + [\mathbf{x}_1]$ and $[b] \cdot [\mathbf{x}]$. They take $([b], [b] \cdot [\mathbf{x}])$ as the result.

8.2 General Linear Systems Protocol

Let $[A]$, $[\mathbf{y}]$ be sharings, where $A \in K^{n,n}$ and $\mathbf{y} \in K^n$. If $\mathbf{x}$ is a solution of $A\mathbf{x} = \mathbf{y}$, then the complete set of solutions is given by $\mathbf{x} + \mathrm{Ker}(A)$.

Using the Random Sampling Protocol it is now an easy task for the network to securely solve a system of linear equations efficiently in constant rounds.

Assume for the moment that the system is solvable. The network first securely generates $[\mathbf{u}_1], \ldots, [\mathbf{u}_n]$, where the $\mathbf{u}_i$ are independently random samples from Ker A. With high probability, these actually generate Ker A. The network defines $[B]$ such that the i-th column of B is $\mathbf{u}_i$. Next, they securely compute $[\mathbf{x}]$, where $\mathbf{x}$ is an arbitrary solution of the linear system. The result of the protocol is $([\mathbf{x}], [B])$.

To deal with the general case, where the system may not be solvable, we first have the network securely compute $[b]$ using the Sub-Space Membership Protocol, where b is the bit that indicates whether it is solvable. After $[\mathbf{x}], [B]$ has been securely computed, they securely compute $([b] \cdot [\mathbf{x}], [b] \cdot [B])$, and take $([b], ([b] \cdot [\mathbf{x}], [b] \cdot [B]))$ as the result.

9 Achieving Perfect Correctness and Privacy

By inspection of our protocols, non-zero error probabilities arise when the network happens to select zeroes of "hidden" polynomials. Since upper-bounds on their degree are known, such errors can be avoided altogether by passing to an extension field and having the network select elements with sufficiently large algebraic degree instead. This, together with some other minor modifications, leads to protocols with perfect correctness in all cases. In [11] we study efficient alternatives with perfect privacy, thereby avoiding the need for large fields.

Acknowledgements. We 'd like to thank Donald Beaver, Vanessa Daza, Yuval Ishai, Erich Kaltofen, Jesper Buus Nielsen, and the anonymous referees for useful comments. Thanks to Michael Rabin for suggesting an approach based on Leverier's Lemma in Section 6.5.

References

1. J. Bar-Ilan, D. Beaver: *Non-cryptographic fault-tolerant computing in constant number of rounds of interaction*, Proc. ACM PODC '89, pp. 201-209, 1989.
2. D. Beaver, S. Micali, P. Rogaway: *The Round Complexity of Secure Protocols*, Proc. 22nd ACM STOC, pp. 503–513, 1990.
3. D. Beaver: *Minimal Latency Secure Function Evaluation*, Proc. EUROCRYPT '00, Springer Verlag LNCS, vol. 1807, pp. 335–350.
4. A. Beimel, A. Gál: *On Arithmetic Branching Programs*, J. Comp. & Syst. Sc., 59, pp. 195–220, 1999.
5. M. Ben-Or, S. Goldwasser, A. Wigderson: *Completeness theorems for non-cryptographic fault-tolerant distributed computation*, Proc. ACM STOC '88, pp. 1–10, 1988.
6. R. Canetti, U. Feige, O. Goldreich, M. Naor: *Adaptively secure multi-party computation*, Proc. ACM STOC '96, pp. 639–648, 1996.
7. D. Chaum, C. Crépeau, I. Damgård: *Multi-party unconditionally secure protocols*, Proc. ACM STOC '88, pp. 11–19, 1988.
8. R. Cramer, I. Damgård, J. Buus Nielsen: *Multiparty Computation from Threshold Homomorphic Encryption*, Proc. EUROCRYPT '01, Springer Verlag LNCS, vol. 2045, pp. 280–300., 2001.
9. R. Cramer, I. Damgård, U. Maurer: *General Secure Multi-Party Computation from any Linear Secret-Sharing Scheme*, Proc. EUROCRYPT '00, Springer Verlag LNCS, vol 1807, pp. 316–334. Full version available from IACR eprint archive, 2000.
10. R. Cramer, I. Damgård, S. Dziembowski, M. Hirt and T. Rabin: *Efficient multiparty computations secure against an adaptive adversary*, Proc. EUROCRYPT '99, Springer Verlag LNCS, vol. 1592, pp. 311–326, 1999.
11. R. Cramer, I. Damgård, V. Daza: work in progress, 2001.
12. U. Feige, J. Kilian, M. Naor: *A Minimal Model for Secure Computation*, Proc. ACM STOC '94, pp. 554–563, 1994.
13. R. Gennaro, M. Rabin, T. Rabin: *Simplified VSS and fast-track multi-party computations with applications to threshold cryptography*, Proc. ACM PODC'98, pp. 101–111, 1998.
14. O. Goldreich, S. Micali and A. Wigderson: *How to play any mental game or a completeness theorem for protocols with honest majority*, Proc. ACM STOC '87, pp. 218–229, 1987.
15. M. Hirt, U. Maurer: *Player simulation and general adversary structures in perfect multi-party computation*, Journal of Cryptology, vol. 13, no. 1, pp. 31–60, 2000. (Preliminary version in Proc. ACM PODC'97, pp. 25–34, 1997)
16. Y. Ishai, E. Kushilevitz: *Private Simultaneous Messages Protocols with Applications*, Proc. 5th Israel Symposium on Theoretical Comp. Sc. (ISTCS '97), pp. 174–183, 1997.
17. Y. Ishai, E. Kushilevitz: *Randomizing Polynomials: A New Paradigm for Round-Efficient Secure Computation*, Proc. of FOCS '00, 2000.
18. M. Karchmer, A. Wigderson: *On span programs*, Proc. of Structure in Complexity '93, pp. 102–111, 1993.
19. J. Kilian: *Founding Cryptography on Oblivious Transfer*, Proc. ACM STOC '88, pp. 20-31, 1988.
20. F. T. Leighton: *Introduction to Parallel Algorithms and Architectures: Arrays–Trees–Hypercubes*, Morgan Kaufmann Publishers, 1992.

21. M. Mahajan and V. Vinay: *Determinant: combinatorics, algorithms and complexity*, Chicago J. Theoret. Comput. Sci., Article 5, 1997.
22. K. Mulmuley: *A Fast Parallel Algorithm to Compute the Rank of a Matrix over an Arbitrary Field*, Combinatorica, vol. 7, pp. 101–104, 1987.
23. T. Rabin, M. Ben-Or: *Verifiable secret sharing and multi-party protocols with honest majority*, Proc. ACM STOC '89, pp. 73–85, 1989.
24. T. Rabin: *Robust sharing of secrets when the dealer is honest or cheating*, J. ACM, 41(6):1089-1109, November 1994.
25. A. Yao: *Protocols for Secure Computation*, Proc. IEEE FOCS '82, pp. 160–164, 1982.

Two-Party Generation of DSA Signatures
(Extended Abstract)

Philip MacKenzie and Michael K. Reiter

Bell Labs, Lucent Technologies, Murray Hill, NJ, USA

Abstract. We describe a means of sharing the DSA signature function, so that two parties can efficiently generate a DSA signature with respect to a given public key but neither can alone. We focus on a certain instantiation that allows a proof of security for concurrent execution in the random oracle model, and that is very practical. We also briefly outline a variation that requires more rounds of communication, but that allows a proof of security for sequential execution without random oracles.

1 Introduction

In this paper we present an efficient and provably secure protocol by which alice and bob, each holding a share of a DSA [25] private key, can (and must) interact to generate a DSA signature on a given message with respect to the corresponding public key. As noted in previous work on multiparty DSA signature generation (e.g., [26,7,16]), shared generation of DSA signatures tends to be more complicated than shared generation of many other types of ElGamal-based signatures [10] because (i) a shared secret must be inverted, and (ii) a multiplication must be performed on two shared secrets. One can see this difference by comparing a Harn signature [20] with a DSA signature, say over parameters $<g,p,q>$, with public/secret key pair $<y(=g^x \bmod p), x>$ and ephemeral public/secret key pair $<r(=g^k \bmod p), k>$. In a Harn signature, one computes

$$s \leftarrow x(\mathsf{hash}(m)) - kr \bmod q$$

and returns a signature $<r, s>$, while for a DSA signature, one computes

$$s \leftarrow k^{-1}(\mathsf{hash}(m) + xr) \bmod q,$$

and returns a signature $<r \bmod q, s>$. Obviously, to compute the DSA signature the ephemeral secret key must be inverted, and the resulting secret value must be multiplied by the secret key. For security, all of these secret values must be shared, and thus inversion and multiplication on shared secrets must be performed. Protocols to perform these operations have tended to be much more complicated than protocols for adding shared secrets.

J. Kilian (Ed.): CRYPTO 2001, LNCS 2139, pp. 137–154, 2001.

Of course, protocols for generic secure two-party computation (e.g., [34]) could be used to perform two-party DSA signature generation, but here we explore a more efficient protocol to solve this particular problem. To our knowledge, the protocol we present here is the first practical and provably secure protocol for two-party DSA signature generation. As building blocks, it uses a public key encryption scheme with certain useful properties (for which several examples exist) and efficient special-purpose zero-knowledge proofs. The assumptions under which these building blocks are secure are the assumptions required for security of our protocol. For example, by instantiating our protocol with particular constructions, we can achieve a protocol that is provably secure under the decision composite residuosity assumption (DCRA) [31] and the strong RSA assumption [2] when executed sequentially, or one that is provably secure in the random oracle model [5] under the DCRA and strong RSA assumption, even when arbitrarily many instances of the protocol are run concurrently. The former protocol requires eight messages, while the latter protocol requires only four messages.

Our interest in two-party DSA signature generation stems from our broader research into techniques by which a device that performs private key operations (signatures or decryptions) in networked applications, and whose local private key is activated with a password or PIN, can be immunized against offline dictionary attacks in case the device is captured [27]. Briefly, we achieve this by involving a remote server in the device's private key computations, essentially sharing the cryptographic computation between the device and the server. Our original work [27] showed how to accomplish this for the case of RSA functions or certain discrete-log-based functions other than DSA, using known techniques for sharing those functions between two parties. The important case of DSA signatures is enabled by the techniques of this paper. Given our practical goals, in this paper we focus on the most efficient (four message, random oracle) version of our protocol, which is quite suitable for use in the context of our system.

2 Related Work

Two-party generation of DSA signatures falls into the category of threshold signatures, or more broadly, threshold cryptography. Early work in the field is due to Boyd [4], Desmedt [8], Croft and Harris [6], Frankel [13], and Desmedt and Frankel [9]. Work in threshold cryptography for discrete-log based cryptosystems other than DSA is due to Desmedt and Frankel [9], Hwang [22], Pedersen [33], Harn [20], Park and Kurosawa [32], Herzberg *et al.* [21], Frankel *et al.* [14], and Jarecki and Lysyanskaya [23].

Several works have developed techniques directly for shared generation of DSA signatures. Langford [26] presents threshold DSA schemes ensuring unforgeability against one corrupt player out of $n \geq 3$; of t corrupt players out of n for arbitrary $t < n$ under certain restrictions (see below); and of t corrupt players out of $n \geq t^2 + t + 1$. Cerecedo *et al.* [7] and Gennaro *et al.* [16] present threshold schemes that prevent t corrupt players out of $n \geq 2t + 1$ from forging, and thus require a majority of correct players. Both of these works further develop *robust* solutions, in which the t corrupted players cannot interfere with the other $n-t$ signing a message, provided that stronger conditions on n and t are met (at least $n \geq 3t+1$). However, since we consider the two party case only, robustness is not a goal here.

The only previous proposal that can implement two-party generation of DSA signatures is due to Langford [26, Section 5.1], which ensures unforgeability against t corrupt players out of n for an arbitrary $t < n$. This is achieved, however, by using a trusted center to precompute the ephemeral secret key k for each signature and to share $k^{-1} \bmod q$ and $k^{-1}x \bmod q$ among the n parties. That is, this solution circumvents the primary difficulties of sharing DSA signatures—inverting a shared secret and multiplying shared secrets, as discussed in Section 1—by using a trusted center. Recognizing the significant drawbacks of a trusted center, Langford extends this solution by replacing the trusted center with three centers (that protect k^{-1} and $k^{-1}x$ from any one) [26, Section 5.2], thereby precluding this solution from being used in the two-party case. In contrast, our solution suffices for the two-party case without requiring the players to store precomputed, per-signature values. Since our motivating application naturally admits a trusted party for initializing the system (see [27]), for the purposes of this extended abstract we assume a trusted party to initialize alice and bob with shares of the private signing key. In the full version of this paper, we will describe the additional machinery needed to remove this assumption.

3 Preliminaries

Security parameters. Let κ be the main cryptographic security parameter used for, e.g., hash functions and discrete log group orders; a reasonable value today may be $\kappa = 160$. We will use $\kappa' > \kappa$ as a secondary security parameter for public key modulus size; reasonable values today may be $\kappa' = 1024$ or $\kappa' = 2048$.

Signature schemes. A *digital signature scheme* is a triple (G_{sig}, S, V) of algorithms, the first two being probabilistic, and all running in expected polynomial time. G_{sig} takes as input $1^{\kappa'}$ and outputs a public key pair (pk, sk), i.e., $(pk, sk) \leftarrow G_{sig}(1^{\kappa'})$. S takes a message m and a secret key sk as input and outputs a signature σ for m, i.e., $\sigma \leftarrow S_{sk}(m)$. V takes a message m, a public key pk, and a candidate signature σ' for m and returns the bit $b = 1$ if σ' is a valid signature for m, and otherwise returns the bit $b = 0$. That is, $b \leftarrow V_{pk}(m, \sigma')$. Naturally, if $\sigma \leftarrow S_{sk}(m)$, then $V_{pk}(m, \sigma) = 1$.

DSA. The Digital Signature Algorithm [25] was proposed by NIST in April 1991, and in May 1994 was adopted as a standard digital signature scheme in the U.S. [12]. It is a variant of the ElGamal signature scheme [10], and is defined as follows, with $\kappa = 160$, κ' set to a multiple of 64 between 512 and 1024, inclusive, and hash function hash defined as SHA-1 [11]. Let "$z \leftarrow_R S$" denote the assignment to z of an element of S selected uniformly at random. Let $\equiv_q$ denote equivalence modulo q.

$G_{DSA}(1^{\kappa'})$: Generate a κ-bit prime q and κ'-bit prime p such that q divides $p-1$. Then generate an element g of order q in $\mathbb{Z}_p^*$. The triple $<g,p,q>$ is public. Finally generate $x \leftarrow_R \mathbb{Z}_q$ and $y \leftarrow g^x \bmod p$, and let $<g,p,q,x>$ and $<g,p,q,y>$ be the secret and public keys, respectively.

$S_{<g,p,q,x>}(m)$: Generate an ephemeral secret key $k \leftarrow_R \mathbb{Z}_q$ and ephemeral public key $r \leftarrow g^k \bmod p$. Compute $s \leftarrow k^{-1}(\mathsf{hash}(m) + xr) \bmod q$. Return $<r \bmod q, s>$ as the signature of m.

$V_{<g,p,q,y>}(m, <r,s>)$: Return 1 if $0 < r < q$, $0 < s < q$, and $r \equiv_q (g^{\mathsf{hash}(m)s^{-1}} y^{rs^{-1}} \bmod p)$ where s^{-1} is computed modulo q. Otherwise, return 0.

Encryption schemes. An *encryption scheme* is a triple (G_{enc}, E, D) of algorithms, the first two being probabilistic, and all running in expected polynomial time. G_{enc} takes as input $1^{\kappa'}$ and outputs a public key pair (pk, sk), i.e., $(pk, sk) \leftarrow G_{enc}(1^{\kappa'})$. E takes a public key pk and a message m as input and outputs an encryption c for m; we denote this $c \leftarrow E_{pk}(m)$. D takes a ciphertext c and a secret key sk and returns either a message m such that c is a valid encryption of m, if such an m exists, and otherwise returns $\perp$.

Our protocol employs a semantically secure encryption scheme with a certain additive homomorphic property. For any public key pk output from the G_{enc} function, let M_{pk} be the space of possible inputs to E_{pk}, and C_{pk} to be the space of possible outputs of E_{pk}. Then we require that there exist an efficient implementation of an additional function $+_{pk} : C_{pk} \times C_{pk} \rightarrow C_{pk}$ such that (written as an infix operator):

$$m_1, m_2, m_1 + m_2 \in M_{pk} \quad \Rightarrow \quad D_{sk}(E_{pk}(m_1) +_{pk} E_{pk}(m_2)) = m_1 + m_2 \quad (1)$$

Examples of cryptosystems for which $+_{pk}$ exist (with $M_{pk} = [-v, v]$ for a certain value v) are due to Naccache and Stern [28], Okamoto and Uchiyama [30], and Paillier [31].[1] Note that (1) further implies the existence of an efficient function $\times_{pk} : C_{pk} \times M_{pk} \rightarrow C_{pk}$ such that

$$m_1, m_2, m_1 m_2 \in M_{pk} \quad \Rightarrow \quad D_{sk}(E_{pk}(m_1) \times_{pk} m_2) = m_1 m_2 \quad (2)$$

In addition, in our protocol, a party may be required to generate a noninteractive zero knowledge proof of a certain predicate P involving decryptions of elements of C_{pk}, among other things. We denote such a proof as $\mathsf{zkp}[P]$. In Section 6.1, we show how these proofs can be accomplished if the Paillier cryptosystem is in use. We emphasize, however, that our use of the Paillier cryptosystem is only exemplary; the other cryptosystems cited above could equally well be used with our protocol.

[1] The cryptosystem of Benaloh [1] also has this additive homomorphic property, and thus could also be used in our protocol. However, it would be less efficient for our purposes.

System model. Our system includes two parties, alice and bob. Communication between alice and bob occurs in *sessions* (or protocol runs), one per message that they sign together. alice plays the role of session initiator in our protocol. We presume that each message is implicitly labeled with an identifier for the session to which it belongs. Multiple sessions can be executed concurrently.

The adversary in our protocol controls the network, inserting and manipulating communication as it chooses. In addition, it takes one of two forms: an alice-compromising adversary learns all private initialization information for alice. A bob-compromising adversary is defined similarly.

We note that a proof of security in this two-party system extends to a proof of security in an n-party system in a natural way, assuming the adversary decides which parties to compromise before any session begins. The basic idea is to guess for which pair of parties the adversary forges a signature, and focus the simulation proof on those two parties, running all other parties as in the real protocol. The only consequence is a factor of roughly n^2 lost in the reduction argument from the security of the signature scheme.

4 Signature Protocol

In this section we present a new protocol called S-DSA by which alice and bob sign a message m.

4.1 Initialization

For our signature protocol, we assume that the private key x is multiplicatively shared between alice and bob, i.e., that alice holds a random private value $x_1 \in \mathbb{Z}_q$ and bob holds a random private value $x_2 \in \mathbb{Z}_q$ such that $x \equiv_q x_1 x_2$. We also assume that along with y, $y_1 = g^{x_1} \bmod p$ and $y_2 = g^{x_2} \bmod p$ are public. In this extended abstract, we do not concern ourselves with this initialization step, but simply assume it is performed correctly, e.g., by a trusted third party. We note, however, that achieving this without a trusted third party is not straightforward (e.g., see [17]), and so we will describe such an initialization protocol in the full version of this paper.

We use a multiplicative sharing of x to achieve greater efficiency than using either polynomial sharing or additive sharing. With multiplicative sharing of keys, inversion and multiplication of shared keys becomes trivial, but addition of shared keys becomes more complicated. For DSA, however, this approach seems to allow a much more efficient two-party protocol.

In addition to sharing x, our protocol assumes that alice holds the private key sk corresponding to a public encryption key pk, and that there is another public encryption key pk' for which alice does not know the corresponding sk'. (As above, we assume that these keys are generated correctly, e.g., by a trusted third party.) Also, it is necessary for our particular zero-knowledge proof constructions that the range of M_{pk} be at least $[-q^8, q^8]$ and the range of $M_{pk'}$ be at least $[-q^6, q^6]$, although we believe a slightly tighter analysis would allow both to have a range of $[-q^6, q^6]$.

4.2 Signing Protocol

The protocol by which alice and bob cooperate to generate signatures with respect to the public key $<g, p, q, y>$ is shown in Figure 1. As input to this protocol, alice receives the message m to be signed. bob receives no input (but receives m from alice in the first message).

Upon receiving m to sign, alice first computes its share k_1 of the ephemeral private key for this signature, computes $z_1 = (k_1)^{-1} \bmod q$, and encrypts both z_1 and $x_1 z_1 \bmod q$ under pk. alice's first message to bob consists of m and these ciphertexts, α and ζ. bob performs some simple consistency checks on α and ζ (though he cannot decrypt them, since he does not have sk), generates his share k_2 of the ephemeral private key, and returns his share $r_2 = g^{k_2} \bmod p$ of the ephemeral public key.

Once alice has received r_2 from bob and performed simple consistency checks on it (e.g., to determine it has order q modulo $\mathbb{Z}_p^*$), she is able to compute the ephemeral public key $r = (r_2)^{k_1} \bmod p$, which she sends to bob in the third message of the protocol. alice also sends a noninteractive zero-knowledge proof Π that there are values η_1 $(= z_1)$ and η_2 $(= x_1 z_1 \bmod q)$ that are consistent with r, r_2, y_1, α and ζ, and that are in the range $[-q^3, q^3]$. This last fact is necessary so that bob's subsequent formation of (a ciphertext of) s does not leak information about his private values.

Upon receiving $<r, \Pi>$, bob verifies Π and performs additional consistency checks on r. If these pass, then he proceeds to compute a ciphertext μ of the value s (modulo q) for the signature, using the ciphertexts α and ζ received in the first message from alice; the values $\mathsf{hash}(m)$, $z_2 = (k_2)^{-1} \bmod q$, $r \bmod q$, and x_2; and the special $\times_{pk}$ and $+_{pk}$ operators of the encryption scheme. In addition, bob uses $+_{pk}$ to "blind" the plaintext value with a random, large multiple of q. So, when alice later decrypts μ, she statistically gains no information about bob's private values. In addition to returning μ, bob computes and returns $\mu' \leftarrow E_{pk'}(z_2)$ and a noninteractive zero-knowledge proof Π' that there are values η_1 $(= z_2)$ and η_2 $(= x_2 z_2 \bmod p)$ consistent with r_2, y_2, μ and μ', and that are in the range $[-q^3, q^3]$. After receiving and checking these values, alice recovers s from μ to complete the signature.

The noninteractive zero-knowledge proofs Π and Π' are assumed to satisfy the usual completeness, soundness, and zero-knowledge properties as defined in [3,29], except using a public random hash function (i.e., a random oracle) instead of a public random string. In particular, we assume in Section 5 that (1) these proofs have negligible simulation error probability, and in fact a simulator exists that generates a proof that is statistically indistinguishable from a proof generated by the real prover, and (2) these proofs have negligible soundness error probability, i.e., the probability that a prover could generate a proof for a false statement is negligible. The implementations of Π and Π' in Section 6 enforce these properties under reasonable assumptions. To instantiate this protocol without random oracles, Π and Π' would need to become interactive zero-knowledge protocols. It is not too difficult to construct four-move protocols for Π and Π', and by overlapping some messages, one can reduce the total number of moves in this instantiation of the S-DSA protocol to eight. For brevity, we omit the full description of this instantiation.

When the zero-knowledge proofs are implemented using random oracles, we can show that our protocol is secure even when multiple instances are executed concurrently. Perhaps the key technical aspect is that we only require proofs of language membership, which can be implemented using random oracles without requiring rewinding in the simulation proof. In particular, we avoid the need for any proofs of knowledge that would require rewinding in knowledge extractors for the simulation proof, even if random oracles are used. The need for rewinding (and particularly, nested rewinding) causes many proofs of security to fail in the concurrent setting (e.g., [24]).

5 Security for S-DSA

In this section we sketch a formal proof of security for our protocol. We begin by defining security for signatures and encryption in Section 5.1 and for S-DSA in Section 5.2. We then state our theorems and proofs in Section 5.3.

5.1 Security for DSA and Encryption

First we state requirements for security of DSA and encryption. For DSA, we specify existential unforgeability versus chosen message attacks [19]. That is, a forger is given $<g,p,q,y>$, where $(<g,p,q,y>, <g,p,q,x>) \leftarrow G_{DSA}(1^{\kappa'})$, and tries to forge signatures with respect to $<g,p,q,y>$. It is allowed to query a signature oracle (with respect to $<g,p,q,x>$) on messages of its choice. It succeeds if after this it can output some (m,σ) where $V_{<g,p,q,y>}(m,\sigma) = 1$ but m was not one of the messages signed by the signature oracle. We say a forger (q,ϵ)-breaks DSA if the forger makes q queries to the signature oracle and succeeds with probability at least ϵ.

For encryption, we specify semantic security [18]. That is, an attacker A is given pk, where $(pk, sk) \leftarrow G_{enc}(1^{\kappa'})$. A generates $X_0, X_1 \in M_{pk}$ and sends these to a test oracle, which chooses $b \leftarrow_R \{0,1\}$, and returns $Y = E_{pk}(X_b)$. Finally A outputs b', and succeeds if $b' = b$. We say an attacker A ϵ-breaks encryption if $2 \cdot \Pr(A \text{ succeeds}) - 1 \geq \epsilon$. Note that this implies $\Pr(A \text{ guesses } 0 \mid b = 0) - \Pr(A \text{ guesses } 0 \mid b = 1) \geq \epsilon$.

5.2 Security for S-DSA

A forger F is given $<g,p,q,y>$, where $(<g,p,q,y>, <g,p,q,x>) \leftarrow G_{DSA}(1^{\kappa'})$, and the public data generated by the initialization procedure for S-DSA, along with the secret data of either alice or bob (depending on the type of forger). As in the security definition for signature schemes, the goal of the forger is to forge signatures with respect to $<g,p,q,y>$. Instead of a signature oracle, there is an alice oracle and a bob oracle.

alice		bob
$k_1 \leftarrow_R \mathbb{Z}_q$		
$z_1 \leftarrow_R (k_1)^{-1} \bmod q$		
$\alpha \leftarrow E_{pk}(z_1)$		
$\zeta \leftarrow E_{pk}(x_1 z_1 \bmod q)$		
	$<m,\alpha,\zeta>$ $\longrightarrow$	
		abort if $(\alpha \notin C_{pk} \ \vee \ \zeta \notin C_{pk})$
		$k_2 \leftarrow_R \mathbb{Z}_q$
		$r_2 \leftarrow g^{k_2} \bmod p$
	r_2 $\longleftarrow$	
abort if $(r_2 \notin \mathbb{Z}_p^* \ \vee \ (r_2)^q \not\equiv_p 1)$		
$r \leftarrow (r_2)^{k_1} \bmod p$		
$\Pi \leftarrow \mathsf{zkp}\left[\begin{array}{lr} \exists \eta_1, \eta_2 : \eta_1, \eta_2 \in [-q^3, q^3] & \\ \wedge & r^{\eta_1} \equiv_p r_2 \\ \wedge & g^{\eta_2/\eta_1} \equiv_p y_1 \\ \wedge & D_{sk}(\alpha) \equiv_q \eta_1 \\ \wedge & D_{sk}(\zeta) \equiv_q \eta_2 \end{array}\right]$		
	$<r,\Pi>$ $\longrightarrow$	
		abort if $(r \notin \mathbb{Z}_p^* \ \vee \ r^q \not\equiv_p 1)$
		abort if (verify(Π) = false)
		$m' \leftarrow \mathsf{hash}(m)$
		$r' \leftarrow r \bmod q$
		$z_2 \leftarrow (k_2)^{-1} \bmod q$
		$c \leftarrow_R \mathbb{Z}_{q^5}$
		$\mu \leftarrow (\alpha \times_{pk} m' z_2) +_{pk} (\zeta \times_{pk} r' x_2 z_2) +_{pk} E_{pk}(cq)$
		$\mu' \leftarrow E_{pk'}(z_2)$
		$\Pi' \leftarrow \mathsf{zkp}\left[\begin{array}{lr} \exists \eta_1, \eta_2 : & \eta_1, \eta_2 \in [-q^3, q^3] \\ \wedge & (r_2)^{\eta_1} \equiv_p g \\ \wedge & g^{\eta_2/\eta_1} \equiv_p y_2 \\ \wedge & D_{sk'}(\mu') \equiv_q \eta_1 \\ \wedge \ D_{sk}(\mu) \equiv_q & D_{sk}((\alpha \times_{pk} m'\eta_1) \\ & +_{pk} (\zeta \times_{pk} r'\eta_2)) \end{array}\right]$
	$<\mu,\mu',\Pi'>$ $\longleftarrow$	
abort if $(\mu \notin C_{pk} \ \vee \ \mu' \notin C_{pk'})$		
abort if (verify(Π') = false)		
$s \leftarrow D_{sk}(\mu) \bmod q$		
publish $<r \bmod q, s>$		

Fig. 1. S-DSA shared signature protocol

F may query the alice oracle by invoking *aliceInv1*(m), *aliceInv2*(r_2), or *aliceInv3*($<\mu, \mu', \Pi'>$) for input parameters of F's choosing. (These invocations are also accompanied by a session identifier, which is left implicit.) These invocations correspond to a request to initiate the protocol for message m and the first and second messages received ostensibly from bob, respectively. These return outputs of the form $<m, \alpha, \zeta>$, $<r, \Pi>$, or a signature for the message m from the previous *aliceInv1* query in the same session, respectively, or abort. Analagously, F may query the bob oracle by invoking *bobInv1*($<m, \alpha, \zeta>$) or *bobInv2*($<r, \Pi>$) for arguments of the F's choosing. These return messages of the form r_2 or $<\mu, \mu', \Pi'>$, respectively, or abort. F may invoke these queries in any order, arbitrarily many times.

An alice-compromising forger F *succeeds* if after gaining access to the private initialization state of alice, and invoking the alice and bob oracles as it chooses, it can output (m, σ) where $V_{<g,p,q,y>}(m, \sigma) = 1$ and m is not one of the messages sent to bob in a *bobInv1* query. Similarly, a bob-compromising forger F *succeeds* if after gaining access to the private initialization state of bob, and invoking the alice and bob oracles as it chooses, it can output (m, σ) where $V_{<g,p,q,y>}(m, \sigma) = 1$ and m is not one of the messages sent to alice in a *aliceInv1* query.

Let q_{alice} be the number of *aliceInv1* queries to alice. Let q_{bob} be the number of *bobInv1* queries. Let q_o be the number of other oracle queries. Let $\overline{q} = <q_{\mathsf{alice}}, q_{\mathsf{bob}}, q_o>$. In a slight abuse of notation, let $|\overline{q}| = q_{\mathsf{alice}} + q_{\mathsf{bob}} + q_o$, i.e., the total number of oracle queries. We say a forger $(\overline{q}, \epsilon)$-breaks S-DSA if it makes $|\overline{q}|$ oracle queries (of the respective type and to the respective oracles) and succeeds with probability at least ϵ.

5.3 Theorems

Here we state theorems and provide proof sketches showing that if a forger breaks the S-DSA system with non-negligible probability, then either DSA or the underlying encryption scheme used in S-DSA can be broken with non-negligible probability. This implies that if DSA and the underlying encryption scheme are secure, our system will be secure.

We prove security separately for alice-compromising and bob-compromising forgers. The idea behind each proof is a simulation argument. Assuming that a forger F can break the S-DSA system, we then construct a forger F^* that breaks DSA. Basically F^* will run F over a simulation of the S-DSA system, and when F succeeds in forging a signature in the simulation of S-DSA, then F^* will succeed in forging a DSA signature.

In the security proof against an alice-compromising forger F, there is a slight complication. If F were able to break the encryption scheme (G_{enc}, E, D), an attacker F^* as described above may not be able to simulate properly. Thus we show that either F forges signatures in a simulation where the encryptions are of strings of zeros, and thus we can construct a forger F^* for DSA, or F does not forge signatures in that simulation, and thus it must be able to distinguish the true encryptions from the zeroed encryptions. Then we can construct an attacker A that breaks the underlying encryption scheme. A similar complication arises in the security proof against a bob-compromising forger F, and the simulation argument is modified in a similar way.

Theorem 1 below states that an alice-compromising forger that breaks S-DSA with a non-negligible probability can break either DSA or (G_{enc}, E, D) with non-negligible probability. Theorem 2 makes a similar claim for a bob-compromising forger. In these theorems, we use "$\approx$" to indicate equality to within negligible factors. Moreover, in our simulations, the forger F is run at most once, and so the times of our simulations are straightforward and omitted from our theorem statements.

Theorem 1. *Suppose an* alice*-compromising forger* $(\overline{q}, \epsilon)$*-breaks* S-DSA. *Then either there exists an attacker that* ϵ'*-breaks* (G_{enc}, E, D) *with* $\epsilon' \approx \frac{\epsilon}{2q_{\mathsf{bob}}}$, *or there exists a forger that* $(q_{\mathsf{bob}}, \epsilon'')$*-breaks DSA with* $\epsilon'' \approx \frac{\epsilon}{2}$.

Proof. Assume an alice-compromising forger F $(\overline{q}, \epsilon)$-breaks the S-DSA scheme. Then consider a simulation SIM of the S-DSA scheme that takes as input a DSA public key $<g, p, q, y>$, a corresponding signature oracle, and a public key pk' for the underlying encryption scheme. SIM generates the initialization data for alice: $x_1 \leftarrow_R \mathbb{Z}_q$ and $(pk, sk) \leftarrow G_{enc}(1^{\kappa'})$, and gives these to F. The public data y, $y_2 = g^{(x_1^{-1} \bmod q)} \bmod p$, and pk' are also revealed to F. Then SIM responds to alice queries as a real alice oracle would, and to bob queries using the help of the DSA signature oracle, since SIM does not know the x_2 value used by a real bob oracle. Specifically SIM answers as follows:

1. *bobInv1*($<m, \alpha, \zeta>$): Set $z_1 \leftarrow D_{sk}(\alpha)$. Query the DSA signature oracle with m to get a signature $<\hat{r}, \hat{s}>$, and compute $r \leftarrow g^{\mathsf{hash}(m)\hat{s}^{-1}} y^{\hat{r}\hat{s}^{-1}} \bmod p$ where $\hat{s}^{-1}$ is computed modulo q. Compute $r_2 \leftarrow r^{z_1} \bmod p$, and return r_2.
2. *bobInv2*($<r, \Pi>$): Reject if Π is invalid, $r \notin \mathbb{Z}_p^*$ or $r^q \not\equiv_p 1$. Else, choose $c \leftarrow_R \mathbb{Z}_{q^5}$ and set $\mu \leftarrow E_{pk}(\hat{s} + cq)$. Set $\mu' \leftarrow E_{pk'}(0)$, and generate Π' using the simulator for the zkp []. Return $<\mu, \mu', \Pi'>$.

Notice that SIM sets μ' to an encryption of zero, and simulates the proof of consistency Π'. In fact, disregarding the negligible statistical difference between the simulated Π' proofs and the real Π' proofs, the only way SIM and the real S-DSA scheme differ (from F's viewpoint) is with respect to the μ' values, i.e., the (at most) q_{bob} ciphertexts generated using pk'.

Now consider a forger F^* that takes as input a DSA public key $<g, p, q, y>$ and corresponding signature oracle, generates a public key pk' using $<pk', sk'> \leftarrow G_{enc}(1^{\kappa'})$, runs SIM using these parameters as inputs, and outputs whatever F outputs. If F produces a forgery with probability at least $\frac{\epsilon}{2}$ in SIM, F^* produces a forgery in the underlying DSA signature scheme with probability at least $\frac{\epsilon}{2}$.

Otherwise F produces a forgery with probability less than $\frac{\epsilon}{2}$ in SIM. Then using a standard hybrid argument, we can construct an attacker A that ϵ'-breaks the semantic security of the underlying encryption scheme for pk', where $\epsilon' \approx \frac{\epsilon}{2q_{\mathsf{bob}}}$. Specifically, A takes a public key pk' and corresponding test oracle as input, generates a DSA public/private key pair $(<g,p,q,y>, <g,p,q,x>) \leftarrow G_{DSA}(1^{\kappa'})$, and runs a slightly modified SIM using $<g,p,q,y>$ as the DSA public key parameter, simulating the DSA signature oracle with $<g,p,q,x>$, and using pk' as the public encryption key parameter. SIM is modified only in the *bobInv2* query, as follows:

1. A computes the value $z_2 \leftarrow (kz_1)^{-1} \bmod q$, where k was computed in the simulation of the DSA signature oracle in the corresponding *bobInv1* query,
2. A chooses to produce the first j ciphertexts under pk' as in the real protocol (i.e., $\mu' \leftarrow E_{pk'}(z_2)$), for a random $j \in \{0, \ldots, q_{\mathsf{bob}}\}$, and
3. A produces the next ciphertext under pk' by using the response from the test oracle with input $X_0 = z_2$ and $X_1 = 0$.

Finally A outputs 0 if F produces a forgery, and 1 otherwise. Since the case of $j = 0$ corresponds to SIM, and the case of $j = q_{\mathsf{bob}}$ corresponds to the real protocol, an averaging argument can be used to show that A ϵ'-breaks the semantic security of the underlying encryption scheme for pk' with probability $\epsilon' \approx \frac{\epsilon}{2q_{\mathsf{bob}}}$.

Theorem 2. *Suppose a* bob-*compromising forger* $(\overline{q}, \epsilon)$*-breaks* S-DSA. *Then either there exists an attacker that* ϵ'*-breaks* (G_{enc}, E, D) *with* $\epsilon' \approx \frac{\epsilon}{4q_{\mathsf{alice}}}$, *or there exists a forger that* $(q_{\mathsf{alice}}, \epsilon'')$*-breaks DSA, with* $\epsilon'' \approx \frac{\epsilon}{2}$.

Proof. Assume a bob-compromising forger F $(\overline{q}, \epsilon)$-breaks the S-DSA scheme. Then consider a simulation SIM of the S-DSA scheme that takes as input a DSA public key $<g,p,q,y>$, a corresponding signature oracle, and a public key pk for the underlying encryption scheme. SIM generates the initialization data for bob: $x_2 \leftarrow_R \mathbb{Z}_q$ and $(pk', sk') \leftarrow G_{enc}(1^{\kappa'})$, and gives these to F. The public data y, $y_1 = g^{(x_2^{-1} \bmod q)} \bmod p$, and pk are also revealed to F. Then SIM responds to bob queries as a real bob oracle would, and to alice queries using the help of the DSA signature oracle, since SIM does not know the x_1 value used by a real alice oracle. Specifically SIM answers as follows:

1. *aliceInv1*(m): Set $\alpha \leftarrow E_{pk}(0)$ and $\zeta \leftarrow E_{pk}(0)$, and return $<m, \alpha, \zeta>$.
2. *aliceInv2*(r_2): Reject if $r_2 \notin \mathbb{Z}_p^*$ or $(r_2)^q \not\equiv_p 1$. Call the DSA signature oracle with m, let $(\hat{r}, \hat{s})$ be the resulting signature, and compute $r \leftarrow g^{\mathsf{hash}(m)\hat{s}^{-1}} y^{\hat{r}\hat{s}^{-1}} \bmod p$ where $\hat{s}^{-1}$ is computed modulo q. Construct Π using the simulator for the zkp []. Store $<\hat{r}, \hat{s}>$ and return $<r, \Pi>$.
3. *aliceInv3*$(<\mu, \mu', \Pi'>)$: Reject if $\mu \notin C_{pk}$, $\mu' \notin C_{pk'}$, or the verification of Π' fails. Otherwise, return $<\hat{r}, \hat{s}>$.

Notice that SIM sets α and ζ to encryptions of zero, and simulates the proof of consistency Π. In fact, disregarding the negligible statistical difference between the simulated Π proofs and the real Π proofs, the only way SIM and the real S-DSA scheme differ (from F's viewpoint) is with respect to the α and ζ values, i.e., the (at most) $2q_{\mathsf{alice}}$ ciphertexts generated using pk.

Now consider a forger F^* that takes as input a DSA public key $<g, p, q, y>$ and a corresponding signature oracle, generates a public key pk using $<pk, sk> \leftarrow G_{enc}(1^{\kappa'})$, runs SIM using these parameters as inputs, and outputs whatever F outputs. If F produces a forgery with probability at least $\frac{\epsilon}{2}$ in SIM, F^* produces a forgery in the underlying DSA signature scheme with probability at least $\frac{\epsilon}{2}$.

Otherwise F produces a forgery with probability less than $\frac{\epsilon}{2}$ in SIM. Then using a standard hybrid argument, we can construct an attacker A that ϵ'-breaks the semantic security of the underlying encryption scheme for pk, where $\epsilon' \approx \frac{\epsilon}{4q_{\mathsf{alice}}}$. Specifically, A takes a public key pk and corresponding test oracle as input, generates a DSA public/private key pair $(<g, p, q, y>, <g, p, q, x>) \leftarrow G_{DSA}(1^{\kappa'})$, and runs a slightly modified SIM using $<g, p, q, y>$ as the DSA public key parameter, and using pk as the public encryption key parameter. SIM is modified only in the alice oracle queries, as follows:

1. In *aliceInv1*,
   a) A chooses to produce the first j ciphertexts under pk as in the real protocol (i.e., either $\alpha \leftarrow E_{pk}(z_1)$ or $\zeta \leftarrow E_{pk}(x_1 z_1 \bmod q)$), for a random $j \in \{0, \ldots, 2q_{\mathsf{alice}}\}$,
   b) A produces the next ciphertext under pk by using the response from the test oracle with input X_0 being the plaintext from the real protocol (i.e., either $X_0 = z_1$ or $X_0 = x_1 z_1 \bmod q$, depending on whether j is even or odd) and $X_1 = 0$.
2. In *aliceInv2*, A computes r as in the real protocol, without calling the DSA signature oracle.
3. In *aliceInv3*, instead of returning the result of calling the DSA signature oracle, A computes $z_2 \leftarrow D_{sk'}(\mu')$ and $k_2 \leftarrow (z_2)^{-1} \bmod q$, sets $k \leftarrow k_1 k_2 \bmod q$, and returns the DSA signature for m using DSA secret key $<g, p, q, x>$ with k as the ephemeral secret key.

Finally A outputs 0 if F produces a forgery, and 1 otherwise. Since the case of $j = 0$ corresponds to SIM (in particular, notice that the distribution of r is identical), and the case of $j = 2q_{\mathsf{alice}}$ corresponds to the real protocol, an averaging argument can be used to show that A ϵ'-breaks the semantic security of the underlying encryption scheme for pk with probability $\epsilon' \approx \frac{\epsilon}{4q_{\mathsf{alice}}}$.

6 Proofs Π and Π'

In this section we provide an example of how alice and bob can efficiently construct and verify the noninteractive zero-knowledge proofs Π and Π'. The form of these proofs naturally depends on the encryption scheme (G_{enc}, E, D), and the particular encryption scheme for which we detail Π and Π' here is that due to Paillier [31]. We reiterate, however, that our use of Paillier is merely exemplary, and similar proofs Π and Π' can be constructed with other cryptosystems satisfying the required properties (see Section 3).

We caution the reader that from this point forward, our use of variables is not necessarily consistent with their prior use in the paper; rather, it is necessary to replace certain variables or reuse them for different purposes.

6.1 The Paillier Cryptosystem

A specific example of a cryptosystem that has the homomorphic properties required for our protocol is the first cryptosystem presented in [31]. It uses the facts that $w^{\lambda(N)} \equiv_N 1$ and $w^{N\lambda(N)} \equiv_{N^2} 1$ for any $w \in \mathbb{Z}^*_{N^2}$, where $\lambda(N)$ is the Carmichael function of N. Let L be a function that takes input elements from the set $\{u < N^2 | u \equiv 1 \bmod N\}$ and returns $L(u) = \frac{u-1}{N}$. We then define the Paillier encryption scheme (G_{Pai}, E, D) as follows. This definition differs from that in [31] only in that we define the message space M_{pk} for public key $pk = <N, g>$ as $M_{<N,g>} = [-(N-1)/2, (N-1)/2]$ (versus $\mathbb{Z}_N$ in [31]).

$G_{Pai}(1^{\kappa'})$: Choose $\kappa'/2$-bit primes p, q, set $N = pq$, and choose a random element $g \in \mathbb{Z}^*_{N^2}$ such that $\gcd(L(g^{\lambda(N)} \bmod N^2), N) = 1$. Return the public key $<N, g>$ and the private key $<N, g, \lambda(N)>$.

$E_{<N,g>}(m)$: Select a random $x \in \mathbb{Z}^*_N$ and return $c = g^m x^N \bmod N^2$.

$D_{<N,g,\lambda(N)>}(c)$: Compute $m = \frac{L(c^{\lambda(N)} \bmod N^2)}{L(g^{\lambda(N)} \bmod N^2)} \bmod N$. Return m if $m \leq (N-1)/2$, and otherwise return $m - N$.

$c_1 +_{<N,g>} c_2$: Return $c_1 c_2 \bmod N^2$.

$c \times_{<N,g>} m$: Return $c^m \bmod N^2$.

Paillier [31] shows that both $c^{\lambda(N)} \bmod N^2$ and $g^{\lambda(N)} \bmod N^2$ are elements of the form $(1+N)^d \equiv_{N^2} 1 + dN$, and thus the L function can be easily computed for decryption. The security of this cryptosystem relies on the *Decision Composite Residuosity Assumption*, DCRA.

6.2 Proof Π

In this section we show how to efficiently implement the proof Π in our protocol when the Paillier cryptosystem is used. Π' is detailed in Section 6.3. Both proofs rely on the following assumption:

Strong RSA Assumption. Given an RSA modulus generator G_{RSA} that takes as input $1^{\kappa'}$ and produces a value N that is the product of two random primes of length $\kappa'/2$, the Strong RSA assumption states that for any probabilistic polynomial-time attacker A:

$$\Pr[N \leftarrow G_{RSA}(1^{\kappa'}); y \leftarrow_R \mathbb{Z}^*_N; (x, e) \leftarrow A(N, y) : (e \geq 3) \wedge (y \equiv_N x^e)]$$

is negligible.

In our proofs, it is assumed that there are public values $\tilde{N}$, h_1 and h_2. Soundness requires that $\tilde{N}$ be an RSA modulus that is the product of two strong primes and for which the factorization is unknown to the prover, and that the discrete logs of h_1 and h_2 relative to each other modulo $\tilde{N}$ are unknown to the prover. Zero knowledge requires that discrete logs of h_1 and h_2 relative to each other modulo $\tilde{N}$ exist (i.e., that h_1 and h_2 generate the same group). As in Section 4.1, here we assume that these parameters are distributed to alice and bob by a trusted third party. In the full paper, we will describe how this assumption can be eliminated.

Now consider the proof Π. Let p and q be as in a DSA public key, $pk = <N, g>$ be a Paillier public key, and $sk = <N, g, \lambda(N)>$ be the corresponding private key, where $N > q^6$. For public values c, d, w_1, w_2, m_1, m_2, we construct a zero-knowledge proof Π of:

$$P = \begin{bmatrix} \exists x_1, x_2 : x_1, x_2 \in [-q^3, q^3] \\ \wedge \qquad c^{x_1} \equiv_p w_1 \\ \wedge \qquad d^{x_2/x_1} \equiv_p w_2 \\ \wedge \qquad D_{sk}(m_1) = x_1 \\ \wedge \qquad D_{sk}(m_2) = x_2 \end{bmatrix}$$

The proof is constructed in Figure 2, and its verification procedure is given in Figure 3. We assume that $c, d, w_1, w_2 \in \mathbb{Z}_p^*$ and are of order q, and that $m_1, m_2 \in \mathbb{Z}_{N^2}^*$. (The prover should verify this if necessary, and abort if not true.) We assume the prover knows $x_1, x_2 \in \mathbb{Z}_q$ and $r_1, r_2 \in \mathbb{Z}_N^*$ such that $c^{x_1} \equiv_p w_1$, $d^{x_2/x_1} \equiv_p w_2$, $m_1 \equiv_{N^2} g^{x_1}(r_1)^N$ and $m_2 \equiv_{N^2} g^{x_2}(r_2)^N$. The prover need not know sk, though a malicious prover might. If necessary, the verifier should verify that $c, d, w_1, w_2 \in \mathbb{Z}_p^*$ and are of order q, and that $m_1, m_2 \in \mathbb{Z}_{N^2}^*$.

Intuitively, the proof works as follows. Commitments z_1 and z_2 are made to x_1 and x_2 over the RSA modulus $\tilde{N}$, and these are proven to fall in the desired range using proofs as in [15]. Simultaneously, it is shown that the commitment z_1 corresponds to the decryption of m_1 and the discrete log of w_1. Also simultaneously, it is shown that the commitment z_2 corresponds to the decryption of m_2, and that the discrete log of w_2 is the quotient of the two commitments. The proof is shown in two columns, the left column used to prove the desired properties of x_1, w_1 and m_1, and the right column used to prove the desired properties of x_2, w_2 and m_2. The proof of the following lemma will appear in the full version of this paper.

Lemma 1. *Π is a noninteractive zero-knowledge proof of P.*

6.3 Proof Π'

Now we look at the proof Π'. Let p and q be as in a DSA public key, $pk = <N, g>$ and $sk = <N, g, \lambda(N)>$ be a Paillier key pair with $N > q^8$, and $pk' = <N', g'>$ and $sk' = <N', g', \lambda(N')>$ be a Paillier key pair with $N' > q^6$. For values c, d, w_1, w_2, m_1, m_2, m_3, m_4 such that for some $n_1, n_2 \in [-q^4, q^4]$, $D_{sk}(m_3) = n_1$ and $D_{sk}(m_4) = n_2$, we construct a zero-knowledge proof Π' of:

$$P' = \left[\begin{array}{lr} \exists x_1, x_2, x_3 : & x_1, x_2 \in [-q^3, q^3] \\ \wedge & x_3 \in [-q^7, q^7] \\ \wedge & c^{x_1} \equiv_p w_1 \\ \wedge & d^{x_2/x_1} \equiv_p w_2 \\ \wedge & D_{sk'}(m_1) = x_1 \\ \wedge\ D_{sk}(m_2) = n_1 x_1 + n_2 x_2 + q x_3 & \end{array}\right]$$

We note that P' is stronger than what is needed as shown in Figure 1. The proof is constructed in Figure 4, and the verification procedure for it is given in Figure 5. We assume that $c, d, w_1, w_2 \in \mathbb{Z}_p^*$ and are of order q, and that $m_1 \in \mathbb{Z}_{(N')^2}^*$ and $m_2 \in \mathbb{Z}_{N^2}^*$. (The prover should verify this if necessary.) We assume the prover knows $x_1, x_2 \in \mathbb{Z}_q$, $x_3 \in \mathbb{Z}_{q^5}$, and $r_1, r_2 \in \mathbb{Z}_N^*$, such that $c^{x_1} \equiv_p w_1$, $d^{x_2/x_1} \equiv_p w_2$, $m_1 \equiv_{(N')^2} (g')^{x_1}(r_1)^{N'}$ and $m_2 \equiv_{N^2} (m_3)^{x_1}(m_4)^{x_2} g^{q x_3}(r_2)^N$. The

$$\begin{array}{ll}
\alpha \leftarrow_R \mathbb{Z}_{q^3} & \delta \leftarrow_R \mathbb{Z}_{q^3} \\
\beta \leftarrow_R \mathbb{Z}_N^* & \mu \leftarrow_R \mathbb{Z}_N^* \\
\gamma \leftarrow_R \mathbb{Z}_{q^3 \tilde{N}} & \nu \leftarrow_R \mathbb{Z}_{q^3 \tilde{N}} \\
\rho_1 \leftarrow_R \mathbb{Z}_{q\tilde{N}} & \rho_2 \leftarrow_R \mathbb{Z}_{q\tilde{N}} \\
 & \rho_3 \leftarrow_R \mathbb{Z}_q \\
 & \epsilon \leftarrow_R \mathbb{Z}_q \\
 & \\
z_1 \leftarrow (h_1)^{x_1}(h_2)^{\rho_1} \bmod \tilde{N} & z_2 \leftarrow (h_1)^{x_2}(h_2)^{\rho_2} \bmod \tilde{N} \\
u_1 \leftarrow c^{\alpha} \bmod p & y \leftarrow d^{x_2+\rho_3} \bmod p \\
u_2 \leftarrow g^{\alpha}\beta^N \bmod N^2 & v_1 \leftarrow d^{\delta+\epsilon} \bmod p \\
u_3 \leftarrow (h_1)^{\alpha}(h_2)^{\gamma} \bmod \tilde{N} & v_2 \leftarrow (w_2)^{\alpha} d^{\epsilon} \bmod p \\
 & v_3 \leftarrow g^{\delta}\mu^N \bmod N^2 \\
 & v_4 \leftarrow (h_1)^{\delta}(h_2)^{\nu} \bmod \tilde{N} \\
 & \\
\multicolumn{2}{l}{e \leftarrow \mathsf{hash}(c, w_1, d, w_2, m_1, m_2, z_1, u_1, u_2, u_3, z_2, y, v_1, v_2, v_3, v_4)} \\
 & \\
s_1 \leftarrow e x_1 + \alpha & t_1 \leftarrow e x_2 + \delta \\
s_2 \leftarrow (r_1)^e \beta \bmod N & t_2 \leftarrow e\rho_3 + \epsilon \bmod q \\
s_3 \leftarrow e\rho_1 + \gamma & t_3 \leftarrow (r_2)^e \mu \bmod N^2 \\
 & t_4 \leftarrow e\rho_2 + \nu \\
 & \\
\multicolumn{2}{l}{\Pi \leftarrow <z_1, u_1, u_2, u_3, z_2, y, v_1, v_2, v_3, v_4, s_1, s_2, s_3, t_1, t_2, t_3, t_4>}
\end{array}$$

Fig. 2. Construction of Π

$$\begin{array}{ll}
\multicolumn{2}{c}{<z_1, u_1, u_2, u_3, z_2, y, v_1, v_2, v_3, v_4, s_1, s_2, s_3, t_1, t_2, t_3, t_4> \leftarrow \Pi} \\
\text{Verify } s_1, t_1 \in \mathbb{Z}_{q^3}. & \text{Verify } d^{t_1+t_2} \equiv_p y^e v_1. \\
\text{Verify } c^{s_1} \equiv_p (w_1)^e u_1. & \text{Verify } (w_2)^{s_1} d^{t_2} \equiv_p y^e v_2. \\
\text{Verify } g^{s_1}(s_2)^N \equiv_{N^2} (m_1)^e u_2. & \text{Verify } g^{t_1}(t_3)^N \equiv_{N^2} (m_2)^e v_3. \\
\text{Verify } (h_1)^{s_1}(h_2)^{s_3} \equiv_{\tilde{N}} (z_1)^e u_3. & \text{Verify } (h_1)^{t_1}(h_2)^{t_4} \equiv_{\tilde{N}} (z_2)^e v_4.
\end{array}$$

Fig. 3. Verification of Π

$$
\begin{array}{ll}
\alpha \leftarrow_R \mathbb{Z}_{q^3} & \delta \leftarrow_R \mathbb{Z}_{q^3} \\
\beta \leftarrow_R \mathbb{Z}^*_{N'} & \mu \leftarrow_R \mathbb{Z}^*_N \\
\gamma \leftarrow_R \mathbb{Z}_{q^3 \tilde{N}} & \nu \leftarrow_R \mathbb{Z}_{q^3 \tilde{N}} \\
\rho_1 \leftarrow_R \mathbb{Z}_{q\tilde{N}} & \rho_2 \leftarrow_R \mathbb{Z}_{q\tilde{N}} \\
 & \rho_3 \leftarrow_R \mathbb{Z}_q \\
 & \rho_4 \leftarrow_R \mathbb{Z}_{q^5 \tilde{N}} \\
 & \epsilon \leftarrow_R \mathbb{Z}_q \\
 & \sigma \leftarrow_R \mathbb{Z}_{q^7} \\
 & \tau \leftarrow_R \mathbb{Z}_{q^7 \tilde{N}} \\
 & \\
z_1 \leftarrow (h_1)^{x_1}(h_2)^{\rho_1} \bmod \tilde{N} & z_2 \leftarrow (h_1)^{x_2}(h_2)^{\rho_2} \bmod \tilde{N} \\
u_1 \leftarrow c^{\alpha} \bmod p & y \leftarrow d^{x_2+\rho_3} \bmod p \\
u_2 \leftarrow (g')^{\alpha}\beta^{N'} \bmod (N')^2 & v_1 \leftarrow d^{\delta+\epsilon} \bmod p \\
u_3 \leftarrow (h_1)^{\alpha}(h_2)^{\gamma} \bmod \tilde{N} & v_2 \leftarrow (w_2)^{\alpha} d^{\epsilon} \bmod p \\
 & v_3 \leftarrow (m_3)^{\alpha}(m_4)^{\delta} g^{q\sigma}\mu^N \bmod N^2 \\
 & v_4 \leftarrow (h_1)^{\delta}(h_2)^{\nu} \bmod \tilde{N} \\
 & z_3 \leftarrow (h_1)^{x_3}(h_2)^{\rho_4} \bmod \tilde{N} \\
 & v_5 \leftarrow (h_1)^{\sigma}(h_2)^{\tau} \bmod \tilde{N}
\end{array}
$$

$$
e \leftarrow \mathsf{hash}(c, w_1, d, w_2, m_1, m_2, z_1, u_1, u_2, u_3, z_2, z_3, y, v_1, v_2, v_3, v_4, v_5)
$$

$$
\begin{array}{ll}
s_1 \leftarrow e x_1 + \alpha & t_1 \leftarrow e x_2 + \delta \\
s_2 \leftarrow (r_1)^e \beta \bmod N' & t_2 \leftarrow e\rho_3 + \epsilon \bmod q \\
s_3 \leftarrow e\rho_1 + \gamma & t_3 \leftarrow (r_2)^e \mu \bmod N \\
 & t_4 \leftarrow e\rho_2 + \nu \\
 & t_5 \leftarrow e x_3 + \sigma \\
 & t_6 \leftarrow e\rho_4 + \tau
\end{array}
$$

$$
\Pi' \leftarrow \langle z_1, u_1, u_2, u_3, z_2, z_3, y, v_1, v_2, v_3, v_4, v_5, s_1, s_2, s_3, t_1, t_2, t_3, t_4, t_5, t_6 \rangle
$$

Fig. 4. Construction of Π'

$$
\langle z_1, u_1, u_2, u_3, z_2, z_3, y, v_1, v_2, v_3, v_4, v_5, s_1, s_2, s_3, t_1, t_2, t_3, t_4, t_5, t_6 \rangle \leftarrow \Pi'
$$

Verify $s_1, t_1 \in \mathbb{Z}_{q^3}$.	Verify $d^{t_1+t_2} \equiv_p y^e v_1$.
Verify $t_5 \in \mathbb{Z}_{q^7}$.	Verify $(w_2)^{s_1} d^{t_2} \equiv_p y^e v_2$.
Verify $c^{s_1} \equiv_p (w_1)^e u_1$.	Verify $(m_3)^{s_1}(m_4)^{t_1} g^{q t_5}(t_3)^N \equiv_{N^2} (m_2)^e v_3$.
Verify $(g')^{s_1}(s_2)^{N'} \equiv_{(N')^2} (m_1)^e u_2$.	Verify $(h_1)^{t_1}(h_2)^{t_4} \equiv_{\tilde{N}} (z_2)^e v_4$.
Verify $(h_1)^{s_1}(h_2)^{s_3} \equiv_{\tilde{N}} (z_1)^e u_3$.	Verify $(h_1)^{t_5}(h_2)^{t_6} \equiv_{\tilde{N}} (z_3)^e v_5$.

Fig. 5. Verification of Π'

prover need not know sk or sk', though a malicious prover might know sk'. We assume the verifier knows n_1 and n_2. If necessary, the verifier should verify that $c, d, w_1, w_2 \in \mathbb{Z}_p^*$ and are of order q, and that $m_1 \in \mathbb{Z}^*_{(N')^2}$ and $m_2 \in \mathbb{Z}^*_{N^2}$. The proof of the following lemma will appear in the full version of this paper.

Lemma 2. *Π' is a noninteractive zero-knowledge proof of P'.*

References

1. J. Benaloh. Dense probabilistic encryption. In *Workshop on Selected Areas of Cryptography*, pages 120–128, 1994.
2. N. Barić and B. Pfitzmann. Collision-free accumulators and fail-stop signature schemes without trees. In *EUROCRYPT '96* (LNCS 1233), pages 480–494, 1997.
3. M. Blum, A. DeSantis, S. Micali, and G. Persiano. Noninteractive zero-knowledge. *SIAM Journal of Computing* 20(6):1084–1118, 1991.
4. C. Boyd. Digital multisignatures. In H. J. Beker and F. C. Piper, editors, *Cryptography and Coding*, pages 241–246. Clarendon Press, 1986.
5. M. Bellare and P. Rogaway. Random oracles are practical: A paradigm for designing efficient protocols. In 1st *ACM Conference on Computer and Communications Security*, pages 62–73, November 1993.
6. R. A. Croft and S. P. Harris. Public-key cryptography and reusable shared secrets. In H. Baker and F. Piper, editors, *Cryptography and Coding*, pages 189–201, 1989.
7. M. Cerecedo, T. Matsumoto, H. Imai. Efficient and secure multiparty generation of digital signatures based on discrete logarithms. *IEICE Trans. Fundamentals of Electronics Communications and Computer Sciences* E76A(4):532–545, April 1993.
8. Y. Desmedt. Society and group oriented cryptography: a new concept. In *CRYPTO '87* (LNCS 293), pages 120–127, 1987.
9. Y. Desmedt and Y. Frankel. Threshold cryptosystems. In *CRYPTO '89* (LNCS 435), pages 307–315, 1989.
10. T. ElGamal. A public key cryptosystem and a signature scheme based on discrete logarithms. *IEEE Transactions on Information Theory*, 31:469–472, 1985.
11. FIPS 180-1. Secure hash standard. Federal Information Processing Standards Publication 180-1, U.S. Dept. of Commerce/NIST, National Technical Information Service, Springfield, Virginia, 1995.
12. FIPS 186. Digital signature standard. Federal Information Processing Standards Publication 186, U.S. Dept. of Commerce/NIST, National Technical Information Service, Springfield, Virginia, 1994.
13. Y. Frankel. A practical protocol for large group oriented networks. In *EUROCRYPT '89* (LNCS 434), pages 56–61, 1989.
14. Y. Frankel, P. MacKenzie, and M. Yung. Adaptively-secure distributed threshold public key systems. In *European Symposium on Algorithms* (LNCS 1643), pages 4–27, 1999.
15. E. Fujisaki and T. Okamoto. Statistical zero-knowledge protocols to prove modular polynomial relations. In *CRYPTO '97* (LNCS 1294), pages 16–30, 1997.
16. R. Gennaro, S. Jarecki, H. Krawczyk, and T. Rabin. Robust threshold DSS signatures. In *EUROCRYPT '96* (LNCS 1070), pages 354–371, 1996.
17. R. Gennaro, S. Jarecki, H. Krawczyk, and T. Rabin. Secure distributed key generation for discrete-log based cryptosystems. In *EUROCRYPT '99* (LNCS 1592), pages 295–310, 1999.
18. S. Goldwasser and S. Micali. Probabilistic encryption. *Journal of Computer and System Sciences* 28:270–299, 1984.
19. S. Goldwasser, S. Micali, and R. L. Rivest. A digital signature scheme secure against adaptive chosen-message attacks. *SIAM Journal of Computing* 17(2):281–308, April 1988.
20. L. Harn. Group oriented (t, n) threshold digital signature scheme and digital multisignature. *IEE Proc.-Comput. Digit. Tech.* 141(5):307–313, 1994.

21. A. Herzberg, M. Jakobsson, S. Jarecki, H. Krawczyk, and M. Yung. Proactive public-key and signature schemes. In 4th *ACM Conference on Computer and Communications Security*, pages 100–110, 1997.
22. T. Hwang. Cryptosystem for group oriented cryptography. In *EUROCRYPT '90* (LNCS 473), pages 352–360, 1990.
23. S. Jarecki and A. Lysyanskaya. Adaptively secure threshold cryptography: introducing concurrency, removing erasures. In *EUROCRYPT 2000* (LNCS 1807), pages 221–242, 2000.
24. J. Kilian, E. Petrank, and C. Rackoff. Lower bounds for zero knowledge on the internet. In 39th *IEEE Symposium on Foundations of Computer Science*, pages 484–492, 1998.
25. D. W. Kravitz. Digital signature algorithm. U.S. Patent 5,231,668, 27 July 1993.
26. S. Langford. Threshold DSS signatures without a trusted party. In *CRYPTO '95* (LNCS 963), pages 397–409, 1995.
27. P. MacKenzie and M. K. Reiter. Networked cryptographic devices resilient to capture. DIMACS Technical Report 2001-19, May 2001. Extended abstract in *2001 IEEE Symposium on Security and Privacy*, May 2001.
28. D. Naccache and J. Stern. A new public-key cryptosystem. In *EUROCRYPT '97* (LNCS 1233), pages 27–36, 1997.
29. M. Naor and M. Yung. Public-key cryptosystems provably secure against chosen ciphertext attacks. In 22nd *ACM Symposium on Theory of Computing*, pages 427–437, 1990.
30. T. Okamoto and S. Uchiyama. A new public-key cryptosystem, as secure as factoring. In *EUROCRYPT '98* (LNCS 1403), pages 308–318, 1998.
31. P. Paillier. Public-key cryptosystems based on composite degree residuosity classes. In *EUROCRYPT '99* (LNCS 1592), pages 223–238, 1999.
32. C. Park and K. Kurosawa. New ElGamal type threshold digital signature scheme. *IEICE Trans. Fundamentals of Electronics Communications and Computer Sciences* E79A(1):86–93, January, 1996.
33. T. Pedersen. A threshold cryptosystem without a trusted party. In *EUROCRYPT '91* (LNCS 547), pages 522–526, 1991.
34. A. Yao. Protocols for secure computation. In 23rd *IEEE Symposium on Foundations of Computer Science*, pages 160–164, 1982.

Oblivious Transfer in the Bounded Storage Model

Yan Zong Ding

DEAS, Harvard University, Cambridge MA 02138, USA
zong@deas.harvard.edu

Abstract. Building on a previous important work of Cachin, Crépeau, and Marcil [15], we present a provably secure and more efficient protocol for $\binom{2}{1}$-Oblivious Transfer with a *storage-bounded* receiver. A public random string of n bits long is employed, and the protocol is secure against any receiver who can store γn bits, $\gamma < 1$. Our work improves the work of CCM [15] in two ways. First, the CCM protocol requires the sender and receiver to store $O(n^c)$ bits, $c \sim 2/3$. We give a similar but more efficient protocol that just requires the sender and receiver to store $O(\sqrt{kn})$ bits, where k is a security parameter. Second, the basic CCM Protocol was proved in [15] to guarantee that a dishonest receiver who can store $O(n)$ bits succeeds with probability at most $O(n^{-d})$, $d \sim 1/3$, although repitition of the protocol can make this probability of cheating exponentially small [20]. Combining the methodologies of [24] and [15], we prove that in our protocol, a dishonest storage-bounded receiver succeeds with probability only $2^{-O(k)}$, without repitition of the protocol. Our results answer an open problem raised by CCM in the affirmative.

1 Introduction

Oblivious Transfer (OT) was introduced by Rabin [47] in 1981, and has since then become one of the most fundamental and powerful tools in cryptography. An important generalization, known as one-out-of-two oblivious transfer and denoted $\binom{2}{1}$-OT, was introduced by Even, Goldreich, and Lempel [28] in 1982. Informally speaking, in a $\binom{2}{1}$-OT, a sender Alice has two secret bits $M_0, M_1 \in \{0,1\}$, and a receiver Bob has a secret bit $\delta \in \{0,1\}$. Alice sends M_0, M_1 in such a way that Bob receives M_δ, but does not learn both M_0 and M_1, and Alice learns nothing about δ. Crépeau proved in 1987 that OT and $\binom{2}{1}$-OT are equivalent [19]. In 1988, Kilian proved that every secure two-party and multiparty computation can be reduced to OT [33].

Traditionally, protocols for OT have been based on unproven complexity assumptions that certains problems, such as integer factorization, are computationally hard, or that trapdoor permutations exist. The solutions so obtained, although significant, have a drawback. Namely, they do not guarantee *everlasting security.* A dishonest player can store the entire conversation during the protocol, and attempt to subvert the security of the protocol later, when enabled by breakthroughs in computing technology and/or code-breaking algorithms. While

J. Kilian (Ed.): CRYPTO 2001, LNCS 2139, pp. 155–170, 2001.

determining the computational complexity of factorization, or proving the existence of trapdoor permutations, is still beyond the reach of complexity theory, continuing advances in factoring algorithms will jeopardize the security of protocols based on factoring. In addition, these protocols will become insecure if quantum computers become available [50]. Similar threats exist for protocols based on other hardness assumptions. We thus seek protocols that are provably secure in face of any future advances in algorithms and computing technology.

The ground breaking work of Cachin, Crépeau, and Marcil [15] in 1998 gave the first provably secure protocol for $\binom{2}{1}$-OT in the *Bounded Storage Model*, without any complexity assumption. The bounded storage model, introduced by Maurer [37], imposes a bound B on the adversary's *storage* capacity only. A public random string of n bits long, $n > B$, is employed in order to defeat the adversary. Although a trusted third party is not necessary in principle, in a practical implementation, the string α may be one in a steady flow of random strings $\alpha_1, \alpha_2, \ldots$, each of length n, broadcast from a satellite at a very high rate, and available to all. When α is broadcast, the adversary is allowed to compute an *arbitrary* function f on α, provided that the length $|f(\alpha)| \leq B$.

In the context of OT, the storage bound is placed on one of the two parties, WLOG say the receiver. By the reversibility of OT [21], the case where the storage bound is placed on the sender, is equivalent. The CCM protocol [15] guarantees provable security against any dishonest sender who is unbounded in every way, and against any *computationally unbounded* dishonest receiver who stores no more than $B = \gamma n$ bits, $\gamma < 1$. Furthermore, the security against a dishonest receiver is preserved regardless of future increases in storage capacity. Together with the completeness of OT [33], a fundamental implication of [15] is that every information-theoretically secure two-party and multi-party computation, in principle, is feasible in the bounded storage model.

The work of CCM [15], however, has two undesirable aspects. First, while providing security against a dishonest receiver who stores $B = O(n)$ bits, the CCM protocol also requires honest sender and receiver to store $O(n^c)$ bits, $c \sim 2/3$. Since n is very large, this requirement could be rather excessive. Second, the CCM protocol was proved in [15] to guarantee that a receiver who stores $O(n)$ bits succeeds with probability at most $O(n^{-d})$, $d \sim 1/3$. Note that this probability is usually not as small as desired. Of course, repitition of the protocol can make this probability of cheating exponentially small [20].

Our Results. Building on the work of Cachin, Crépeau, and Marcil [15], we give a similar but more efficient protocol for $\binom{2}{1}$-OT in the bounded storage model. The major difference between our protocol and the CCM Protocol is that the CCM Protocol uses an extra distillation step, which involves many bits divided into polynomially large blocks, and the extraction of a nearly random bit from each block. Getting rid of this distillation step, we reduce the storage requirement to $O(\sqrt{kn})$, where k is a security parameter. Combining the methodologies of [24] and [15], we prove that in our protocol, any dishonest receiver who stores $O(n)$ bits succeeds with probability at most $2^{-O(k)}$, without repetition of the protocol. Our results answer positively an open problem raised in [15].

1.1 Related Work

OT and $\binom{2}{1}$-OT were introduced by Rabin [47] and Even *et al* [28] respectively. Their equivalence was established by Crépeau [19]. There is a vast literature on the relationships between OT and other cryptographic primitives, and between OT variants. OT can be used to construct protocols for secret key agreement [47], [8], [52], contract signing [28], bit commitment and zero-knowledge proof [33], and general secure multi-party computation [52], [30], [31], [33], [32], [35], [36], [22]. It was proved by Kilian that every secure two-party and multi-party computation reduces to OT [33]. Information-theoretic reductions between OT variants were studied in [10], [11], [19], [20], [21], [12], [9], [14], [25].

In traditional cryptography, protocols for OT have been designed under the assumptions that factoring is hard [47], discret log is hard [6], and trapdoor permutations exist [28], [52], [30], [31]. OT has also been studied in the quantum model [7], and the noisy channel model [20]. Recently OT has been extended to various distributed and concurrent settings [5], [49], [29], [44], and these protocols are either based on complexity assumption, or information-theoretically secure using private channels and auxilliary servers. Cachin, Crépeau, and Marcil [15] gave the first secure two-party protocol for $\binom{2}{1}$-OT in the bounded storage and public random string model, without any complexity assumption, and without private channels or auxilliary servers.

The public random string model was introduced by Rabin [48]. The bounded storage model was introduced by Maurer [37]. Secure encryption in the bounded storage model was first studied in [37], [16], but later significantly stronger results appeared in [1], [2], [24]. Information-theoretically secure key agreement was investigated in [38], [39], [16], [40], [41], [42].

The bounded *space* model for zero-knowledge proof was studied in [18], [17], [34], [23], [26], [27], [3]. Pseudorandomness in the bounded space model was studied in [45], [46]. However, note the important difference between the bounded space model and the bounded storage model: the bounded *space* model imposes a bound on the computation space of the adversary, whereas in the bounded *storage* model the adversary can compute an function with arbitrarily high complexity, provided that the length of the output is bounded.

2 Preliminaries

This section provides the building blocks for our protocol and analysis. Throughout the paper, k is a security parameter, n is the length of a public random string, and $B = \gamma n$, $\gamma < 1$, is the storage bound on the receiver Bob. For simplicity and WLOG, we consider $B = n/6$ (i.e. $\gamma = 1/6$). Similar results hold for any $\gamma < 1$.

Definition 1. *Denote* $[n] = \{1, \ldots, n\}$. *Let* $\mathcal{K} \stackrel{d}{=} \{s \subset [n] : |s| = k\}$ *be the set of all* k*-element subsets of* $[n]$.

Definition 2. *For* $s = \{\sigma_1, \ldots, \sigma_k\} \in \mathcal{K}$ *and* $\alpha \in \{0,1\}^n$, *define* $s(\alpha) \stackrel{d}{=} \bigoplus_{i=1}^{k} \alpha[\sigma_i]$, *where* $\oplus$ *denotes XOR, and* $\alpha[\sigma_i]$ *is the* σ_i*-th bit of* α.

Definition 3. *Let $H \subset \{0,1\}^n$. Let $s \in \mathcal{K}$. We say that s is* good *for H if*

$$\left| \frac{|\{\alpha \in H : s(\alpha) = 0\}|}{|H|} - \frac{|\{\alpha \in H : s(\alpha) = 1\}|}{|H|} \right| < 2^{-k/3}. \quad (1)$$

Thus, if s is good for H, then $\{s(\alpha) : \alpha \in H\}$ is well balanced between 0's and 1's.

Definition 4. *Let $H \subset \{0,1\}^n$. We say that H is* fat *if $|H| \geq 2^{0.813n}$.*

The following Lemma 1 says that if H is *fat*, then *almost all* $s \in \mathcal{K}$ are *good* for H. The lemma follows directly from Main Lemma 1 of [24], by considering k-tuples in $[n]^k$ with *distinct* coordinates.

Lemma 1. *Let $H \subset \{0,1\}^n$. Denote*

$$B_H \stackrel{d}{=} \{s \in \mathcal{K} : s \text{ is not } good\ for\ H\}. \quad (2)$$

If H is fat, *and $k < \sqrt{n}$* [1], *then*

$$|B_H| < |\mathcal{K}| \cdot 2^{-k/3} = \binom{n}{k} \cdot 2^{-k/3}. \quad (3)$$

In Appendix A we will give a proof lemma 1 from Main Lemma 1 of [24].

Notation: Let F be a finite set. The notation $x \stackrel{R}{\longleftarrow} F$ denotes choosing x uniformly from F.

Lemma 2. *Let $0 < \gamma, \nu < 1$ and $\nu < 1 - \gamma$. For any function $f : \{0,1\}^n \longrightarrow \{0,1\}^{\gamma n}$, for $\alpha \stackrel{R}{\longleftarrow} \{0,1\}^n$,*

$$\Pr\left[\left|f^{-1}(f(\alpha))\right| \geq 2^{(1-\gamma-\nu)n}\right] > 1 - 2^{-\nu n}.$$

Proof. Any function $f : \{0,1\}^n \longrightarrow \{0,1\}^{\gamma n}$ partitions $\{0,1\}^n$ into $2^{\gamma n}$ disjoint subsets $\Omega_1, \ldots, \Omega_{2^{\gamma n}}$, one for each $\eta \in \{0,1\}^{\gamma n}$, such that for each i, $\forall \alpha, \beta \in \Omega_i$, $f(\alpha) = f(\beta) = \eta_i \in \{0,1\}^{\gamma n}$. Let $\mu = 1 - \gamma - \nu$. We now bound the number of $\alpha \in \{0,1\}^n$ s.t. $\left|f^{-1}(f(\alpha))\right| < 2^{\mu n}$. Since there are at most $2^{\gamma n}$ j's such that $|\Omega_j| < 2^{\mu n}$, it follows that

$$\begin{aligned} \left|\{\alpha \in \{0,1\}^n : |f^{-1}(f(\alpha))| < 2^{\mu n}\}\right| &= \sum_{j:|\Omega_j|<2^{\mu n}} |\Omega_j| \\ &< 2^{\gamma n} \cdot 2^{\mu n} = 2^{(1-\nu)n}. \end{aligned}$$

[1] The condition $k < \sqrt{n}$ in Lemma 1 is valid, because k, the security parameter (e.g. $k = 1000$), is negligbly small compared to n (e.g. $n = 10^{15}$), which is larger than the adversary's storage capacity.

Therefore, for $\alpha \xleftarrow{R} \{0,1\}^n$,

$$\Pr\left[\left|f^{-1}(f(\alpha))\right| < 2^{\mu n}\right] = \frac{\left|\{\alpha \in \{0,1\}^n : |f^{-1}(f(\alpha))| < 2^{\mu n}\}\right|}{2^n} < \frac{2^{(1-\nu)n}}{2^n} = 2^{-\nu n}.$$

□

Corollary 1. *For any function $f : \{0,1\}^n \longrightarrow \{0,1\}^{n/6}$, for $\alpha \xleftarrow{R} \{0,1\}^n$,*

$$\Pr\left[f^{-1}(f(\alpha)) \text{ is fat}\right] > 1 - 2^{-0.02n}.$$

Proof. Let $\gamma = 1/6$ and $\nu = 0.02$ in Lemma 2. □

The rest of this section is devoted to the crucial tools employed by the CCM Protocol and our protocol.

2.1 Birthday Paradox

Lemma 3. *Let $\mathcal{A}, \mathcal{B} \subset [n]$ be two independent random subsets of $[n]$ with $|\mathcal{A}| = |\mathcal{B}| = u$. Then the expected size $E[|\mathcal{A} \cap \mathcal{B}|] = u^2/n$.*

Corollary 2. *Let $\mathcal{A}, \mathcal{B} \subset [n]$ be two independent random subsets of $[n]$ with $|\mathcal{A}| = |\mathcal{B}| = \sqrt{kn}$. Then the expected size $E[|\mathcal{A} \cap \mathcal{B}|] = k$.*

We now wish to bound the probability that $|\mathcal{A} \cap \mathcal{B}|$ deviates from the expectation. Note that standard Chernoff-Hoeffding bounds do not directly apply, since elements of the subsets $\mathcal{A}$ and $\mathcal{B}$ are chosen without replacement. We use the following version of Chernoff-Hoeffding from [4].

Lemma 4. [4] *Let $Z_1, \ldots, Z_u$ be Bernoulli trials (not necessarily independent), and let $0 \leq p_i \leq 1$, $1 \leq i \leq u$. Assume that $\forall\, i$ and $\forall\, (e_1, \cdots, e_{i-1}) \in \{0,1\}^{i-1}$,*

$$\Pr\left[Z_i = 1 \mid Z_1 = e_1, \ldots, Z_{i-1} = e_{i-1}\right] \geq p_i.$$

Let $W = \sum_{i=1}^{u} p_i$. Then for $\delta < 1$,

$$\Pr\left[\sum_{i=1}^{u} Z_i < W \cdot (1-\delta)\right] < e^{-\delta^2 W/2}. \tag{4}$$

Corollary 3. *Let $\mathcal{A}, \mathcal{B} \subset [n]$ be two independent random subsets of $[n]$ with $|\mathcal{A}| = |\mathcal{B}| = 2\sqrt{kn}$. Then*

$$\Pr\left[|\mathcal{A} \cap \mathcal{B}| < k\right] < e^{-k/4}. \tag{5}$$

Proof. Let $u = 2\sqrt{kn}$. Consider any fixed u-subset $\mathcal{B} \subset [n]$, and a randomly chosen u-subset $\mathcal{A} = \{\mathcal{A}_1, \ldots, \mathcal{A}_u\} \subset [n]$. For $i = 1, \ldots, u$, let Z_i be the Bernoulli trial such that $Z_i = 1$ if and only if $\mathcal{A}_i \in \mathcal{B}$. Then clearly

$$\Pr[Z_i = 1 \mid Z_1 = e_1, \ldots, Z_{i-1} = e_{i-1}] \geq \frac{u - (i-1)}{n - (i-1)} > \frac{u - (i-1)}{n}. \quad (6)$$

Let $p_i = \frac{u - (i-1)}{n}$. Let $W = \sum_{i=1}^{u} p_i$. Then by (6),

$$W > \frac{1}{n} \cdot \sum_{i=1}^{u} i > \frac{u^2}{2n} = 2k. \quad (7)$$

Therefore, (5) follows from (4) and (7), with $\delta = 1/2$. □

2.2 Interactive Hashing

Interactive Hashing is a protocol introduced by M. Noar, Ostrovsky, Venkatesan, and Yung in the context of bit commitment and zero-knowledge proof [43]. Cachin, Crépeau, and Marcil [15] gave a new elegant analysis of interactive hashing. The protocol involves two parties, Alice and Bob. Bob has a secret t-bit string $\chi \in T \subset \{0,1\}^t$, where $|T| \leq 2^{t-k}$ and T is unknown to Alice. The protocol is defined to be correct and secure if

1. Bob sends χ in such a way that Alice receives two strings $\chi_0, \chi_1 \in \{0,1\}^t$, one of which is χ, but Alice does not know which one is χ.
2. Bob cannot force both χ_0 and χ_1 to be in T.

The following interactive hashing protocol is due to [43]. The same idea involving taking inner products over $GF(2)$, was first introduced by Valiant and V. Vazirani earlier in the complexity of UNIQUE SATISFIABILITY [51].

NOVY Protocol: Alice *randomly* chooses $t-1$ *linearly independent* vectors $a_1, \ldots, a_{t-1} \in \{0,1\}^t$. The protocol then proceeds in $t-1$ rounds. In Round i, for each $i = 1, \ldots, t-1$,

1. Alice sends a_i to Bob.
2. Bob computes $b_i = a_i \cdot \chi$, where $\cdot$ denotes inner product, and sends b_i to Alice.

After the $t-1$ rounds, both Alice and Bob have the same system of linear equations $a_i \cdot x = b_i$ over $GF(2)$. Since the vectors $a_1, \ldots, a_{t-1} \in \{0,1\}^t$ are linearly independent, the system of $t-1$ linear equations over $GF(2)$ with t unknowns has exactly two solutions, one of which is χ. Therefore, by solving the systems of equations $a_i \cdot x = b_i$, Alice receives two strings χ_0, χ_1, one of which is χ. It is clear that information-theoretically, Alice does not know which solution is χ. Thus Condition 1 of interactive hashing is satisfied.

The following important lemma, regarding Condition 2 of interactive hashing, was proved in [15]. The same result in a non-adversarial setting, more precisely in the case that the Bob is honest, was proved in [51].

Lemma 5. [15] *Suppose Alice and Bob engage in interactive hashing of a t-bit string, $\lg t \leq k \leq t$, by the NOVY protocol. Let $T \subset \{0,1\}^t$ be any subset with $|T| \leq 2^{t-k}$. Then the probability that Bob can answer Alice's queries in such a way that T contains both strings χ_0, χ_1 received by Alice, is at most $2^{-O(k)}$.*

Corollary 4. *Let Alice and Bob engage in interactive hashing of a t-bit string as above. Let $T_0, T_1 \subset \{0,1\}^t$ be any two subsets with $|T_0|, |T_1| \leq 2^{t-k}$. Then the probability that Bob can answer Alice's queries in such a way that either $\chi_0 \in T_0 \wedge \chi_1 \in T_1$, or $\chi_0 \in T_1 \wedge \chi_1 \in T_0$, is at most $2^{-O(k)}$.*

Proof. Let $T = T_0 \cup T_1$ in Lemma 5. □

3 Protocol for $\binom{2}{1}$-OT

Recall that in a $\binom{2}{1}$-OT, the sender Alice has two secret bits $M_0, M_1 \in \{0,1\}$, and the receiver Bob has a secret bit $\delta \in \{0,1\}$. By definition, a $\binom{2}{1}$-OT protocol is correct and secure if the following three conditions are all satisfied:

1. Bob receives M_δ.
2. Bob learns nothing about $M_{1\oplus\delta}$, except with a small probability $\nu(k)$, where k is a security parameter.
3. Alice learns nothing about δ.

3.1 Outline of Basic Ideas

We first outline the basic ideas underling our protocol for $\binom{2}{1}$-OT. First, Alice chooses random $\mathcal{A} \subset [n]$, and Bob chooses random $\mathcal{B} \subset [n]$, with $|\mathcal{A}| = |\mathcal{B}| = u = 2\sqrt{kn}$. Public random string $\alpha \xleftarrow{R} \{0,1\}^n$ is broadcast. Alice retains $\alpha[i]$ $\forall\, i \in \mathcal{A}$, and Bob retains $\alpha[j]$ $\forall\, j \in \mathcal{B}$. Alice then sends her subset $\mathcal{A}$ to Bob, and Bob computes $\mathcal{A} \cap \mathcal{B}$. By the birthday paradox (Corollary 3), with very high probability, $|\mathcal{A} \cap \mathcal{B}| \geq k$.

Fact 1 (Encoding of Subsets) [15] *Each of the $\binom{u}{k}$ k-element subsets of $[u] = \{1,\ldots,u\}$ can be uniquely encoded as a $\lg\binom{u}{k}$-bit string. See* [15] *for an efficient method of encoding and decoding.*

Next, Bob encodes a random k-subset $s \subset \mathcal{A} \cap \mathcal{B}$ as a $\lg\binom{u}{k}$-bit string, and sends s to Alice via the NOVY interactive hashing protocol. By the end of interactive hashing, Alice and Bob will have created two "keys", a good key $S_G = s$, and a bad key S_B, each a k-subset of $\mathcal{A}$, such that: Bob knows $S_G(\alpha)$, but learns nothing about $S_B(\alpha)$, and Alice knows both $S_G(\alpha)$ and $S_B(\alpha)$, but does not know which key is good and which key is bad.

Once the keys S_G and S_B are created, the rest of the protocol is trivial. If Bob wants to read M_δ, then he simply asks Alice to encrypt M_δ with the good key S_G, and $M_{1\oplus\delta}$ with the good key S_B, i.e. Bob ask Alice to send $M_\delta \oplus S_G(\delta)$ and $M_{1\oplus\delta} \oplus S_B(\delta)$. The correctness and security of the protocol follow from the properties of S_B and S_G described above.

3.2 The Protocol, and Main Results

Notation: For a bit $Y \in \{0,1\}$, denote $\overline{Y} \stackrel{d}{=} 1 \oplus Y$.

Definition 5. *Let $\mathcal{X} = \{x_1, \ldots, x_u\}$ be an u-element set. For each subset $J \subset [u]$, define $\mathcal{X}_J \stackrel{d}{=} \{x_i : i \in J\}$.*

Notation: From now on, let $u = 2\sqrt{kn}$.

Our protocol for $\binom{2}{1}$-OT, Protocol A, is described below. Protocol A uses two public random strings $\alpha_0, \alpha_1 \stackrel{R}{\longleftarrow} \{0,1\}^n$. In each of Steps 2 and 3, Alice and Bob each store $u = 2\sqrt{kn}$ bits. In the interactive hashing of Step 4, Alice transmits and Bob stores t^2 bits, where $t = \lg \binom{u}{k} < k \cdot (\lg u - \lg k/e)$. Since $k << n$, the storage requirement is dominated by $O(u) = O(\sqrt{kn})$.

Protocol A:

1. Alice randomly chooses $\mathcal{A}^{(0)} = \{\mathcal{A}_1^{(0)}, \ldots, \mathcal{A}_u^{(0)}\}$, $\mathcal{A}^{(1)} = \{\mathcal{A}_1^{(1)}, \ldots, \mathcal{A}_u^{(1)}\} \subset [n]$, with $|\mathcal{A}^{(0)}| = |\mathcal{A}^{(1)}| = u$. Bob also chooses random $\mathcal{B}^{(0)} = \{\mathcal{B}_1^{(0)}, \ldots, \mathcal{B}_u^{(0)}\}$, $\mathcal{B}^{(1)} = \{\mathcal{B}_1^{(1)}, \ldots, \mathcal{B}_u^{(1)}\} \subset [n]$, with $|\mathcal{B}^{(0)}| = |\mathcal{B}^{(1)}| = u$.
2. The first public random string $\alpha_0 \stackrel{R}{\longleftarrow} \{0,1\}^n$ is broadcast. Alice stores the u bits $\alpha_0[\mathcal{A}_1^{(0)}], \ldots, \alpha_0[\mathcal{A}_u^{(0)}]$, and Bob stores $\alpha_0[\mathcal{B}_1^{(0)}], \ldots, \alpha_0[\mathcal{B}_u^{(0)}]$.
3. After a short pause, the second public random string $\alpha_1 \stackrel{R}{\longleftarrow} \{0,1\}^n$ is broadcast. Alice stores $\alpha_1[\mathcal{A}_1^{(1)}], \ldots, \alpha_1[\mathcal{A}_u^{(1)}]$, and Bob stores $\alpha_1[\mathcal{B}_1^{(1)}], \ldots, \alpha_1[\mathcal{B}_u^{(1)}]$.
4. Alice sends $\mathcal{A}^{(0)}, \mathcal{A}^{(1)}$ to Bob. Bob flips a coin $c \stackrel{R}{\longleftarrow} \{0,1\}$, and computes $\mathcal{A}^{(c)} \cap \mathcal{B}^{(c)}$. If $|\mathcal{A}^{(c)} \cap \mathcal{B}^{(c)}| < k$, then $\mathcal{R}$ aborts. Otherwise, Bob chooses a random k-subset $s = \left\{\mathcal{A}_{i_1}^{(c)}, \ldots, \mathcal{A}_{i_k}^{(c)}\right\} \subset \mathcal{A}^{(c)} \cap \mathcal{B}^{(c)}$, and sets $I = \{i_1, \ldots, i_k\}$. Thus by Definition 5, $s = \mathcal{A}_I^{(c)}$.
5. Bob encodes I as a t-bit string, where $t = \lg \binom{u}{k}$, and sends I to Alice via the NOVY interactive hashing protocol in $t-1$ rounds. Alice receives two k-subsets $I_0 < I_1 \subset [u]$. For some $b \in \{0,1\}$, $I = I_b$, but Alice does *not* know b. Bob also computes I_0, I_1 by solving the same system of linear equations, and knows b.
6. Bob sends $\varepsilon = b \oplus c$ and $\tau = \delta \oplus c$ to Alice, where c and b are defined in Steps 4 and 5 respectively.
7. Alice sets $s_0 = \mathcal{A}_{I_\varepsilon}^{(0)}$, $X_0 = s_0(\alpha_0)$, $s_1 = \mathcal{A}_{I_{\overline{\varepsilon}}}^{(1)}$, and $X_1 = s_1(\alpha_1)$. Alice then computes $C_0 = X_\tau \oplus M_0$, and $C_1 = X_{\overline{\tau}} \oplus M_1$, and sends C_0, C_1 to Bob.
8. Bob reads $M_\delta = C_\delta \oplus X_c = C_\delta \oplus \bigoplus_{j=1}^k \alpha_c[\mathcal{A}_{i_j}^{(c)}]$. (Note that an honest Bob following the protocol has stored $\alpha_c[\mathcal{A}_{i_j}^{(c)}]$ $\forall\, 1 \le j \le k$. Recall from Step 4 that $\forall\, 1 \le j \le k$, $\mathcal{A}_{i_j}^{(c)} \in s \subset \mathcal{B}^{(c)}$).

Remark: Each of $\mathcal{A}^{(0)}, \mathcal{A}^{(1)}, \mathcal{B}^{(0)}, \mathcal{B}^{(1)}$, as described in Protocol A, consists of u independently chosen elements of $[n]$, resulting in $u \lg n$ bits each. However, as noted in [15], we can reduce the number of bits for describing the sets to $O(k \log n)$, by choosing the elements with $O(k)$-wise independence, without significantly affecting the results.

Lemma 6. *The probability that an honest receiver Bob aborts in Step 4 of the protocol, is at most* $e^{-k/4}$.

Proof. By Corollary 3, $\Pr\left[\left|\mathcal{A}^{(c)} \cap \mathcal{B}^{(c)}\right| < k\right] \; < \; e^{-k/4}$. □

The following two lemmas about Protocol A are immediate.

Lemma 7. *The receiver Bob can read* M_δ *simply by following the protocol.*

Lemma 8. *The sender Alice learns nothing about* δ.

Proof. Because Alice does not learn c (defined in Step 4) and b (defined in Step 5) in Protocol A. □

Therefore, Conditions 1 and 2 for a correct and secure $\binom{2}{1}$-OT, are satisfied. We now come to the most challenging part, namely, Condition 3 regarding the security against a dishonest receiver Bob, who can store $B = n/6$ bits, and whose goal is to learn both M_0 and M_1. While α_0 is broadcast in Step 2, Bob computes an *arbitrary* function $\eta_0 = A_0(\alpha_0)$ using unlimited computing power, provided that $|\eta_0| = n/6$; and while α_1 is broadcast in Step 3, Bob computes an arbitrary function $\eta_1 = A_1(\eta_0, \alpha_1)$, $|\eta_1| = n/6$. In Steps 4 - 6, using η_1 and $\mathcal{A}^{(0)}, \mathcal{A}^{(1)}$, Bob employs an arbitrary strategy in interacting with Alice. At the end of the protocol, Bob attempts to learn both M_0 and M_1, using his information η_1 on (α_0, α_1), C_0, C_1 received from Alice in Step 7, and all information $\mathcal{I}$ he obtains in Steps 4 - 6. Thus in particular, $\mathcal{I}$ includes $\mathcal{A}^{(0)}, \mathcal{A}^{(1)}$ received from Alice in Step 4, and I_0, I_1 obtained in Step 5.

Theorem 1. *For any* $A_0 : \{0,1\}^n \longrightarrow \{0,1\}^{n/6}$ *and* $A_1 : \{0,1\}^{n/6} \times \{0,1\}^n \longrightarrow \{0,1\}^{n/6}$, *for any strategy Bob employs in Steps 4 - 6 of Protocol* A, *with probability at least* $1 - 2^{-O(k)} - 2^{-0.02n+1}$, $\exists\, \beta \in \{0,1\}$ *such that for any distinguisher* $\mathcal{D}$,

$$\left|\Pr\left[\mathcal{D}(\eta_1, \mathcal{I}, X_{\overline{\beta}}, X_\beta) = 1\right] - \Pr\left[\mathcal{D}(\eta_1, \mathcal{I}, X_{\overline{\beta}}, 1 \oplus X_\beta) = 1\right]\right| \; < \; 2^{-k/3}, \quad (8)$$

where $\eta_1 = A_1(\eta_0, \alpha_1)$, $\eta_0 = A_0^{(0)}(\alpha_0)$, $\mathcal{I}$ *denotes all the information Bob obtains in Steps 4 - 6, and* X_0, X_1 *are defined in Step 7 of Protocol* A.

Theorem 1 says that using all the information he has in his bounded storage, Bob is not able to distinguish between $(X_{\overline{\beta}}, X_\beta)$ and $(X_{\overline{\beta}}, 1 \oplus X_\beta)$, for some $\beta \in \{0,1\}$, where X_0, X_1 are defined in Step 7 of Protocol A. From Theorem 1, we obtain:

Theorem 2. *For any* $A_0 : \{0,1\}^n \longrightarrow \{0,1\}^{n/6}$ *and* $A_1 : \{0,1\}^{n/6} \times \{0,1\}^n \longrightarrow \{0,1\}^{n/6}$*, for any strategy Bob employs in Steps 4 - 6 of Protocol* A*, with probability at least* $1 - 2^{-O(k)} - 2^{-0.02n+1}$, $\exists\, \beta \in \{0,1\}$ *such that* $\forall\, M_0, M_1 \in \{0,1\}$, $\forall\, \delta \in \{0,1\}$*, for any distinguisher* $\mathcal{D}$,

$$\begin{aligned} \Big| &\Pr\left[\mathcal{D}(\eta_1, \mathcal{I}, X_{\overline{\beta}} \oplus M_\delta, X_\beta \oplus M_{\overline{\delta}}) = 1\right] \\ &- \Pr\left[\mathcal{D}(\eta_1, \mathcal{I}, X_{\overline{\beta}} \oplus M_\delta, X_\beta \oplus \overline{M_{\overline{\delta}}}) = 1\right]\Big| \;<\; 2^{-k/3}, \end{aligned} \tag{9}$$

where X_0, X_1, η_1 *and* $\mathcal{I}$ *are as above. Therefore, the VIEW of Bob is essentially the same if* $M_{\overline{\delta}}$ *is replaced by* $\overline{M_{\overline{\delta}}} = 1 \oplus M_{\overline{\delta}}$*. Hence, in Protocol* A*, Bob learns essentially nothing about any non-trivial function or relation involving* both M_0 *and* M_1.

Proof. It is clear that (9) follows from (8). Therefore, Theorem 2 follows from Theorem 1. □

4 Proof of Theorem 1

In this section, we consider a *dishonest* receiver Bob, and prove Theorem 1.

We first note that it suffices to prove the theorem in the case that Bob's recording functions A_0, A_1 are deterministic. This does not detract from the generality of our results for the following reason. By definition, a randomized algorithm is an algorithm that uses a random help-string r for computing its output. A randomized algorithm A with each fixed help-string r gives rise to a *deterministic* algorithm A^r. Therefore, that Theorem 1 holds for any deterministic recording algorithm implies that for any randomized recording algorithm A, *for each fixed* help-string r, A using r cannot succeed. Hence, by an averaging argument, A using a randomly chosen r does not help. The reader might notice that the help-string r could be arbitrarily long since Bob has unlimited computing power. In particular, it could be that $|r| > B$, thereby giving rise to a deterministic recording algorithm with length $|A^r| = |A| + |r| > B$. But our model imposes *no restriction* on the *program size* of the recording algorithm. The only restriction is that the length of the output $|A^r(\alpha)| = B$ for each r. In the formal model, A is an unbounded non-uniform Turing Machine whose output tape is bounded by B bits.

We prove a slightly stronger result, namely, Theorem 1 holds even if Bob stores not only η_1, but also η_0, where $\eta_0 = A_0(\alpha_0)$ and $\eta_1 = A_1(\eta_0, \alpha_1)$, A_0, A_1 are Bob's recording functions, and α_0, α_1 are the public random strings used in Steps 2 and 3 of Protocol A. Let

$$\begin{aligned} H_0 &\stackrel{d}{=} A_0^{-1}(\eta_0) = \{\alpha \in \{0,1\}^n : A_0(\alpha) = \eta_0\}\,; \\ H_1 &\stackrel{d}{=} \{\alpha \in \{0,1\}^n : A_1(\eta_0, \alpha) = \eta_1\}\,. \end{aligned}$$

After η_0 and η_1 are computed in Steps 2 and 3 of Protocol A, the receiver Bob can compute H_0 and H_1, using unlimited computing power and space. But given

η_0 and η_1, all Bob knows about (α_0, α_1) is that it is *uniformly random* in $H_0 \times H_1$, i.e. each element of $H_0 \times H_1$ is equally likely to be (α_0, α_1) .

Recall from Definition 4 that $H \subset \{0,1\}^n$ is *fat* if $|H| > 2^{0.813n}$. By Corollary 1 and a union bound, for $\alpha_0, \alpha_1 \xleftarrow{R} \{0,1\}^n$, for any recording functions A_0, A_1,

$$\Pr[\textit{Both } H_0 \text{ and } H_1 \text{ are } \textit{fat}] \;>\; 1 \;-\; 2^{-0.02n+1}. \tag{10}$$

Thus, consider the case that both H_0 and H_1 are fat. By Lemma 1, for any fat $H \subset \{0,1\}^n$,

$$|B_H| \;<\; |\mathcal{K}| \cdot 2^{-k/3} \;=\; \binom{n}{k} \cdot 2^{-k/3}, \tag{11}$$

where B_H is defined in (2), i.e. almost all k-subsets of $[n]$ are *good* for H (See Definition 3 for the definition of goodness). Next, we show that if H is fat, then for a uniformly random $\mathcal{A} \subset [n]$ s.t. $|\mathcal{A}| = u$, with overwhelming probability, almost all k-subsets of $\mathcal{A}$ are *good* for H.

Definition 6. *For $\mathcal{A} \subset [n]$, define $\mathcal{K}_\mathcal{A} \stackrel{d}{=} \{s \subset \mathcal{A} : |s| = k\}$, i.e. $\mathcal{K}_\mathcal{A}$ is the set of all k-subsets of $\mathcal{A}$.*

Definition 7. *For $\mathcal{A} \subset [n]$ and $H \subset \{0,1\}^n$, define*

$$B_H^\mathcal{A} \stackrel{d}{=} \{s \in \mathcal{K}_\mathcal{A} : s \textit{ is } \text{not } \textit{good for } H\}.$$

Lemma 9. *Let $H \subset \{0,1\}^n$ be fat. For a uniformly random $\mathcal{A} \subset [n]$ with $|\mathcal{A}| = u$,*

$$\Pr\left[|B_H^\mathcal{A}| \;<\; |\mathcal{K}_\mathcal{A}| \cdot 2^{-k/6} \;=\; \binom{u}{k} \cdot 2^{-k/6}\right] \;>\; 1 \;-\; 2^{-k/6}.$$

In other words, for almost all $\mathcal{A} \subset [n]$ with $|\mathcal{A}| = u$, almost all k-subsets of $\mathcal{A}$ are good for any fat H.

Proof. Let $\mathcal{U}$ be the set of all the $\binom{n}{u}$ u-subsets of $[n]$. For each $\mathcal{A} \in \mathcal{U}$, let $W_\mathcal{A} \stackrel{d}{=} |B_H^\mathcal{A}|$, i.e. $W_\mathcal{A}$ is the number of k-subsets of $\mathcal{A}$ that are *bad* for H. Let $W \stackrel{d}{=} \sum_{\mathcal{A} \in \mathcal{U}} W_\mathcal{A}$. Since each k-subset of $[n]$ is contained in exactly $\binom{n-k}{u-k}$ u-subsets, in the sum W each bad k-subset of $[n]$ for H, i.e. every element of B_H (defined in (2)), is counted exactly $\binom{n-k}{u-k}$ times. Together with (11), we have

$$W \;=\; \sum_{\mathcal{A} \in \mathcal{U}} W_\mathcal{A} \;=\; |B_H| \cdot \binom{n-k}{u-k} \;<\; \binom{n-k}{u-k}\binom{n}{k} \cdot 2^{-k/3}. \tag{12}$$

Fact 2 *For $k \le u \le n$,*

$$\binom{n}{k}\binom{n-k}{u-k} \;=\; \binom{n}{u}\binom{u}{k}. \tag{13}$$

Therefore, by (12) and (13),

$$W = \sum_{\mathcal{A}\in\mathcal{U}} W_{\mathcal{A}} < \binom{n}{u}\binom{u}{k}\cdot 2^{-k/3}. \quad (14)$$

It follows that there can be at most a $2^{-k/6}$ fraction of u-subsets $\mathcal{A}$ such that $\left|B_H^{\mathcal{A}}\right| \geq \binom{u}{k}\cdot 2^{-k/6}$, for otherwise we would have $W \geq \binom{u}{k}\cdot 2^{-k/6}\cdot\binom{n}{u}\cdot 2^{-k/6} = \binom{n}{u}\binom{u}{k}\cdot 2^{-k/3}$, contradicting (14). The lemma thus follows. □

Again let $\mathcal{A}^{(0)}, \mathcal{A}^{(1)}$ be the random u-subsets of $[n]$ Alice chooses in Step 1 of Protocol A. By (10), Lemma 9 and a union bound, for $\alpha_0, \alpha_1 \xleftarrow{R} \{0,1\}^n$, and uniformly random $\mathcal{A}^{(0)}, \mathcal{A}^{(1)} \subset [n]$ with $|\mathcal{A}^{(0)}| = |\mathcal{A}^{(1)}| = u$, with probability at least $1 - 2^{-k/6+1} - 2^{-0.02n+1}$,

$$\left|B_{H_0}^{\mathcal{A}^{(0)}}\right|, \left|B_{H_1}^{\mathcal{A}^{(1)}}\right| < \binom{u}{k}\cdot 2^{-k/6}. \quad (15)$$

Thus consider the case that both $B_{H_0}^{\mathcal{A}^{(0)}}, B_{H_1}^{\mathcal{A}^{(1)}}$ satisfy (15).

For each $c \in \{0,1\}$, denote $\mathcal{A}^{(c)} = \left\{\mathcal{A}_1^{(c)}, \ldots, \mathcal{A}_u^{(c)}\right\}$. Recall from Definition 5 that for $J = \{j_1, \ldots, j_k\} \subset [u]$, $\mathcal{A}_J^{(c)} \stackrel{d}{=} \left\{\mathcal{A}_{j_1}^{(c)}, \ldots, \mathcal{A}_{j_k}^{(c)}\right\}$. By Definition 6, $\mathcal{A}_J^{(c)} \in \mathcal{K}_{\mathcal{A}^{(c)}}$. Define

$$T_0 \stackrel{d}{=} \left\{J \subset [u] : |J| = k \ \wedge\ \mathcal{A}_J^{(0)} \in B_{H_0}^{\mathcal{A}^{(0)}}\right\},$$
$$T_1 \stackrel{d}{=} \left\{J \subset [u] : |J| = k \ \wedge\ \mathcal{A}_J^{(1)} \in B_{H_1}^{\mathcal{A}^{(1)}}\right\}.$$

Clearly $|T_0| = \left|B_{H_0}^{\mathcal{A}^{(0)}}\right|$, and $|T_1| = \left|B_{H_1}^{\mathcal{A}^{(1)}}\right|$. Thus by (15), we have

$$|T_0|, |T_1| < \binom{u}{k}\cdot 2^{-k/6}. \quad (16)$$

Consider I_0, I_1 defined in Step 5 of Protocol A. Let ε be the first bit Bob sends Alice in Step 6 of Protocol A. Then by (10), (15), (16), and Corollary 4 of Lemma 5 on interactive hashing, for any strategy Bob uses in Steps 4 - 6, with probability at least $1 - 2^{-O(k)} - 2^{-0.02n+1}$, $I_\varepsilon \notin T_0 \ \vee\ I_{\overline{\varepsilon}} \notin T_1$, where $\overline{\varepsilon} = 1 \oplus \varepsilon$. WLOG, say $I_{\overline{\varepsilon}} \notin T_1$. Let $s_0 = \mathcal{A}_{I_\varepsilon}^{(0)}$, $X_0 = s_0(\alpha_0)$, $s_1 = \mathcal{A}_{I_{\overline{\varepsilon}}}^{(1)}$, and $X_1 = s_1(\alpha_1)$, as defined in Step 7 of Protocol A. Since $I_{\overline{\varepsilon}} \notin T_1$, by definition $s_1 \notin B_{H_1}^{\mathcal{A}^{(1)}}$, i.e. s_1 is *good* for H_1. Note again that given η_0 and η_1, and thus H_0 and H_1, all Bob knows about (α_0, α_1) is that (α_0, α_1) is *uniformly random* in $H_0 \times H_1$. Since s_1 is good for H_1, by (1) for the definition of goodness, for $\alpha_1 \xleftarrow{R} H_1$,

$$|\Pr[X_1 = 0] - \Pr[X_1 = 1]| < 2^{-k/3}. \quad (17)$$

For $(\alpha_0, \alpha_1) \xleftarrow{R} H_0 \times H_1$, X_0 and X_1 are *independent*. Thus together with (17), for $(\alpha_0, \alpha_1) \xleftarrow{R} H_0 \times H_1$, for any $b_0 \in \{0,1\}$,

$$|\Pr[X_1 = 0 \mid X_0 = b_0] - \Pr[X_1 = 1 \mid X_0 = b_0]| < 2^{-k/3}. \quad (18)$$

Thus, from (18) and all the above, Theorem 1 follows (with $\beta = 1$).

5 Discussion

Building on the work of Cachin, Crépeau, and Marcil [15], we have given a similar but more efficient protocol for $\binom{2}{1}$-OT in the bounded storage model, and provided a stronger security analysis.

Having proved a stronger result than that of [15], we note that the model of [15] is slightly stronger than ours in the following sense. In [15], the dishonest receiver Bob computes an arbitrary function on *all* public random bits, and stores B bits of output. In our model, α_0 is first broadcast, Bob computes and stores $\eta_0 = A_0(\alpha_0)$, which is a function of α_0. Then α_0 disappears. After a short pause, α_1 is broadcast, and Bob computes and stores $\eta_1 = A_1(\eta_0, \alpha_1)$, which is a function of η_0 and α_1. However, we claim that our model is reasonable, as with limited storage, in practice it is impossible for Bob to compute a function on all of α_0 and α_1, with $|\alpha_0| = |\alpha_1| > B$, that are broadcast one after another, with a pause in between. Furthermore, we believe that by a more detailed analysis, it is possible to show that our results hold even in the stronger model, where Bob computes an arbitrary function $A(\alpha_0, \alpha_1)$ on all bits of α_0 and α_1.

As the CCM Protocol, our protocol employs interactive hashing, resulting in an inordinate number of interactions. Further, the communication complexity of the NOVY protocol is quadratic in the size of the string to be transmitted. It thus remains a most important open problem to make this part of the protocol non-interactive and more communication efficient.

Can the storage requirement of our protocol be further improved? For very large n, $\Omega(\sqrt{kn})$ may not be small enough to be practical. It becomes another important open problem to investigate the feasibility of reducing the storage requirement for OT in the bounded storage model, and establish lower bounds.

We also note that the constant hidden by $O(\cdot)$ in our results is not optimized. We believe that this can be improved by refining the analysis of Lemma 9, as well as the analysis of interactive hashing in [15].

References

1. Y. Aumann and M. O. Rabin. Information Theoretically Secure Communication in the Limited Storage Space Model. In *Advances in Cryptology - CRYPTO '99*, pages 65-79, 1999.
2. Y. Aumann, Y. Z. Ding, and M. O. Rabin. Everlasting Security in the Bounded Storage Model. Accepted to *IEEE Transactions on Information Theory*, 2001.
3. Y. Aumann and U. Feige. One message proof systems with known space verifier. In *Advances in Cryptology - CRYPTO '93*, pages 85-99, 1993.
4. Y. Aumann and M. O. Rabin. Clock Construction in Fully Asynchronous Parallel Systems and PRAM Simulation. *TCS*, 128(1):3-30, 1994.
5. D. Beaver. Commoditiy-Based Cryptography. In *Proc. 29th ACM Symposium on Theory of Computing*, pages 446-455, 1997.
6. M. Bellare and S. Micali. Non-interactive oblivious transfer and applications. In *Advances in Cryptology - CRYPTO '89*, pages 200-215, 1989.
7. C. H. Bennett, G. Brassard, C. Crépeau, and M.H. Skubiszewska. Practical quantum oblivious transfer protocols. In *Advances in Cryptology - CRYPTO '91*, pages 351-366, 1991.

8. M. Blum. How to exchange (secret) keys. *ACM Transactions of Computer Systems*, 1(20): 175-193, 1983.
9. G. Brassard and C. Crépeau. Oblivious transfers and privacy amplification. In *Advances in Cryptology - EUROCRYPT '97*, pages 334-347, 1997.
10. G. Brassard, C. Crépeau, and J-M. Roberts. Information theoretic reductions among disclosure problems. In *Proc. 27th IEEE Symposium on the Foundations of Computer Science*, pages 168-173, 1986.
11. G. Brassard, C. Crépeau, and J-M. Roberts. All-or-nothing disclosure of secrets. In *Advances in Cryptology - CRYPTO '86*, pages 234-238, 1986.
12. G. Brassard, C. Crépeau, and M. Sántha. Oblivious transfers and intersecting codes. *IEEE Transactions on Information Theory*, 42(6): 1769-80, 1996.
13. C. Cachin. *Entropy Measures and Unconditional Security in Cryptography*, volume 1 of *ETH Series in Information Security and Cryptography.* Hartun-Gorre Verlag, Konstanz, Germany, 1997.
14. C. Cachin. On the foundations of oblivious transfer. In *Advances in Cryptology - EUROCRYPT '98*, pages 361-374, 1998.
15. C. Cachin, C. Crépeau, and J. Marcil. Oblivious transfer with a memory-bounded receiver. In *Proc. 39th IEEE Symposium on Foundations of Computer Science*, pages 493-502, 1998.
16. C. Cachin and U. Maurer. Unconditional security against memory bounded adversaries. In *Advances in Cryptology - CRYPTO '97*, pages 292-306, 1997.
17. A. Condon. Bounded Space Probabilistic Games. *JACM*, 38(2):472-494, 1991.
18. A. Condon, and R. Ladner. Probabilistic Game Automata. *JCSS*, 36(3):452-489, 1987.
19. C. Crépeau. Equivalence between two flavours of oblivious transfer. In *Advances in Cryptology - CRYPTO '87*, pages 351-368, 1987.
20. C. Crépeau and J. Kilian. Achieving oblivious transfer using weakened security assumptions. In *Proc. 29th IEEE Symposium on the Foundations of Computer Science*, 42-52, 1988.
21. C. Crépeau and M. Sántha. On the reversibility of oblivious transfer. In *Advances in Cryptology - EUROCRYPT '91*, pages 106-113, 1991.
22. C. Crépeau, J. van de Graff, and A. Tapp. Committed oblivious transfer and private multi-party computations. In *Advances in Cryptology - CRYPTO '95*, pages 110-123, 1995.
23. A. De-Santis, G. Persiano, and M. Yung. One-message statistical zero-knowledge proofs with space-bounded verifier. In *Proc. 19th ICALP*, pages 28-40, 1992.
24. Y. Z. Ding and M. O. Rabin. Provably Secure and Non-Malleable Encryption. To appear, 2001.
25. Y. Dodis and S. Micali. Lower bounds for oblivious transfer reductions. In *Advances in Cryptology - EUROCRYPT '99*, pages 42-55, 1999.
26. C. Dwork and L. J. Stockmeyer. Finite State Verifiers I: The Power of Interaction. *JACM* 39(4): 800-828, 1992
27. C. Dwork and L. J. Stockmeyer. Finite State Verifiers II: Zero Knowledge. *JACM* 39(4): 829-858, 1992.
28. S. Even, O. Goldreich, and A. Lempel. A randomized protocol for signing contracts. In *Advances in Cryptology - CRYPTO '82*, pages 205-210, 1982.
29. J. A. Garay and P. Mackenzie. Concurrent Oblivious Transfer. In *Proc. 41th IEEE Symposium on the Foundations of Computer Science*, pages 314-324, 2000.
30. O. Goldreich, S. Micali, and A. Wigderson. How to play any mental game or a completeness theorem for protocols with honest majority. In *Proc. 19th ACM Symposium on Theory of Computing*, pages 218-229, 1987.

31. O. Goldreich and R. Vainish. How to solve any protocol problem - an efficiency improvement. In *Advances in Cryptology - CRYPTO '87*, pages 73-86, 1987.
32. S. Goldwasser and L. Levin. Fair Computation of General Functions in Presence of Immoral Majority. In *Advances in Cryptology - CRYPTO '90*, pages 77-93, 1990.
33. J. Kilian. Founding cryptography on oblivious transfer. In *Proc. 20th ACM Symposium on Theory of Computing*, pages 20-31, 1988.
34. J. Kilian. Zero-knowledge with Log-Space Verifiers. In *Proc. 29th IEEE Symposium on the Foundations of Computer Science*, pages 25-35, 1988.
35. J. Kilian. A general completeness theorem for two-party games. In *Proc. 23th ACM Symposium on Theory of Computing*, pages 553-560, 1991.
36. J. Kilian, E. Kushilevitz, S. Micali, and R. Ostrovsky. Reducibility and completeness in private computations. *SIAM Journal on Computing*, 29(4): 1189-1208, 2000.
37. U. Maurer. Conditionally-perfect secrecy and a provably-secure randomized cipher. *Journal of Cryptology*, 5(1):53-66, 1992.
38. U. Maurer. Secret key agreement by public discussion from common information. *IEEE Transactions on Information Theory*, 39(3):733-742, 1993.
39. U. Maurer and S. Wolf. Toward characterizing when information-theoretic secret key agreement is possible. In *Advances in Cryptology - ASIACRYPT'96*, pages 196-209, 1996.
40. U. Maurer. Information-theoretically secure secret-key agreement by NOT authenticated public discussion. *Advances in Cryptology - EUROCRYPT '97*, pages 209-225, 1997.
41. U. Maurer and S. Wolf. Unconditional secure key agreement and the intrinsic conditional information. *IEEE Transaction on Information Theory*, 45(2): 499-514, 1999.
42. U. Maurer and S. Wolf. Information-Theoretic Key Agreement: From Weak to Strong Secrecy for Free. In *Advances in Cryptology - EUROCRYPT '00*, pages 351-368, 2000.
43. M. Naor, R. Ostrovsky, R. Venkatesan, and M. Yung. Perfect zero-knowledge arguments for NP using any one-way function. *Journal of Cryptology*, 11(2): 87-108, 1998.
44. M. Naor and B. Pinkas. Distributed Oblivious Transfer. In *Advances in Cryptology - ASIACRYPT '00*, pages 205-219, 2000.
45. N. Nisan. Pseudorandom generators for space-bounded computation. In *Proc. 22rd ACM Symposium on Theory of Computing*, pages 204-212, 1990.
46. N. Nisan and D. Zuckerman. Randomness is linear in space. *JCSS* 52(1): 43-52, 1996.
47. M. O. Rabin. How to exchange secrets by oblivious transfer. Technical Report TR-81, Harvard University, 1981.
48. M. O. Rabin. Transaction Protection by Beacons. *JCSS* 27(2): 256-267, 1983.
49. R. Rivest. Unconditionally Secure Commitment and Oblivious Transfer Schemes Using Private Channels and a Trusted Initializer. Manuscript, 1999.
50. P. W. Shor. Polynomial-time algorithms for prime factorization and discrete logarithms on a quantum computer. *SIAM J. Computing*, 26(5): 1484-1509, 1997.
51. L. G. Valiant and V. V. Vazirani. NP is as easy as detecting unique solutions. In *Proc. ACM Symposium on Theory of Computing*, pages 458-463, 1985.
52. A. C. Yao. How to generate and exchange secrets. In *Proc. 27th IEEE Symposium on the Foundations of Computer Science*, pages 162-167, 1986.

Appendix A: Proof of Lemma 1

Definition 8. *Let $s = (\sigma_1, \ldots, \sigma_k) \in [n]^k$. For $\alpha \in \{0,1\}^n$, define $s(\alpha)$ as in Definition 2, i.e. $s(\alpha) \stackrel{d}{=} \bigoplus_{i=1}^{k} \alpha[\sigma_i]$.*

Definition 9. *Let $s \in [n]^k$. Let $H \subset [n]$. Define the goodness of s for H as in Definition 3, i.e. s is good for H if (1) holds.*

The following main lemma is proved in [24].

Main Lemma 1 [24] *Let $H \subset \{0,1\}^n$. Denote*

$$\hat{B}_H \stackrel{d}{=} \left\{ s \in [n]^k : s \textit{ is } \text{not } \textit{good for } H \right\}. \tag{19}$$

If H is fat, *then*

$$|\hat{B}_H| < n^k \cdot 2^{-k/3-1}. \tag{20}$$

We now prove Lemma 1 from Main Lemma 1. Let $\tilde{B}_H \subset \hat{B}_H$ be the subset of bad k-tuples with k *distinct* coordinates, i.e.

$$\tilde{B}_H \stackrel{d}{=} \left\{ s = (\sigma_1, \ldots, \sigma_k) \in \hat{B}_H : \ \sigma_i \neq \sigma_j \ \forall \ i \neq j \right\}. \tag{21}$$

Then clearly

$$|\tilde{B}_H| = |B_H| \cdot k!, \tag{22}$$

where B_H is defined in (2). By way of contradiction, suppose that Lemma 1 does not hold, i.e.

$$|B_H| \geq \binom{n}{k} \cdot 2^{-k/3}. \tag{23}$$

Then by (22) and (23), and the fact that $\tilde{B}_H \subset \hat{B}_H$, we have

$$|\hat{B}_H| \geq |\tilde{B}_H| = |B_H| \cdot k! \geq \binom{n}{k} \cdot k! \cdot 2^{-k/3}. \tag{24}$$

Observe that

$$\binom{n}{k} \cdot k! = n^k \cdot \left(1 - \frac{1}{n}\right) \cdots \left(1 - \frac{k-1}{n}\right) > n^k \cdot \left(1 - \frac{\sum_{i=1}^{k-1} i}{n}\right)$$
$$> n^k \cdot \left(1 - \frac{k^2}{2n}\right) > \frac{n^k}{2} \quad \text{for } k < \sqrt{n}. \tag{25}$$

Therefore, if Lemma 1 does not hold, i.e. if (23) holds, then by (24) and (25),

$$|\hat{B}_H| > n^k \cdot 2^{-k/3-1}, \tag{26}$$

contradicting (20). Thus, Lemma 1 must hold.

Parallel Coin-Tossing and Constant-Round Secure Two-Party Computation

Yehuda Lindell

Department of Computer Science and Applied Math,
Weizmann Institute of Science, Rehovot, ISRAEL.
lindell@wisdom.weizmann.ac.il

Abstract. In this paper we show that any *two-party* functionality can be securely computed in a *constant number of rounds*, where security is obtained against malicious adversaries that may arbitrarily deviate from the protocol specification. This is in contrast to Yao's constant-round protocol that ensures security only in the face of semi-honest adversaries, and to its malicious adversary version that requires a polynomial number of rounds.

In order to obtain our result, we present a constant-round protocol for secure coin-tossing of polynomially many coins (in parallel). We then show how this protocol can be used in conjunction with other existing constructions in order to obtain a constant-round protocol for securely computing any two-party functionality. On the subject of coin-tossing, we also present a constant-round *perfect* coin-tossing protocol, where by "perfect" we mean that the resulting coins are guaranteed to be statistically close to uniform (and not just pseudorandom).

1 Introduction

1.1 Secure Two-Party Computation

In the setting of two-party computation, two parties, with respective private inputs x and y, wish to jointly compute a functionality $f(x,y) = (f_1(x,y), f_2(x,y))$, such that the first party receives $f_1(x,y)$ and the second party receives $f_2(x,y)$. This functionality may be probabilistic, in which case $f(x,y)$ is a random variable. Loosely speaking, the security requirements are that nothing is learned from the protocol other than the output (*privacy*), and that the output is distributed according to the prescribed functionality (*correctness*). The actual definition [14,1,5] blends these two conditions (see Section 2). This must be guaranteed even when one of the parties is adversarial. Such an adversary may be *semi-honest* in which case it correctly follows the protocol specification, yet attempts to learn additional information by analyzing the transcript of messages received during the execution. On the other hand, an adversary may be *malicious*, in which case it can arbitrarily deviate from the protocol specification.

The first general solutions for the problem of secure computation were presented by Yao [17] for the two-party case (with security against semi-honest adversaries) and Goldreich, Micali and Wigderson [13] for the multi-party case

J. Kilian (Ed.): CRYPTO 2001, LNCS 2139, pp. 171–189, 2001.

(with security even against malicious adversaries). Thus, despite the stringent security requirements placed on such protocols, wide-ranging completeness results were established, demonstrating that *any* probabilistic polynomial-time functionality can be securely computed (assuming the existence of trapdoor permutations).

Yao's protocol. In [17], Yao presented a *constant-round* protocol for securely computing any functionality, where the adversary may be semi-honest. Denote Party 1 and Party 2's respective inputs by x and y and let f be the functionality that they wish to compute (for simplicity, assume that both parties wish to receive $f(x, y)$). Loosely speaking, Yao's protocol works by having one of the parties (say Party 1) first generate an "encrypted" circuit computing $f(x, \cdot)$ and send it to Party 2. The circuit is such that it reveals nothing in its encrypted form and therefore Party 2 learns nothing from this stage. However, Party 2 can obtain the output $f(x, y)$ by "decrypting" the circuit. In order to ensure that nothing is learned beyond the output itself, this decryption must be "partial" and must reveal $f(x, y)$ only. Without going into unnecessary details, this is accomplished by Party 2 obtaining a series of keys corresponding to its input y such that given these keys and the circuit, the output value $f(x, y)$ (and only this value) may be obtained. Of course, Party 2 must obtain these keys without revealing anything about y and this can be done by running $|y|$ instances of a (semi-honest) secure 2-out-of-1 Oblivious Transfer protocol [7], which is constant-round. By running the Oblivious Transfer protocols in parallel, this protocol requires only a constant number of rounds.

Now consider what happens if Yao's protocol is run when the adversary may be malicious. Firstly, we have no guarantee that Party 1 constructed the circuit so that it correctly computes $f(x, \cdot)$. Thus, *correctness* may be violated (intuitively, this can be solved using zero-knowledge proofs). Secondly, the Oblivious Transfer protocol must satisfy the requirements for secure computation (in the face of malicious adversaries), and must maintain its security when run in parallel. We note that we know of no such (highly secure) oblivious transfer protocol that runs in a constant number of rounds. Finally, if the functionality f is *probabilistic*, then Party 1 must be forced to input a truly random string into the circuit. Thus, some type of coin-tossing protocol is also required.

Secure protocol compilation. As we have mentioned, Goldreich, Micali and Wigderson [12,13] showed that assuming the existence of trapdoor permutations, there exist protocols for securely computing any multi-party functionality, where the adversary may be malicious. They achieve this in two stages. First, they show a protocol for securely computing any functionality in the semi-honest adversarial model. Next, they construct a *protocol compiler* that takes any semi-honest protocol and "converts" it into a protocol that is secure in the malicious model. As this compiler is generic, it can be applied to *any* semi-honest protocol and in particular, to the constant-round two-party protocol of Yao. However, due to the nature of their compilation, the output protocol is no longer constant-round.

1.2 Our Results

The focus of this paper is to construct a protocol compiler such that the round-complexity of the compiled protocol is of the same order as that of the original protocol. We observe that the only component of the GMW compiler for which there does *not* exist a constant-round construction is that of coin-tossing in the well [3]. Therefore, our technical contribution is in constructing a constant-round protocol for coin-tossing in the well *of polynomially many coins.* That is, we obtain the following theorem (informally stated):

Theorem 1 (constant-round coin-tossing): *Assuming the existence of one-way functions, there exists a* constant-round *protocol for the coin-tossing functionality (as required by the* GMW *compiler).*

In order to construct such a constant-round protocol we introduce a technique relating to the use of commitment schemes, which we believe may be useful in other settings as well. Commitment schemes are a basic building block and are used in the construction of many protocols. Consider, for example, Blum's protocol for coin-tossing a single bit [3]. In this protocol, Party 1 sends a commitment to a random-bit; then, Party 2 replies with its own random bit and finally Party 1 decommits. The difficulty in simulating such protocols is that the simulator only knows the correct value to commit to *after* the other party sends its message. However, since the simulator is bound to its commitment, it must somehow guess the correct value *before* this message is sent. In case the messages are long (say n bits rather than a single bit or $\log n$ bits), this may be problematic. Thus, rather than decommitting, we propose to have the party reveal the committed value and then prove (in zero-knowledge) the validity of this revealed value. In a real execution, this is equivalent to decommitting, since the committing party is effectively bound to the committed value by the zero-knowledge proof. However, the simulator is able to provide a *simulated* zero-knowledge proof (rather than a real one). Furthermore, this proof remains indistinguishable from a real proof even if the revealed value is incorrect (and thus the statement is false). Therefore, the simulator can effectively "decommit" to any value it wishes and is not bound in any way by the original commitment that it sends.

Combining the constant-round protocol of Theorem 1 with other known constructions, we obtain the following theorem:

Theorem 2 *Assume the existence of one-way functions. Then, there exists a protocol compiler that given a two-party protocol Π for securely computing f in the semi-honest model produces a two-party protocol Π' that securely computes f in the malicious model, so that the number of rounds of communication in Π' is within a* constant factor *of the number of rounds of communication in Π.*

We stress that, when ignoring the "round preservation" clause, the existence of a protocol compiler is not new and has been shown in [12,13] (in fact, as we have mentioned, we use most of the components of their compiler). Our contribution is in reducing the overhead of the compiler, in terms of the round-complexity, to a constant. The main result, stated in the following theorem, is obtained by applying the compiler of Theorem 2 to the constant-round protocol of Yao.

Theorem 3 *Assuming the existence of trapdoor permutations, any two-party functionality can be securely computed in the malicious model in a* constant number of rounds.

On the subject of coin-tossing, we also present a constant-round protocol for "perfect" coin-tossing (of polynomially many coins) that guarantees that the output of the coin-tossing protocol is statistically close to uniform, and not just computationally indistinguishable.

1.3 Related Work

In the setting of multi-party computation with an honest majority, Beaver, Micali and Rogaway [2] showed that any functionality can be securely computed in a constant number of rounds, where the adversary may be malicious. Unfortunately, their technique relies heavily on the fact that a majority of the parties are honest and as such cannot be applied to the case of two-party protocols. As we have described, in this paper we establish the analogous result for the setting of *two*-party computation.

1.4 Organization

In Section 2 we present the definition of secure two-party computation. Then, in Section 3 we discuss the protocol compiler of GMW and observe that in order to achieve "round-preserving" compilation, one needs only to construct a constant-round coin-tossing protocol. Our technical contribution in this paper thus begins in Section 4 where we present such a constant-round coin-tossing protocol. Finally, in Section 5 we show how perfect coin-tossing can be achieved.

2 Definitions – Secure Computation

In this section we present the definition of secure two-party computation. Our presentation is based on [9], which in turn follows [14,1,5]. We first introduce the following notation: U_n denotes the uniform distribution over $\{0,1\}^n$; for a set S we denote $s \in_R S$ when s is chosen uniformly from S; finally, computational indistinguishability is denoted by $\stackrel{c}{\equiv}$ and statistical closeness by $\stackrel{s}{\equiv}$.

Two-party computation. A two-party protocol problem is cast by specifying a random process that maps pairs of inputs to pairs of outputs (one for each party). We refer to such a process as a functionality and denote it $f : \{0,1\}^* \times \{0,1\}^* \rightarrow \{0,1\}^* \times \{0,1\}^*$, where $f = (f_1, f_2)$. That is, for every pair of inputs (x, y), the output-pair is a random variable $(f_1(x, y), f_2(x, y))$ ranging over pairs of strings. The first party (with input x) wishes to obtain $f_1(x, y)$ and the second party (with input y) wishes to obtain $f_2(x, y)$. We often denote such a functionality by $(x, y) \mapsto (f_1(x, y), f_2(x, y))$. Thus, for example, the basic coin-tossing functionality is denoted by $(1^n, 1^n) \mapsto (U_n, U_n)$.

Adversarial behavior. Loosely speaking, the aim of a secure two-party protocol is to protect an honest party against dishonest behavior by the other party. This "dishonest behavior" can manifest itself in a number of ways; in particular, we focus on what are known as *semi-honest* and *malicious* adversaries. A semi-honest adversary follows the prescribed protocol, yet attempts to learn more information than "allowed" from the execution. Specifically, the adversary may record the entire message transcript of the execution and attempt to learn something beyond the protocol output. On the other hand, a malicious adversary may arbitrarily deviate from the specified protocol. When considering malicious adversaries, there are certain undesirable actions that cannot be prevented. Specifically, a party may refuse to participate in the protocol, may substitute its local input (and enter with a different input) and may abort the protocol prematurely.

Security of protocols (informal). The security of a protocol is analyzed by comparing what an adversary can do in the protocol to what it can do in an ideal scenario that is secure by definition. This is formalized by considering an *ideal* computation involving an incorruptible *trusted third party* to whom the parties send their inputs. The trusted party computes the functionality on the inputs and returns to each party its respective output. Loosely speaking, a protocol is secure if any adversary interacting in the real protocol (where no trusted third party exists) can do no more harm than if it was involved in the above-described ideal computation.

Execution in the ideal model. The ideal model differs for semi-honest and malicious parties. First, for semi-honest parties, an ideal execution involves each party sending their respective input to the trusted party and receiving back their prescribed output. An honest party then outputs this output, whereas a semi-honest party may output an arbitrary (probabilistic polynomial-time computable) function of its initial input and the message it obtained from the trusted party. (See [9] for a formal definition of the ideal and real models for the case of semi-honest adversaries.)

We now turn to the ideal model for malicious parties. Since some malicious behavior cannot be prevented (for example, early aborting), the definition of the ideal model in this case is somewhat more involved. An ideal execution proceeds as follows:

Inputs: Each party obtains an input, denoted z.

Send inputs to trusted party: An honest party always sends z to the trusted party. A malicious party may, depending on z, either abort or sends some $z' \in \{0,1\}^{|z|}$ to the trusted party.

Trusted party answers first party: In case it has obtained an input pair, (x, y), the trusted party (for computing f), first replies to the first party with $f_1(x, y)$. Otherwise (i.e., in case it receives only one input), the trusted party replies to both parties with a special symbol, $\perp$.

Trusted party answers second party: In case the first party is malicious it may, depending on its input and the trusted party's answer, decide to *stop* the trusted party. In this case the trusted party sends $\perp$ to the second party.

Otherwise (i.e., if not stopped), the trusted party sends $f_2(x,y)$ to the second party.

Outputs: An honest party always outputs the message it has obtained from the trusted party. A malicious party may output an arbitrary (probabilistic polynomial-time computable) function of its initial input and the message obtained from the trusted party.

Let $f : \{0,1\}^* \times \{0,1\}^* \mapsto \{0,1\}^* \times \{0,1\}^*$ be a functionality, where $f = (f_1, f_2)$, and let $\overline{M} = (M_1, M_2)$ be a pair of families of non-uniform probabilistic *expected polynomial-time* machines (representing parties in the ideal model). Such a pair is admissible if for at least one $i \in \{1,2\}$ we have that M_i is honest. Then, the joint execution of f under $\overline{M}$ in the ideal model (on input pair (x,y)), denoted $\text{ideal}_{f,\overline{M}}(x,y)$, is defined as the output pair of M_1 and M_2 from the above ideal execution. For example, in the case that M_1 is malicious and always aborts at the outset, the joint execution is defined as $(M_1(x,\perp),\perp)$. Whereas, in case M_1 never aborts, the joint execution is defined as $(M_1(x, f_1(x',y)), f_2(x',y))$ where $x' = M_1(x)$ is the input that M_1 gives to the trusted party.

Execution in the real model. We next consider the real model in which a real (two-party) protocol is executed (and there exists no trusted third party). In this case, a malicious party may follow an arbitrary feasible strategy; that is, any strategy implementable by non-uniform expected polynomial-time machines. In particular, the malicious party may abort the execution at any point in time (and when this happens prematurely, the other party is left with no output).

Let f be as above and let Π be a two-party protocol for computing f. Furthermore, let $\overline{M} = (M_1, M_2)$ be a pair of families of non-uniform probabilistic *expected polynomial-time* machines (representing parties in the real model). Such a pair is admissible if for at least one $i \in \{1,2\}$ we have that M_i is honest (i.e., follows the strategy specified by Π). Then, the joint execution of Π under $\overline{M}$ in the real model (on input pair (x,y)), denoted $\text{real}_{\Pi,\overline{M}}(x,y)$, is defined as the output pair of M_1 and M_2 resulting from the protocol interaction.

Security as emulation of a real execution in the ideal model. Having defined the ideal and real models, we can now define security of protocols. Loosely speaking, the definition asserts that a secure two-party protocol (in the real model) emulates the ideal model (in which a trusted party exists). This is formulated by saying that admissible pairs in the ideal model are able to simulate admissible pairs in an execution of a secure real-model protocol.

Definition 4 (security in the malicious model): *Let f and Π be as above. Protocol Π is said to* securely compute f (in the malicious model) *if there exists a probabilistic polynomial-time computable transformation of pairs of admissible families of non-uniform probabilistic expected polynomial-time machines $\overline{A} = (A_1, A_2)$ for the real model into pairs of admissible families of non-uniform probabilistic expected polynomial-time machines $\overline{B} = (B_1, B_2)$ for the ideal model such that*

$$\{\text{ideal}_{f,\overline{B}}(x,y)\}_{x,y \text{ s.t. } |x|=|y|} \stackrel{c}{\equiv} \{\text{real}_{\Pi,\overline{A}}(x,y)\}_{x,y \text{ s.t. } |x|=|y|}$$

Remark: The above definition is different from the standard definition in that the adversary (in both the ideal and real models) is allowed to run in *expected* polynomial-time (rather than *strict* polynomial-time). This seems to be inevitable given that currently known constant-round zero-knowledge proofs require *expected* polynomial-time simulation. We stress that an honest party always runs in *strict* polynomial time.

3 Two-Party Computation Secure Against Malicious Adversaries

3.1 The Compiler of Goldreich, Micali, and Wigderson [13]

Goldreich, Micali and Wigderson [13] showed that assuming the existence of trapdoor permutations, there are secure protocols (in the malicious model) for any multi-party functionality. Their methodology works by first presenting a protocol secure against semi-honest adversaries. Next, a *compiler* is applied that transforms *any* protocol secure against semi-honest adversaries into a protocol secure against malicious adversaries. Thus, their compiler can also be applied to the constant-round two-party protocol of Yao [17] (as it is secure against semi-honest adversaries). However, as we shall see, the output protocol itself is *not* constant-round. In this section, we describe the [13] compiler and show what should be modified in order to obtain a *constant-round* compiler instead.

Enforcing semi-honest behavior. The GMW compiler takes for input a protocol secure against semi-honest adversaries; from here on we refer to this as the "basic protocol". Recall that this protocol is secure in the case that each party follows the protocol specification exactly, using its input and uniformly chosen random tape. Thus, in order to obtain a protocol secure against malicious adversaries, we need to enforce potentially malicious parties to behave in a semi-honest manner. First and foremost, this involves forcing the parties to follow the prescribed protocol. However, this only makes sense relative to a *given* input and random tape. Furthermore, a malicious party must be forced into using a *uniformly chosen* random tape. This is because the security of the basic protocol may depend on the fact that the party has no freedom in setting its own randomness.

An informal description of the GMW compiler. In light of the above discussion, the compiler begins by having each party commit to its input. Next, the parties run a coin-tossing protocol in order to fix their random tapes (clearly, this protocol must be secure against malicious adversaries). A regular coin-tossing protocol in which both parties receive the same uniformly distributed string does not help us here. This is because the parties' random tapes must remain secret. This is solved by augmenting the coin-tossing protocol so that one party receives a uniformly distributed string (to be used as its random tape) and the other party receives a commitment to that string. Now, following these two steps, each party holds its own uniformly distributed random-tape and a commitment to the other party's input and random-tape. Therefore, each party can be "forced" into working consistently with this specific input and random-tape.

We now describe how this behavior is enforced. A protocol specification is a deterministic function of a party's view consisting of its input, random tape and messages received so far. As we have seen, each party holds a commitment to the input and random tape of the other party. Furthermore, the messages sent so far are public. Therefore, the assertion that a new message is computed according to the protocol is of the $\mathcal{NP}$ type (and the party sending the message knows an adequate $\mathcal{NP}$-witness to it). Thus, the parties can use zero-knowledge proofs to show that their steps are indeed according to the protocol specification. As the proofs used are zero-knowledge, they reveal nothing. On the other hand, due to the soundness of the proofs, even a malicious adversary cannot deviate from the protocol specification without being detected. We thus obtain a reduction of the security in the malicious case to the given security of the basic protocol against semi-honest adversaries.

In summary, the components of the compiler are as follows (from here on "secure" refers to security against malicious adversaries):

1. **Input Commitment:** In this phase the parties execute a secure protocol for the following functionality:
$$((x, r), 1^n) \mapsto (\lambda, C(x; r))$$
where x is the party's input string (and r is the randomness chosen by the committing party).
A secure protocol for this functionality involves the committing party sending $C(x; r)$ to the other party followed by a zero-knowledge proof of knowledge of (x, r). Note that this functionality ensures that the committing party knows the value being committed to.
2. **Coin Generation:** The parties generate t-bit long random tapes (and corresponding commitments) by executing a secure protocol in which one party receives a commitment to a uniform string of length t and the other party receives the string itself (to be used as its random tape) and the decommitment (to be used later for proving "proper behavior"). That is, the parties compute the functionality:
$$(1^n, 1^n) \mapsto ((U_t, U_{t \cdot n}), C(U_t; U_{t \cdot n}))$$
(where we assume that to commit to a t-bit string, C requires $t \cdot n$ random bits).
3. **Protocol Emulation:** In this phase, the parties run the basic protocol whilst proving (in zero-knowledge) that their steps are consistent with their input string, random tape and prior messages received.

A detailed description of each phase of the compiler and a full proof that the resulting protocol is indeed secure against malicious adversaries can be found in [9].

3.2 Achieving Round-Preserving Compilation

As we have mentioned, our aim in this work is to show that the GMW compiler can be implemented so that the number of rounds in the resulting compiled protocol is within a *constant factor* of the number of rounds in the original semi-honest protocol. We begin by noting that using currently known constructions, Phases 1 and 3 of the GMW compiler can be implemented in a constant number of rounds. That is,

Proposition 5 *Assuming the existence of one-way functions, both the* input-commitment *and* protocol-emulation *phases can be securely implemented in a* constant number of rounds.

First consider the input-commitment phase. As mentioned above, this phase can be securely implemented by having the committing party send a perfectly binding commitment of its input to the other party, followed by a zero-knowledge proof of knowledge of the committed value. Both constant-round commitment schemes and constant-round zero-knowledge arguments of knowledge are known to exist by the works of Naor [15] and Feige and Shamir [8], respectively (these constructions can also be based on any one-way function). Thus the input-commitment phase can be implemented as required for Proposition 5.[1] Next, we recall that a secure implementation of the protocol emulation phase requires zero-knowledge proofs for $\mathcal{NP}$ only. Thus, once again, using the argument system of [8], this can be implemented in a constant number of rounds (using any one-way function).

Constant-round coin tossing. In contrast to the input-commitment and protocol-emulation phases of the GMW compiler, known protocols for tossing polynomially many coins do not run in a constant number of rounds. Rather, single coins are tossed sequentially (and thus $poly(n)$ rounds are needed). In particular, the proof of [9] does *not* extend to the case that many coins are tossed in parallel. Thus, in order to obtain a round-preserving compiler, it remains to present a secure protocol for the coin-generation functionality that works in a constant number of rounds. Furthermore, it is preferable to base this protocol on the existence of one-way functions only (so that this seemingly minimal assumption is all that is needed for the entire compiler). In the next section we present such a coin-tossing protocol, thereby obtaining Theorem 2 (as stated in the introduction).

3.3 Constant-Round Secure Computation

Recall that by Yao [17], assuming the existence of trapdoor permutations, any two-party functionality can be securely computed in the *semi-honest* model in a constant-number of rounds. Thus, applying the constant-round compiler of Theorem 2 to Yao's protocol, we obtain a constant-round protocol that is secure

[1] We note that the protocol for the commit-functionality, as described in [9], is for a single-bit only (and thus the compiler there runs this protocol sequentially for each bit of the input). However, the proof for the commit-functionality remains almost identical when the functionality is extended to commitments of $poly(n)$-bit strings (rather than for just a single-bit).

in the *malicious* model, and prove Theorem 3. That is, assuming the existence of trapdoor permutations, any two-party functionality can be securely computed in the *malicious* model in a *constant-number of rounds.*

4 The Augmented Coin-Tossing Protocol

4.1 The Augmented Coin-Tossing Functionality

In this section we present our coin-tossing protocol, thus proving Theorem 1. In a basic coin-tossing functionality, both parties receive identical uniformly distributed strings. That is, the functionality is defined as: $(1^n, 1^n) \mapsto (U_m, U_m)$ for some $m = poly(n)$. This basic coin-tossing is augmented as follows. Let F be any deterministic function. Then, define the augmented coin-tossing functionality by:

$$(1^n, 1^n) \mapsto (U_m, F(U_m))$$

That is, the first party indeed receives a uniformly distributed string. However, the second party receives F applied to that string (rather than the string itself). Setting F to the *identity function*, we obtain basic coin-tossing. However, recall that the coin-generation component of the GMW compiler requires the following functionality:

$$(1^n, 1^n) \mapsto ((U_t, U_{t \cdot n}), C(U_t; U_{t \cdot n}))$$

where C is a commitment scheme (and we assume that C requires n random bits for every bit committed to). Then, this functionality can be realized with our augmentation by setting $m = t + t \cdot n$ and $F(U_m) = C(U_t; U_{t \cdot n})$. Thus, the second party receives a commitment to a uniformly distributed string of length t and the first party receives the string and its decommitment. Recall that in the compiler, the party uses the t-bit string as its random tape and the decommitment in order to prove in zero-knowledge that it is acting consistently with this random tape (and its input).

4.2 Motivating Discussion

In order to motivate our construction of a constant-round coin-tossing protocol, we consider the special case of basic coin-tossing (i.e., where F is the identity function). A natural attempt at a coin-tossing protocol follows:

Protocol 1 (Attempt at Basic Coin-Tossing):

1. *Party 1 chooses a random string* $s_1 \in_R \{0,1\}^m$ *and sends* $c = \text{Commit}(s_1) = C(s_1; r)$ *(where* r *is randomly chosen).*
2. *Party 2 chooses a random string* $s_2 \in_R \{0,1\}^m$ *and sends it to Party 1.*
3. *Party 1 decommits to* s_1 *sending the pair* (s_1, r)*.*

Party 1 always outputs $s \stackrel{\text{def}}{=} s_1 \oplus s_2$*, whereas Party 2 outputs* $s_1 \oplus s_2$ *if Party 1's decommitment is correct and* $\perp$ *otherwise.*

We note that when $m = 1$ (i.e., a single bit), the above protocol is the basic coin-tossing protocol of Blum [3] (a formal proof of the security of this protocol can be found in [9]). However, here we are interested in a parallelized version where the parties attempt to simultaneously generate an m-bit random string (for any $m = poly(n)$). Intuitively, due to the secrecy of the commitment scheme, the string s_2 chosen by (a possibly malicious) Party 2 cannot be dependent on the value of s_1. Thus if s_1 is chosen uniformly, the resulting string $s = s_1 \oplus s_2$ is close to uniform. On the other hand, consider the case that Party 1 may be malicious. Then, by the protocol, Party 1 is committed to s_1 before Party 2 sends s_2. Thus, if s_2 is chosen uniformly, the string $s = s_1 \oplus s_2$ is uniformly distributed. We note that due to the binding property of the commitment scheme, Party 1 cannot alter the initial string committed to. We conclude that neither party is able to bias the output string.

However, the infeasibility of either side to bias the resulting string is not enough to show that the protocol is secure. This is because the definition of secure computation requires that the protocol simulate an *ideal execution* in which a trusted third party chooses a random string s and gives it to both parties. Loosely speaking, this means that there exists a simulator that works in the ideal model and simulates an execution with a (possibly malicious) party such that the joint output distribution (in this ideal scenario) is indistinguishable from when the parties execute the real protocol.

Protocol 1 seems not to fulfill this more stringent requirement. That is, our problem in proving the security of Protocol 1 is with constructing the required simulator. The main problem that occurs is regarding the simulation of Party 2.

Simulating a malicious Party 2: The simulator receives a uniformly distributed string s and must generate an execution consistent with s. That is, the commitment $c = C(s_1)$ given by the simulator to Party 2 must be such that $s_1 \oplus s_2 = s$ (where s_2 is the string sent by Party 2 in Step 2 of the protocol). However, s_1 is chosen and fixed (via a perfectly binding commitment) *before* s_2 is chosen by Party 2. Since the commitment is perfectly binding, even an all-powerful simulator cannot "cheat" and decommit to a different value. This problem is compounded by the fact that Party 2 may choose s_2 based on the commitment received to s_1 (by say invoking a pseudorandom function on c). Therefore, rewinding Party 2 and setting s_1 to equal $s \oplus s_2$ will not help (as s_2 will change and thus once again $s_1 \oplus s_2$ will equal s with only negligible probability). We note that this problem does not arise in the single-bit case as there are only two possible values for s_2 and thus the simulator succeeds with probability $1/2$ each time.

A problem relating to abort: The above problem arises even when the parties *never* abort. However, another problem in simulation arises due to the ability of the parties to abort. In particular, simulation of Party 1 in Protocol 1 is easy assuming that Party 1 never aborts. On the other hand, when Party 1's abort probability is unknown (and specifically when it is neither negligible nor noticeable), we do not know how to construct a simulator that does not skew the real probability of abort in the simulated execution. Once again, this problem is considerably easier in the single-bit case since Party 1's decision of whether or not to abort is based on only a single bit sent by Party 2 in Step 2 of the protocol (and so there are only three possible probabilities).

We note that basic coin-tossing is a special case of the augmented coin-tossing functionality. Thus, the same problems (and possibly others) must be solved in order to obtain an augmented coin-tossing protocol. As we will show, our solutions for these problems are enough for the augmented case as well.

Evidence that Protocol 1 is not secure: One may claim that the above *3-round* protocol may actually be secure and that the above-described difficulties are due to our proof technique. However, it can be shown that if there exists a secure 3-round protocol for coin-tossing (where the simulation uses black-box access to the malicious party), then there exist 3-round black-box zero-knowledge arguments for $\mathcal{NP}$. By [11], this would imply that $\mathcal{NP} \subseteq \mathcal{BPP}$. We note that all known simulations of secure protocols are indeed black-box.

4.3 The Actual Protocol

Before presenting the protocol itself, we discuss how we solve the problems described in the above motivating discussion.

- *Party 1 is malicious:* As described, when Party 1 is malicious, the problem that arises is that of aborting. In particular, Party 1 may decide to abort depending on the string s_2 sent to it by Party 2. This causes a problem in ensuring that the probability of abort in the simulation is negligibly close to that in a real execution. This is solved by having Party 1 send a proof of knowledge of s_1 after sending the commitment. Then, the simulator can extract s_1 from the proof of knowledge and can send $s_2 = s_1 \oplus s$ (where s is the string chosen by the trusted party) without waiting for Party 1 to decommit in a later step.
- *Party 2 is malicious:* As described, the central problem here is that Party 1 must commit itself to s_1 before s_2 is known (yet $s_1 \oplus s_2$ must equal s). This cannot be solved by rewinding because Party 2 may choose s_2 based on the commitment to s_1 that it receives (and thus changing the commitment changes the value of s_2). We solve this problem by not having Party 1 decommit at all; rather, it sends $s = s_1 \oplus s_2$ (or $F(s_1 \oplus s_2)$ in the augmented case) and proves in zero-knowledge that the value sent is consistent with its commitment and s_2. Thus, the simulator (who can generate proofs to false statements of this type) is able to "cheat" and send s (or $F(s)$) irrespective of the *real* value committed to in Step 1.[2]

  This technique of not decommitting, but rather revealing the committed value and proving (in zero-knowledge) that this value is correct, is central to

[2] In general, nothing can be said about a simulated proof of a false statement. However, in the specific case of statements regarding commitment values, proofs of false statements are indistinguishable from proofs of valid statements. This is due to the hiding property of the commitment scheme.

our simulation strategy. Specifically, it enables us to "decommit" to a value that is unknown at the time of the commitment. (As we have mentioned, in order for the simulation to succeed, Party 2 must be convinced that the commitment of Step 1 is to s_1, where $s_1 \oplus s_2 = s$. However, the correct value of s_1 is only known to the simulator after Step 2.)

We now present our constant-round protocol for the augmented secure coin-tossing functionality: $(1^n, 1^n) \mapsto (U_m, F(U_m))$, for $m = poly(n)$. For the sake of simplicity, our presentation uses a non-interactive commitment scheme (which is easily constructed given any 1–1 one-way function). However, the protocol can easily be modified so that an interactive commitment scheme is used instead (in particular, the two-round scheme of Naor [15]).

Protocol 2 (Augmented Parallel Coin-Tossing):

1. *Party 1 chooses $s_1 \in_R \{0,1\}^m$ and sends $c = C(s_1; r)$ for a random r to Party 2 (using a perfectly binding commitment scheme).*
2. *Party 1 proves knowledge of (s_1, r) with a (constant round) zero-knowledge argument of knowledge with negligible error. If the proof fails, then Party 2 aborts with output $\perp$.*
3. *Party 2 chooses $s_2 \in_R \{0,1\}^m$ and sends s_2 to Party 1.*
4. *If until this point Party 1 received an invalid message from Party 2, then Party 1 aborts, outputting $\perp$.*
   Otherwise, Party 1 sends $y = F(s_1 \oplus s_2)$.
5. *Party 1 proves to Party 2 using a (constant round) zero-knowledge argument that there exists a pair (s_1, r) such that $c = C(s_1; r)$ and $y = F(s_1 \oplus s_2)$ (that is, Party 1 proves that y is consistent with c and s_2).[3] If the proof fails, then Party 2 aborts with output $\perp$.*
6. Output:
   - *Party 1 outputs $s_1 \oplus s_2$ (even if Party 2 fails to correctly complete the verification of the proof in Step 5).*
   - *Party 2 outputs y.*

Round complexity: Using the constant-round zero-knowledge argument system of Feige and Shamir [8] and the constant-round commitment scheme of Naor [15], Protocol 2 requires a constant number of rounds only. We note that the proof system of [8] is also a proof of knowledge.

[3] It may appear that the reason that Party 1 does not decommit to c is due to the fact that Party 2 should only learn $F(s)$, and not s itself (if Party 1 decommits, then s is clearly revealed). Following this line of thinking, if F was the identity function, then Steps 4 and 5 could be replaced by Party 1 sending the actual decommitment. However, we stress that we do not know how to prove the security of such a modified protocol. The fact that Party 1 does not decommit, even when F is the identity function, is crucial to our proof of security.

Sufficient assumptions: All the components of Protocol 2 can be implemented using any one-way function. In particular the string commitment of Naor [15] can be used (this requires an additional pre-step in which Party 2 sends a random string to Party 1; however this step is of no consequence to the proof). Furthermore, the zero-knowledge argument of knowledge of [8] can be used in both Steps 2 and 5. Since both the [15] and [8] protocols only assume the existence of one-way functions, this is the only assumption required for the protocol.

Theorem 6 *Assuming the existence of one-way functions, Protocol 2 is a secure protocol for augmented parallel coin-tossing.*

Proof: We need to show how to efficiently transform any admissible pair of machines (A_1, A_2) for the real model into an admissible pair of machines (B_1, B_2) for the ideal model. We denote the trusted third party by T, the coin-tossing functionality by f and Protocol 2 by Π. We first consider the case that A_1 is adversarial.

Lemma 7 *Let (A_1, A_2) be an admissible pair of probabilistic expected polynomial-time machines for the real model in which A_2 is* honest. *Then, there exists an efficient transformation of (A_1, A_2) into an admissible pair of probabilistic expected polynomial-time machines (B_1, B_2) for the ideal model such that*

$$\{\mathsf{ideal}_{f,\overline{B}}(1^n, 1^n)\}_{n \in \mathsf{N}} \stackrel{c}{\equiv} \{\mathsf{real}_{\Pi,\overline{A}}(1^n, 1^n)\}_{n \in \mathsf{N}}$$

Proof Sketch: In this case the second party is honest and thus B_2 is determined. We now briefly describe the transformation of the real-model adversary A_1 into an ideal-model adversary B_1. Machine B_1 emulates an execution of A_1 with A_2 by playing the role of (an honest party) A_2 in most of the execution. (In particular, B_1 verifies the zero-knowledge proofs provided by A_1 and "checks" that A_1 is not cheating.) However, instead of randomly choosing the string s_2 in Step 3 (as A_2 would), machine B_1 first obtains the value s_1 (committed to by A_1) by running the extractor for the proof of knowledge of Step 2. Then, B_1 sets $s_2 = s_1 \oplus s$ where s is the output provided by the trusted third party.

It is easy to see that if A_1 follows the instructions of Protocol 2, then the output distributions in the ideal and real models are identical. This is because A_1's view in the ideal-model emulation with B_1 is identical to that of a real execution with A_2. Furthermore, since $s_1 \oplus s_2 = s$, the result of the execution is consistent with the outputs chosen by the trusted third party. However, A_1 may not follow the instructions of the protocol and nevertheless we must show that the real and ideal output distributions remain computationally indistinguishable (in fact, they are even *statistically close*). This can be seen by noticing that differences between the ideal and real executions can occur only if the extraction fails even though A_1 succeeded in proving the proof of Step 2, or if A_1 successfully cheats in the zero-knowledge proof of Step 5. Since both of these events occur with at most negligible probability, we have that the distributions are statistically close. ■

We now consider the case that A_2 is adversarial.

Lemma 8 *Let (A_1, A_2) be an admissible pair of probabilistic expected polynomial-time machines for the real model in which A_1 is* honest. *Then, there exists an efficient transformation of (A_1, A_2) into an admissible pair of probabilistic expected polynomial-time machines (B_1, B_2) for the ideal model such that*

$$\{\mathsf{ideal}_{f,\overline{B}}(1^n, 1^n)\}_{n\in\mathsf{N}} \stackrel{\mathsf{c}}{\equiv} \{\mathsf{real}_{\Pi,\overline{A}}(1^n, 1^n)\}_{n\in\mathsf{N}}$$

Proof Sketch: In this case the first party is honest and thus B_1 is determined. We now transform the real-model adversary A_2 into an ideal-model adversary B_2, where the transformation is such that B_2 uses black-box access to A_2. Specifically, B_2 chooses a uniform random tape, denoted R, for A_2 and invokes A_2 on input 1^n and random tape R. Once the input and random tape are fixed, A_2 is a deterministic function of messages received during a protocol execution. Thus $A_2(1^n, R, \overline{m})$ denotes the message sent by A_2 with input 1^n, random-tape R and sequence $\overline{m}$ of incoming messages to A_2.

The transformation works by having B_2 emulate an execution of A_2, while playing A_1's role. Machine B_2 does this when interacting with the trusted third party T and its aim is to obtain an execution with A_2 that is consistent with the output received from T. Therefore, B_2 has both external communication with T and "internal", emulated communication with A_2. Machine B_2 works as follows:

1. The ideal adversary B_2 chooses a uniformly distributed random tape R for the real adversary A_2, invokes $A_2(1^n, R)$ and (internally) passes to A_2 the commitment $c = C(0^m; r)$ for a random r (recall that in a real execution, A_2 expects to receive $C(s_1; r)$ for a random s_1).
2. B_2 invokes the *simulator* for the zero-knowledge argument of knowledge of the decommitment of c, using $A_2(1^n, R, c)$ as the verifier. (That is, this is a simulation of the proof of knowledge that A_1 is supposed to prove to A_2 in a real execution.)
3. B_2 obtains s_2 from A_2. (Recall that this is formally stated by having B_2 compute the function $A_2(1^n, R, c, t_{pok})$, where t_{pok} is the resulting transcript from the zero-knowledge proof of knowledge simulation).
   If at any point until here A_2 sent an invalid message, then B_2 aborts and outputs $A_2(1^n, R, c, t_{pok})$.
4. The ideal adversary B_2 sends 1^n to the (external) trusted third party T and receives the output $F(s)$.
   Next, B_2 (internally) passes to A_2 the string $y = F(s)$.
5. B_2 invokes the simulator for the zero-knowledge proof of Step 5 of the Protocol with the verifier role being played by $A_2(1^n, R, c, t_{pok}, y)$. Denote the transcript from the simulation of the zero-knowledge proof by t_{pf}.
6. B_2 outputs $A_2(1^n, R, c, t_{pok}, y, t_{pf})$.

We need to show that

$$\{\mathsf{ideal}_{f,\overline{B}}(1^n, 1^n)\}_{n\in\mathsf{N}} \stackrel{\mathsf{c}}{\equiv} \{\mathsf{real}_{\Pi,\overline{A}}(1^n, 1^n)\}_{n\in\mathsf{N}}$$

The following differences are evident between the ideal and real executions:

- The commitment received by A_2 (in the internal emulation by B_2) is to 0^m, rather than to a random string consistent with $y = F(s)$ and s_2 (as is the case in a real execution). However, by the indistinguishability of commitments, this should not make a difference.
- In the internal emulation by B_2, the zero-knowledge proofs are simulated and not real proofs. However, by the indistinguishability of simulated proofs, this should also not make a difference. As mentioned above, this holds even though the statement "proved" by B_2 in Step 5 is false with overwhelming probability.

The natural way to proceed at this point would be to define a hybrid experiment in which the commitment given by B_2 to A_2 is to s_1 and yet the zero-knowledge proofs are simulated. (In this hybrid experiment, s_1 must be such that $y = F(s_1 \oplus s_2)$.) However, such a hybrid experiment is problematic because the value of s_1 that is consistent with both y (from T) and s_2 is *unknown* at the point that B_2 generates the commitment. We must therefore bypass this problem before defining the hybrid experiment. We do this by defining the following *mental experiment* with a modified party B_2' (replacing Step 4 only of B_2 above):

4′. B_2' chooses $s_1 \in_R \{0,1\}^m$ (independently of what it has previously seen) and computes $y = F(s_1 \oplus s_2)$ (rather than obtaining $y = F(s)$ from T).
Next, B_2' (internally) passes A_2 the string y.

Notice that B_2' *does not* interact with any trusted third party at all. Rather, it chooses a uniformly distributed s, and computes $F(s)$ itself (observe that choosing s_1 uniformly and setting $s = s_1 \oplus s_2$ is equivalent to uniformly choosing s). We stress that B_2' does not work in the ideal model, but is rather a mental experiment. Despite this, since B_2' chooses s_1 independently of what it has seen, the distribution generated by B_2' is identical to that of the ideal model (where s is chosen by the trusted party).

Next, notice that if we move the step in which s_1 is chosen to before the first step of B_2', then this makes no difference to the output distribution. Having done this, it is possible for B_2' to send a commitment to s_1 rather than to 0^m. Thus, the above-described hybrid experiment can be defined. That is, we define a hybrid setting (with a party B_2'') in which B_2'' initially sends a commitment to s_1 (rather than to 0^m). Thus, in terms of the commitment, the hybrid experiment is identical to a real execution (and different to the mental experiment and ideal model). On the other hand, the zero-knowledge proofs in the hybrid experiment are simulated (as in the mental experiment), rather than being actual proofs (as in the real model). Then, the indistinguishability of the mental experiment from the real model is demonstrated by first showing the indistinguishability of the the hybrid and mental experiments (where the only difference is regarding the initial commitment) and then showing the indistinguishability of the hybrid and real executions (where the only difference is regarding the simulated zero-knowledge proofs). Since the output of an ideal-model execution is identically distributed to the output from the mental experiment, this completes the proof. ■

This completes the proof of Theorem 6. ■

4.4 Comparing Protocol 2 to the Protocol of [9]

The protocol for augmented coin-tossing presented by Goldreich [9] is for tossing a single bit only (i.e., where $m = 1$). Thus, in order to toss polynomially many coins, Goldreich suggests running the single-bit protocol many times sequentially. However, the only difference between Protocol 2 and the protocol of [9] is that here m can be any value polynomial in n and there m is fixed at 1 (i.e., by setting $m = 1$ in our protocol, we obtain the exact protocol of [9]). Despite this, our *proof* is different and works for any $m = poly(n)$ whereas the *proof* of [9] relies heavily on $m = 1$ (or at the most $m = O(\log n)$).[4] Furthermore, there is a conceptual difference in the role of the two zero-knowledge proofs in the protocol. In [9], these proofs are used in order to obtain *augmented* coin-tossing (and are not needed for the case that F is the identity function). However, here these proofs are used for obtaining coin-tossing of $m = poly(n)$ coins in parallel, even when F is the identity function.

5 Perfect Coin-Tossing

In this section we present a constant-round protocol for *perfect* coin tossing. By perfect coin tossing, we mean that the output distribution of a real execution is *statistically close* to the output distribution of the ideal process (rather than the distributions being only computationally indistinguishable as required by secure computation); see Theorem 9. As in the previous section, the functionality we consider is that of augmented coin tossing: $(1^n, 1^n) \mapsto (U_m, F(U_m))$.

The protocol is almost identical to Protocol 2 except that the commitment scheme used is perfectly hiding and the zero-knowledge arguments are perfect. These primitives are known to exist assuming the existence of families of clawfree functions or collision-resistant hash functions. Thus we rely here on a (seemingly) stronger assumption than merely the existence of one-way functions. We note that Protocol 1 is a protocol for *perfect coin tossing* of a single bit and thus perfect coin tossing of m coins can be achieved in $O(m)$ rounds (see the proof in [9] which actually demonstrates statistical closeness). In this section we show that perfect coin tossing of polynomially many coins can also be achieved in a *constant number of rounds.*

Protocol 3 (Augmented Perfect Coin-Tossing):
An augmented perfect coin-tossing protocol is constructed by taking Protocol 2 and making the following modifications:

- *The commitment sent by Party 1 in Step 1 is* perfectly hiding.
- *The proof of knowledge provided by Party 1 in Step 2 is* perfect *zero-knowledge.*
- *The proof provided by Party 1 in Step 5 is a* perfect *zero-knowledge* proof of knowledge. *(Recall that in Protocol 2, this proof need not be a proof of knowledge.)*

[4] In private communication, Goldreich stated that he did not know whether or not his protocol [9] can be parallelized.

Constant-round perfect zero-knowledge arguments of knowledge are known to exist assuming the existence of constant-round perfectly hiding commitment schemes [4,8]. Furthermore, constant-round perfectly-hiding commitment schemes can be constructed using families of clawfree [10] or collision-resistant hash functions [16,6]. These commitment schemes work by having the receiver first uniformly choose a function f from the family designated in the protocol. The receiver then sends f to the sender who uses it to commit to a string by sending a single message. Thus, using such a scheme, Protocol 3 begins by Party 2 choosing a function f from a clawfree or collision-resistant family and sending it to Party 1. Then, Party 1 commits using f.

We stress the use of arguments of knowledge for *both* proofs here, whereas in Protocol 2 the proof of Step 5 need not be a proof of knowledge. The reason for this is that since the commitment is perfectly hiding, c is essentially a valid commitment to *every* $s_1 \in \{0,1\}^m$. Thus, every y is "consistent" with c and s_2. Therefore, what we need to ensure is that y is consistent with s_2 and the decommitment of c that are *known* to Party 1. This can be accomplished using a proof of knowledge.

Theorem 9 *Assuming the existence of perfectly-hiding commitment schemes, Protocol 3 is a secure protocol for augmented* perfect *coin-tossing. That is, there exists an efficient transformation of every admissible pair of probabilistic expected polynomial-time machines for the real model (A_1, A_2) into an admissible pair of probabilistic expected polynomial-time machines for the ideal model (B_1, B_2), such that*

$$\{\mathsf{ideal}_{f,\overline{B}}(1^n, 1^n)\} \stackrel{s}{\equiv} \{\mathsf{real}_{\Pi_2,\overline{A}}(1^n, 1^n)\}$$

where f is the augmented coin-tossing functionality and Π_2 denotes Protocol 3.

The proof of this theorem is very similar to the proof of Theorem 6; the main difference being with respect to the fact that the initial commitment is not perfectly binding.

Acknowledgements. We would like to thank Oded Goldreich for his invaluable contribution to all aspects of this work. We would also like to thank Moni Naor for his suggestion that we look at the question of perfect coin-tossing as well.

References

1. D. Beaver. Foundations of Secure Interactive Computing. In *Crypto91*, Springer-Verlag LNCS Vol. 576, pages 377–391, 1991.
2. D. Beaver, S. Micali and P. Rogaway. The Round Complexity of Secure Protocols. In *22nd STOC*, pages 503–513, 1990.
3. M. Blum. Coin Flipping by Phone. *IEEE Spring COMPCOM*, pages 133–137, February 1982.
4. G. Brassard, C. Crepeau and M. Yung. Constant-round perfect zero-knowledge computationally convincing protocols. In *Theoretical Computer Science*, Vol. 84 (1), pp. 23–52, 1991.

5. R. Canetti. Security and Composition of Multi-party Cryptographic Protocols. *Journal of Cryptology*, Vol. 13, No. 1, pages 143–202, 2000.
6. I. Damgard, T. Pederson and B. Pfitzmann. On the Existence of Statistically Hiding Bit Commitment Schemes and Fail-Stop Signatures. In *Crypto93*, Springer-Verlag LNCS Vol. 773, pages 250–265, 1993.
7. S. Even, O. Goldreich and A. Lempel. A Randomized Protocol for Signing Contracts. *Communications of the ACM* **28**, pp. 637–647, 1985.
8. U. Feige and A. Shamir. Zero-Knowledge Proofs of Knowledge in Two Rounds. In *Crypto89*, Springer-Verlag LNCS Vol. 435, pp 526-544.
9. O. Goldreich. *Secure Multi-Party Computation.* Manuscript. Preliminary version, 1998. Available from: `www.wisdom.weizmann.ac.il/~oded/pp.html`.
10. O. Goldreich. *Foundations of Cryptography: Volume 1 – Basic Tools.* Cambridge University Press, 2001.
11. O. Goldreich and H. Krawczyk. On the Composition of Zero-Knowledge Proof Systems. *SIAM Journal on Computing*, Vol. 25, No. 1, February 1996, pages 169-192.
12. O. Goldreich, S. Micali and A. Wigderson. Proofs that Yield Nothing but their Validity or All Languages in NP Have Zero-Knowledge Proof Systems. *JACM*, Vol. 38, No. 1, pages 691–729, 1991.
13. O. Goldreich, S. Micali and A. Wigderson. How to Play any Mental Game – A Completeness Theorem for Protocols with Honest Majority. In *19th STOC*, pages 218–229, 1987. For details see [9].
14. S. Micali and P. Rogaway. Secure Computation. Unpublished manuscript, 1992. Preliminary version in *Crypto'91*, Springer-Verlag (LNCS 576), 1991.
15. M. Naor. Bit Commitment using Pseudorandom Generators. *Journal of Cryptology*, Vol. 4, pages 151–158, 1991.
16. M. Naor and M. Yung. Universal One-Way Hash Functions and their Cryptographic Applications. In *21st STOC*, pages 33–43, 1989.
17. A.C. Yao. How to Generate and Exchange Secrets. In *27th FOCS*, pages 162–167, 1986.

Faster Point Multiplication on Elliptic Curves with Efficient Endomorphisms

Robert P. Gallant[1], Robert J. Lambert[1], and Scott A. Vanstone[1,2]

[1] Certicom Research, Canada
{rgallant,rlambert,svanstone}@certicom.com
[2] University of Waterloo, Canada

Abstract. The fundamental operation in elliptic curve cryptographic schemes is the multiplication of an elliptic curve point by an integer. This paper describes a new method for accelerating this operation on classes of elliptic curves that have efficiently-computable endomorphisms. One advantage of the new method is that it is applicable to a larger class of curves than previous such methods. For this special class of curves, a speedup of up to 50% can be expected over the best general methods for point multiplication.

1 Introduction

Let E be an elliptic curve defined over a finite field $\mathbb{F}_q$. The dominant cost operation in elliptic curve cryptographic schemes is *point multiplication*, namely computing kQ where Q is an elliptic curve point and k is an integer. This operation is the additive analogue of the exponentiation operation α^k in a general (multiplicative-written) finite group. The basic technique for exponentiation is the repeated square-and-multiply algorithm. Numerous methods for speeding up exponentiation and point multiplication have been discussed in the literature; for a survey, see [11,12,17]. These methods can be categorized as follows:

1. Generic methods which can be applied to speed up exponentiation in any finite abelian group, including:
   a) Comb techniques (e.g. [15]) which precompute tables which depend on Q. Such techniques are applicable when the base point Q is fixed and known a priori, for example in ECDSA signature generation.
   b) Addition chains which are useful when k is fixed, for example in RSA decryption.
   c) Windowing techniques which are useful when the base point Q is not known a priori, for example in Diffie-Hellman key agreement.
   d) Simultaneous multiple exponentiation techniques for computing expressions $k_1Q_1 + k_2Q_2 + \cdots + k_tQ_t$, for example in ECDSA signature verification.
2. Exponent recoding techniques which replace the binary representation of k with a representation which has fewer non-zero terms (e.g, [10,19]).
3. Methods which are particular to elliptic curve point multiplication such as:

J. Kilian (Ed.): CRYPTO 2001, LNCS 2139, pp. 190–200, 2001.

a) Selection of an underlying finite field which enables faster field arithmetic. For example, selection of a prime field $\mathbb{F}_p$ where p is a Mersenne prime or a Mersenne-like prime [31], or an optimal field extension [2].
b) Selection of a representation of the underlying finite field which enables faster field arithmetic. For example, selection of an irreducible trinomial as the reduction polynomial for binary extension fields.
c) Selection of a point representation which enables faster elliptic curve arithmetic [6].
d) Selection of an elliptic curve with special properties, for example Koblitz curves [13].

Koblitz curves are elliptic curves defined over $\mathbb{F}_2$, and were first proposed for cryptographic use in [13]. The primary advantage of Koblitz curves is that the Frobenius endomorphism can be exploited to devise fast point multiplication algorithms that do not use any point doublings [30,32]. These techniques can be generalized to use arbitrary endomorphisms but are generally not efficient.

The contribution of this paper is a new technique for speeding up point multiplication of elliptic curves having an efficiently-computable endomorphism. While the technique is not as efficient as the methods of Solinas [30,32] for Koblitz curves, they are useful for speeding up point multiplication on a larger class of elliptic curves, for example certain curves over prime fields. Such elliptic curves over prime fields have been included in the WAP WTLS (Wireless Transport Layer Security) standard [33]. We believe the ideas discussed in this paper are new (though not difficult). In particular, we believe that the approach of decomposing k modulo n, and applying just one application of the endomorphism is different than the methods of previous papers. The result is a technique which works on a wider class of curves (in particular, curves defined over prime fields), and works with endomorphisms whose computational cost is not necessarily cheaper than a point operation. For this class of curves, a speedup of up to 50% can be expected over the best general methods for point multiplication.

The remainder of this paper is organized as follows. §2 defines an endomorphism and reviews how the Frobenius endomorphism can be used to speed up point multiplication on Koblitz curves. Our new work for speeding up point multiplication on elliptic curves which have efficiently-computable endomorphisms is described in §3 and §4. The security of the new method is considered in §5. Finally, we draw our conclusions and discuss avenues for future work in §6.

2 Endomorphisms

Let E be an elliptic curve defined over the finite field $\mathbb{F}_q$. The point at infinity is denoted by $\mathcal{O}$. For any $n \geq 1$, the group of $\mathbb{F}_{q^n}$-rational points on E is denoted by $E(\mathbb{F}_{q^n})$.

An *endomorphism* of E is a rational map $\phi : E \to E$ satisfying $\phi(\mathcal{O}) = \mathcal{O}$ [27]. If the rational map is defined over $\mathbb{F}_q$, then the endomorphism ϕ is also said to be defined over $\mathbb{F}_q$. In this case, ϕ is a group homomorphism of $E(\mathbb{F}_q)$, and also of $E(\mathbb{F}_{q^n})$ for any $n \geq 1$.

Example 1. Let E be an elliptic curve defined over $\mathbb{F}_q$. For each $m \in \mathbb{Z}$ the *multiplication by m* map $[m] : E \to E$ defined by $P \mapsto mP$ is an endomorphism defined over $\mathbb{F}_q$. A special case is the *negation* map defined by $P \mapsto -P$.

Example 2. Let E be an elliptic curve defined over $\mathbb{F}_q$. Then the qth power map $\phi : E \to E$ defined by $(x, y) \mapsto (x^q, y^q)$ and $\mathcal{O} \mapsto \mathcal{O}$ is an endomorphism defined over $\mathbb{F}_q$, called the *Frobenius* endomorphism. Since exponentiating to the qth power is a linear operation in $\mathbb{F}_{q^n}$, computation of $\phi(P)$ is normally quite fast. For example, if a normal basis of $\mathbb{F}_{q^n}$ over $\mathbb{F}_q$ is used, this computation can be implemented as a cyclic shift of the vector representation.

Example 3 (§7.2.3 of [5]). Let $p \equiv 1 \pmod 4$ be a prime, and consider the elliptic curve

$$E_1 \ : \ y^2 = x^3 + ax \tag{1}$$

defined over $\mathbb{F}_p$. Let $\alpha \in \mathbb{F}_p$ be an element of order 4. Then the map $\phi : E_1 \to E_1$ defined by $(x, y) \mapsto (-x, \alpha y)$ and $\mathcal{O} \mapsto \mathcal{O}$ is an endomorphism defined over $\mathbb{F}_p$. If $P \in E(\mathbb{F}_p)$ is a point of prime order n, then ϕ acts on $\langle P \rangle$ as a multiplication map $[\lambda]$, i.e., $\phi(Q) = \lambda Q$ for all $Q \in \langle P \rangle$, where λ is an integer satisfying $\lambda^2 \equiv -1 \pmod n$. Note that $\phi(Q)$ can be computed using only one multiplication in $\mathbb{F}_p$.

Example 4 (§7.2.3 of [5]). Let $p \equiv 1 \pmod 3$ be a prime, and consider the elliptic curve

$$E_2 \ : \ y^2 = x^3 + b \tag{2}$$

defined over $\mathbb{F}_p$. Let $\beta \in \mathbb{F}_p$ be an element of order 3. Then the map $\phi : E_2 \to E_2$ defined by $(x, y) \mapsto (\beta x, y)$ and $\mathcal{O} \mapsto \mathcal{O}$ is an endomorphism defined over $\mathbb{F}_p$. If $P \in E(\mathbb{F}_p)$ is a point of prime order n, then ϕ acts on $\langle P \rangle$ as a multiplication map $[\lambda]$, where λ is an integer satisfying $\lambda^2 + \lambda \equiv -1 \pmod n$. Note that $\phi(Q)$ can be computed using only one multiplication in $\mathbb{F}_p$.

Example 5 (§7.2.3 of [5]). Let $p > 3$ be a prime such that -7 is a perfect square in $\mathbb{F}_p$, and let $\omega = (1 + \sqrt{-7})/2$, and let $a = (\omega - 3)/4$. Consider the elliptic curve

$$E_3 \ : \ y^2 = x^3 - \frac{3}{4}x^2 - 2x - 1 \tag{3}$$

defined over $\mathbb{F}_p$. Then the map $\phi : E_3 \to E_3$ defined by

$$(x, y) \ \mapsto \ \left(\omega^{-2}\frac{x^2 - \omega}{x - a}, \ \omega^{-3}y\frac{x^2 - 2ax + \omega}{(x - a)^2} \right)$$

and $\mathcal{O} \mapsto \mathcal{O}$ is an endomorphism defined over $\mathbb{F}_p$. Computing the endomorphism is a little harder than doubling a point.

Example 6 (§14B of [7]). Let $p > 3$ be a prime such that -2 is a perfect square in $\mathbb{F}_p$, and consider the elliptic curve

$$E_4 \; : \; y^2 = 4x^3 - 30x - 28 \tag{4}$$

defined over $\mathbb{F}_p$. Then the map $\phi : E_4 \to E_4$ defined by

$$(x, y) \;\mapsto\; \left(-\frac{2x^2 + 4x + 9}{4(x+2)}, \; -\frac{2x^2 + 8x - 1}{4\sqrt{-2}(x+2)^2} y \right)$$

and $\mathcal{O} \mapsto \mathcal{O}$ is an endomorphism defined over $\mathbb{F}_p$. Computing the endomorphism is a little harder than doubling a point.

The existing methods [13,14,20,29,32] for point multiplication which exploit efficiently-computable endomorphisms all use the Frobenius endomorphism. Let E be an elliptic curve defined over a small field $\mathbb{F}_q$, and let ϕ be the Frobenius endomorphism. To compute kP, where $P \in E(\mathbb{F}_{q^n})$, these methods first compute $k' = k \bmod (\phi^n - 1)$ in the ring $\mathbb{Z}[\phi]$. Then, one computes a ϕ-adic expansion $k' = \sum_{i=0}^{t} c_i \phi^i$, where the c_i are elements of a small set, e.g., $\{-q/2, \ldots, q/2\}$, and $t \approx n$. Finally, kP can be efficiently computed as follows:

$$kP \;=\; k'P \;=\; \sum_{i=0}^{t} c_i \phi^i(P). \tag{5}$$

The expression (5) can be evaluated using traditional windowing techniques. Observe that the (slow) point doublings in traditional repeated add-and-double algorithms have been replaced by (fast) evaluations of the Frobenius map.

The methods based on Frobenius map expansions can in principle be extended to an arbitrary endomorphism ψ. However, these techniques will no longer be efficient if computing ψ is more expensive than a point doubling. Furthermore, one may not have $\psi^n - 1 = 0$, so the ψ-adic expansion of k may be significantly longer than the binary expansion of k. Finally, the existing techniques do not apply when $\mathrm{Norm}(\psi) = 1$ (as is the case in Examples 3 and 4) since these techniques require a division operation by ψ which yield a nontrivial remainder having norm less than $\mathrm{Norm}(\psi)$.

In the next section, we present a new method that exploits efficiently-computable endomorphisms such as the ones in Examples 3, 4, 5 and 6 to speed up point multiplication.

3 Using Efficient Endomorphisms

Let E be an elliptic curve defined over $\mathbb{F}_q$, and let $P \in E(\mathbb{F}_q)$ be a point of prime order n. Let ϕ be an endomorphism defined over $\mathbb{F}_q$, and suppose that the characteristic polynomial of ϕ has a root λ modulo n—since the characteristic polynomial of an endomorphism has degree two we expect that roughly half of all curves will have a root modulo n. The map ϕ acts on $\langle P \rangle$ as a multiplication map $[\lambda]$.

The methods described will be advantageous if computing ϕ costs less than computing about $(\log_2 n)/3$ point doublings. In practice, we expect the algorithm to be applied when the cost of ϕ is less than (say) 5 point doubles.

The problem we consider is that of computing kP for k selected uniformly at random from the interval $[1, n-1]$. The basic idea of the paper is as follows. Suppose that we can efficiently write $k = k_1 + k_2\lambda \bmod n$, where $k_1, k_2 \in [0, \lceil\sqrt{n}\rceil]$ (see §4). Then we have

$$\begin{aligned} kP &= (k_1 + k_2\lambda)P \\ &= k_1P + k_2(\lambda P) \\ &= k_1P + k_2\phi(P). \end{aligned} \tag{6}$$

Now (6) can be computed using any of the 'simultaneous multiple exponentiation' type algorithms[1], the simplest of which we review below. In the following, $(u_{t-1}, \ldots, u_1, u_0)_2$ denotes the binary representation of the integer u, and w is the window width.

Algorithm 1. Simultaneous multiple point multiplication

INPUT: w, $u = (u_{t-1}, \ldots, u_1, u_0)_2$, $v = (v_{t-1}, \ldots, v_1, v_0)_2$, P, Q.
OUTPUT: $uP + vQ$.

1. *Compute* $iP + jQ$ *for all* $i, j \in [0, 2^w - 1]$.
2. *Write* $u = (u^{d-1}, \ldots, u^1, u^0)$ *and* $v = (v^{d-1}, \ldots, v^1, v^0)$ *where each* u^i *and* v^i *is a bitstring of length* w, *and* $d = \lceil t/w \rceil$.
3. $R \leftarrow \mathcal{O}$.
4. *For* i *from* $d-1$ *downto* 0 *do*
   4.1 $R \leftarrow 2^w R$.
   4.2 $R \leftarrow R + (u^iP + v^iQ)$.
5. *Return(R).*

Analysis. Since the bitlengths of k_1 and k_2 in (6) are half the bitlength of k, we might expect to obtain a significant speedup because we have eliminated a significant number of point doublings at the expense of a few point additions. A precise analysis is complicated due to the large number of point multiplication techniques available. Nevertheless, the following provides some indication of the relative benefits of our method.

Assume that k is a randomly selected t-bit integer. When $t = 160$, Algorithm 2 of [18] (an exponent recoding and sliding window algorithm) is among the best algorithms for computing kP. This method costs approximately 157 point doubles and 34 point additions using windows of size 4 [18]. To compare this traditional method with the proposed method, we need an algorithm for computing $k_1P + k_2Q$ (where in our case $Q = \phi(P)$). The following is straightforward and useful for our purposes, but we cannot find a reference for it.

[1] These are also known as 'exponentiation using vector-addition chains', 'Shamir's trick', or 'multi-exponentiation'.

Algorithm 2 of [18] can be combined with the simultaneous multiple exponentiation technique of Algorithm 1 to give an algorithm which is among the best [24] for computing $k_1P + k_2Q$. Essentially, this combined algorithm computes ϵP and ϵQ for the integers ϵ corresponding to allowable windows, then writes each of k_1 and k_2 in signed windowed-NAF form as in [18]. Finally a left-to-right algorithm is used to iteratively double a common accumulator and add in an ϵP or ϵQ as appropriate. After $\max\{\log_2 k_1, \log_2 k_2\}$ iterations, the accumulator holds the desired $k_1P + k_2Q$.

Using this algorithm in the proposed method to compute $kP = k_1P + k_2(\lambda P)$ costs approximately 79 point doubles and 38 point additions (when using windows of size 3 [18]) plus 1 evaluation of the map ϕ. If the cost of a point doubling is 8 field multiplications and the cost of a point addition is 11 field multiplications (as is the case with Jacobian coordinates [4]), then the ratio of the running times of the proposed method to the traditional method is ≈ 0.66. Thus the new method for point multiplication is roughly 50% faster than the traditional method when $t = 160$. As the bitlength of k increases, the ratio essentially decreases[2] and so the relative performance of the new method gets better. For example, with a bitlength of $t = 512$, the ratio is about .62.

Remark. If computing ϕ is cheaper than a point addition, then a few additions can be saved as follows. In the above 'simultaneous windowed-NAF' method for computing $k_1P + k_2Q$, we initially compute and store points ϵP and ϵQ for small values of ϵ. If $Q = \phi(P)$, and computing ϕ is cheaper than a point addition, then we can instead compute $\epsilon Q = \epsilon\phi(P) = \phi(\epsilon P)$. For example, in the width-3 windowed method of [18], computing $k_1P + k_2\phi(P)$ saves 3 additions at the expense of 3 additional applications of ϕ.

Example 7. An example of an elliptic curve for which our new method is applicable is

$$E\ :\ y^2 = x^3 + 3$$

over the prime field $\mathbb{F}_p$, where

$$p = 1461501637330902918203684832716283019655932313743$$

is a 160-bit prime, and

$$\#E(\mathbb{F}_p) = 1461501637330902918203687013445034429194588307251$$

is prime. This curve is included in the WAP specification of the WTLS protocol [33].

[2] There are occasional minor bumps corresponding to window size changes.

4 Decomposing k

In this section we describe an algorithm which takes as input integers n, λ and $k \in_R [1, n-1]$, and returns integers k_1 and k_2 such that $k \equiv k_1 + k_2\lambda \pmod{n}$. The integers k_1 and k_2 returned are distinguished in that they are both small or, equivalently, the vector $(k_1, k_2) \in \mathbb{Z} \times \mathbb{Z}$ has small Euclidean norm. The term "small" will be made precise below.

Let $G = \mathbb{Z} \times \mathbb{Z}$ and consider the homomorphism $f : G \to \mathbb{Z}_n$ defined by $(i, j) \mapsto (i+\lambda j) \bmod n$. We wish to find a short vector $u \in G$ such that $f(u) = k$; the components of u can then be used as the required k_1 and k_2. Note that it is easy to find a vector $v \in G$ such that $f(v) = k$; $v = (k, 0)$ is such a vector. The problem is in finding a vector that is also short.

Our approach is the following. We first find linearly independent short vectors $v_1, v_2 \in G$ such that $f(v_1) = f(v_2) = 0$. We then find a vector v in the integer lattice generated by v_1 and v_2 that is close to $(k, 0)$. It then follows that $u = (k, 0) - v$ is a short vector with $f(u) = f((k, 0)) - f(v) = k$. Note that both subproblems can be solved using lattice basis reduction algorithms. However, the direct methods presented here are far less cumbersome to implement.

Finding v_1 and v_2. The problem of finding two independent short vectors v_1, v_2 such that $f(v_1) = f(v_2) = 0$ can be solved using the extended Euclidean algorithm. We apply the extended Euclidean algorithm to find the greatest common divisor of n and λ. (This gcd is 1 since n is prime.) The algorithm produces a sequence of equations

$$s_i n + t_i \lambda = r_i, \text{ for } i = 0, 1, 2, \dots, \tag{7}$$

where $s_0 = 1$, $t_0 = 0$, $r_0 = n$, $s_1 = 0$, $t_1 = 1$, $r_1 = \lambda$, and $r_i \geq 0$ for all i. The following properties of the extended Euclidean algorithm are well-known and can be easily proven by induction.

Lemma 1. *Let s_i, t_i, r_i be the sequence of variables in (7) produced by an application of the extended Euclidean algorithm to positive integers n and λ.*

(i) $r_i > r_{i+1} \geq 0$ *for all* $i \geq 0$.
(ii) $|s_i| < |s_{i+1}|$ *for* $i \geq 1$.
(iii) $|t_i| < |t_{i+1}|$ *for* $i \geq 0$.
(iv) $r_{i-1}|t_i| + r_i|t_{i-1}| = n$ *for all* $i \geq 1$.

Let m be the greatest index for which $r_m \geq \sqrt{n}$. Then $r_m|t_{m+1}| + r_{m+1}|t_m| = n$, and $|t_{m+1}| < \sqrt{n}$. We take $v_1 = (r_{m+1}, -t_{m+1})$. By (7) we have $f(v_1) = 0$. Also, since $|t_{m+1}| < \sqrt{n}$ and $|r_{m+1}| < \sqrt{n}$, we have $||v_1|| \leq \sqrt{2n}$. We also take v_2 to be the shorter of $(r_m, -t_m)$ and $(r_{m+2}, -t_{m+2})$. Again by (7), we have $f(v_2) = 0$. Heuristically one expects that v_2 is also short.[3] Observe that v_1 and v_2 are linearly independent since otherwise if $v_2 = (r_m, -t_m)$ (say), then

[3] Experiments with various values of λ also validate this assumption. It is impossible to prove this without further restrictions; for example consider $\lambda = n - 1$.

$$\frac{r_{m+1}}{r_m} = \frac{-t_{m+1}}{-t_m} = \frac{t_{m+1}}{t_m};$$

but $r_{m+1}/r_m < 1$ by Lemma 1(i) and $|t_{m+1}/t_m| > 1$ by Lemma 1(iii).

Notice that since v_1 and v_2 only depend on n and λ (and not on k), they can be precomputed if n and λ are shared domain parameters.

Finding v. A vector v in the integer lattice generated by v_1 and v_2 that is close to $(k, 0)$ can be easily found using elementary linear algebra[1]. By considering $(k, 0)$, v_1 and v_2 as vectors in $\mathbb{Q} \times \mathbb{Q}$, we can write $(k, 0) = \beta_1 v_1 + \beta_2 v_2$, where $\beta_1, \beta_2 \in \mathbb{Q}$. Then round β_1, β_2 to the nearest integers: $b_1 = \lfloor \beta_1 \rceil$, $b_2 = \lfloor \beta_2 \rceil$. Finally, let $v = b_1 v_1 + b_2 v_2$.

The following proves that the vector u is indeed short.

Lemma 2. *The vector $u = (k, 0) - v$, where v is constructed as above, has norm at most $max(||v_1||, ||v_2||)$.*

Proof. We have

$$\begin{aligned} u &= (k, 0) - v \\ &= (\beta_1 v_1 + \beta_2 v_2) - (b_1 v_1 + b_2 v_2) \\ &= (\beta_1 - b_1) v_1 + (\beta_2 - b_2) v_2. \end{aligned}$$

Finally, since $|\beta_1 - b_1| \leq \frac{1}{2}$ and $|\beta_2 - b_2| \leq \frac{1}{2}$, by the Triangle Inequality we have

$$\begin{aligned} ||u|| &\leq \tfrac{1}{2}||v_1|| + \tfrac{1}{2}||v_2|| \\ &\leq \max(||v_1||, ||v_2||). \end{aligned}$$

□

5 Security Considerations

Elliptic curves having efficiently-computable endomorphisms should be regarded as "special" elliptic curves. Using "special" instances of cryptographic schemes is sometimes done for efficiency reasons (for example the use of low encryption-exponent RSA, or the use of small subgroups hidden in a larger group as with DSA). However in any instance of a cryptographic scheme, there is always the chance that an attack will be forthcoming that applies to the special instance and significantly weakens the security. Such is the case here as well.

When selecting an elliptic curve E over $\mathbb{F}_q$ for cryptographic use, one must ensure that the order $\#E(\mathbb{F}_q)$ of the elliptic curve is divisible by a large prime number n (say $n \geq 2^{160}$) in order to prevent the Pohlig-Hellman [22] and Pollard's rho [23,21] attacks. In addition, one must ensure that $\#E(\mathbb{F}_q) \neq q$ in order to prevent the Semaev-Satoh-Araki-Smart attack [26,25,28], and that n does not divide $q^i - 1$ for all $1 \leq i \leq 20$ in order to prevent the Weil pairing [16] and Tate pairing attacks [8]. Given a curve satisfying these conditions, there is no attack known that significantly reduces the time required to compute elliptic curve discrete logarithms. Many such curves having efficient endomorphisms exist and hence appear suitable for cryptographic use. One attack on the elliptic curve discrete logarithm problem on such curves is along the lines of [9] and [34]. The application of such ideas does not reduce the time to compute a logarithm by more than a small factor.

The number of curves for which this technique applies seems to be reasonably large. For instance, one of the Examples 3, 4, 5 and 6 provide a candidate for most primes p.

6 Conclusions and Further Work

We described a new method for accelerating point multiplication on classes of elliptic curves that have efficiently-computable endomorphisms. The new method for point multiplication is roughly 50% faster than the best general methods. One advantage of the new method is that it is applicable to a larger class of curves than previous such methods. For example, the method is applicable to classes of curves over prime fields and, in particular, is well suited to two curves over prime fields included in the WAP WTLS specification.

One direction in which our method can be generalized is to use higher powers of the endomorphism. For example, one could write $k \equiv k_1 + k_2\lambda + k_3\lambda^2 \bmod n$ for $t/3$-bit integers k_1, k_2, k_3. This could be done by first finding three linearly independent vectors v_1, v_2, v_3 in $\mathbb{Z} \times \mathbb{Z} \times \mathbb{Z}$ each of length roughly $n^{1/3}$, and lying in the kernel of the homomorphism $f : \mathbb{Z} \times \mathbb{Z} \times \mathbb{Z}$ defined by $(x, y, z) \mapsto x + y\lambda + z\lambda^2 \bmod n$. Experimentally, we found that if λ satisfies $\lambda^2 - T\lambda + N \equiv 0 \bmod n$ for a random prime n, random T (Trace) and random N (Norm), then the application of LLL to the lattice generated by $\{(\lambda^2, 0, -1), (\lambda, -1, 0), (0, \lambda, -1), (n, 0, 0), (0, n, 0), (0, 0, n)\}$ results in 3 independent vectors of length about $n^{1/3}$, *provided* at least one of N, T has magnitude at least $n^{1/3}$. In this case the application of 'simultaneous multiple exponentiation' type techniques yield an even better improvement over traditional algorithms, with the relevant ratio around 1/2.

We warn that generating k by simply choosing k_1, k_2, k_3 first requires care: for example, if $\lambda^2 + \lambda + 1 \equiv 0 \bmod n$ (as in Example 4) then $k_1 + k_2\lambda + k_3\lambda^2 \equiv (k_1 - k_3) + (k_2 - k_3)\lambda \bmod n$. Thus simply choosing k_1, k_2, k_3 randomly in $[0, n^{1/3}]$ and setting $k \equiv k_1 + k_2\lambda + k_3\lambda^2 \bmod n$ will result in a k having a considerable bias, and consequently the resulting cryptographic scheme may be susceptible to an attack like Bleichenbacher's attack [3] on the DSA as specified in FIPS 186.

Acknowledgements. The authors would like to thank Charles Lam, Alfred Menezes, and John Proos for several very helpful comments and suggestions.

References

1. L. Babai, "On Lovász' Lattice Reduction and the Nearest Lattice Point Problem", Combinatorica **6** (1986), 1-13

2. D. Bailey and C. Paar, "Optimal extension fields for fast arithmetic in public-key algorithms", *Advances in Cryptology - Crypto '98*, 1998, 472-485.
3. D. Bleichenbacher, "On the generation of DSA one-time keys", preprint, November 2000.
4. D. Chudnovsky and G. Chudnovsky, "Sequences of numbers generated by addition in formal groups and new primality and factoring tests", *Advances in Applied Mathematics*, **7** (1987), 385-434.
5. H. Cohen, *A Course in Computational Algebraic Number Theory*, Springer-Verlag, 3rd printing, 1996.
6. H. Cohen, A. Miyaji and T. Ono, "Efficient elliptic curve exponentiation using mixed coordinates", *Advances in Cryptology-Asiacrypt '98*, 1998, 51-65.
7. D. Cox, *Primes of the Form* $x^2 + ny^2$. *Fermat, Class Field Theory and Complex Multiplication*, Wiley, 1989.
8. G. Frey and H. Rück, "A remark concerning m-divisibility and the discrete logarithm in the divisor class group of curves", *Mathematics of Computation*, **62** (1994), 865-874.
9. R. Gallant, R. Lambert and S. Vanstone, "Improving the parallelized Pollard lambda search on anomalous binary curves", *Mathematics of Computation*, **69** (2000), 1699-1705.
10. D. Gollmann, Y. Han and C. Mitchell, "Redundant integer representations and fast exponentiation", *Designs, Codes and Cryptography*, **7** (1996), 135-151.
11. D. Gordon, "A survey of fast exponentiation methods", *Journal of Algorithms*, **27** (1998), 129-146.
12. D. Hankerson, J. Hernandez and A. Menezes, "Software implementation of elliptic curve cryptography over binary fields", *Proceedings of CHES 2000*, LNCS **1965** (2000), 1-24.
13. N. Koblitz, "CM-curves with good cryptographic properties", *Advances in Cryptology - Crypto '91*, 1992, 279-287.
14. N. Koblitz, "An elliptic curve implementation of the finite field digital signature algorithm", *Advances in Cryptology - Crypto '98*, 1998, 327-337.
15. C. Lim and P. Lee, "More flexible exponentiation with precomputation", *Advances in Cryptology - Crypto '94*, 1994, 95-107.
16. A. Menezes, T. Okamoto and S. Vanstone, "Reducing elliptic curve logarithms to logarithms in a finite field", *IEEE Transactions on Information Theory*, **39** (1993), 1639-1646.
17. A. Menezes, P. van Oorschot and S. Vanstone, *Handbook of Applied Cryptography*, CRC Press, 1996.
18. A. Miyaji, T. Ono and H. Cohen, "Efficient elliptic curve exponentiation", *Proceedings of ICICS '97*, 1997, 282-290.
19. F. Morain and J. Olivos, "Speeding up the computations on an elliptic curve using addition-subtraction chains", *Informatique Théorique et Applications*, **24** (1990), 531-544.
20. V. Müller, "Fast multiplication in elliptic curves over small fields of characteristic two", *Journal of Cryptology*, **1** (1998), 219-234.
21. P. van Oorschot and M. Wiener, "Parallel collision search with cryptanalytic applications", *Journal of Cryptology*, **12** (1999), 1-28.
22. S. Pohlig and M. Hellman, "An improved algorithm for computing logarithms over $GF(p)$ and its cryptographic significance", *IEEE Transactions on Information Theory*, **24** (1978), 106-110.
23. J. Pollard, "Monte Carlo methods for index computation mod p", *Mathematics of Computation*, **32** (1978), 918-924.

24. J. Proos, personal communication, March 2000.
25. T. Satoh and K. Araki, "Fermat quotients and the polynomial time discrete log algorithm for anomalous elliptic curves", *Commentarii Mathematici Universitatis Sancti Pauli*, **47** (1998), 81-92.
26. I. Semaev, "Evaluation of discrete logarithms in a group of p-torsion points of an elliptic curve in characteristic p", *Mathematics of Computation*, **67** (1998), 353-356.
27. J. Silverman, *The Arithmetic of Elliptic Curves*, Springer-Verlag, 1986.
28. N. Smart, "The discrete logarithm problem on elliptic curves of trace one", *Journal of Cryptology*, **12** (1999), 193-196.
29. N. Smart, "Elliptic curve cryptosystems over small fields of odd characteristic", *Journal of Cryptology*, **12** (1999), 141-151.
30. J. Solinas, "An improved algorithm for arithmetic on a family of elliptic curves", *Advances in Cryptology – Crypto '97*, 1997, 357-371.
31. J. Solinas, "Generalized Mersenne numbers", Technical Report CORR 99-39, Dept. of C&O, University of Waterloo, 1999.
32. J. Solinas, "Efficient arithmetic on Koblitz curves", *Designs, Codes and Cryptography*, **19** (2000), 195-249.
33. WAP WTLS, *Wireless Application Protocol Wireless Transport Layer Security Specification*, Wireless Application Protocol Forum, February 1999. Drafts available at http://www.wapforum.org
34. M. Wiener and R. Zuccherato, "Faster attacks on elliptic curve cryptosystems", *Selected Areas in Cryptography*, LNCS **1556** (1999), 190-200.

On the Unpredictability of Bits of the Elliptic Curve Diffie–Hellman Scheme

Dan Boneh[1]* and Igor E. Shparlinski[2]**

[1] Department of Computer Science, Stanford University, CA, USA
dabo@cs.stanford.edu
[2] Department of Computing, Macquarie University, Sydney, NSW 2109, Australia
igor@comp.mq.edu.au

Abstract. Let $\mathbb{E}/\mathbb{F}_p$ be an elliptic curve, and $G \in \mathbb{E}/\mathbb{F}_p$. Define the Diffie–Hellman function as $\mathsf{DH}_{\mathbb{E},G}(aG, bG) = abG$. We show that if there is an efficient algorithm for predicting the LSB of the x or y coordinate of abG given $\langle \mathbb{E}, G, aG, bG \rangle$ for a certain family of elliptic curves, then there is an algorithm for computing the Diffie–Hellman function on all curves in this family. This seems stronger than the best analogous results for the Diffie–Hellman function in $\mathbb{F}_p^*$. Boneh and Venkatesan showed that in $\mathbb{F}_p^*$ computing approximately $(\log p)^{1/2}$ of the bits of the Diffie–Hellman secret is as hard as computing the entire secret. Our results show that just predicting one bit of the Elliptic Curve Diffie–Hellman secret in a family of curves is as hard as computing the entire secret.

1 Introduction

We recall how the Diffie–Hellman key exchange scheme works in an arbitrary finite cyclic group $\mathcal{G}$ of order T. Let g be a generator g of $\mathcal{G}$. Then to establish a common key, two communicating parties, *Alice* and *Bob* execute the following protocol, see [15,25]: *Alice* chooses a random integer $x \in [1, T-1]$, computes and sends $X = g^x$ to *Bob*. *Bob* chooses a random integer $y \in [1, T-1]$, computes and sends $Y = g^y$ to *Alice*. Now both *Alice* and *Bob* can compute the common *Diffie–Hellman secret*

$$K = Y^x = X^y = g^{xy}.$$

The Computational Diffie–Hellman assumption (CDH) in the group $\mathcal{G}$ states that no efficient algorithm can compute g^{xy} given g, g^x, g^y. However, this does not mean that one cannot compute a few bits of g^{xy} or perhaps predict some bits of g^{xy}. In fact, to use the Diffie–Hellman protocol in an efficient system one usually relies on the stronger Decision Diffie–Hellman assumption (DDH) [3]. Ideally, one would like to show than an algorithm for DDH in the group $\mathcal{G}$ implies an algorithm for CDH in $\mathcal{G}$. As a first step we show that, in the group of points of an elliptic curve over a finite field, predicting the least significant bit (LSB)

* Supported by NSF and the Packard Foundation.
** Supported in part by ARC

J. Kilian (Ed.): CRYPTO 2001, LNCS 2139, pp. 201–212, 2001.

of the Diffie–Hellman secret, for many curves in a family of curves, is as hard as computing the entire secret. Such results were previously known for the RSA function [1,7] but not for Diffie-Hellman.

Let p be prime and let $\lfloor s \rfloor_p$ denote the remainder of an integer s on division by p. We also use $\log z$ to denote the binary logarithm of $z > 0$. In the classical settings $\mathcal{G}$ is selected as the multiplicative group $\mathbb{F}_p^*$ of a finite field of p elements (and thus g is a primitive root of $\mathbb{F}_p$). In this case, Boneh and Venkatesan [5] showed that about $\log^{1/2} p$ most significant bits of $\lfloor g^{xy} \rfloor_p$ are as hard to find as $\lfloor g^{xy} \rfloor_p$ itself. The result is based on lattice reduction techniques. A similar result holds for the least significant bits as well. González Vasco and Shparlinski [10] used exponential sums to extend this result to subgroups $\mathcal{G}$ of $\mathbb{F}_p^*$. It has turned out that the lattice reduction technique used in [5] coupled with the exponential sum technique lead to a series of new results about the bits security of some cryptographic constructions [11,14,22,23] as well as to attacks on some of them [6, 13,17,18].

However the case where $\mathcal{G}$ is the point group of an elliptic curve has turned out to be much harder for applications of the lattice reduction based technique of [5] because of the inherited nonlinearity of the problem. Although some results have recently been obtained in [4] they are much weaker that those known for subgroups of $\mathbb{F}_p^*$. Here, using a very different technique, we show that working with a certain family of isomorphic curves (rather than with one fixed curve) allows to obtain results that are stronger than those known for subgroups of $\mathbb{F}_p^*$. By using certain twists of the given curve we show that predicting the least significant bit of the elliptic curve Diffie–Hellman secret in a family of curves is as hard as computing the entire secret. Since our techniques work with many curves at once they do not extend to the case of subgroups of $\mathbb{F}_p^*$.

2 Elliptic Curve Diffie–Hellman Scheme

Throughout the paper we let p be a prime and let $\mathbb{F}_p$ be the finite field of size p. Let $\mathbb{E}$ be an elliptic curve over $\mathbb{F}_p$, given by an affine Weierstrass equation of the form

$$Y^2 = X^3 + AX + B, \qquad 4A^3 + 27B^2 \neq 0 \tag{1}$$

It is known [24] that the set $\mathbb{E}(\mathbb{F}_p)$ of $\mathbb{F}_p$-rational points of $\mathbb{E}$ form an Abelian group under an appropriate composition rule and with the point at infinity $\mathcal{O}$ as the neutral element. We also recall that

$$|N - p - 1| \le 2p^{1/2},$$

where $N = |\mathbb{E}(\mathbb{F}_p)|$ is the number of $\mathbb{F}_p$-rational points, including the point at infinity $\mathcal{O}$.

Let $G \in \mathbb{E}$ be a point of order q, that is, q is the size of the cyclic group generated by G. Then the common key established at the end of the Diffie–Hellman protocol with respect to the curve $\mathbb{E}$ and the point G is $abG = (x, y) \in \mathbb{E}$ for some integers $a, b \in [1, q-1]$.

Throughout the paper we use the fact that the representation of $\mathbb{E}$ contains the field of definition of $\mathbb{E}$. With this convention, an algorithm given the representation of $\mathbb{E}/\mathbb{F}_p$ as input does not need to also be given p. The algorithm obtains p from the representation of $\mathbb{E}$.

Diffie-Hellman Function: Let $\mathbb{E}$ be an elliptic curve over $\mathbb{F}_p$ and let $G \in \mathbb{E}$ be a point of prime order q. We define the Diffie-Hellman function as:

$$\mathsf{DH}_{\mathbb{E},G}(aG, bG) = abG$$

where a, b are integers in $[1, q-1]$. The Diffie-Hellman problem on $\mathbb{E}$ is to compute $\mathsf{DH}_{\mathbb{E},G}(P, Q)$ given $\mathbb{E}, G, P, Q$. Clearly we mostly focus on curves in which the Diffie-Hellman problem is believed to be hard. Throughout we say that a randomized algorithm $\mathcal{A}$ computes the Diffie-Hellman function if $\mathcal{A}(\mathbb{E}, G, aG, bG) = abG$ holds with probability at least $1 - 1/p$. The probability is over the random bits used by $\mathcal{A}$.

Twists on elliptic curves: Let $\mathbb{E}$ be an elliptic curve over $\mathbb{F}_p$ given by the *Weierstrass equation* $y^2 = x^3 + Ax + B$. Our proofs rely on using certain *twists* of the elliptic curve. For $\lambda \in \mathbb{F}_p^*$ define $\phi_\lambda(\mathbb{E})$ to be the (twisted) elliptic curve:

$$Y^2 = X^3 + A\lambda^4 X + B\lambda^6. \tag{2}$$

We remark that $4(A\lambda^4)^3 + 27(B\lambda^6)^2 = (4A^3 + 27B^2)\lambda^{12} \neq 0$ for $\lambda \in \mathbb{F}_p^*$. Hence, $\phi_\lambda(\mathbb{E})$ is an elliptic curve for any $\lambda \in \mathbb{F}_p^*$. Throughout the paper we are working with the family of curves $\{\phi_\lambda(\mathbb{E}_0)\}_{\lambda \in \mathbb{F}_p^*}$ associated with a given curve $\mathbb{E}_0$.

It is easy to verify that for any point $P = (x, y) \in \mathbb{E}$ and any $\lambda \in \mathbb{F}_p^*$ the point $P_\lambda = (x\lambda^2, y\lambda^3) \in \phi_\lambda(\mathbb{E})$. Moreover, from the explicit formulas for the group law on $\mathbb{E}$ and $\phi_\lambda(\mathbb{E})$, see [2,24], we derive that for any points $P, Q, R \in \mathbb{E}$ with $P + Q = R$ we also have $P_\lambda + Q_\lambda = R_\lambda$. In particular, for any $G \in \mathbb{E}$ we have:

$$xG_\lambda = (xG)_\lambda, \qquad yG_\lambda = (yG)_\lambda, \qquad xyG_\lambda = (xyG)_\lambda.$$

Hence, the map $\phi_\lambda : \mathbb{E} \to \phi_\lambda(\mathbb{E})$ mapping $P \in \mathbb{E}$ to $P_\lambda \in \phi_\lambda(\mathbb{E})$ is a homomorphism. In fact, it is easy to verify that ϕ_λ is an isomorphism of groups. This means that

$$\mathsf{DH}_{\phi_\lambda(\mathbb{E}),G_\lambda}(P_\lambda, Q_\lambda) = \phi_\lambda[\mathsf{DH}_{\mathbb{E},G}(P, Q)].$$

Hence, if the Diffie-Hellman function is hard to compute in $\mathbb{E}$ then it is also hard to compute for all curves in $\{\phi_\lambda(\mathbb{E})\}_{\lambda \in \mathbb{F}_p^*}$.

3 Main Results

We denote by $\mathsf{LSB}(z)$ the *least significant bit* of an integer $z \geq 0$. When $z \in \mathbb{F}_p$ we let $\mathsf{LSB}(z)$ be $\mathsf{LSB}(x)$ for the unique integer $x \in [0, p-1]$ such that $x \equiv z \bmod p$.

Let p be a prime, and let $\mathbb{E}$ be an elliptic curve over $\mathbb{F}_p$. Let $G \in \mathbb{E}$ be a point of order q, for some prime q. We say that an algorithm $\mathcal{A}$ has advantage ϵ in predicting the LSB of the x-coordinate of the Diffie-Hellman function on $\mathbb{E}$ if:

$$\mathrm{Adv}^X_{\mathbb{E},G}(\mathcal{A}) = \left|\Pr_{a,b}[A(\mathbb{E}, G, aG, bG) = \mathsf{LSB}(x)] - \frac{1}{2}\right| > \varepsilon$$

where $abG = (x, y) \in \mathbb{E}$ and a, b are chosen uniformly at random in $[1, q-1]$. We write $\mathrm{Adv}^X_{\mathbb{E},G}(\mathcal{A}) > \varepsilon$. Similarly, we say that algorithm $\mathcal{A}$ has advantage ϵ in predicting the LSB of the y-coordinate of the Diffie-Hellman function if:

$$\mathrm{Adv}^Y_{\mathbb{E},G}(\mathcal{A}) = \left|\Pr_{a,b}[A(\mathbb{E}, G, aG, bG) = \mathsf{LSB}(y)] - \frac{1}{2}\right| > \varepsilon$$

where $abG = (x, y) \in \mathbb{E}$. We write $\mathrm{Adv}^Y_{\mathbb{E},G}(\mathcal{A}) > \varepsilon$.

The following result shows that no algorithm can have a non-negligible advantage in predicting the LSB of the x or y coordinates of the Diffie-Hellman secret for many curves in $\{\phi_\lambda(\mathbb{E}_0)\}_{\lambda\in\mathbb{F}_p^*}$, unless the Diffie–Hellman problem is easy on $\mathbb{E}_0$.

Theorem 1. *Let $\epsilon, \delta \in (0, 1)$. Let p be a prime, and let $\mathbb{E}_0$ be an elliptic curve over $\mathbb{F}_p$. Let $G \in \mathbb{E}_0$ be a point of prime order. Suppose there is a t-time algorithm $\mathcal{A}$ such that either:*

1. *$Adv^X_{\phi_\lambda(\mathbb{E}_0),\phi_\lambda(G)}(\mathcal{A}) > \varepsilon$ for at least a δ-fraction of the $\lambda \in \mathbb{F}_p^*$, or*
2. *$Adv^Y_{\phi_\lambda(\mathbb{E}_0),\phi_\lambda(G)}(\mathcal{A}) > \varepsilon$ for at least a δ-fraction of the $\lambda \in \mathbb{F}_p^*$.*

Then the Diffie–Hellman function $\mathsf{DH}_{\mathbb{E}_0,G}(P, Q)$ can be computed in expected time $t \cdot T(\log p, \frac{1}{\varepsilon\delta})$ where T is some fixed polynomial independent of p and $\mathbb{E}_0$.

Theorems 1 shows that, if the Diffie-Hellman problem is hard in $\mathbb{E}_0$, then no efficient algorithm can predict the least significant bit of the X or Y coordinates of the Diffie–Hellman function for a non-negligible fraction of the curves in $\{\phi_\lambda(\mathbb{E}_0)\}_{\lambda\in\mathbb{F}_p^*}$. The proof of Theorem 1 is given in Section 6. Note the theorem does not give a curve in $\{\phi_\lambda(\mathbb{E}_0)\}_{\lambda\in\mathbb{F}_p^*}$ for which the LSB of the X coordinate is a hard-core bit — it can still be the case that for every curve $\mathbb{E} \in \{\phi_\lambda(\mathbb{E}_0)\}_{\lambda\in\mathbb{F}_p^*}$ there is an efficient algorithm that predicts the LSB of $\mathsf{DH}_{\mathbb{E},G}$ for that curve only. However, there cannot be a single efficient algorithm that predicts this LSB for a non-negligible fraction of the curves in $\{\phi_\lambda(\mathbb{E}_0)\}_{\lambda\in\mathbb{F}_p^*}$.

An immediate corollary of Theorem 1 gives a hard core predicate for a simple extension of the Diffie-Hellman function. Let $\overline{\mathsf{DH}}_{\mathbb{E},G}$ be the function:

$$\overline{\mathsf{DH}}_{\mathbb{E},G}(P, Q, \lambda) = \mathsf{DH}_{\phi_\lambda(\mathbb{E}),G_\lambda}(P_\lambda, Q_\lambda)$$

where $G_\lambda = \phi_\lambda(G)$ and similarly P_λ, Q_λ. Note that this function basically uses λ as an index indicating in which group to execute the Diffie-Hellman protocol. Then the LSB of the X or Y coordinates is a hard-core bit of this function assuming the Diffie-Hellman problem is hard in $\mathbb{E}$.

Corollary 1. *Let $\mathbb{E}$ be an elliptic curve over $\mathbb{F}_p$ and let $G \in \mathbb{E}$ be of prime order q. Suppose there is a t-time algorithm $\mathcal{A}$ such that*

$$\Pr_{a,b,\lambda}\left[\mathcal{A}(\mathbb{E}, G, aG, bG, \lambda) = \mathsf{LSB}(x)\right] > \frac{1}{2} + \varepsilon$$

where $\overline{\mathsf{DH}}_{\mathbb{E},G}(aG, bG, \lambda) = (x, y) \in \phi_\lambda(\mathbb{E})$. Here a, b are uniformly chosen in $[1, q-1]$ and $\lambda \in \mathbb{F}_p^*$. Then the Diffie–Hellman function $\mathsf{DH}_{\mathbb{E}_0,G}$ can be computed in expected time $t \cdot T(\log p, \frac{1}{\varepsilon})$ where T is some fixed polynomial independent of p and $\mathbb{E}_0$.

We note that there are other ways of extending the Diffie-Hellman function to obtain a hard-core bit [8,12].

4 Review of the ACGS Algorithm

The proof of Theorem 1 uses an algorithm due to Alexi, Chor, Goldreich, and Schnorr [1]. We refer to this algorithm as the ACGS algorithm. For completeness, we briefly review the algorithm here. First, we define the following variant of the Hidden Number Problem (HNP) presented in [5].

HNP-CM*:* Fix an $\varepsilon > 0$. Let p be a prime. For an $\alpha \in \mathbb{F}_p$ let $L : \mathbb{F}_p^* \to \{0, 1\}$ be a function satisfying

$$\Pr_{t \in \mathbb{F}_p^*}\left[L(t) = \mathsf{LSB}\left(\lfloor \alpha \cdot t \rfloor_p\right)\right] \geq \frac{1}{2} + \varepsilon. \tag{3}$$

The HNP-CM problem is: given an oracle for $L(t)$, find α in polynomial time (in $\log p$ and $1/\epsilon$). Clearly we wish to show an algorithm for this problem that works for the smallest possible ε. For small ϵ there might be multiple α satisfying condition (3) (polynomially many in ε^{-1}). In this case the list-HNP-CM problem is to find the list of all such $\alpha \in \mathbb{F}_p$. Note that it is easy to verify that a given α belongs to the list of solutions by picking polynomially many random samples $x \in \mathbb{F}_p$ (say, $O(1/\varepsilon^2)$ samples suffice) and testing that $L(x) = \mathsf{LSB}(\lfloor \alpha x \rfloor_p)$ holds sufficiently often.

We refer to the above problem as HNP-CM to denote the fact that we are free to evaluate $L(t)$ at any multiplier t of our choice (the CM stands for Chosen Multiplier). In the original HNP studied in [5] one is only given samples $(t, L(t))$ for random t. The following theorem shows how to solve the HNP-CM for any $\varepsilon > 0$. The proof of the theorem (using different terminology) can be found in [1] and [7].

Theorem 2 (ACGS). *Let p be an n-bit prime and let $\varepsilon > 0$. Then, given ε, the list-*HNP-CM *problem can be solved in expected polynomial time in n and $1/\varepsilon$.*

Proof Sketch For $\alpha \in \mathbb{F}_p^*$ let $f_\alpha(t) : \mathbb{F}_p \to \{0, 1\}$ be a function such that $f_\alpha(t) = \mathsf{LSB}\left(\lfloor \alpha t \rfloor_p\right)$ for all $t \in \mathbb{F}_p$. It is well known that given an oracle for

$f_\alpha(t)$ it is possible to recover α using polynomially many queries (polynomial in $\log p$). See [1,7] or Theorem 7 of [5]. In fact, using the method of [1], it suffices to make queries only at t for which $\lfloor t\alpha \rfloor_p < p \cdot \varepsilon/2$ (as a result the run time is polynomial in $\log p$ and $1/\varepsilon$). Hence, the main challenge is in building an oracle for $f_\alpha(t)$ from an oracle for $L(t)$. The ACGS algorithm constructs an oracle for $f_\alpha(t)$ for every $\alpha \in \mathbb{F}_p^*$ that satisfies the condition (3). This construction is at the heart of the ACSG algorithm.

Let $m = n \cdot \frac{1}{\varepsilon^2}$. We show how to evaluate $f_\alpha(t)$ given an oracle for the function $L(t)$. We first pick random $u, v \in \mathbb{F}_p$. We use the same u, v to answer all queries to $f_\alpha(t)$. We assume that we know the $2\log m$ most significant bits and the least significant bit of $\lfloor u\alpha \rfloor_p$, $\lfloor v\alpha \rfloor_p$. This assumption is valid since we intend to run the ACGS algorithm with all possible values for these $2 + \lceil 4 \log m \rceil$ bits. In one of these iterations we obtain the correct values for the $2 + \lceil 4 \log m \rceil$ most significant bits and least significant bit of $\lfloor u\alpha \rfloor_p$, $\lfloor v\alpha \rfloor_p$. Note that different guesses for these bits will lead to oracles for $f_\alpha(t)$ for different values of α.

For $i = 1, \ldots, m$ let $r_i = \lfloor iu + v \rfloor_p$. Then $r_1, \ldots, r_m$ are pair wise independent values in $\mathbb{F}_p$ (over the choice of u, v). One can easily show (as in [1,7]) that using the knowledge of the most significant bits of $u\alpha, v\alpha \bmod p$ and the least significant bit, it is easy to determine $b_i = \mathsf{LSB}(\lfloor r_i \alpha \rfloor_p)$ for $i = 1, \ldots, m$. Therefore, to evaluate $f_\alpha(t)$ do the following:

1. Evaluate $a_i = L(t + r_i)$. Set $f_i = a_i \oplus b_i$, for $i = 1, \ldots, m$, where $\oplus$ denotes addition modulo 2.
2. Respond with $f_\alpha(t) = \text{Majority}(f_1, \ldots, f_m)$.

For a given $i \in [1, m]$ we say that a_i is correct if $a_i = \mathsf{LSB}(\lfloor \alpha(t + r_i) \rfloor_p)$. Recall that we only make $f_\alpha(t)$ queries at t satisfying $\lfloor t\alpha \rfloor_p < p \cdot \varepsilon/2$. Therefore, $\lfloor \alpha(t + r_i) \rfloor_p = \lfloor \alpha t \rfloor_p + \lfloor \alpha r_i \rfloor_p$, as integers, with probability at least $1 - \varepsilon/2$. Then $\mathsf{LSB}(\lfloor \alpha(t + r_i) \rfloor_p) = \mathsf{LSB}(\lfloor \alpha t \rfloor_p) \oplus \mathsf{LSB}(\lfloor \alpha r_i \rfloor_p)$. It follows that if a_i is correct then $f_i = \mathsf{LSB}(\lfloor t\alpha \rfloor_p)$ with probability at least $1 - \varepsilon/2$.

Since each r_i is uniformly distributed in $\mathbb{F}_p$ (over the choice of u, v) it follows that each a_i is correct with probability at least $\frac{1}{2} + \epsilon$. Since the r_i's are pair wise independent it follows that the f_i's are pair wise independent. Therefore, by Chebychev's inequality we obtain the correct value of $f_\alpha(t)$ with probability $1 - 1/n$. The exact analysis is given in [1]. Since we are able to construct an almost perfect subroutine for $f_\alpha(t)$ for all α satisfying the condition (3) the ACGS algorithm will produce a polynomial (in $\log p$) length list of candidates containing all required α. Note that it is easy to verify that a given α in the resulting list satisfies the condition (3) by picking polynomially many random samples $x \in \mathbb{F}_p$ and testing that $L(x) = \mathsf{LSB}(\lfloor \alpha x \rfloor_p)$ holds sufficiently often. □

We note that Fischlin and Schnorr [7] presented a more efficient algorithm for the HNP-CM. They rely on sub-sampling in Step 2 above to reduce the number of queries to the oracle for L.

5 Quadratic and Cubic Hidden Number Problems

To prove the main results of Section 3 we actually need an algorithm for the following variant of the HNP-CM problem.

HNP-CM^d*:* Fix an integer $d > 0$ and an $\varepsilon > 0$. Let p be a prime. For an $\alpha \in \mathbb{F}_p^*$ let $L^{(d)} : \mathbb{F}_p^* \to \{0, 1\}$ be a function satisfying

$$\Pr_{t \in \mathbb{F}_p^*} \left[L^{(d)}(t) = \mathsf{LSB}\left(\lfloor \alpha t^d \rfloor_p \right) \right] \geq \frac{1}{2} + \varepsilon. \tag{4}$$

The HNP-CM^d problem is: given an oracle for $L^{(d)}(t)$, find α in polynomial time. For small ε there might be multiple α satisfying condition (4) (polynomially many in ε^{-1}). In this case the list-HNP-CM^d problem is to find all such $\alpha \in \mathbb{F}_p^*$. We prove the following simple result regarding the list-HNP-CM^d problem. We use this theorem for $d = 2$ and $d = 3$.

Theorem 3. *Fix an integer $d > 1$. Let p be a n-bit prime and let $\varepsilon > 0$. Then, given ε, the* HNP-CM^d *problem can be solved in expected polynomial time in* $\log p$ *and d/ε.*

Proof. Let $L^{(d)}$ be a function satisfying the condition (4). Let $R : \mathbb{F}_p \to \{0, 1\}$ be a random function chosen uniformly from the set of all functions from $\mathbb{F}_p$ to $\{0, 1\}$. Let $S : \mathbb{F}_p^d \to \mathbb{F}_p$ be a function satisfying $S(x)^d \equiv x \bmod p$ for all $x \in \mathbb{F}_p^d$ and chosen at random from the set of such functions. Here $\mathbb{F}_p^d$ is the set of d'th powers in $\mathbb{F}_p$. The function S is simply a function mapping a d'th power $x \in \mathbb{F}_p^d$ to a randomly chosen d'th root of x. Next, define the following function $L(t)$:

$$L(t) = \begin{cases} L^{(d)}(S(t)) & \text{if } t \in \mathbb{F}_p^d, \\ R(t) & \text{otherwise.} \end{cases}$$

We claim that for any $\alpha \in \mathbb{F}_p^*$ satisfying the condition (4) we have that $L(t)$ satisfies

$$\Pr_{t,R,S} \left[L(t) = \mathsf{LSB}\left(\lfloor \alpha \cdot t \rfloor_p \right) \right] \geq \frac{1}{2} + \varepsilon/d.$$

To see this, fix an $\alpha \in \mathbb{F}_p$ satisfying the condition (4). Let $\mathcal{B}_t$ be the event that $L(t) = \mathsf{LSB}\left(\lfloor \alpha \cdot t \rfloor_p \right)$. Let $\mathcal{B}_t^d$ be the event that $L^{(d)}(t) = \mathsf{LSB}\left(\lfloor \alpha \cdot t^d \rfloor_p \right)$. Observe that if t is uniform in $\mathbb{F}_p^d \setminus \{0\}$ then $S(t)$ is uniform in $\mathbb{F}_p^*$. Let $e = \gcd(p-1, d)$.

If $e = 1$ then $\mathbb{F}_p = \mathbb{F}_p^d$ and therefore:

$$\Pr_{t,R,S} [\mathcal{B}_t] = \Pr_{t,R,S} \left[\mathcal{B}_{S(t)}^d \right] = \Pr_{x \in \mathbb{F}_p^*} \left[\mathcal{B}_x^d \right] \geq \frac{1}{2} + \varepsilon.$$

Hence, in this case the claim is correct. When $e > 1$ then the size of $\mathbb{F}_p^d \setminus \{0\} = \mathbb{F}_p^e \setminus \{0\}$ is $\frac{p-1}{e}$. Therefore:

$$\begin{aligned}\Pr_{t,R,S}[\mathcal{B}_t] &= \frac{1}{e} \Pr_{t,R,S}\left[\mathcal{B}_t \mid t \in \mathbb{F}_p^d\right] + \left(1 - \frac{1}{e}\right) \Pr_{t,R,S}\left[\mathcal{B}_t \mid t \notin \mathbb{F}_p^d\right] \\ &= \frac{1}{e} \Pr_{t,R,S}\left[\mathcal{B}_{S(t)}^d \mid t \in \mathbb{F}_p^e\right] + \left(1 - \frac{1}{e}\right) \cdot \frac{1}{2} \\ &\geq \frac{1}{e}\left(\frac{1}{2} + \varepsilon\right) + \left(1 - \frac{1}{e}\right) \cdot \frac{1}{2} = \frac{1}{2} + \frac{\varepsilon}{e} \geq \frac{1}{2} + \frac{\varepsilon}{d}\end{aligned}$$

and hence the claim holds in this case as well.

We see that an oracle for $L^{(d)}$ with advantage ϵ immediately gives rise to an oracle for L with advantage ϵ/d. Hence, we can use the ACGS algorithm to find the list of solutions to the given $\mathsf{HNP\text{-}CM}^d$ problem. When the ACGS algorithm runs we build the functions R and S as they are needed to respond to ACGS's queries to L. The ACGS algorithm will produce a super set of the solution set to the list-$\mathsf{HNP\text{-}CM}^d$ within the required time bound. Note that we may need to prune some of the solutions produced by the ACGS algorithm: we only output the α's for which the condition (4) holds. □

6 Proof of Main Results

We are now ready to prove Theorem 1. The proof reduces the problem of computing the Diffie–Hellman function to the Hidden Number Problem described in Section 5. We also use the following two simple lemmas. For a curve $\mathbb{E}/\mathbb{F}_p$ and $G \in \mathbb{E}$ of order q define:

$$F_{\mathbb{E},G,\lambda}(\mathcal{B}) = \Pr_{a,b}[\mathcal{B}(\phi_\lambda(\mathbb{E}), \phi_\lambda(G), \phi_\lambda(aG), \phi_\lambda(bG)) = \mathsf{LSB}(x_\lambda)]$$

where $\phi_\lambda(abG) = (x_\lambda, y_\lambda) \in \phi_\lambda(\mathbb{E})$ and a, b are uniform in $[1, q-1]$. Note that the probability space includes the random bits used by $\mathcal{B}$.

Lemma 1. *Let p be a prime, and let $\mathbb{E}$ be an elliptic curve over $\mathbb{F}_p$. Let $G \in \mathbb{E}$. Suppose there is a t-time algorithm $\mathcal{A}$ such that $Adv^X_{\phi_\lambda(\mathbb{E}),\phi_\lambda(G)}(\mathcal{A}) > \varepsilon$ for at least a δ-fraction of the $\lambda \in \mathbb{F}_p^*$.*
Then, given ϵ, δ, there is a t'-time algorithm $\mathcal{B}$ such that:
  (1) for at least a δ-fraction of the $\lambda \in \mathbb{F}_p^*$ we have that: $F_{\mathbb{E},G,\lambda}(\mathcal{B}) > \frac{1}{2} + \epsilon/2$, and
  (2) for the remaining $\lambda \in \mathbb{F}_p^*$ we have that: $F_{\mathbb{E},G,\lambda}(\mathcal{B}) > \frac{1}{2} - \frac{\varepsilon\delta}{4}$
Furthermore, $t' = t \cdot T(1/\epsilon\delta)$ for some fixed polynomial T independent of $p, \mathbb{E}$.

Proof. On input $\langle \mathbb{E}, G, P, Q \rangle$ algorithm $\mathcal{B}$ works as follows:
  1. Pick $u = (4/\epsilon\delta)^3$ random $a, b \in [1, q-1]$ pairs and run $\mathcal{A}$ on all tuples $\langle \mathbb{E}, G, aG, bG \rangle$.

2. let v be the number of runs in which $\mathcal{A}$ correctly outputs $\mathsf{LSB}((abG)_x)$.
3. if $v > u/2$ then $\mathcal{B}$ outputs $\mathcal{A}(\mathbb{E}, G, P, Q)$, otherwise $\mathcal{B}$ output the complement of $\mathcal{A}(\mathbb{E}, G, P, Q)$.

Let $\tau \geq \epsilon\delta/4$. For all $\lambda \in \mathbb{F}_p^*$ for which $\mathrm{Adv}^X_{\phi_\lambda(\mathbb{E}),\phi_\lambda(G)}(\mathcal{A}) > \tau$ we have that $\mathcal{B}$ satisfies: $F_{\mathbb{E},G,\lambda}(\mathcal{B}) > \frac{1}{2} + \tau/2$. This follows directly from Chebychev's inequality. For all other λ's, by definition of $\mathrm{Adv}(\mathcal{A})$ we have $F_{\mathbb{E},G,\lambda}(\mathcal{B}) > \frac{1}{2} - \epsilon\delta/4$. Hence, both conditions 1 and 2 are satisfied. □

Lemma 2. *Let $\mathcal{B}$ be an algorithm satisfying the two conditions of Lemma 1. Then*

$$\Pr_{\lambda \in \mathbb{F}_p^*}[\mathcal{B}(\phi_\lambda(\mathbb{E}), \phi_\lambda(G), \phi_\lambda(aG), \phi_\lambda(bG)) = \mathsf{LSB}(x_\lambda)] \geq \frac{1}{2} + \frac{\varepsilon\delta}{4}$$

holds with probability at least $\frac{\varepsilon\delta}{8}$ over the choice of $a, b \in [1, q-1]$, where $\phi_\lambda(abG) = (x_\lambda, y_\lambda)$.

Proof. The proof uses a standard counting argument. Algorithm $\mathcal{B}$ induces a matrix M whose entries are real numbers in $[0,1]$. There is a column in M for every $\lambda \in \mathbb{F}_p^*$ and a row for every $(a,b) \in [1, q-1]^2$. The entry at the λ'th column and (a,b)'th row is simply

$$\Pr[\mathcal{B}(\phi_\lambda(\mathbb{E}), \phi_\lambda(G), \phi_\lambda(aG), \phi_\lambda(bG)) = \mathsf{LSB}(x_\lambda)].$$

The probability is over the random bits used by $\mathcal{B}$. Suppose the matrix M has n columns and m rows. Since $\mathcal{B}$ satisfies the two condition of Lemma 1 we know that the sum of all the entries in M, which we call the weight of M denote by weight(M) is at least

$$\mathrm{weight}(M) > nm\left[\delta\left(\frac{1}{2} + \frac{\varepsilon}{2}\right) + (1-\delta)\left(\frac{1}{2} - \frac{\varepsilon\delta}{4}\right)\right] > nm\left(\frac{1}{2} + \frac{\delta\varepsilon}{4}\right).$$

Let R be the number of the rows in M must have weight at least $n[\frac{1}{2} + \frac{\varepsilon\delta}{8}]$ (the weight of a row is the sum of the entries in that row). We have

$$Rn + (m-R)n\left[\frac{1}{2} + \frac{\varepsilon\delta}{8}\right] \geq \mathrm{weight}(M) > nm\left(\frac{1}{2} + \frac{\delta\varepsilon}{4}\right).$$

Therefore

$$R\left[\frac{1}{2} - \frac{\varepsilon\delta}{8}\right] > \frac{\varepsilon\delta}{8}m.$$

The result now follows. □

We also need to review a theorem due to Shoup (Theorem 7 of [21]). The theorem shows that an algorithm that outputs a list of candidates for the Diffie-Hellman function can be easily converted into an algorithm that computes the Diffie-Hellman function. For concreteness we state the theorem as it applies to elliptic curves over $\mathbb{F}_p$.

Theorem 4 (Shoup). *Let $\mathbb{E}$ be an elliptic curve over $\mathbb{F}_p$ and let $G \in \mathbb{E}$ be an element of prime order q. Suppose there is a t-time algorithm $\mathcal{A}$ that given $aG, bG \in \mathbb{E}$ outputs a set of size m satisfying $\mathsf{DH}_{\mathbb{E},G}(aG, bG) \in A(\mathbb{E}, G, aG, bG)$ with probability at least 7/8. Then there is an algorithm $\mathcal{B}$ that computes the Diffie-Hellman function in $\mathbb{E}$ in time $t' = t(\log p) + T(m, \log p)$. Here T is a fixed polynomial independent of p and $\mathbb{E}$.*

Proof of Theorem 1: Let $\mathbb{E}$ be a curve over $\mathbb{F}_p$ and $G \in \mathbb{E}$ of prime order q. Suppose there is an expected t-time algorithm $\mathcal{A}$ such that $\mathrm{Adv}^X_{\phi_\lambda(\mathbb{E}),\phi_\lambda(G)}(\mathcal{A}) > \varepsilon$ for at least a δ-fraction of the $\lambda \in \mathbb{F}_p^*$. We show how to compute the Diffie–Hellman function $\mathsf{DH}_{\mathbb{E},G}$.

We are given $A = aG$ and $B = bG$ in $\mathbb{E}$. We wish to compute the point $C = abG \in \mathbb{E}$. We first randomize the problem by computing $A' = a_0 A$ and $B' = b_0 B$ for random $a_0, b_0 \in [1, q-1]$. If $C' = \mathsf{DH}_{\mathbb{E},G}(A', B')$ then $C = c_0 C'$ where $c_0 \equiv (a_0 b_0)^{-1} \bmod q$. Hence, it suffices to find C'. Write $C' = (x_0, y_0)$.

Since $\phi_\lambda : \mathbb{E} \to \phi_\lambda(\mathbb{E})$ is an isomorphism it follows that

$$\mathsf{DH}_{\phi_\lambda(\mathbb{E}),\phi_\lambda(G)}(\phi_\lambda(A'), \phi_\lambda(B')) = \phi_\lambda(C') = (\lambda^2 x_0, \lambda^3 y_0).$$

Since A', B' are uniformly distributed in the group generated by G (excluding $\mathcal{O}$) we can apply both Lemma 1 and Lemma 2 to obtain an algorithm $\mathcal{B}$ satisfying:

$$\Pr_\lambda[\mathcal{B}(\phi_\lambda(\mathbb{E}), \phi_\lambda(G), \phi_\lambda(A'), \phi_\lambda(B')) = \mathsf{LSB}(\lambda^2 x_0)] > \frac{1}{2} + \frac{\varepsilon\delta}{8} \qquad (5)$$

is true with probability at least $\varepsilon\delta/8$ over the choice of a_0, b_0 in $[1, q-1]$.

For now we assume that (5) holds. We obtain an $\mathsf{HNP\text{-}CM}^2$ problem where x_0 is the hidden number. To see this, define:

$$L^{(2)}(\lambda) = \mathcal{A}(\phi_\lambda(\mathbb{E}), \phi_\lambda(G), \phi_\lambda(A'), \phi_\lambda(B')).$$

Then the condition 5 implies that $\Pr_\lambda[L^{(2)}(\lambda) = \mathsf{LSB}(\lambda^2 x_0)] > \frac{1}{2} + \frac{\varepsilon\delta}{8}$. We can therefore use the algorithm of Theorem 3 to find a list of candidates $x_1, \ldots, x_n \in \mathbb{F}_p$ containing the desired x_0.

To ensure that condition (5) holds, we repeat this process $\lceil 8/\varepsilon\delta \rceil$ times and build a list of candidates of size $O(n/\delta\varepsilon)$. Then condition (5) holds with constant probability during one of these iterations. Therefore, the list of candidates contains the correct x_0 with constant probability. By solving for y we obtain a list of candidates for C'. That is, we obtain a set S' such that $C' \in S' \subseteq \mathbb{E}$. This list S' can be easily converted to a list of candidates S for C by setting $S = \{c_0 P \mid P \in S'\}$.

Therefore, we just constructed a polynomial time algorithm (in $\log p$ and $\frac{1}{\varepsilon\delta}$) that for any $aG, bG \in \mathbb{E}$ outputs a polynomial size list containing C with constant probability. Using Theorem 4 this algorithm gives an algorithm for computing the Diffie-Hellman function in $\mathbb{E}$ in the required time bound.

To complete the proof of the theorem we also need to consider an algorithm predicting the LSB of the y-coordinates. That is, suppose there is an expected

t-time algorithm $\mathcal{A}$ such that $\mathrm{Adv}^{Y}_{\phi_\lambda(\mathbb{E}),\phi_\lambda(G)}(\mathcal{A}) > \varepsilon$ for a δ-fraction of $\lambda \in \mathbb{F}_p^*$. We show how to compute the Diffie–Hellman function $\mathsf{DH}_{\mathbb{E},G}$. The proof in this case is very similar to the proof for the x-coordinate. The only difference is that since we are using the Y coordinate we obtain an $\mathsf{HNP\text{-}CM}^3$ problem. We use Lemma 1 and Lemma 2 to obtain an $\mathsf{HNP\text{-}CM}^3$ oracle with advantage $\varepsilon\delta/8$ in predicting $\mathsf{LSB}(\lambda^3 y_0)$. The theorem now follows from the algorithm for $\mathsf{HNP\text{-}CM}^3$ given in Theorem 3.

□

7 Conclusions

We have showed that no algorithm can predict the LSB of the X and Y coordinates of the elliptic curve Diffie–Hellman secret for a non-negligible fraction of the curves in $\{\phi_\lambda(\mathbb{E}_0)\}_{\lambda\in\mathbb{F}_p^*}$, assuming the Diffie–Hellman problem is hard on some curve $\mathbb{E}_0 \in \{\phi_\lambda(\mathbb{E}_0)\}_{\lambda\in\mathbb{F}_p^*}$. Our proofs use reductions between many curves by randomly twisting the curve $\mathbb{E}_0$. We hope these techniques will eventually lead to a proof that if CDH is hard on a certain curve $\mathbb{E}$ then the LSB of Diffie-Hellman is a hard core predicate on that curve.

References

1. W. Alexi, B. Chor, O. Goldreich, and C. Schnorr. 'RSA and Rabin functions: Certain parts are as hard as the whole', *SIAM J. Computing*, **17**(1988), 194–209, Nov. 1988.
2. I. Blake, G. Seroussi, and N. Smart, *Elliptic Curves in Cryptography*, London Mathematical Society, Lecture Notes Series, **265**, Cambridge University Press, 1999.
3. D. Boneh, 'The decision Diffie–Hellman problem', *In Proc. 3rd Algorithmic Number Theory Symposium*, Lect. Notes in Comp. Sci., Springer-Verlag, Berlin, **1423** (1998), 48–63.
4. D. Boneh, S. Halevi and N. A. Howgrave-Graham, 'The modular inversion hidden number problem', *Preprint*, 2001.
5. D. Boneh and R. Venkatesan, 'Hardness of computing the most significant bits of secret keys in Diffie–Hellman and related schemes', *In Proc. Crypto '96*, Lect. Notes in Comp. Sci., Springer-Verlag, Berlin, **1109** (1996), 129–142. Recent version available at `http://crypto.stanford.edu/~dabo/`.
6. E. El Mahassni, P. Q. Nguyen and I. E. Shparlinski, 'The insecurity of Nyberg–Rueppel and other DSA-like signature schemes with partially known nonces', *Proc. Workshop on Lattices and Cryptography*, Boston, MA, 2001 (to appear).
7. R. Fischlin, C. Schnorr, 'Stronger security proofs for RSA and Rabin bits', *J. Cryptology*, **13** (2000), 221–244.
8. O. Goldreich, L. Levin, 'A hard core predicate for any one way function', *In Proc. 21st ACM Symp. on Theory of Comput.*, 1989, 25–32.
9. M. I. González Vasco and M. Näslund, 'A survey of hard core functions', *In Proc. Workshop on Cryptography and Computational Number Theory*, Singapore 1999, Birkhäuser, 2001, 227–256.

10. M. I. González Vasco and I. E. Shparlinski, 'On the security of Diffie–Hellman bits', *Proc. Workshop on Cryptography and Computational Number Theory*, Singapore 1999, Birkhäuser, 2001, 257–268.
11. M. I. González Vasco and I. E. Shparlinski, 'Security of the most significant bits of the Shamir message passing scheme', *Math. Comp.* (to appear).
12. M. Naslund, 'All bits in $ax+b$ mod p are hard', *In Proc. Crypto '96*, Lect. Notes in Comp. Sci., Springer-Verlag, Berlin, **1109** (1996), 114–128.
13. N. A. Howgrave-Graham and N. P. Smart, 'Lattice attacks on digital signature schemes', *Designs, Codes and Cryptography* (to appear).
14. N. A. Howgrave-Graham, P. Q. Nguyen and I. E. Shparlinski, 'Hidden number problem with hidden multipliers, timed-release crypto and noisy exponentiation', *Preprint*, 2000, 1–26.
15. A. J. Menezes, P. C. van Oorrschot and S. A. Vanstone, *Handbook of applied cryptography*, CRC Press, Boca Raton, FL, 1996.
16. P. Q. Nguyen, 'The dark side of the hidden number problem: Lattice attacks on DSA', *Proc. Workshop on Cryptography and Computational Number Theory*, Singapore 1999, Birkhäuser, 2001, 321–330.
17. P. Q. Nguyen and I. E. Shparlinski, 'The insecurity of the Digital Signature Algorithm with partially known nonces', *Preprint*, 2000, 1–26.
18. P. Q. Nguyen and I. E. Shparlinski, 'The insecurity of the elliptic curve Digital Signature Algorithm with partially known nonces', *Preprint*, 2000, 1–24.
19. P. Q. Nguyen and J. Stern, 'Lattice reduction in cryptology: An update', *In Proc. 4th Algorithmic Number Theory Symposium*, Lect. Notes in Comp. Sci., Springer-Verlag, Berlin, **1838** (2000), 85–112.
20. H. Niederreiter, *Random number generation and quasi–Monte Carlo methods*, SIAM, Philadelphia, 1992.
21. V. Shoup, 'Lower bounds for discrete logarithms and related problems', *In Proc. Eurocrypt '97*, Lect. Notes in Comp. Sci., Springer-Verlag, Berlin, **1233** (1997), 256–266.
22. I. E. Shparlinski, 'Sparse polynomial approximation in finite fields', *Proc. 33rd ACM Symp. on Theory of Comput.*, Crete, Greece, July 6-8, 2001 (to appear).
23. I. E. Shparlinski, 'On the generalized hidden number problem and bit security of XTR', *In Proc. the 14th Symp. on Appl. Algebra, Algebraic Algorithms, and Error-Correcting Codes*, Lect. Notes in Comp. Sci., Springer-Verlag, Berlin (to appear).
24. J. H. Silverman, *The arithmetic of elliptic curves*, Springer-Verlag, Berlin, 1995.
25. D. R. Stinson, *Cryptography: Theory and practice*, CRC Press, Boca Raton, FL, 1995.

Identity-Based Encryption from the Weil Pairing

Dan Boneh[1]* and Matt Franklin[2]**

[1] Computer Science Department, Stanford University, Stanford CA 94305-9045
dabo@cs.stanford.edu

[2] Computer Science Department, University of California, Davis CA 95616-8562
franklin@cs.ucdavis.edu

Abstract. We propose a fully functional identity-based encryption scheme (IBE). The scheme has chosen ciphertext security in the random oracle model assuming an elliptic curve variant of the computational Diffie-Hellman problem. Our system is based on the Weil pairing. We give precise definitions for secure identity based encryption schemes and give several applications for such systems.

1 Introduction

In 1984 Shamir [27] asked for a public key encryption scheme in which the public key can be an arbitrary string. In such a scheme there are four algorithms: (1) setup generates global system parameters and a master-key, (2) extract uses the master-key to generate the private key corresponding to an arbitrary public key string $\mathsf{ID} \in \{0,1\}^*$, (3) encrypt encrypts messages using the public key ID, and (4) decrypt decrypts messages using the corresponding private key.

Shamir's original motivation for identity-based encryption was to simplify certificate management in e-mail systems. When Alice sends mail to Bob at `bob@hotmail.com` she simply encrypts her message using the public key string "`bob@hotmail.com`". There is no need for Alice to obtain Bob's public key certificate. When Bob receives the encrypted mail he contacts a third party, which we call the Private Key Generator (PKG). Bob authenticates himself to the PKG in the same way he would authenticate himself to a CA and obtains his private key from the PKG. Bob can then read his e-mail. Note that unlike the existing secure e-mail infrastructure, Alice can send encrypted mail to Bob even if Bob has not yet setup his public key certificate. Also note that key escrow is inherent in identity-based e-mail systems: the PKG knows Bob's private key. We discuss key revocation, as well as several new applications for IBE schemes in the next section.

Since the problem was posed in 1984 there have been several proposals for IBE schemes (e.g., [7,29,28,21]). However, none of these are fully satisfactory. Some solutions require that users not collude. Other solutions require the PKG to spend a long time for each private key generation request. Some solutions

* Supported by DARPA contract F30602-99-1-0530 and the Packard Foundation.
** Supported by an NSF Career Award.

J. Kilian (Ed.): CRYPTO 2001, LNCS 2139, pp. 213–229, 2001.

require tamper resistant hardware. It is fair to say that constructing a usable IBE system is still an open problem. Interestingly, the related notions of identity-based signature and authentication schemes, also introduced by Shamir [27], do have satisfactory solutions [11,10].

In this paper we propose a fully functional identity-based encryption scheme. The performance of our system is comparable to the performance of ElGamal encryption in $\mathbb{F}_p^*$. The security of our system is based on a natural analogue of the computational Diffie-Hellman assumption on elliptic curves. Based on this assumption we show that the new system has chosen ciphertext security in the random oracle model. Using standard techniques from threshold cryptography [14,15] the PKG in our scheme can be distributed so that the master-key is never available in a single location. Unlike common threshold systems, we show that robustness for our distributed PKG is free.

Our IBE system can be built from any bilinear map $e : \mathbb{G}_1 \times \mathbb{G}_1 \to \mathbb{G}_2$ between two groups $\mathbb{G}_1, \mathbb{G}_2$ as long as a variant of the Computational Diffie-Hellman problem in $\mathbb{G}_1$ is hard. We use the Weil pairing on elliptic curves as an example of such a map. Until recently the Weil pairing has mostly been used for attacking elliptic curve systems [22,13]. Joux [17] recently showed that the Weil pairing can be used for "good" by using it in a protocol for three party one round Diffie-Hellman key exchange. Using similar ideas, Verheul [30] recently constructed an ElGamal encryption scheme where each public key has two corresponding private keys. In addition to our identity-based encryption scheme, we show how to construct an ElGamal encryption scheme with "built-in" key escrow, i.e., where one global escrow key can decrypt ciphertexts encrypted under any public key.

To argue about the security of our IBE system we define chosen ciphertext security for identity-based encryption. Our model is slightly stronger than the standard model for chosen ciphertext security [25,1]. While mounting a chosen ciphertext attack on the public key ID, the attacker could ask the PKG for the private key of some public key $\mathsf{ID}' \neq \mathsf{ID}$. This private key might help the attacker. Hence, during the chosen ciphertext attack we allow the attacker to obtain the private key for any public key of her choice other than the one on which the attacker is being challenged. Even with the help of such queries the attacker should have negligible advantage in defeating the semantic security of the system.

The rest of the paper is organized as follows. Several applications of identity-based encryption are discussed in Section 1.1. We then give precise definitions and security models in Section 2. Basic properties of the Weil pairing – sufficient for an understanding of our constructions – are discussed in Section 3. Our main identity-based encryption scheme is presented in Section 4. Some extensions and variations (efficiency improvements, distribution of the master-key) are considered in Section 5. Our construction for ElGamal encryption with a global escrow key is described in Section 6. Conclusions and open problems are discussed in Section 7.

1.1 Applications for Identity-Based Encryption

The original motivation for identity-based encryption is to help the deployment of a public key infrastructure. In this section, we show several other unrelated applications.

Revocation of Public Keys. Public key certificates contain a preset expiration date. In an IBE system key expiration can be done by having Alice encrypt e-mail sent to Bob using the public key: "`bob@hotmail.com` ‖ `current-year`". In doing so Bob can use his private key during the current year only. Once a year Bob needs to obtain a new private key from the PKG. Hence, we get the effect of annual private key expiration. Note that unlike the existing PKI, Alice does not need to obtain a new certificate from Bob every time Bob refreshes his certificate.

One could potentially make this approach more granular by encrypting e-mail for Bob using "`bob@hotmail.com` ‖ `current-date`". This forces Bob to obtain a new private key every day. This might be feasible in a corporate PKI where the PKG is maintained by the corporation. With this approach key revocation is quite simple: when Bob leaves the company and his key needs to be revoked, the corporate PKG is instructed to stop issuing private keys for Bob's e-mail address. The interesting property is that Alice does not need to communicate with any third party to obtain Bob's daily public key. This approach enables Alice to send messages into the future: Bob will only be able to decrypt the e-mail on the date specified by Alice (see [26,8] for methods of sending messages into the future using a stronger security model).

Delegation of Decryption Keys. Another application for IBE systems is delegation of decryption capabilities. We give two example applications. In both applications the user Bob plays the role of the PKG. Bob runs the setup algorithm to generate his own IBE system parameters params and his own master-key. Here we view params as Bob's public key. Bob obtains a certificate from a CA for his public key params. When Alice wishes to send mail to Bob she first obtains Bob's public key params and public key certificate.

1. **Delegation to a laptop.** Suppose Alice encrypts mail to Bob using the current date as the IBE encryption key (she uses Bob's params as the IBE system parameters). Since Bob has the master-key he can extract the private key corresponding to this IBE encryption key and then decrypt the message. Now, suppose Bob goes on a trip for seven days. Normally, Bob would put his private key on his laptop. If the laptop is stolen the private key is compromised. When using the IBE system Bob could simply install on his laptop the seven private keys corresponding to the seven days of the trip. If the laptop is stolen, only the private key for those seven days are compromised. The master-key is unharmed. This is analogous to the delegation scenario for *signature schemes* considered by Goldreich et al. [16].

2. Delegation of duties. Suppose Alice encrypts mail to Bob using the subject line as the IBE encryption key. Bob can decrypt mail using his master-key. Now, suppose Bob has several assistants each responsible for a different task (e.g. one is 'purchasing', another is 'human-resources', etc.). Bob gives one private key to each of his assistants corresponding to the assistant's responsibility. Each assistant can then decrypt messages whose subject line falls within its responsibilities, but it cannot decrypt messages intended for other assistants. Note that Alice only obtains a single public key from Bob (params), and she uses that public key to send mail with any subject line of her choice. The mail can only be read by the assistant responsible for that subject.

More generally, IBE can simplify various systems that manage a large number of public keys. Rather than storing a big database of public keys the system can either derive these public keys from usernames, or simply use the integers $1, \ldots, n$ as distinct public keys.

2 Definitions

Bilinear Map. Let $\mathbb{G}_1$ and $\mathbb{G}_2$ be two cyclic groups of order q for some large prime q. In our system, $\mathbb{G}_1$ is the group of points of an elliptic curve over $\mathbb{F}_p$ and $\mathbb{G}_2$ is a subgroup of $\mathbb{F}_{p^2}^*$. Therefore, we view $\mathbb{G}_1$ as an additive group and $\mathbb{G}_2$ as a multiplicative group. A map $\hat{e} : \mathbb{G}_1 \times \mathbb{G}_1 \to \mathbb{G}_2$ is said to be *bilinear* if $\hat{e}(aP, bQ) = \hat{e}(P, Q)^{ab}$ for all $P, Q \in \mathbb{G}_1$ and all $a, b \in \mathbb{Z}$. As we will see in Section 3, the Weil pairing is an example of an efficiently computable non-degenerate bilinear map.

Weil Diffie-Hellman Assumption (WDH). Our IBE system can be built from any bilinear map $\hat{e} : \mathbb{G}_1 \times \mathbb{G}_1 \to \mathbb{G}_2$ for which the following assumption holds: there is no efficient algorithm to compute $\hat{e}(P, P)^{abc} \in \mathbb{G}_2$ from $P, aP, bP, cP \in \mathbb{G}_1$ where $a, b, c \in \mathbb{Z}$. This assumption is precisely defined in Section 3. We note that this WDH assumption implies that the Diffie-Hellman problem is hard in the group $\mathbb{G}_1$.

Identity-Based Encryption. An identity-based encryption scheme is specified by four randomized algorithms: Setup, Extract, Encrypt, Decrypt:

Setup: takes a security parameter k and returns params (system parameters) and master-key. The system parameters include a description of a finite message space $\mathcal{M}$, and a description of a finite ciphertext space $\mathcal{C}$. Intuitively, the system parameters will be publicly known, while the master-key will be known only to the "Private Key Generator" (PKG).

Extract: takes as input params, master-key, and an arbitrary $\mathsf{ID} \in \{0, 1\}^*$, and returns a private key d. Here ID is an arbitrary string that will be used as a public key, and d is the corresponding private decryption key. The Extract algorithm extracts a private key from the given public key.

Encrypt: takes as input params, ID, and $M \in \mathcal{M}$. It returns a ciphertext $C \in \mathcal{C}$.

Decrypt: takes as input params, ID, $C \in \mathcal{C}$, and a private key d. It return $M \in \mathcal{M}$.

These algorithms must satisfy the standard consistency constraint, namely when d is the private key generated by algorithm Extract when it is given ID as the public key, then

$$\forall M \in \mathcal{M} : \mathsf{Decrypt}(\mathsf{params}, \mathsf{ID}, C, d) = M \quad \text{where} \quad C = \mathsf{Encrypt}(\mathsf{params}, \mathsf{ID}, M)$$

Chosen ciphertext security. Chosen ciphertext security (IND-CCA) is the standard acceptable notion of security for a public key encryption scheme [25,1,9]. Hence, it is natural to require that an identity-based encryption scheme also satisfy this strong notion of security. However, the definition of chosen ciphertext security must be strengthened a bit. The reason is that when an attacker attacks a public key ID in an identity-based system, the attacker might already possess the private keys of users $\mathsf{ID}_1, \ldots, \mathsf{ID}_n$ of her choice. The system should remain secure under such an attack. Hence, the definition of chosen ciphertext security must allow the attacker to obtain the private key associated with any identity ID_i of her choice (other than the public key ID being attacked). We refer to such queries as private key extraction queries. Another difference is that the attacker is challenged on a public key ID of her choice (as opposed to a random public key).

We say that an identity-based encryption scheme is semantically secure against an adaptive chosen ciphertext attack (IND-ID-CCA) if no polynomially bounded adversary $\mathcal{A}$ has a non-negligible advantage against the Challenger in the following game:

Setup: The challenger takes a security parameter k and runs the Setup algorithm. It gives the adversary the resulting system parameters params. It keeps the master-key to itself.

Phase 1: The adversary issues queries $q_1, \ldots, q_m$ where query q_i is one of:

- Extraction query $\langle \mathsf{ID}_i \rangle$. The challenger responds by running algorithm Extract to generate the private key d_i corresponding to the public key $\langle \mathsf{ID}_i \rangle$. It sends d_i to the adversary.
- Decryption query $\langle \mathsf{ID}_i, C_i \rangle$. The challenger responds by running algorithm Extract to generate the private key d_i corresponding to ID_i. It then runs algorithm Decrypt to decrypt the ciphertext C_i using the private key d_i. It sends the resulting plaintext to the adversary.

These queries may be asked adaptively, that is, each query q_i may depend on the replies to $q_1, \ldots, q_{i-1}$.

Challenge: Once the adversary decides that Phase 1 is over it outputs two plaintexts $M_0, M_1 \in \mathcal{M}$ and an identity ID on which it wishes to be challenged. The only constraint is that ID did not appear in any private key extraction query in Phase 1.

The challenger picks a random bit $b \in \{0, 1\}$ and sets $C = \mathsf{Encrypt}(\mathsf{params}, \mathsf{ID}, M_b)$. It sends C as the challenge to the adversary.

Phase 2: The adversary issues more queries $q_{m+1}, \ldots, q_n$ where query q_i is one of:

- Extraction query $\langle \mathsf{ID}_i \rangle$ where $\mathsf{ID}_i \neq \mathsf{ID}$. Challenger responds as in Phase 1.
- Decryption query $\langle \mathsf{ID}_i, C_i \rangle \neq \langle \mathsf{ID}, C \rangle$. Challenger responds as in Phase 1.

These queries may be asked adaptively as in Phase 1.

Guess: Finally, the adversary outputs a guess $b' \in \{0, 1\}$. The adversary wins the game if $b = b'$.

We refer to such an adversary $\mathcal{A}$ as an IND-ID-CCA attacker. We define adversary $\mathcal{A}$'s advantage in attacking the scheme as: $\mathrm{Adv}(\mathcal{A}) = \left|\Pr[b = b'] - \frac{1}{2}\right|$. The probability is over the random bits used by the challenger and the adversary. We say that the IBE system is semantically secure against an adaptive chosen ciphertext attack (IND-ID-CCA) if no polynomially bounded adversary has a non-negligible advantage in attacking the scheme. As usual, "non-negligible" should be understood as larger than $1/f(k)$ for some polynomial f (recall k is the security parameter). Note that the standard definition of chosen ciphertext security (IND-CCA) [25,1] is the same as above except that there are no private key extraction queries and the attacker is challenged on a random public key (rather than a public key of her choice).

Private key extraction queries are related to the definition of chosen ciphertext security in the multiuser settings [4]. After all, our definition involves multiple public keys belonging to multiple users. In [4] the authors show that that multiuser IND-CCA is reducible to single user IND-CCA using a standard hybrid argument. This does not hold in the identity-based settings, IND-ID-CCA, since the attacker gets to choose which public keys to corrupt during the attack. To emphasize the importance of private key extraction queries we note that our IBE system can be easily modified (by removing one of the hash functions) into a system which has chosen ciphertext security when private extraction queries are disallowed. However, the scheme is completely insecure when extraction queries are allowed.

One way identity-based encryption. The proof of security for our IBE system makes use of a weak notion of security called one-way encryption (OWE) [12]. OWE is defined for standard public key encryption schemes (not identity based) as follows: the attacker $\mathcal{A}$ is given a random public key K_{pub} and a ciphertext C which is the encryption of a random message M using K_{pub}. The attacker's goal is to recover the corresponding plaintext. It has advantage ϵ in attacking the system if $\Pr[\mathcal{A}(K_{pub}, C) = M] = \epsilon$. We say that the public key scheme is a one-way encryption scheme (OWE) if no polynomial time attacker has non-negligible advantage in attacking the scheme. See [12] for precise definitions.

For identity-based encryption, we strengthen the definition as follows. We say that an IBE scheme is a one-way identity-based encryption scheme (ID-OWE) if no polynomially bounded adversary $\mathcal{A}$ has a non-negligible advantage against the Challenger in the following game:

Setup: The challenger takes a security parameter k and runs the Setup algorithm. It gives the adversary the resulting system parameters params. It keeps the master-key to itself.

Phase 1: The adversary issues private key extraction queries $\mathsf{ID}_1, \ldots, \mathsf{ID}_m$. The challenger responds by running algorithm Extract to generate the private key d_i corresponding to the public key ID_i. It sends d_i to the adversary. These queries may be asked adaptively.

Challenge: Once the adversary decides that Phase 1 is over it outputs a public key $\mathsf{ID} \neq \mathsf{ID}_1, \ldots, \mathsf{ID}_m$ on which it wishes to be challenged. The challenger picks a random $M \in \mathcal{M}$ and encrypts M using ID as the public key. It then sends the resulting ciphertext C to the adversary.

Phase 2: The adversary issues more extraction queries $\mathsf{ID}_{m+1}, \ldots, \mathsf{ID}_n$. The only constraint is that $\mathsf{ID}_i \neq \mathsf{ID}$. The challenger responds as in Phase 1.

Guess: Finally, the adversary outputs a guess $M' \in \mathcal{M}$. The adversary wins the game if $M = M'$.

We refer to such an attacker $\mathcal{A}$ as an ID-OWE attacker. We define adversary's $\mathcal{A}$'s advantage in attacking the scheme as: $\mathrm{Adv}(\mathcal{A}) = \Pr[M = M']$. The probability is over the random bits used by the challenger and the adversary. Note that the definitions of OWE is the same as ID-OWE except that there are no private key extraction queries and the attacker is challenged on a random public key (rather than a public key of her choice).

3 Properties of the Weil Pairing

The bilinear map $\hat{e} : \mathbb{G}_1 \times \mathbb{G}_1 \to \mathbb{G}_2$ discussed in Section 2 is implemented via the Weil pairing. In this section we describe the basic properties of this pairing and the complexity assumption needed for the security of our system. To make the presentation concrete we consider a specific supersingular elliptic curve. In Section 5 we describe several extensions and observations for our approach. The complete definition and algorithm for computing the pairing are given in the full version of the paper [2].

Let p be a prime satisfying $p = 2 \bmod 3$ and $p = 6q - 1$ for some prime q. Let E be the elliptic curve defined by the equation $y^2 = x^3 + 1$ over $\mathbb{F}_p$. We state a few elementary facts about this curve:

Fact 1: Since $x^3 + 1$ is a permutation on $\mathbb{F}_p$ it easily follows that $E/\mathbb{F}_p$ contains $p+1$ points. We let O denote the point at infinity. Let $P \in E/\mathbb{F}_p$ be a generator of the group of points of order $q = (p+1)/6$. We denote this group by G_q.

Fact 2: For any $y_0 \in \mathbb{F}_p$ there is a unique point (x_0, y_0) on $E/\mathbb{F}_p$. Hence, if (x, y) is a random non-zero point on $E/\mathbb{F}_p$ then y is uniform in $\mathbb{F}_p$. We use this property to simplify the proof of security.

Fact 3: Let $1 \neq \zeta \in \mathbb{F}_{p^2}$ be a solution of $x^3 - 1 = 0 \bmod p$. Then the map $\phi(x, y) = (\zeta x, y)$ is an automorphism of the group of points on the curve E. Note that when $P = (x, y) \in E/\mathbb{F}_p$ we have that $\phi(P) \in E/\mathbb{F}_{p^2}$, but $\phi(P) \notin E/\mathbb{F}_p$. Hence, $P \in E/\mathbb{F}_p$ is linearly independent of $\phi(P) \in E/\mathbb{F}_{p^2}$.

Fact 4: Since the points P and $\phi(P)$ are linearly independent they generate a group isomorphic to $\mathbb{Z}_q \times \mathbb{Z}_q$. We denote this group of points by $E[q]$.

Let μ_q be the subgroup of $\mathbb{F}_{p^2}^*$ containing all elements of order $q = (p+1)/6$. The Weil pairing on the curve $E/\mathbb{F}_{p^2}$ is a mapping $e : E[q] \times E[q] \to \mu_q$. We define the modified Weil pairing $\hat{e} : G_q \times G_q \to \mu_q$ to be:

$$\hat{e}(P, Q) = e(P, \phi(Q))$$

The modified Weil pairing satisfies the following properties:

1. Bilinear: For all $P, Q \in G_q$ and for all $a, b \in \mathbb{Z}$ we have $\hat{e}(aP, bQ) = \hat{e}(P, Q)^{ab}$.
2. Non-degenerate: $\hat{e}(P, P) \in \mathbb{F}_{p^2}$ is an element of order q, and in fact a generator of μ_q.
3. Computable: Given $P, Q \in G_q$ there is an efficient algorithm, due to Miller, to compute $\hat{e}(P, Q)$. This algorithm is described in [2]. Its run time is comparable to a full exponentiation in $\mathbb{F}_p$.

3.1 Weil Diffie-Hellman Assumption

Joux and Nguyen [18] point out that although the Computational Diffie-Hellman problem (CDH) appears to be hard in the group G_q, the Decisional Diffie-Hellman problem (DDH) is easy in G_q. Observe that given $P, aP, bP, cP \in G_q$ we have

$$c = ab \bmod q \quad \Longleftrightarrow \quad \hat{e}(P, cP) = \hat{e}(aP, bP)$$

Hence, the modified Weil pairing provides an easy test for Diffie-Hellman tuples. Consequently, one cannot use the DDH assumption to build cryptosystems in the group G_q. The security of our system is based on the following natural variant of the Computational Diffie-Hellman assumption.

Weil Diffie-Hellman Assumption (WDH): Let $p = 2 \bmod 3$ be a k-bit prime and $p = 6q - 1$ for some prime q. Let $E/\mathbb{F}_p$ be the curve $y^2 = x^3 + 1$ and let $P \in E/\mathbb{F}_p$ be a point of order q. The WDH problem is as follows: Given $\langle P, aP, bP, cP \rangle$ for random $a, b, c \in \mathbb{Z}_q^*$ compute $W = \hat{e}(P, P)^{abc} \in \mathbb{F}_{p^2}$. The WDH Assumption states that when p is a random k-bit prime there is no probabilistic polynomial time algorithm for the WDH problem. An algorithm $\mathcal{A}$ has advantage ϵ in solving WDH if $\Pr\left[\mathcal{A}(P, aP, bP, cP) = \hat{e}(P, P)^{abc}\right] \geq \epsilon$. Joux [17] previously used an analogue of the WDH assumption to construct a one-round three party Diffie-Hellman protocol. Verheul [30] recently used a related hardness assumption.

To conclude this section we point out that the discrete log problem in G_q is easily reducible to the discrete log problem in $\mathbb{F}_{p^2}^*$ (see [22,13]). To see this observe that given $P \in G_q$ and $Q = aP$ we can define $g = \hat{e}(P, P)$ and $h = \hat{e}(Q, P)$. Then $h = g^a$ and $h, g \in \mathbb{F}_{p^2}^*$. Hence, computing discrete log in $\mathbb{F}_{p^2}^*$ is sufficient for computing discrete log in G_q. For proper security of discrete log in $\mathbb{F}_p^*$ one often uses primes p that are 1024-bits long. Since we need discrete log in G_q to be difficult our system also uses primes p that are at least 1024-bits long.

4 Our Identity-Based Encryption Scheme

We describe our scheme in stages. First we give a basic identity-based encryption scheme which is not secure against an adaptive chosen ciphertext attack. The only reason for describing the basic scheme is to make the presentation easier to follow. Our full scheme, described in Section 4.3, extends the basic scheme to get security against an adaptive chosen ciphertext attack (IND-ID-CCA) in the random oracle model.

4.1 MapToPoint

Let p be a prime satisfying $p = 2 \bmod 3$ and $p = 6q-1$ for some prime $q > 3$. Let E be the elliptic curve $y^2 = x^3+1$ over $\mathbb{F}_p$. Our IBE scheme makes use of a simple algorithm for converting an arbitrary string $\mathsf{ID} \in \{0,1\}^*$ to a point $Q_{\mathsf{ID}} \in E/\mathbb{F}_p$ of order q. We refer to this algorithm as MapToPoint. We describe one of several ways of doing so. Let G be a cryptographic hash function $G : \{0,1\}^* \to \mathbb{F}_p$ (in the security analysis we view G as a random oracle). Algorithm $\mathsf{MapToPoint}_G$ works as follows:
1. Compute $y_0 = G(\mathsf{ID})$ and $x_0 = (y_0^2 - 1)^{1/3} = (y_0^2 - 1)^{(2p-1)/3} \bmod p$.
2. Let $Q = (x_0, y_0) \in E/\mathbb{F}_p$. Set $Q_{\mathsf{ID}} = 6Q$. Then Q_{ID} has order q as required.
This completes the description of MapToPoint. We note that there are 5 values of $y_0 \in \mathbb{F}_p$ for which $6Q = (x_0, y_0) = \mathcal{O}$ (these are the non-$\mathcal{O}$ points of order dividing 6). When $G(\mathsf{ID})$ is one of these 5 values Q_{ID} will not have order q. Since it is extremely unlikely for $G(\mathsf{ID})$ to hit one of these five points, for simplicity we say that such ID's are invalid. It is easy to extend algorithm MapToPoint to handle these five y_0 values as well.

4.2 BasicIdent

To explain the basic ideas underlying our IBE system we describe the following simple scheme, called BasicIdent. We present the scheme by describing the four algorithms: Setup, Extract, Encrypt, Decrypt. We let k be the security parameter given to the setup algorithm.

Setup: The algorithm works as follows:

Step 1: Choose a large k-bit prime p such that $p = 2 \bmod 3$ and $p = 6q - 1$ for some prime $q > 3$. Let E be the elliptic curve defined by $y^2 = x^3 + 1$ over $\mathbb{F}_p$. Choose an arbitrary $P \in E/\mathbb{F}_p$ of order q.

Step 2: Pick a random $s \in \mathbb{Z}_q^*$ and set $P_{pub} = sP$.

Step 3: Choose a cryptographic hash function $H : \mathbb{F}_{p^2} \to \{0,1\}^n$ for some n. Choose a cryptographic hash function $G : \{0,1\}^* \to \mathbb{F}_p$. The security analysis will view H and G as random oracles.

The message space is $\mathcal{M} = \{0,1\}^n$. The ciphertext space is $\mathcal{C} = E/\mathbb{F}_p \times \{0,1\}^n$. The system parameters are params $= \langle p, n, P, P_{pub}, G, H \rangle$. The master-key is $s \in \mathbb{Z}_q$.

Extract: For a given string $\mathsf{ID} \in \{0,1\}^*$ the algorithm builds a private key d as follows:
Step 1: Use $\mathsf{MapToPoint}_G$ to map ID to a point $Q_{\mathsf{ID}} \in E/\mathbb{F}_p$ of order q.
Step 2: Set the private key d_{ID} to be $d_{\mathsf{ID}} = sQ_{\mathsf{ID}}$ where s is the master key.

Encrypt: To encrypt $M \in \mathcal{M}$ under the public key ID do the following: (1) use $\mathsf{MapToPoint}_G$ to map ID into a point $Q_{\mathsf{ID}} \in E/\mathbb{F}_p$ of order q, (2) choose a random $r \in \mathbb{Z}_q$, and (3) set the ciphertext to be

$$C = \langle rP, \;\; M \oplus H(g_{\mathsf{ID}}^r) \rangle \quad \text{where} \quad g_{\mathsf{ID}} = \hat{e}(Q_{\mathsf{ID}}, P_{pub}) \in \mathbb{F}_{p^2}$$

Decrypt: Let $C = \langle U, V \rangle \in \mathcal{C}$ be a ciphertext encrypted using the public key ID. If $U \in E/\mathbb{F}_p$ is not a point of order q reject the ciphertext. Otherwise, to decrypt C using the private key d_{ID} compute:

$$V \oplus H(\hat{e}(d_{\mathsf{ID}}, U)) = M$$

This completes the description of BasicIdent. We first verify consistency. When everything is computed as above we have:
1. During encryption M is Xored with the hash of: g_{ID}^r.
2. During decryption V is Xored with the hash of: $\hat{e}(d_{\mathsf{ID}}, U)$.
These masks used during encryption and decryption are the same since:

$$\hat{e}(d_{\mathsf{ID}}, U) = \hat{e}(sQ_{\mathsf{ID}}, rP) = \hat{e}(Q_{\mathsf{ID}}, P)^{sr} = \hat{e}(Q_{\mathsf{ID}}, P_{pub})^r = g_{\mathsf{ID}}^r$$

Thus, applying decryption after encryption produces the original message M as required. We note that there is no need to devise attacks against this basic scheme since it is only presented for simplifying the exposition. The next section describes the full scheme.

Performance. Algorithms Setup and Extract are very simple algorithms. At the heart of both algorithms is a standard multiplication on the curve $E/\mathbb{F}_p$. Algorithm Encrypt requires that the encryptor compute the Weil pairing of Q_{ID} and P_{pub}. Note that this computation is independent of the message, and hence can be done once and for all. Once g_{ID} is computed the performance of the system is almost identical to standard ElGamal encryption. We also note that the ciphertext length is the same as in regular ElGamal encryption in $\mathbb{F}_p$. Decryption is a simple Weil pairing computation.

Security. Next, we study the security of this basic scheme. The following theorem shows that the scheme is a one-way identity based encryption scheme (ID-OWE) assuming WDH is hard.

Theorem 1. *Let the hash functions H, G be random oracles. Suppose there is an* ID-OWE *attacker $\mathcal{A}$ that has advantage ϵ against the scheme* BasicIdent. *Suppose $\mathcal{A}$ make at most $q_E > 0$ private key extraction queries and $q_H > 0$ hash queries. Then there is an algorithm $\mathcal{B}$ for computing WDH with advantage at least* $\frac{\epsilon}{e(1+q_E)\cdot q_H} - \frac{1}{q_H \cdot 2^n}$. *Here $e \approx 2.71$ is the base of the natural logarithm. The running time of $\mathcal{B}$ is $O(time(\mathcal{A}))$.*

To prove the theorem we need to define a related Public Key Encryption scheme (not an identity scheme), called PubKeyEnc. PubKeyEnc is described by three algorithms: keygen, encrypt, decrypt.

keygen: The algorithm works as follows:

Step 1: Choose a large k-bit prime p such that $p = 2 \bmod 3$ and $p = 6q - 1$ for some prime $q > 3$. Let E be the elliptic curve defined by $y^2 = x^3 + 1$ over $\mathbb{F}_p$. Choose an arbitrary $P \in E/\mathbb{F}_p$ of order q.

Step 2: Pick a random $s \in \mathbb{Z}_q^*$ and set $P_{pub} = sP$.
Pick a random point $Q_{\mathsf{ID}} \in E/\mathbb{F}_p$ of order q. Then Q_{ID} is in the group generated by P.

Step 3: Choose a cryptographic hash function $H : \mathbb{F}_{p^2} \to \{0,1\}^n$ for some n.

Step 4: The public key is $\langle p, n, P, P_{pub}, Q_{\mathsf{ID}}, H \rangle$. The private key is $d_{\mathsf{ID}} = sQ_{\mathsf{ID}}$.

encrypt: To encrypt $M \in \{0,1\}^n$ choose a random $r \in \mathbb{Z}_q$ and set the ciphertext to be:

$$C = \langle rP,\ \ M \oplus H(g^r) \rangle \quad \text{where} \quad g = \hat{e}(Q_{\mathsf{ID}}, P_{pub}) \in \mathbb{F}_{p^2}$$

decrypt: Let $C = \langle U, V \rangle \in \mathcal{C}$ be a ciphertext encrypted using the public key $\langle p, n, P, P_{pub}, Q_{\mathsf{ID}}, H \rangle$. To decrypt C using the private key d_{ID} compute:

$$V \oplus H(\hat{e}(d_{\mathsf{ID}}, U)) = M$$

This completes the description of PubKeyEnc. We now prove Theorem 1 in two steps. We first show that an ID-OWE attack on BasicIdent can be converted to a OWE attack on PubKeyEnc. This step shows that private key extraction queries do not help the attacker. We then show that PubKeyEnc is OWE if the WDH assumption holds. The proofs of these two lemmas appear in the full version of the paper [2].

Lemma 1. *Let G be a random oracle from $\{0,1\}^*$ to $\mathbb{F}_p$. Let $\mathcal{A}$ be an* ID-OWE *attacker that has advantage ϵ against* BasicIdent*. Suppose $\mathcal{A}$ makes at most $q_E > 0$ private key extraction queries. Then there is a* OWE *attacker $\mathcal{B}$ that has advantage $\epsilon/e(1 + q_E)$ against* PubKeyEnc*. Its running time is $O(time(\mathcal{A}))$.*

Lemma 2. *Let H be a random oracle from $\mathbb{F}_{p^2}$ to $\{0,1\}^n$. Let $\mathcal{A}$ be a* OWE *attacker that has advantage ϵ against* PubKeyEnc*. Suppose $\mathcal{A}$ makes a total of $q_H > 0$ queries to H. Then there is an algorithm $\mathcal{B}$ that solves the WDH problem with advantage at least $(\epsilon - \frac{1}{2^n})/q_H$ and a running time $O(time(\mathcal{A}))$.*

Proof of Theorem 1. The theorem follows directly from Lemma 1 and Lemma 2. Composing both reductions shows that an ID-OWE attacker on BasicIdent with advantage ϵ gives an algorithm for WDH with advantage $(\epsilon/e(1 + q_E) - 1/2^n)/q_H$, as required. □

4.3 Identity-Based Encryption with Chosen Ciphertext Security

We use a technique due to Fujisaki-Okamoto [12] to convert the BasicIdent scheme of the previous section into a chosen ciphertext secure IBE system (in the sense of Section 2) in the random oracle model. Let $\mathcal{E}$ be a public key encryption scheme. We denote by $\mathcal{E}_{pk}(M;r)$ the encryption of M using the random bits r under the public key pk. Fujisaki-Okamoto define the hybrid scheme $\mathcal{E}^{hy}$ as:

$$\mathcal{E}^{hy}_{pk}(M) \quad = \quad \mathcal{E}_{pk}(\sigma; H_1(\sigma, M)) \, \Big\| \, G_1(\sigma) \oplus M$$

Here σ is generated at random and H_1, G_1 are cryptographic hash functions. Fujisaki-Okamoto show that if $\mathcal{E}$ is a one-way encryption scheme then $\mathcal{E}^{hy}$ is a chosen ciphertext secure system (IND-CCA) in the random oracle model (assuming $\mathcal{E}_{pk}$ satisfies some natural constraints).

We apply this transformation to BasicIdent and show that the resulting IBE system is IND-ID-CCA. We obtain the following IBE scheme which we call FullIdent. Recall that n is the length of the message to be encrypted.

Setup: As in the BasicIdent scheme. In addition, we pick a hash function $H_1 : \{0,1\}^n \times \{0,1\}^n \to \mathbb{F}_q$, and a hash function $G_1 : \{0,1\}^n \to \{0,1\}^n$.

Extract: As in the BasicIdent scheme.

Encrypt: To encrypt $M \in \{0,1\}^n$ under the public key ID do the following: (1) use algorithm $\mathsf{MapToPoint}_G$ to convert ID into a point $Q_{\mathsf{ID}} \in E/\mathbb{F}_p$ of order q, (2) choose a random $\sigma \in \{0,1\}^n$, (3) set $r = H_1(\sigma, M)$, and (4) set the ciphertext to be

$$C = \langle rP, \ \sigma \oplus H(g^r_{\mathsf{ID}}), \ M \oplus G_1(\sigma) \rangle \quad \text{where} \quad g_{\mathsf{ID}} = \hat{e}(Q_{\mathsf{ID}}, P_{pub}) \in \mathbb{F}_{p^2}$$

Decrypt: Let $C = \langle U, V, W \rangle \in \mathcal{C}$ be a ciphertext encrypted using the public key ID. If $U \in E/\mathbb{F}_p$ is not a point of order q reject the ciphertext. To decrypt C using the private key d_{ID} do:

1. Compute $\quad V \oplus H(\hat{e}(d_{\mathsf{ID}}, U)) = \sigma$.
2. Compute $\quad W \oplus G_1(\sigma) = M$.
3. Set $r = H_1(\sigma, M)$. Test that $U = rP$. If not, reject the ciphertext.
4. Output M as the decryption of C.

This completes the description of FullIdent. Note that M is encrypted as $W = M \oplus G_1(\sigma)$. This can be replaced by $W = E_{G_1(\sigma)}(M)$ where E is a semantically secure symmetric encryption scheme (see [12]).

Security. The following theorem shows that FullIdent is a chosen ciphertext secure IBE (i.e. IND-ID-CCA), assuming WDH is hard.

Theorem 2. *Let $\mathcal{A}$ be a t-time* IND-ID-CCA *attacker on* FullIdent *that achieves advantage ϵ. Suppose $\mathcal{A}$ makes at most q_E extraction queries, at most q_D decryption queries, and at most q_H, q_{G1}, q_{H1} queries to the hash functions H, G_1, H_1*

respectively. Then there is a t_1*-time algorithm for WDH that achieves advantage* ϵ_1 *where*

$$t_1 = FO_{time}(t, q_{G_1}, q_{H_1})$$
$$\epsilon_1 = \left(FO_{adv}(\epsilon(\tfrac{1}{eq_E} - \tfrac{q_D}{q}), q_{G_1}, q_{H_1}, q_D) - 1/2^n\right)/q_H$$

where the functions FO_{time} *and* FO_{adv} *are defined in Theorem 3.*

The proof of the theorem is based on the theorem below due to Fujisaki and Okamoto (Theorem 14 in [12]). We state their theorem as it applies to the public key encryption scheme PubKeyEnc of the previous section. Let $\mathsf{PubKeyEnc}^{hy}$ be the result of applying the Fujisaki-Okamoto transformation to PubKeyEnc.

Theorem 3 (FO). *Suppose there is a* $(t, q_{G_1}, q_{H_1}, q_D)$ IND-CCA *attacker that achieves advantage* ϵ *when attacking* $\mathsf{PubKeyEnc}^{hy}$*. Then there is a* (t_1, ϵ_1) OWE *attacker on* PubKeyEnc *where*

$$t_1 = FO_{time}(t, q_{G_1}, q_{H_1}) = t + O((q_{G_1} + q_{H_1}) \cdot n), \quad \text{and}$$
$$\epsilon_1 = FO_{adv}(\epsilon, q_{G_1}, q_{H_1}, q_D) = \frac{1}{2(q_{G_1} + q_{H_1})}\left[(\epsilon + 1)(1 - 2/q)^{q_D} - 1\right]$$

We also need the following lemma to translate between an IND-ID-CCA chosen ciphertext attack on FullIdent and an IND-CCA chosen ciphertext attack on $\mathsf{PubKeyEnc}^{hy}$. The proof appears in the full version of the paper [2].

Lemma 3. *Let* $\mathcal{A}$ *be an* IND-ID-CCA *attacker that has advantage* ϵ *against the IBE scheme* FullIdent*. Suppose* $\mathcal{A}$ *makes at most* $q_E > 0$ *private key extraction queries and at most* q_D *decryption queries. Then there is an* IND-CCA *attacker* $\mathcal{B}$ *that has advantage at least* $\epsilon(\frac{1}{eq_E} - \frac{q_D}{q})$ *against* $\mathsf{PubKeyEnc}^{hy}$*. Its running time is* $O(time(\mathcal{A}))$.

Proof of Theorem 2. By Lemma 3 an IND-ID-CCA attacker on FullIdent implies an IND-CCA attacker on $\mathsf{PubKeyEnc}^{hy}$. By Theorem 3 an IND-CCA attacker on $\mathsf{PubKeyEnc}^{hy}$ implies a OWE attacker on PubKeyEnc. By Lemma 2 a OWE attacker on PubKeyEnc implies an algorithm for WDH. Composing all these reductions gives the required bounds. □

5 Extensions and Observations

Tate pairing and other curves. Our IBE system has some flexibility in terms of the curves being used and the definition of the pairing. For example, one could use the curve $y^2 = x^3 + x$ with its endomorphism $\phi : (x, y) \to (-x, iy)$ where $i^2 = -1$. We do not explore this here, but note that both encryption and decryption in FullIdent can be made faster by using the Tate pairing. In general, one can use any efficiently computable bilinear pairing $\hat{e} : \mathbb{G}_1 \times \mathbb{G}_1 \to \mathbb{G}_2$ between two groups $\mathbb{G}_1, \mathbb{G}_2$ as long as the WDH assumption holds. One would also need a way to map identities in $\{0,1\}^*$ uniformly onto $\mathbb{G}_1$.

Distributed PKG. In the standard use of an IBE in an e-mail system the master-key stored at the PKG must be protected in the same way that the private key of a CA is protected. One way of protecting this key is by distributing it among different sites using techniques of threshold cryptography [14]. Our IBE system supports this in a very efficient and robust way. Recall that the master-key is some $s \in \mathbb{F}_q$. in order to generate a private key the PKG computes $Q_{priv} = sQ_{\mathsf{ID}}$, where Q_{ID} is derived from the user's public key ID. This can easily be distributed in a t-out-of-n fashion by giving each of the n PKGs one share s_i of a Shamir secret sharing of s mod q. When generating a private key each of the t chosen PKGs simply responds with $Q^{(i)}_{priv} = s_i Q_{\mathsf{ID}}$. The user can then construct Q_{priv} as $Q_{priv} = \sum \lambda_i Q^{(i)}_{priv}$ where the λ_i's are the appropriate Lagrange coefficients.

Furthermore, it is easy to make this scheme robust against dishonest PKGs using the fact that DDH is easy in G_q (the group generated by P). During setup each of the n PKGs publishes $P^{(i)}_{pub} = s_i P$. During a key generation request the user can verify that the response from the i'th PKG is valid by testing that:

$$\hat{e}(Q^{(i)}_{priv}, P) = \hat{e}(Q_{\mathsf{ID}}, P^{(i)}_{pub})$$

Thus, a misbehaving PKG will be immediately caught. There is no need for zero-knowledge proofs as in regular robust threshold schemes. The PKG's master-key can be generated in a distributed fashion using the techniques of [15].

Note that a distributed master-key also enables decryption on a *per-message* basis, without any need to derive the corresponding decryption key. For example, threshold decryption of BasicIdent ciphertext (U, V) is straightforward if each PKG responds with $\hat{e}(s_i Q_{ID}, U)$.

Working in subgroups. The performance of our IBE system can be improved if we work in a small subgroup of the curve. For example, choose a 1024-bit prime $p = 2 \bmod 3$ with $p = aq - 1$ for some 160-bit prime q. The point P is then chosen to be a point of order q. Each public key ID is converted to a group point by hashing ID to a point Q on the curve and then multiplying the point by a. The system is secure if the WDH assumption holds in the group generated by P. The advantage is that Weil computations are done on points of small order, and hence is much faster.

IBE implies signatures. Moni Naor has observed that an IBE scheme can be immediately converted into a public key signature scheme. The intuition is as follows. The private key for the signature scheme is the master key for the IBE scheme. The public key for the signature scheme is the global system parameters for the IBE scheme. The signature on a message M is the IBE decryption key for ID $= M$. To verify a signature, choose a random message M', encrypt M' using the public key ID $= M$, and then attempt to decrypt using the given signature on M as the decryption key. If the IBE scheme is IND-ID-CCA, then the signature scheme is existentially unforgeable against a chosen message attack. Note that, unlike most signature schemes, the signature verification algorithm here is randomized. This shows that secure IBE schemes require both public key encryption and digital signatures. We note that the signature scheme derived from our IBE system has some interesting properties [3].

6 Escrow ElGamal Encryption

In this section we note that the Weil pairing enables us to add a global escrow capability to the ElGamal encryption system. A single escrow key enables the decryption of ciphertexts encrypted under any public key. Paillier and Yung have shown how to add a global escrow capability to the Paillier encryption system [24]. Our ElGamal escrow system works as follows:

Setup: The algorithm works as follows:

Step 1: Choose a large k-bit prime p such that $p = 2 \bmod 3$ and $p = 6q - 1$ for some prime $q > 3$. Let E be the elliptic curve defined by $y^2 = x^3 + 1$ over $\mathbb{F}_p$. Choose an arbitrary $P \in E/\mathbb{F}_p$ of order q.

Step 2: Pick a random $s \in \mathbb{Z}_q$ and set $Q = sP$.

Step 3: Choose a cryptographic hash function $H : \mathbb{F}_{p^2} \to \{0,1\}^n$.

The message space is $\mathcal{M} = \{0,1\}^n$. The ciphertext space is $\mathcal{C} = E/\mathbb{F}_p \times \{0,1\}^n$. The system parameters are $\mathsf{params} = \langle p, n, P, Q, H \rangle$. The escrow key is $s \in \mathbb{Z}_q$.

keygen: A user generates a public/private key pair for herself by picking a random $x \in \mathbb{Z}_q$ and computing $P_{pub} = xP$. Her private key is x, her public key is P_{pub}.

Encrypt: To encrypt $M \in \{0,1\}^n$ under the public key P_{pub} do the following: (1) pick a random $r \in \mathbb{Z}_q$, and (2) set the ciphertext to be:

$$C = \langle rP,\ M \oplus H(g^r) \rangle \quad \text{where} \quad g = \hat{e}(P_{pub}, Q) \in \mathbb{F}_{p^2}$$

Decrypt: Let $C = \langle U, V \rangle$ be a ciphertext encrypted using P_{pub}. If $U \in E/\mathbb{F}_p$ is not a point of order q reject the ciphertext. To decrypt C using the private key x do:

$$V \oplus H(\hat{e}(U, xQ)) = M$$

Escrow-decrypt: To decrypt $C = \langle U, V \rangle$ using the escrow key s do:

$$V \oplus H(\hat{e}(U, sP_{pub})) = M$$

A standard argument shows that assuming WDH the system has semantic security in the random oracle model (recall that since DDH is easy we cannot prove semantic security based on DDH). Yet, the escrow agent can decrypt any ciphertext encrypted using any user's public key. The decryption capability of the escrow agent can be distributed using the PKG distribution techniques described in Section 5.

Using a similar hardness assumption, Verheul [30] has recently described an ElGamal encryption system with non-global escrow. Each user constructs a public key with two corresponding private keys, and gives one of the private keys to the trusted third party. Although both private keys can be used to decrypt, only the user's private key can be used simultaneously as the signing key for a discrete logarithm based signature scheme.

7 Summary and Open Problems

We defined chosen ciphertext security for identity-based systems and proposed a fully functional IBE scheme. The scheme has chosen ciphertext security in the random oracle model assuming WDH, a natural analogue of the computational Diffie-Hellman problem. The WDH assumption deserves further study considering the powerful cryptosystems derived from it. For example, it could be interesting to see whether the techniques of [20] can be used to prove that the WDH assumption is equivalent to the discrete log assumption on the curve for certain primes p.

It is natural to try and build chosen ciphertext secure identity based systems that are secure under standard complexity assumptions (rather than the random oracle model). One might hope to use the techniques of Cramer-Shoup [6] to provide chosen ciphertext security based on DDH. Unfortunately, as mentioned in Section 2 the DDH assumption is false in the group of points on the curve E. However, a natural variant of DDH does seem to hold. In particular, the following two distributions appear to be computationally indistinguishable: $\langle P, aP, bP, cP, abcP \rangle$ and $\langle P, aP, bP, cP, rP \rangle$ where a, b, c, r are random in $\mathbb{Z}_q$. We refer to this assumption as WDDH. It is natural to ask whether there is a chosen ciphertext secure identity-based system strictly based on WDDH. Such a scheme would be the analogue of the Cramer-Shoup system.

References

1. M. Bellare, A. Desai, D. Pointcheval, P. Rogaway, "Relations among notions of security for public-key encryption schemes", Proc. Crypto '98, pp. 26–45, 1998.
2. D. Boneh, M. Franklin, "Identity based encryption from the Weil pairing", Full version available at `http://crypto.stanford.edu/ibe`
3. D. Boneh, B. Lynn, H. Shacham, "Short signatures from the Weil pairing", manuscript.
4. M. Bellare, A. Boldyreva, S. Micali, "Public-key Encryption in a Multi-User Setting: Security Proofs and Improvements", Proc. Eurocrypt 2000, LNCS 1807, 2000.
5. J. Coron, "On the exact security of Full-Domain-Hash", Proc. of Crypto 2000.
6. R. Cramer and V. Shoup, "A practical public key cryptosystem provably secure against adaptive chosen ciphertext attack", in proc. Crypto '98, pp. 13–25.
7. Y. Desmedt and J. Quisquater, "Public-key systems based on the difficulty of tampering", Proc. Crypto '86, pp. 111-117, 1986.
8. G. Di Crescenzo, R. Ostrovsky, and S. Rajagopalan, "Conditional Oblivious Transfer and Timed-Release Encryption", Proc. of Eurocrypt '99.

9. D. Dolev, C. Dwork, M. Naor, "Non-malleable cryptography", SIAM J. of Computing, Vol. 30(2), pp. 391–437, 2000.
10. U. Feige, A. Fiat and A. Shamir, "Zero-knowledge proofs of identity", *J. Cryptology*, vol. 1, pp. 77–94, 1988.
11. A. Fiat and A. Shamir, "How to prove yourself: Practical solutions to identification and signature problems", Proc. Crypto '86, pp. 186–194, 1986.
12. E. Fujisaki and T. Okamoto, "Secure integration of asymmetric and symmetric encryption schemes", Proc. Crypto '99, pp. 537–554, 1999.
13. G. Frey, M. Müller, H. Rück, "The Tate pairing and the discrete logarithm applied to elliptic curve cryptosystems", IEEE Tran. on Info. Th., Vol. 45, pp. 1717–1718, 1999.
14. P. Gemmell, "An introduction to threshold cryptography", in CryptoBytes, a technical newsletter of RSA Laboratories, Vol. 2, No. 7, 1997.
15. R. Gennaro, S. Jarecki, H. Krawczyk, T. Rabin, "Secure Distributed Key Generation for Discrete-Log Based Cryptosystems", *Advances in Cryptology - Eurocrypt '99*, Springer-Verlag LNCS 1592, pp. 295–310, 1999.
16. O. Goldreich, B. Pfitzmann and R. Rivest, "Self-delegation with controlled propagation -or- What if you lose your laptop", proc. Crypto '98, pp. 153–168, 1998.
17. A. Joux, "A one round protocol for tripartite Diffie-Hellman", Proc of ANTS 4, LNCS 1838, pp. 385–394, 2000.
18. A. Joux, K. Nguyen, "Separating Decision Diffie-Hellman from Diffie-Hellman in cryptographic groups", available from `eprint.iacr.org`.
19. S. Lang, "Elliptic functions", Addison-Wesley, Reading, 1973.
20. U. Maurer, "Towards proving the equivalence of breaking the Diffie-Hellman protocol and computing discrete logarithms", Proc. Crypto '94, pp. 271–281.
21. U. Maurer and Y. Yacobi, "Non-interactive public-key cryptography", proc. Eurocrypt '91, pp. 498–507.
22. A. Menezes, T. Okamoto, S. Vanstone, "Reducing elliptic curve logarithms to logarithms in a finite field", IEEE Tran. on Info. Th., Vol. 39, pp. 1639–1646, 1993.
23. V. Miller, "Short programs for functions on curves", unpublished manuscript.
24. P. Paillier and M. Yung, "Self-escrowed public-key infrastructures" in Proc. ICISC, pp. 257–268, 1999.
25. C. Rackoff, D. Simon, "Noninteractive zero-knowledge proof of knowledge and chosen ciphertext attack", in proc. Crypto '91, pp. 433–444, 1991.
26. R. Rivest, A. Shamir and D. Wagner, "Time lock puzzles and timed release cryptography," Technical report, MIT/LCS/TR-684
27. A. Shamir, "Identity-based cryptosystems and signature schemes", Proc. Crypto '84, pp. 47–53.
28. S. Tsuji and T. Itoh, "An ID-based cryptosystem based on the discrete logarithm problem", *IEEE Journal on Selected Areas in Communication*, vol. 7, no. 4, pp. 467–473, 1989.
29. H. Tanaka, "A realization scheme for the identity-based cryptosystem", Proc. Crypto '87, pp. 341–349, 1987.
30. E. Verheul, "Evidence that XTR is more secure than supersingular elliptic curve cryptosystems", Proc. Eurocrypt 2001.

A Chosen Ciphertext Attack on RSA Optimal Asymmetric Encryption Padding (OAEP) as Standardized in PKCS #1 v2.0

James Manger

Telstra Research Laboratories,
Level 7, 242 Exhibition Street, Melbourne 3000, Australia
James.H.Manger@team.telstra.com

Abstract. An adaptive chosen ciphertext attack against PKCS #1 v2.0 RSA OAEP encryption is described. It recovers the plaintext – not the private key – from a given ciphertext in a little over $\log_2 n$ queries of an oracle implementing the algorithm, where n is the RSA modulus. The high likelihood of implementations being susceptible to this attack is explained as well as the practicality of the attack. Improvements to the algorithm to defend against the attack are discussed.

Keywords: chosen ciphertext attack, RSA, OAEP, PKCS

1 Introduction

At CRYPTO '98 Daniel Bleichenbacher presented an adaptive chosen ciphertext attack against PKCS #1 v1.5 RSA block type 2 padding [1]. The attack needs roughly one million oracle queries to succeed for a 1024-bit RSA key. He concluded that RSA encryption should include an integrity check and that the phase between decryption and integrity verification is crucial, because any information leaking from this phase can present a security risk. Version 2.0 of PKCS #1 introduced a new algorithm RSAES- OAEP that uses Optimal Asymmetric Encryption Padding (OAEP) to counteract this attack [2][5]. It says, "a chosen ciphertext attack is ineffective against a plaintext-aware encryption scheme such as RSAES-OAEP". However, the design of RSAES-OAEP makes it highly likely that implementations will leak information between the decryption and integrity check operations making them susceptible to a chosen ciphertext attack that requires many orders of magnitude less effort than similar attacks against PKCS #1 v1.5 block type 2 padding. The attack needs roughly one thousand oracle queries to succeed for a 1024-bit RSA key.

Section 2 summarizes RSA Optimal Asymmetric Encryption Padding as defined in PKCS #1 v2.0 . [1] Section 3 describes a chosen ciphertext against this algorithm. Section 4 explores the practicality of the assumptions necessary for

[1] The same algorithm is standardized in IEEE 1363, where the relevant message encoding method for encryption is called EME1 [4]

J. Kilian (Ed.): CRYPTO 2001, LNCS 2139, pp. 230–238, 2001.

the attack to proceed. Section 5 discusses approaches for changing the algorithm or its implementation to prevent the attack and restore the intended security properties.

2 RSAES-OAEP

RSAES-OAEP encryption starts by encoding a seed, a hash, padding octets and the secret (typically a session key) into an octet string. Masking operations effectively randomize these octets before they are treated as the unsigned binary representation of an integer – the integer used in the RSA modular exponentiation operation. The number of padding octets is chosen so that the encoding consumes one less octet than required for a unsigned binary representation of the modulus. This ensures the integer is less than the modulus as required in RSA. Alteratively, the encoded messages can be considered as an octet string the same length as the modulus, but with the most significant octet set to '00'h.

Figure 1 shows the RSAES-OAEP decryption and decoding process. The ciphertext is converted to the plaintext by modular exponentiation with the private exponent followed by integer-to-octet translation. A mask generation function (MGF) uses the least significant portion of the plaintext to unmask the seed. A mask generated from the seed unmasks a hash, padding and the confidential message. The integrity of the ciphertext is verified by comparing the unmasked hash to an independently calculated hash of the parameters (and by checking the padding).

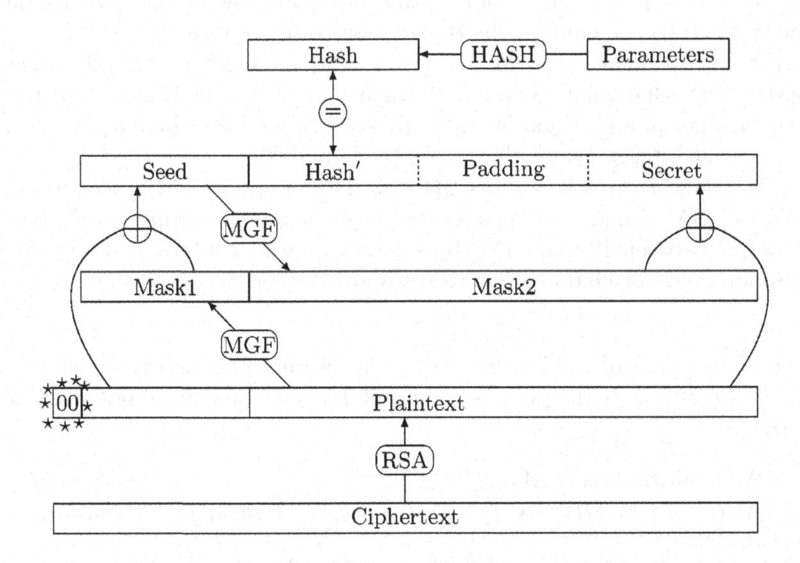

Fig. 1. RSAES-OAEP Decoding

After the private key operation the decryption operation can fail in the integer-to-octet translation (e.g. the integer is too large to fit in one fewer octets than the modulus) or in the OAEP-decoding (e.g. integrity check fails). In both instances PKCS #1 v2.0 says to output "`decryption error`" and stop.

3 Chosen Ciphertext Attack

Let n be an RSA modulus, with e and d the public and private exponents respectively. Let $k = \lceil \log_{256} n \rceil$ be the byte length of n and let $B = 2^{8(k-1)}$. [2]

Assume an attacker knows the public key (n, e) and has access to an oracle that for any chosen ciphertext x indicates whether the corresponding plaintext $y \equiv x^d \pmod n$ is less than B or not — returning "$y < B$" or "$y \geq B$". For the last assumption to hold it is sufficient for the oracle to distinguish a failure in the integer-to-octets conversion (in which case "$y \geq B$" is returned) from any subsequent failure, e.g. of the integrity check.

The attacker wishes to determine the plaintext $m \equiv c^d \pmod n$ corresponding to a captured ciphertext c. The basic step is to choose a multiple f and send $f^e \cdot c \pmod n$ to the oracle. This ciphertext corresponds to the plaintext $f \cdot m$. [3] The oracle indicates if this is in the range $[0, B)$ or (B, n) modulo n, thus providing a mathematical relationship about m that reduces the range (or ranges) in which it must lie. The aim is to reduce this range with successive oracle queries until just one value is left — m.

The approach of the attack described in this paper is to choose values of f such that the range where $f \cdot m$ could lie spans exactly one boundary between a region where $f \cdot m < B \pmod n$ and a region where $f \cdot m \geq B \pmod n$. The oracle response narrows the range to one of these regions.

Initially we know $m \in [0, B)$, as all valid messages are in this range by construction. One point to note is that since $m < B$ there is always a multiple of m that lies in any region of width B. For instance, for any integer i there is always some integer f such that $f \cdot m \in [in, in + B)$.

The following attack assumes $2B < n$. This assumption will usually be satisfied as RSA moduli are typically chosen to be exact multiples of 8 bits long making n between 128 and 256 times larger than B. Situations where this assumption does not hold are discussed toward the end of this section.

Step 1: Try multiples of $2, 4, 8, \ldots 2^i, \ldots$ in turn until the oracle returns "$\geq B$". For each multiple f_1 the possible values of $f_1 \cdot m$ span a single boundary point at B.

1.1 We know $m \in [0, B)$. Let $f_1 = 2$.
1.2 So $f_1 \cdot m \in [0, 2B)$. Try f_1 with the oracle, i.e. send $f_1^e \cdot c \pmod n$.

[2] Any number less than B encoded into k octets will start with a '00'h octet.
[3] $(f^e \cdot c)^d \equiv f^{ed} \cdot c^d \equiv f \cdot m \pmod n$

1.3a If the oracle indicates "$< B$":
This implies $f_1 \cdot m \in [0, B)$, so $2f_1 \cdot m \in [0, 2B)$.
Set $f_1 \leftarrow 2f_1$ and go back to step 1.2.
1.3b If the oracle indicates "$\geq B$:
This implies $f_1 \in [B, 2B)$ for a known (even) multiple f_1. Rephrasing this gives $\frac{f_1}{2} \cdot m \in [\frac{B}{2}, B)$ for a known multiple $\frac{f_1}{2}$. Now move to the next step.

Step 2: Start with a multiple f_2 such that $f_2 \cdot m$ is just less than $n + B$ for the maximum possible m. Keep increasing this multiple until the oracle returns "$< B$". For each multiple f_2 the possible values of $f_2 \cdot m$ span a single boundary point at n.

2.1 We have $\frac{f_1}{2} \cdot m \in [\frac{B}{2}, B)$. Let $f_2 = \lfloor \frac{n+B}{B} \rfloor \cdot \frac{f_1}{2}$.
2.2 So $f_2 \cdot m \in [\frac{n}{2}, n + B)$. Try f_2 with the oracle.
2.3a If the oracle indicates "$\geq B$":
This implies $f_2 \cdot m \in [\frac{n}{2}, n)$, so $(f_2 + \frac{f_1}{2}) \cdot m \in [\frac{n}{2}, n + B)$.
Set $f_2 \leftarrow f_2 + \frac{f_1}{2}$ and go back to step 2.2.
2.3b If the oracle indicates "$< B$":
This implies $f_2 \cdot m \in [n, n + B)$ for a known multiple f_2. Now move to the next step.

As f_2 increases at iterations through step 2.3a the lower bound on $f_2 \cdot m$ increases, eventually exceeding n when $f_2 = \lceil \frac{2n}{B} \rceil \cdot \frac{f_1}{2}$. Branch 2.3b must occur at or before this multiple. That is, step 2 will always terminate — taking at most $\lceil \frac{n}{B} \rceil$ oracle queries.

Step 3: Try multiples f_3 that give a range for $f_3 \cdot m$ about $2B$ integers wide and spanning a single boundary point. Each oracle response will half the range back to a width of about B integers, so the next multiple is approximately twice the previous value.

3.1 We have $f_2 \cdot m \in [n, n + B)$.
Rephrasing, we have a multiple f_2 and a range $[m_{\text{min}}, m_{\text{max}})$ of possible m values, where $m_{\text{min}} = \lceil \frac{n}{f_2} \rceil$, $m_{\text{max}} = \lfloor \frac{n+B}{f_2} \rfloor$ and $f_2 \cdot (m_{\text{max}} - m_{\text{min}}) \approx B$.
3.2 Choose a multiple f_{tmp} such that the width of $f_{\text{tmp}} \cdot m$ is approximately $2B$.
$f_{\text{tmp}} = \lfloor \frac{2B}{m_{\text{max}} - m_{\text{min}}} \rfloor$. This value is about double the previous multiple.
3.3 Select a boundary point, $in + B$, near the range of $f_{\text{tmp}} \cdot m$.
$i = \lfloor \frac{f_{\text{tmp}} \cdot m_{\text{min}}}{n} \rfloor$.
3.4 Choose a multiple f_3 such that $f_3 \cdot m$ spans a single boundary point at $in + B$.
$f_3 = \lceil \frac{in}{m_{\text{min}}} \rceil$. This gives $f_3 \cdot m \in [in, in + 2B)$ (though the upper bound is only approximate). f_3 is approximately equal to f_{tmp}. Try f_3 with the oracle.
3.5a If the oracle indicates "$\geq B$":
This implies $f_3 \cdot m \in [in + B, in + 2B)$.
Set $m_{\text{min}} \leftarrow \lceil \frac{in+B}{f_3} \rceil$ and go back to step 3.2.

3.5b If the oracle indicates "$< B$":
This implies $f_3 \cdot m \in [in, in + B)$.
Set $m_{\max} \leftarrow \lfloor \frac{in+B}{f_3} \rfloor$ and go back to step 3.2.

Each answer from the oracle in step 3 selects either the top or bottom half (approximately) of the $f_3 \cdot m$ range, halving the range of possible m values. Eventually the range in which m lies narrows to a single number, which is the desired plaintext. At this point $f_3 \approx B = 2^{8(k-1)}$.

The description of step 3 above does not provide a proof that those particular choices of multiples, boundary points and interval widths will always work for any key or message. Minor variations on these choices can make the attack algorithm marginally more efficient. See [1] for a more mathematically rigorous analysis of a closely related problem.

3.1 Complexity

Steps 1 and 3 approximately halve the range of possible m values with each iteration so between them they take about $\log_2 B = 8(k-1)$ oracle queries. [4] Step 2 takes at most $\lceil \frac{n}{B} \rceil$ oracle queries (which must be ≤ 256), and half this number on average.

RSA moduli are typically chosen to be exact multiples of 8 bits long, e.g. 1024-bit moduli are far more prevalent than, say, 1021-bit moduli. Hence, for typical keys $\lceil \frac{n}{B} \rceil$ is in the range (128, 256], so step 2 will typically take on the order of 100 oracle queries.

For a 1024-bit RSA key the attack requires about 1100 oracle queries, for a 2048-bit key about 2200.

3.2 When $n < 2B$

The attack procedure described above assumes $2B < n$. If this is not the case, an indication from the oracle of "$< B$" when $f = 2$ narrows the range in which $f \cdot m$ lies not to a single region, but to a pair of regions: $f \cdot m \in [0, B) \bigcup [n, 2B)$. The range in which m is known to lie is reduced is total size, but is no longer confined to a single interval. This somewhat complicates the decision about which multiples to try but an adaptive chosen ciphertext attack will still work. The chosen ciphertext attack against RSA block type 2 padding had a similar issue — see [1] for a full analysis.

3.3 Comparison to the RSA Block Type 2 Attack

Analysis in [1] of the number of oracle queries required for a chosen ciphertext attack found an expression with two terms: the first term inversely proportional

[4] Reduction of the range of possible m values in step 2 slightly reduces the number of oracle queries required during steps 1 and 3, but this number also slightly increases (by a few percent) as the ranges in step 3 not being exactly centred on boundary points.

to the probability that a random integer from $[0, n)$ conforms to the encoding format; the second term proportional to $\log_2 n$. The first term dominates for RSA block type 2 padding (making the number of required queries quite dependent on various implementation issues, i.e. how the encoding format is checked). For RSAES-OAEP the first term corresponds to the number of oracle queries in step 2, which is an order of magnitude less than the second term.

4 Likelihood of Susceptibility

The chosen ciphertext attack described in the previous section starts with an assumption that the attacker can distinguish a failure in the integer-to-octets conversion from any subsequent failure, e.g. of the integrity check during OAEP-decoding. PKCS #1 v2.0, however, recognizes this risk by explicitly stating "it is important that the error messages output in steps 4 [integer-to- octets conversion] and 5 [OAEP decoding] be the same". [5] This section investigates why, in spite of this statement, it is likely that many RSAES-OAEP implementations will be susceptible to chosen ciphertext attack.

4.1 Spelling

Simply misspelling a word, including a full-stop or starting with a capital letter at one point is sufficient to distinguish two error messages that are otherwise the same. Having to relying for security on the absence of any such trivial occurrence in an implementation should not be necessary.

4.2 Logs

Even when a system avoids revealing error details in, say, its protocol response it is likely to reveal more detailed error descriptions in its logs. [6] "`Integer too large`" and "`decoding error`" – included in PKCS #1 v2.0 as error messages from sub-routines used by RSAES-OAEP – are just the sort of details a log may contain yet their presence is sufficient for the attack to proceed. Requiring access to system logs clearly lessens the risk of an attack but it is still an attack that must be considered. Logs are typically available to a much larger set of people than have direct access to a private key and logs will be given less protection (and should not be required to have the same protection as a private key).

[5] PKCS #1 v2.0, section 7.1.2 *Decryption operation*, last paragraph.

[6] Divulging less detail and only very general error indications is a well-known security technique, but it does come at a cost. Less information for an attacker also means less information for developers, support staff and users to understand the state of a system and respond appropriately.

4.3 Other Error Conditions

There are many possible errors that are not mentioned in the definition of RSAES-OAEP in PKCS #1 v2.0. This seems sensible as most are implementation issues but it becomes problematic when, due to the algorithm's design, these errors can have serious security implications. Consider what could happen when an unsupported mask generation function (MGF) is specified (by the attacker, along with his chosen ciphertext). Though not explicitly considered in PKCS #1, some sort of error must result, say "`unsupported algorithm`", and it may not be detected until the MGF is first used – in the OAEP-decoding stage. Any indication that the OAEP-decoding stage has been reached, however, is sufficient for the attack to proceed as it implies the previous integer-to-octet conversion stage was successful, i.e. plaintext $< B$.

4.4 Timing

Even identical error responses can be distinguished if they take different amounts of time to occur. For instance, detecting an integrity error during OAEP-decoding takes at least the time of two mask generation operations longer than detecting an error in the integer-to-octet conversion. Though this time difference may be small compared to the total response time (e.g. the modular exponentiation is likely to take much longer) it is still likely to be measurable, even if extra oracle queries and statistical analysis have to be employed.

RSAES-OAEP offers an even bigger target for a timing attack. The integrity check compares a hash from the OAEP-decoding to a locally calculated hash of the parameters. The parameters can be an octet string of arbitrary length chosen by the attacker. The hash is only needed in the OAEP-decoding stage and it is reasonable to assume many implementations would calculate it during this stage (as the standard suggests), but this point is after the integer-to-octet conversion. An attacker can achieve whatever time difference he or she requires to distinguish the relevant error sources by using a sufficiently large octet string for the parameters — set the parameters to be 10MB long and do the attack with a wristwatch.

This use of the hash operation to attack RSAES-OAEP illustrates the algorithm's fragile nature. The hash does not involve the private key or the secret in the plaintext at all, so even a diligent implementer is unlikely to expect its operation to impact the security. Performing the hash operation before the integer-to-octet conversion eliminates its usefulness in a timing attack.

4.5 Summary

An algorithm that relies on identical responses to errors (despite their disparate sources), no access to logs, a specific (undocumented and not obvious) order of sub-tasks and attention to timing must be considered quite fragile. Though it is possible some implementations of RSAES-OAEP will be immune, it is quite likely that many others will be susceptible to the chosen ciphertext attack described in this paper. To some degree RSAES-OAEP achieves security through obscurity — obscurity of the source of errors, of implementation details and of timing information. Obscurity, however, is widely recognized as a poor principle for designing an algorithm.

5 Directions towards a Solution

The attack relies on distinguishing different actions of the oracle resulting from a decision about the structure of the plaintext. This suggests two possible approaches for a solution: ensure the actions are indistinguishable; or avoid any decision based on the structure of the plaintext. The former approach uses obscurity to achieve security, while the latter approach offers better hope of reducing the security dependence on seemingly innocuous implementation choices.

PKCS #1 v2.0 makes a basic effort at obscurity by outputting the same error message for all identified errors. PKCS #1 v2.1 draft 2 enhances this effort by noting that errors from integer-to-octet conversion and OAEP-decoding must be indistinguishable and, importantly, that execution time must not reveal which error occurred [3]. [7]

A naive solution for avoiding a decision about the structure of the plaintext is to simply ignore its structure, i.e. ignore its most significant octet (after converting integer m to k octets). Ignoring this octet during decryption allows it to be set to any value (e.g. a random value) during encryption (subject to the restriction $m < n$). As it stands, however, this is not a good solution because these modifications mean the algorithm is no longer plaintext-aware — destroying the security proof that OAEP offered. An operation on a ciphertext that only altered the most significant octet of the corresponding plaintext would produce a different, but still valid, ciphertext without requiring knowledge of the plaintext. How to perform such an operation is an open question (at least to the author), as is the question of how such an ability would affect security in practice.

Another open question is how to modify RSAES-OAEP to eliminate the last vestige of structure from the plaintext, yet retain a proof of its security against chosen ciphertext attack in the random oracle model. Not only would such a solution avoid decisions based on plaintext structure - it would ensures no such decision could reasonably be made (even inadvertently) as there is no structure upon which to make it. [8]

5.1 Best Practise

Though the check that $m < B$ is the basis of the attack, it is other details (such the time a hash operation takes) that allow the attack to proceed. This rein-

[7] PKCS #1 v2.1 draft 2, section 7.1.2 *[RSAES-OAEP] Decryption operation*, see the note at the bottom of page 18.

[8] Such an inadvertent decision (i.e. a software bug) has been noticed by the author in one RSAES-OAEP implementation. It never explicitly checked if the plaintext "`integer [was] too large`", but just assumed it would fit in $k - 1$ octets and suffered buffer overflow problems when this was not the case.

forces Bleichenbacher's conclusion that the "integrity check must be performed in the correct step of the protocol – preferably immediately after decryption" [1]. Moving any processing that does not have to occur between the decryption and integrity check to another location is a practical step towards satisfying this criterion, hence lessening the exposure to chosen ciphertext attacks (though it does not, by itself, eliminate the threat). Processes that could be performed before the decryption operation in RSAES-OAEP include hashing the parameters, confirming relevant MGF and hash algorithms are supported and allocating memory required during mask generation and OAEP-decoding. Rearranging these processes should occur in implementations and also in standards defining algorithms, as the latter are the specification from which implementations are built.

6 Conclusion

Optimal Asymmetric Encryption Padding adds an integrity check and masks the structure of the message being encrypted to achieve plaintext-awareness and consequent protection against chosen ciphertext attack. However, translating the octet-aligned OAEP process into integers modulo n in RSAES-OAEP reintroduced sufficient structure to make an adaptive chosen ciphertext attack possible, with a high likelihood, in many implementations.

Acknowledgements. I thank the Director of Research, Telstra Research Laboratories, for supporting this work. I also thank the reviewers of this paper for highlighting the risks of simply ignoring the structure in RSAES-OAEP.

References

1. D. Bleichenbacher: Chosen Ciphertext Attacks Against Protocols Based on the RSA Encryption Standard PKCS #1. In Hugo Krawczyk (ed.), *Advances in Cryptology - CRYPTO '98*, pages 1–12, Berlin, Springer, 1998 (Lecture Notes in Computer Science, vol. 1462).
2. PKCS #1 v2.0: *RSA Cryptography Standard*, 1 October 1998. `http://www.rsasecurity.com/rsalabs/pkcs/`
3. PKCS #1 v2.1 draft 2: *RSA Cryptography Standard*, 5 January 2001. `http://www.rsasecurity.com/rsalabs/pkcs/`
4. IEEE 1363 draft 13: *Standard Specifications for Public Key Cryptography*, 12 November 1999. `http://grouper.ieee.org/groups/1363/`
5. M. Bellare and P. Rogaway: Optimal Asymmetric Encryption Padding — How to Encrypt with RSA. In *Advances in Cryptology — EUROCRYPT '94*, pages 92–111, Springer-Verlag, 1994.

OAEP Reconsidered
(Extended Abstract)

Victor Shoup

IBM Zurich Research Lab, Säumerstr. 4, 8803 Rüschlikon, Switzerland
sho@zurich.ibm.com

Abstract. The OAEP encryption scheme was introduced by Bellare and Rogaway at Eurocrypt '94. It converts any trapdoor permutation scheme into a public-key encryption scheme. OAEP is widely believed to provide resistance against adaptive chosen ciphertext attack. The main justification for this belief is a supposed proof of security in the random oracle model, assuming the underlying trapdoor permutation scheme is one way.

This paper shows conclusively that this justification is invalid. First, it observes that there appears to be a non-trivial gap in the OAEP security proof. Second, it proves that this gap cannot be filled, in the sense that there can be no standard "black box" security reduction for OAEP. This is done by proving that there exists an oracle relative to which the general OAEP scheme is insecure.

The paper also presents a new scheme OAEP+, along with a complete proof of security in the random oracle model. OAEP+ is essentially just as efficient as OAEP, and even has a tighter security reduction.

It should be stressed that these results do not imply that a particular instantiation of OAEP, such as RSA-OAEP, is insecure. They simply undermine the original justification for its security. In fact, it turns out—essentially by accident, rather than by design—that RSA-OAEP is secure in the random oracle model; however, this fact relies on special algebraic properties of the RSA function, and not on the security of the general OAEP scheme.

1 Introduction

It is generally agreed that the "right" definition of security for a public key encryption scheme is *security against adaptive chosen ciphertext attack*, as defined in [RS91]. This notion of security is equivalent to other useful notions, such as the notion of *non-malleability*, as defined in [DDN91,DDN00].

[DDN91] proposed a scheme that is provably secure in this sense, based on standard intractability assumptions. While this scheme is useful as a proof of concept, it is quite impractical. [RS91] also propose a scheme that is also provably secure; however, it too is also quite impractical, and moreover, it has special "public key infrastructure" requirements.

In 1993, Bellare and Rogaway proposed a method for converting any trapdoor permutation scheme into an encryption scheme [BR93]. They proved that this

J. Kilian (Ed.): CRYPTO 2001, LNCS 2139, pp. 239–259, 2001.

scheme is secure against adaptive chosen ciphertext attack in the *random oracle model*, provided the underlying trapdoor permutation scheme is one way.

In the random oracle model, one analyzes the security of the scheme by *pretending* that a cryptographic hash function is really a *random oracle.*

The encryption scheme in [BR93] is very efficient from the point of view of computation time. However, it has a "message expansion rate" that is not as good as some other encryption schemes.

In 1994, Bellare and Rogaway proposed another method for converting any trapdoor permutation scheme into an encryption scheme [BR94]. This scheme goes by the name OAEP. The scheme when instantiated with the RSA function [RSA78] goes by the name RSA-OAEP, and is the industry-wide standard for RSA encryption (PKCS#1 version 2, IEEE P1363). It is just as efficient computationally as the scheme in [BR93], but it has a better message expansion rate. With RSA-OAEP, one can encrypt messages whose bit-length is up to just a few hundred bits less than the number of bits in the RSA modulus, yielding a ciphertext whose size is the same as that of the RSA modulus.

Besides its efficiency in terms of both time and message expansion, and its compatibility with more traditional implementations of RSA encryption, perhaps one of the reasons that OAEP is so popular is the widespread *belief* that the scheme is provably secure in the random oracle model, provided the underlying trapdoor permutation scheme is one way.

In this paper we argue that this belief is unjustified. Specifically, we argue that in fact, no *complete* proof of the general OAEP method has ever appeared in the literature. Moreover, we prove that no proof is attainable using standard "black box" reductions (even in the random oracle model). Specifically, we show that there exists an oracle relative to which the general OAEP scheme is insecure. We then present a variation, OAEP+, and a complete proof of security in the random oracle model. OAEP+ is essentially just as efficient as OAEP.

There is one more twist to this story: we observe that RSA-OAEP with encryption exponent 3 actually *is* provably secure in the random oracle model; the proof, of course, is not a "black box" reduction, but exploits special algebraic properties of the RSA function. These observations were subsequently extended in [FOPS00,FOPS01] to RSA-OAEP with arbitrary encryption exponent.

Note that although the precise specification of standards (PKCS#1 version 2, IEEE P1363) differ in a few minor points from the scheme described in [BR94], none of these minor changes affect the arguments we make here.

1.1 A Missing Proof of Security

[BR94] contains a valid proof that OAEP satisfies a certain technical property which they call "plaintext awareness." Let us call this property *PA1*. However, it is claimed *without proof* that PA1 implies security against chosen ciphertext attack and non-malleability. Moreover, it is not even clear if the authors mean adaptive chosen ciphertext attack (as in [RS91]) or *indifferent* (a.k.a. *lunchtime*) chosen ciphertext attack (as in [NY90]).

Later, in [BDPR98], a new definition of "plaintext awareness" is given. Let us call this property *PA2*. It is claimed in [BDPR98] that OAEP is "plaintext aware." It is not clear if the authors mean to say that OAEP is PA1 or PA2; in any event, they certainly do not prove anything new about OAEP in [BDPR98]. Furthermore, [BDPR98] contains a valid proof that PA2 implies security against adaptive chosen ciphertext attack.

Notice that nowhere in this chain of reasoning is a proof that OAEP is secure against adaptive chosen ciphertext attack. What is missing is a proof that either OAEP is PA2, or that PA1 implies security against adaptive chosen ciphertext attack.

We should point out, however, that PA1 is trivially seen to imply security against *indifferent* chosen ciphertext attack, and thus OAEP is secure against indifferent chosen ciphertext attack. However, this is a strictly weaker and much less useful notion of security than security against adaptive chosen ciphertext attack.

1.2 Our Contributions

In §4, we give a rather informal argument that there is a non-trivial obstruction to obtaining a complete proof of security for OAEP against adaptive chosen ciphertext attack (in the random oracle model).

In §5, we give more formal and compelling evidence for this. Specifically, we prove that if one-way trapdoor permutation schemes with an additional special property exist, then OAEP when instantiated with such a one-way trapdoor permutation scheme is in fact *insecure.* We do not know how to prove the existence of such special one-way trapdoor permutation schemes (assuming, say, that one-way trapdoor permutation schemes exist at all). However, we prove that there exists an oracle, relative to which such special one-way trapdoor permutation schemes exists. It follows that relative to an oracle, the OAEP construction is not secure.

Actually, our proofs imply something slightly stronger: relative to an oracle, OAEP is *malleable* with respect to a *chosen plaintext* attack.

Of course, such relativized results do not necessarily imply anything about the ordinary, unrelativized security of OAEP. But they do imply that standard proof techniques, in which the adversary and the trapdoor permutation are treated as "black boxes," cannot possibly yield a proof of security, since they would relativize. Certainly, all of the arguments in [BR94] and [BDPR98] involve only "black box" reductions, and so they cannot possibly be modified to yield a proof of security.

In §6, we present a new scheme, called OAEP+. This is a variation of OAEP that is essentially just as efficient in all respects as OAEP, but for which we provide a complete, detailed proof of security against adaptive chosen ciphertext attack. Moreover, the security reduction for OAEP+ is somewhat tighter than for OAEP.

We conclude the paper in §7 on a rather ironic note. After considering other variations of OAEP, we sketch a proof that RSA-OAEP with encryption exponent 3 actually *is* secure in the random oracle model. This fact, however, makes essential use of Coppersmith's algorithm [Cop96] for solving low-degree modular equations. This proof of security does not generalize to large encryption exponents, and in particular, it does not cover the popular encryption exponent $2^{16}+1$.

Part of the irony of this observation is that Coppersmith viewed his own result as a reason *not* to use exponent 3, while here, it ostensibly gives one reason why one perhaps *should* use exponent 3.

It is also worth noting here that by using Coppersmith's algorithm, one gets a fairly tight security reduction for exponent-3 RSA-OAEP, and an even tighter reduction for exponent-3 RSA-OAEP+. These reductions are much more efficient than either the (incorrect) reduction for OAEP in [BR94], or our general reduction for OAEP+. Indeed, these general reductions are so inefficient that they fail to provide any truly meaningful security guarantees for, say, 1024-bit RSA, whereas with the use of Coppersmith's algorithm, the security guarantees are much more meaningful.

Subsequent to the distribution of the original version of this paper [Sho00], it was shown in [FOPS00] that RSA-OAEP with an arbitrary encryption exponent is indeed secure against adaptive chosen ciphertext attack in the random oracle model. We remark, however, that the reduction in [FOPS00] is significantly less efficient than our general reduction for OAEP+, and so it provides a less meaningful security guarantee for typical choices of security parameters. This may be a reason to consider using RSA-OAEP+ instead of RSA-OAEP.

We also mention the subsequent work of [Bon01], which considers OAEP-like variations of RSA as well as Rabin encryption.

Let us be clear about the implications of our results. They do not imply an attack on RSA-OAEP. They only imply that the original *justification* for the *belief* that OAEP in general—and hence RSA-OAEP in particular—is resistant against adaptive chosen ciphertext attack was *invalid*. As it turns out, our observations on exponent-3 RSA-OAEP, and the more general results of [FOPS00] on arbitrary-exponent RSA-OAEP, imply that RSA-OAEP is indeed secure against adaptive chosen ciphertext attack in the random oracle model. However, the security of RSA-OAEP does not follow from the security of OAEP in general, but rather, relies on specific algebraic properties of the RSA function.

Before moving ahead, we recall some definitions in §2, and the OAEP scheme itself in §3.

2 Preliminaries

2.1 Security against Chosen Ciphertext Attack

We recall the definition of security against adaptive chosen ciphertext attack.

We begin by describing the attack scenario.

Stage 1. The key generation algorithm is run, generating the public key and private key for the cryptosystem. The adversary, of course, obtains the public key, but not the private key.

Stage 2. The adversary makes a series of arbitrary queries to a *decryption oracle*. Each query is a ciphertext y that is decrypted by the decryption oracle, making use of the private key of the cryptosystem. The resulting decryption is given to the adversary. The adversary is free to construct the ciphertexts in an arbitrary way—it is certainly *not* required to compute them using the encryption algorithm.

Stage 3. The adversary prepares two messages x_0, x_1, and gives these to an *encryption oracle*. The encryption oracle chooses $b \in \{0,1\}$ at random, encrypts x_b, and gives the resulting "target" ciphertext y^* to the adversary. The adversary is free to choose x_0 and x_1 in an arbitrary way, except that if message lengths are not fixed by the cryptosystem, then these two messages must nevertheless be of the same length.

Stage 4. The adversary continues to submit ciphertexts y to the decryption oracle, subject only to the restriction that $y \neq y^*$.

Stage 5. The adversary outputs $\hat{b} \in \{0,1\}$, representing its "guess" of b.

That completes the description of the attack scenario.

The adversary's *advantage* in this attack scenario is defined to be $|\Pr[\hat{b} = b] - 1/2|$.

A cryptosystem is defined to be *secure against adaptive chosen ciphertext attack* if for any efficient adversary, its advantage is negligible.

Of course, this is a complexity-theoretic definition, and the above description suppresses many details, e.g., there is an implicit security parameter which tends to infinity, and the terms "efficient" and "negligible" are technical terms, defined in the usual way. Also, we shall work in a *uniform* model of computation (i.e., Turing machines).

The definition of security we have presented here is from [RS91]. It is called *IND-CCA2* in [BDPR98]. It is known to be equivalent to other notions, such as non-malleability [DDN91,BDPR98,DDN00], which is called *NM-CCA2* in [BDPR98].

It is fairly well understood and accepted that this notion of security is the "right" one, in the sense that a general-purpose cryptosystem that is to be deployed in a wide range of applications should satisfy this property. Indeed, with this property, one can typically establish the security of larger systems that use such a cryptosystem as a component.

There are other, weaker notions of security against chosen ciphertext attack. For example, [NY90] define a notion that is sometimes called *security against indifferent chosen ciphertext attack*, or *security against lunchtime attack*. This definition of security is exactly the same as the one above, except that Stage 4 is omitted—that is, the adversary does not have access to the decryption oracle after it obtains the target ciphertext. While this notion of security may seem natural, it is actually not sufficient in many applications. This notion is called *IND-CCA1* in [BDPR98].

2.2 One-Way Trapdoor Permutations

We recall the notion of a trapdoor permutation scheme. This consists of a probabilistic *permutation generator* algorithm that outputs (descriptions of) two algorithms f and g, such that the function computed by f is a permutation on the set of k-bit strings, and the function computed by g is its inverse.

An attack on a trapdoor permutation scheme proceeds as follows. First the generator is run, yielding f and g. The adversary is given f, but not g. Additionally, the adversary is given a random $y \in \{0,1\}^k$. The adversary then computes and outputs a string $w \in \{0,1\}^k$.

The adversary's *success probability* is defined to $\Pr[f(w) = y]$.

The scheme is called a *one-way* trapdoor permutation scheme if for any efficient adversary, its success probability is negligible. As above, this is a complexity theoretic definition, and we have suppressed a number of details, including a security parameter, which is input to the permutation generator; the parameter k, as well as the running times of f and g, should be bounded by a polynomial in this security parameter.

2.3 The Random Oracle Model

The random oracle model was introduced in [BR93] as a means of heuristically analyzing a cryptographic primitive or protocol. In this approach, one equips all of the algorithms associated with the primitive or protocol (including the adversary's algorithms) with oracle access to one or more functions. Each of these functions is a map from $\{0,1\}^a$ to $\{0,1\}^b$, for some specified values a and b. One then reformulates the definition of security so that in the attack game, each of these functions is chosen at random from the set of all functions mapping $\{0,1\}^a$ to $\{0,1\}^b$.

In an actual implementation, one typically instantiates these random oracles as cryptographic hash functions.

Now, a proof of security in the random oracle model does not necessarily imply *anything* about security in the "real world" where actual computation takes place (see [CGH98]). Nevertheless, it seems that designing a scheme so that it is provably secure in the random oracle model is a good engineering principle, at least when all known schemes that are provably secure without the random oracle heuristic are too impractical. Subsequent to [BR93], many other papers have proposed and analyzed cryptographic schemes in the random oracle model.

3 OAEP

We now describe the OAEP encryption scheme, as described in §6 of [BR94].

The general scheme makes use of a one-way trapdoor permutation. Let f be the permutation, acting on k-bit strings, and g its inverse. The scheme also makes use of two parameters k_0 and k_1, which should satisfy $k_0 + k_1 < k$. It should also be the case that 2^{-k_0} and 2^{-k_1} are negligible quantities. The scheme encrypts messages $x \in \{0,1\}^n$, where $n = k - k_0 - k_1$.

The scheme also makes use of two functions, $G : \{0,1\}^{k_0} \to \{0,1\}^{n+k_1}$, and $H : \{0,1\}^{n+k_1} \to \{0,1\}^{k_0}$. These two functions will be modeled as random oracles in the security analysis.

We describe the key generation, encryption, and decryption algorithms of the scheme.

Key generation. This simply runs the generator for the one-way trapdoor permutation scheme, obtaining f and g. The public key is f, and the private key is g.

Encryption. Given a plaintext x, the encryption algorithm randomly chooses $r \in \{0,1\}^{k_0}$, and then computes

$$s \in \{0,1\}^{n+k_1},\ t \in \{0,1\}^{k_0},\ w \in \{0,1\}^k,\ y \in \{0,1\}^k$$

as follows:

$$s = G(r) \oplus (x \,\|\, 0^{k_1}), \tag{1}$$

$$t = H(s) \oplus r, \tag{2}$$

$$w = s \,\|\, t, \tag{3}$$

$$y = f(w). \tag{4}$$

The ciphertext is y.

Decryption. Given a ciphertext y, the decryption algorithm computes

$$w \in \{0,1\}^k,\ s \in \{0,1\}^{n+k_1},\ t \in \{0,1\}^{k_0},\ r \in \{0,1\}^{k_0},$$
$$z \in \{0,1\}^{n+k_1},\ x \in \{0,1\}^n,\ c \in \{0,1\}^{k_1}$$

as follows:

$$w = g(y), \tag{5}$$

$$s = w[0 \ldots n + k_1 - 1], \tag{6}$$

$$t = w[n + k_1 \ldots k], \tag{7}$$

$$r = H(s) \oplus t, \tag{8}$$

$$z = G(r) \oplus s, \tag{9}$$

$$x = z[0 \ldots n - 1], \tag{10}$$

$$c = z[n \ldots n + k_1 - 1]. \tag{11}$$

If $c = 0^{k_1}$, then the algorithm outputs the cleartext x; otherwise, the algorithm *rejects* the ciphertext, and does not output a cleartext.

4 An Informal Argument that OAEP Cannot Be Proven Secure

In this section, we discuss the gap in the proof in [BR94]. The reader may safely choose to skip this section upon first reading.

We first recall the main ideas of the proof in [BR94] that OAEP is "plaintext aware" in the random oracle model, where G and H are modeled as random oracles.

The argument shows how a simulator that has access to a table of input/output values for the points at which G and H were queried can simulate the decryption oracle without knowing the private key. As we shall see, one must distinguish between random oracle queries made by the adversary and random oracle queries made by the encryption oracle. This is a subtle point, but the failure to make this distinction is really at the heart of the flawed reasoning in [BR94].

To make our arguments clearer, we introduce some notational conventions. First, any ciphertext y implicitly defines values w, s, t, r, z, x, c via the decryption equations (5)-(11). Let y^* denote the target ciphertext, and let $w^*, s^*, t^*, r^*, z^*, x^*, c^*$ be the corresponding implicitly defined values for y^*. Note that $x^* = x_b$ and $c^* = 0^{k_1}$.

Let S_G the set of values r at which G was queried *by the adversary*. Also, let S_H be the set of values s at which H was queried *by the adversary*. Further, let $S_G^* = S_G \cup \{r^*\}$ and $S_H^* = S_H \cup \{s^*\}$, where r^*, s^* are the values implicitly defined by y^*, as described above. We view these sets as growing incrementally as the adversary's attack proceeds—elements are added to these only when a random oracle is queried by the adversary or by the encryption oracle.

Suppose the simulator is given a ciphertext y to decrypt. One can show that if $r \notin S_G^*$, then with overwhelming probability the actual decryption algorithm would reject y; this is because in this case, s and $G(r)$ are independent, and so the probability that $c = 0^{k_1}$ is 2^{-k_1}. Moreover, if $s \notin S_H^*$, then with overwhelming probability, $r \notin S_G^*$; this is because in this case, t and $H(s)$ are independent, and so r is independent of the adversary's view. From this argument, it follows that the actual decryption algorithm would reject with overwhelming probability, unless $r \in S_G^*$ and $s \in S_H^*$.

If the decryption oracle simulator (a.k.a., plaintext extractor) has access to S_G^* and S_H^*, as well as the corresponding outputs of G and H, then it can effectively simulate the decryption without knowing the secret key, as follows. It simply enumerates all $r' \in S_G^*$ and $s' \in S_H^*$, and for each of these computes

$$t' = H(s') \oplus r', \ w' = s' \,\|\, t', \ y' = f(w').$$

If y' is equal to y, then it computes the corresponding x' and c' values, via the equations (10) and (11); if $c' = 0^{k_1}$, it outputs x', and otherwise rejects. If no y' equals y, then it simply outputs reject.

Given the above arguments, it is easy to see that this simulated decryption oracle behaves exactly like the actual decryption oracle, except with negligible probability. Certainly, if some $y' = y$, the simulator's response is correct, and if no $y' = y$, then the above arguments imply that the real decryption oracle would have rejected y with overwhelming probability.

From this, one would like to conclude that the decryption oracle does not help the adversary. But this reasoning is invalid. Indeed, the adversary in the actual attack has access to S_G and S_H, along with the corresponding outputs of G and H, but does not have direct access to $r^*, G(r^*), s^*, H(s^*)$. Thus, the above decryption simulator has more power than does the adversary. Moreover, if we give the decryption simulator access to $r^*, G(r^*), s^*, H(s^*)$, then the proof that x^* is well hidden, unless the adversary can invert f, is doomed to failure: if the simulator needs to "know" r^* and s^*, then it must already "know" w^*, and so one can not hope use the adversary to compute something that the simulator did not already know.

On closer observation, it is clear that the decryption simulator does not need to know $s^*, G(s^*)$: if $s = s^*$, then it must be the case that $t \neq t^*$, which implies that $r \neq r^*$, and so $c = 0^{k_1}$ with negligible probability. Thus, it is safe to reject all ciphertexts y such that $s = s^*$.

If one could make an analogous argument that the decryption simulator does not need to know $r^*, G(r^*)$, we would be done. This is unfortunately not the case, as the following example illustrates.

The arguments in [BR94] simply do not take into account the random oracle queries made by the decryption oracle. All these arguments really show is that OAEP is secure against indifferent chosen ciphertext attack.

4.1 An Example

Suppose that we have an algorithm that actually can invert f. Now of course, in this case, we will not be able to construct a counter-example to the security of OAEP, but we will argue that the proof technique fails. In particular, we show how to build an adversary that uses the f-inverting algorithm to break the cryptosystem, but it does so in such a way that no simulator given black box access to the adversary and its random oracle queries can use our adversary to compute $f^{-1}(y^*)$ for a given value of y^*.

We now describe adversary. Upon obtaining the target ciphertext y^*, the adversary computes w^* using the algorithm for inverting f, and then extracts the corresponding values s^* and t^*. The adversary then chooses an arbitrary, non-zero $\Delta \in \{0,1\}^n$, and computes:

$$s = s^* \oplus (\Delta \,\|\, 0^{k_1}), \;\; t = t^* \oplus H(s^*) \oplus H(s), \;\; w = s \,\|\, t, \;\; y = f(w).$$

It is easily verified that y is a valid encryption of $x = x^* \oplus \Delta$, and clearly $y \neq y^*$. So if the adversary submits y to the decryption oracle, he obtains x, from which he can then easily compute x^*.

This adversary clearly breaks the cryptosystem—in fact, its advantage is 1/2. However, note in this attack, the adversary only queries the oracle H at the points s and s^*. It never queries the oracle G at all. In fact $r = r^*$, and the attack succeeds just where the gap in the proof was identified above.

What information has a simulator learned by interacting with the adversary as a black box? It has only learned s^* and s (and hence Δ). So it has learned the first $n + k_1$ bits of the pre-image of y^*, but the last k_0 remain a complete mystery to the simulator, and in general, they will not be easily computable from the first $n + k_1$ bits. The simulator also has seen the value y submitted to the decryption oracle, but it does not seem likely that this can be used by the simulator to any useful effect.

5 Formal Evidence that the OAEP Construction Is Not Sound

In this section, we present strong evidence that the OAEP construction is not sound. First, we show that if a special type of one-way trapdoor permutation f_0 exists, then in fact, we can construct another one-way trapdoor permutation f such that OAEP using f is *insecure*. Although we do not know how to explicitly construct such a special f_0, we can show that there is an oracle relative to which one exists. Thus, there is an oracle relative to which OAEP is insecure. This in turn implies that there is no standard "black box" security reduction for OAEP.

Definition 1. *We call a permutation generator* XOR-malleable *if the following property holds. There exists an efficient algorithm U, such that for infinitely many values of the security parameter, $U(f_0, f_0(t), \delta) = f_0(t \oplus \delta)$ with non-negligible probability. Here, the probability is taken over the random bits of the permutation generator, and random bit strings t and δ in the domain $\{0,1\}^{k_0}$ of the generated permutation f_0.*

Theorem 1. *If there exists an XOR-malleable one-way trapdoor permutation scheme, then there exists a one-way trapdoor permutation scheme such that when OAEP is instantiated with this scheme, the resulting encryption scheme is insecure (in the random oracle model).*

We now prove this theorem, which is based on the example presented in §4.1.

Let f_0 be the given XOR-malleable one-way trapdoor permutation on k_0-bit strings. Let U be the algorithm that computes $f_0(t \oplus \delta)$ from $(f_0, f_0(t), \delta)$. Choose $n > 0$, $k_1 > 0$, and set $k = n + k_0 + k_1$. Let f be the permutation on k-bit strings defined as follows: for $s \in \{0,1\}^{n+k_1}, t \in \{0,1\}^{k_0}$, let $f(s \,\|\, t) = s \,\|\, f_0(t)$.

It is clear that f is a one-way trapdoor permutation.

Now consider the OAEP scheme that uses this f as its one-way trapdoor permutation, and uses the parameters k, n, k_0, k_1 for the padding scheme.

Recall our notational conventions: any ciphertext y implicitly defines values w, s, t, r, z, x, c, and the target ciphertext y^* implicitly defines $w^*, s^*, t^*, r^*, z^*, x^*, c^*$.

We now describe the adversary. Upon obtaining the target ciphertext y^*, the adversary decomposes y^* as $y^* = s^* \,\|\, f_0(t^*)$. The adversary then chooses an arbitrary, non-zero $\Delta \in \{0,1\}^n$, and computes:

$$s = s^* \oplus (\Delta \,\|\, 0^{k_1}), \ v = U(f_0, f_0(t^*), H(s^*) \oplus H(s)), \ y = s \,\|\, v.$$

It is easily verified that y is a valid encryption of $x = x^* \oplus \Delta$, provided $v = f_0(t^* \oplus H(s^*) \oplus H(s))$, which by our assumption of XOR-malleability occurs with non-negligible probability. Indeed, we have

$$t = t^* \oplus H(s^*) \oplus H(s),$$
$$r = H(s) \oplus t = H(s^*) \oplus t^* = r^*,$$
$$z = G(r) \oplus s = G(r^*) \oplus s^* \oplus (\Delta \,\|\, 0^{k_1}) = (x^* \oplus \Delta) \,\|\, 0^{k_1}.$$

So if the adversary submits y to the decryption oracle, he obtains x, from which he can then easily compute x^*.

This adversary clearly breaks the cryptosystem. That completes the proof of the theorem.

Note that in the above attack, $r = r^*$ and the adversary never explicitly queried G at r, but was able to "hijack" $G(r)$ from the encryption oracle—this is the essence of the problem with OAEP.

Note that this also attack shows that the scheme is malleable with respect to chosen plaintext attack.

Of course, one might ask if it is at all reasonable to believe that XOR-malleable one-way trapdoor permutations exist at all. First of all, note that the standard RSA function is a one-way trapdoor permutation that is not XOR-malleable, but is still malleable in a very similar way: given $\alpha = (a^e \bmod N)$ and $(b \bmod N)$, we can compute $((ab)^e \bmod N)$ as $(\alpha \cdot (b^e \bmod N))$. Thus, we can view the RSA function itself as a kind of malleable one-way trapdoor permutation, but where XOR is replaced by multiplication mod N. In fact, one could modify the OAEP scheme so that $t, H(s)$ and r are numbers mod N, and instead of the relation $t = H(s) \oplus r$, we would use the relation $t = H(s) \cdot r \bmod N$. It would seem that if there were a proof of security for OAEP, then it should go through for this variant of OAEP as well. But yet, this variant of OAEP is clearly insecure, even though the underlying trapdoor permutation is presumably one way.

Another example is exponentiation in a finite abelian group. For a group element g, the function mapping a to g^a is malleable with respect to both addition *and* multiplication modulo the order of g. Although for appropriate choices of groups this function is a reasonable candidate for a one-way permutation, it does not have a trapdoor.

Beyond this, we prove a relativized result.

Theorem 2. *There exists an oracle, relative to which XOR-malleable one-way trapdoor permutations exist.*

This theorem provides some evidence that the notion of an XOR-malleable one-way trapdoor permutation scheme is not *a priori* vacuous.

Also, Theorems 1 and 2 imply the following.

Corollary 1. *There exists an oracle, relative to which the OAEP construction is insecure.*

We should stress the implications of this corollary.

Normally, to prove the security of a cryptographic system, one proves this via a "black box" security reduction from solving the underlying "hard" problem to breaking the cryptographic system. Briefly, such a reduction for a cryptosystem based on a general trapdoor permutation scheme would be an efficient, probabilistic algorithm that inverts a permutation f on a random point, given oracle access to an adversary that successfully breaks cryptosystem (instantiated with f) and the permutation f. It should work for *all* adversaries and *all* permutations, even ones that are not efficiently computable, or even computable at all. Whatever the adversary's advantage is in breaking the cryptosystem, the success probability of the inversion algorithm should not be too much smaller.

We do not attempt to make a more formal or precise definition of a black-box security reduction, but it should be clear that any such reduction would imply security relative to any oracle. So Corollary 1 implies that there is no black-box security reduction for OAEP.

For lack of space, we do not present the proof of Theorem 2 in this extended abstract. The reader is referred to the full-length version of this paper [Sho00].

6 OAEP+

We now describe the OAEP+ encryption scheme, which is just a slight modification of the OAEP scheme.

The general scheme makes use of a one-way trapdoor permutation. Let f be the permutation, acting on k-bit strings, and g its inverse. The scheme also makes use of two parameters k_0 and k_1, which should satisfy $k_0 + k_1 < k$. It should also be the case that 2^{-k_0} and 2^{-k_1} are negligible quantities. The scheme encrypts messages $x \in \{0,1\}^n$, where $n = k - k_0 - k_1$.

The scheme also makes use of three functions:

$$G : \{0,1\}^{k_0} \to \{0,1\}^n, \; H' : \{0,1\}^{n+k_0} \to \{0,1\}^{k_1}, \; H : \{0,1\}^{n+k_1} \to \{0,1\}^{k_0}.$$

These three functions will be modeled as independent random oracles in the security analysis.

We describe the key generation, encryption, and decryption algorithms of the scheme.

Key generation. This simply runs the generator for the one-way trapdoor permutation scheme, obtaining f and g. The public key is f, and the private key is g.

Encryption. Given a plaintext x, the encryption algorithm randomly chooses $r \in \{0,1\}^{k_0}$, and then computes

$$s \in \{0,1\}^{n+k_1}, \; t \in \{0,1\}^{k_0}, \; w \in \{0,1\}^k, \; y \in \{0,1\}^k$$

as follows:

$$s = (G(r) \oplus x) \,\|\, H'(r \,\|\, x), \tag{12}$$
$$t = H(s) \oplus r, \tag{13}$$
$$w = s \,\|\, t, \tag{14}$$
$$y = f(w). \tag{15}$$

The ciphertext is y.

Decryption. Given a ciphertext y, the decryption algorithm computes

$$w \in \{0,1\}^k,\ s \in \{0,1\}^{n+k_1},\ t, r \in \{0,1\}^{k_0},\ x \in \{0,1\}^n,\ c \in \{0,1\}^{k_1}$$

as follows:

$$w = g(y), \tag{16}$$
$$s = w[0 \ldots n + k_1 - 1], \tag{17}$$
$$t = w[n + k_1 \ldots k], \tag{18}$$
$$r = H(s) \oplus t, \tag{19}$$
$$x = G(r) \oplus s[0 \ldots n - 1], \tag{20}$$
$$c = s[n \ldots n + k_1 - 1]. \tag{21}$$

If $c = H'(r \,\|\, x)$, then the algorithm outputs the cleartext x; otherwise, the algorithm *rejects* the ciphertext, and does not output a cleartext.

Theorem 3. *If the underlying trapdoor permutation scheme is one way, then OAEP+ is secure against adaptive chosen ciphertext attack in the random oracle model.*

We start with some notations and conventions.

Let A be an adversary, and let $\mathbf{G}_0$ be the original attack game. Let b and $\hat{b}$ be as defined in §2.1, and let S_0 be the event that $b = \hat{b}$.

Let q_G, q_H, and $q_{H'}$ bound the number of queries made by A to the oracles G, H, and H' respectively, and let q_D bound the number of decryption oracle queries.

We assume without loss of generality that whenever A makes a query of the form $H'(r \,\|\, x)$, for any $r \in \{0,1\}^{k_0}, x \in \{0,1\}^n$, then A has previously made the query $G(r)$.

We shall show that

$$|\Pr[S_0] - 1/2| \le \mathit{InvAdv}(A') + (q_{H'} + q_D)/2^{k_1} + (q_D + 1)q_G/2^{k_0}, \tag{22}$$

where $\mathit{InvAdv}(A')$ is the success probability that a particular adversary A' has in breaking the one-way trapdoor permutation scheme on k-bit inputs. The time and space requirements of A' are related to those of A as follows:

$$\mathit{Time}(A') = O(\mathit{Time}(A) + q_G q_H T_f + (q_G + q_{H'} + q_H + q_D)k); \tag{23}$$
$$\mathit{Space}(A') = O(\mathit{Space}(A) + (q_G + q_{H'} + q_H)k). \tag{24}$$

Here, T_f is the time required to compute f, and space is measured in bits of storage. These complexity estimates assume a standard random-access model of computation.

Any ciphertext y implicitly defines values w, s, t, r, x, c via the decryption equations (16)-(21). Let y^* denote the target ciphertext, and let $w^*, s^*, t^*, r^*, x^*, c^*$ be the corresponding implicitly defined values for y^*. Note that $x^* = x_b$ and $c^* = H'(r^* \| x^*)$.

We define sets S_G and S_H, as in §4, as follows. Let S_G be the set of values r at which G was queried by A. Also, let S_H be the set of values s at which H was queried by A. Additionally, define $S_{H'}$ to be the set of pairs (r, x) such that H' was queried at $r \| x$ by A. We view these sets as growing incrementally as A's attack proceeds—elements are added to these only when a random oracle is queried by A.

We also define A's *view* as the sequence of random variables

$$View = \langle X_0, X_1, \ldots, X_{q_G+q_{H'}+q_H+q_D+1} \rangle,$$

where X_0 consists of A's coin tosses and the public key of the encryption scheme, and where each X_i for $i \geq 1$ consists of a response to either a random oracle query, a decryption oracle query, or the encryption oracle query. The ith such query is a function of $\langle X_0, \ldots, X_{i-1} \rangle$. The adversary's final output $\hat{b}$ is a function of *View*. At any fixed point in time, A has made some number, say m, queries, and we define

$$CurrentView = \langle X_0, \ldots, X_m \rangle.$$

Our overall strategy for the proof is as follows. We shall define a sequence $\mathbf{G}_1, \mathbf{G}_2, \ldots, \mathbf{G}_5$ of modified attack games. Each of the games $\mathbf{G}_0, \mathbf{G}_1, \ldots, \mathbf{G}_5$ operate on the same underlying probability space. In particular, the public key and private key of the cryptosystem, the coin tosses of A, the values of the random oracles G, H', H, and the hidden bit b take on *identical* values across all games. Only some of the rules defining how the view is computed differ from game to game. For any $1 \leq i \leq 5$, we let S_i be the event that $b = \hat{b}$ in game $\mathbf{G}_i$. Our strategy is to show that for $1 \leq i \leq 5$, the quantity $|\Pr[S_{i-1}] - \Pr[S_i]|$ is negligible. Also, it will be evident from the definition of game $\mathbf{G}_5$ that $\Pr[S_5] = 1/2$, which will imply that $|\Pr[S_0] - 1/2|$ is negligible.

In games $\mathbf{G}_1$, $\mathbf{G}_2$, and $\mathbf{G}_3$, we incrementally modify the decryption oracle, so that in game $\mathbf{G}_3$, the modified decryption oracle operates without using the trapdoor for f at all. In games $\mathbf{G}_4$ and $\mathbf{G}_5$, we modify the encryption oracle, so that in game $\mathbf{G}_5$, the hidden bit b is completely independent of *View*.

To make a rigorous and precise proof, we state following very simple, but useful lemma, which we leave to the reader to verify.

Lemma 1. *Let E, E', and F be events defined on a probability space such that* $\Pr[E \wedge \neg F] = \Pr[E' \wedge \neg F]$. *Then we have* $|\Pr[E] - \Pr[E']| \leq \Pr[F]$.

Game $\mathbf{G}_1$. Now we modify game $\mathbf{G}_0$ to define a new game $\mathbf{G}_1$.

We modify the decryption oracle as follows. Given a ciphertext y, the new decryption oracle computes w, s, t, r, x, c as usual. If the old decryption oracle rejects, so does the new one. But the new decryption oracle also rejects if $(r, x) \notin S_{H'}$. More precisely, if the new decryption oracle computes r via equation (19), and finds that $r \notin S_G$, then it rejects right away, without ever querying $G(r)$; if $r \in S_G$, then x is computed, but if $(r, x) \notin S_{H'}$, it rejects without querying

$H'(r \| x)$. Recall that by convention, if A queried $H'(r \| x)$, it already queried $G(r)$. One sees that in game $\mathbf{G}_1$, the decryption oracle never queries G or H' at points other than those at which A did.

Let F_1 be the event that a ciphertext is rejected in $\mathbf{G}_1$ that would not have been rejected under the rules of game $\mathbf{G}_0$.

Consider a ciphertext $y \neq y^*$ submitted to the decryption oracle. If $r = r^*$ and $x = x^*$, then we must have $c \neq c^*$; in this case, however, we will surely reject under the rules of game $\mathbf{G}_0$. So we assume that $r \neq r^*$ or $x \neq x^*$. Now, the encryption oracle has made the query $H'(r^* \| x^*)$, but not $H'(r \| x)$, since $(r, x) \neq (r^*, x^*)$. So if A has not made the query $H'(r \| x)$, the value of $H'(r \| x)$ is independent of *CurrentView*, and hence, is independent of c, which is a function of *CurrentView* and H. Therefore, the probability that $c = H'(r \| x)$ is $1/2^{k_1}$.

From the above, it follows that $\Pr[F_1] \leq q_D/2^{k_1}$. Moreover, it is clear by construction that $\Pr[S_0 \wedge \neg F_1] = \Pr[S_1 \wedge \neg F_1]$, since the two games proceed identically unless the event F_1 occurs; that is, the value of *View* is the same in both games, provided F_1 does not occur. So applying Lemma 1 with (S_0, S_1, F_1), we have

$$|\Pr[S_0] - \Pr[S_1]| \leq q_D/2^{k_1}. \tag{25}$$

Game $\mathbf{G}_2$. Now we modify game $\mathbf{G}_1$ to obtain a new game $\mathbf{G}_2$. In this new game, we modify the decryption oracle yet again. Given a ciphertext y, the new decryption oracle computes w, s, t, r, x, c as usual. If the old decryption oracle rejects, so does the new one. But the new decryption oracle also rejects if $s \notin S_H$. More precisely, if the new decryption oracle computes s via equation (17), and finds that $s \notin S_H$, then it rejects right away, without ever querying $H(s)$. Thus, in game $\mathbf{G}_2$, the decryption oracle never queries G, H', or H at points other than those at which A did.

Let F_2 be the event that a ciphertext is rejected in $\mathbf{G}_2$ that would not have been rejected under the rules of game $\mathbf{G}_1$.

Consider a ciphertext $y \neq y^*$ with $s \notin S_H$ submitted to the decryption oracle. We consider two cases.

Case 1: $s = s^*$. Now, $s = s^*$ and $y \neq y^*$ implies $t \neq t^*$. Moreover, $s = s^*$ and $t \neq t^*$ implies that $r \neq r^*$. If this ciphertext is rejected in game $\mathbf{G}_2$ but would not be under the rules in game $\mathbf{G}_1$, it must be the case that $H'(r^* \| x^*) = H'(r \| x)$. The probability that such a collision can be found over the course of the attack is $q_{H'}/2^{k_1}$. Note that r^* is fixed by the encryption oracle, and so "birthday attacks" are not possible.

Case 2: $s \neq s^*$. In this case, the oracle H was never queried at s by either A, the encryption oracle, or the decryption oracle. Since $t = H(s) \oplus r$, the value r is independent of *CurrentView*. It follows that the probability that $r \in S_G$ is at most $q_G/2^{k_0}$. Over the course of the entire attack, these probabilities sum to $q_D q_G/2^{k_0}$.

It follows that $\Pr[F_2] \leq q_{H'}/2^{k_1} + q_D q_G/2^{k_0}$. Moreover, it is clear by construction that $\Pr[S_1 \wedge \neg F_2] = \Pr[S_2 \wedge \neg F_2]$, since the two games proceed identically unless F_2 occurs. So applying Lemma 1 with (S_1, S_2, F_2), we have

$$|\Pr[S_1] - \Pr[S_2]| \le q_{H'}/2^{k_1} + q_D q_G/2^{k_0}. \tag{26}$$

Game $\mathbf{G}_3$. Now we modify game $\mathbf{G}_2$ to obtain an equivalent game $\mathbf{G}_3$. We modify the decryption oracle so that it does not make use of the trapdoor for f at all.

Conceptually, this new decryption oracle iterates through all pairs $(r', x') \in S_{H'}$. For each of these, it does the following. First, it sets $s' = (G(r') \oplus x') \,\|\, H'(r' \,\|\, x')$. Note that both G and H' have already been queried at the given points. Second, if $s' \in S_H$, it then computes

$$t' = H(s') \oplus r', \; w' = s' \,\|\, t', \; y' = f(w').$$

If y' is equal to y, it stops and outputs x'.

If the above iteration terminates without having found some $y' = y$, then the new decryption oracle simply rejects.

It is clear that games $\mathbf{G}_3$ and $\mathbf{G}_2$ are identical, and so

$$\Pr[S_3] = \Pr[S_2]. \tag{27}$$

To actually implement this idea, one would build up a table, with one entry for each $(r', x') \in S_{H'}$. Each entry in the table would contain the corresponding value s', along with y' if s' is currently in S_H. If s' is currently not in S_H, we place y' in the table entry if and when A eventually queries $H(s')$. When a ciphertext y is submitted to the decryption oracle, we simply perform a table lookup to see if there is a y' in the table that is equal to y. These tables can all be implemented using standard data structures and algorithms. Using search tries to implement the table lookup, the total running time of the simulated decryption oracle over the course of game $\mathbf{G}_3$ is

$$O(\min(q_{H'}, q_H)T_f + (q_G + q_{H'} + q_H + q_D)k).$$

Note also that the space needed is essentially linear: $O((q_G + q_{H'} + q_H)k)$ bits.

Remark. Let us summarize the modifications made so far. We have modified the decryption oracle so that it does not make use of the trapdoor for f at all; moreover, the decryption oracle never queries G, H', or H at points other than those at which A did.

Game $\mathbf{G}_4$. In this game, we modify the random oracles and slightly modify the encryption oracle. The resulting game $\mathbf{G}_4$ is equivalent to game $\mathbf{G}_3$; however, this rather technical "bridging" step will facilitate the analysis of more drastic modifications of the encryption oracle in games $\mathbf{G}_5$ and $\mathbf{G}'_5$ below.

We introduce random bit strings $r^+ \in \{0,1\}^{k_0}$ and $g^+ \in \{0,1\}^n$. We also introduce a new random oracle

$$h^+ : \{0,1\}^n \to \{0,1\}^{k_1}.$$

Game $\mathbf{G}_4$ is the same as game $\mathbf{G}_3$, except that we apply the following special rules.

R1: In the encryption oracle, we compute

$$y^* = f(s^* \,\|\, (H(s^*) \oplus r^*)),$$

where

$$r^* = r^+ \text{ and } s^* = (g^+ \oplus x_b) \,\|\, h^+(x_b).$$

R2: Whenever the random oracle G is queried at r^+, we respond with the value g^+, instead of $G(r^+)$.

R3: Whenever the random oracle H' is queried at a point $r^+ \,\|\, x$ for some $x \in \{0,1\}^n$, we respond with the value $h^+(x)$, instead of $H'(r^+ \,\|\, x)$.

That completes the description of game $\mathbf{G}_4$. It is a simple matter to verify that the the random variable $\langle \mathit{View}, b \rangle$ has the same distribution in both games $\mathbf{G}_3$ and $\mathbf{G}_4$, since we have simply replaced one set of random variables by a different, but identically distributed, set of random variables. In particular,

$$\Pr[S_4] = \Pr[S_3]. \tag{28}$$

Game $\mathbf{G}_5$. This game is identical to game $\mathbf{G}_4$, except that we drop rules **R2** and **R3**, while retaining rule **R1**.

In game $\mathbf{G}_5$, it will not in general hold that $x^* = x_b$ or that $H(r^* \,\|\, x^*) = c^*$. Moreover, since the value g^+ is not used anywhere else in game $\mathbf{G}_5$ other than to "mask" x_b in the encryption oracle, we have

$$\Pr[S_5] = 1/2. \tag{29}$$

Despite the above differences, games $\mathbf{G}_4$ and $\mathbf{G}_5$ proceed identically unless A queries G at r^* or H' at $r^* \,\|\, x$ for some $x \in \{0,1\}^n$. Recall that by our convention, whenever A queries H' at $r^* \,\|\, x$ for some $x \in \{0,1\}^n$, then G has already been queried at r^*. Let F_5 be the event that in game $\mathbf{G}_5$, A queries G at r^*. We have $\Pr[S_4 \wedge \neg F_5] = \Pr[S_5 \wedge \neg F_5]$, and so by Lemma 1 applied to (S_4, S_5, F_5),

$$|\Pr[S_4] - \Pr[S_5]| \leq \Pr[F_5]. \tag{30}$$

Game $\mathbf{G}'_5$. We introduce an auxiliary game $\mathbf{G}'_5$ in order to bound $\Pr[F_5]$. In game $\mathbf{G}'_5$, we modify the encryption oracle once again. Let $y^+ \in \{0,1\}^k$ be a random bit string. Then in the encryption oracle, we simply set $y^* = y^+$, ignoring the encryption algorithm altogether.

It is not too hard to see that the random variable $\langle \mathit{View}, r^* \rangle$ has the same distribution in both games $\mathbf{G}_5$ and $\mathbf{G}'_5$. Indeed, the distribution of $\langle \mathit{View}, r^* \rangle$ in game $\mathbf{G}_5$ clearly remains the same if we instead choose r^* and s^* at random, and compute $y^* = f(s^* \,\|\, (H(s^*) \oplus r^*))$. Simply choosing y^* at random clearly induces the same distribution on $\langle \mathit{View}, r^* \rangle$. In particular, if we define F'_5 to be the event that in game $\mathbf{G}'_5$ A queries G at r^*, then

$$\Pr[F_5] = \Pr[F'_5]. \tag{31}$$

So our goal now is to bound $\Pr[F'_5]$. To this end, let F''_5 be the event that A queries H at s^* in game $\mathbf{G}'_5$. Then we have

$$\Pr[F'_5] = \Pr[F'_5 \wedge F''_5] + \Pr[F'_5 \wedge \neg F''_5]. \tag{32}$$

First, we claim that

$$\Pr[F_5' \wedge F_5''] \leq \mathit{InvAdv}(A'), \tag{33}$$

where $\mathit{InvAdv}(A')$ is the success probability of an inverting algorithm A' whose time and space requirements are bounded as in (23) and (24). To see this, observe that if A queries G at r^* and H at s^*, then we can easily convert the attack into an algorithm A' that computes $f^{-1}(y^+)$ on input y^+. A' simply runs A against game $\mathbf{G}_5'$. When A terminates, A' enumerates all $r' \in S_G$ and $s' \in S_H$, and for each of these computes

$$t' = H(s') \oplus r', \ w' = s' \,\|\, t', \ y' = f(w').$$

If y' is equal to y^+, then A' outputs w' and terminates.

Although game $\mathbf{G}_5'$ is defined with respect to random oracles, there are no random oracles in A'. To implement A', one simulates the random oracles that appear in game $\mathbf{G}_5'$ in the "natural" way. That is, whenever A queries a random oracle at a new point, A' generates an output for the oracle at random and puts this into a lookup table keyed by the input to the oracle. If A has previously queried the oracle at a point, A' takes the output value from the lookup table. Again, using standard algorithms and data structures, such as search tries, the running time and space complexity of A' are easily seen to be bounded as claimed in (23) and (24).

Unfortunately, the running time of A' is much worse than that of the simulated decryption oracle described in game $\mathbf{G}_3$. But at least the space remains essentially linear in the total number of oracle queries.

We also claim that

$$\Pr[F_5' \wedge \neg F_5''] \leq q_G/2^{k_0}. \tag{34}$$

To see this, consider a query of G at r, prior to which H has not been queried at s^*. Since $t^* = H(s^*) \oplus r^*$, the value r^* is independent of *CurrentView*, and so $\Pr[r = r^*] = 1/2^{k_0}$. The bound (34) now follows.

Equations (32)-(34) together imply

$$\Pr[F_5'] \leq \mathit{InvAdv}(A') + q_G/2^{k_0}. \tag{35}$$

Equations (25), (26), (27), (28), (29), (30), (31), and (35) together imply (22).

That completes the proof of Theorem 3.

Remark. Our reduction from inverting f to breaking OAEP+ is tighter than the corresponding reduction for OAEP in [BR94]. In particular, the OAEP+ construction facilitates a much more efficient "plaintext extractor" than the OAEP construction. The latter apparently requires either

- time proportional to $q_D q_G q_H$ and space linear in the number of oracle queries, or
- time proportional to $q_D + q_G q_H$ and space proportional to $q_G q_H$ (if one builds a look-up table).

For OAEP+, the total time and space complexity of the plaintext extractor in game $\mathbf{G}_3$ is linear in the number of oracle queries. Unfortunately, our inversion algorithm for OAEP+ in game $\mathbf{G}'_5$ still requires time proportional to $q_G q_H$, although its space complexity is linear in the number of oracle queries. We should remark that as things now stand, the reductions for OAEP+ are not tight enough to actually imply that an algorithm that breaks, say, 1024-bit RSA-OAEP+ in a "reasonable" amount of time implies an algorithm that solves the RSA problem in time faster than the best known factoring algorithms. However, as we shall see in §7.2, for exponent-3 RSA-OAEP+, one can in fact get a very tight reduction. An interesting open problem is to get a tighter reduction for OAEP+ or a variant thereof.

7 Further Observations

7.1 Other Variations of OAEP

Instead of modifying OAEP as we did, one could also modify OAEP so that instead of adding the data-independent redundancy 0^{k_1} in (1), one added the data-dependent redundancy $H''(x)$, where H'' is a hash function mapping n-bit strings to k_1-bit strings. This variant of OAEP—call it OAEP′—suffers from the same problem from which OAEP suffers. Indeed, Theorem 1 holds also for OAEP′.

7.2 RSA-OAEP with Exponent 3 Is Provably Secure

Consider RSA-OAEP. Let N be the modulus and e the encryption exponent. Then this scheme actually *is* secure in the random oracle model, provided $k_0 \leq \log_2 N/e$. This condition is satisfied by typical implementations of RSA-OAEP with $e = 3$.

We sketch very briefly why this is so.

We first remind the reader of the attempted proof of security of OAEP in §4, and we adopt all the notation specified there.

Suppose an adversary submits a ciphertext y to the decryption oracle. We observed in §4 that if the adversary never explicitly queried $H(s)$, then with overwhelming probability, the actual decryption oracle would reject. The only problem was, we could not always say the same thing about $G(r)$ (specifically, when $r = r^*$).

For a bit string v, let $I(v)$ denote the unique integer such that v is a binary representation of $I(v)$.

If a simulated decryption oracle knows s (it will be one of the adversary's H-queries), then $X = I(t)$ is a solution to the equation

$$(X + 2^{k_0} I(s))^e \equiv y \pmod{N}.$$

To find $I(t)$, we can apply Coppersmith's algorithm [Cop96]. This algorithm works provided $I(t) < N^{1/e}$, which is guaranteed by our assumption that $k_0 \leq \log_2 N/e$.

More precisely, for all $s' \in S_H$, the simulated decryption oracle tries to find a corresponding solution t' using Coppersmith's algorithm. If all of these attempts fail, then the simulator rejects y. Otherwise, knowing s and t, it decrypts y in the usual way.

We can also apply Coppersmith's algorithm in the step of the proof where we use the adversary to help us to extract a challenge instance of the RSA problem.

Not only does this prove security, but we get a more efficient reduction—the implied inverting algorithm has a running time roughly equal to that of the adversary, plus $O(q_D q_H T_C)$, where T_C is the running time of Coppersmith's algorithm.

We can also use the same observation to speed up the reduction for exponent-3 RSA-OAEP+. The total running time of the implied inversion algorithm would be roughly equal to that of the adversary, plus $O(q_H T_C)$; that is, a factor of q_D faster than the inversion algorithm implied by RSA-OAEP. Unlike the generic security reduction for OAEP+, this security reduction is essentially tight, and so it has much more meaningful implications for the security of the scheme when used with a typical, say, 1024-bit RSA modulus.

7.3 RSA-OAEP with Large Exponent

In our example in §4.1, as well as in our proof of Theorem 1, the adversary is able to create a valid ciphertext y without ever querying $G(r)$. However, this adversary queries both $H(s)$ *and* $H(s^*)$. As we already noted, the adversary must query $H(s)$. But it turns out that if the adversary avoids querying $G(r)$, he must query $H(s^*)$. This observation was made by [FOPS00], who then further observed that this implies the security of RSA-OAEP with arbitrary encryption exponent in the random oracle model. We remark, however, that the reduction in [FOPS00] is significantly less efficient than our general reduction for OAEP+. In particular, their reduction only implies that if an adversary has advantage ϵ in breaking RSA-OAEP, then there is an algorithm that solves the RSA inversion problem with probability about ϵ^2. Moreover, their inversion algorithm is even somewhat slower than that of the (incorrect) inversion algorithm for OAEP in [BR94]. There is still the possibility, however, that a more efficient reduction for RSA-OAEP can be found.

Acknowledgments. Thanks to Jean-Sebastien Coron for pointing out an error in a previous draft. Namely, it was claimed that the the variant OAEP′ briefly discussed in §7.1 could also be proven secure, but this is not so.

References

[BDPR98] M. Bellare, A. Desai, D. Pointcheval, and P. Rogaway. Relations among notions of security for public-key encryption schemes. In *Advances in Cryptology–Crypto '98*, pages 26–45, 1998.

[Bon01] D. Boneh. Simplified OAEP for the RSA and Rabin functions. In *Advances in Cryptology–Crypto 2001*, 2001.

[BR93] M. Bellare and P. Rogaway. Random oracles are practical: a paradigm for designing efficient protocols. In *First ACM Conference on Computer and Communications Security*, pages 62–73, 1993.

[BR94] M. Bellare and P. Rogaway. Optimal asymmetric encryption. In *Advances in Cryptology—Eurocrypt '94*, pages 92–111, 1994.

[CGH98] R. Canetti, O. Goldreich, and S. Halevi. The random oracle model, revisted. In *30th Annual ACM Symposium on Theory of Computing*, 1998.

[Cop96] D. Coppersmith. Finding a small root of a univariate modular equation. In *Advances in Cryptology-Eurocrypt '96*, pages 155–165, 1996.

[DDN91] D. Dolev, C. Dwork, and M. Naor. Non-malleable cryptography. In *23rd Annual ACM Symposium on Theory of Computing*, pages 542–552, 1991.

[DDN00] D. Dolev, C. Dwork, and M. Naor. Non-malleable cryptography. *SIAM J. Comput.*, 30(2):391–437, 2000.

[FOPS00] E. Fujisaki, T. Okamoto, D. Pointcheval, and J. Stern. RSA-OAEP is still alive! Cryptology ePrint Archive, Report 2000/061, 2000. `http://eprint.iacr.org`.

[FOPS01] E. Fujisaki, T. Okamoto, D. Pointcheval, and J. Stern. RSA-OAEP is secure under the RSA assumption. In *Advances in Cryptology-Crypto 2001*, 2001.

[NY90] M. Naor and M. Yung. Public-key cryptosystems provably secure against chosen ciphertext attacks. In *22nd Annual ACM Symposium on Theory of Computing*, pages 427–437, 1990.

[RS91] C. Rackoff and D. Simon. Noninteractive zero-knowledge proof of knowledge and chosen ciphertext attack. In *Advances in Cryptology-Crypto '91*, pages 433–444, 1991.

[RSA78] R. L. Rivest, A. Shamir, and L. M. Adleman. A method for obtaining digital signatures and public-key cryptosystems. *Communications of the ACM*, pages 120–126, 1978.

[Sho00] V. Shoup. OAEP reconsidered. Cryptology ePrint Archive, Report 2000/060, 2000. `http://eprint.iacr.org`.

RSA–OAEP Is Secure under the RSA Assumption

Eiichiro Fujisaki[1], Tatsuaki Okamoto[1], David Pointcheval[2], and Jacques Stern[2]

[1] NTT Labs, 1-1 Hikarino-oka, Yokosuka-shi, 239-0847 Japan.
{fujisaki,okamoto}@isl.ntt.co.jp.
[2] Dépt d'Informatique, ENS – CNRS, 45 rue d'Ulm, 75230 Paris Cedex 05, France.
{David.Pointcheval,Jacques.Stern}@ens.fr
http://www.di.ens.fr/users/{pointche,stern}.

Abstract. Recently Victor Shoup noted that there is a gap in the widely-believed security result of OAEP against adaptive chosen-ciphertext attacks. Moreover, he showed that, presumably, OAEP cannot be proven secure from the *one-wayness* of the underlying trapdoor permutation. This paper establishes another result on the security of OAEP. It proves that OAEP offers semantic security against adaptive chosen-ciphertext attacks, in the random oracle model, under the *partial-domain* one-wayness of the underlying permutation. Therefore, this uses a formally stronger assumption. Nevertheless, since partial-domain one-wayness of the RSA function is equivalent to its (full-domain) one-wayness, it follows that the security of RSA–OAEP can actually be proven under the sole RSA assumption, although the reduction is not tight.

1 Introduction

The OAEP conversion method [3] was introduced by Bellare and Rogaway in 1994 and was believed to provide semantic security against adaptive chosen-ciphertext attacks [7,12], based on the one-wayness of a trapdoor permutation, using the (corrected) definition of plaintext-awareness [1].

Victor Shoup [15] recently showed that it is quite unlikely that such a security proof exists — at least for non-malleability — under the one-wayness of the permutation. He also proposed a slightly modified version of OAEP, called OAEP+, which can be proven secure, under the one-wayness of the permutation.

Does Shoup's result mean that OAEP is insecure or that it is impossible to prove the security of OAEP? This is a totally misleading view: the result only states that it is highly unlikely to find any proof, under just the one-wayness assumption. In other words, Shoup's result does not preclude the possibility of proving the security of OAEP from stronger assumptions.

This paper uses such a stronger assumption. More precisely, in our reduction, a new computational assumption is introduced to prove the existence of a simulator of the decryption oracle. Based on this idea, we prove that OAEP is semantically secure against adaptive chosen-ciphertext attack in the random

J. Kilian (Ed.): CRYPTO 2001, LNCS 2139, pp. 260–274, 2001.

oracle model [3], under the *partial-domain* one-wayness of the underlying permutation, which is stronger than the original assumption.

Since partial-domain one-wayness of the RSA function [13] is equivalent to the (full-domain) one-wayness, the security of RSA-OAEP can actually be proven under the one-wayness of the RSA function.

The rest of this paper is organized as follows. Section 2 recalls the basic notions of asymmetric encryption and the various security notions. Section 3 reviews the OAEP conversion [3]. Sections 4 and 5 present our new security result together with a formal proof for general OAEP applications. In Section 6, we focus on the RSA application of OAEP, RSA-OAEP.

2 Public-Key Encryption

The aim of public-key encryption is to allow anybody who knows the public key of Alice to send her a message that only she will be able to recover it through her private key.

2.1 Definitions

A public-key encryption scheme is defined by the three following algorithms:

- The *key generation algorithm* $\mathcal{K}$. On input 1^k, where k is the security parameter, the algorithm $\mathcal{K}$ produces a pair $(\mathsf{pk}, \mathsf{sk})$ of matching public and secret keys. Algorithm $\mathcal{K}$ is probabilistic.
- The *encryption algorithm* $\mathcal{E}$. Given a message m and a public key pk, $\mathcal{E}$ produces a ciphertext c of m. This algorithm may be probabilistic.
- The *decryption algorithm* $\mathcal{D}$. Given a ciphertext c and the secret key sk, $\mathcal{D}$ returns the plaintext m. This algorithm is deterministic.

2.2 Security Notions

The first security notion that one would like for an encryption scheme is *one-wayness*: starting with just public data, an attacker cannot recover the complete plaintext of a given ciphertext. More formally, this means that for any adversary $\mathcal{A}$, her success in inverting $\mathcal{E}$ without the secret key should be negligible over the probability space $\mathcal{M} \times \Omega$, where $\mathcal{M}$ is the message space and Ω is the space of the random coins r used for the encryption scheme, and the internal random coins of the adversary:

$$\mathsf{Succ}^{\mathsf{ow}}(\mathcal{A}) = \Pr_{m,r}[(\mathsf{pk}, sk) \leftarrow \mathcal{K}(1^k) : \mathcal{A}(\mathsf{pk}, \mathcal{E}_{\mathsf{pk}}(m; r)) = m].$$

However, many applications require more from an encryption scheme, namely *semantic security* (*a.k.a. polynomial security* or *indistinguishability of encryptions* [7], denoted IND): if the attacker has some information about the plaintext, for example that it is either "yes" or "no" to a crucial query, no adversary should learn more with the view of the ciphertext. This security notion requires computational impossibility to distinguish between two messages, chosen by the adversary, one of which has been encrypted, with a probability significantly better than one half: her advantage $\mathsf{Adv}^{\mathsf{ind}}(\mathcal{A})$, where the adversary $\mathcal{A}$ is seen as a 2-stage Turing machine $(\mathcal{A}_1, \mathcal{A}_2)$, should be negligible, where $\mathsf{Adv}^{\mathsf{ind}}(\mathcal{A})$ is formally defined as.

$$2 \times \Pr_{b,r}\left[\begin{array}{l}(\mathsf{pk}, \mathsf{sk}) \leftarrow \mathcal{K}(1^k), (m_0, m_1, s) \leftarrow \mathcal{A}_1(\mathsf{pk}),\\ c = \mathcal{E}_{\mathsf{pk}}(m_b; r) : \mathcal{A}_2(m_0, m_1, s, c) = b\end{array}\right] - 1.$$

Another notion was defined thereafter, the so-called *non-malleability* [6], in which the adversary tries to produce a new ciphertext such that the plaintexts are meaningfully related. This notion is stronger than the above one, but it is equivalent to semantic security in the most interesting scenario [1].

On the other hand, an attacker can use many kinds of attacks: since we are considering asymmetric encryption, the adversary can encrypt any plaintext of her choice with the public key, hence *chosen-plaintext attack*. She may, furthermore, have access to more information, modeled by partial or full access to some oracles: a plaintext-checking oracle which, on input of a pair (m, c), answers whether c encrypts the message m. This attack has been named the *Plaintext-Checking Attack* [11]; a validity-checking oracle which, on input of a ciphertext c, just answers whether it is a valid ciphertext. This weak oracle (involved in the reaction attacks [8]) had been enough to break some famous encryption schemes [4,9], namely PKCS #1 v1.5; or the decryption oracle itself, which on the input of any ciphertext, except the challenge ciphertext, responds with the corresponding plaintext (*non-adaptive/adaptive chosen-ciphertext attacks* [10,12]). The latter, the adaptive chosen-ciphertext attack denoted CCA2, is clearly the strongest one.

A general study of these security notions and attacks was given in [1], we therefore refer the reader to this paper for more details. However, the by now expected security level for public-key encryption schemes is semantic security against adaptive chosen-ciphertext attacks (IND-CCA2) – where the adversary just wants to distinguish which plaintext, between two messages of her choice, had been encrypted; she can ask any query she wants to a decryption oracle (except the challenge ciphertext). This is the strongest scenario one can define.

3 Review of OAEP

3.1 The Underlying Problems

Consider permutation $f : \{0,1\}^k \longrightarrow \{0,1\}^k$, which can also be seen as

$$f : \{0,1\}^{n+k_1} \times \{0,1\}^{k_0} \longrightarrow \{0,1\}^{n+k_1} \times \{0,1\}^{k_0},$$

with $k = n + k_0 + k_1$. In the original description of OAEP from [3], it is only required that f is a trapdoor one-way permutation. However, in the following, we consider two additional related problems: the partial-domain one-wayness and the set partial-domain one-wayness of permutation f:

– (τ, ε)-One-Wayness of f, means that for any adversary $\mathcal{A}$ whose running time is bounded by τ, the success probability $\mathsf{Succ}^{\mathsf{ow}}(\mathcal{A})$ is upper-bounded by ε, where
$$\mathsf{Succ}^{\mathsf{ow}}(\mathcal{A}) = \Pr_{s,t}[\mathcal{A}(f(s,t)) = (s,t)];$$
– (τ, ε)-Partial-Domain One-Wayness of f, means that for any adversary $\mathcal{A}$ whose running time is bounded by τ, the success probability $\mathsf{Succ}^{\mathsf{pd-ow}}(\mathcal{A})$ is upper-bounded by ε, where
$$\mathsf{Succ}^{\mathsf{pd-ow}}(\mathcal{A}) = \Pr_{s,t}[\mathcal{A}(f(s,t)) = s];$$
– $(\ell, \tau, \varepsilon)$-Set Partial-Domain One-Wayness of f, means that for any adversary $\mathcal{A}$ that outputs a set of ℓ elements within time bound τ, the success probability $\mathsf{Succ}^{\mathsf{s-pd-ow}}(\mathcal{A})$ is upper-bounded by ε, where
$$\mathsf{Succ}^{\mathsf{s-pd-ow}}(\mathcal{A}) = \Pr_{s,t}[s \in \mathcal{A}(f(s,t))].$$

We denote by $\mathsf{Succ}^{\mathsf{ow}}(\tau)$, (resp. $\mathsf{Succ}^{\mathsf{pd-ow}}(\tau)$ and $\mathsf{Succ}^{\mathsf{s-pd-ow}}(\ell, \tau)$) the maximal success probability $\mathsf{Succ}^{\mathsf{ow}}(\mathcal{A})$ (resp. $\mathsf{Succ}^{\mathsf{pd-ow}}(\mathcal{A})$ and $\mathsf{Succ}^{\mathsf{s-pd-ow}}(\mathcal{A})$). The maximum ranges over all adversaries whose running time is bounded by τ. In the third case, there is an obvious additional restriction on this range from the fact that $\mathcal{A}$ outputs sets with ℓ elements. It is clear that for any τ and $\ell \geq 1$,
$$\mathsf{Succ}^{\mathsf{s-pd-ow}}(\ell, \tau) \geq \mathsf{Succ}^{\mathsf{pd-ow}}(\tau) \geq \mathsf{Succ}^{\mathsf{ow}}(\tau).$$

Note that, by randomly selecting an element in the set returned by an adversary to the Set Partial-Domain One-Wayness, one breaks Partial-Domain One-Wayness with probability $\mathsf{Succ}^{\mathsf{s-pd-ow}}(\mathcal{A})/\ell$. This provides the following inequality $\mathsf{Succ}^{\mathsf{pd-ow}}(\tau) \geq \mathsf{Succ}^{\mathsf{s-pd-ow}}(\ell, \tau)/\ell$. However, for specific choices of f, more efficient reductions may exist. Also, in some cases, all three problems are polynomially equivalent. This is the case for the RSA permutation [13], hence the results in section 6.

3.2 The OAEP Cryptosystem

We briefly describe the OAEP cryptosystem $(\mathcal{K}, \mathcal{E}, \mathcal{D})$ obtained from a permutation f, whose inverse is denoted by g. We need two hash functions G and H:
$$G : \{0,1\}^{k_0} \longrightarrow \{0,1\}^{k-k_0} \text{ and } H : \{0,1\}^{k-k_0} \longrightarrow \{0,1\}^{k_0}.$$

Then,

– $\mathcal{K}(1^k)$: specifies an instance of the function f, and of its inverse g. The public key pk is therefore f and the secret key sk is g.
– $\mathcal{E}_{\mathsf{pk}}(m; r)$: given a message $m \in \{0,1\}^n$, and a random value $r \stackrel{R}{\leftarrow} \{0,1\}^{k_0}$, the encryption algorithm $\mathcal{E}_{\mathsf{pk}}$ computes
$$s = (m \| 0^{k_1}) \oplus G(r) \text{ and } t = r \oplus H(s),$$
and outputs the ciphertext $c = f(s,t)$.

– $\mathcal{D}_{\mathsf{sk}}(c)$: thanks to the secret key, the decryption algorithm $\mathcal{D}_{\mathsf{sk}}$ extracts

$$(s,t) = g(c), \text{ and next } r = t \oplus H(s) \text{ and } M = s \oplus G(r).$$

If $[M]_{k_1} = 0^{k_1}$, the algorithm returns $[M]^n$, otherwise it returns "Reject".

In the above description, $[M]_{k_1}$ denotes the k_1 least significant bits of M, while $[M]^n$ denotes the n most significant bits of M.

4 Security Result

In their paper [3], Bellare and Rogaway provided a security analysis, which proved that the OAEP construction together with any trapdoor one-way permutation is semantically security and (weakly) plaintext-aware. Unfortunately, this just proves semantic security against non-adaptive chosen-ciphertext attacks (*a.k.a.* lunchtime attacks [10] or IND-CCA1). Even if the achieved security was believed to be stronger (namely IND-CCA2), it had never been proven. Thus, Shoup [15] recently showed that it is quite unlikely that such a security proof exists, for any trapdoor one-way permutation. However, he provided a specific proof for RSA with public exponent 3.

In the following, we provide a general security analysis, but under a stronger assumption about the underlying permutation. Indeed, we prove that the scheme is IND-CCA2 in the random oracle model [2], relative to the *partial-domain* one-wayness of function f. More precisely, the following exact security result holds.

Theorem 1. *Let $\mathcal{A}$ be a CCA2–adversary against the "semantic security" of the OAEP conversion $(\mathcal{K}, \mathcal{E}, \mathcal{D})$, with advantage ε and running time t, making q_D, q_G and q_H queries to the decryption oracle, and the hash functions G and H respectively. Then,* $\mathsf{Succ}^{\mathsf{pd-ow}}(t')$ *is greater than*

$$\frac{1}{q_H} \cdot \left(\frac{\varepsilon}{2} - \frac{2q_D q_G + q_D + q_G}{2^{k_0}} - \frac{2q_D}{2^{k_1}} \right),$$

where $t' \leq t + q_G \cdot q_H \cdot (T_f + \mathcal{O}(1))$, and T_f denotes the time complexity of function f.

In order to prove this theorem relative to the partial-domain one-wayness of the permutation, one can use the related notion of set partial-domain one-wayness. The theorem follows from the inequalities of the previous section together with the lemma stated below.

Lemma 2. *Let $\mathcal{A}$ be a CCA2–adversary against the "semantic security" of the OAEP conversion $(\mathcal{K}, \mathcal{E}, \mathcal{D})$, with advantage ε and running time t, making q_D, q_G and q_H queries to the decryption oracle, and the hash functions G and H respectively. Then,* $\mathsf{Succ}^{\mathsf{s-pd-ow}}(q_H, t')$ *is greater than*

$$\frac{\varepsilon}{2} - \frac{2q_D q_G + q_D + q_G}{2^{k_0}} - \frac{2q_D}{2^{k_1}},$$

where $t' \leq t + q_G \cdot q_H \cdot (T_f + \mathcal{O}(1))$, *and* T_f *denotes the time complexity of function* f.

The next section is devoted to proving this lemma. Hereafter, we will repeatedly use the following simple result:

Lemma 3. *For any probability events* E, F *and* G

$$\Pr[\mathsf{E} \wedge \mathsf{F} \mid \mathsf{G}] \leq \begin{cases} \Pr[\mathsf{E} \mid \mathsf{F} \wedge \mathsf{G}] \\ \Pr[\mathsf{F} \mid \mathsf{G}]. \end{cases}$$

5 Proof of Lemma 2

We prove lemma 2 in three stages. The first presents the reduction of IND-CCA2 adversary $\mathcal{A}$ to algorithm $\mathcal{B}$ for breaking the partial-domain one-wayness of f. Note that, in the present proof, we are just interested in security under the partial-domain one-wayness of f, and not under the full-domain one-wayness of f as in the original paper [3]. The second shows that the decryption oracle simulation employed in this reduction works correctly with overwhelming probability under the partial-domain one-wayness of f. This latter part differs from the original proof [3], and corrects the recently spotted flaw [15]. Finally, we analyze the success probability of our reduction in total, through the incorporation of the above-mentioned analysis of the decryption oracle simulation.

5.1 Description of the Reduction

In this first part, we recall how reduction operates. Let $\mathcal{A} = (A_1, A_2)$ be an adversary against the semantic security of $(\mathcal{K}, \mathcal{E}, \mathcal{D})$, under chosen-ciphertext attacks. Within time bound t, $\mathcal{A}$ asks q_D, q_G and q_H queries to the decryption oracle and the random oracles G and H respectively, and distinguishes the right plaintext with an advantage greater than ε. Let us describe the reduction $\mathcal{B}$.

Top Level Description of the Reduction

1. $\mathcal{B}$ is given a function f (defined by the public key) and $c^\star \leftarrow f(s^\star, t^\star)$, for $(s^\star, t^\star) \stackrel{R}{\leftarrow} \{0,1\}^{k-k_0} \times \{0,1\}^{k_0}$. The aim of $\mathcal{B}$ is to recover the partial pre-image $s^\star$ of $c^\star$.
2. $\mathcal{B}$ runs A_1 on the public data, and gets a pair of messages $\{m_0, m_1\}$ as well as state information st. It chooses a random bit b, and then gives $c^\star$ to A_1, as the ciphertext of m_b. $\mathcal{B}$ simulates the answers to the queries of A_1 to the decryption oracle and random oracles G and H respectively. See the description of these simulations below.
3. $\mathcal{B}$ runs $A_2(c^\star, st)$ and finally gets answer b'. $\mathcal{B}$ simulates the answers to the queries of A_2 to the decryption oracle and random oracles G and H respectively. See the description of these simulations below. $\mathcal{B}$ then outputs the partial pre-image $s^\star$ of $c^\star$, if one has been found among the queries asked to H (see below), or the list of queries asked to H.

Simulation of Random Oracles G and H. The random oracle simulation has to simulate the random oracle answers, managing query/answer lists G-List and H-List for the oracles G and H respectively, both are initially set to empty lists:

- for a fresh query γ to G, one looks at the H-List, and for any query δ asked to H with answer H_δ, one builds $z = \gamma \oplus H_\delta$, and checks whether $c^\star = f(\delta, z)$. If for some δ, that relation holds, function f has been inverted, and we can still correctly simulate G, by answering $G_\gamma = \delta \oplus (m_b \| 0^{k_1})$. Note that G_γ is then a uniformly distributed value since $\delta = s^\star$, and the latter is uniformly distributed. Otherwise, one outputs a random value G_γ. In both cases, the pair (γ, G_γ) is concatenated to the G-List.
- for a fresh query δ to H, one outputs a random value H_δ, and the pair (δ, H_δ) is concatenated to the H-List. Note that, once again, for any $(\gamma, G_\gamma) \in$ G-List, one may build $z = \gamma \oplus H_\delta$, and check whether $c^\star = f(\delta, z)$. If for some γ that relation holds, we have inverted the function f.

Simulation of the Decryption Oracle. On query $c = f(s,t)$ to the decryption oracle, decryption oracle simulation $\mathcal{DS}$ looks at each query-answer $(\gamma, G_\gamma) \in$ G-List and $(\delta, H_\delta) \in$ H-List. For each pair taken from both lists, it defines

$$\sigma = \delta, \tau = \gamma \oplus H_\delta, \mu = G_\gamma \oplus \delta,$$

and checks whether

$$c = f(\sigma, \tau) \text{ and } [\mu]_{k_1} = 0^{k_1}.$$

As soon as both equalities hold, $\mathcal{DS}$ outputs $[\mu]^n$. If no such pair is found, "Reject" is returned.

Remarks. When we have found the pre-image of $c^\star$, and thus inverted f, we could output the expected result $s^\star$ and stop the reduction. But for this analysis, we assume the reduction goes on and that $\mathcal{B}$ only outputs it, or the list of queries asked to H, once A_2 has answered b' (or after a time limit).

Even if no answer is explicitly specified, except by a random value for new queries, some are implicitly defined. Indeed, $c^\star$ is defined to be a ciphertext of m_b with random tape $r^\star$:

$$r^\star \leftarrow H(s^\star) \oplus t^\star \text{ and } G(r^\star) \leftarrow s^\star \oplus (m_b \| 0^{k_1}).$$

Since $H(s^\star)$ is randomly defined, $r^\star$ can be seen as a random variable. Let us denote by AskG the event that query $r^\star$ has been asked to G, and by AskH the event that query $s^\star$ has been asked to H. Let us furthermore denote by GBad the event that $r^\star$ has been asked to G, but the answer is something other than $s^\star \oplus (m_b \| 0^{k_1})$ (bit b is fixed in the reduction scenario). Note that the event GBad implies AskG. As seen above, GBad is the only event that makes the random oracle simulation imperfect, in the chosen-plaintext attack scenario. In the chosen-ciphertext attack scenario, we described a decryption simulator that may sometimes fail. Such an event of decryption failure will be denoted by DBad. We thus denote Bad = GBad $\vee$ DBad.

5.2 Notations

In order to proceed to the analysis of the success probability of the above-mentioned reduction, one needs to set up notations. First, we still denote with a star ($\star$) all variables related to the challenge ciphertext $c^\star$, obtained from the encryption oracle. Indeed, this ciphertext, of either m_0 or m_1, implicitly defines hash values, but the corresponding pairs may not appear in the G or H lists. All other variables refer to the decryption query c, asked by the adversary to the decryption oracle, and thus to be decrypted by this simulation. We consider several further events about a ciphertext queried to the decryption oracle:

- CBad denotes the union of the bad events, CBad = RBad ∨ SBad, where
  - SBad denotes the event that $s = s^\star$;
  - RBad denotes the event that $r = r^\star$, and thus $H(s) \oplus t = H(s^\star) \oplus t^\star$;
- AskRS denotes the intersection of both events about the oracle queries, AskRS = AskR ∧ AskS, which means that both r and s have been asked to G and H respectively, since
  - AskR denotes the event that r $(= H(s) \oplus t)$ has been asked to G;
  - AskS denotes the event that s has been asked to H;
- Fail denotes the event that the above decryption oracle simulator outputs a wrong decryption answer to query c. (More precisely, we may denote Fail_i for event Fail on the i-th query c_i $(i = 1, \ldots, q_D)$. For our analysis, however, we can evaluate probabilities regarding event Fail_i in a uniform manner for any i. Hence, we just employ notation Fail.) Therefore, in the global reduction, the event DBad will be set to true as soon as one decryption simulation fails.

Note that the Fail event is limited to the situation in which the plaintext-extractor rejects a ciphertext whereas it would be accepted by the actual decryption oracle. Indeed, as soon as it accepts, we see that the ciphertext is actually valid and corresponds to the output plaintext.

5.3 Analysis of the Decryption Oracle Simulation

We analyze the success probability of decryption oracle simulator $\mathcal{DS}$.

Security Claim. We claim the following, which repairs the previous proof [3], based on the new computational assumption. More precisely, we show that additional cases to consider, due to the corrected definition of plaintext-awareness [1], are very unlikely under the partial-domain one-wayness of the permutation f:

Lemma 4. *When at most one ciphertext $c^\star = f(s^\star, t^\star)$ has been directly obtained from the encryption oracle, but $s^\star$ has not been asked to H, the decryption oracle simulation $\mathcal{DS}$ can correctly produce the decryption oracle's output on query (ciphertext) c $(\neq c^\star)$ with probability greater than ε', within time bound t', where*

$$\varepsilon' \geq 1 - \left(\frac{2}{2^{k_1}} + \frac{2q_G + 1}{2^{k_0}}\right) \quad \text{and } t' \leq q_G \cdot q_H \cdot (T_f + \mathcal{O}(1)).$$

Before we start the analysis, we recall that the decryption oracle simulator is given the ciphertext c to be decrypted, as well as the ciphertext $c^\star$ obtained from the encryption oracle and both the G-List and H-List resulting from the interactions with the random oracles G and H. Let us first see that the simulation uniquely defines a possible plaintext, and thus can output the first one it finds. Indeed, with the above definition, several pairs could satisfy the equalities. However, since function f is a permutation, and thus one-to-one, the value of $\sigma = s$ is uniquely defined, and thus δ and H_δ. Similarly, $\tau = t$ is uniquely defined, and thus γ and G_γ: at most one μ may be selected. Then either $[\mu]_{k_1} = 0^{k_1}$ or not.

In the above, one should keep in mind that the G-List and H-List correspond to input-output pairs for the functions G and H. Thus, at most one output is related to a given input.

If the ciphertext has been correctly built by the adversary (r has been asked to G and s to H), the simulation will output the correct answer. However, it will output "Reject" in any other situation, whereas the adversary may have built a valid ciphertext without asking both queries to the random oracles G and H.

Success Probability. Since our goal is to prove the security relative to the partial-domain one-wayness of f, we are only interested in the probability of the event Fail, while $\neg$AskH occurred, which may be split according to other events. Granted $\neg$CBad $\wedge$ AskRS, the simulation is perfect, and cannot fail. Thus, we have to consider the complementary events:

$$\Pr[\mathsf{Fail} \mid \neg\mathsf{AskH}] = \Pr[\mathsf{Fail} \wedge \mathsf{CBad} \mid \neg\mathsf{AskH}] + \Pr[\mathsf{Fail} \wedge \neg\mathsf{CBad} \wedge \neg\mathsf{AskRS} \mid \neg\mathsf{AskH}].$$

Concerning the latter contribution to the right hand side, we first note that both

$$\begin{aligned} \neg\mathsf{AskRS} &= \neg\mathsf{AskR} \vee \neg\mathsf{AskS} = (\neg\mathsf{AskR}) \vee (\neg\mathsf{AskS} \wedge \mathsf{AskR}) \\ \neg\mathsf{CBad} &= \neg\mathsf{RBad} \wedge \neg\mathsf{SBad}. \end{aligned}$$

Forgetting $\neg$AskH for a while, using lemma 3, one gets that $\Pr[\mathsf{Fail} \wedge \neg\mathsf{CBad} \wedge \neg\mathsf{AskRS}]$ is less than

$$\begin{aligned} &\Pr[\mathsf{Fail} \wedge \neg\mathsf{RBad} \wedge \neg\mathsf{AskR}] + \Pr[\mathsf{Fail} \wedge \neg\mathsf{SBad} \wedge (\mathsf{AskR} \wedge \neg\mathsf{AskS})] \\ &\leq \Pr[\mathsf{Fail} \mid \neg\mathsf{AskR} \wedge \neg\mathsf{RBad}] + \Pr[\mathsf{AskR} \mid \neg\mathsf{AskS} \wedge \neg\mathsf{SBad}]. \end{aligned}$$

But without having asked r to G, taking into account the further event $\neg$RBad, $G(r)$ is unpredictable, and thus the probability that $[s \oplus G(r)]_{k_1} = 0^{k_1}$ is less than 2^{-k_1}. On the other hand, the probability of having asked r to G, without any information about $H(s)$ and thus about r ($H(s)$ not asked, and $s \neq s^\star$, which both come from the conditioning $\neg$AskS $\wedge$ $\neg$SBad), is less than $q_G \cdot 2^{-k_0}$. Furthermore, this event is independent of AskH, which yields

$$\Pr[\mathsf{Fail} \wedge \neg\mathsf{CBad} \wedge \neg\mathsf{AskRS} \mid \neg\mathsf{AskH}] \leq 2^{-k_1} + q_G \cdot 2^{-k_0}.$$

We now focus on the former term, Fail $\wedge$ CBad, while $\neg$AskH, which was missing in the original proof [3] based on a weaker notion of plaintext-awareness.

It can be split according to the disjoint sub-cases of CBad, which are SBad and $\neg\mathsf{SBad} \wedge \mathsf{RBad}$. Then again using lemma 3,

$$\Pr[\mathsf{Fail} \wedge \mathsf{CBad} \mid \neg\mathsf{AskH}] \leq \Pr[\mathsf{Fail} \mid \mathsf{SBad} \wedge \neg\mathsf{AskH}] + \Pr[\mathsf{RBad} \mid \neg\mathsf{SBad} \wedge \neg\mathsf{AskH}].$$

The latter event means that RBad occurs provided $s \neq s^\star$ and the adversary has not queried $s^\star$ from H. When $s^\star$ has not been asked to H and $s \neq s^\star$, $H(s^\star)$ is unpredictable and independent of $H(s)$ as well as t and $t^\star$. Then, event RBad, $H(s^\star) = H(s) \oplus t \oplus t^\star$, occurs with probability at most 2^{-k_0}.

The former event can be further split according to AskR, and, using once again lemma 3, it is upper-bounded by

$$\Pr[\mathsf{AskR} \mid \mathsf{SBad} \wedge \neg\mathsf{AskH}] + \Pr[\mathsf{Fail} \mid \neg\mathsf{AskR} \wedge \mathsf{SBad} \wedge \neg\mathsf{AskH}].$$

The former event means that r is asked to G whereas $s = s^\star$ and $H(s^\star)$ is unpredictable, thus $H(s)$ is unpredictable. Since r is unpredictable, the probability of this event is at most $q_G \cdot 2^{-k_0}$ (the probability of asking r to G). On the other hand, the latter event means that the simulator rejects the valid ciphertext c whereas $H(s)$ is unpredictable and r is not asked to G. From the one-to-one property of the Feistel network, it follows from $s = s^\star$ that $r \neq r^\star$, and thus $G(r)$ is unpredictable. Then the redundancy cannot hold with probability greater than 2^{-k_1}. To sum up, $\Pr[\mathsf{Fail} \mid \mathsf{SBad} \wedge \neg\mathsf{AskH}] \leq 2^{-k_1} + q_G \cdot 2^{-k_0}$, thus $\Pr[\mathsf{Fail} \wedge \mathsf{CBad} \mid \neg\mathsf{AskH}] \leq 2^{-k_1} + (q_G + 1) \cdot 2^{-k_0}$.

As a consequence,

$$\Pr[\mathsf{Fail} \mid \neg\mathsf{AskH}] \leq \frac{2}{2^{k_1}} + \frac{2q_G + 1}{2^{k_0}}.$$

The running time of this simulator includes just the computation of $f(\sigma, \tau)$ for all possible pairs and is thus bounded by $q_G \cdot q_H \cdot (T_f + \mathcal{O}(1))$.

5.4 Success Probability of the Reduction

This subsection analyzes the success probability of our reduction with respect to the advantage of the IND-CCA2 adversary. The goal of the reduction is, given $c^\star = f(s^\star, t^\star)$, to obtain $s^\star$. Therefore, the success probability is obtained by the probability that event AskH occurs during the reduction (*i.e.*, $\Pr[\mathsf{AskH}] \leq \mathsf{Succ}^{\mathsf{s-pd-ow}}(q_H, t')$, where t' is the running time of the reduction).

We thus evaluate $\Pr[\mathsf{AskH}]$ by splitting event AskH according to event Bad.

$$\Pr[\mathsf{AskH}] = \Pr[\mathsf{AskH} \wedge \mathsf{Bad}] + \Pr[\mathsf{AskH} \wedge \neg\mathsf{Bad}].$$

First let us evaluate the first term.

$$\begin{aligned}
\Pr[\mathsf{AskH} \wedge \mathsf{Bad}] &= \Pr[\mathsf{Bad}] - \Pr[\neg\mathsf{AskH} \wedge \mathsf{Bad}] \\
&\geq \Pr[\mathsf{Bad}] - \Pr[\neg\mathsf{AskH} \wedge \mathsf{GBad}] - \Pr[\neg\mathsf{AskH} \wedge \mathsf{DBad}] \\
&\geq \Pr[\mathsf{Bad}] - \Pr[\mathsf{GBad} \mid \neg\mathsf{AskH}] - \Pr[\mathsf{DBad} \mid \neg\mathsf{AskH}]
\end{aligned}$$

$$\geq \Pr[\mathsf{Bad}] - \Pr[\mathsf{AskG} \mid \neg\mathsf{AskH}] - \Pr[\mathsf{DBad} \mid \neg\mathsf{AskH}]$$
$$\geq \Pr[\mathsf{Bad}] - \frac{q_G}{2^{k_0}} - q_D\left(\frac{2}{2^{k_1}} + \frac{2q_G+1}{2^{k_0}}\right)$$
$$\geq \Pr[\mathsf{Bad}] - \frac{2q_Dq_G + q_D + q_G}{2^{k_0}} - \frac{2q_D}{2^{k_1}}.$$

Here, $\Pr[\mathsf{DBad} \mid \neg\mathsf{AskH}] \leq q_D\left(2 \cdot 2^{-k_1} + (2q_G+1)\cdot 2^{-k_0}\right)$ is directly obtained from lemma 4, and $\Pr[\mathsf{GBad} \mid \neg\mathsf{AskH}] \leq \Pr[\mathsf{AskG} \mid \neg\mathsf{AskH}]$ is obtained from the fact that event GBad implies AskG. When $\neg$AskH occurs, $H(s^\star)$ is unpredictable, and $r^\star = t^\star \oplus H(s^\star)$ is also unpredictable. Hence $\Pr[\mathsf{AskG} \mid \neg\mathsf{AskH}] \leq q_G \cdot 2^{-k_0}$.

We then evaluate the second term.

$$\begin{aligned}\Pr[\mathsf{AskH} \wedge \neg\mathsf{Bad}] &= \Pr[\neg\mathsf{Bad}] \cdot \Pr[\mathsf{AskH} \mid \neg\mathsf{Bad}]\\ &\geq \Pr[\neg\mathsf{Bad}] \cdot \Pr[\mathcal{A} = b \wedge \mathsf{AskH} \mid \neg\mathsf{Bad}]\\ &\geq \Pr[\neg\mathsf{Bad}] \cdot (\Pr[\mathcal{A} = b \mid \neg\mathsf{Bad}] - \Pr[\mathcal{A} = b \wedge \neg\mathsf{AskH} \mid \neg\mathsf{Bad}]).\end{aligned}$$

Here, when $\neg$AskH occurs, $H(s^\star)$ is unpredictable, thus $r^\star = t^\star \oplus H(s^\star)$ is unpredictable, and so is b as well. This fact is independent from event $\neg$Bad. Hence $\Pr[\mathcal{A} = b \wedge \neg\mathsf{AskH} \mid \neg\mathsf{Bad}] \leq \Pr[\mathcal{A} = b \mid \neg\mathsf{AskH} \wedge \neg\mathsf{Bad}] = 1/2$. Furthermore,

$$\frac{\varepsilon}{2} + \frac{1}{2} \leq \Pr[\mathcal{A} = b] \leq \Pr[\mathcal{A} = b \mid \neg\mathsf{Bad}] \cdot \Pr[\neg\mathsf{Bad}] + \Pr[\mathsf{Bad}].$$

Therefore,

$$\Pr[\mathsf{AskH} \wedge \neg\mathsf{Bad}] \geq \left(\frac{\varepsilon}{2} + \frac{1}{2} - \Pr[\mathsf{Bad}]\right) - \frac{\Pr[\neg\mathsf{Bad}]}{2} = \frac{\varepsilon - \Pr[\mathsf{Bad}]}{2}.$$

Combining the evaluation for the first and second terms, and from the fact that $\Pr[\mathsf{Bad}] \geq 0$, one gets

$$\Pr[\mathsf{AskH}] \geq \frac{\varepsilon}{2} - \frac{2q_Dq_G + q_D + q_G}{2^{k_0}} - \frac{2q_D}{2^{k_1}}.$$

5.5 Complexity Analysis

Note that during the execution of $\mathcal{B}$, for any new G-query γ, one has to look at all query-answer pairs (δ, H_δ) in the H-List, and to compute $s = \delta$, $t = \gamma \oplus H_\delta$ as well as $f(s,t)$.

Apparently, one should perform this computation again to simulate the decryption of any ciphertext. Proper bookkeeping allows the computation to be done once for each pair, when the query is asked to the hash functions. Thus, the time complexity of the overall reduction is $t' = t + q_G \cdot q_H \cdot (T_f + \mathcal{O}(1))$, where T_f denotes the time complexity for evaluating function f.

6 Application to RSA–OAEP

The main application of OAEP is certainly the famous RSA–OAEP, which has been used to update the PKCS #1 standard [14]. In his paper [15], Shoup was able to repair the security result for a small exponent, $e = 3$, using Coppersmith's algorithm from [5]. However, our result can be applied to repair RSA–OAEP, regardless of the exponent; thanks to the random self-reducibility of RSA, the partial-domain one-wayness of RSA is equivalent to that of the whole RSA problem, as soon as a constant fraction of the most significant bits (or the least significant bits) of the pre-image can be recovered.

We note that, in the original RSA–OAEP [3], the most significant bits are involved in the H function, but in PKCS #1 standards v2.0 and v2.1 [14] and RFC2437, the least significant bits are used: the value $\mathsf{maskedSeed}\|\mathsf{maskedDB}$ is the input to f, the RSA function, where $\mathsf{maskedSeed}$ plays the role of t, and $\mathsf{maskedDB}$ the role of s. But we insist on the fact that the following result holds in both situations (and can be further extended).

One may also remark that the following argument can be applied to any random (multiplicatively) self-reducible problem, such as the Rabin function. Before presenting the final reduction, let us consider the problem of finding small solutions for a linear modular equation.

Lemma 5. *Consider an equation $t + \alpha u = c \bmod N$ which has solutions t and u smaller than 2^{k_0}. For all values of α, except a fraction $2^{2k_0+6}/N$ of them, (t, u) is unique and can be computed within time bound $\mathcal{O}((\log N)^3)$.*

Proof. Consider the lattice

$$L(\alpha) = \{(x, y) \in \mathbb{Z}^2 \mid x - \alpha y = 0 \bmod N\}.$$

We say that $L(\alpha)$ is an ℓ-good lattice (and that α is an ℓ-good value) if there is no non-zero vector of length at most ℓ (with respect to the Euclidean norm). Otherwise, we use the wording ℓ-bad lattices (and ℓ-bad values respectively). It is clear that there are approximately less than $\pi\ell^2$ such ℓ-bad lattices, which we bound by $4\ell^2$. Indeed, each bad value for α corresponds to a point with integer coordinates in the disk of radius ℓ. Thus, the proportion of bad values for α is less than $4\ell^2/N$.

Given an ℓ-good lattice, one applies the Gaussian reduction algorithm. One gets within time $\mathcal{O}((\log N)^3)$ a basis of $L(\alpha)$ consisting of two non-zero vectors U and V such that

$$\|U\| \leq \|V\| \text{ and } |(U, V)| \leq \|U\|^2/2.$$

Let T be the point (t, u), where (t, u) is a solution of the equation $t + \alpha u = c \bmod N$, with both t and u less than 2^{k_0}:

$$T = \lambda U + \mu V, \text{ for some real } \lambda, \mu.$$

$$\|T\|^2 = \lambda^2\|U\|^2 + \mu^2\|V\|^2 + 2\lambda\mu(U,V) \geq (\lambda^2 + \mu^2 - \lambda\mu) \times \|U\|^2$$
$$\geq ((\lambda - \mu/2)^2 + 3\mu^2/4) \times \|U\|^2 \geq 3\mu^2/4 \times \|U\|^2 \geq 3\mu^2\ell^2/4.$$

Since furthermore we have $\|T\|^2 \leq 2 \times 2^{2k_0}$,

$$|\mu| \leq \frac{2\sqrt{2} \cdot 2^{k_0}}{\sqrt{3} \cdot \ell}, \text{ and } |\lambda| \leq \frac{2\sqrt{2} \cdot 2^{k_0}}{\sqrt{3} \cdot \ell} \text{ by symmetry.}$$

Assuming that we have set from the beginning $\ell = 2^{k_0+2} > 2^{k_0+2}\sqrt{2/3}$, then

$$-\frac{1}{2} < \lambda, \mu < \frac{1}{2}.$$

Choose any integer solution $T_0 = (t_0, u_0)$ of the equation simply by picking a random integer u_0 and setting $t_0 = c - \alpha u_0 \bmod N$. Write it in the basis (U, V): $T_0 = \rho U + \sigma V$ using real numbers ρ and σ. These coordinates can be found, so $T - T_0$ is a solution to the homogeneous equation, and thus indicate a lattice point: $T - T_0 = aU + bV$, with unknown integers a and b. But,

$$T = T_0 + aU + bV = (a + \rho)U + (b + \sigma)V = \lambda U + \mu V,$$

with $-1/2 \leq \lambda, \mu \leq 1/2$. As a conclusion, a and b are the closest integers to $-\rho$ and $-\sigma$ respectively. With a, b, ρ and σ, one can easily recover λ and μ and thus t and u, which are necessarily unique. □

Lemma 6. *Let $\mathcal{A}$ be an algorithm that outputs a q-set containing $k - k_0$ of the most significant bits of the e-th root of its input (partial-domain RSA, for any $2^{k-1} < N < 2^k$, with $k > 2k_0$), within time bound t, with probability ε. There exists an algorithm $\mathcal{B}$ that solves the RSA problem (N, e) with success probability ε', within time bound t' where*

$$\varepsilon' \geq \varepsilon \times (\varepsilon - 2^{2k_0 - k + 6}),$$
$$t' \leq 2t + q^2 \times \mathcal{O}(k^3).$$

Proof. Thanks to the random self-reducibility of RSA, with part of the bits of the e-th root of $X = (x \cdot 2^{k_0} + r)^e \bmod N$, and the e-th root of $Y = X\alpha^e = (y \cdot 2^{k_0} + s)^e \bmod N$, for a randomly chosen α, one gets both x and y. Thus,

$$(y \cdot 2^{k_0} + s) = \alpha \times (x \cdot 2^{k_0} + r) \bmod N$$
$$\alpha r - s = (y - x\alpha) \times 2^{k_0} \bmod N$$

which is a linear modular equation with two unknowns r and s which is known to have small solutions (smaller than 2^{k_0}). It can be solved using lemma 5.

Algorithm $\mathcal{B}$ just runs twice $\mathcal{A}$, on inputs X and $X\alpha^e$ and next runs the Gaussian reduction on all the q^2 pairs of elements coming from both sets. If the partial pre-images are in the sets, they will be found, unless the random α is bad (*cf.* the Gaussian reduction in lemma 5.) □

Remark 7. The above lemma can be extended to the case where a constant fraction Θ of the leading or trailing bits of the e-th root is found. The reduction runs $1/\Theta$ times the adversary $\mathcal{A}$, and the success probability decreases to approximately $\varepsilon^{1/\Theta}$. Extensions to any constant fraction of consecutive bits are also possible. Anyway, in PKCS #1 v2.0, k_0 is much smaller than $k/2$.

Theorem 8. *Let $\mathcal{A}$ be a CCA2–adversary against the "semantic security" of RSA–OAEP (with a k-bit long modulus, with $k > 2k_0$), with running time bounded by t and advantage ε, making q_D, q_G and q_H queries to the decryption oracle, and the hash functions G and H respectively. Then, the RSA problem can be solved with probability ε' greater than*

$$\frac{\varepsilon^2}{4} - \varepsilon \cdot \left(\frac{2q_Dq_G + q_D + q_G}{2^{k_0}} + \frac{2q_D}{2^{k_1}} + \frac{32}{2^{k-2k_0}} \right)$$

within time bound $t' \leq 2t + q_H \cdot (q_H + 2q_G) \times \mathcal{O}(k^3)$.

Proof. Lemma 2 states that

$$\mathsf{Succ}^{\mathsf{s-pd-ow}}(q_H, t'') \geq \frac{\varepsilon}{2} - \frac{2q_Dq_G + q_D + q_G}{2^{k_0}} - \frac{2q_D}{2^{k_1}},$$

with $t'' \leq t + q_G \cdot q_H \cdot (T_f + \mathcal{O}(1))$, and $T_f = \mathcal{O}(k^3)$. Using the previous results relating q_H-set partial-domain–RSA and RSA, we easily conclude. □

7 Conclusion

Our conclusion is that one can still trust the security of RSA–OAEP, but the reduction is more costly than the original one. However, for other OAEP applications, more care is needed, since the security does not actually rely on the one-wayness of the permutation, only on its partial-domain one-wayness.

Acknowledgments. We thank Victor Shoup, Don Coppersmith and Dan Boneh for fruitful comments.

References

1. M. Bellare, A. Desai, D. Pointcheval, and P. Rogaway. Relations among Notions of Security for Public-Key Encryption Schemes. In *Crypto '98*, LNCS 1462, pages 26–45. Springer-Verlag, Berlin, 1998.
2. M. Bellare and P. Rogaway. Random Oracles Are Practical: a Paradigm for Designing Efficient Protocols. In *Proc. of the 1st CCS*, pages 62–73. ACM Press, New York, 1993.
3. M. Bellare and P. Rogaway. Optimal Asymmetric Encryption – How to Encrypt with RSA. In *Eurocrypt '94*, LNCS 950, pages 92–111. Springer-Verlag, Berlin, 1995.

4. D. Bleichenbacher. A Chosen Ciphertext Attack against Protocols based on the RSA Encryption Standard PKCS #1. In *Crypto '98*, LNCS 1462, pages 1–12. Springer-Verlag, Berlin, 1998.
5. D. Coppersmith. Finding a Small Root of a Univariate Modular Equation. In *Eurocrypt '96*, LNCS 1070, pages 155–165. Springer-Verlag, Berlin, 1996.
6. D. Dolev, C. Dwork, and M. Naor. Non-Malleable Cryptography. *SIAM Journal on Computing*, 30(2):391–437, 2000.
7. S. Goldwasser and S. Micali. Probabilistic Encryption. *Journal of Computer and System Sciences*, 28:270–299, 1984.
8. C. Hall, I. Goldberg, and B. Schneier. Reaction Attacks Against Several Public-Key Cryptosystems. In *Proc. of ICICS'99*, LNCS, pages 2–12. Springer-Verlag, 1999.
9. M. Joye, J. J. Quisquater, and M. Yung. On the Power of Misbehaving Adversaries and Security Analysis of the Original EPOC. In *CT – RSA '2001*, LNCS 2020, pages 208–222. Springer-Verlag, Berlin, 2001.
10. M. Naor and M. Yung. Public-Key Cryptosystems Provably Secure against Chosen Ciphertext Attacks. In *Proc. of the 22nd STOC*, pages 427–437. ACM Press, New York, 1990.
11. T. Okamoto and D. Pointcheval. REACT: Rapid Enhanced-security Asymmetric Cryptosystem Transform. In *CT – RSA '2001*, LNCS 2020, pages 159–175. Springer-Verlag, Berlin, 2001.
12. C. Rackoff and D. R. Simon. Non-Interactive Zero-Knowledge Proof of Knowledge and Chosen Ciphertext Attack. In *Crypto '91*, LNCS 576, pages 433–444. Springer-Verlag, Berlin, 1992.
13. R. Rivest, A. Shamir, and L. Adleman. A Method for Obtaining Digital Signatures and Public Key Cryptosystems. *Communications of the ACM*, 21(2):120–126, February 1978.
14. RSA Data Security, Inc. Public Key Cryptography Standards – PKCS.
15. V. Shoup. OAEP Reconsidered. In *Crypto '2001*, LNCS. Springer-Verlag, Berlin, 2001.

Simplified OAEP for the RSA and Rabin Functions

Dan Boneh*

Computer Science Department, Stanford University
dabo@cs.stanford.edu

Abstract. Optimal Asymmetric Encryption Padding (OAEP) is a technique for converting the RSA trapdoor permutation into a chosen ciphertext secure system in the random oracle model. OAEP padding can be viewed as two rounds of a Feistel network. We show that for the Rabin and RSA trapdoor functions a much simpler padding scheme is sufficient for chosen ciphertext security in the random oracle model. We show that only one round of a Feistel network is sufficient. The proof of security uses the algebraic properties of the RSA and Rabin functions.

1 Introduction

In an influential paper Bellare and Rogaway [2] introduced the Optimal Asymmetric Encryption Padding (OAEP) system. OAEP is most commonly used for strengthening the RSA and Rabin encryption schemes. OAEP is widely deployed and appears in several standards. Shoup [11] recently described a modification to OAEP called OAEP+ that provably converts any trapdoor permutation into a chosen ciphertext secure system in the random oracle model. Shoup also showed that applying OAEP to the RSA permutation with public exponent $e = 3$ gives a chosen ciphertext secure system in the random oracle model. Fujisaki et al.[8] were able to extend the result and prove that the same holds for the RSA permutation with any RSA public exponent e.

We show that for the RSA and Rabin systems, much simpler padding schemes can be shown to be chosen ciphertext secure in the random oracle model. We introduce two simple padding schemes. The first is called Simple-OAEP, or SAEP for short. The second is called SAEP$^+$. We note that simplifying the padding scheme makes the system easier to describe and easier to implement, and thus is more elegant. Simplifying the padding scheme has little bearing on performance since padding time is negligible compared to public key operations.

We begin by describing SAEP and SAEP$^+$ padding (see Figure 1). Let M be a message $M \in \{0,1\}^m$ and let r be a random string $r \in \{0,1\}^{s_1}$. Let H be a hash function from $\{0,1\}^{s_1}$ to $\{0,1\}^{m+s_0}$. Let G be a hash function from $\{0,1\}^{m+s_1}$ to $\{0,1\}^{s_0}$. Define the new padding schemes SAEP and SAEP$^+$ as follows:

* Supported by NSF and the Packard Foundation.

J. Kilian (Ed.): CRYPTO 2001, LNCS 2139, pp. 275–291, 2001.

$$\mathsf{SAEP}(M,r) = ((M \parallel \quad 0^{s_0} \quad) \oplus H(r)) \parallel r$$
$$\mathsf{SAEP}^{+}(M,r) = ((M \parallel G(M\|r)\ ) \oplus H(r)) \parallel r$$

These padding schemes are to be used as preprocessing functions with the Rabin or RSA trapdoor functions. To encrypt a message $M \in \{0,1\}^m$ first pick a random $r \in \{0,1\}^{s_1}$, compute $y = \mathsf{SAEP}(M,r)$, and set $C = y^2 \bmod N$ or $C = y^e \bmod N$ for some RSA exponent e.

Both schemes provide security against an adaptive chosen ciphertext attack in the random oracle model for appropriate values of m, s_0, s_1. Let N be an n-bit modulus. We prove the following results for the Rabin and RSA functions:

SAEP: Let Rabin-SAEP be the encryption scheme resulting from combining SAEP with the Rabin trapdoor function, $f(x) = x^2 \bmod N$ (as described in the next section). We show that Rabin-SAEP provides chosen ciphertext security whenever $m+s_0 < n/2$ and $m < n/4$. Security is based on the hardness of factoring large RSA composites. The reduction is very efficient. It is based entirely on applying Coppersmith's algorithm [6] to quadratic and quartic polynomials. SAEP works well with the Rabin function, but is hard to use with RSA, as explained in Section 4.

SAEP$^+$: Both RSA-SAEP$^+$ (for any RSA exponent e) and Rabin-SAEP$^+$ can be shown to be chosen ciphertext secure whenever $m + s_0 < n/2$. The reduction to factoring for Rabin-SAEP$^+$ is extremely efficient. The proof is based on Coppersmith's algorithm. For RSA-SAEP$^+$ the reduction to breaking RSA is less efficient. Its running time is similar to the running time of the reduction in the proof of security for RSA-OAEP [8].

SAEP$^+$ is more flexible than SAEP in a number of ways. First, SAEP$^+$ can be used with both Rabin and RSA (although Rabin is preferred). Second, SAEP$^+$ can encrypt messages of longer size. For example, when using a 1024 bit modulus ($n = 1024$) one often takes $s_0 = 128$ for proper security. In this case, the maximum message length in SAEP is 256 bits. In SAEP$^+$ the maximum length is 384 bits. Note that since a 1024-bit modulus is often used for transporting a 128-bit session-key, both SAEP and SAEP$^+$ are adequate for this purpose.

In some cases it might be desirable to allow for longer messages to be encrypted with SAEP$^+$. In Section 5 we note that the proof of security for RSA-SAEP$^+$ can be extended so that the scheme is secure whenever $m+s_0 < n(1-\delta)$ for any fixed $\delta > 0$. This means M could be almost as long as the modulus. However, the efficiency of the reduction to breaking RSA degrades exponentially in $\frac{1}{\delta}$. Hence, throughout the paper we stick with $\delta = 1/2$. The extended proof is based on solutions to the Hidden Number Problem [4] modulo a composite.

Both SAEP and SAEP$^+$ work best with the Rabin function. The resulting systems are better than their RSA counterparts in all aspects: (1) encryption is slightly faster, (2) the reduction given in the security proof is more efficient, and (3) security relies on the difficulty of factoring rather than the difficulty of inverting the RSA permutation.

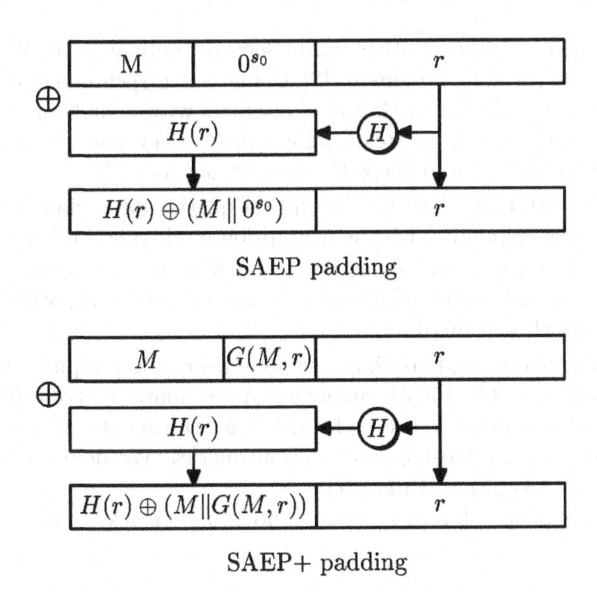

Fig. 1. SAEP and SAEP$^+$ padding

Comparison of OAEP *and* SAEP. OAEP, presented by Bellare and Rogaway, and OAEP+, presented by Shoup, both provide chosen ciphertext security for the RSA trapdoor permutation (although OAEP+ has a more efficient security proof). These padding schemes are defined as follows:

$$\mathsf{OAEP}(M, r) = (\quad (M \| 0^{s_0}) \oplus H(r) \quad) \; \| \; (r \oplus G(\quad (M \| 0^{s_0}) \oplus H(r) \quad))$$

$$\mathsf{OAEP+}(M, r) = (\,(M \oplus H(r)) \,\|\, W(M, r)\,) \; \| \; (r \oplus G(\,(M \oplus H(r)) \,\|\, W(M, r)\,))$$

where H, G, W are hash functions. Schematically both OAEP and OAEP+ look like two rounds of a Feistel network. Clearly the new padding schemes, SAEP and SAEP$^+$ are simpler. These new schemes are only a single round of a Feistel network.

Although the new padding schemes are simpler than OAEP, they are slightly more restrictive. Using OAEP and OAEP+ one can encrypt messages that are almost as long as the modulus. For example, for a 1024-bit modulus it is safe to encrypt messages that are 768-bits long. In contrast, using the same modulus size, SAEP$^+$ can only encrypt 384-bit messages. This difference is irrelevant for common applications (e.g. key transport), but is worth pointing out.

1.1 Chosen Ciphertext Security

Adaptive chosen ciphertext security is the accepted notion for secure encryption. We have confidence in this notion since it captures a wide range of attacks, and is equivalent to several other useful security notions [7,3]. We present the definition due to Rackoff and Simon [12]. Define a (t, q_D) chosen ciphertext attack algorithm $\mathcal{A}$ as a t-time algorithm that interacts with a challenger as follows:

Setup: The challenger generates a public/private key pair. It gives the public key to the attacker $\mathcal{A}$ and keeps the private key to itself.

Phase I: The attacker $\mathcal{A}$ issues decryption queries for various ciphertexts C. The challenger responds with the decryption of all valid ciphertexts.

Challenge: At some point algorithm $\mathcal{A}$ outputs two messages M_0, M_1. The challenger responds with a ciphertext C^* which is the encryption of M_b where b is randomly chosen in $\{0, 1\}$.

Phase II: The attacker $\mathcal{A}$ continues to issue decryption requests C, subject to the constraint $C \neq C^*$. Finally algorithm $\mathcal{A}$ terminates and outputs $b' \in \{0, 1\}$.

We say that the attacker is successful if $b = b'$. During the attack the attacker is allowed to make at most q_D decryption queries. We define the adversary's advantage as: $\text{adv}(\mathcal{A}) = \left|\Pr[b = b'] - \frac{1}{2}\right|$

We say that a system is (t, ϵ, q_D) secure if no (t, q_D) attacker has advantage more than ϵ.

Random oracles: To analyze the security of certain natural constructions Bellare and Rogaway introduced an idealized world called the random oracle model [1]. A system that has chosen ciphertext security in this idealized world is said to be chosen ciphertext secure in the random oracle model. Security in the random oracle model does not imply security in the real world [5]. Nevertheless, the random oracle model is a useful tool for validating natural constructions. Given an encryption scheme using hash functions $H_1, \ldots, H_n$ we use $(t, q_D, q_{H_1}, \ldots, q_{H_n})$ to denote a (t, q_D) chosen ciphertext attacker that makes at most q_{H_i} queries to the hash function H_i.

1.2 Coppersmith's Algorithm

The proofs of security for SAEP and SAEP$^+$ are based on an important result due to Coppersmith [6]. Coppersmith proved the following theorem:

Theorem 1 (Coppersmith). *Let N be an integer and let $f(x) \in \mathbb{Z}_N[x]$ be a monic polynomial of degree d. Then there is an efficient algorithm to find all $x_0 \in \mathbb{Z}$ such that $f(x_0) = 0 \bmod N$ and $|x_0| < N^{1/d}$.*

We denote by $T_C(N, d)$ the running time of Coppersmith's algorithm when finding roots of a polynomial $f \in \mathbb{Z}[x]$ of degree d. In our proofs we only apply Coppersmith's algorithm to quadratic and quartic polynomials.

2 Full Description of SAEP and SAEP$^+$

We now give a full description of the SAEP and SAEP$^+$ systems for RSA and Rabin. We first describe these schemes as they apply to the Rabin function.

In doing so we deal with complications that arise from the fact that $f(x) = x^2 \bmod N$ is not a permutation of $\mathbb{Z}_N^*$. Let m, s_0, s_1 be security parameters. Set $n = m + s_0 + s_1$. We will make use of a hash function $H : \{0,1\}^{s_1} \to \{0,1\}^{m+s_0}$. The Rabin-SAEP system is composed of three algorithms: key-gen, encrypt, decrypt. We describe each of these algorithms in turn:

key-gen: The key generation algorithm takes a security parameter n and produces an $(n+2)$-bit RSA modulus $N = pq$ where p and q are $(n/2+1)$-bit primes. We require that $p = q = 3 \bmod 4$. We also require that $N \in [2^{n+1}, 2^{n+1} + 2^n)$, i.e. that the two most significant bits of N are '10'. Any of the standardized methods can be used to generate p and q [9]. The public key is N. The private key is the factorization of N, namely $\langle p, q \rangle$.

encrypt: We wish to encrypt a message $M \in \{0,1\}^m$:

Step 1: Pick a random $r \in \{0,1\}^{s_1}$.

Step 2: Set $t = 0^{s_0}$.

Step 3: Set $v = M \| t \in \{0,1\}^{m+s_0}$.

Step 4: Set $x = v \oplus H(r)$.

Step 5: Set $y = x \| r \in \{0,1\}^n$. We view y as an n-bit integer. Note that $y < N/2$.

Step 6: Define the ciphertext C as $C = y^2 \bmod N$.

decrypt: Given a ciphertext $C \in \mathbb{Z}_N$ we decrypt using the steps below. We let A and B be the Chinese Remainder coefficients, i.e. A is $1 \bmod p$ and $0 \bmod q$, and B is $0 \bmod p$ and $1 \bmod q$.

Step 1: Compute $z_p = C^{\frac{p+1}{4}} \bmod p$ and $z_q = C^{\frac{q+1}{4}} \bmod q$. Since $p = q = 3 \bmod 4$ it follows that z_p, z_q are square roots of C in $\mathbb{Z}_p, \mathbb{Z}_q$ respectively.

Step 2: Test that $z_p^2 = C \bmod p$ and $z_q^2 = C \bmod q$. If either condition does not hold, then C is not a quadratic residue in $\mathbb{Z}_N$. Reject this C as an invalid ciphertext.

Step 3: Set $y_1 = A \cdot z_p + B \cdot z_q \bmod N$ and $y_2 = A \cdot z_p - B \cdot z_q \bmod N$. The four square roots of $C \bmod N$ are $\pm y_1$ and $\pm y_2$. Two of these four roots must be greater than $N/2$ and hence can be discarded. Let y_1, y_2 be the two remaining square roots. If neither of y_1, y_2 is in $[0, 2^n)$ then reject C as an invalid ciphertext. Without loss of generality we assume both y_1, y_2 are in $[0, 2^n)$.

Step 4: View both y_1 and y_2 as strings in $\{0,1\}^n$. Write $y_1 = x_1 \| r_1$ and $y_2 = x_2 \| r_2$ with $x_1, x_2 \in \{0,1\}^{m+s_0}$ and $r_1, r_2 \in \{0,1\}^{s_1}$.

Step 5: Set $v_1 = x_1 \oplus H(r_1)$ and $v_2 = x_2 \oplus H(r_2)$.

Step 6: Write $v_1 = M_1 \| t_1$ and $v_2 = M_2 \| t_2$ where $M_1, M_2 \in \{0,1\}^m$ and $t_1, t_2 \in \{0,1\}^{s_0}$.

Step 7: For $i = 1, 2$ test if t_i is equal to 0^{s_0}. If this condition holds for either none or both of v_1, v_2 then reject C as an invalid ciphertext.

Step 8: Let $i \in \{1, 2\}$ be the unique i for which the condition of Step 7 holds. Output M_i as the decryption of C.

Note that in Step 7, if both t_1 and t_2 are equal to 0^{s_0} the decryptor cannot choose between them. Hence, in this case the ciphertext is rejected. This means that with very low probability, namely 2^{-s_0}, a valid ciphertext might be rejected by the decryptor (recall that typically $s_0 \geq 128$). For most applications such low error probabilities can be ignored. One concern is whether a malicious encryptor can create a valid ciphertext that will be rejected by the decryptor in Step 7. It is easy to show that in the random oracle model the encryptor would have to spend expected time $O(2^{s_0})$ to create such a ciphertext. This is sufficient for most applications. We note that if a negligible error probability is unacceptable then the encryptor could keep choosing random r's until y has Jacobi symbol 1. This enables the decryptor to select the correct square root by choosing the unique root $y_i \in [0, 2^n)$ with Jacobi symbol 1. However, this is unnecessary and makes the scheme less efficient.

During decryption invalid ciphertexts can be rejected in Steps 2 and 3 as well as in Step 7. Manger [10] points out the importance of preventing an attacker from distinguishing between rejections at the various steps, say, using timing analysis. Implementors must ensure that the reason a ciphertext is rejected is hidden from the outside world. Indeed, our proof of security fails if this is not the case.

*Description of Rabin-*SAEP$^+$*:* The description of Rabin-SAEP$^+$ is very similar to Rabin-SAEP. SAEP$^+$ makes use of an additional hash function $G : \{0,1\}^{m+s_1} \to \{0,1\}^{s_0}$. Key generation for Rabin-SAEP$^+$ is identical to key generation for Rabin-SAEP. Encryption differs only in Step 2 where t is defined as $t = G(M, r) \in \{0,1\}^{s_0}$. Decryption differs only in Step 7 where the condition tested is whether t_i is equal to $G(M_i, r_i)$.

The description of RSA-SAEP$^+$ is analogous to the one given above. Decryption is a bit simpler since one does not have to worry about multiple preimages to the RSA trapdoor permutation.

2.1 Complexity Assumptions

Throughout the paper we use the following standard complexity assumptions:

Factoring assumption: We say that a t-time algorithm $\mathcal{B}$ is an (n, t) factoring algorithm with advantage ϵ if $\mathcal{B}$ succeeds with probability at least ϵ in factoring n-bit integers generated by the key-gen algorithm. The probability is over the random bits used by algorithms key-gen and $\mathcal{B}$. We write $\text{adv}(\mathcal{B}) = \epsilon$. We say that the (n, t, ϵ) factoring assumption holds if there is no (n, t) factoring algorithm with advantage ϵ.

RSA assumption: We say that a t-time algorithm $\mathcal{B}$ is an (n, e, t) algorithm for computing e'th roots in $\mathbb{Z}_N$ with advantage ϵ if $\mathcal{B}$ succeeds with probability at least ϵ in computing $x^{1/e} \bmod N$ for an n-bit integer N generated by the key-gen algorithm and a random $x \in \mathbb{Z}_N$. The probability is over x and the random bits used by algorithms key-gen, $\mathcal{B}$. We write $\text{adv}(\mathcal{B}) = \epsilon$. We say that the (n, e, t, ϵ) RSA assumption holds if there is no (n, e, t) algorithm with advantage ϵ.

3 Two Simple Facts

We state two simple facts that will be useful in the proof of security.

Fact 2. *Let $N = pq$ be an $n+2$-bit integer generated by the* key-gen *algorithm, i.e. $N \in [2^{n+1}, 2^{n+1} + 2^n)$. Let α be a random integer in $[0, 2^n)$ and $C^* = \alpha^2$. Then with probability at least 1/3 (over the choice of α) there exist two distinct integers $y_1^*, y_2^* \in [0, 2^n)$ such that $(y_1^*)^2 = (y_2^*)^2 = C^* \bmod N$.*

Proof. The condition $N \in [2^{n+1}, 2^{n+1} + 2^n)$ implies that $2^n < N/2$ and that $N/2^{n+1} < 3/2$. Let $\alpha \in [0, 2^n)$. Since $2^n < N/2$ we know that $C^* = \alpha^2 \bmod N$ always has either one or two square roots in $[0, 2^n)$. Let A be the number of $\alpha \in [0, 2^n)$ so that $\alpha^2 \bmod N$ has one root in $[0, 2^n)$. Let B be the number of $\alpha \in [0, 2^n)$ so that $\alpha^2 \bmod N$ has two roots in $[0, 2^n)$. We know $A + B = 2^n$. Furthermore, we know that for every $\alpha \in [0, 2^n)$ relatively prime to N we have that $\alpha^2 \bmod N$ has exactly two roots in $[0, N/2)$. The number of α not relatively prime to N is at most $p + q$. Therefore, $A < (N/2 - 2^n) + p + q$ and hence $B > 2^n - (N/2 - 2^n) - p - q = 2^{n+1} - N/2 - p - q$. We get that:

$$\frac{B}{2^n} > 2 - \frac{N}{2^{n+1}} - \frac{p+q}{2^n} > \frac{1}{2} - \frac{p+q}{2^n} > \frac{1}{3}$$

□

Fact 3. *Let $N = pq$ be an $n+2$-bit integer generated by the* key-gen *algorithm. Let α be a random integer in $[0, 2^n)$ and set $C^* = \alpha^2$. Let $y_1^*, y_2^* \in [0, 2^n)$ be two integers such that $(y_1^*)^2 = (y_2^*)^2 = C^* \bmod N$. When C^* has two distinct roots in $[0, 2^n)$ we assume $y_1^* \neq y_2^*$, otherwise set $y_1^* = y_2^*$. Let c be a random bit in $\{1, 2\}$. Then y_c^* is a uniform random variable in $[0, 2^n)$ over the choice of (α, c).*

The proof of Fact 3 is immediate.

4 Proof of Security of Rabin-SAEP

We show that an attacker capable of mounting a successful adaptive chosen ciphertext attack on Rabin-SAEP in the random oracle model can be used to efficiently factor large integers. We use m, s_0, s_1 as the security parameters of SAEP and set $n = m + s_0 + s_1$. Recall that the SAEP key-gen algorithm generates an $(n + 2)$-bit modulus N.

Theorem 4. *Let $N = pq$ be an integer generated by the Rabin-*SAEP key-gen *algorithm given the security parameter n. We assume $m < n/4$ and $m + s_0 < n/2$. Let $\mathcal{A}$ be a (t, q_D, q_H) chosen ciphertext attack algorithm in the random oracle model. Suppose $\mathcal{A}$ has advantage ϵ when attacking Rabin-*SAEP *modulo N. Then there is a uniform algorithm $\mathcal{B}$ for factoring N with the following parameters:*

$$time(\mathcal{B}) = time(\mathcal{A}) + O(q_D q_H T_C + q_D T_C')$$

$$adv(\mathcal{B}) \geq \tfrac{1}{6} \cdot adv(\mathcal{A}) \cdot (1 - \frac{2q_D}{2^{s_0}} - \frac{2q_D}{2^{s_1}})$$

Here $T_C = T_C(n, 2)$ and $T_C' = T_C(n, 4)$.

Proof of Theorem 4. Let N be an $(n+2)$-bit integer generated by the keygen algorithm. To factor N algorithm $\mathcal{B}$ begins by picking a random $\alpha \in [0, 2^n)$ and computing $C^* = \alpha^2 \bmod N$. We show an algorithm that takes C^* as input, interacts with $\mathcal{A}$, and outputs a square root $\alpha' \in [0, 2^n)$ of $C^* \bmod N$ with probability at least $\epsilon' = \epsilon \cdot (1 - 2q_D/2^{s_0} - 2q_D/2^{s_1})$. By Fact 2 we know that C^* has two distinct square roots $\gamma, \gamma' \in [0, 2^n)$ with probability at least 1/3. Therefore, $\alpha \neq \alpha'$ with probability 1/6. When this happens, we can factor N by computing $\gcd(N, \alpha - \alpha')$. Since both $0 \leq \alpha, \alpha' < N/2$ this is guaranteed to give a non-trivial factor of N. Overall, we succeed in factoring N with probability at least $\frac{1}{6}\epsilon'$ as required.

The rest of the proof focuses on computing a square root α' of C^*. We construct a simulator that given C^* interacts with algorithm $\mathcal{A}$ and produces a root. The simulator responds to $\mathcal{A}$'s decryption queries and H hash queries, and provides algorithm $\mathcal{A}$ with the challenge ciphertext. We first give a high level description of the simulator (the simulator is described in detail below). The simulator gives C^* as the challenge ciphertext to the attacker $\mathcal{A}$. Suppose $C^* = (y_1^*)^2 = (y_2^*)^2 \bmod N$ for some $y_1^*, y_2^* \in [0, 2^n)$ (unknown to the simulator). For $i = 1, 2$ write $y_i^* = x_i^* \| r_i^*$ with $r_i^* \in \{0,1\}^{s_1}$ and $x_i^* \in \{0,1\}^{m+s_0}$. If $\mathcal{A}$ is to have any information about the decryption of C^* we will show that it must either query the function H at a point r_i^* or issue a decryption query involving one of r_1^*, r_2^* as described below. First, we show that once the simulator receives a query for one of $H(r_1^*)$ or $H(r_2^*)$ it can easily deduce a square root of C^*. Given r_i^* we know that x_i^* is a root of $f(x) = (2^{s_1}x + r_i^*)^2 - C^* \bmod N$. Since $x_i^* < 2^{m+s_0} < \sqrt{N}$, the simulator can use Coppersmith's algorithm to find x_i^*. Then $y^* = x_i^* \| r_i^*$ is a square root of C^* as required.

Next, we give a high level description of how the simulator responds to $\mathcal{A}$'s decryption queries. Suppose the attacker issues a decryption query for the ciphertext C. Let $C = y^2 \bmod N$ for some $y \in [0, 2^n)$ and let r be the s_1 least significant bits of y. We will show that if C is a valid ciphertext, then $H(r)$ must already be defined (otherwise, with high probability, the string 0^{s_0} will not be found when unpadding y). Hence, the r used to create C must satisfy one of the following: (1) the attacker queried $H(r)$ prior to issuing the decryption query, or (2) $r = r_1^*$ or $r = r_2^*$. Suppose method (1) is used. Then when the decryption query is issued, the simulator already has r, which enables it to find the square root of C, as above. Suppose method (2) is used, i.e. $r = r_i^*$ for some $i \in \{1, 2\}$. In this case, assuming C is a valid ciphertext, we know that $y = y_i^* + 2^{s_0+s_1}\Delta$ for some $|\Delta| < 2^m < N^{1/4}$. Hence, define the two polynomials:

$$f(z) = z^2 - C^* \quad \text{and} \quad g(z, \Delta) = (z + 2^{s_0+s_1}\Delta)^2 - C$$

Then $f(y_i^*) = g(y_i^*, \Delta) = 0 \bmod N$. Therefore, Δ must be a root of the resultant $h = \mathrm{Res}_z(f, g)$ which is a quartic polynomial in Δ. Since $|\Delta| < N^{1/4}$ we can use Coppersmith's algorithm to find Δ. Using Δ the simulator easily finds y_i^* which is a square root of C^* as required. Hence, decryption queries for valid ciphertexts are either correctly answered or they lead directly to a square root of C^*.

We are now ready to describe the complete simulator for computing square roots. It works as follows:

Setup: The simulator gives $\mathcal{A}$ the value N as the public key to be attacked. It also gives $\mathcal{A}$ the security parameters m, s_0, s_1.

***H*-queries:** At any time $\mathcal{A}$ can query H at $r \in \{0,1\}^{s_1}$. The simulator needs to respond with $H(r)$. To respond to such H-queries the simulator maintains a list, called the H_{list}. The H_{list} is a list of tuples of the form $\langle z, H(z)\rangle$ that records all responses to previous H-queries. The H_{list} is initially empty. To respond to the query r the simulator works as follows:

Step 1: If r already appears as the left hand side of some tuple $\langle z, H(z)\rangle$ in the H_{list} then respond to $\mathcal{A}$ with $H(r) = H(z)$.

Step 2: Consider the polynomial $f(x) = (2^{s_1}x + r)^2 - C^*$. The simulator runs Coppersmith's algorithm to try to find a solution $|x_0| < 2^{m+s_0} < \sqrt{N}$ satisfying $f(x_0) = 0 \bmod N$. If a solution is found, the simulator outputs $2^{s_1}x_0 + r$ as the square root of C^* and terminates the simulation.

Step 3: Otherwise, the simulator picks a random $w \in \{0,1\}^{m+s_0}$ and sets $H(r) = w$. It adds the tuple $\langle r, w\rangle$ to the H_{list} and responds to $\mathcal{A}$ by saying $H(r) = w$.

Challenge: At some point $\mathcal{A}$ produces two plaintexts $M_0, M_1 \in \{0,1\}^m$ where it wishes to be challenged. The simulator responds with C^* as the challenge ciphertext.

Decryption queries: Let $C \in \mathbb{Z}_N$ be a ciphertext output by $\mathcal{A}$. The simulator must decrypt C or reject it as an invalid ciphertext. We construct a plaintext extractor to decrypt C. The plaintext extractor takes C, H_{list}, C^* as input and works as follows:

Step 1: For each tuple $\langle r, H(r)\rangle$ on the H_{list} consider the polynomial $f_r(X) = (2^{s_1}x + r)^2 - C$. The simulator runs Coppersmith's algorithm on each $f_r(x)$ to try to find an $|x_0| < \sqrt{N}$ satisfying $f_r(x_0) = 0 \bmod N$. Suppose an x_0 is found for some r_0 on the H_{list}. In this case, the simulator found a square root of C, namely $2^{s_1}x_0 + r_0$. Using $H(r_0)$ from the H_{list} the simulator checks that x_0 is a properly padded SAEP message. If so, it gives $\mathcal{A}$ the plaintext. If not, the simulator rejects C as an invalid ciphertext.

Step 2: Suppose no r_0 on the H_{list} is found. Consider the two polynomials

$$f(z) = z^2 - C^* \quad \text{and} \quad g(z, \Delta) = (z + 2^{s_0+s_1}\Delta)^2 - C$$

Let $h(\Delta)$ be the resultant of the two polynomials with respect to z. Then $h(\Delta)$ is a quartic polynomial. Use Coppersmith's algorithm to try to find a $\Delta_0 < 2^m < N^{1/4}$ such that $h(\Delta_0) = 0 \bmod N$. If such a Δ_0 is found then we know $f(y^*) = g(y^*, \Delta_0) = 0 \bmod N$ where y^* is some square root of C^*. Then the simulator can easily find y^* by computing the gcd of the univariate polynomials $f(z)$ and $g(z, \Delta_0)$. Since these two monic quadratic polynomials must be different (since $C \neq C^*$) their gcd must be a linear polynomial having y^* as a root. The simulator outputs y^* as the square root of C^* and terminates the simulation.

Step 3: If both Step 1 and Step 2 fail to resolve the decryption query, the ciphertext C is rejected as an invalid ciphertext. Note that Step 2 is only done in Phase 2 of the attack.

This completes the description of the simulator. The simulator's running time is as stated in the statement of Theorem 4. It remains to calculate the success probability of computing a square root of C^*. Let y_1^*, y_2^* be the two square roots of C^* mod N in $[0, 2^n)$. If C^* only has one such square root then set $y_1^* = y_2^*$. Let r_1^*, r_2^* be the s_1 least significant bits of y_1^*, y_2^* respectively. We are successful if during the simulation either: (1) $\mathcal{A}$ issues a query for one of $H(r_1^*), H(r_2^*)$, or (2) $\mathcal{A}$ issues a decryption query for a valid ciphertext $C \neq C^*$ where the s_1 least significant bits of some $\sqrt{C} \in [0, 2^n)$ equal r_1^* or r_2^*. If either one of these queries occurs during the attack we say that $\mathcal{A}$ issued an r^* query. We denote by $\mathcal{A}(r^*)$ the event that $\mathcal{A}$ issues an r^* query during the attack. Our goal is to show that during the simulation $\Pr_{sim}[\mathcal{A}(r^*)]$ is non-negligible.

Lemma 1. *Let $\mathcal{A}$ be a (t, q_D, q_H) chosen ciphertext attacker with adv(A) $\geq \epsilon$. Then* $\Pr_{sim}[\mathcal{A}(r^*)] \geq \epsilon(1 - \frac{2q_D}{2^{s_0}} - \frac{2q_D}{2^{s_1}})$.

Proof. We first note that during the real attack we have $\Pr_{real}[\mathcal{A}(r^*)] \geq \epsilon$. To see this observe that if $\mathcal{A}$ does not issue an r^* query during the real attack then the decryption of the challenge C^* is independent of $\mathcal{A}$'s view (since $H(r_1^*), H(r_2^*)$ are independent of $\mathcal{A}$'s view). Hence, since adv($\mathcal{A}$) $\geq \epsilon$, it follows that in the real attack $\mathcal{A}$ must make an r^* query with probability at least ϵ, i.e. $\Pr_{real}[\mathcal{A}(r^*)] \geq \epsilon$.

Next, we show that with high probability $\mathcal{A}$ cannot distinguish the real attack from the simulation until it issues an r^* query. We say that the event GoodSim occurred if the following two events happen:

- The simulator never rejects a valid decryption query issued by $\mathcal{A}$ (the validity of a query is determined relative to the oracle H at the end of the simulation), and
- During phase I of the attack (i.e. prior to being given the challenge) algorithm $\mathcal{A}$ did not issue a decryption query for C where $C = y^2 \bmod N$ and the s_1 least significant bits of $y \in [0, 2^n)$ are equal to r_1^* or r_2^*.

We show that when GoodSim occurs the simulation and the real attack are indistinguishable. We then show that GoodSim occurs with high probability.

Claim 1: $\Pr_{real}[\mathcal{A}(r^*)] = \Pr_{sim}[\mathcal{A}(r^*)|\text{GoodSim}]$.

Proof: We show that when GoodSim occurs $\mathcal{A}$'s view during the simulation is sampled from the same distribution as $\mathcal{A}$'s view during the real attack. By construction, all responses to H queries are as in a real attack. Similarly, when GoodSim occurs all responses to decryption queries are as in a real attack. Hence, the only thing to show is that the challenge C^* given by the simulator is sampled from the same distribution as in a real attack. Recall that C^* is generated by picking a random $\alpha \in [0, 2^n)$ and computing $C^* = \alpha^2 \bmod N$. For C^* to be an encryption of M_0 or M_1 we must introduce an implicit constraint on H, namely $H(r^*) = w^*$ for some $\langle r^*, w^* \rangle$. We show that w^* is uniform in $\{0,1\}^{m+s_0}$ and that $w^*, H(r^*)$ are both independent of the attacker's view at the end of phase I. Hence, setting $H(r^*) = w^*$ is consistent with a real attack. Proving this requires some care for the Rabin function.

Let $c \in \{1, 2\}$ be a random bit. If C^* has two square roots in $[0, 2^n)$ we use the bit c to pick one of them at random. Let y^* be the chosen square root (unknown to the simulator). By Fact 3 we know that y^* is uniform in $\{0,1\}^n$ (over the probability space induced by $\langle \alpha, c \rangle$). Write $y^* = x^* \| r^*$ with $x^* \in \{0,1\}^{m+s_0}$

and $r^* \in \{0,1\}^{s_1}$. Choose a random $b \in \{0,1\}$ and set $v^* = M_b \| 0^{s_0}$. The random bit b indicates whether C^* is an encryption of M_0 or M_1. Finally, set $H(r^*) = v^* \oplus x^*$. Since y^* is uniform in $[0, 2^n)$ we know that x^* is uniformly distributed in $\{0,1\}^{m+s_0}$. Hence, $v^* \oplus x^* \in \{0,1\}^{m+s_0}$ is a uniform random string. It is independent of $\mathcal{A}$'s view at the end of phase I as required since at that time C^* has not yet been used to answer any queries.

Next, we show that at the end of phase I (just before $\mathcal{A}$ receives the challenge) $H(r^*)$ is independent of $\mathcal{A}$'s view (otherwise we cannot set $H(r^*) = v^* \oplus x^*$). This is immediate by the following facts: (1) we may assume that during phase I the attacker does not issue a query for $H(r^*)$ since otherwise the event $\mathcal{A}(r^*)$ has already occurred and there is nothing more to prove. (2) the second part of GoodSim implies that during phase I the attacker did not issue a decryption query that restricts $H(r^*)$. Hence, at the end of phase I we know that $H(r^*)$ is independent of the attacker's view. This completes the proof of Claim 1.

Claim 2: $\Pr[\mathsf{GoodSim}] \geq 1 - \frac{2q_D}{2^{s_0}} - \frac{2q_D}{2^{s_1}}$.
Proof: Let C be a decryption query issued by the attacker and rejected by the simulator (i.e. C fails steps 1 and 2 of response to decryption queries). We show that the probability that C is valid is at most $2/2^{s_0}$. Let y_1, y_2 be the square roots of C in $[0, 2^n)$. Let M_1, r_1, x_1, t_1, v_1 and M_2, r_2, x_2, t_2, v_2 be the unpadding of y_1, y_2 as defined in Section 2. Then C is a valid ciphertext only if either $t_1 = 0^{s_0}$ or $t_2 = 0^{s_0}$. Since C failed to satisfy the condition of Step 1 we know that $\mathcal{A}$ has not yet issued a query for $H(r_1)$ or $H(r_2)$. Since C failed to satisfy Step 2 we know that $r_1, r_2 \neq r_1^*$ and $r_1, r_2 \neq r_2^*$. Hence, $H(r_1)$ and $H(r_2)$ are independent of the attacker's current view. Therefore, the probability that $t_1 = 0^{s_0}$ or $t_2 = 0^{s_0}$ is at most $2/2^{s_0}$. Since the attacker makes at most q_D queries, the probability that any of these queries are incorrectly rejected is at most $2q_D/2^{s_0}$.

To bound the probability for the second part of GoodSim observe that during phase I the challenge C^* is independent of the attacker's view. Therefore, the probability that a decryption query during phase I happened to use r_1^* or r_2^* is at most $2/2^{s_1}$. Therefore, the probability that any of the queries during phase I use r_1^* or r_2^* is at most $2q_D/2^{s_1}$. To conclude we have that $\Pr[\mathsf{GoodSim}] \geq 1 - 2q_D/2^{s_0} - 2q_D/2^{s_1}$ as required. This completes the proof of Claim 2.
The proof of the lemma now follows from Claims 1 and 2:

$$\Pr_{sim}\left[\mathcal{A}(r^*)\right] \geq \Pr_{sim}\left[\mathcal{A}(r^*)\middle|\mathsf{GoodSim}\right] \cdot \Pr\left[\mathsf{GoodSim}\right] =$$

$$\Pr_{real}[\mathcal{A}(r^*)] \cdot \Pr\left[\mathsf{GoodSim}\right] \geq \epsilon(1 - \frac{2q_D}{2^{s_0}} - \frac{2q_D}{2^{s_1}})$$

As required. This concludes the proof of Lemma 1 and Theorem 4. □

Extensions. SAEP is not known to be secure for the general RSA trapdoor permutation, $f(x) = x^e \bmod N$. For very small RSA exponents one can show some limited security. For example, for $e = 3$ SAEP has chosen ciphertext security whenever $m + s_0 < n/3$ and $m < n/9$. For typical RSA modulus sizes, these restrictions on the message length make it difficult to use this system.

5 Proof of Security for RSA-SAEP+ and Rabin-SAEP+

The proof of security for SAEP+ holds in a more general settings than the proof of SAEP. As in the previous section, we use m, s_0, s_1 as the security parameters of SAEP+ and set $n = m + s_0 + s_1$.

Let $f(x,r)$ be a trapdoor permutation acting on strings in $\{0,1\}^{m+s_0} \times \{0,1\}^{s_1}$. As usual we assume f is selected from a family $\mathcal{F}$ of such trapdoor permutations. Following the notation of [8] we define the set partial one-wayness problem as follows:
Set partial one-wayness: We say that an algorithm $\mathcal{A}$ solves the (f,k) partial one-wayness problem if given $f(x,r)$ the algorithm produces a set $S = \{r_1,\ldots,r_k\} \subseteq \{0,1\}^{s_1}$ such that $r \in S$. More precisely, we say that $\mathcal{A}$ has advantage ϵ if

$$\mathsf{adv}^{p-ow}(\mathcal{A}) = Pr_{x,r}[r \in \mathcal{A}(f(x,r))] \geq \epsilon$$

Consider the $f-$SAEP+ cryptosystem obtained by padding the message M with SAEP+ prior to encrypting with f. We first show that a successful chosen ciphertext attacker on $f-$SAEP+ can be used to solve the set partial one-wayness problem for f. We then discuss the applications to the RSA and Rabin functions.

Theorem 5. *Let $\mathcal{A}$ be a (t, q_D, q_H, q_G) chosen ciphertext attack algorithm in the random oracle model. Suppose $\mathcal{A}$ has advantage ϵ when attacking $f -$ SAEP+. Then there is a uniform algorithm $\mathcal{B}$ for solving the (f, q_H) set partial one-wayness problem with the following parameters:*

$$time(\mathcal{B}) \leq time(\mathcal{A}) + O(q_H + q_G + q_D)$$
$$adv^{p-ow}(\mathcal{B}) \geq adv(\mathcal{A})(1 - q_D/2^{s_0} - q_D/2^{s_1})$$

Proof. Algorithm $\mathcal{B}$ is given $C^* = f(x^*, r^*)$ for some random $x^*\|r^* \in \{0,1\}^n$. Our goal is to output a list of size q_H containing r^*. We construct a simulator that interacts with algorithm $\mathcal{A}$ and produces the required output. Note that since f is a permutation, $x^*\|r^*$ is unique given C^*.

We first give a high level description of the simulator. During the simulation, $\mathcal{A}$ outputs two plaintexts M_0, M_1 where it wishes to be challenged. The simulator responds with C^* as the challenge ciphertext. We view C^* as the encryption of M^*, where M^* is one of the two challenge plaintexts M_0, M_1. We will show that if $\mathcal{A}$ is to have any information about the decryption of C^* it must query the function H at the point r^*. Therefore, if we place all of $\mathcal{A}$'s queries to H in a list, called the H_{list}, then with non-negligible probability the H_{list} is a solution to the set partial one-wayness problem.

Next, we show how to respond to decryption queries. Say the attacker wishes to decrypt the ciphertext C. Suppose C is a valid ciphertext, and is the encryption of some message M. Furthermore, let $C = f(x,r)$. We will show that if C is a valid ciphertext, then both $G(M,r)$ and $H(r)$ are already defined. Hence, the r used to create C must satisfy one of the following: (1) the attacker queried $G(M,r)$ and $H(r)$ prior to issuing the decryption query, or (2) $r = r^*$ and $M = M^*$. Suppose method (1) is used. Then when the decryption query is issued, the simulator has already been queried on $G(M,r)$. Hence, to decrypt C

the simulator simply checks to see which pair $\langle M, r\rangle$ on the list of queries to G is the decryption of C. Suppose method (2) is used, i.e. $r = r^*$ and $M = M^*$. In this case $C = C^*$ and hence this is an invalid decryption query since it matches the challenge ciphertext. Consequently, all decryption queries can be correctly answered.

We now give the detailed description of the simulator $\mathcal{B}$.

Setup: The simulator gives $\mathcal{A}$ the security parameters m, s_0, s_1, and identifies the function f within the family of trapdoor permutations $\mathcal{F}$.

H-queries: At any time $\mathcal{A}$ can query H at $r \in \{0,1\}^{s_1}$. The simulator needs to respond with $H(r)$. To respond to such H-queries the simulator maintains a list, called the H_{list}. The H_{list} is a list of tuples of the form $\langle z, H(z)\rangle$ that records all responses to previous H-queries. The H_{list} is initially empty. To respond to the query r the simulator works as follows:

Step 1: If r already appears as the left hand side of some tuple $\langle z, H(z)\rangle$ in the H_{list} then respond to $\mathcal{A}$ with $H(r) = H(z)$.

Step 2: Otherwise, the simulator picks a random $w \in \{0,1\}^{m+s_0}$ and sets $H(r) = w$. It adds the tuple $\langle r, w\rangle$ to the H_{list} and responds to $\mathcal{A}$ by saying $H(r) = w$.

G-queries: At any time $\mathcal{A}$ can query G at $G(M_0, r_0)$ where $M_0 \in \{0,1\}^m$ and $r_0 \in \{0,1\}^{s_1}$. The simulator needs to produce $G(M_0, r_0)$. To respond to such G-queries the simulator maintains a list, called the G_{list}. It is a list of tuples of the form $\langle M, r, G(M,r), C\rangle$ that records all responses to previous G-queries. The last entry, C, is the ciphertext that results from encrypting M using the random string r (see Step 2 below). The G_{list} is initially empty. To respond to the query (M_0, r_0) the simulator works as follows:

Step 1: If (M_0, r_0) appears as the left hand side of some tuple $\langle M_0, r_0, u, C\rangle$ in the G_{list} then respond to $\mathcal{A}$ with $G(M_0, r_0) = u$.

Step 2: Otherwise, the simulator picks a random $u \in \{0,1\}^{s_0}$ and sets $G(M_0, r_0) = u$. It then runs the algorithm for responding to an H query to obtain the value of $H(r_0)$. The simulator then computes $C_0 = f(\mathsf{SAEP}^+(M_0, r_0))$, which is the ciphertext obtained from encrypting M_0 using r_0. Note that at this point $H(r_0)$ and $G(M_0, r_0)$ are well defined, so that C_0 is well defined. The simulator adds $\langle M_0, r_0, u, C_0\rangle$ to the G_{list} and responds to $\mathcal{A}$ by saying $G(M_0, r_0) = u$.

Challenge: At some point $\mathcal{A}$ produces two plaintexts $M_0, M_1 \in \{0,1\}^m$ where it wishes to be challenged. The simulator responds with C^* as the challenge ciphertext.

Decryption queries: Let $C \in \mathbb{Z}_N$ be a ciphertext output by $\mathcal{A}$. The simulator must decrypt C or reject it as an invalid ciphertext. We construct a plaintext extractor to decrypt C. The plaintext extractor is very simple: search the G_{list} to see if it contains a tuple $\langle M, r, u, C\rangle$ with C as the last entry. If so, respond with M as the decryption of C. Otherwise, reject the ciphertext as an invalid ciphertext.

This completes the description of the simulator. Algorithm $\mathcal{B}$ outputs the H_{list} at the end of the simulation as its solution to the given set partial one-wayness problem. One can easily verify that the running time of $\mathcal{B}$ is as stated in the statement of the theorem. We are assuming that searching the H_{list} and G_{list} takes constant time.

It remains to calculate the probability that r^* is contained in one of the tuples on the final H_{list}. This happens if $\mathcal{A}$ issues a query for $H(r^*)$ or a query for $G(-, r^*)$. We denote the probability of this event by $\Pr_{sim}[r^* \in H_{list}]$. We note that once the attacker queries $H(r^*)$ it can easily distinguish the simulation from a real attack: the simulator defines $H(r^*)$ to be a random string, but then C^* is unlikely to be the encryption of M_0 or M_1. Hence, the attacker may choose to abort the attack. However, at that point r^* is already in the H_{list} as required. The next lemma shows that $\Pr_{sim}[r^* \in H_{list}]$ is sufficiently large.

Lemma 2. *Let $\mathcal{A}$ be a (t, q_D, q_H, q_G) chosen ciphertext attacker for $f-\mathsf{SAEP}^+$ with advantage ϵ.*
Then $\Pr_{sim}[r^* \in H_{list}] \geq \epsilon(1 - q_D/2^{s_0} - q_D/2^{s_1})$.

Proof As in the proof of Lemma 1 we have that in the real attack $\Pr_{real}[r^* \in H_{list}] \geq \epsilon$. It remains to show that with high probability $\mathcal{A}$ cannot distinguish the simulation from the real attack until it issues a query for $H(r^*)$ or $G(-, r^*)$. Let $\mathsf{GoodSim}$ be the event defined as in the proof of Lemma 1, namely we say that the event $\mathsf{GoodSim}$ occurred if the following two events happen:

- The simulator never rejects a valid decryption query issued by $\mathcal{A}$ (the validity of a query is determined relative to the oracle H at the end of the simulation), and
- During phase I of the attack (i.e. prior to being given the challenge) algorithm $\mathcal{A}$ did not issue a decryption query for C where $C = f(x, r^*)$ for some $x \in \{0,1\}^{m+s_0}$.

Claim 1: $\Pr_{real}[r^* \in H_{list}] = \Pr_{sim}[r^* \in H_{list} | \mathsf{GoodSim}]$.
Proof: We show that when $\mathsf{GoodSim}$ occurs $\mathcal{A}$'s view during the simulation is sampled from the same distribution as $\mathcal{A}$'s view during the real attack. Observe that the simulator provides a perfect simulation of the H and G oracles. Also, when $\mathsf{GoodSim}$ occurs all decryption queries are answered correctly. Next we show that the challenge ciphertext C^* given to $\mathcal{A}$ is distributed as in the real attack. Recall that x^*, r^* are chosen at random. Let M_0, M_1 be the messages on which $\mathcal{A}$ wishes to be challenged. Pick a random $b \in \{0,1\}$. We make C^* be the encryption of M_b. To do so, pick a random $t^* \in \{0,1\}^{s_0}$ and define $G(M_b, r^*) = t^*$. Set $v^* = M_b \| t^*$ and define $H(r^*) = v^* \oplus x^*$. Then C^* is the encryption of M_b. Furthermore, t^* and $v^* \oplus x^*$ are random strings independent of $\mathcal{A}$'s view at the end of phase I as required. To complete the proof we need to argue that at the end of phase I the hash values $G(M_b, r^*)$ and $H(r^*)$ are independent of the attacker's view (otherwise we cannot set $G(M_b, r^*) = t^*$ and $H(r^*) = v^* \oplus x^*$). We do so in the same way as at the end of Claim 1 of Lemma 1.
Claim 2: $\Pr[\mathsf{GoodSim}] \geq 1 - \frac{q_D}{2^{s_0}} - \frac{q_D}{2^{s_1}}$.
Proof: Let C be a decryption query issued by the attacker and rejected by the simulator. Let $C = f(x, r)$, and let M, t, v be the unpadding of $x \| r$ as described in Section 2. Then C is a valid ciphertext only if $t = G(M, r)$. Since C is rejected by the simulator we know that the attacker did not issue a query for $G(M, r)$. Similarly, since $C \neq C^*$ we know that $\langle M, r \rangle$ is not equal to $\langle M_b, r^* \rangle$. Hence, $G(M, r)$ is independent of the attacker's current view. Therefore, the probability

that $t = G(M, r)$ is $1/2^{s_0}$. Since the attacker makes at most q_D queries, the probability that any decryption query is incorrectly rejected is at most $q_D/2^{s_0}$. We bound the probability for the second part of GoodSim as we did in the proof of Claim 2 of Lemma 1. Overall, we get that $\Pr[\mathsf{GoodSim}] \geq 1 - q_D/2^{s_0} - q_D/2^{s_1}$ as required. This concludes the proof of Claim 2.
The proof of the lemma now follows from Claims 1 and 2 as in the calculation at the end of Lemma 1. This concludes the proof of Theorem 5. □

We now describe how Theorem 5 applies to the Rabin and RSA functions. For the Rabin function we obtain an extremely efficient reduction to factoring. For the RSA permutation we obtain a reduction to breaking RSA, but the reduction is not as efficient. Since the Rabin function is not a permutation on $\mathbb{Z}_N^*$ one needs to extend the proof of Theorem 5 to this case. The extension is done using the same techniques as in Theorem 4. Theorem 5 remains unchanged.

Corollary 1 (Rabin-SAEP$^+$). *Consider the Rabin-SAEP$^+$ scheme, with $m + s_0 < n/2$. Suppose the (n, t, ϵ) factoring assumption holds. Then Rabin-SAEP$^+$ is $(t', \epsilon', q_D, q_H, q_G)$ chosen ciphertext secure in the random oracle model for t', ϵ' satisfying:*

$$t' \leq t - O(q_D + q_G + q_H T_C), \quad \textit{and}$$
$$\tfrac{1}{6}\epsilon' \geq \epsilon + q_D/2^{s_0} + q_D/2^{s_1}$$

where $T_C = T_C(n, 2)$.

Proof Suppose $\mathcal{A}$ is a (t', q_D, q_H, q_G) chosen ciphertext attacker on Rabin-SAEP$^+$ with advantage ϵ'. Let f_N be the function $f_N(x) = x^2 \bmod N$ for some N generated by the Rabin-SAEP$^+$ key-gen algorithm. By Theorem 5 there exists a t_0-time algorithm $\mathcal{B}$ that solves the (f_N, q_H) set partial one-wayness problem with advantage ϵ_0 for some t_0, ϵ_0.

We construct an algorithm $\mathcal{C}$ for factoring N. The algorithm starts by picking a random $\alpha \in [0, 2^n)$ and computing $C^* = \alpha^2 \bmod N$. It then runs $\mathcal{B}$ on input C^*. With probability at least ϵ_0 we obtain a set $S = \{r_1, \ldots, r_{q_H}\} \subseteq \{0,1\}^{s_1}$ of size q_H with the following property: there exists an integer $x \in [0, 2^{m+s_0})$ and $r \in S$ such that $(2^{s_1}x + r)^2 = C^* \bmod N$. Since $x < \sqrt{N}$ we can then find x, r by running Coppersmith's algorithm on all q_H candidates for r. Once x, r are found, we obtain a square root $\alpha' \in [0, 2^n)$ of $C^* \bmod N$. Then the factorization of N is revealed with probability at least $1/6$ by computing $\gcd(N, \alpha - \alpha')$. To see this observe that by Fact 2, $C^* \bmod N$ has two square roots in $[0, 2^n)$ with probability at least $1/3$. Therefore, $\alpha \neq \alpha'$ with probability $1/6$. Since $0 \leq \alpha, \alpha' < N/2$ the GCD gives a non-trivial factor of N. The resulting factoring algorithm $\mathcal{C}$ has running time: $\mathrm{time}(\mathcal{C}) = t_0 + q_H T_C = t' + O(q_D + q_G + q_H T_C)$ and success probability at least $\mathrm{adv}(\mathcal{C}) = \frac{1}{6}\epsilon_0 = \frac{1}{6}\epsilon'(1 - q_D/2^{s_0} - q_D/2^{s_1})$. The corollary now follows. □

Corollary 2 (RSA-SAEP$^+$). *Consider the RSA-SAEP$^+$ scheme, with $m + s_0 < n/2$. Suppose the (n, e, t, ϵ) RSA assumption holds for some $e > 0$. Then RSA-SAEP$^+$ is $(t', \epsilon', q_D, q_H, q_G)$ chosen ciphertext secure in the random oracle model for t', ϵ' satisfying:*

$$t' \le t/2 - O(q_D + q_G + q_H^2), \quad and$$
$$\epsilon' \ge \epsilon^{1/2} + q_D/2^{s_0} + q_D/2^{s_1}$$

Proof Suppose $\mathcal{A}$ is a (t', q_D, q_H, q_G) chosen ciphertext attacker with advantage ϵ'. Let f_N be the function $f_N(x) = x^e \bmod N$ for some N generated by the RSA-SAEP+ key-gen algorithm. By Theorem 5 there exists a t_0-time algorithm $\mathcal{B}$ that solves the (f_N, q_H) set partial one-wayness problem with advantage ϵ_0 for some t_0, ϵ_0. Fujisaki et al. [8] show that, when $m + s_0 < n/2$, such an algorithm can be used to compute the e'th root of C^* modulo N. They do so by running algorithm $\mathcal{B}$ on both C^* and αC^* for a random $\alpha \in \mathbb{Z}_N$. The resulting sets S and S_α expose the e'th root of C^* in time $O(q_H^2)$. Hence, we obtain an algorithm for breaking RSA in time $2t_0 = 2t' + O(q_D + q_G + q_H^2)$ and success probability $\epsilon_0^2 = (\epsilon'(1 - q_D/2^{s_0} - q_D/2^{s_1}))^2 \ge (\epsilon' - q_D/2^{s_0} - q_D/2^{s_1})^2$. The corollary now follows. □

Note that the reduction time for RSA-SAEP+ is quadratic in q_H and the success probability is quadratic in ϵ. This is not as efficient as the reduction for Rabin-SAEP+ which is linear time.

Accommodating large messages in RSA-SAEP+. Note that in Corollary 2 the message length must satisfy $m+s_0 < n/2$. We briefly note that the corollary remains true even if $m+s_0 < (1-\delta)n$ for any fixed $\delta > 0$. To do so run algorithm $\mathcal{B}$ on $c = 1/\delta$ random values $\alpha_1 C^*, \dots, \alpha_c C^*$. We obtain c lists of size q_H each. Suppose we find a c-tuple $c^* = \langle r_1^*, \dots r_c^* \rangle$ (one entry from each list) that is the correct solution to these c partial one-wayness problems. Then we obtain the δn least significant bits of each $\alpha_i C^* \bmod N$ where the α_i are random in $\mathbb{Z}_N$. Finding C^* from this tuple is a standard Hidden Number Problem (HNP) modulo N. We can use the algorithm in [4] to efficiently find C^*. The analysis in [4], which applies to HNP modulo primes, extends to handle RSA composites $N = pq$ as well. The resulting algorithm for breaking RSA has a running time of $O(q_H^c)$, since we must try all c-tuples c^*, and a success probability of $O(\epsilon^c)$, since $\mathcal{B}$ must succeed on all c iterations. Consequently, this reduction becomes very inefficient for small δ.

6 Conclusions

We showed that OAEP can be simplified significantly when applied to the Rabin and RSA functions. OAEP can be viewed as two rounds of a Feistel network. The simplified schemes, SAEP and SAEP+, require only one round of Feistel. The proof of security for the two schemes is based on the algebraic properties of the Rabin and RSA functions. When using an n-bit modulus Rabin-SAEP is secure whenever $m + s_0 < n/2$ and $m < n/4$. SAEP+ is secure whenever $m + s_0 < n/2$. The proof of security for RSA-SAEP+ has the same efficiency as the proof for RSA-OAEP [8]. For Rabin-SAEP+ the proof is as efficient as the proof for Rabin-OAEP+ [11].

The padding SAEP+ is superior to SAEP both in terms of the reduction efficiency and in terms of the weaker restriction on the message length. For practical purposes one is most likely to use SAEP$^+$ rather than SAEP. Nevertheless, it is useful to know that Rabin-SAEP, which is a slightly simpler construction, also provides chosen ciphertext security when appropriate parameters are used.

Acknowledgments. The author thanks David Pointcheval, Jacques Stern, Victor Shoup, and Phong Nguyen for helpful discussions.

References

1. M. Bellare, P. Rogaway, "Random oracles are practical: a paradigm for designing efficient protocols", In ACM conference on Computers and Communication Security, pp. 62–73, 1993.
2. M. Bellare, P. Rogaway, "Optimal asymmetric encryption", Eurocrypt '94, pp. 92–111, 1994.
3. M. Bellare, A. Desai, D. Pointcheval, P. Rogaway, "Relations among notions of security for public-key encryption schemes", in proc. Crypto '98, pp. 26–45, 1998.
4. D. Boneh, R. Venkatesan, "Hardness of computing the most significant bits of secret keys in Diffie-Hellman and related schemes", in proc. Crypto '96, pp. 129–142, 1996.
5. R. Canetti, O. Goldreich, S. Halevi, "The random oracle model, revisited", in proc. STOC '98.
6. D. Coppersmith. Small solutions to polynomial equations, and low exponent RSA vulnerabilities. *Journal of Cryptology*, vol. 10, pp. 233–260, 1997.
7. D. Dolev, C. Dwork, M. Naor, "Non-malleable cryptography", SIAM J. of Computing, Vol. 30(2), pp. 391–437, 2000.
8. E. Fujisaki, T. Okamoto, D. Pointcheval, J. Stern, "RSA-OAEP is secure under the RSA assumption", In proc. Crypto '2001, Springer-Verlag, 2001.
9. A. Menezes, P. van Oorschot and S. Vanstone, *Handbook of Applied Cryptography*, CRC Press, 1996.
10. J. Manger, "A chosen ciphertext attack on RSA Optimal Asymmetric Encryption Padding (OAEP) as standardized in PKCS #1", In proc. Crypto '2001.
11. V. Shoup, "OAEP reconsidered", In proc. Crypto '2001, Springer-Verlag, 2001.
12. C. Rackoff, D. Simon, "Non-interactive zero-knowledge proof of knowledge and chosen ciphertext attack", in proc. Crypto '91, pp. 433–444, 1991.

Online Ciphers and the Hash-CBC Construction

Mihir Bellare[1], Alexandra Boldyreva[1], Lars Knudsen[2], and Chanathip Namprempre[1]

[1] Department of Computer Science & Engineering
University of California, San Diego
La Jolla, California 92093
{mihir,aboldyre,meaw}@cs.ucsd.edu
http://www-cse.ucsd.edu/users/{mihir,aboldyre,cnamprem}
[2] Department of Informatics
PB 7800, N-5020 Bergen, Norway
lars@ramkilde.com
http://www.ramkilde.com

Abstract. We initiate a study of on-line ciphers. These are ciphers that can take input plaintexts of large and varying lengths and will output the ith block of the ciphertext after having processed only the first i blocks of the plaintext. Such ciphers permit length-preserving encryption of a data stream with only a single pass through the data. We provide security definitions for this primitive and study its basic properties. We then provide attacks on some possible candidates, including CBC with fixed IV. Finally we provide a construction called HCBC which is based on a given block cipher E and a family of AXU functions. HCBC is proven secure against chosen-plaintext attacks assuming that E is a PRP secure against chosen-plaintext attacks.

1 Introduction

We begin by saying what we mean by on-line ciphers. We then describe a notion of security for them, and discuss constructions and analyses. Finally, we discuss usage, applications, and related work.

1.1 Online Ciphers

A *cipher* over domain D is a function $F\colon \{0,1\}^k \times D \to D$ such that for each key K the map $F(K,\cdot)$ is a length-preserving permutation on D, and possession of K enables one to both compute and invert $F(K,\cdot)$. The most popular examples are block ciphers, where $D = \{0,1\}^n$ for some n called the block length; these are fundamental tools in cryptographic protocol design. However, one might want to encipher data of large size, in which case one needs a cipher whose domain D is appropriately large. (A common choice, which we make, is to set the domain to $D_{d,n}$, the set of all strings having a length that is at most some large value d, and is also divisible by n.) Matyas and Meyer refer to these as "general" ciphers [10].

J. Kilian (Ed.): CRYPTO 2001, LNCS 2139, pp. 292–309, 2001.

In this paper, we are interested in general ciphers that are computable in an on-line manner. Specifically, cipher F is said to be *on-line* if the following is true. View the input plaintext $M = M[1] \ldots M[l]$ to an instance $F(K, \cdot)$ of the cipher as a sequence of n-bit blocks, and similarly for the output ciphertext $F(K, M) = C[1] \ldots C[l]$. Then, given the key K, for all i, it should be possible to compute output block $C[i]$ after having seen input blocks $M[1] \ldots M[i]$. That is, $C[i]$ does not depend on blocks $i + 1, \ldots, l$ of the plaintext.

An on-line cipher permits real-time, length-preserving encryption of a data stream without recourse to buffering, which can be attractive in some practical settings.

The intent of this paper is to find efficient, proven secure constructions of on-line ciphers and to further explore the applications. Let us now present the relevant security notions and our results.

1.2 A Notion of Security for Online Ciphers

A commonly accepted notion of security to target for a cipher is that it be a pseudorandom permutation (PRP), as defined by Luby and Rackoff [9]. Namely, for a cipher F to be a PRP, it should be computationally infeasible, given an oracle g, to have non-negligible advantage in distinguishing between the case where g is a random instance of F and the case where g is a randomly-chosen, length-preserving permutation on the domain of the cipher. However, if a cipher is on-line, then the ith block of the ciphertext does not depend on blocks $i + 1, i + 2, \ldots$ of the plaintext. This is necessary, since otherwise it would not be possible to output the ith ciphertext block having seen only the first i plaintext blocks. Unfortunately, this condition impacts security, since a cipher with this property certainly cannot be a PRP. An easy distinguishing test is to ask the given oracle g the two-block queries AB and AC, getting back outputs WX and YZ respectively, and if $W = Y$ then bet that g is an instance of the cipher. This test has a very high advantage since the condition being tested fails with high probability for a random length-preserving permutation.

For an on-line cipher, then, we must give up on the requirement that it meet the security property of being a PRP. Instead, we define and target an appropriate alternative notion of security. This is quite natural; we simply ask that the cipher behave "as randomly as possible" subject to the constraint of being on-line. We say that a length-preserving permutation π is *on-line* if for all i the ith output block of π depends only on the first i input blocks to π, and let $\mathsf{OPerm}_{d,n}$ denote the set of all length-preserving permutations π on domain $D_{d,n}$. The rest is like for a PRP, with members of this new set playing the role of the "ideal" objects to which cipher instances are compared: it should be computationally infeasible, given an oracle g, to have non-negligible advantage in distinguishing between the case where g is a random instance of F and the case where g is a random member of $\mathsf{OPerm}_{d,n}$. A cipher secure in this sense is called an on-line-PRP.

The fact that an on-line-PRP meets a notion of security that is relatively weak compared to a PRP might at first lead one to question the introduction of such a notion. However, finding appropriate balances between security and practical constraints is an impactful and active research endeavor where the goal is not necessarily to achieve some strong notion of security but to have the "best possible" security under given practical constraints, so that weaker notions of security are useful. Furthermore, we will see that in this case, even this weak primitive, if properly used, can provide strong security.

1.3 Candidates for Online Ciphers

To the best of our knowledge, the problem of designing on-line ciphers with security properties as strong as those required by our definition has not been explicitly addressed before. When one comes to consider this problem, however, it is natural to test first some existing candidate ciphers or natural constructions from the literature. We consider some of them and present attacks that are helpful to gather intuition about the kinds of security properties we are seeking.

It is natural to begin with standard modes of operation of a block cipher, such as CBC. However, CBC is an encryption scheme, not a cipher; each invocation chooses a new random initial vector as a starting point and makes this part of the ciphertext. In particular, it is not length-preserving. The natural way to modify it to be a cipher is to fix the initial vector. There are a couple of choices: make it a known public value, or, hopefully better for security, make it a key that will be part of the secret key of the cipher. The resulting ciphers are certainly on-line, but they do not meet the notion of security we have defined. In other words, the CBC cipher with fixed IV, whether public or private, can be easily distinguished from a random on-line permutation. Attacks demonstrating this are provided in Section 4.

We then consider the Accumulated Block Chaining (ABC) mode proposed by Knudsen in [7], which is a generalization of the Infinite Garble Extension mode proposed by Campbell [5]. It was designed to have "infinite error propagation," a property that intuitively seems necessary for a secure on-line cipher but which, as we will see, is not sufficient. In Section 4, we present attacks demonstrating that this is not a secure on-line cipher.

1.4 The HCBC Online Cipher and Its Security

We seek a construction of a secure on-line cipher based on a given block cipher $E\colon \{0,1\}^{ek} \times \{0,1\}^n \to \{0,1\}^n$. We provide a construction called HCBC that uses a family $H\colon \{0,1\}^{hk} \times \{0,1\}^n \to \{0,1\}^n$ of Almost-XOR-Universal (AXU) hash functions [8]. The key $eK\|hK$ for an instance $\mathsf{HCBC}(eK\|hK, \cdot)$ of the cipher consists of a key eK for the block cipher and a key hK specifying a member $H(hK, \cdot)$ of the family H. The construction is just like CBC, except that a ciphertext block is first hashed via $H(hK, \cdot)$ before being XORed with the next plaintext block. (The initial vector is fixed to 0^n.) A picture is in Figure 3, and a full description of the construction is in Section 6. It is easy to see that this cipher is on-line.

We stress that the hash functions map n bits to n bits, meaning work on inputs of the block length, as does the given block cipher. Numerous designs of fast AXU families are known, so that our construction is quite efficient. For an overview of the state-of-the-art of AXU families refer to [12].

We prove that HCBC meets the notion of security for an on-line cipher that we discussed above, assuming that the underlying block cipher E is a PRP. The proof involves finding and exploiting a way of looking at an on-line cipher as a 2^n-ary tree of permutations on n bits, and then going through a hybrid argument involving a sequence of different games that "move" from $\mathsf{OPerm}_{d,n}$ to HCBC.

1.5 Security against Chosen Ciphertext Attacks

The notions of PRPs and on-line PRPs that we have discussed above represent security under chosen-plaintext attack. A stronger requirement is security under chosen-ciphertext attack. For a PRP this means that the adversary has an oracle not just for the challenge permutation, but also for its inverse. (An object secure in this sense was called a strong PRP in [11] and a super-PRP in [9].) This notion is easily adapted to yield a notion of on-line PRPs secure against chosen-ciphertext attack. We provide an attack showing that HCBC is not secure against chosen-ciphertext attack. The question of finding a construction of an on-line PRP secure against chosen ciphertext attack, based on a block cipher assumed to be a PRP secure against chosen-ciphertext attack, is open. In the full version of this paper [1] we report on some efforts to this end.

1.6 Usage and Application of Online Ciphers

There are settings in which the input plaintext is being streamed to a device that has limited memory for buffering and wants to produce output at the same rate at which it is getting input. The on-line property becomes desirable in these settings. The most direct usage of an on-line cipher will be in settings where, additionally, there is a constraint requiring the length of the ciphertext to equal the length of the plaintext. (Otherwise, one can use a standard mode of encryption like CBC, since it has the on-line property. But it is length expanding in the sense that the length of the ciphertext exceeds that of the plaintext, due to the changing initial vector.) This type of constraint occurs when one is dealing with fixed packet formats or legacy code.

However, an on-line cipher is more generally useful, via the "encode-then-encipher" paradigm discussed in [4]. This paradigm was presented for ciphers that are PRPs, and says that enciphering yields an IND-CPA secure encryption scheme if the message space has enough entropy, and provides integrity (meaning achieves INT-CTXT) if the message space contains enough redundancy. (The privacy requires that the PRP be secure against chosen-plaintext attack, while the integrity requires security against chosen-ciphertext attack.) Entropy and redundancy might be present in the data, as often happens when enciphering structured data like packets, which have fixed formats and often contain counters. Or, entropy and redundancy can be explicitly added, for example by inserting a random value and a constant string in the message. (This will of course increase the size of the plaintext, so is only possible when data expansion is permitted.)

Claims similar to those made in [4] remain true even if the cipher is an on-line-PRP rather than a PRP. Specifically, the requirement on the message space must be strengthened to require not just that entropy be present, but that it be in the first blocks of the message; and similarly, that redundancy not just be present, but be at the end of the data. Again, one might already have data of such structure, in which case the encryption will be length preserving yet provide semantic security and integrity, or one can prepend a random number and append a constant to the message, getting the same properties but at the cost of data expansion.

1.7 Related Work

The problem addressed by our Hash-CBC construction is that of building a general cipher from a block cipher. Naor and Reingold [11] consider this problem for the case where the general cipher is to be a PRP or strong PRP, while we want the general cipher to be an on-line-PRP or strong-on-line-PRP. The constructions of [11, Section 7] are not on-line; indeed, they cannot be, since they achieve the stronger security notion of a PRP. Our construction, however, follows that of [11] in using hash functions in combination with block ciphers. A problem that has received a lot of attention is to take a PRP and produce another having twice the input block length of the original [9,11]. We are, however, interested in allowing inputs of varying and very large size, not merely twice the block size.

2 Definitions

We recall basic definitions of families of functions and ciphers following [2].

NOTATION. A *string* is a member of $\{0,1\}^*$. If x is a string, then $|x|$ denotes its length. The empty string is denoted ε. If $x, y \in \{0,1\}^*$ are strings, then we denote by $\mathsf{LCP}_n(x,y)$ the *longest common n-prefix* of x, y. This is the longest string s such that $|s|$ is a multiple of n, and s is a prefix of both x and y. A map $f\colon D \to R$ is a *permutation* if $D = R$ and f is a bijection (i.e. one-to-one and onto). A map $f\colon D \to R$ is *length-preserving* if $|f(x)| = |x|$ for all $x \in D$. If $n \geq 1, d \geq 1$ are integers, then $D_{d,n}$ denotes the set of all strings whose length is a positive multiple of n bits and at most dn bits. If $P \in D_{d,n}$, then $P[i]$ denotes its ith block, meaning $P = P[1] \dots P[l]$ where $l = |P|/n$ and $|P[i]| = n$ for all $i = 1, \dots, l$. We will typically consider functions whose inputs and outputs are in $D_{d,n}$, so that both are viewed as sequences of blocks where each block is n bits long. We let $f^{(i)}$ denote the function which on input M returns the ith block of $f(M)$. (Or ε if $|f(M)| < ni$.)

FUNCTION FAMILIES AND CIPHERS. A *family of functions* is a map $F\colon\ Keys(F) \times Dom(F) \to Ran(F)$ where $Keys(F)$ is the *key space* of F; $Dom(F)$ is the *domain* of F; and $Ran(F)$ is the *range* of F. If $Keys(F) = \{0,1\}^k$, then we refer to k as the key-length. The two-input function F takes a key $K \in \mathit{Keys}(F)$ and an input $x \in \mathit{Dom}(F)$ to return a point $F(K,x) \in \mathit{Ran}(F)$. For each key $K \in \mathit{Keys}(F)$, we define the map $F_K\colon \mathit{Dom}(F) \to \mathit{Ran}(F)$ by $F(K,x)$ for all $x \in \mathit{Dom}(F)$. Thus, F specifies a collection of maps from $\mathit{Dom}(F)$ to $\mathit{Ran}(F)$, each map being associated with a key. (That is why F is called a family of functions.) We refer to $F(K,\cdot)$ as an *instance* of F. The operation of choosing a key at random from the key space is denoted $K \stackrel{R}{\leftarrow} \mathit{Keys}(F)$. We write $f \stackrel{R}{\leftarrow} F$ for the operation $K \stackrel{R}{\leftarrow} \mathit{Keys}(F)\,;\, f \leftarrow F(K,\cdot)$. That is, $f \stackrel{R}{\leftarrow} F$ denotes the operation of selecting at random a function from the family F. When f is so selected it is called a *random instance* of F. Let $\mathsf{Rand}_{n,n}$ be the family of all functions mapping $\{0,1\}^n$ to $\{0,1\}^n$ so that $f \stackrel{R}{\leftarrow} \mathsf{Rand}_{n,n}$ denotes the operation of selecting at random a function from $\{0,1\}^n$ to $\{0,1\}^n$. Similarly, let Perm_n be the family of all permutations mapping $\{0,1\}^n$ to $\{0,1\}^n$ so that $\pi \stackrel{R}{\leftarrow} \mathsf{Perm}_n$ denotes the operation of selecting at random a permutation on $\{0,1\}^n$. We say that F is a *cipher* if $\mathit{Dom}(F) = \mathit{Ran}(F)$ and each instance $F(K,\cdot)$ of F is a length-preserving permutation. A *block cipher* is a cipher whose domain and range equal $\{0,1\}^n$ for some integer n called the *block size*. (For example, the AES has block size 128.) If F is a cipher, then F^{-1} is the *inverse cipher*, defined by $F^{-1}(K,x) = F(K,\cdot)^{-1}(x)$ for all $K \in \mathit{Keys}(F)$ and $x \in \mathit{Dom}(F)$.

Pseudorandomness of ciphers. A "secure" cipher is one that approximates a family of random permutations; the "better" the approximation, the more secure the cipher. This is formalized following [6,9]. A *distinguisher* is an algorithm that has access to one or more oracles and outputs a bit. Let F: $\mathit{Keys}(F) \times \{0,1\}^n \to \{0,1\}^n$ be a family of functions with domain and range $\{0,1\}^n$. Let A_1 be a distinguisher with one oracle and A_2 a distinguisher with two oracles. Let

$$\mathbf{Adv}_F^{\text{prp-cpa}}(A_1) = \Pr\left[g \stackrel{R}{\leftarrow} F \;:\; A_1^g = 1\right] - \Pr\left[g \stackrel{R}{\leftarrow} \mathsf{Perm}_n \;:\; A_1^g = 1\right].$$

If $F\colon \mathit{Keys}(F) \times \{0,1\}^n \to \{0,1\}^n$ is a cipher, then we also let

$$\mathbf{Adv}_F^{\text{prp-cca}}(A_2) = \Pr\left[g \stackrel{R}{\leftarrow} F \;:\; A_2^{g,g^{-1}} = 1\right] - \Pr\left[g \stackrel{R}{\leftarrow} \mathsf{Perm}_n \;:\; A_2^{g,g^{-1}} = 1\right].$$

These capture the *advantage* of the distinguisher in question in the task of distinguishing a random instance of F from a random permutation on D. In the first case, the distinguisher gets to query the challenge instance. In the second, it also gets to query the inverse of the challenge instance. For any integers $t, q_e, q_d, \mu_e, \mu_d$, we now let

$$\mathbf{Adv}_F^{\text{prp-cpa}}(t, q_e, \mu_e) = \max_{A_1}\left\{\, \mathbf{Adv}_F^{\text{prp-cpa}}(A_1) \,\right\}$$

$$\mathbf{Adv}_F^{\text{prp-cca}}(t, q_e, \mu_e, q_d, \mu_d) = \max_{A_2}\left\{\, \mathbf{Adv}_F^{\text{prp-cca}}(A_2) \,\right\}.$$

The maximum is over all distinguishers having time-complexity t, making to the g oracle at most q_e queries totaling at most μ_e bits, and, in the second case, also making to the g^{-1} oracle at most q_d queries totaling at most μ_d bits. We say that a PRP F is *secure against chosen-plaintext attacks* if the function $\mathbf{Adv}_F^{\text{prp-cpa}}(t, q_e)$ grows "slowly." Similarly, we say that a PRP F is *se-*

cure against chosen-ciphertext attacks if the function $\mathbf{Adv}_F^{\text{prp-cca}}(t, q_e, q_d)$ grows "slowly." Time complexity includes the time to reply to oracle calls by computation of $F(K,\cdot)$ or $F(K,\cdot)^{-1}$.

3 Online Ciphers and Their Basic Properties

We say that a function $f\colon D_{d,n} \to D_{d,n}$ is n-on-line if the i-th block of the output is determined completely by the first i blocks of the input. A more formal definition follows. We refer the reader to Section 2 for the definition of $f^{(i)}$.

Definition 1. Let $n, d \geq 1$ be integers, and let $f\colon D_{d,n} \to D_{d,n}$ be a length-preserving function. We say that f is *n-on-line* if there exists a function $X\colon D_{d,n} \to \{0,1\}^n$ such that for every $M \in D_{d,n}$ and every $i \in \{1,\ldots,|M|/n\}$ it is the case that

$$f^{(i)}(M) = X(M[1]\ldots M[i]) .$$

A cipher F having domain and range a subset of $D_{d,n}$ is said to be n-on-line if for every $K \in \mathit{Keys}(F)$ the function $F(K,\cdot)$ is on-line. ∎

Definition 2. Let f be an n-on-line function. Let $i \geq 1$. Fix $M[1],\ldots,M[i-1] \in \{0,1\}^n$. Define the function $\Pi^f_{M[1]\ldots M[i-1]}\colon \{0,1\}^n \to \{0,1\}^n$ by

$$\Pi^f_{M[1]\ldots M[i-1]}(x) = f^{(i)}(M[1]\ldots M[i-1]x)$$

for all $x \in \{0,1\}^n$. ∎

Proposition 1. *If f is an n-on-line permutation, $i \geq 1$ and $M[1],\ldots,M[i-1] \in \{0,1\}^n$, then the map $\Pi^f_{M[1]\ldots M[i-1]}$ is a permutation on $\{0,1\}^n$.*

The proof of proposition 1 is in the full version of this paper [1].

PSEUDORANDOMNESS OF ON-LINE CIPHERS. Let $\mathsf{OPerm}_{d,n}$ denote the family of all n-on-line, length-preserving permutations on $D_{d,n}$. A "secure" on-line cipher is one that closely approximates $\mathsf{OPerm}_{d,n}$; the "better" the approximation, the more "secure" the on-line cipher. This formalization is analogous to the previously presented formalization of the pseudorandomness of ciphers. Let $F\colon \mathit{Keys}(F) \times D_{d,n} \to D_{d,n}$ be a family of functions with domain and range $D_{d,n}$. Let A_1 be a distinguisher with one oracle and A_2 a distinguisher with two oracles. Let

$$\mathbf{Adv}_F^{\text{oprp-cpa}}(A_1) = \Pr\left[g \stackrel{R}{\leftarrow} F \;:\; A_1^g = 1\right] - \Pr\left[g \stackrel{R}{\leftarrow} \mathsf{OPerm}_{d,n} \;:\; A_1^g = 1\right] .$$

If $F\colon \mathit{Keys}(F) \times \{0,1\}^n \to \{0,1\}^n$ is a cipher, then we also let

$$\mathbf{Adv}_F^{\text{oprp-cca}}(A_2) = \Pr\left[g \stackrel{R}{\leftarrow} F \;:\; A_2^{g,g^{-1}} = 1\right] - \Pr\left[g \stackrel{R}{\leftarrow} \mathsf{OPerm}_{d,n} \;:\; A_2^{g,g^{-1}} = 1\right] .$$

These capture the *advantage* of the distinguisher in question in the task of distinguishing a random instance of F from a random, length-preserving, n-on-line permutation on $D_{d,n}$. In the first case, the distinguisher gets to query the challenge instance. In the second, it also gets to query the inverse of the challenge instance. For any integers $t, q_e, \mu_e, q_d, \mu_d$, we now let

$$\mathbf{Adv}_F^{\text{oprp-cpa}}(t, q_e, \mu_e) = \max_{A_1} \{ \mathbf{Adv}_F^{\text{oprp-cpa}}(A_1) \}$$

$$\mathbf{Adv}_F^{\text{oprp-cca}}(t, q_e, \mu_e, q_d, \mu_d) = \max_{A_2} \{ \mathbf{Adv}_F^{\text{oprp-cca}}(A_2) \} .$$

The maximum is over all distinguishers having time-complexity t, making to the oracle g at most q_e queries totaling at most μ_e bits, and, in the second case, also making to the g^{-1} oracle at most q_d queries totaling at most μ_d bits. We say that an online PRP (OPRP) F is secure against chosen plaintext attacks if the function $\mathbf{Adv}_F^{\text{oprp-cpa}}(t, q_e, \mu_e)$ grows "slowly." Similarly, we say that an OPRP F is secure against chosen ciphertext attacks if the function $\mathbf{Adv}_F^{\text{oprp-cca}}(t, q_e, \mu_e, q_d, \mu_d)$ grows "slowly." Time complexity includes the time to reply to oracle calls by computation of $F(K, \cdot)$ or $F(K, \cdot)^{-1}$.

TREE-BASED CHARACTERIZATION. We present a tree-based characterization of n-on-line ciphers that is useful to gain intuition and to analyze constructs. Let $N = 2^n$. An N-ary tree of functions is an N-ary tree T each node of which is labeled by a function mapping $\{0,1\}^n$ to $\{0,1\}^n$. We label each edge in the tree in a natural way via a string in $\{0,1\}^n$. Then, each node in the tree is described by a sequence of edge labels defining the path from the root to the node in question. The function labeling node x in the tree, where x is a string of length ni for some $0 \le i \le d$, is then denoted T_x. A tree defines a function T from $D_{d,n}$ to $D_{d,n}$ as described below. If the nodes in the tree are labeled with permutations, then the tree also defines an inverse function T^{-1}.

$T(M[1] \ldots M[l])$	$T^{-1}(C[1] \ldots C[l])$
$x \leftarrow \varepsilon$	$x \leftarrow \varepsilon$
For $i = 1, \ldots, l$ do	For $i = 1, \ldots, l$ do
$C[i] \leftarrow T_x(M[i])$	$M[i] \leftarrow T_x^{-1}(C[i])$
$x \leftarrow x \| C[i]$	$x \leftarrow x \| C[i]$
EndFor	EndFor
Return $C[1] \ldots C[l]$	Return $M[1] \ldots M[l]$

Here, $1 \le l \le d$. Let $G : \mathit{Keys}(G) \times \{0,1\}^n \to \{0,1\}^n$ be a function family. (We are most interested in the case where G is Perm_n or $\mathsf{Rand}_{n,n}$.) We let $\mathsf{Tree}(n, G, d)$ denote the set of all 2^n-ary trees of functions in which each function is an instance of G and the depth of the tree is d. This set is viewed as equipped with a distribution under which each node of the tree is assigned a random instance of G, and the assignments to the different nodes are independent. We claim that a tree-based construction defined above is a valid characterization of on-line ciphers, as stated in the following proposition and proven in [1].

Proposition 2. *There is a bijection between* $\mathsf{Tree}(n, \mathsf{Perm}_n, d)$ *and* $\mathsf{OPerm}_{d,n}$.

INVERSION. It turns out that the inverse of an on-line permutation is itself on-line, as stated below and proven in [1].

Proposition 3. *Let $f\colon D_{d,n} \to D_{d,n}$ be an n-on-line permutation, and let $g = f^{-1}$. Then g is an n-on-line permutation.*

We note that the proof does not tell us anything about the computational complexity of function f^{-1}, meaning it could be the case that f is efficiently computable, but the f^{-1} given by Proposition 3 is not. However, whenever we design a cipher F, we will make sure that both $F(K, \cdot)$ and $F^{-1}(K, \cdot)$ are efficiently computable given K, and will explicitly specify F^{-1} in order to make this clear.

4 Analysis of Some Candidate Ciphers

We consider several candidates for on-line ciphers. First, we consider one based on the basic CBC mode. Then, we consider the Accumulated Block Chaining (ABC) proposed by Knudsen in [7], which is a generalization of the Infinite Garble Extension mode proposed by Campbell [5]. In this section, we let $E\colon \{0,1\}^{ek} \times \{0,1\}^n \to \{0,1\}^n$ be a given block cipher with key size ek and block size n.

CBC AS AN ON-LINE CIPHER. In CBC encryption based on E, one usually uses a new, random IV for every message. This does not yield a cipher, let alone an on-line one. To get an on-line cipher, we fix the IV. We can, however, make it secret; this can only increase security. In more detail, the *CBC cipher associated to E*, denoted OCBC, has key space $\{0,1\}^{ek+n}$. For $M, C \in D_{d,n}$, $eK \in \{0,1\}^{ek}$ and $C[0] \in \{0,1\}^n$, we define

$\mathsf{OCBC}(eK \| C[0], M)$	$\mathsf{OCBC}^{-1}(eK \| C[0], C)$
Parse M as $M[1] \ldots M[l]$ with $l \geq 1$	Parse C as $C[1] \ldots C[l]$ with $l \geq 1$
For $i = 1, \ldots, l$ do	For $i = 1, \ldots, l$ do
$\quad C[i] \leftarrow E(eK, M[i] \oplus C[i-1])$	$\quad M[i] \leftarrow E^{-1}(eK, C[i]) \oplus C[i-1]$
Return $C[1] \ldots C[l]$	Return $M[1] \ldots M[l]$

Here, $C[0]$ is the IV. The key is the pair $eK \| C[0]$, consisting of a key eK for the block cipher, and the IV. It is easy to check that the above cipher is on-line. For clarity, we have also shown the inverse cipher. We now present the attack. The adversary A shown in Figure 1 gets an oracle g where g is either an instance of OCBC or an instance of $\mathsf{OPerm}_{d,n}$. We claim that

$$\mathbf{Adv}^{\text{oprp-cpa}}_{\mathsf{OCBC}}(A) \geq 1 - 2^{-n} . \tag{1}$$

We justify Equation (1) in the full version of this paper [1]. Since A made only 3 oracle queries, this shows that the CBC mode with a fixed IV is not a secure on-line cipher.

The idea of the attack is to gather some input-output pairs for the cipher. Then we use these values to construct a new sequence of input blocks so that one of the input blocks to E collides with one of the previous input blocks to E. This enables us to predict an output block of the cipher. If our prediction is correct, then we know that the oracle is an instance of OCBC with overwhelming probability.

```
Distinguisher A^g
    Let M[2],...,M[l] be any n-bit strings
    Let M_1 = 0^n M[2]...M[l] and let M_2 = 1^n M[2]...M[l]
    Let C_1[1]...C_1[l] <- g(M_1) and let C_2[1]...C_2[l] <- g(M_2)
    Let M_3[2] = M[2]⊕C_1[1]⊕C_2[1] and let M_3 = 1^n M_3[2]M[3]...M[l]
    Let C_3[1]...C_3[l] <- g(M_3)
    If C_3[2] = C_1[2] then return 1 else return 0
```

Fig. 1. Attack on the CBC based on-line cipher.

ABC AS AN ON-LINE CIPHER. Knudsen in [7] proposes the Accumulated Block Chaining (ABC) mode of operation for block ciphers. This is an on-line cipher that is a natural starting point in the problem of finding a secure on-line cipher because it has the property of "infinite error propagation." We formalize and analyze ABC with regard to meeting our security requirements.

The mode is parameterized by initial values $P[0], C[0] \in \{0,1\}^n$ and also by a public function h: $\{0,1\}^n \to \{0,1\}^n$. (Instantiations for h suggested in [7] include the identity function, the constant function always returning 0^n, and the function which rotates its input by one bit.) We are interested in the security of the mode across various settings and choices of these parameters. (In particular, we want to consider the case where the initial values are public and also the case where they are secret, and see how the choice of h impacts security in either case.) Accordingly, it is convenient to first introduce auxiliary functions EABC and DABC. For $M, C \in D_{d,n}$ and $eK \in \{0,1\}^k$, we define

EABC($eK, P[0], C[0], M$)	DABC($eK, P[0], C[0], C$)
Parse M as $M[1] \ldots M[l]$ with $l \geq 1$	Parse C as $C[1] \ldots C[l]$ with $l \geq 1$
For $i = 1, \ldots, l$ do	For $i = 1, \ldots, l$ do
$P[i] \leftarrow M[i] \oplus h(P[i-1])$	$P[i] \leftarrow E^{-1}(eK, C[i] \oplus P[i-1])$
$C[i] \leftarrow E(eK, P[i] \oplus C[i-1])$	$\oplus C[i-1]$
$\oplus P[i-1]$	$M[i] \leftarrow P[i] \oplus h(P[i-1])$
EndFor	EndFor
Return $C[1] \ldots C[l]$	Return $M[1] \ldots M[l]$

We now define two versions of the ABC cipher. The first uses public initial values, while the second uses secret initial values. The *ABC cipher with public initial values* associated to E, denoted PABC, has key space $\{0,1\}^k$ and domain and range $D_{d,n}$. We fix values $P[0], C[0] \in \{0,1\}^n$ which are known to all parties including the adversary. We then define the cipher and the inverse cipher as follows:

PABC(eK, M)	PABC^{-1}(eK, C)
Return EABC($eK, P[0], C[0], M$)	Return DABC($eK, P[0], C[0], C$)

The *ABC cipher with secret initial values* associated to E, denoted SABC, has key space $\{0,1\}^{k+2n}$ and domain and range $D_{d,n}$. The key is $eK \| P[0] \| C[0]$. We then define the cipher and the inverse cipher as follows:

Distinguisher A^g
  Let $M[2],\ldots,M[l]$ be any n-bit strings
  Let $M_1 = 0^n M[2]\ldots M[l]$ and let $M_2 = 1^n M[2]\ldots M[l]$
  Let $C_1[1]\ldots C_1[l] \leftarrow g(M_1)$ and let $C_2[1]\ldots C_2[l] \leftarrow g(M_2)$
  Let $M_3[2] = M[2] \oplus C_1[1] \oplus C_2[1] \oplus h(0^n \oplus h(P[0])) \oplus h(1^n \oplus h(P[0]))$
  Let $M_3 = 1^n M_3[2] M[3]\ldots M[l]$
  Let $C_3[1]\ldots C_3[l] \leftarrow g(M_3)$
  If $C_3[2] = C_1[2] \oplus 1^n$, then return 1 else return 0

Fig. 2. Attack on the ABC based on-line cipher.

$\mathsf{SABC}(eK\|P[0]\|C[0],M)$	$\mathsf{SABC}^{-1}(eK\|P[0]\|C[0],C)$
Return $\mathrm{EABC}(eK,P[0],C[0],M)$	Return $\mathrm{DABC}(eK,P[0],C[0],C)$

It is easy to check that both the above ciphers are n-on-line.

We show that the ABC cipher with public initial values is not a secure OPRP for *all* choices of the function h. The attack is shown in Figure 2. The adversary A gets an oracle g where g is either an instance of PABC or an instance of $\mathsf{OPerm}_{d,n}$. The adversary can mount this attack because the function h as well as the value $P[0]$ are public. We claim that

$$\mathbf{Adv}^{\text{oprp-cpa}}_{\mathsf{PABC}}(A) \geq 1 - 2 \cdot 2^{-n} . \tag{2}$$

Since A made only three oracle queries, this means that PABC is not a secure on-line cipher.

We show that the ABC cipher with secret initial values is not a secure OPRP for a class of functions h that includes the ones suggested in [7]. Specifically, let us say that a function $h\colon \{0,1\}^n \to \{0,1\}^n$ is *linear* if $h(x \oplus y) = h(x) \oplus h(y)$ for all $x, y \in \{0,1\}^n$. (Notice that the identity function, the constant function always returning 0^n, and the function which rotates its input by one bit are all linear.) For any linear hash function h, we simply note that the above attack applies. This is because the fourth line of the adversary's code can be replaced by

  Let $M_3[2] = M[2] \oplus C_1[1] \oplus C_2[1] \oplus h(0^n) \oplus h(1^n)$

The adversary can compute $M_3[2]$ because h is public. The fact that h is linear means that the value $M_3[2]$ is the same as before, so the attack has the same success probability. The analysis for the attacks against both PABC and SABC appear in the full version of this paper [1].

5 Lemmas about AXU Families

Our constructions of on-line ciphers will use the families of AXU (Almost Xor Universal) functions as defined by Krawczyk [8]. We recall the definition, and then prove some lemmas that will be helpful in our analyses.

Definition 3. Let $n, hk \geq 1$ be integers, and let $H\colon \{0,1\}^{hk} \times \{0,1\}^n \to \{0,1\}^n$ be a family of functions. Let

$$\mathbf{Adv}_H^{\mathrm{axu}} = \max_{x_1,x_2,y} \left\{ \Pr\left[K \stackrel{R}{\leftarrow} \{0,1\}^{hk} \; : \; H(K,x_1) \oplus H(K,x_2) = y \right] \right\}$$

where the maximum is over all *distinct* $x_1, x_2 \in \{0,1\}^n$ and all $y \in \{0,1\}^n$. ∎

The "advantage function" based notation we are introducing is novel: previous works used instead the term "ϵ-AXU" family to refer to a family H that, in our notation, has $\mathbf{Adv}_H^{\mathrm{axu}} \leq \epsilon$. We find the "advantage function" based notation more convenient, and more consistent with the rest of our security definitions.

The definition is information-theoretic, talking of the maximum value of some probability. We will find it convenient to think in terms of an adversary attacking the scheme, and will use the following lemma. We stress that below there are no limits on the running time of the adversary. This lemma is standard, and follows easily from Definition 3, so we omit the proof.

Lemma 1. *Let $n, hk \geq 1$ be integers, and let $H\colon \{0,1\}^{hk} \times \{0,1\}^n \to \{0,1\}^n$ be a family of functions. Let A be any possibly probabilistic algorithm that takes no inputs and returns a triple (x_1, x_2, y) of n-bit strings. Then*

$$\Pr\left[(x_1,x_2,y) \stackrel{R}{\leftarrow} A \,;\, K \stackrel{R}{\leftarrow} \{0,1\}^{hk} \; : \; H(K,x_1) \oplus H(K,x_2) = y\right] \;\leq\; \mathbf{Adv}_H^{\mathrm{axu}}.$$

In the formulation of Lemma 1, it is important that the adversary is constrained to pick x_1, x_2, y before the K is chosen. In our upcoming analyses, we will, in contrast, be considering an adversary that obtains some partial information regarding $H(K,\cdot)$ in the course of its search for a certain kind of "collision," and uses this to guide its search. Specifically, our adversary B can be viewed as having access to an oracle that knows a key K. The adversary functions in stages. In stage i, it produces a pair (x_i, y_i) of values which it submits to the oracle. The latter responds with a bit indicating whether or not there exists some $j \in \{1, \ldots, i-1\}$ such that $H(K,x_j) \oplus H(K,x_i) = y_j \oplus y_i$. (The oracle is stateful because it has to remember the adversary queries from previous stages in order to be able to answer the current query.) We wish to argue that the partial information about $H(K,\cdot)$ that is obtained by the adversary via this process is not too large. Specifically, we argue that the probability that the adversary ever gets back a positive response from the oracle is $O(q^2) \cdot \mathbf{Adv}_H^{\mathrm{axu}}$.

In the formal definition that follows, we first describe an algorithm that serves as a stateful oracle discussed above. Then, we describe an experiment in which the adversary B with oracle access to the algorithm is executed.

Definition 4. Let $H\colon \{0,1\}^{hk} \times \{0,1\}^n \to \{0,1\}^n$ be a family of hash functions, and let hK be a string of length hk. We define the following stateful algorithm D. It maintains a counter i and arrays X, Y, and takes n-bit strings x, y as inputs. Then, we let B be an adversary with oracle access to D_{hK} and define an experiment in which B executes.

Algorithm $D_{hK}(x,y)$
  $i \leftarrow i+1 \,;\, r \leftarrow 0 \,;\, X[i] \leftarrow x \,;\, Y[i] \leftarrow y$

For $j = 1, \ldots, i-1$ do
  If $(H(hK, X[j]) \oplus Y[j] = H(hK, X[i]) \oplus Y[i])$ and $(X[j] \neq X[i])$ then $r \leftarrow j$
EndFor
Return r

Experiment $\mathbf{Exp}_H^{\text{axu-cr}}(B)$
  $hK \stackrel{R}{\leftarrow} \{0,1\}^{hk}$
  Initialize D_{hK} with $i = 0$ and X, Y empty
  Run $B^{D_{hk}(\cdot,\cdot)}$ until it halts
  If B made some oracle query that received a non-zero response,
    then return 1, else return 0.

We define the *advantage* of the adversary B and the *AXU-Collision advantage function* of H as follows. For any integer q,

$$\mathbf{Adv}_H^{\text{axu-cr}}(B) = \Pr[\,\mathbf{Exp}_H^{\text{axu-cr}}(B) = 1\,]$$
$$\mathbf{Adv}_H^{\text{axu-cr}}(q) = \max_B \{\, \mathbf{Adv}_H^{\text{axu-cr}}(B) \,\}$$

where the maximum is taken over all adversaries making q queries. ▮

The following lemma states the relationship between Definition 3 and Definition 4. The proof is presented in the full version of this paper [1].

Lemma 2. *Let* H: $\{0,1\}^{hk} \times \{0,1\}^n \to \{0,1\}^n$ *be a family of hash functions. Then,*

$$\mathbf{Adv}_H^{\text{axu-cr}}(q) \;\leq\; q(q-1) \cdot \mathbf{Adv}_H^{\text{axu}} \,.$$

6 The HCBC Cipher

In this section, we suggest a construction of an on-line cipher. We call it HCBC and prove its security against chosen-plaintext attacks. This construction is similar to the CBC mode of encryption. The only difference is that each output block passes through a keyed hash function before getting exclusive-or-ed with the next input block. The key of the hash function is kept secret.

Construction 1. Let $n, d \geq 1$ be integers, and let E: $\{0,1\}^{ek} \times \{0,1\}^n \to \{0,1\}^n$ be a block cipher. Let H: $\{0,1\}^{hk} \times \{0,1\}^n \to \{0,1\}^n$ be a family of hash functions. We associate to them a cipher HCBC: $\{0,1\}^{ek+hk} \times D_{d,n} \to D_{d,n}$. A key for it is a pair $eK \| hK$ where eK is a key for E and hK is a key for H. The cipher and its inverse are defined as follows for $M, C \in D_{d,n}$. Figure 3 illustrates the cipher.

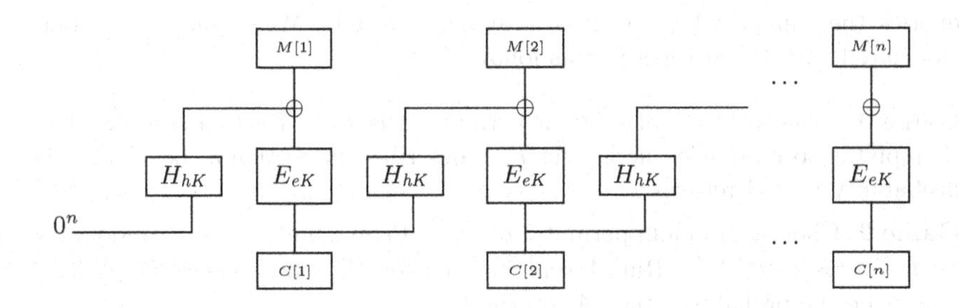

Fig. 3. The HCBC cipher.

$\mathsf{HCBC}(eK\|hK, M)$	$\mathsf{HCBC}^{-1}(eK\|hK, C)$
Parse M as $M[1]\ldots M[l]$ with $l \geq 1$	Parse C as $C[1]\ldots C[l]$ with $l \geq 1$
$C[0] \leftarrow 0^n$	$C[0] \leftarrow 0^n$
For $i = 1, \ldots, l$ do	For $i = 1, \ldots, l$ do
$P[i] \leftarrow H(hK, C[i-1]) \oplus M[i]$	$P[i] \leftarrow E^{-1}(eK, C[i])$
$C[i] \leftarrow E(eK, P[i])$	$M[i] \leftarrow H(hK, C[i-1]) \oplus P[i]$
EndFor	EndFor
Return $C[1]\ldots C[l]$	Return $M[1]\ldots M[l]$ ∎

The following theorem implies that, if E is a PRP secure against chosen-plaintext attacks and H is an AXU family of hash functions, then HCBC is an OPRP secure against chosen-plaintext attacks.

Theorem 1. *Let E: $\{0,1\}^{ek} \times \{0,1\}^n \to \{0,1\}^n$ be a block cipher, and let H: $\{0,1\}^{hk} \times \{0,1\}^n \to \{0,1\}^n$ be a family of hash functions. Let* HCBC *be the n-on-line cipher associated to them as per Construction 1. Then, for any integers $t, q_e, \mu_e \geq 0$ such that $\mu_e/n \leq 2^{n-1}$, we have*

$$\mathbf{Adv}^{\text{oprp-cpa}}_{\mathsf{HCBC}}(t, q_e, \mu_e) \leq$$

$$\mathbf{Adv}^{\text{prp-cpa}}_{E}(t, \mu_e/n, \mu_e) + \left(\frac{\mu_e^2 - n\mu_e}{n^2}\right) \cdot \mathbf{Adv}^{\text{axu}}_{H} + \frac{\mu_e^2 + 2n(q_e+1)\mu_e}{n^2 \cdot 2^n} .$$

HCBC is not secure against chosen-ciphertext attacks. We present an attack in the full version of this paper [1].

A complete proof of Theorem 1 can be found in the full version of this paper [1]. In the rest of this section, we provide an overview of this proof.

We introduce the notation $\mathsf{HCBC}_\pi(hK, \cdot)$ to denote an instance of a cipher defined by Construction 1 where a permutation π and π^{-1} are used in place of a permutation from the family E and its inverse, respectively. The proof looks at an on-line cipher as a 2^n-ary tree of permutations on $\{0,1\}^n$, and goes through a hybrid argument involving a sequence of different games that "move" from $\mathsf{OPerm}_{d,n}$ to HCBC. Let A be an adversary that has oracle access to a length-preserving function f: $D_{d,n} \to D_{d,n}$. We assume that A makes at most q_e oracle queries the sum of whose lengths is at most μ_e bits. We define three games associated with the adversary A as follows.

Game 1. Choose a tree of random permutations $T \xleftarrow{R} \mathsf{Tree}(n, \mathsf{Perm}_n, d)$. Run A, replying to its oracle queries via T as described in Section 3. Let P_1 be the probability that A returns 1.

Game 2. Choose a random permutation, $\pi \xleftarrow{R} \mathsf{Perm}_n$, and choose a random key for H via $hK \xleftarrow{R} \{0,1\}^{hk}$. Run A, replying to its oracle queries via $\mathsf{HCBC}_\pi(hK, \cdot)$. Let P_2 be the probability that A returns 1.

Game 3. Choose random keys for E and H via $eK \xleftarrow{R} \{0,1\}^{ek}$ and $hK \xleftarrow{R} \{0,1\}^{hk}$, respectively. Run A, replying to its oracle queries via $\mathsf{HCBC}(eK \| hK, \cdot)$. Let P_3 be the probability that A returns 1.

By the definition of $\mathbf{Adv}^{\text{oprp-cpa}}_{\mathsf{HCBC}}(A)$, we have

$$\mathbf{Adv}^{\text{oprp-cpa}}_{\mathsf{HCBC}}(A) = P_3 - P_1 = (P_3 - P_2) + (P_2 - P_1)\,. \tag{3}$$

We bound the difference terms via the following lemmas:

Lemma 3. $P_3 - P_2 \leq \mathbf{Adv}^{\text{prp-cpa}}_{E}(t, \mu_e/n, \mu_e)$

Lemma 4. $P_2 - P_1 \leq \dfrac{\mu_e^2 + 2n(q_e+1)\mu_e}{n^2 \cdot 2^n} + \mathbf{Adv}^{\text{axu-cr}}_{H}(\mu_e/n)$

Equation (3), Lemma 2, and the above lemmas imply the statement of the theorem. We proceed to discuss the proofs of the lemmas.

The proof of Lemma 3 is a standard simulation argument, detailed in [1]. The rest of this section is devoted to an overview of the proof of Lemma 4. We let $M_1, \ldots, M_{q_e}$ denote A's queries, where $M_j = M_j[1] \ldots M_j[l_j]$ for $j = 1, \ldots, q_e$. Let hK denote the key of the hash function, and π the choice of permutation from Perm_n, that underly Game 2. Then we introduce the following notation in this game:

For each $j = 1, \ldots, q_e$
  Let $C_j[0] = 0^n$
  For $i = 1, \ldots, l_j$
    Let $P_j[i] = H(hK, C_j[i-1]) \oplus M_j[i]$ and let $C_j[i] = \pi(P_j[i])$

We now define some events in Game 2:

Event ZO_2 : There exist (i,j) such that $1 \leq j \leq q_e$, $1 \leq i \leq l_j$ and $C_j[i] = 0^n$

Event HC : There exist $(i,j), (i',j')$ such that $1 \leq j < j' \leq q_e$, $1 \leq i \leq l_j$, $1 \leq i' \leq l_{j'}$ and $P_j[i] = P_{j'}[i']$, but $C_j[i-1] \neq C_{j'}[i'-1]$

Event B_2 : $\mathsf{ZO}_2 \vee \mathsf{HC}$.

Now let T denote the random choice of tree from $\mathsf{Tree}(n, \mathsf{Perm}_n, d)$ that underlies Game 1 and introduce the following notation in this game:

For each $j = 1, \ldots, q_e$
Let $x_j[0] = \varepsilon$
For $i = 1, \ldots, l_j$
Let $C_j[i] = T_{x_j[i-1]}(M_j[i])$ and let $x_j[i] = x_j[i-1] \| C_j[i]$

We now define some events in Game 1:

Event ZO_1 : There exist (i, j) such that $1 \le j \le q_e$, $1 \le i \le l_j$ and $C_j[i] = 0^n$

Event OC : There exist $(i, j), (i', j')$ such that $1 \le j < j' \le q_e$, $1 \le i \le l_j$, $1 \le i' \le l_{j'}$ and $x_j[i-1] \ne x_{j'}[i'-1]$ but $C_j[i] = C_{j'}[i']$

Event B_1 : $\mathsf{ZO}_1 \vee \mathsf{OC}$

Let $\Pr_1[\cdot]$ denote the probability function underlying Game 1, namely that created by the random choice $T \stackrel{R}{\leftarrow} \mathsf{Tree}(n, \mathsf{Perm}_n, d)$, and let $\Pr_2[\cdot]$ denote the probability function underlying Game 2, namely that created by the random choices of π and hK. Let F denote $\mathsf{HCBC}_\pi(hK, \cdot)$.

Claim. $\Pr_2[A^F = 1 \mid \overline{\mathsf{B}}_2] = \Pr_1[A^T = 1 \mid \overline{\mathsf{B}}_1]$

Given this claim, a conditioning argument can be used to show that

$$P_2 - P_1 \le \Pr_2[\mathsf{HC}] + \Pr_2[\mathsf{ZO}_2] + \Pr_1[\mathsf{B}_1] .$$

The terms are bounded via the following claims:

Claim. $\Pr_2[\mathsf{HC}] \le \mathbf{Adv}_H^{\text{axu-cr}}(\mu_e/n)$ ∎

Claim. $\Pr_2[\mathsf{ZO}_2] \le \dfrac{2\mu_e}{n \cdot 2^n}$ ∎

Claim. $\Pr_1[\mathsf{B}_1] \le \dfrac{\mu_e^2 + 2nq_e\mu_e}{n^2 \cdot 2^n}$ ∎

The proofs of the four claims above can be found in [1]. We conclude this sketch by providing some intuition regarding the choice of the "bad" events, beginning with the following definition.

Definition. Suppose $1 \le j, j' \le q$, $1 \le i \le l_j$ and $1 \le i' \le l_{j'}$. We say that $(i, j) \prec (i', j')$ if: either $j = j'$ and $i < i'$, or $j < j'$. We say that (i', j') is *trivial* if there exists some $j < j'$ such that $ni' \le |\mathsf{LCP}_n(M_j, M_{j'})|$.

We claim that the bad event B_2 has been chosen so that, in its absence, the following is true for every non-trivial (i', j'): If $(i, j) \prec (i', j')$ then $P_j[i] \ne P_{j'}[i']$. In other words, any two input points to the function π are unequal unless they are equal for the trivial reason that the corresponding message prefixes are equal. This means that in the absence of the bad event, ciphertext blocks whose value is not "forced" by message prefix conditions are random but distinct, being outputs of a random permutation. We have choosen event B_1 in Game 1 so that the output distribution here, conditioned on the absence of this event, is the same.

7 Usage of Online Ciphers

The use of an on-line ciphers can provide strong privacy and authenticity properties, even though the cipher itself is weak compared to a standard one, if the plaintext space has appropriate properties. This follows via the the encode-then-encipher paradigm of [4], under which we imagine an explicit encoding step applied to the raw data before enciphering. While [4] say that randomness and redundancy anywhere in the message suffices, we have to be more constrained: we *prepend* randomness and *append* redundancy.

Construction 2. Let n, d be integers, and let $F\colon Keys(F) \times D_{d,n} \to D_{d,n}$ be a cipher. We associate to them the following symmetric encryption scheme $\mathcal{SE} = (\mathcal{K}, \mathcal{E}, \mathcal{D})$:

Algorithm $\mathcal{K}$	Algorithm $\mathcal{E}(K, M)$	Algorithm $\mathcal{D}(K, C)$
$K \stackrel{R}{\leftarrow} Keys(F)$	$r \stackrel{R}{\leftarrow} \{0,1\}^n$	$x \leftarrow F^{-1}(K, C)$
Return K	$x \leftarrow r \| M \| 0^n$	If $\lvert x \rvert < 3n$ then return $\perp$
	$C \leftarrow F(K, x)$	Parse x as $r \| M \| \tau$ with $\lvert r \rvert = \lvert \tau \rvert = n$
	Return C	If $\tau = 0^n$ then return M
		Else return $\perp$ ∎

We want to show that this scheme provides privacy, when F is an n-on-line cipher secure against chosen-plaintext attacks, and authenticity, when F is an n-on-line cipher secure against chosen-ciphertext attacks. Definitions for these privacy and authenticity notions are standard (see for example [3]). Briefly, the symmetric encryption scheme achieves privacy and is called IND-CPA-secure if no polynomial time adversary, which gets to see ciphertexts for plaintexts of its choice and is given a challenge ciphertext, can get "any" information about the underlying plaintext. The symmetric encryption scheme achieves integrity and is called INT-CTXT-secure if no polynomial time adversary, which gets to see ciphertexts of plaintexts of its choice, can create a "new" valid ciphertext. The following claims state our results.

Proposition 4. *Let $F\colon Keys(F) \times D_{d,n} \to D_{d,n}$ be an n-on-line cipher, and let $\mathcal{SE} = (\mathcal{K}, \mathcal{E}, \mathcal{D})$ be the symmetric encryption scheme defined in Construction 2. Then, for any integers $t, q_e, \mu_e \geq 0$,*

$$\mathbf{Adv}^{\text{ind-cpa}}_{\mathcal{SE}}(t, q_e, \mu_e) \leq 2\mathbf{Adv}^{\text{oprp-cpa}}_{F}(t, q_e, \mu_e) + \frac{q_e^2}{2^n} .$$

Also, for any integers $t, q_e, q_d, \mu_e, \mu_d \geq 0$,

$$\mathbf{Adv}^{\text{int-ctxt}}_{\mathcal{SE}}(t, q_e, q_d, \mu_e, \mu_d) \leq 2\mathbf{Adv}^{\text{oprp-cca}}_{F}(t, q_e, \mu_e, q_d, \mu_d) + \frac{q_d}{2^n} .$$

That is, if F is an n-on-line cipher secure against chosen-plaintext attacks, then $\mathcal{SE}$ is IND-CPA secure, and if F is also secure against chosen-ciphertext attacks, then $\mathcal{SE}$ is INT-CTXT secure.

The proof of Proposition 4 is simple and follows [4]. We present it in [1]. Note that if n-on-line ciphers are used to encrypt messages which by their nature start with at least n random bits and end with some fixed sequence of n bits than we get a symmetric encryption scheme that achieves privacy and integrity and, moreover, is length-preserving.

Acknowledgments. The UCSD authors are supported in part by Bellare's 1996 Packard Foundation Fellowship in Science and Engineering. We thank Anand Desai, Bogdan Warinschi, and the Crypto 2001 program committee for their helpful comments.

References

1. M. Bellare, A. Boldyreva, L. Knudsen, C. Namprempre. On-line ciphers and the Hash-CBC construction. Full version of this paper, available via `http://www-cse.ucsd.edu/users/mihir`.
2. M. Bellare, J. Kilian, and P. Rogaway. The security of the cipher block chaining message authentication code. In *Journal of Computer and System Sciences*, volume 61, No. 3, pages 362-399, Dec 2000. Academic Press.
3. M. Bellare and C. Namprempre. Authenticated encryption: Relations among notions and analysis of the generic composition paradigm. In T. Okamoto, editor, *Advances in Cryptology — ASIACRYPT ' 00*, volume 1976 of *Lecture Notes in Computer Science*, pages 531–545, Berlin, Germany, Dec. 2000. Springer-Verlag.
4. M. Bellare and P. Rogaway. Encode-then-encipher encryption: How to exploit nonces or redundancy in plaintexts for efficient cryptography. In T. Okamoto, editor, *Advances in Cryptology — ASIACRYPT ' 00*, volume 1976 of *Lecture Notes in Computer Science*, pages 317–330, Berlin, Germany, Dec. 2000. Springer-Verlag.
5. C. Campbell. Design and specification of cryptographic capabilities. In D. Brandstad, editor, *Computer Security and the Data Encryption Standard,* National Bureau of Standards Special Publications 500-27, U.S. Department of Commerce, pages 54-66, February 1978.
6. O. Goldreich, S. Goldwasser, and S. Micali. How to construct random functions, *Journal of the ACM,* Vol. 33, No. 4, 1986, pp. 210–217.
7. L. Knudsen. Block chaining modes of operation. Reports in Informatics, Report 207, Dept. of Informatics, University of Bergen, October 2000.
8. H. Krawczyk. LFSR-based hashing and authenticating. In Y. Desmedt, editor, *Advances in Cryptology — CRYPTO '94*, volume 839 of *Lecture Notes in Computer Science*, pages 129–139, Berlin, Germany, 1994. Springer-Verlag.
9. M. Luby and C. Rackoff. How to construct pseudo-random permutations from pseudo-random functions. *SIAM Journal of Computing,* Vol. 17, No. 2, pp. 373–386, April 1988.
10. C. Meyer and Matyas. *A new direction in Computer Data Security.* John Wiley & Sons, 1982.
11. M. Naor and O.Reingold. On the construction of pseudorandom permutations: Luby-Rackoff Revisited. In J. Feigenbaum, editor, *Journal of Cryptology,* Volume 12, Number 1, Winter 1999. Springer-Verlag.
12. W. Nevelsteen and B. Preneel. Software performance of universal hash functions. In J. Stern, editor, *Advances in Cryptology — EUROCRYPT '99*, volume 1592 of *Lecture Notes in Computer Science*, pages 24–41, Berlin, Germany, 1999. Springer-Verlag.

The Order of Encryption and Authentication for Protecting Communications (or: How Secure Is SSL?)*

Hugo Krawczyk

EE Department,
Technion, Haifa, Israel.
hugo@ee.technion.ac.il

Abstract. We study the question of how to generically compose *symmetric* encryption and authentication when building "secure channels" for the protection of communications over insecure networks. We show that any secure channels protocol designed to work with any combination of secure encryption (against chosen plaintext attacks) and secure MAC must use the encrypt-then-authenticate method. We demonstrate this by showing that the other common methods of composing encryption and authentication, including the authenticate-then-encrypt method used in SSL, are not generically secure. We show an example of an encryption function that provides (Shannon's) perfect secrecy but when combined with any MAC function under the authenticate-then-encrypt method yields a totally insecure protocol (for example, finding passwords or credit card numbers transmitted under the protection of such protocol becomes an easy task for an active attacker). The same applies to the encrypt-and-authenticate method used in SSH.
On the positive side we show that the authenticate-then-encrypt method is secure if the encryption method in use is either CBC mode (with an underlying secure block cipher) or a stream cipher (that xor the data with a random or pseudorandom pad). Thus, while we show the generic security of SSL to be broken, the current practical implementations of the protocol that use the above modes of encryption are safe.

1 Introduction

The most widespread application of cryptography in the Internet these days is for implementing a *secure channel* between two end points and then exchanging information over that channel. Typical implementations first call a key-exchange protocol for establishing a shared key between the parties, and then use this key to authenticate and encrypt the transmitted information using (efficient) symmetric-key algorithms. The three most popular protocols that follow this approach are SSL [11] (or TLS [9]), IPSec [18,19] and SSH [27]. In particular, SSL is used to protect a myriad of passwords, credit card numbers, and other

* A full version of this paper can be found in [21].

J. Kilian (Ed.): CRYPTO 2001, LNCS 2139, pp. 310–331, 2001.

sensitive data transmitted between Web clients and servers, and is used to secure many other applications. IPSec is the standard for establishing a secure channel between any two IP entities for protecting information at the network layer.

As said, all these protocols apply both symmetric authentication (MAC) and encryption to the transmitted data. Interestingly, each of these three popular protocols have chosen a *different* way to combine authentication and encryption. We describe these three methods (here x is a message; $\mathcal{E}nc(\cdot)$ is a symmetric encryption function; $\mathcal{A}uth(\cdot)$ is a message authentication code; and ',' denotes concatenation — in this notation the secret keys to the algorithms are implicit):

SSL: $a = \mathcal{A}uth(x)$, $C = \mathcal{E}nc(x, a)$, transmit C
IPSec: $C = \mathcal{E}nc(x)$, $a = \mathcal{A}uth(C)$, transmit (C, a)
SSH: $C = \mathcal{E}nc(x)$, $a = \mathcal{A}uth(x)$, transmit (C, a).

We refer to these three methods as *authenticate-then-encrypt* (abbreviated *AtE*), *encrypt-then-authenticate* (*EtA*), and *encrypt-and-authenticate* (*E&A*), respectively.

This disparity of choices reflects lack of consensus in the cryptography and security communities as for the right way to apply these functions. But is there a "right way", or are all equally secure? Clearly, the answer to this question depends on the assumptions one makes on the encryption and authentication functions. However, since protocols like the above are usually built using cryptographic functions as replaceable modules, the most useful form of this question is obtained by considering both functionalities, encryption and authentication, as *generic cryptographic primitives* with well defined (and independent from each other) properties. Moreover, we want these properties to be commonly achieved by the known efficient methods of symmetric encryption and authentication, and expected to exist in future practical realizations of these functions as well.

Specifically, we consider generic MAC functions secure against *chosen-message attacks* and generic symmetric encryption functions secure against *chosen-plaintext attacks.* These security properties are the most common notions used to model the security of these cryptographic primitives. In particular, chosen-message security of the authentication function allows to use the MAC in the above protocols independently of the encryption in cases where only integrity protection is required but not secrecy. As for encryption, chosen-plaintext security is the most common property under which encryption modes are designed and analyzed. We note that a stronger property of encryption is resistance to chosen-ciphertext attacks; while this property is important against active attacks it is NOT present in the prevalent modes of symmetric encryption (such as in stream ciphers or CBC mode even when the underlying block cipher is chosen-ciphertext secure) and therefore assuming this strong property as the basic secrecy requirement of the encryption function would exclude the use of such standard efficient mechanisms.

Rather than just studying the above ways of composing encryption and authentication as a stand-alone composed primitive, our focus is on the more comprehensive question of whether these methods provide for truly secure communications (i.e., secrecy and integrity) when embedded in a protocol that runs in a real adversarial network setting (where links are controlled by the attacker, where some of the parties running the protocol may be corrupted, where multiple security sessions are run simultaneously and maliciously interleaved, etc.).

Recent results. In a recent work, Canetti and Krawczyk [8] describe a model of secure channels that encompasses both the initial exchange of a key between pairs of communicating parties and the use of the resultant shared key for the application of symmetric encryption and authentication on the transmitted data. The requirements made from secure channels in this model include protecting the data's integrity (in the sense of simulating ideally authenticated channels) and secrecy (in the sense of plaintext indistinguishability) in the presence of a network attacker with powerful and realistic abilities of the type mentioned above. A main result in [8] is that if the key is shared securely then applying to the data the encrypt-then-authenticate method achieves secure channels provided that the encryption function is semantically secure (or plaintext-indistinguishable) under a chosen-plaintext attack and the authentication function is a MAC that resists chosen message attacks. This provides one important answer to the questions raised above: *it proves that encrypt-then-authenticate is a generically secure method for implementing secure channels.*

Our results. In this paper we complement the above result on the encrypt-then-authenticate method with contrasting results on the other two methods.

THE GENERIC INSECURITY OF *AtE*. We show that the authenticate-then-encrypt method (as in SSL) *is not generically secure* under the sole assumption that the encryption function is secure against chosen plaintext attacks and the MAC secure against chosen message attacks. We show an example of a simple encryption function that enjoys perfect (in the sense of Shannon) secrecy against chosen plaintext attacks and when combined under the *AtE* method with any MAC (even a perfect one) results in a *totally breakable implementation of secure channels.* To illustrate the insecurity of the resultant scheme we show how passwords (and credit card numbers, etc) transmitted under such a method can be easily discovered by an active attacker that modifies some of the information on the links. A major issue to highlight here is that the attack is not against the authenticity of information but against its secrecy! This result is particularly unfortunate in the case of SSL where protection of this form of sensitive information is one of the most common uses of the protocol.

THE GENERIC INSECURITY OF *E&A*. The above example is used also to demonstrate the insecurity of the encrypt-and-authenticate method (as in SSH) where the same attack (and consequences) is possible. It is worth noting that the *E&A* is obviously insecure if one uses a MAC function that leaks information on the data. However, what our attack shows is that the method is not generically secure even if one assumes a stronger MAC function with secrecy properties as commonly used in practice (e.g. a MAC realized via a pseudorandom family or if the MAC's tag itself is encrypted).

THE SECURITY OF *AtE* WITH SPECIFIC ENCRYPTION MODES. This paper does not bring just bad news. We also show that the authenticate-then-encrypt method *is secure* under two very common forms of encryption: CBC mode (with an underlying secure block cipher) and stream ciphers (that xor the data with a random or pseudorandom pad). We provide a (near optimal) quantified security analysis of these methods. While these positive results do not resolve the "generic weakness" of the authenticate-then-encrypt method (and of SSL), they do show that the common implementations currently in use do result in a secure channels protocol.

In conjunction, these results show a quite complete picture of the security (and lack of security) of these methods. They point to the important conclusion that any secure channels protocol designed to work with any combination of secure encryption (against chosen plaintext attacks) and secure MAC must use the encrypt-then-authenticate method. On the other hand, protocols that use the authenticate-then-encrypt method with encryption in either stream cipher or CBC modes are safe. However, we note the fragility of this last statement: very simple (seemingly innocuous) changes to the encryption function, including changes that do not influence the secrecy protection provided by the encryption when considered as a stand-alone primitive, can be fatal for the security of the implemented channels. This is illustrated by our example of a perfect cipher where the sole use of a simple encoding before encryption compromises the security of the transmitted data, or by the case of CBC encryption where the join encryption of message and MAC results in a secure protocol but separate encryption of these elements is insecure. Thus, when using a non-generically secure method one has to be very careful with *any* changes to existing functions or with the introduction of new encryption mechanisms (even if these mechanisms are secure as stand-alone functions).

Open question. Our results demonstrate that chosen-plaintext security is not a sufficient condition for an encryption scheme to guarantee a secure authenticate-then-encrypt composition even if the MAC is secure. An interesting open question is to find a stronger property that is enjoyed by common modes of encryption but at the same time is sufficient to ensure the security of the authenticate-then-encrypt method when combined with a secure MAC. Note that we are looking for a property that is significantly weaker than chosen-ciphertext security since the latter is *not* achieved by most symmetric encryption modes, but also because our results show that this condition is not really necessary.

Related work. While the interaction between symmetric encryption and authentication is a fundamental issue in the design of cryptographic protocols, this question seems to have received surprisingly little explicit attention in the cryptographic literature until very recently. In contrast, in the last year we have seen a significant amount of work dealing with this and related questions.

We already mentioned the work by Canetti and Krawczyk [8] that establishes the security of the encrypt-then-authenticate method for building secure channels. Here, we use this result (and some extensions of it) as a basis to derive some of our positive results. In particular, we borrow from that paper the formalization of the notion of secure channels; a short outline of this model is presented in Section 2.3 but the reader is referred directly to [8] for the (many missing) details.

A recent, independent, work that deals directly with the ordering of generic encryption and authentication is Bellare and Namprempre [5]. They study the same three forms of composition as in this paper but focus on the properties of the composed function as a stand-alone composed primitive rather than in the context of its application to secure channels as we do. The main contribution of [5] is in providing careful quantitative relations and reductions between different methods and security notions related to these forms of composition. These results, however, are insufficient in general for claiming the security, or demonstrating the insecurity, of channels that use these methods for protecting data. For example, while [5] show that authenticate-then-encrypt is not necessarily CCA-secure, it turns out (by results in [8] and here) that the lack of this property is no reason to consider insecure the channels that use such a method (moreover, even the specific non-CCA example in [5] does provide secure channels). This demonstrates that the consideration of secure channels requires a finer treatment of the question of encryption/authentication composition (see discussion at the beginning of Section 4.2). In particular, none of our results is claimed or implied by [5].

A related subject that received much attention recently is the construction of encryption modes that provide integrity in addition to secrecy. Katz and Yung [16] suggest a mode of operation for block ciphers that provides such functional combination; for their analysis (and for its independent interest) they introduce the notion of "unforgeable encryption". A very similar notion is also introduced in [5] and called there "integrity of ciphertexts" (INT-CTXT). We use this notion in our work too (see Section 3) as a tool in some of our proofs. In another recent work, An and Bellare [1] study the use of redundancy functions (with and without secret keys) as a method for adding authentication to encryption functions. They show several positive and negative results about the type of redundancy functions that are required in combination with different forms of encryption and security notions. Our results concerning the authenticate-then-encrypt method with stream ciphers and CBC modes contribute also to this research direction since these results provide sufficient and necessary conditions on the redundancy functions (viewed as MAC functions) required for providing integrity to these important modes of encryption. Of particular interest is our proof that a secure *AtE* composition that uses CBC encryption requires a strong underlying MAC; this contradicts a common intuition that (since the message and MAC are encrypted) weaker "redundancy functions" could replace the full-fledge MAC.

Recently, Jutla [15] devised an elegant CBC-like scheme that provides integrity at little cost beyond the traditional CBC method, as well as a parallel mode of encryption with integrity guarantee (a related scheme is presented in [26]). We note that while schemes such as [15] can be used to efficiently implement secure channels that provide secrecy and authenticity, generic schemes like encrypt-then-authenticate have several design and analysis advantages due to their modularity and the fact that the encryption and authentication components can be designed, analyzed and replaced independently of each other. In particular, generic schemes can allow for faster implementations than the specific ones; even today the combination of fast stream ciphers with a fast MAC function such as UMAC [6] under the encrypt-then-authenticate method would result in a faster mechanism than the one proposed in [15] which requires the use of block ciphers. Also, having a separate MAC from encryption allows for much more efficient authentication in the cases where secrecy is not required.

2 Preliminaries

We informally outline some well-known notions of security for MAC and encryption functions as used throughout the paper, and introduce some notation. References are given below for formal treatment of these notions. We also sketch the model of "secure channels" from [8].

2.1 Secure Message Authentication

Functions that provide a way to verify the integrity of information (for example, against unauthorized changes over a communications network) and which use a shared secret key are called *MAC* (*message authentication codes*). The notion of a MAC and its security definition is well understood [4]. Here we outline the main ingredients of this definition as used later in the paper.

A MAC scheme is described as a family of (deterministic) functions over a given domain and range. (We will usually assume the domain to be $\{0,1\}^*$ and the range $\{0,1\}^n$ for fixed size n.). The key shared by the parties that use the MAC scheme determines a specific function from this family. This specific function is used to compute an *authentication tag* on each transmitted message and the tag is appended to the message. A recipient of the information that knows the MAC key can re-compute the tag on the received message and compare to the received tag. Security of a MAC scheme is defined through the inability of an attacker to produce a *forgery*, namely, to generate a message, not transmitted between the legitimate parties, with its valid authentication tag. The formal definition of security provides the attacker with access to a *MAC oracle* $\mathcal{O}_{\text{MAC}}$ that on input a message x outputs the authentication tag corresponding to that message. The oracle uses for its responses a key that is generated according to the probability distribution of keys defined by the MAC scheme. The attacker succeeds if after this interaction with the oracle it is able to find a forgery (for a message not previously queried). To quantify security we say that *a MAC scheme has security* $\mathcal{E}_M(q,Q,T)$ *if any attacker that works time* T *and asks* q *queries from* $\mathcal{O}_{\text{MAC}}$ *involving a total of* Q *bits has probability at most* $\mathcal{E}_M(q,Q,T)$ *to produce a forgery.*

Remark. In the case of MAC functions (e.g., randomized ones) where there may be multi-valued valid tags for the same message, we extend the definition of security as follows. If the messages queried to $\mathcal{O}_{\text{MAC}}$ are $x_1, x_2, \ldots, x_q$ and the responses from $\mathcal{O}_{\text{MAC}}$ are $t_1, t_2, \ldots, t_q$ then a forgery (x,t) output by the attacker is considered valid if $(x,t) \neq (x_i,t_i)$ for all $i = 1,\ldots,q$. (Namely, we consider the attacker successful even in case its forgery includes a queried message as long as the tag t was not generated by the oracle for that message.) This technical strengthening of the definition is used in some of our results. This notion appears (due to similar reasons) also in [5].

2.2 Secure Symmetric Encryption

We do not develop a formal definition of encryption security here as the subject is well established and treated extensively in the literature. Yet, we summarize informally the main aspects of the security notions of symmetric encryption that are relevant to our work and establish some notation. For formal and precise definitions see the references mentioned below.

An encryption scheme is a triple of (probabilistic) algorithms (*KEYGEN*, *ENC*, *DEC*) where *KEYGEN* defines the process (and resultant probability distribution) by which keys are generated, while *ENC* and *DEC* are the encryption and decryption operations with the usual inverse properties. To simplify notation we use *ENC* to denote the encryption operation itself but also as representing the whole scheme (i.e., a triple as above). The main notion behind the common definitions of security of encryption is *semantic security* [13], or its (usually) equivalent formulation via *plaintext indistinguishability*. In this formulation an attacker against a scheme *ENC* is given a *target ciphertext* y and two candidate plaintexts x_1, x_2 such that $y = ENC(x_i)$, $i \in_R \{0,1\}$.[1] The encryption scheme has the indistinguishability property if the attacker cannot guess the right value of i with probability significantly better than 1/2. The security of the scheme is quantified via the time invested by the attacker and the probability beyond 1/2 to guess correctly.

The above describes the goal of the attacker but not the ways of attack it is allowed to use. Two common models of attack are CPA (chosen plaintext attack) and CCA (chosen ciphertext attack). In the first the attacker has access to an *encryption oracle* $\mathcal{O}_{\mathrm{ENC}}$ to which it can present plaintexts and receive the ciphertexts resulting from the encryption of these plaintexts. In the second model the attacker can, in addition to the above queries to the encryption oracle, also ask for decryptions of arbitrary ciphertexts (except for the target ciphertext y) from a *decryption oracle* $\mathcal{O}_{\mathrm{DEC}}$. We note that both $\mathcal{O}_{\mathrm{ENC}}$ and $\mathcal{O}_{\mathrm{DEC}}$ use the same key for their responses which is also the key under which the target ciphertext y, as described above, is produced. In both cases the queries to the oracles can be generated adaptively by the attacker, i.e. as a function of previous responses from the oracles and of the target ciphertext y (actually, also the candidate plaintexts x_1, x_2 on which the target ciphertext y is computed can be chosen by the attacker). Under these formulations two new parameters enter the quantification of security: the number of queries to $\mathcal{O}_{\mathrm{ENC}}$ and the number of queries to $\mathcal{O}_{\mathrm{DEC}}$ (the latter is 0 in the case of CPA). A finer quantification would also consider the total number of bits in these queries.

[1] We use the notation $a \in_R A$ to denote that the element a is chosen with uniform probability from the set A.

As it is customary we denote the above two notions of encryption security as IND-CPA and IND-CCA. Extensive treatment of these notions can be found among other works in [13,12,2] and [24,3,17], respectively. A notion strongly related to IND-CCA is non-malleability of ciphertexts [10] which we do not use directly here; a weaker notion of CCA security was introduced earlier in [23]. We also note that we are only concerned with symmetric encryption; asymmetric encryption shares many of the same aspects but there are some important differences as well (in particular, in the asymmetric case encryption oracles are meaningless since everyone can encrypt at will any plaintext).

2.3 Secure Channels

In order to claim our positive results, i.e. that a certain combination of encryption and authentication provides secure communications, we need to define what is meant by such "secure communications". For this we use the model of secure channels introduced by Canetti and Krawczyk [8] and which is intended to capture the standard network-security practice in which communications over public networks are protected through "sessions" between pairs of communicating parties, and where each session consists of two stages. First, the two parties run a key-exchange protocol that establishes an authenticated and secret session key shared between the parties. Then, in the second stage, this session key is used, together with symmetric-key cryptographic functions, to protect the integrity and/or secrecy of the transmitted data. The formalism of [8] involves the definition of a key-exchange protocol for implementation of the session and key establishment stage, as well as of two functions, snd and rcv, that define the actions applied to transmitted data for protection over otherwise insecure links. A protocol that follows this formalism is called in [8] a "network channels protocol", and its security is defined in terms of authentication and secrecy.

These notions are defined in [8] in the context of communications controlled by an attacker with full control of the information sent over the links and with the capability of corrupting sessions and parties. We refer to the full version of [8] for a full description of the adversarial model and security definitions. Here we only mention briefly the main elements in this definition concerning the functions snd and rcv. The function snd represents the operations and transformations applied to a message by its sender in order to protect it from adversarial action over the communication links. Namely, when a message m is to be transmitted from party P to party Q under a session s established between these parties, the function snd is applied to m and, possibly, to additional information such as a message identifier. The definition of snd typically consists of the application of some combination of a MAC and symmetric encryption keyed via the session key. The function rcv describes the action at the receiving end for "decoding" and verifying incoming messages, and it typically involves the verification of a MAC and/or the decryption of an incoming ciphertext.

Roughly speaking, [8] define that authentication is achieved by the protocol if any message decoded and accepted as valid by the receiving party to a session was indeed sent by the partner to that session. (That is, any modification of messages produced by the attacker over the communications links, including the injection or replay of messages, should be detected and rejected by the recipient; in [8] this is formalized as the "emulation" of an ideally-authenticated channel.) Secrecy is formalized in the tradition of semantic security: among the many messages exchanged in a session the attacker chooses a pair of "test messages" of which only one is sent; the attacker's goal is to guess which one was sent. Security is obtained if the attacker cannot guess correctly with probability significantly greater than 1/2. A network channels protocol is called a secure channels protocol if it achieves both authentication and secrecy in the sense outlined above.

In this paper we focus on the way the functions snd and rcv are to be defined to achieve secure channels, i.e. to provide both authentication and secrecy in the presence of an attacker as above. We say that any of the combinations *EtA*, *AtE*, *E&A* implements secure channels if when used as the specification of the snd and rcv functions the resultant protocol is a "secure channels protocol". Note that we are not concerned here with a specific key-exchange mechanism, but rather assume a secure key-exchange protocol [8], and may even assume an "ideally shared" session key.

3 CUF-CPA: Ciphertext Unforgeability

In addition to the traditional notions of security for an encryption scheme outlined in Section 2.2 we use the following notion of security that we call ciphertext unforgeability. A similar notion has been recently (and independently) used in [16,5] where it is called "existential unforgeability of encryption" and "integrity of ciphertexts (INT-CTXT)", respectively.

Let *ENC* be a symmetric encryption scheme, and k be a key for *ENC*. Let $P(k)$ be the set of plaintexts on which ENC_k is defined, and $C(k)$ be the set of ciphertexts $\{y : \exists x \in P(k) \text{ s.t. } y = ENC_k(x)\}$ (note that if *ENC* is not deterministic then by $y = ENC_k(x)$ we mean that there is a run of *ENC* on x that outputs y). We call $C(k)$ the set of valid ciphertexts under key k. For example, under a block cipher only strings of the block length are valid ciphertexts while in the basic CBC mode only strings that are multiples of the block length can be valid ciphertexts. We assume that the decryption oracle $\mathcal{O}_{\text{DEC}}$ outputs a special "invalidity symbol" $\perp$ when queried with an invalid ciphertext (and otherwise outputs the unique decrypted plaintext x).

We say that an encryption scheme is ciphertext unforgeable, and denote it CUF-CPA, if it is infeasible for any attacker $\mathcal{F}$ (called a *"ciphertext forger"*) that has access to an encryption oracle $\mathcal{O}_{\text{ENC}}$ with key k to produce a valid ciphertext under k not generated by $\mathcal{O}_{\text{ENC}}$ as response to one of the queries by $\mathcal{F}$. More precisely, we quantify ciphertext unforgeability by the function $\mathcal{E}_U(q, Q, T)$ defined as the maximal probability of success for any ciphertext forger $\mathcal{F}$ that queries q plaintexts totalling Q bits and spends time T in the attack. We stress that this definition does not involve access to a decryption oracle and thus its name CUF-CPA (this is consistent with other common notations of the form

X-Y where X represents the goal of the attacker and Y the assumed abilities of the attacker).

Our main use of the CUF-CPA notion is for proving (see Section 5) that under certain conditions the *AtE* composition is secure, i.e., it implements secure channels. However, the notion of CUF-CPA while sufficient for our purposes is actually stronger than needed. For example, any scheme *ENC* that allows for arbitrary padding of ciphertexts to a length-boundary (e.g., to a multiple of 8-bits) will not be CUF-CPA (since given a ciphertext with padded bits any change to these bits will result in a different yet valid ciphertext). However, such a scheme may be perfectly secure in the context of implementing secure channels (see [8]); moreover, schemes of this type are common in practice. Thus, in order to avoid an artificial limitation of the schemes that we identify as secure for implementing secure channels we present next a relaxation of the CUF-CPA notion that is still sufficient for our purposes (we stress that this is not necessarily the weakest relaxation for this purpose and other weakenings of the CUF-CPA notion are possible).

Let ρ be a *polynomial-time computable* relation on pairs of ciphertexts computed under the encryption function *ENC* with the property that $\rho(c, c')$ implies that c and c' decrypt to the same plaintext. Then we say that the encryption scheme *ENC* is CUF_ρ-CPA if for any valid ciphertext c that the attacker can feasibly produce there exists a ciphertext c' output by the encryption oracle such that $\rho(c, c')$. When the relation ρ is not explicitly described we will refer to this notion as loose ciphertext unforgeability.

For instance, in the above example of a scheme that allows for arbitrary padding of ciphertexts, if one defines $\rho(c, c')$ to hold if c and c' differ only on the padding bits, then the scheme can achieve CUF_ρ-CPA. We note that while CUF-CPA implies CCA-security, loose CUF-CPA does not (as the above "padding example" shows). Indeed, as we pointed out in the introduction (see also Section 4.2) CCA-security is not a necessary condition for a MAC/encryption combination to implement secure channels.

4 Generic Composition of Encryption and Authentication

In this section we study the security of the three methods, *EtA*, *AtE*, *E&A*, under generic symmetric encryption and MAC functions where the only assumption is that the encryption is IND-CPA and the MAC is secure against chosen message attacks. Our focus is on the appropriateness of these methods to provide security to transmitted data in a realistic setting of adversarially-controlled networks. In other words, we are interested in whether each one of these methods when applied to adversarially-controlled communication channels achieve the goals of information secrecy and integrity. As we will see only the encrypt-then-authenticate method is generically secure.

4.1 The Known Security of Encrypt-then-Authenticate

The results in this subsection are from [8] and we present them briefly for completeness. We refer the reader to that paper for details. In particular, in the statement of the next theorem we use the notion of "secure channels" as introduced in the above paper and sketched in Section 2.3.

Theorem 1. [8] *If ENC is a symmetric encryption scheme secure in the sense of IND-CPA and MAC is a secure MAC family then method EtA(ENC, MAC) implements secure channels.*

Following our terminology from Section 2.3, the meaning of the above theorem is that if in the network channels model of [8] one applies to each transmitted message the composed function *EtA(ENC, MAC)* (as the snd function) then the secrecy and authenticity of the resultant network channels is guaranteed. More precisely, in proving the above theorem, [8] specify the snd function as follows. First, a pair of (computationally independent) keys, κ_a and κ_e, are derived from each session key. Then, for each transmitted message, m, a *unique* message identifier *m-id* is chosen (e.g., a sequence number). Finally, the function snd produces a triple (x, y, z) where $x = m\text{-}id$, $y = ENC_{\kappa_e}(m)$, $z = MAC_{\kappa_a}(m\text{-}id, y)$. On an incoming message (x', y', z') the rcv function verifies the uniqueness of message identifier x' and the validity of the *MAC* tag z (computed on (x', y')); if the checks succeeds y' is decrypted under key κ_e and the resultant plaintext accepted as a valid message.[2]

A main contribution of the present paper is in showing (see next subsections) that a *generic* result as in Theorem 1 cannot hold for any of the other two methods, *AtE* and *E&A* (even if the used keys are shared with perfect security). Therefore, any secure channels protocol designed to work with *any* combination of secure encryption (against chosen plaintext attacks) and secure MAC must use the encrypt-then-authenticate method. However, we note in Section 5 that the above theorem can be extended in the setting of method *AtE* if one assumes a stronger property on the encryption function; in particular, we show two important cases that satisfy the added security requirement.

Remark. Note that the authentication of the ciphertext provides plaintext integrity as long as the encryption and decryption keys used at the sender and receiver, respectively, are the same. While this key synchrony is implicit in our analytical models [8], a key mismatch can happen in practice. A system concerned with detecting such cases can check the plaintext for redundancy information (such redundancy exists in most applications: e.g., message formats, non-cryptographic checksums, etc.). If the redundancy entropy is significant then a key mismatch will corrupt this redundancy with high probability.

[2] Protocols that use a synchronized counter as the message identifier, e.g. SSL, do not need to transmit this value; yet they must include it under the *MAC* computation and verification. If transmitted, identifiers are not encrypted under ENC_{κ_e} since they are needed for verifying the *MAC* value before the decryption is applied.

4.2 Authenticate-then-Encrypt Is Not Generically Secure

Here we show that the authenticate-then-encrypt method *AtE*(*ENC*, *MAC*) is not guaranteed to be secure for implementing secure channels even if the function *ENC* is IND-CPA and *MAC* provides message unforgeability against chosen message attacks. First, however, we discuss shortly why this result does not follow from [5] where it is shown that the *AtE* composition (viewed as an encryption scheme) does not necessarily provide IND-CCA. The reason is simple: as demonstrated in [8] IND-CCA *is not a necessary condition for a combination of encryption and MAC functions to implement secure channels.* An example is provided by the main construction of secure channels in [8] (see Theorem 1): if the MAC used in this scheme enjoys regular MAC security, rather than the strengthened notion described in the last remark of Section 2.1, then this construction guarantees secure channels but not necessarily CCA security. (For example, if the *MAC* function has the property that flipping the last bit of an authentication tag does not change the validity of the tag, then the scheme in [8] is not IND-CCA yet it suffices for implementing secure channels. For a similar example, see remark on "multi-valued MAC" following our Theorem 3.) Moreover, the specific example from [5] of a non-CCA *AtE*(*ENC*, *MAC*) scheme[3] can by itself be used to show an example of a non-CCA scheme that provably provides secure channels. Therefore, the result in [5] does not say anything about the suitability of *AtE*(*ENC*, *MAC*) for implementing secure channels; it rather points out to the fact that while CCA security is a useful security notion it is certainly too strong for some (fundamental) applications such as secure channels.

Thus if we want to establish the *insecurity* of authenticate-then-encrypt channels under generic composition we need to show an explicit example and a successful attack. We provide such example now. In this example the encryption scheme is IND-CPA (actually, it enjoys "perfect secrecy" in the sense of Shannon) but when combined with *any* MAC function under the *AtE* method the secrecy of the composed scheme breaks completely under an active attack.

The encryption function ENC^*. We start by defining an encryption scheme ENC^* that can be based on any stream cipher *ENC* (i.e. any encryption function that uses a random or pseudorandom pad to xor with the data). The scheme ENC^* preserves the IND-CPA security of the underlying scheme *ENC*. In particular, if *ENC* has perfect secrecy (i.e., uses a perfect one-time pad encryption) so does ENC^*. Next, we define ENC^*.

Given an n-bit plaintext x (for any n), ENC^* first applies an encoding of x into a $2n$-bit string x' obtained by representing each bit x_i, $i = 1, \ldots, n$, in x with two bits in x' as follows:

1. if bit $x_i = 0$ then the pair of bits (x'_{2i-1}, x'_{2i}) is set to $(0, 0)$;
2. if bit $x_i = 1$ then the pair of bits (x'_{2i-1}, x'_{2i}) is set to $(0, 1)$ or to $(1, 0)$ (by arbitrary choice of the encrypting party).

[3] Just append an arbitrary one-bit pad to the ciphertext and discard the bit before decryption.

The encryption function *ENC* is then applied to x'. For decrypting $y = ENC^*(x)$ one first applies the decryption function of *ENC* to obtain x' which is then decoded into x by mapping a pair $(0,0)$ into 0 and either pair $(0,1)$ or $(1,0)$ into 1. If x' contains a pair (x'_{2i-1}, x'_{2i}) that equals $(1,1)$ the decoding outputs the invalidity sign $\perp$.

The attack when only encryption is used. For the sake of presentation let's first assume that *only* ENC^* is applied to the transmitted data (we will then treat the *AtE* case where a MAC is applied to the data before encryption). In this case when an attacker $\mathcal{A}$ sees a transmitted ciphertext $y = ENC^*(x)$ it can learn the first bit x_1 of x as follows. It intercepts y, flips (from 0 to 1 and from 1 to 0) the first two bits (y_1, y_2) of y, and sends the modified ciphertext y' to its destination. If A can obtain the information of whether the decryption output a valid or invalid plaintext then A learns the first bit of x. This is so since, as it can be easily seen, the modified y' is valid if and only if $x_1 = 1$. (Remember that we are using a stream cipher to encrypt x'.) Clearly, this breaks the secrecy of the channel (note that the described attack can be applied to any of the bits of the plaintext). One question that arises is whether it is realistic to assume that the attacker learns the validity or invalidity of the ciphertext. The answer is that this is so for many practical applications that will show an observable change of behavior if the ciphertext is invalid (in particular, many applications will return an error message in this case).

To make the point even clearer consider a protocol that transmits passwords and uses ENC^* to protect passwords over the network (this is, for example, one of the very common uses of SSL). The above attack if applied to one of the bits of the password (we assume that the attacker knows the placement of the password field in the transmitted data) will work as follows. If the attacked bit is 1 then the password authentication will succeed in spite of the change in the ciphertext. If it is 0 the password authentication will fail. In this case success or failure is reported back to the remote machine and then learned by the attacker. In applications where the same password is used multiple times (again, as in many applications protected by SSL) the attacker can learn the password bit-by-bit. The same can be applied to other sensitive information such as to credit card numbers where a mistake in this number will be usually reported back and the validity/invalidity information will be learned by A.

The attack against the $AtE(ENC^*, MAC)$ **scheme.** Consider now the case of interest for us in which the encryption is applied not just to the data but also to a MAC function computed on this data. Does the above attack applies? The answer is YES. The MAC is applied to the data before encoding and encryption and therefore if the original bit is 1 the change in ciphertext will result in the same decrypted plaintext and then the MAC check will succeed. Similarly, if the original bit is 0 the decrypted plaintext will have a 1 instead and the MAC will fail. All the attacker needs now is the information of whether the MAC succeeded or not. Note that in a sense *the MAC just makes things worse* since regardless of the semantics of the application a failure of authentication is easier to learn by the attacker: either via returned error messages, or by other effects on the application that can be observed by the attacker.

Discussion: what have we learned? The example using ENC^* is certainly sufficient to show that the method *AtE* can be insecure even if the encryption function is IND-CPA secure and the MAC unforgeable (note that this conclusion does not depend on any specific formalization of secure communications; any reasonable definition of security must label the above protocol as insecure). Therefore, if one wants to claim the security of $AtE(ENC, MAC)$ for particular functions *ENC* and *MAC* one needs to analyze the combination as a whole or use stronger or specific properties of the encryption function (see Section 5). An interesting issue here is how plausible it is that people will ever use an encryption scheme such as ENC^*. We note that although this scheme does not appear to be the most natural encryption mechanism some (equally insecure) variants of it may arise in practice. First the application of an encoding to a plaintext before encryption is used many times for padding and other purposes and is a particularly common practice in public key encryption algorithms. Second, encodings of this type can be motivated by *stronger* security requirements: e.g. to prevent an attacker from learning the exact length of transmitted messages or other traffic analysis information. In this case one could use an encoding similar to ENC^* but with variable size codes. (Just to make the point: note that a good example of traffic analysis arises in the above examples where the attacker has a lot to learn from error-reporting messages; even in cases where this information is encrypted it can usually be learned through the analysis of packet lengths, etc.) Another setting where plaintext encoding is introduced in order to improve security is for combating timing and power analysis attacks.

The bottom line is that it is highly desirable to have schemes that are robust to generic composition and are not vulnerable when seemingly innocuous changes are made to an algorithm (or when a new more secure or more efficient algorithm or mode is adopted)[4].

4.3 Encrypt-and-Authenticate Is Not Generically Secure

The first observation to make regarding the encrypt-and-authenticate method is that under the common requirements from a MAC function this method cannot guarantee the protection of secrecy (even against a passive eavesdropper). This is so since a MAC can be secure against forgeries but still leak information on the plaintext. Thus, the really interesting question is whether the method becomes secure if we avoid this obvious weakness via the use of a "secrecy protecting" MAC such as one implemented via a pseudorandom function or when the MAC tag is encrypted (most, if not all, MAC functions used in practice are believed to protect secrecy). Unfortunately, however, the attack from the previous section applies here too, thus showing the (generic) insecurity of the *E&A* method even under the above stronger forms of MAC. (See also last remark in Section 5.2.)

[4] See the last remark in Section 5.2 for another example where seemingly harmless changes transform a secure protocol into an insecure one.

5 Authenticate-then-Encrypt with CBC and OTP Modes

In Section 4.2 we saw that authenticate-then-encrypt cannot guarantee secure channels under the sole assumption that the encryption function is IND-CPA, even if the MAC function is perfectly secure. In this section we prove that for two common modes of encryption, CBC (with a secure underlying block cipher) and OTP (stream ciphers that xor data with a (pseudo) random pad), the *AtE* mode does work for implementing secure channels.

5.1 A Sufficient Condition for the Security of *AtE*

We start by pointing out to the following Theorem that can be proven in the security model of [8] (see Section 2.3).

Theorem 2. (derived from [8]) *Let ENC be an IND-CPA encryption function and MAC a MAC function. If the composed function AtE(ENC, MAC), considered as an encryption scheme, is (loose) CUF-CPA, then AtE(ENC, MAC) implements secure channels.*

That is, under the assumptions on the *ENC* and *MAC* functions as stated in the Theorem, applying the function *AtE*(*ENC*, *MAC*) to information transmitted over adversarially-controlled links protects the secrecy and integrity of this information. More specifically, the Theorem implies the following definition of the function snd in the network channels model of [8] (see Section 2.3). For each transmitted message m with unique message identifier *m-id* the function snd produces a pair (x, y) where $x = m\text{-}id$ and $y = ENC_{\kappa_e}(m, MAC_{\kappa_a}(m\text{-}id, m))$, where the keys κ_e and κ_a are computationally independent keys derived from the session key. On an incoming message (x', y') the rcv function verifies the uniqueness of message identifier x', decrypts y' under key κ_e, verifies the validity of the decrypted *MAC* tag, and if all tests succeed the recipient accepts the decrypted message as valid. We note that if the message identifier is maintained in synchrony by sender and receiver (as in SSL) then there is no need to send its value over the network. On the other hand, if sent, the message identifier can be encrypted too. The above Theorem holds in either case.

We stress that the Theorem holds for strict CUF-CPA as well as for the relaxed "loose" version (see Section 3).

Based on this Theorem, and on the fact that OTP and CBC are IND-CPA [2], we can prove the security of *AtE* under OTP and CBC by showing that in this case the resultant *AtE* scheme is CUF-CPA. The rest of this section is devoted to prove these facts.

5.2 *AtE* with OTP

The OTP scheme. Let F be a family of functions with domain $\{0,1\}^\ell$ and range $\{0,1\}^{\ell'}$. We define the encryption scheme $OTP(F)$ to work on messages of length at most ℓ' as follows. A key in the encryption scheme is a description of a member f of the family F. The OTP encryption under f of plaintext x is performed by choosing $r \in_R \{0,1\}^\ell$ and computing $c = f(r) \oplus x$ where $f(r)$ is truncated to the length of x. The ciphertext is the pair (r, c). Decryption works in the obvious way. If F is the set of *all* functions with the above domain and range and f is chosen at random from this family we get perfect secrecy against chosen-plaintext attacks as long as there are no repetitions in the values r chosen by the encryptor (after encrypting q different messages a repetition happens with probability $q^2/2^\ell$); we denote this scheme by $OTP_\$$. If F is a family of pseudorandom functions then the same security is achieved but in a computational sense, i.e., up to the "indistinguishability distance" between the pseudorandom family and a truly random function. A formal and exact-security treatment of this mode of encryption can be found in [2].

The $AtE(OTP_\$, MAC)$ composition. Let MAC be a MAC family with n-bit outputs, and k a key to a member of that family. Let f be a random function with domain and range as defined above. The $AtE(OTP_\$, MAC)$ function with f and k acts as follows: (i) it receives as input a message x of length at most $\ell' - n$, (ii) computes $t = MAC_k(x)$, (iii) appends t to x, (iv) outputs the OTP encryption under f of the concatenated message (x, t).

The following theorem establishes the CUF-CPA security of $AtE(OTP_\$, MAC)$ as a function of the security $\mathcal{E}_M(\cdot,\cdot,\cdot)$ of MAC.

Theorem 3. *If MAC is a MAC family that resists one-query attacks then $AtE(OTP_\$, MAC)$ is CUF-CPA (and then by Theorem 2 it implements secure channels). More precisely, any ciphertext forger $\mathcal{F}$ against $AtE(OTP_\$, MAC)$ that runs time T has success probability $\mathcal{E}_U$ of at most $q^2/2^\ell + \mathcal{E}_M(1, p, T')$, where ℓ is a parameter of $OTP_\$$, q is the number of queries $\mathcal{F}$ makes during the attack, p is an upper bound on the length of each such query and on the length of the output forgery, and $T' = T + cqp$ for some constant c.*

For a proof of the Theorem see [21].

Using standard techniques one can show that the theorem holds also for a OTP scheme realized via a family of pseudorandom functions if we add to the above probability bound the distinguishability distance between the pseudorandom family and a truly random function. Also, the $q^2/2^\ell$ component can be eliminated if one uses non-repeating nonces instead of random r's (such as in counter mode or via a stateful pseudorandom generator used to generate a pseudorandom pad).

Remark (*Tightness: one-query resistance is necessary*). Here is an example of a MAC that *does not* resist one-queries and with which valid ciphertext can be forged against $AtE(OTP_\$, MAC)$. Assume MAC allows for finding two same-length messages with the same MAC tag. (For example, MAC first zeros the last bit of the message and then applies a secure MAC function on the resultant message. Thus, MAC resists zero-queries but fails to one-queries: ask for a MAC on a message, then forge for the message with last bit flipped.) The strategy of the ciphertext forger against $AtE(OTP_\$, MAC)$ is to find such pair of messages x_1, x_2. Then, it queries the first one and gets the ciphertext (r, c).

Finally, it outputs the forgery (r, c') where c' is obtained from c by xor-ing x_2 to the first $|x_2|$ bits of c. It is easy to see that (r, c') decrypts to $(x_2, MAC(x_2))$.

Remark (*Multi-valued MAC*). In Section 2.1 we strengthened the regular security definition of a MAC function in the case that the function allows for different valid authentication tags for the same message. This extended definiton is used (explicitly) in the proof of Theorem 3 and is essential for ensuring the CUF-CPA property of $AtE(OTP_\$, MAC)$. To see this, let MAC be a secure single-valued MAC function and define MAC' to be the same as MAC except that an additional arbitrary bit is appended to each authentication tag. The verification procedure will just ignore this bit. It is easy to see that in this case $AtE(OTP_\$, MAC')$ will not be CUF-CPA. However, if one examines the proof of Theorem 3 it can be seen that $AtE(OTP_\$, MAC')$ achieves loose CUF-CPA (see Section 3) and then it is sufficient for implementing secure channels (which is what we care about). So can we dispense of the strengthened notion of MAC when multi-valued MACs are used? The answer is no. It is possible to build a multi-valued function MAC' that satisfies the regular MAC definition, but not the strengthened version, for which $AtE(OTP_\$, MAC')$ is *insecure* for building secure channels (see [21]).

Remark (*Sufficiency of redundancy functions*). In [1] An and Bellare investigate the question of whether simple redundancy functions (such as combinatorial hash functions) applied to a plaintext before encryption suffice for providing ciphertext unforgeability. In the case of AtE with OTP it seems natural to assume that a simple combinatorial property of the redundancy function such as AXU [20,25] should suffice. (In particular, this seems so since such a property is sufficient [20] if one only considers *plaintext integrity* where only the output of the redundancy function is encrypted under an OTP scheme.) However, this turns out not to be true in the case of ciphertext unforgeability. We can show an example of an $\mathcal{E}$-AXU (and also $\mathcal{E}$-balanced [20]) MAC family for which $AtE(OTP_\$, MAC)$ is not CUF-CPA. It seems plausible, however, that a more involved combinatorial property (involving the length of messages) of the MAC function could suffice to guarantee ciphertext unforgeability in the case of AtE with OTP. Actually, it is interesting to note that if the authentication tag is positioned *before* the message, instead of at the end as defined above, the AXU property is indeed sufficient (assuming fixed-length and single-valued valid authentication tags).

Remark (*Beware of "slight changes"*). To highlight the "fragility" of the result in Theorem 3 we note that the proof of this theorem uses in an essential way the fact that the encryption is applied as a whole on the concatenated message and MAC tag. If we were to encrypt these two values *separately* (i.e., using separate IVs for the encryption of the message and of the MAC) even under a truly random function we would not get CUF or CCA security. More significantly, such separate encryption results in *insecure channels.* Indeed, under this method an active attacker can get to learn whether two transmitted messages, possibly with different message identifiers, are the same, something clearly unwanted in a secure protocol. (This weakness allows for actual attacks on practical applications, in particular several forms of "dictionary attacks"[5])

In addition, this observation shows another weakness of the encrypt-and-authenticate method (Section 4.3) since it exhibits the insecurity of this method even under the use of a standard stream cipher for encryption and even when the MAC tag is encrypted.

5.3 *AtE* with CBC

The CBC scheme. Let ℓ be a positive integer and F be a family of permutations over $\{0,1\}^\ell$. We define the encryption scheme $CBC(F)$ to work on messages of length a multiple of ℓ. A key in the encryption scheme is a description of a member f of the family F. The CBC encryption under f of plaintext x is performed by partitioning x into blocks $x[1],\ldots,x[p]$ of length ℓ each, then choosing $r \in_R \{0,1\}^\ell$ (called the IV) and computing the ciphertext $c = c[0], c[1], \ldots, c[p]$ as $c[0] = r, c[i] = f(c[i-1] \oplus x[i]), i = 1,\ldots,p$. Decryption works in the obvious inverse way. If F is the set of *all* permutations over $\{0,1\}^\ell$ and f is chosen at random from F then we denote the scheme by $CBC_\$$. A formal and exact-security treatment of this mode of encryption can be found in [2] who in particular prove it to be IND-CPA also in the case where F is a pseudorandom family (in this case the security depends on the "indistinguishability distance" between the pseudorandom family and a truly random function).

The $AtE(CBC_\$, MAC)$ composition. Let MAC be a MAC family with ℓ-bit outputs, and k a key to a member of that family. Let f be a random permutation over $\{0,1\}^\ell$. The $AtE(CBC_\$, MAC)$ function with f and k acts as follows: (i) it receives as input a message x of length multiple of ℓ, (ii) computes $t = MAC_k(x)$, (iii) appends t to x, (iv) outputs the CBC encryption under f of the concatenated message (x,t) (note that the resultant output is two blocks longer than x due to the added block t and the prepended IV r).

The following theorem establishes the CUF-CPA security of $AtE(CBC_\$, MAC)$ as a function of the security $\mathcal{E}_M(\cdot,\cdot,\cdot)$ of MAC.

Theorem 4. *If MAC is a secure MAC family then $AtE(CBC_\$, MAC)$ is CUF-CPA (and then by Theorem 2 it implements secure channels). More precisely, any ciphertext forger $\mathcal{F}$ against $AtE(CBC_\$, MAC)$ that runs time T has success probability $\mathcal{E}_U$ of at most*

$$Q^2/2^\ell + 2q\mathcal{E}_M(0,0,T') + \mathcal{E}_M(1,p\ell,T') + 2\mathcal{E}_M(q^*,q^*p\ell,T')$$

where q is the number of plaintexts queried by $\mathcal{F}$, p is an upper bound on the number of blocks in each of these queries, p^* is the length in blocks of the forgery y^* output by $\mathcal{F}$, $q^* = \min\{q,p^*\}$, Q is the total number of blocks in the responses to $\mathcal{F}$'s queries plus p^*, and $T' = T + cQ$ for constant c.

For a proof of the Theorem see [21].

[5] One such example would be finding passwords sent in the `telnet` protocol even if the protocol is run over a secure channel protected as above; this is particularly facilitated by the fact that in this case *individual password characters* are transmitted separately, and thus a dictionary attack can be mounted on individual characters.

Using standard techniques one can show that the theorem holds also for a CBC scheme realized via a family of pseudorandom permutations if we add to the above probability bound the distinguishability distance between the pseudorandom family and a truly random function. However, we note, that in this case the distinguisher not only gets access to an oracle that computes the function but also to an oracle that computes the inverse function (that is, we need to assume the family of permutations to be "super pseudorandom" [22]).

Remark (*Tightness: the necessity of the bound* $\mathcal{E}_M(q^*)$). The most "expensive" term in MAC security in the expression of the theorem is the value $\mathcal{E}_M(q^*)$ since other terms only require protection against one-query or zero-query. Since an attacker $\mathcal{F}$ does not get to see any of the MAC values one could wonder why such a strong security from the MAC is required. We show here that, in contrast to the *AtE*($OTP_\$$, *MAC*) case, this requirement is unavoidable. Specifically, we present for any $i = 0, 1, 2, \ldots$, an example of a MAC function *MAC* that is secure against i queries but yields an insecure *AtE*($CBC_\$$, *MAC*) scheme with $q = i + 1$ (and $p^* = 2i + 4$). We describe the example for $i = 1$, the extension to other values is straightforward.

Let $\{g_k\}_k$ be a family of pseudorandom functions from $(\{0,1\}^\ell)^*$ to $\{0,1\}^{\ell/2}$. Define a MAC family MAC' on the same domain as $\{g_k\}_k$, and with ℓ-bit outputs as follows: $MAC'_{(k_1,k_2)}(x) = (g_{k_1}(x), g_{k_2}(g_{k_1}(x)))$. Define a second MAC family *MAC* that uses the same set of keys as MAC' and such that on key (k_1, k_2):

1. if the input x contains two ℓ-bit blocks b_i and b_j, $i < j$, such that $b_i \neq b_j$ and both have the property that applying g_{k_2} to the first half of the block yields the second half of the block then output b_i as the *MAC* value for x.
2. otherwise, output $MAC'_{(k_1,k_2)}(x)$

It is easy to see that the so defined *MAC* has security of roughly $2^{\ell/2}$ against single queries (but is totally insecure after two queries since the output of *MAC* provides the block format that makes the authentication tag "trivial"). We show that it yields a *AtE*($CBC_\$$, *MAC*) scheme whose ciphertexts are forgeable after two queries even if the encryption permutation f is purely random. The ciphertext forger $\mathcal{F}$ against *AtE*($CBC_\$$, *MAC*) proceeds as follows:

1. Choose two arbitrary one-block long plaintexts x_1, x_2 as the two queries.
2. Let the responses y_1, y_2 be the triples: $(r_1, c_1 = f(r_1 \oplus x_1), m_1 = f(c_1 \oplus MAC(x_1)))$ and $(r_2, c_2 = f(r_2 \oplus x_2), m_2 = f(c_2 \oplus MAC(x_2)))$.
3. Output forgery $y^* = (c_1, m_1, c_2, m_2, c_1, m_1)$.

A simple examination shows that y^* is a valid ciphertext.

One consequence of the above lower bound on the required security of *MAC* is that, somewhat surprisingly, the MAC function cannot be replaced by a simple combinatorial hash function, such as one enjoying AXU (see remark on "redundancy functions" in Section 5.2). Indeed, had AXU been sufficient then one-query resistant MACs would suffice too (since one-query resistance implies AXU). We note that a modified CBC-like mode for which AXU is sufficient is presented in [1].

In contrast to the above lower bound, we do not know if the term $q\mathcal{E}_M(0)$ in the bound of the theorem is necessary or not; we do not have so far an example that shows this term to be unavoidable. Thus, it may well be the case that a more careful analysis could lower the factor q (actually, even with the current analysis it is possible to replace the factor q with q^* by a slightly more involved argument).

Remark (*Non-adaptive security of MAC suffices*). It is interesting to note that the requirement from the security of the MAC in Theorem 4 is for *non-adaptive* queries only. This can be seen by inspecting the proof of the theorem, where the MAC forger $\mathcal{G}$ that we build makes non-adaptive queries only.

Remark (*Beware of "slight changes"*). Similarly to the case of $AtE(OTP_\$, MAC)$ the proof of Theorem 4 uses in an essential way the fact that the encryption is done as a whole on the concatenated message and *MAC*. It is easy to build a ciphertext forgery attack in case the encryption of the plaintext and of the *MAC* tag are done separately (i.e. with independently chosen IVs).

Acknowledgment. I would like to thank Yaron Scheffer for motivating conversations on this topic and for "forcing" me to find an explicit counter-example for the *AtE* method; Yaron also helped in simplifying a previous example. I also thank Mihir Bellare for interesting conversations and for highlighting some of the subtleties related to the subject of this paper, and to Ran Canetti and Jonathan Katz for valuable comments on earlier drafts of the paper.
This research is supported by an Irwin and Bethea Green & Detroit Chapter Career Development Chair, and by the Fund for the Promotion of Research at the Technion.

References

1. J. An, M. Bellare, "Does encryption with redundancy provide authenticity?", *Advances in Cryptology – EUROCRYPT 2001 Proceedings*, Lecture Notes in Computer Science, Vol. 2045, Springer-Verlag, B. Pfitzmann, ed, 2001.
2. M. Bellare, A. Desai, E. Jokipii, and P. Rogaway, "A concrete security treatment of symmetric encryption: Analysis of the DES modes of operation", *Proceedings of the 38th Symposium on Foundations of Computer Science*, IEEE, 1997.
3. M. Bellare, A. Desai, D. Pointcheval, and P. Rogaway, "Relations Among Notions of Security for Public-Key Encryption Schemes", *Advances in Cryptology - CRYPTO'98 Proceedings*, Lecture Notes in Computer Science Vol. 1462, H. Krawczyk, ed., Springer-Verlag, 1998, pp. 26–45.
4. M. Bellare, J. Kilian and P. Rogaway, " The security of cipher block chaining", *Advances in Cryptology – CRYPTO'94 Proceedings*, Lecture Notes in Computer Science Vol. 839, Y. Desmedt, ed., Springer-Verlag, 1994. pp. 341-358.
5. M. Bellare and C. Namprempre, "Authenticated encryption: Relations among notions and analysis of the generic composition paradigm", *Advances in Cryptology - ASIACRYPT'00 Proceedings*, Lecture Notes in Computer Science Vol. 1976, T. Okamoto, ed., Springer-Verlag, 2000.
6. Black, J., Halevi, S., Krawczyk, H., Krovetz, T., and Rogaway, P., "UMAC: Fast and Secure Message Authentication", *Advances in Cryptology – CRYPTO'99 Proceedings*, Lecture Notes in Computer Science, Vol. 1666, Springer-Verlag, M. Wiener, ed, 1999, pp. 216–233.

7. Bleichenbacher, D., "Chosen Ciphertext Attacks against Protocols Based on RSA Encryption Standard PKCS #1", *Advances in Cryptology - CRYPTO'98 Proceedings*, Lecture Notes in Computer Science Vol. 1462, H. Krawczyk, ed., Springer-Verlag, 1998, pp. 1–12.
8. Canetti, R., and Krawczyk, H., "Analysis of Key-Exchange Protocols and Their Use for Building Secure Channels", *Advances in Cryptology – EUROCRYPT 2001 Proceedings*, Lecture Notes in Computer Science, Vol. 2045, Springer-Verlag, B. Pfitzmann, ed, 2001, pp. 453–474. Full version in: *Cryptology ePrint Archive* (`http://eprint.iacr.org/`), Report 2001/040.
9. T. Dierks and C. Allen, "The TLS Protocol – Version 1", *Request for Comments 2246*, 1999.
10. D. Dolev, C. Dwork, and M. Naor. "Non-malleable cryptography". *Proceedings of the 23rd Annual ACM Symposium on Theory of Computing*, pages 542-552, 1991.
11. A. Frier, P. Karlton, and P. Kocher, "The SSL 3.0 Protocol", Netscape Communications Corp., Nov 18, 1996. `http://home.netscape.com/eng/ssl3/ssl-toc.html`
12. O. Goldreich, *"Foundations of Cryptography (Fragments of a book)"*, Weizmann Inst. of Science, 1995. `http://www.wisdom.weizmann.ac.il/ oded/frag.html`
13. S. Goldwasser, and S. Micali. "Probabilistic Encryption", *Journal of Computer and System Sciences*, Vol. 28, 1984, pp. 270-299.
14. Halevi, S., and Krawczyk H., "Public-Key Cryptography and Password Protocols", *ACM Transactions on Information and System Security*, Vol. 2, No. 3, August 1999, pp. 230–268.
15. C. Jutla, "Encryption Modes with Almost Free Message Integrity", *Advances in Cryptology – EUROCRYPT 2001 Proceedings*, Lecture Notes in Computer Science, Vol. 2045, Springer-Verlag, B. Pfitzmann, ed, 2001.
16. J. Katz and M. Yung, "Unforgeable encryption and adaptively secure modes of operations", *Fast Software Encryption'00*, 2000.
17. J. Katz and M. Yung, "Complete characterization of security notions for probabilistic private-key encryption", *Proceedings of the 32nd Annual ACM Symposium on Theory of Computing*, 2000.
18. S. Kent and R. Atkinson, "Security Architecture for the Internet Protocol", *Request for Comments 2401*, Nov. 1998.
19. S. Kent and R. Atkinson, "IP Encapsulating Security Payload (ESP)", *Request for Comments 2406*, Nov. 1998.
20. H. Krawczyk, "LFSR-based Hashing and Authentication", *Proceedings of CRYPTO '94*, Lecture Notes in Computer Science, vol. 839, Y. Desmedt, ed., Springer-Verlag, 1994, pp. 129-139.
21. H. Krawczyk, "The order of encryption and authentication for protecting communications (Or: how secure is SSL?)". Full version: `http://eprint.iacr.org/2001.`
22. M. Luby and C. Rackoff, "How to construct pseudorandom permutations from pseudorandom functions", *SIAM J. on Computing*, Vol 17, Number 2, April 1988, pp. 373–386.
23. M. Naor and M. Yung, "Public key cryptosystems provably secure against chosen ciphertext attacks". *Proceedings of the 22nd Annual ACM Symposium on Theory of Computing*, 1990.
24. C. Rackoff and D. Simon, "Non-interactive zero-knowledge proof of knowledge and chosen ciphertext attack", *Advances in Cryptology - CRYPTO'91 Proceedings*, Lecture Notes in Computer Science Vol. 576, J. Feigenbaum ed, Springer-Verlag.
25. P. Rogaway. " Bucket Hashing and its application to Fast Message Authentication", *Proceedings of CRYPTO '95*, Lecture Notes in Computer Science, vol. 963, D. Coppersmith, ed., Springer-Verlag, 1995, pp. 15-25.

26. P. Rogaway, M. Bellare, J. Black, and T. Krovetz, "OCB Mode", *Cryptology ePrint Archive*, Report 2001/026.
27. T. Ylonen, T. Kivinen, M. Saarinen, T. Rinne, and S. Lehtinen, "SSH Transport Layer Protocol", January 2001, `draft-ietf-secsh-transport-09.txt`.

Forward-Secure Signatures with Optimal Signing and Verifying

Gene Itkis[1] and Leonid Reyzin[2]

[1] Boston University Computer Science Dept.
111 Cummington St.
Boston, MA 02215, USA
itkis@bu.edu

[2] Laboratory for Computer Science
Massachusetts Institute of Technology
Cambridge, MA 02139 USA
reyzin@theory.lcs.mit.edu
http://theory.lcs.mit.edu/~reyzin

Abstract. We propose the first forward-secure signature scheme for which both signing and verifying are as efficient as for one of the most efficient ordinary signature schemes (Guillou-Quisquater [GQ88]), each requiring just two modular exponentiations with a short exponent. All previously proposed forward-secure signature schemes took significantly longer to sign and verify than ordinary signature schemes.

Our scheme requires only fractional increases to the sizes of keys and signatures, and no additional public storage. Like the underlying [GQ88] scheme, our scheme is provably secure in the random oracle model.

1 Introduction

THE PURPOSE OF FORWARD SECURITY. Ordinary digital signatures have a fundamental limitation: if the secret key of a signer is compromised, all the signatures (past and future) of that signer become worthless. This limitation undermines, in particular, the non-repudiation property that digital signatures are often intended to provide. Indeed, one of the easiest ways for Alice to repudiate her signatures is to post her secret key anonymously somewhere on the Internet and claim to be a victim of a computer break-in. In principle, various revocation techniques can be used to prevent users from accepting signatures with compromised keys. However, even with these techniques in place, the users who had accepted signatures *before* the keys were compromised are now left at the mercy of the signer, who could (and, if honest, would) re-issue the signatures with new keys.

Forward-secure signature schemes, first proposed by Anderson in [And97] and formalized by Bellare and Miner in [BM99], are intended to address this limitation. Namely, the goal of a forward-secure signature scheme is to preserve the validity of *past* signatures even if the current secret key has been compromised. This is accomplished by dividing the total time that given public key is

J. Kilian (Ed.): CRYPTO 2001, LNCS 2139, pp. 332–354, 2001.

valid into T *time periods*, and using a different secret key in each time period (while the public key remains fixed). Each subsequent secret key is computed from the current secret key via a *key update* algorithm. The time period during which a message is signed becomes part of the signature. Forward security property means that even if the current secret key is compromised, a forger cannot forge signatures for past time periods.

PRIOR SCHEMES. Prior forward-secure signature schemes can be divided into two categories: those that use arbitrary signature schemes in a black-box manner, and those that modify specific signature scheme.

In the first category, the schemes use some method in which a master public key is used to certify (perhaps via a chain of certificates) the current public key for a particular time period. Usually, these schemes require increases in storage space by noticeable factors in order to maintain the current (public) certificates and the (secret) keys for issuing future certificates. They also require longer verification times than ordinary signatures do, because the verifier needs to verify the entire certificate chain in addition to verifying the actual signature on the message. There is, in fact, a trade-off between storage space and verification time. The two best such schemes are the tree-based scheme of Bellare and Miner [BM99][1] (requiring storage of about $\log_2 T$ secret keys and non-secret certificates, and verification of about $\log_2 T$ ordinary signatures) and the scheme of Krawczyk [Kra00] (requiring storage of T non-secret certificates, and verification of only 2 ordinary signatures).

In the second category, there have been two schemes proposed so far (both in the random oracle model): the scheme of Bellare and Miner [BM99] based on the Fiat-Shamir scheme [FS86], and the scheme of Abdalla and Reyzin [AR00] based the 2^t-th root scheme [OO88,OS90,Mic94]. While needing less space than the schemes in the first category, both [BM99] and [AR00] require signing and verification times that are linear in T.

OUR RESULTS. We propose a scheme in the second category, based on one of the most efficient ordinary signature schemes, due to Guillou-Quisquater [GQ88]. It uses just two modular exponentiations with short exponents for both signing and verifying.

Ours is the first forward-secure scheme where *both signing and verifying are as efficient as the underlying ordinary signature scheme.* Moreover, in our scheme *the space requirements for keys and signatures are nearly the same as those in the underlying signature scheme* (for realistic parameter values, less than 50% more).

The price of such efficient signing and verifying and storage is in the running times of our key generation and update routines: both are linear in T (however, so is the key generation and non-secret storage in the scheme of [Kra00]; as well as the key generation, signing and verifying in the Fiat-Shamir-based scheme of [BM99] and the scheme of [AR00]). However, key generation and update are

[1] Some improvements to tree-based scheme of [BM99] (not affecting this discussion) have been proposed in [AR00] and [MI].

(presumably) performed much less frequently than signing and verifying, and can be performed off-line as long in advance as necessary. Moreover, we show that, if we are willing to tolerate secret storage of $1 + \log_2 T$ values, we can reduce the running time of the key update algorithm to be logarithmic in T without affecting the other components (this, rather unexpectedly, involves an interesting application of pebbling). For realistic parameter values, the total storage requirements, even with these additional secrets, are still less than in all prior schemes; the only exception is the [AR00] scheme, which has very inefficient signing and verifying.

Our scheme is provably secure in the random oracle model based on a variant of the strong RSA assumption (precisely defined in Section 2.2).

2 Background

2.1 Definitions

This section closely follows the first formal definition of forward-secure signatures proposed by Bellare and Miner [BM99]. Their definition, in turn, is based on the Goldwasser, Micali and Rivest's [GMR88] definition of (ordinary) digital signatures secure against adaptive chosen message attacks.

KEY EVOLUTION. The approach taken by forward-secure schemes is to change the secret key periodically (and require the owner to properly destroy the old secret key[2]). Thus we consider time to be divided into time periods; at the end of each time period, a new secret key is produced and the old one is destroyed. The number of the time period when a signature was generated is part of the signature and is input to the verification algorithm; signatures with incorrect time periods should not verify.

Of course, while modifying the secret key, one would like to keep the public key fixed. This can, for example, be achieved by use of a "master" public key, which is somehow used to certify a temporary public key for the current time period (note however, than one needs to be careful not to keep around the corresponding "master" secret key—its presence would defeat the purpose of forward security) . The first simple incarnation of this approach was proposed by [And97]; a very elegant tree-based solution was proposed by [BM99]; another approach, based on generating all of the certificates in advance, was put forward by [Kra00]. However, in general, one can conceive of schemes where the public

[2] Obviously, if the key owner does not properly destroy her old keys, an attacker can obtain them and thus forge the "old" signatures. Moreover, if the key owner does not detect that the current key was leaked, the attacker may hold on to the compromised key for a few time periods, and forge "old" signatures then. Indeed, proper deletion of the old keys and proper intrusion detection are non-trivial tasks. However, it is reasonable to insist that the key owner perform such deletion and intrusion detection—certainly more reasonable than insisting that she guarantee the secrecy of her active keys through resistance to any intrusion attack.

key stays fixed but no such certificates of per-period public keys are present (and, indeed, such schemes are proposed in [BM99,AR00], as well as in this paper).

The notion of a *key-evolving* signature scheme captures, in full generality, the idea of a scheme with a fixed public key and a varying secret key. It is, essentially, a regular signature scheme with the additions of time periods and the key update algorithm. Note that this notion is purely functional: security is addressed separately, in the definition of forward security (which is the appropriate security notion for key-evolving signature schemes).

Thus, a *key-evolving digital signature scheme* is a quadruple of algorithms, $\mathsf{FSIG} = (\mathsf{FSIG.key}, \mathsf{FSIG.sign}, \mathsf{FSIG.ver}, \mathsf{FSIG.update})$, where:

- $\mathsf{FSIG.key}$, the *key generation* algorithm, is a probabilistic algorithm which takes as input a security parameter $k \in \mathsf{N}$ (given in unary as 1^k) and the total number of periods T and returns a pair (SK_1, PK), the initial secret key and the public key;
- $\mathsf{FSIG.sign}$, the (possibly probabilistic) *signing* algorithm, takes as input the secret key $SK_j = \langle S_j, j, T\rangle$ for the time period $j \leq T$ and the message M to be signed and returns the signature $\langle j, sign\rangle$ of M for time period j;
- $\mathsf{FSIG.ver}$, the (deterministic) *verification* algorithm, takes as input the public key PK, a message M, and a candidate signature $\langle j, sign\rangle$, and returns 1 if $\langle j, sign\rangle$ is a *valid* signature of M or 0, otherwise. It is required that $\mathsf{FSIG.ver}(PK, M, \mathsf{FSIG.sign}(SK_j, M)) = 1$ for every message M and time period j.
- $\mathsf{FSIG.update}$, the (possibly probabilistic) *secret key update* algorithm, takes as input the secret key SK_j for the current period $j < T$ and returns the new secret key SK_{j+1} for the next period $j+1$.

We adopt the convention that SK_{T+1} is the empty string and $\mathsf{FSIG.update}(SK_T)$ returns SK_{T+1}.

When we work in the random oracle model, all the above-mentioned algorithms would have an additional security parameter, 1^l, and oracle access to a public hash function $H : \{0,1\}^* \rightarrow \{0,1\}^l$, which is assumed to be random in the security analysis.

Forward Security. Forward security captures the notion that it should be computationally infeasible for any adversary to forge a signature for any past time period even in the event of exposure of the current secret key. Of course, since the update algorithm is public, nothing can be done with respect to future secret keys, except for revoking the public key (thus invalidating all signatures for the time period of the break-in and thereafter). To define forward security formally, the notion of a secure digital signature of [GMR88] is extended in [BM99] to take into account the ability of the adversary to obtain a key by means of a break-in.

Intuitively, in this new model, the forger first conducts an adaptive chosen message attack (cma), requesting signatures on messages of its choice for as many time periods as he desires. Whenever he chooses, he "breaks in": requests the secret key SK_b for the current time period b and then outputs an (alleged) signature on a message M of his choice for a time period $j < b$. The forger is considered to be successful if the signature is valid and the pair (M, j) was not queried during cma.

Formally, let the forger $F = \langle F.\mathsf{cma}, F.\mathsf{forge}\rangle$. For a key pair $(PK, SK_0) \stackrel{R}{\leftarrow}$ $\mathsf{FSIG.key}(k, \ldots, T)$, $F.\mathsf{cma}$, given PK and T, outputs (CM, b), where b is the break-in time period and CM is a set of adaptively chosen message-period pairs (the set of signatures $sign(CM)$ of the current set CM is available to F at all times, including during the construction of CM)[3]. Finally, $F.\mathsf{forge}$ outputs $\langle M, j, sig\rangle \leftarrow F.\mathsf{forge}(CM, sign(CM), SK_b)$. We say that F is successful if $\langle M, j\rangle \notin CM, j < b$, and $\mathsf{FSIG.ver}_{PK}(M, \langle j, sig\rangle) = 1$. (Note: formally, the components of F can communicate all the necessary information, including T and b, via CM.)

Define $\mathbf{Succ}^{\mathrm{fwsig}}(\mathsf{FSIG}[k, T], F)$ to be the probability (over coin tosses of F and FSIG) that F is successful. Let the function $\mathbf{InSec}^{\mathrm{fwsig}}(\mathsf{FSIG}[k, T], t, q_{\mathrm{sig}})$ (the *insecurity* function) be the maximum, over all algorithms F that are restricted to running time t and q_{sig} signature queries, of $\mathbf{Succ}^{\mathrm{fwsig}}(\mathsf{FSIG}[k, T], F)$.

The insecurity function above follows the "concrete security" paradigm and gives us a measure of how secure or insecure the scheme really is. Therefore, we want its value to be as small as possible. Our goal in a security proof will be to find an upper bound for it.

The above definition can be translated to the random oracle model in a standard way [BR93]: by introducing an additional security parameter 1^l, allowing all algorithms the access to the random oracle $H : \{0, 1\}^* \to \{0, 1\}^l$, and considering q_{hash}, the number of queries to the random oracle, as one more parameter for the forger.

2.2 Assumption

We use a variant of the strong RSA assumption (to the best of our knowledge, first introduced independently in [BP97] and [FO97]), which postulates that it is to compute *any* root of a fixed value modulo a composite integer. More precisely, the strong RSA assumption states that it is intractable, given n that is a product of two primes and a value α in Z_n^*, to find $\beta \in Z_n^*$ and $r > 1$ such that $\beta^r = \alpha$.

However, we modify the assumption in two ways. First, we restrict ourselves to the moduli that are products of so-called "safe" primes (a safe prime is one of the form $2q + 1$, where q itself is a prime). Note that, assuming safe primes

[3] Note that the [BM99] definition, which captures what F can do in practice, allows the messages-period pairs to be added to CM only in the order of increasing time periods and without knowledge of any secret keys. However, allowing the forger to construct CM in arbitrary order, and even to obtain SK_b in the middle of the CM construction (so that some messages be constructed by the forger *with* the knowledge of SK_b) would not affect our (and their) results. Similarly, the forger can be allowed to obtain more than one secret key — we only care about the earliest period b for which the secret key is given to the forger. So, the forger may adaptively select some messages which are signed for him, then request some period's secret key; then adaptively select more messages and again request a key, etc.

are frequent, this restriction does not strengthen the assumption. Second, we upperbound the permissible values or r by 2^{l+1}, where l is a security parameter for our scheme (in an implementation, l will be significantly shorter than the length k of the modulus n).

More formally, let A be an algorithm. Consider the following experiment.

Experiment Break-Strong-RSA(k, l, A)
Randomly choose two primes q_1 and q_2 of length $\lceil k/2 \rceil - 1$ each
such that $2q_1 + 1$ and $2q_2 + 1$ are both prime.
$p_1 \leftarrow 2q_1 + 1$; $p_2 \leftarrow 2q_2 + 1$; $n \leftarrow p_1 p_2$
Randomly choose $\alpha \in Z_n^*$.
$(\beta, r) \leftarrow A(n, \alpha)$
`If` $1 < r \leq 2^{l+1}$ `and` $\beta^r \equiv \alpha \pmod{n}$ `then return` 1 `else return` 0

Let $\mathsf{Succ}(A, k, l) = \Pr[Break\text{-}Strong\text{-}RSA(k, l, A) = 1]$. Let $\mathbf{InSec}^{\mathrm{SRSA}}(k, l, t)$ be the maximum of $\mathsf{Succ}(A, k, l)$ over all the adversaries A who run in time at most t. Our assumption is that $\mathbf{InSec}^{\mathrm{SRSA}}(k, l, t)$, for t polynomial in k, is negligible in k. The smaller the value of l, of course, the weaker the assumption.

In fact, for a sufficiently small l, our assumption follows from a variant of the fixed-exponent RSA assumption. Namely, assume that there exists a constant ϵ such that, for every r, the probability of computing, in time t, an r-th root of a random integer modulo a k-bit product of two safe primes, is at most 2^{-k^ϵ}. Then, $\mathbf{InSec}^{\mathrm{SRSA}}(k, l, t) < 2^{l+1-k^\epsilon}$, which is negligible if $l = o(k^\epsilon)$.

2.3 Mathematical Tools

The following two simple statements will be helpful later. They were first pointed out by Shamir [Sha83] in the context of generation of pseudorandom sequences based on the RSA function.

Proposition 1. *Let G be a group. Suppose $e_1, e_2 \in Z$ are such that $\gcd(e_1, e_2) = 1$. Given $a, b \in G$ such that and $a^{e_1} = b^{e_2}$, one can compute c such that $c^{e_2} = a$ in $O(\log(e_1 + e_2))$ group and arithmetic operations.*

Proof. Using Euclid's extended gcd algorithm, within $O(\log(e_1 + e_2))$ arithmetic operations compute f_1, f_2, such that $e_1 f_1 + e_2 f_2 = 1$. Compute $c = a^{f_2} b^{f_1}$, with $O(\log(f_1 + f_2)) = O(\log(e_1 + e_2))$ group operations. Then $c^{e_2} = a^{e_2 f_2} b^{e_2 f_1} = a^{e_2 f_2} a^{e_1 f_1} = a$. □

Lemma 1. *Let G be a finite group. Suppose $e_1 \in Z$ and $e_2 \in Z$ are such that $\gcd(e_1, e_2) = g$ and $\gcd(g, |G|) = 1$. Given $a, b \in G$, such that $a^{e_1} = b^{e_2}$, one can compute c such that $c^{e_2/g} = a$ in $O(\log \frac{e_1 + e_2}{g})$ group and arithmetic operations.*

Proof. Since $\gcd(g, |G|) = 1$, $(z^g = 1) \Rightarrow (z = 1)$ for any $z \in G$. Let $e_1' = e_1/g$, $e_2' = e_2/g$. Then $(a^{e_1'}/b^{e_2'})^g = 1$, so $a^{e_1'} = b^{e_2'}$, so we can apply and Proposition 1 to get c such that $c^{e_2'} = a$. □

2.4 The Guillou-Quisquater Signature Scheme

In [GQ88], Guillou and Quisquater propose the following three-round identification scheme, summarized in Figure 1. Let k and l be two security parameters. The prover's secret key consists of a k-bit modulus n (a product of two random primes p_1, p_2), an $(l+1)$-bit exponent e that is relatively prime to $\phi(n) = (p_1 - 1)(p_2 - 1)$, and a random $s \in Z_n^*$. The public key consists of n, e and v where $v \equiv 1/s^e \pmod{n}$.

In the first round, the prover generates a random $r \in Z_n^*$, computes the commitment $y = r^e \pmod{n}$ and sends y to the the verifier. In the second round, the verifier sends a random l-bit challenge σ to the prover. In the third round, the prover computes and sends to the verifier $z = rs^\sigma$. To check, the verifier computes $y' = z^e v^\sigma$ and checks if $y = y'$ (and $y \not\equiv 0 \pmod{n}$).

The scheme's security is based on the assumption that computing roots modulo composite n is infeasible without knowledge of its factors (the precise assumption varies depending on how e is chosen), and can be proven using Lemma 1. Informally, if the prover can answer two different challenges, σ and τ, for the same y, then it can provide z_σ and z_τ such that $z_\sigma^e v^\sigma = z_\tau^e v^\tau$. Hence, $v^{\sigma-\tau} = (z_\sigma/z_\tau)^e$. Note that e is $l+1$-bits long, hence $e > |\sigma - \tau|$, hence $g = \gcd(\sigma - \tau, e) < e$, so $r = e/g > 1$. By Lemma 1, knowing $v, \sigma - \tau, z_\sigma/z_\tau$ and e allows one to efficiently compute the r-th root of v (to apply the lemma, we need to have g relatively prime with the order $\phi(n)$ of the multiplicative group Z_n^*, which is the case by construction, because e is picked to be relatively prime with $\phi(n)$). Thus, the prover must know at least some root of v (in fact, if e is picked to be prime, then the prover must know precisely the e-th root of v, because $g = 1$ and $r = e$). Note that it is crucial to the proof that $e > 2^l$ and e is relatively prime with $\phi(n)$.

The standard transformation of [FS86] can be applied to this identification scheme to come up with the GQ signature scheme, presented in Figure 1. Essentially, the interactive verifier's l-bit challenge σ is now computed using a random oracle (hash function) $H : \{0,1\}^* \to \{0,1\}^L$ applied to the message M and the commitment y.

3 Our Forward-Secure Scheme

3.1 Main Ideas for Forward Security

The main idea for our forward-secure scheme is to combine the GQ scheme with Shamir's observation (Lemma 1). Namely, let $e_1, e_2, \ldots, e_T$ be distinct integers, all greater than 2^l, all pairwise relatively prime and relatively prime with $\phi(n)$. Let $s_1, s_2, \ldots, s_T$ be such that $s_i^{e_i} \equiv 1/v \pmod{n}$ for $1 \le i \le T$. In time period i, the signer will simply use the GQ scheme with the secret key (n, s_i, e_i) and the verifier will use the GQ scheme with the public key (n, v, e_i). Intuitively, this will be forward-secure because of the relative primality of the e_i's: if the forger breaks-in during time period b and learns the e_b-th, e_{b+1}-th, $\ldots$, e_T-th roots of v, this will not help it compute e_j-th root of v for $j < b$ (nor, more generally, the r-th root of v, where $r | e_j$).

```
algorithm GQ.key(k, l)                    algorithm GQ.sign(M, (n, s, e))
  Generate random ⌈k/2⌉-bit                 r ←R Z*_n
    primes p1, p2                           y ← r^e mod n
  n ← p1 p2                                 σ ← H(y, M)
  s ←R Z*_n                                 z ← r s^σ mod n
  e ←R [2^l, 2^(l+1))                       return (z, σ)
    s.t. gcd(e, φ(n)) = 1
  v ← 1/s^e mod n                         algorithm GQ.ver(M, (n, v, e), (z, σ))
  SK ← (n, s, e)                            if z ≡ 0 (mod n) then return 0
  PK ← (n, v, e)                            y' ← z^e v^σ mod n
  return (SK, PK)                           if σ = H(y', M) then return 1
                                                              else return 0
```

Fig. 1. *The GQ Signature Scheme*

This idea is quite simple. However, we still need to address the following two issues: (i) how the signer computes the s_i's, and (ii) how both the signer and the verifier obtain the e_i's.

COMPUTING s_i'S. Notice that if the signer were required to store all the s_i's, this scheme would require secret storage that is linear in T. However, this problem can be easily resolved. Let $f_i = e_i \cdot e_{i+1} \cdot \ldots \cdot e_T$. Let t_i be such that $t_i^{f_i} \equiv 1/v \pmod{n}$. During the j-th time period, the signer stores s_j and t_{j+1}. At update time, the signer computes $s_{j+1} = t_{j+1}^{f_{j+2}} \bmod n$ and $t_{j+2} = t_{j+1}^{e_{j+1}} \bmod n$. This allows secret storage that is independent of T: only two values modulo n are stored at any time (the f_i and e_i values are not stored—see below). It does, however, require computation linear in T at each update, because of the high cost of computing s_{j+1} from t_{j+1}.

We can reduce the computation at each update to be only logarithmic in T by properly utilizing precomputed powers of t_{j+1}. This will require us, however, to store $1 + \log_2 T$ secrets instead of just two. This optimization concerns only the efficiency of the update algorithm and affects neither the other components of the scheme nor the proof of security, and is therefore presented separately in Section 4.2.

OBTAINING e_i'S. In order for the scheme to be secure, the e_i's need to be relatively prime with each other[4] and with $\phi(n)$, and greater than 2^l. The signer can therefore generate the e_i's simply as distinct $(l+1)$-bit primes. Of course,

[4] In fact, this requirement can be relaxed. We can allow the e_i's not to be pairwise relatively prime, as long as we redefine f_i as $f_i = \mathrm{lcm}(e_i, e_{i+1}, \ldots, e_T)$, and require that e_i be relatively prime with $\phi(n)$ and $e_i/\gcd(e_i, f_{i+1}) > 2^l$. However, we see no advantages in allowing this more general case; the disadvantage is that the e_i's will have to be longer to satisfy the last requirement, and thus the scheme will be less efficient.

to store all the e_i's would require linear in T (albeit public) storage. However, the signer need only store e_j for the current time period j, and generate anew the other e_i's for $i > j$ during key update. This works as long as the signer uses a deterministic algorithm for generating primes: either pseudorandom search or sequential search from fixed starting points. The fact that e_i's are not stored but rather recomputed each time slows down the update algorithm only (and, as we show in Section 3.3, not by much). Note that the way we currently described the update algorithm, for the update at time period j the signer will need to compute $e_{j+1}, \ldots, e_T$. With the optimization of Section 4.2, however, only at most $\log_2 T$ of the e_i's will need to be computed at each update.

We have not yet addressed the issue of how the verifier gets the e_i's. Of course, it could simply generate them the same way that the signer does during each key update. However, this will slow down verification, which is undesirable. The solution is perhaps surprising: the verifier need not know the "true" e_i's at all! The value of e_j can be simply included by the signer in every signature for time period j. Of course, a forger is under no obligation to include the true e_j. Therefore, to avoid ambiguity, we will denote by e the value included in a signature. It may or may not actually equal e_j.

For the security of the scheme, we require that e satisfy the following requirements:

1. e should be included as an argument to the hash function H, so that the forger cannot decide on e after seeing the challenge σ;
2. e should be greater than 2^l, for the same reasons as in the GQ scheme;
3. e should be relatively prime with $\phi(n)$, for the same reasons as in the GQ scheme; and
4. e should be relatively prime with the $e_b, \ldots, e_T$ (where b is the break-in time period), so that the knowledge of the root of v of degree $e_b \cdot e_{b+1} \cdot \ldots \cdot e_T$ does not help the forger compute any root of v of degree $r|e$.

The first two conditions can be easily enforced by the verifier. The third condition can be enforced by having n be a product of two "safe" primes (primes p_1, p_2 that are of the form $p_i = 2q_i + 1$, where q is prime). Then the verifier simply needs to check that e is odd (then it must be relatively prime with $\phi(n)$—otherwise, it would be divisible by q_1, q_2 or q_1q_2, which would imply that the forger could factor n).

It is the fourth condition that presents difficulties. How can the verifier check the that e is relatively prime with $e_b, \ldots, e_T$ without knowing b and the actual values of $e_b, \ldots, e_T$? We accomplish this by splitting the entire interval between 2^l and 2^{l+1} into T consecutive buckets of size $2^l/T$ each, and having each e_i be a prime from the i-th bucket. Then the verifier knows that the actual values $e_{j+1}, \ldots, e_T$ are all at least $2^l(1 + j/T)$ and prime. Thus, as long as e in the signature for time period j is less than $2^l(1+j/T)$, it is guaranteed to be relatively prime with $e_{j+1}, \ldots, e_T$, and hence with $e_b, \ldots, e_T$ (because $b > j$).

Thus, to enforce the above four conditions, the verifier needs to check is that e is odd, is between 2^l and $2^l(1+j/T)$ and is included in the hash computation.

3.2 The Scheme

Our scheme (denoted IR) based on the above ideas is presented in Figure 2. As in the GQ scheme, let $H : \{0,1\}^* \rightarrow \{0,1\}^l$ be a hash function.

3.3 Efficiency

SIGNING AND VERIFYING. The distinghuishing feature of our scheme is the efficiency of the signing and verification algorithms. Both are the same as the already efficient ordinary GQ scheme (verifying has the additional, negligible component of testing whether e is in the right range and odd). Namely, they each take two modular exponentiations, one modular multiplication and an application of H, for a total time of $O(k^2 l)$ plus the time required to evaluate H. (Note that, just like the GQ scheme, one of the two modular exponentiations for signing can be done off-line, before the message is known; also, one of the two modular exponentiations for verifying is of a fixed base v, and can benefit from precomputation.)

KEY GENERATION. We need to make strong assumptions on the distributions of primes in order to estimate efficiency of key generation. First, we assume that at least one in $O(k)$ $\lceil k/2 \rceil$-bit numbers is a prime, and that at least one in $O(k)$ of those is of the form $2q+1$, where q is prime. Then, generating n takes $O(k^2)$ primality tests. Each primality test can be done in $O(k^3)$ bit operations [BS96]. Thus, the modulus n is generated in $O(k^5)$ bit operations (a factor k slower than an RSA modulus, because of the need for safe primes). Similarly, we will assume that at least one in $O(l)$ integers in each bucket $[2^l(1+(i-1)/T), 2^l(1+i/T))$ is a prime, so generating each e_i takes $O(l^4)$ bit operations.

In addition to generating n and the e_i's, key generation needs to compute the product of the e_i's modulo $\phi(n)$, which takes $O(Tkl)$ bit operations, and three modular exponentiations, each taking $O(k^2 l)$ bit operations. Therefore, key generation takes $O(k^5 + l^4 T + k^2 l + klT))$ bit operations.

Note that, similarly to the GQ scheme, n and e_i's may be shared among users if n is generated by a trusted party, because each user need not know the factors of n. Each user can simply generate its own t_1 and v.

KEY UPDATE. Key update cannot multiply all the relevant e_i's modulo $\phi(n)$, because $\phi(n)$ is not available (otherwise, the scheme would not be forward-secure). Therefore, it has to perform $O(T)$ modular exponentiations separately, in addition to regenerating all the e_i's. Thus, it takes $O(k^2 lT + l^4 T)$ bit operations.

Note that the $l^4 T$ component is present in the running time for the update algorithm because of the need to regenerate the e_i's each time. However, for practical values of l (on the order of 100) and k (on the order of 1000), $l^4 T$ is roughly the same as $k^2 lT$, so this only slows down the key update algorithm by a small constant factor. Moreover, in Section 4.1 we show how to reduce the $l^4 T$ component in both key generation and update to $(l^2 + \log^4 T)T$ (at a very slight expense to signing and verifying).

Finally, as shown in Section 4.2, if we are willing to increase secret storage from $2k$ bits (for s_j and t_{j+1}) to $(1+\log_2 T)k$ bits, then we can replace the factor

```
algorithm IR.key(k, l, T)
```
Generate random $(\lceil k/2 \rceil - 1)$-bit primes q_1, q_2 s.t. $p_i = 2q_i + 1$ are both prime
$n \leftarrow p_1 p_2$
$t_1 \stackrel{R}{\leftarrow} Z_n^*$
Generate primes e_i s.t. $2^l(1 + (i-1)/T) \leq e_i < 2^l(1 + i/T)$ for $i = 1, 2, \ldots, T$.
(This generation is done either deterministically or using a small seed *seed* and H as a pseudorandom function.)
$f_2 \leftarrow e_2 \cdot \ldots \cdot e_T \bmod \phi(n)$, where $\phi(n) = 4q_1q_2$
$s_1 \leftarrow t_1^{f_2} \bmod n$
$v \leftarrow 1/s_1^{e_1} \bmod n$
$t_2 \leftarrow t_1^{e_1} \bmod n$
$SK_1 \leftarrow (1, T, n, s_1, t_2, e_1, seed)$
$PK \leftarrow (n, v, T)$
return (SK_1, PK)

```
algorithm IR.update(SK_j)
```
Let $SK_j = (j, T, n, s_j, t_{j+1}, e_j, seed)$
if $j = T$ **then return** ϵ
Regenerate $e_{j+1}, \ldots, e_T$ using *seed*
$s_{j+1} \leftarrow t_{j+1}^{e_{j+2} \cdot \ldots \cdot e_T} \bmod n$; $t_{j+2} \leftarrow t_{j+1}^{e_{j+1}} \bmod n$
return $SK_{j+1} = (j+1, T, n, s_{j+1}, t_{j+2}, e_{j+1}, seed)$

```
algorithm IR.sign(SK_j, M)
```
Let $SK_j = (j, T, n, s_j, t_{j+1}, e_j, seed)$
$r \stackrel{R}{\leftarrow} Z_n^*$
$y \leftarrow r^{e_j} \bmod n$
$\sigma \leftarrow H(j, e_j, y, M)$
$z \leftarrow rs^\sigma \bmod n$
return (z, σ, j, e_j)

```
algorithm IR.ver(PK, M, (z, σ, j, e))
```
Let $PK = (n, v)$
if $e \geq 2^l(1 + j/T)$ **or** $e < 2^l$ **or** e is even **then return** 0
if $z \equiv 0 \pmod n$ **then return** 0
$y' \leftarrow z^e v^\sigma \bmod n$
if $\sigma = H(j, e, y', M)$ **then return** 1 **else return** 0

Fig. 2. *Our forward-secure signature scheme (without efficiency improvements)*

of T in the cost of update by the factor of $\log_2 T$, to get update at the cost of $O((l^4 + k^2 l) \log T)$ (or, if optimization of Section 4.1 is additionally applied, $O((k^2 l + l^2 + \log^4 T) \log T)$).

SIZES. All the key and signature sizes are comparable to those in the ordinary GQ scheme.

The public key has $l+1$ fewer bits than the GQ public key, and the signatures have $l+1$ more bits, because e is included in the signature rather than in the public key. In addition, both the public key and the signature have $\log_2 T$ more bits in order to accommodate T in the public key and the current time period in the signature (this is necessary in any forward-secure scheme). Thus, the total public key length is $2k + \log_2 T$ bits, and signature length is $k + 2l + 1 + \log_2 T$ bits. Optimization of Section 4.1 shortens the signatures slightly, replacing $l+1$ of the signature bits with about $\log_2 T$ bits.

The secret key is $k + 2\log_2 T + |seed|$ bits longer than in the GQ scheme in order to accommodate the current time period j, the total time periods T, the value t_{j+1} necessary to compute future keys and the seed necessary to regenerate the e_i's for $i > j$. Thus, the total secret key length is $3k + l + 1 + |seed| + 2\log_2 T$ bits (note that only $2k$ of these bits need to be kept secret). If the optimization of Section 4.2 is used, then the secret contains an additional $k(\log_2 T - 1)$ bits, all of which need to be kept secret.

3.4 Security

The exact security of our scheme (in the random oracle model) is close to the exact security of the schemes of [BM99,AR00]. The proof is also similar: it closely follows the one in [AR00], combining ideas from [PS96,BM99,MR99].

First, we state the following theorem that will allow us to upper-bound the insecurity function. The full proof of the theorem is very similar to the one in [AR00] and is contained in Appendix A.

Theorem 1. *Given a forger F for $\mathsf{IR}[k,l,T]$ that runs in time at most t, asking q_{hash} hash queries and q_{sig} signing queries, such that $\mathbf{Succ}^{\text{fwsig}}(\mathsf{IR}[k,l,T],F) \geq \varepsilon$, we can construct an algorithm A that, on input n (a product of two safe primes), $\alpha \in Z_n^*$ and l, runs in time t' and outputs (β, r) such that $1 < r \leq 2^{l+1}$ and $\beta^r \equiv \alpha \pmod n$ with probability ε', where*

$$t' = 2t + O(lT(l^2T^2 + k^2))$$
$$\varepsilon' = \frac{\left(\varepsilon - 2^{2-k} q_{\text{sig}}(q_{\text{hash}}+1)\right)^2}{T^2(q_{\text{hash}}+1)} - \frac{\varepsilon - 2^{2-k} q_{\text{sig}}(q_{\text{hash}}+1)}{2^l T}.$$

Proof Outline. A will use F as a subroutine. (Note that A gets to provide the public key for F and to answer its signing and hashing queries.) A bases the public key v on α as follows: it randomly guesses j between 1 and T, hoping that F's eventual forgery will be for the j-th time period. It then generates $e_1, \ldots, e_T$ just like the real signer, sets $t_{j+1} = \alpha$ and computes v as $v = 1/t_{j+1}^{f_{j+1}} \bmod n$, where, as above, $f_{j+1} = e_{j+1} \cdot \ldots \cdot e_T$.

Then A runs F. Answering F's hash and signature queries is easy, because A fully controls the random oracle H. If A's guess for j was correct, and F indeed will output a forgery for the j-th time period, then F's break-in query will be for the secret of a time period $b > j$. A can compute the answer as follows: $t_{b+1} = t_{j+1}^{f_{j+1}/f_b} = \alpha^{e_{j1} \cdot \ldots \cdot e_b}$ and $s_b = t_b^{f_{b+1}} = \alpha^{e_{j1} \cdot \ldots \cdot e_{b-1} \cdot e_{b+1} \cdot \ldots \cdot e_T}$ (the other components of SK_b are not secret, anyway). Suppose A's guess was correct, and in the end F outputs a signature (z, σ, j, e) on some message M. We will assume that F asked a hash query on (j, e, y, M) where $y = z^e v^\sigma \bmod n$ (F can always be modified to do so.)

Then, A runs F the second time with the same random tape, giving the same answers to all the oracle queries before the query (j, e, y, M). For (j, e, y, M), A gives a new answer τ. If F again forges a signature (z', τ, j, e) using the same hash query, we will have that $y \equiv z^e v^\sigma \equiv z'^e v^\tau \pmod n$, so $(z/z')^e \equiv v^{\tau - \sigma} \equiv \alpha^{f_{j+1}(\sigma - \tau)} \pmod n$. Note that because e is guaranteed to be relatively prime with f_{j+1}, and $\sigma - \tau$ has at least one fewer bit than e, $\gcd(f_{j+1}(\sigma - \tau), e) = \gcd(\sigma - \tau, e) < e$ (as long as $\sigma \neq \tau$). Thus, $r = e / \gcd(f_{j+1}(\tau - \sigma), e) > 1$ and, by Lemma 1, A will be able to efficiently compute the r-th root of α.

Please refer to Appendix A for further details. □

This allows us to state the following theorem about the insecurity function of our scheme.

Theorem 2. *For any t, q_{sig}, and q_{hash},*

$$\mathbf{InSec}^{\text{fwsig}}(\mathsf{IR}[k, l, T]; t, q_{\text{sig}}, q_{\text{hash}}) \leq$$
$$T\sqrt{(q_{\text{hash}} + 1)\mathbf{InSec}^{\text{SRSA}}(k, l, t')} + 2^{-l+1} T (q_{\text{hash}} + 1) + 2^{2-k} q_{\text{sig}} (q_{\text{hash}} + 1) \,,$$

where $t' = 2t + O(lT(l^2 T^2 + k^2))$.

Proof. To compute the insecurity function, simply solve for $(\varepsilon - 2^{2-k} q_{\text{sig}}(q_{\text{hash}} + 1))/T$ the quadratic equation in Theorem 1 that expresses ε' in terms of ε to get

$$\begin{aligned}
&(\varepsilon - 2^{2-k} q_{\text{sig}}(q_{\text{hash}} + 1))/T \\
&\qquad = 2^{-l}(q_{\text{hash}} + 1) + \sqrt{2^{-2l}(q_{\text{hash}} + 1)^2 + \varepsilon'(q_{\text{hash}} + 1)} \\
&\qquad \leq 2^{-l}(q_{\text{hash}} + 1) + \sqrt{2^{-2l}(q_{\text{hash}} + 1)^2} + \sqrt{\varepsilon'(q_{\text{hash}} + 1)} \\
&\qquad = 2^{-l+1}(q_{\text{hash}} + 1) + \sqrt{\varepsilon'(q_{\text{hash}} + 1)},
\end{aligned}$$

and then solve the resulting inequality for ε. □

4 Further Improving Efficiency

4.1 Finding the e_i's Faster

Finding e_i's takes time because they need to be $l+1$-bit primes. If we were able to use small primes instead, we could search significantly faster, both because small primes are more frequent and because primality tests are faster for shorter lengths.[5]

We cannot use small primes directly because, as already pointed out, the e_i's must have at least $l+1$ bits. However, we can use powers of small primes that are at least $l+1$ bits. That is, we let ϵ_i be a small prime, $\pi(\epsilon_i)$ be such that $\epsilon_i^{\pi(\epsilon_i)} > 2^l$ and $e_i = \epsilon_i^{\pi(\epsilon_i)}$. As long as π is a deterministic function of its input ϵ (for example, $\pi(\epsilon) = l/\lfloor \log_2 \epsilon \rfloor$), we can replace e in the signature by ϵ, and have the verification algorithm compute $e = \epsilon^{\pi(\epsilon)}$.

Of course, the verification algorithm still needs to ensure that e is relatively prime to $\phi(n)$ and to $e_b, \ldots, e_T$. This is accomplished essentially the same way as before: we divide a space of small integers into T consecutive buckets of some size S each, and have each ϵ_i come from the i-th bucket: $\epsilon_i \in [(i-1)S, iS)$. Then, when verifying a signature for time period j, it will suffice to check that ϵ is odd and comes from a bucket no greater than the j-th: $\epsilon < jS$. It will be then relatively prime to $\epsilon_b, \ldots, \epsilon_T$, and therefore $e = \epsilon^{\pi(\epsilon)}$ will be relatively prime to $e_b, \ldots, e_T$.

When we used large primes, we simply partitioned the space of $(l+1)$-bit integers into large buckets, of size $2^l/T$ each. We could have used smaller buckets, but this offered no advantages. However, now that we are using small primes, it is advantageous to make the bucket size S as small as possible, so that even the largest prime (about TS) is still small.

Thus, to see how much this optimization speeds up the search for the e_i's, we need to upper-bound S. S needs to be picked so that there is at least one prime in each interval $[(i-1)S, iS)$ for $1 \leq i \leq T$. It is reasonable to conjecture that the distance between two consecutive primes P_n and P_{n+1} is at most $(\ln^2 P_n)$ [BS96]. Therefore, because the largest prime we are looking for is smaller than TS, S should be such that $S > \ln^2 TS$. It is easy to see that $S = 4\ln^2 T$ will work for $T \geq 75$. (As a practical matter, computation shows that, for any reasonable value of T, the value of S will be quite small: $S = 34$ will work for $T = 1000$, because the largest gap between the first 1000 primes is 34; by the same reasoning, $S = 72$ will work for $T = 10^4$, $S = 114$ will work for $T = 10^5$, and $S = 154$ will work for $T = 10^6$.) Thus, the ϵ_i's are all less than $4T\ln^2 T$, and therefore the size of each ϵ_i is $O(\log T)$ bits. Thus, finding and testing the primality of the ϵ_i's and then computing the e_i's takes $O(T(\log^4 T + l^2))$ time, as opposed to $O(Tl^4)$ without this optimization.

The resulting scheme will slightly increase verification time: the verifier needs to compute e from ϵ. This takes time $O(l^2)$ (exponentiating any quantity to obtain an $(l+1)$-bit quantity takes time $O(l^2)$), which is lower order than $O(k^2 l)$ verification time. Moreover, it will be impossible to get e_i to be exactly $l+1$ bits (it will be, on average, about $l + (\log_2 T)/2$ bits). This will slow down both verification and signing, albeit by small amounts. Therefore, whether to use the optimization in practice depends on the relative importance of the speeds of signing and verifying vs. the speeds of key generation and update.

[5] In fact, when a table of small primes is readily available (as it often is for reasonably small T), no searching or primality tests are required at all.

4.2 Optimizing Key Update

The key update in our scheme requires computing s_i such that $s_i^{e_i} \equiv 1/v \bmod n$. Knowledge of s_{i-1}, such that $s_{i-1}^{e_{i-1}} \equiv 1/v \bmod n$, does not help, because e_i and e_{i-1} are relatively prime. The easiest way to compute s_i requires knowledge of $\phi(n)$: $s_i \leftarrow 1/v^{1/e_i \bmod \phi(n)} \bmod n$. However, the signer cannot store $\phi(n)$—otherwise the forger would obtain it during a break-in, and thus be able to factor n and produce the past periods' secrets (and signatures). The value of $\phi(n)$ can be used only during the initial key generations stage, after which it should be securely deleted.

To enable generation of current and future s_i's without compromising the past ones, we had defined (in Section 3) a secret t_i for time period i, from which it was possible to derive all future periods' secrets $s_{j \geq i}$. The update of t_i to t_{i+1} can be implemented efficiently (1 exponentiation). However, in this approach the computation of each s_i from t_i requires $\Theta(T-i)$ exponentiations. This computation can be reduced dramatically if the storage is increased slightly.

Specifically, in this section we demonstrate how replacing the single secret t_i with $\log_2 T$ secrets can reduce the complexity of the update algorithm to only $\log_2 T$ exponentiations.

Abstracting the Problem. Consider all subsets of $Z_T = \{1, 2, \ldots, T\}$. Let each such subset S correspond to the secret value $t_S = t_1^{\prod_{i \notin S} e_i}$. For example, t_1 corresponds to Z_T, t_i corresponds to $\{i, i+1, \ldots, T\}$, v^{-1} corresponds to the empty set, and each s_i corresponds to the singleton set $\{i\}$. Raising some secret value t_S to power e_i corresponds to dropping i from S.

Thus, instead of secrets and the exponentiation operation, we can consider sets and the operation of removing an element. Our problem, then, can be reformulated as follows: design an algorithm that, given Z_T, outputs (one-by-one, in order) the singleton sets $\{i\}$ for $1 \leq i \leq T$. The only way to create new sets is to remove elements from known sets. The algorithm should minimize the number of element-removal operations (because they correspond to the expensive exponentiation operations).

Fairly elementary analysis quickly demonstrates that the most efficient solution for this problem (at least for T that is a power of 2) is the following divide-and-conquer algorithm:

Input: An ordered non-empty set A.
Output: Singleton sets $\{x\}$, for $x \in A$, in order.
Steps: If A has one element, output A and return.
Remove the second half of A's elements to get B.
Recurse on B.
Remove the first half of A's elements to get C.
Recurse on C.

This algorithm takes exactly $T \log_2 T$ element-removal operations to output all the singletons. Moreover, the recursion depth is $1 + \log_2 T$, so only $1 + \log_2 T$ sets need to be stored at any time (each set is just a consecutive interval, so the bookkeeping about what each set actually contains is simple).

This recursive algorithm can essentially be the update algorithm for our scheme: at every call to update, we run the recursive algorithm a little further, until it produces the next output. We then stop the recursive algorithm, save its stack (we need to save only $\log_2 T$ secrets, because the remaining one is the output of the algorithm), and run it again at the next call to update. A little more care needs to be taken to ensure forward security: none of the sets stored at time period i should contain elements less than i. This can be done by simply removing i from all sets that still contain in (and that are still needed) during the i-th update. The total amount of work still does not change.

Because there are T calls to update (if we include the initial key generation), the amortized amount of work per update is exactly $\log_2 T$ exponentiations. However, some updates will be more expensive than others, and update will still cost $\Theta(T)$ exponentiations in the worst case. We thus want to improve the worst-case running time of our solution without increasing the (already optimal) total running time. This can be done through pebbling techniques, described below.

PEBBLING. Let each subset of Z_T correspond to a node in a graph. Connect two sets by a directed edge if the destination can be obtained from the source by dropping a single element. The resulting graph is the T-dimensional hypercube, with directions on the edges (going from higher-weight nodes to lower-weight nodes). We can traverse the graph in the direction given by the edges. We start at the node corresponding to Z_T, and need to get to all the nodes corresponding to the singleton sets $\{i\}$.

One way to accomplish this task is given by the above recursive algorithm, which has the minimal total number of steps. However, we would like to minimize not only the total number of steps, but also the number of steps taken *between* any two "consecutive" nodes $\{i\}$ and $\{i+1\}$, while keeping the memory usage low. We will do this by properly arranging different branches of the recursive algorithm to run in parallel.

To help visualize the algorithm, we will represent each set stored as a pebble at the corresponding node in a graph. Then removing an element from a set corresponds to moving the corresponding pebble down the corresponding directed edge. The original set may be preserved, in which case a "clone" of a pebble is left at the original node, or it may be discarded, in which case no such clone is left. Our goal can be reformulated as follows in terms of pebbles: find a pebbling strategy that, starting at the node Z_T, reaches every node $\{i\}$ in order, while minimizing the number of pebbles used at any given time (this corresponds to total secret storage needed), the total number of pebble moves (this corresponds to total number of exponentiations needed), and the number of pebble moves between any two consecutive hits of a singleton (this corresponds to the worst-case cost of the update algorithm).

THE PEBBLING ALGORITHM. We shall assume that $T > 1$ is a power of 2. The following strategy uses at most $1 + \log_2 T$ pebbles, takes $T \log_2 T$ total moves (which is the minimum possible), and requires at most $\log_2 T$ moves per update.

Each pebble has the following information associated with it:

1. its current position, represented by a set $P \subseteq Z_T$ (P will always be a set of consecutive integers $\{P_{\min}, \ldots, P_{\max}\}$);

2. its "responsibility," represented by a set $R \subseteq P$ (R will also always be a set of consecutive integers $\{R_{\min}, \ldots, R_{\max}\}$; moreover $|R|$ will always be a power of 2).

Each pebble's goal is to ensure that it (together with its clones, their clones, etc.) reaches every singleton in its set P. If $R \subsetneq P$, then the pebble can move towards this goal by removing an element from P. If, however, $R = P$, then the pebble has to clone (unless $|P| = |R| = 1$, in which case it has reached its singleton, and can be removed from the graph). Namely, it creates a new pebble with the same P, and responsibility set R' containing only the second half of R. It then changes its own R to $R - R'$ (thus dividing its responsibility evenly between itself and its clone). Now both the pebble and the clone can move towards their disjoint sets of singletons.

We start with a single pebble with $P = R = Z_T$. The above rules for moving and cloning ensure that the combined moves of all the pebbles will be the same as in the recursive algorithm. Thus, the steps of the pebbles are already determined. We now have to specify the timing rules: namely, when the pebbles take their steps. A careful specification is important: if a pebble moves too fast, then it can produce more clones than necessary, thus increasing the total memory; if a pebble moves too slowly, then it may take longer to reach its destination singletons, thus increasing the worst-case cost of update.

In order to specify the timing rules, we will imagine having a clock. The clock "ticks" consecutive integer values, starting with $-T/2+1$. After each clock tick, each pebble will decide whether to move and, if so, for how many moves, as follows:

1. The original pebble always makes two moves per clock tick, until it reaches the singleton $\{1\}$. After reaching the singleton it stops, and then removes itself from the graph on the next clock tick.
2. After a new pebble is cloned with responsibility set R, it stays still for $\lceil |R|/2 \rceil$ clock ticks. After $\lceil |R|/2 \rceil$-th clock-tick following its birth, it starts moving at one move per clock tick. After $|R|$ such moves, it starts moving a two moves per clock tick, until it reaches its leftmost singleton. After reaching the singleton it stops, and then removes itself from the graph on the next clock tick.

We remark that the above rules may seem a bit complex. Indeed, simpler rules can be envisioned: for example, allowing each pebble at most one move per clock tick, and specifying that each pebbles moves following a given clock tick only if it absolutely has to move in order to reach its leftmost singleton on time. However, this set of rules will require $(\log_2 T) - 2$ pebbles (even though at most $\log_2 T$ of them will be moving at any given time). Having pebbles move at variable speeds allows us to delay their cloning, and thus reduces the total number of pebbles, as shown by the following theorem.

Theorem 3. *Suppose $T > 1$ is a power of two. If i is the value most recently ticked by the clock, then the total number of pebbles under the above rules never exceeds $1 + \lfloor \log_2(T - i) \rfloor$ (if $i \geq 0$) or $(\log_2 T) - \lfloor \log_2 -i \rfloor$ (if $-T < i < 0$). The number of moves occurring immediately following the clock tick i also never* *exceeds this quantity. For each i, $1 \leq i \leq T$, a pebble reaches the singleton $i+1$ immediately before the clock ticks the value $i+1$, and is removed before the clock ticks $i+2$.*

Proof. The proof is by induction on $\log_2 T$.

For $T = 2$, we start with a single pebble with $P = R = \{1, 2\}$. After the clock ticks 0, this pebble clones the pebble with $R' = 2$, and itself moves to $P = \{1\}$. The clone waits for one clock tick and then, after the clock ticks 1, the clone moves to $P = \{2\}$.

Suppose the statement is true for some T that is a power of two. We will now prove it for $T' = 2T$. After clock tick $-T + 1$, we have two pebbles: one responsible for $\{1, \ldots, T\}$, and the other responsible for $\{T+1, \ldots, 2T\}$. For the next $T/2 - 1$ clock ticks, the first pebble will move at two steps per tick, and the second one will stay put (thus, the number of moves does not exceed the number of pebbles). After the clock ticks $-T/2$, the first pebble will arrive at position $P = \{1, \ldots, T\}$. Thus, starting at $t = -T + 1$, the inductive hypothesis applies to the all the pebbles that will cover the first half of the singletons: there is a single pebble until $t = -T/2 + 1$ and it is in position $P = \{1, \ldots, T\}$ after clock tick $-T/2 + 1$.

The second pebble will reach the position $P' = \{2, \ldots, T\}$ after the clock ticks $T/2$. Thus, again, after the clock ticks 1, the inductive hypothesis applies to all the pebbles that will cover the second half of the singletons, except that time is shifted forward by T. That is, if $1 \leq i < T$, then the number of pebbles in the second half does not exceed $(\log_2 T) - \lfloor \log_2(T - i) \rfloor$, and if $t \geq T$, then the number of pebbles in the second half does not exceed $1 + \lfloor \log_2(2T - i) \rfloor$.

The key to finishing the proof is to realize that the first half will lose a pebble just as the second half gains one. To be precise, we can consider the following four cases.

- For $-T < i < 0$, we have $(\log_2 T) - \lfloor \log_2 -i \rfloor$ pebbles in the first half (by the inductive hypothesis), and one pebble in the second half, so we have a total of $(\log_2 2T) - \lfloor \log_2 -i \rfloor$ pebbles, as required.
- For $i = 0$, we have $1 + \log_2 T = \log_2 2T$ pebbles in the first half (by the inductive hypothesis), and one pebble in the second half, for a total of $1 + \log_2 2T$ pebbles, as required.
- For $0 < i \leq T$, we have $1 + \lfloor \log_2(T - i) \rfloor$ pebbles in the first half and $(\log_2 T) - \lfloor \log_2(T - i) \rfloor$ pebbles in the second half (both by the inductive hypothesis), for a total of $1 + \log_2 T = 1 + \lfloor \log_2(2T - i) \rfloor$ pebbles, as required.
- For $i > T$, we have no pebbles in the first half and $\lfloor \log_2(2T - i) \rfloor$ pebbles in the second half (by the inductive hypothesis), as required.

It is easy to see that in each of the above four cases, the number of moves does not exceed the number of pebbles (because for every pebble moving at two steps per clock tick, there exists a pebble that is standing still—namely, its most recent clone). □

SECURITY. It is, of course, crucial to ensure that the above changes to the update algorithm do not compromise the security of our scheme. It suffices to prove that every secret stored following the clock tick i can be derived in polynomial time from t_{i+1}. In other words, it suffices to prove that, following the clock tick i, no pebble's position P satisfies $i \in P$. This can be easily done by induction, as long as each pebble moves towards its goal by removing the *smallest* possible element from its position P (the inductive step is proved as follows: if $2T$ is the total number of time periods, then the single pebble responsible for the second half of the singletons will have removed $\{1, \ldots, T/2\}$ from its position following the clock tick 1, and will have removed $\{1, \ldots, T\}$ following the clock tick $T/2+1$).

Acknowledgements. We thank Anna Lysyanskaya and Silvio Micali for helpful discussions about our complexity assumptions; Ron Rivest for sharing his insights on pebbling algorithms; and the anonymous referees for helpful comments.

References

[And97] Ross Anderson. Invited lecture. Fourth Annual Conference on Computer and Communications Security, ACM, 1997.

[AR00] Michel Abdalla and Leonid Reyzin. A new forward-secure digital signature scheme. In *Advances in Cryptology—ASIACRYPT 2000*, Springer-Verlag 2000. Full version available from the Cryptology ePrint Archive, record 2000/002, `http://eprint.iacr.org/`.

[BM99] Mihir Bellare and Sara Miner. A forward-secure digital signature scheme. In *Advances in Cryptology—CRYPTO '99*, Springer-Verlag, 1999. Revised version is available from `http://www.cs.ucsd.edu/mihir/`.

[BP97] Niko Barić and Birgit Pfitzmann. Collision-free accumulators and fail-stop signature schemes without trees. In *Advances in Cryptology—EUROCRYPT 97*, Springer-Verlag, 1997.

[BR93] Mihir Bellare and Phillip Rogaway. Random oracles are practical: A paradigm for designing efficient protocols. In *Proceedings of the 1st ACM Conference on Computer and Communication Security*, pages 62–73, November 1993. Revised version appears in `http://www-cse.ucsd.edu/users/mihir/papers/crypto-papers.html`.

[BS96] Eric Bach and Jeffrey Shallit. *Algorithmic Number Theory.* MIT Press, Cambridge, MA, 1996.

[FO97] Eiichiro Fujisaki and Tatsuaki Okamoto. Statistical zero knowledge protocols to prove modular polynomial relations. In Burton S. Kaliski Jr., editor, *Advances in Cryptology—CRYPTO '97*, volume 1294 of *Lecture Notes in Computer Science*, pages 16–30. Springer-Verlag, 17–21 August 1997.

[FS86] Amos Fiat and Adi Shamir. How to prove yourself: Practical solutions to identification and signature problems. In Andrew M. Odlyzko, editor, *Advances in Cryptology—CRYPTO '86*, volume 263 of *Lecture Notes in Computer Science*, pages 186–194. Springer-Verlag, 1987, 11–15 August 1986.

[GMR88] Shafi Goldwasser, Silvio Micali, and Ronald L. Rivest. A digital signature scheme secure against adaptive chosen-message attacks. *SIAM Journal on Computing*, 17(2):281–308, April 1988.

[Gol88] Shafi Goldwasser, editor. *Advances in Cryptology—CRYPTO '88*, volume 403 of *Lecture Notes in Computer Science.* Springer-Verlag, 1990, 21–25 August 1988.

[GQ88] Louis Claude Guillou and Jean-Jacques Quisquater. A "paradoxical" indentity-based signature scheme resulting from zero-knowledge. In Goldwasser [Gol88], pages 216–231.

[Kra00] Hugo Krawczyk. Simple forward-secure signatures from any signature scheme. In *Seventh ACM Conference on Computer and Communication Security*. ACM, November 1–4 2000.

[MI] Silvio Micali and Gene Itkis. Private Communication.

[Mic94] Silvio Micali. A secure and efficient digital signature algorithm. Technical Report MIT/LCS/TM-501, Massachusetts Institute of Technology, Cambridge, MA, March 1994.

[MR99] Silvio Micali and Leonid Reyzin. Improving the exact security of Fiat-Shamir signature schemes. In R. Baumgart, editor, *Secure Networking — CQRE [Secure] '99*, volume 1740 of *Lecture Notes in Computer Science*, pages 167–182. Springer-Verlag, 1999.

[OO88] Kazuo Ohta and Tatsuaki Okamoto. A modification of the Fiat-Shamir scheme. In Goldwasser [Gol88], pages 232–243.

[OS90] H. Ong and Claus P. Schnorr. Fast signature generation with a Fiat Shamir-like scheme. In I. B. Damgård, editor, *Advances in Cryptology—EUROCRYPT 90*, volume 473 of *Lecture Notes in Computer Science*, pages 432–440. Springer-Verlag, 1991, 21–24 May 1990.

[PS96] David Pointcheval and Jacques Stern. Security proofs for signature schemes. In Ueli Maurer, editor, *Advances in Cryptology—EUROCRYPT 96*, volume 1070 of *Lecture Notes in Computer Science*, pages 387–398. Springer-Verlag, 12–16 May 1996.

[Sha83] Adi Shamir. On the generation of cryptographically strong pseudorandom sequences. *ACM Transactions on Computer Systems*, 1(1):38–44, February 1983.

A Details of the Proof of Theorem 1

First, we assume that if F outputs (z, σ, j, e) as a forgery, then the hashing oracle has been queried on (j, e, y, M), where $y = z^e v^\sigma \bmod n$ (any adversary can be modified to do that; this may raise the number of hash queries to $q_{\text{hash}} + 1$.) We will also assume that F performs the necessary bookkeeping and does not ask the same hash query twice.[6] Note that F may ask the same signature query twice, because the answers will most likely be different.

Recall that A's job, given α and n, is to find (with F's help) β and $r > 1$ such that $\beta^r \equiv \alpha \pmod n$. First, A has to guess the time period for which F will output the forgery: it randomly selects j, $1 < j \le T$ (sometimes A may also succeed if the forgery is for a time period $i < j$, but this not necessary for our argument). A then generates $e_1, \ldots, e_T$ just like the real signer, sets $t_{j+1} = \alpha$ and computes v as $v = 1/t_{j+1}^{f_{j+1}} \bmod n$, where, as above, $f_{j+1} = e_{j+1} \cdot \ldots \cdot e_T$.

A then comes up with a random tape for F, remembers it, and runs F on that tape and the input public key (n, v, T). If F breaks in at time period b, then A can provide F with the secret key as long as $b > j$: knowing t_{j+1} will

[6] This may slightly increase the running time of F, but we will ignore costs of simple table look-up for the purposes of this analysis.

allow A to compute s_b and t_{b+1}. If $b \leq j$, then A aborts (because, in particular, F's forgery cannot be for time period j in that case).

To answer F's signature and hash queries, A maintains two tables: a signature query table and a hash query table.

Signature queries can be answered almost at random, because A controls the hash oracle. In order to answer a signature query number s on a message M_s during time period j_s, A selects a random $z_s \in Z_n^*$ and $\sigma_s \in \{0,1\}^l$, computes $y_s = {z_s}^{e_{j_s}} v^{\sigma_s}$, and checks its signature query table to see if a signature query on M_s during time period j_s has already been asked and if y_s used in answering it. If so, A changes z_s and σ_s to the z and σ that were used in answering that query. Then A adds the entry $(s, j_s, e_{j_s}, y_s, \sigma_s, z_s, M_s)$ to its signature query table and outputs $(z_s, \sigma_s, j_s, e_{j_s})$.

Hash queries are also answered at random. To answer the t-th hash query (j'_t, e'_t, y'_t, M'_t), A first checks its signature query table to see if there is an entry $(s, j_s, e_{j_s}, y_s, \sigma_s, z_s, M_s)$ such that $(j_s, e_{j_s}, y_s, M_s) = (j'_t, e'_t, y'_t, M'_t)$. If so, it just outputs σ_s. Otherwise, it picks a random $\sigma'_t \in \{0,1\}^l$, records in its hash query table the tuple $(t, y'_t, M'_t, j'_t, e'_t, \sigma'_t)$ and outputs σ'_t.

Assume now the break-in query occurs during time period $b > j$, and the valid forgery (z, σ, i, e) is output for a time period $i \leq j$ (if not, or if no valid forgery is output, A fails). Let $y = z^e v^\sigma$. Because we modified F to first ask a hash query on (i, e, y, M), we have that, for some h, $(h, y, M, i, e, \sigma) = (h, y'_h, M'_h, j'_h, e'_h, \sigma'_h)$ in the hash query table (it can't come from the signature query table, because F is not allowed to forge a signature on a message for which it asked a signature query). A finds such an h in its table and remembers it.

A now resets F with the same random tape as the first time, and runs it again, giving the exact same answers to all F's queries before the h-th hash query (it can do so because it has all the answers recorded in the tables). Note that this means that F will be asking the same h-th hash query (i, e, y, M) as the first time. As soon as F asks the h-th hash query, however, A stops giving the answers from the tables and comes up with new answers at random, in the same manner as the first time. Let τ be the new answer given to the h-th hash query, and assume $\tau \neq \sigma$.

Assume again the break-in query occurs during time period $b > j$, and the valid forgery (z', σ', i', e') is output for a time period $i' \leq j$. A again computes $y' = z'^{e'} v^{sigma'}$; by the same reasoning as before, F had to ask a hash query on (i', e', y', M'). Let h' be the number of that query. A finds h' and fails if $h' \neq h$. If, however, $h' = h$, then $(i, e, y, M) = (i', e', y', M')$, simply because the h-th hash query had to be the same in both runs of F. Also then $\sigma' = \tau$. Therefore, $z^e v^\sigma \equiv z'^e v^\tau$, so $(z/z')^e \equiv v^{\tau - \sigma} \equiv \alpha^{f_{j+1}(\sigma - \tau)} \pmod N$.

Note that because e is guaranteed to be relatively prime with f_{j+1} (as long as $i \leq j$), and $\sigma - \tau$ has at least one fewer bit than e, $\gcd(f_{j+1}(\sigma - \tau), e) = \gcd(\sigma - \tau, e) < e$ (as long as $\sigma \neq \tau$). Thus, $r = e / \gcd(f_{j+1}(\sigma - \tau), e) > 1$ and, by Lemma 1, A will be able to efficiently compute the r-th root of α.

Running Time Analysis. A runs F twice. Preparing the public key and answering hashing and signing queries takes A no longer than it would take the real oracles. To find the hashing query corresponding to the forgery and to apply Lemma 1 takes $O(lT(l^2T^2 + k^2))$ bit operations.

PROBABILITY ANALYSIS. We will need the following lemma in our analysis.

Lemma 2. *Let $a_1, a_2, \ldots, a_\lambda$ be real numbers. Let $a = \sum_{\mu=1}^{\lambda} a_\mu$, and let $s = \sum_{\mu=1}^{\lambda} a_\mu^2$. Then $s \geq \frac{a^2}{\lambda}$.*

Proof. Let $b = a/\lambda$ and $b_\mu = b - a_\mu$. Then $\sum_{\mu=1}^{\lambda} b_\mu = \lambda b - \sum_{\mu=1}^{\lambda} a_\mu = 0$. Hence $\sum_{\mu=1}^{\lambda} a_\mu{}^2 = \sum_{\mu=1}^{\lambda}(b - b_\mu)^2 = \lambda b^2 - 2b\sum_{\mu=1}^{\lambda} b_\mu + \sum_{\mu=1}^{\lambda} b_\mu^2 \geq \lambda b^2 = \frac{a^2}{\lambda}$. □

First, consider the probability that A's answers to F's oracle queries are distributed as those of the true oracles that F expects. This is the case unless, for some signature query, the hash value that A needs to define has already been defined through a previous answer to a hash query (call this "A's failure to pretend"). Because z is picked at random from Z_n*, $z^e v^\sigma$ is a random element of Z_n^*. The probability of its collision with a value from a hash query in the same execution of F is at most $(q_{\text{hash}} + 1)/|Z_n^*|$ thus, the probability (taken over only the random choices of A) of A's failure to pretend is at most $q_{\text{sig}}(q_{\text{hash}}+1)/|Z_n^*| \leq q_{\text{sig}}(q_{\text{hash}}+1)2^{2-k}$ (because $|Z_n^*| = 4q_1q_2 > 2^{k-2}$). This is exactly the amount by which F's probability of success is reduced because of interaction with A rather than the real signer. Let $\delta = \varepsilon - q_{\text{sig}}(q_{\text{hash}} + 1)2^{2-k}$.

Let ε_b be the probability that F produces a successful forgery and that its break-in query occurs in time period b. Clearly, $\delta = \sum_{b=2}^{T+1} \varepsilon_b$ (if $b = 1$, then F cannot forge for any time period). Assume now that A picked $j = b - 1$ for some fixed b. The probability of that is $1/T$.

We will now calculate the probability of the event that F outputs a valid forgery based on the same hash query both times and that the hash query was answered differently the second time and that the break-in query was b both times. Let $p_{h,b}$ be the probability that, in one run, F produces a valid forgery based on hash query number h after break-in query in time period b. Clearly,

$$\varepsilon_b = \sum_{h=1}^{q_{\text{hash}}+1} p_{h,b}$$

Let $p_{h,b,S}$ (for a sufficiently long binary string S of length m) be the probability that, in one run, F produces a valid forgery based on hash query number h after break-in query in time period b, given that the string S was used to determine the random tape of F and the responses to all the oracle queries of F until (and not including) the h-th hash query. We have that

$$2^m p_{h,b} = \sum_{S \in \{0,1\}^m} p_{h,b,S}.$$

Given such a fixed string S, the probability that F produces a valid forgery based on the hash query number h after break-in query in time period b in both runs is $p_{h,b,S}^2$ (because the first forgery is now independent of the second forgery). The additional requirement that the answer to the hash query in the second run be different reduces this probability to $p_{h,b,S}(p_{h,b,S} - 2^{-l})$. Thus, the probability $q_{h,b}$ that F produces a valid forgery based on the hash query number h in both runs and that the answer to the hash query is different in the second run and that the break-in query was b in both runs is

$$\begin{aligned} q_{h,b} &= \sum_{S\in\{0,1\}^m} 2^{-m} p_{h,b,S}(p_{h,b,S} - 2^{-l}) \\ &= 2^{-m}\left(\sum_{S\in\{0,1\}^m} p_{h,b,S}^2 - 2^{-l}\sum_{S\in\{0,1\}^m} p_{h,b,S}\right) \\ &\geq \frac{2^{-m}\left(p_{h,b}2^m\right)^2}{2^m} - 2^{-l}p_{h,b} = p_{h,b}^2 - 2^{-l}p_{h,b} \end{aligned}$$

(by Lemma 2).

The probability that F outputs a valid forgery based on the same hash query both times and that the hash query was answered differently in the second run and that the break-in query occurred in time period i is now

$$\sum_{h=1}^{q_{\text{hash}}+1} q_{h,b} \geq \sum_{h=1}^{q_{\text{hash}}+1} p_{h,b}^2 - \sum_{h=1}^{q_{\text{hash}}+1} 2^{-l}p_{h,b} \geq \frac{\varepsilon_b^2}{q_{\text{hash}}+1} - 2^{-l}\varepsilon_b$$

(by Lemma 2).

Note that if this happens, then the forgery occurs in time period $i < b = j+1$ (because the forgery has to occur before the break-in query), so A will be able to take a root of α.

Finally, we again use Lemma 2 to remove the assumption that A picked $j = b-1$ as the time period to get the probability of A's success:

$$\varepsilon' \geq \frac{1}{T}\sum_{i=2}^{T+1}\left(\frac{\varepsilon_b^2}{q_{\text{hash}}+1} - 2^{-l}\varepsilon_b\right) \geq \frac{\delta^2}{T^2(q_{\text{hash}}+1)} - \frac{\delta}{2^l T}. \qquad \square$$

Improved Online/Offline Signature Schemes

Adi Shamir and Yael Tauman

Applied Math. Dept.
The Weizmann Institute of Science
Rehovot 76100, Israel
{shamir,tauman}@wisdom.weizmann.ac.il

Abstract. The notion of on-line/off-line signature schemes was introduced in 1990 by Even, Goldreich and Micali. They presented a general method for converting any signature scheme into an on-line/off-line signature scheme, but their method is not very practical as it increases the length of each signature by a quadratic factor. In this paper we use the recently introduced notion of a trapdoor hash function to develop a new paradigm called *hash-sign-switch*, which can convert any signature scheme into a highly efficient on-line/off-line signature scheme: In its recommended implementation, the on-line complexity is equivalent to about 0.1 modular multiplications, and the size of each signature increases only by a factor of two. In addition, the new paradigm enhances the security of the original signature scheme since it is only used to sign random strings chosen off-line by the signer. This makes the converted scheme secure against adaptive chosen message attacks even if the original scheme is secure only against generic chosen message attacks or against random message attacks.

Keywords: signature schemes, on-line/off-line, trapdoor hash functions.

1 Introduction

Digital signature schemes are among the most fundamental and useful inventions of modern cryptography. In such schemes, each user generates a (private) signing key and a (public) verification key. A user signs a message using his private signing key, and anyone can authenticate the signer and verify the message by using the signer's public verification key. A signature scheme is considered to be secure if signatures on new messages cannot be forged by any attacker who knows the user's public key but not his private key. Many constructions of signature schemes appear in the literature, but most of these schemes have unproven security, and the few schemes that are provably secure (under standard cryptographic assumptions) are not fast enough for many practical applications. Signature schemes that are efficient and provably secure are interesting both from a practical and a theoretical point of view.

In this paper, we introduce a general method for simultaneously improving both the security and the real-time efficiency of any signature scheme by converting it into an efficient *on-line/off-line signature scheme*. This notion was first

J. Kilian (Ed.): CRYPTO 2001, LNCS 2139, pp. 355–367, 2001.

introduced by Even, Goldreich and Micali [1]. The idea is to perform the signature generating procedure in two phases. The first phase is performed off-line (before the message to be signed is given) and the second phase is performed on-line (after the message to be signed is given). On-line/off-line signature schemes are useful, since in many applications the signer has a very limited response time once the message is presented, but he can carry out costly computations between consecutive signing requests. On-line/off-line signature schemes are particularly useful in smart card applications: The off-line phase is implemented either during the card manufacturing process or as a background computation whenever the card is connected to power, and the on-line phase uses the stored result of the off-line phase to sign actual messages. The on-line phase is typically very fast, and hence can be executed efficiently even on a weak processor.

Some signature schemes can be naturally partitioned into off-line and on-line phases. For example, the first step in the Fiat-Shamir, Schnorr, El-Gamal and DSS signature schemes does not depend on the given message, and can thus be carried out off-line. However, these are particular schemes with special structure and specific security assumptions rather than a general and provably secure conversion technique for arbitrary signature schemes.

Even, Goldreich and Micali presented a general method for converting any signature scheme into an on-line/off-line signature scheme. Their method uses a one-time signature scheme, i.e., a scheme which can securely sign only a single message. The essence of their method is to apply (off-line) the ordinary signing algorithm to authenticate a fresh one-time verification key, and then to apply (on-line) the one-time signing algorithm, which is typically very fast. In the basic [1] construction of a one-time bit-oriented signature scheme, the size of each signature is k^2 (where k is the size of the message and the security parameter). Additional constructions were proposed in [1], but they offer a very inefficient tradeoff between the size of the keys and the complexity of the one-time signing algorithm. In this paper, we present a method that increases the length of the signatures by an additive (rather than multiplicative) factor of k bits.

Our method uses a special type of hash functions, called *trapdoor hash functions*. These functions were recently introduced by Krawczyk and Rabin [3], who used them to construct *chameleon signatures*. Chameleon signatures are signatures that commit the signer to the contents of the signed message (as regular signatures do) but do not allow the recipient of the signature to convince third parties that a particular message was signed, since the recipient can change the signed message to any other message of his choice.

A trapdoor hash function is associated with a public key and a private key, referred to as the hash key HK and the trapdoor key TK, respectively. Loosely speaking, a trapdoor hash function is a probabilistic function h, such that collisions are difficult to generate when only HK is known, but easy to generate when TK is also known. More formally, given only HK, it is hard to find two messages m, m' and two auxiliary numbers r, r' such that $h(m;r) = h(m';r')$, but given (HK, TK) and m, m', r', it is easy to find r such that $h(m;r) = h(m';r')$. Note that this requirement is weaker than the requirement of trapdoor permutations, and thus it may be easier to find efficient trapdoor hash functions than to find efficient signature schemes based on trapdoor permutations.

The essence of our method is to hash the given message using a trapdoor hash function (rather than a regular hash function) and then to sign the hashed value using the given signature scheme. The resultant signature scheme can be implemented as an on-line/off-line signature scheme as follows: The off-line phase uses the original signature scheme to sign the hash value $h(m';r')$ of a random message m' and a random auxiliary number r'. Given an actual message m, the on-line phase uses the same precomputed signature of the randomly chosen m' as a signature of the given message m, by using the trapdoor key to find a collision of the form $h(m';r') = h(m;r)$. The signature of m consists of the new auxiliary number r and the precomputed signature of $h(m';r')$. We call this paradigm a *hash-sign-switch scheme.* Notice that the on-line phase is completely independent of the original signature scheme, and consists only of finding a collision of the trapdoor hash function. In particular, we describe a trapdoor hash function in which collisions can be found with time complexity equivalent to about 0.1 modular multiplications. Hence, for *any* signature scheme, its on-line/off-line version can be implemented such that the on-line phase requires only this negligible time complexity, and the size of the signature is only increased by adding r to the original signature.

For any signature scheme, we prove that our on-line/off-line version is at least as secure as the original scheme, provided that the trapdoor hash family is secure. In fact, we prove that the converted scheme is even more secure than the original scheme, since the original scheme is only applied to random messages chosen exclusively by the signer. In particular, we can show that the on-line/off-line signature scheme is secure against adaptive chosen message attacks even if the original signature scheme is secure only against generic chosen message attacks or random message attacks. Note for example, that the Rabin signature scheme [5] and the RSA signature scheme [6] are not secure against adaptive chosen message attacks, but are believed to be secure against random message attacks, and hence we believe that our method enhances the security of these schemes.

2 Definitions and Constructions

In this section, we introduce the basic notations and definitions used in this paper and present some constructions of trapdoor hash functions. For any binary string x, we denote by $|x|$ the length of x. For any finite set V, the notation $x \in_R V$ implies that x is uniformly distributed in V.

We consider the following types of attacks:

- Random message attack: The attacker has access to an oracle that signs (with the unknown signing key SK) random message chosen by the oracle.
- Generic chosen message attack: The attacker is given signatures for a list of messages of his choice. However, this list should be produced before any signature is given, and should be independent of the verification key VK.

- Adaptive chosen message attack: The attacker has access to an oracle that signs any queried message m. In particular, the choice of each query m can depend on the verification key VK and on the signature produced for previous messages.
- Q-adaptive chosen message attack: An adaptive chosen message attack where the attacker can query the oracle at most Q times.

In this work, a signature scheme is considered to be secure (against a certain type of attack) if there does not exist a probabilistic polynomial-time forger that generates a pair consisting of some new message (that was not previously presented to the oracle) and a valid signature, with a probability which is not negligible. This property was called *existential unforgeability* in [2].

In the remaining part of this section, we concentrate on the notion of a trapdoor hash function [3]. A trapdoor hash function is a special type of hash function, whose collision resistance depends on the user's state of knowledge. Every trapdoor hash function is associated with a pair of public key and private key, referred to as the hash key HK and the trapdoor key TK, respectively:

Definition 1. (trapdoor hash family) *A trapdoor hash family consists of a pair $(\mathcal{I}, \mathcal{H})$ such that:*

- *$\mathcal{I}$ is a probabilistic polynomial-time key generation algorithm that on input 1^k outputs a pair (HK, TK), such that the sizes of HK, TK are polynomially related to k.*
- *$\mathcal{H}$ is a family of randomized hash functions. Every hash function in $\mathcal{H}$ is associated with a hash key HK, and is applied to a message from a space $\mathcal{M}$ and a random element from a finite space $\mathcal{R}$. The output of the hash function h_{HK} does not depend on TK.*

A trapdoor hash family $(\mathcal{I}, \mathcal{H})$ has the following properties:

1. Efficiency: *Given a hash key HK and a pair $(m, r) \in \mathcal{M} \times \mathcal{R}$, $h_{HK}(m; r)$ is computable in polynomial time.*
2. Collision resistance: *There is no probabilistic polynomial-time algorithm $\mathcal{A}$ that on input HK outputs, with a probability which is not negligible, two pairs $(m_1, r_1), (m_2, r_2) \in \mathcal{M} \times \mathcal{R}$ that satisfy $m_1 \neq m_2$ and $h_{HK}(m_1; r_1) = h_{HK}(m_2; r_2)$ (the probability is over HK, where $(HK, TK) \leftarrow \mathcal{I}(1^k)$, and over the random coin tosses of algorithm $\mathcal{A}$).* [1]
3. Trapdoor collisions: *There exists a probabilistic polynomial time algorithm that given a pair $(HK, TK) \leftarrow \mathcal{I}(1^k)$, a pair $(m_1, r_1) \in \mathcal{M} \times \mathcal{R}$, and an additional message $m_2 \in \mathcal{M}$, outputs a value $r_2 \in \mathcal{R}$ such that:*
   - *$h_{HK}(m_1; r_1) = h_{HK}(m_2; r_2)$.*
   - *If r_1 is uniformly distributed in $\mathcal{R}$ then the distribution of r_2 is computationally indistinguishable from uniform in $\mathcal{R}$.*

[1] Note that it is not required that given one collision it remains hard to find new collisions. Indeed, all the constructions that we present have the property that given a hash key HK and given a single collision of h_{HK}, one can easily compute a trapdoor key TK such that the pair (HK, TK) is in the range of $\mathcal{I}(1^k)$.

We refer to every member of a trapdoor hash family as a trapdoor hash function. We now present three constructions of trapdoor hash families. The first two constructions were presented in [3], and the third construction is a new one.

1. **A trapdoor hash function based on the Factoring assumption.**
   - The key generation algorithm $\mathcal{I}$: Choose at random two primes $p, q \in \{0,1\}^{k/2}$ such that $p \equiv 3 \ (mod\ 8)$ and $q \equiv 7 \ (mod\ 8)$, and compute $n = pq$. The public hash key is n and the private trapdoor key is (p, q).
   - The hash family $\mathcal{H}$: For a hash key n, h_{HK} is a function from $\mathcal{M} \times QR_n$, where $\mathcal{M}$ is any suffix free subset of $\{0,1\}^*$ and $QR_n \stackrel{\text{def}}{=} \{x \in Z_n^* | (\frac{x}{p}) = (\frac{x}{q}) = 1\}$. Given a message $m = m[1]m[2] \ldots m[|m|]$ and a random value $r \in_R QR_n$, $h_{HK}(m; r) \stackrel{\text{def}}{=} f_{m[1]} \circ f_{m[2]} \circ \cdots \circ f_{m[|m|]}(r)$, where $f_0(x) \stackrel{\text{def}}{=} x^2 \ (mod\ n)$ and $f_1(x) \stackrel{\text{def}}{=} 4x^2 \ (mod\ n)$. (Note that $h(m; r) = 4^m r^{2^{|m|}} \ (mod\ n)$).

   Remark 1. The functions f_0 and f_1 were introduced in [2], who proved that they are claw free permutations, and used this property to construct an (inefficient) provably secure signature scheme.

   Lemma 1. *The pair $(\mathcal{I}, \mathcal{H})$ is a trapdoor hash family, under the Factoring Assumption.*

   A proof of this lemma appears in Appendix A. This trapdoor hash function has the following additional property: There exists a probabilistic polynomial-time algorithm that given a pair (HK, TK) (of hash key and trapdoor key), a message $m \in \mathcal{M}$ and any value c in the image of h_{HK}, outputs $r \in \mathcal{R}$ such that:
   - $h_{HK}(m; r) = c$.
   - If c is uniformly distributed (in the image of h_{HK}) then the distribution of r is computationally indistinguishable from uniform (in $\mathcal{R}$).

   Note that this inversion property is stronger than the ability to generate collisions. We will use it to convert any signature scheme which is provably secure only against random message attacks into a signature scheme which is provably secure against adaptive chosen message attacks.
2. **A trapdoor hash family based on the Discrete Log Assumption**
   - The key generation algorithm $\mathcal{I}$. Choose at random a safe prime $p \in \{0,1\}^k$ (i.e., a prime p such that $q \stackrel{\text{def}}{=} \frac{p-1}{2}$ is prime) and an element $g \in \mathbf{Z}_p^*$ of order q. Choose a random element $\alpha \in_R Z_q^*$ and compute $y = g^\alpha \ (mod\ p)$. The public hash key is (p, g, y) and the private trapdoor key is α.
   - The hash family $\mathcal{H}$. For $HK = (p, g, y)$, $h_{HK} : Z_q \times Z_q \longrightarrow Z_p^*$ is defined as follows: $h_{HK}(m; r) \stackrel{\text{def}}{=} g^m y^r \ (mod\ p)$.

   Lemma 2. *The pair $(\mathcal{I}, \mathcal{H})$ is a trapdoor hash family, under the Discrete Log Assumption.*

   A proof of this lemma appears in Appendix B.

3. **A new trapdoor hash family based on the Factoring Assumption.**
   - The key generation Algorithm $\mathcal{I}$. Choose at random two safe primes $p, q \in \{0,1\}^{k/2}$ (i.e., primes such that $p' \stackrel{\text{def}}{=} \frac{p-1}{2}$ and $q' \stackrel{\text{def}}{=} \frac{q-1}{2}$ are primes) and compute $n = pq$. Choose at random an element $g \in Z_n^*$ of order $\lambda(n)$ ($\lambda(n) \stackrel{\text{def}}{=} lcm(p-1, q-1) = 2p'q'$). The public hash key is (n, g) and the private trapdoor key is (p, q).
   - The hash family $\mathcal{H}$. For $HK = (n, g)$, $h_{HK} : Z_n \times Z_{\lambda(n)} \longrightarrow Z_n^*$ is defined as follows: $h_{HK}(m; r) \stackrel{\text{def}}{=} g^{m \circ r} \ (mod\ n)$ (where $m \circ r$ denotes the concatenation of m and r).

Lemma 3. *The pair $(\mathcal{I}, \mathcal{H})$ is a trapdoor hash family, under the Factoring Assumption.*

A proof of this lemma appears in Appendix C.

We summarize the efficiency analysis of these three constructions of trapdoor hash families in the following table . We assume that the messages in $\mathcal{M}$ and the random seeds in $\mathcal{R}$ are of size $\approx k$.

Construction	Computing h_{HK}	Finding collisions	Inversion prop.	Assumption
1	k mult.	≈ 5 exp.	YES	Factoring
2	1 exp.	≈ 1 mult.	NO	Discrete Log
3	1 exp.	≈ 0.1 mult.	NO	Factoring

Remark 2. The complexity of collision finding in construction 3 is equivalent to about one tenth of a regular modular multiplication, since for 1024 bit keys and 160 bit (hashed) messages, it requires only two additions/subtractions and one reduction of a 1184 bit number modulo a 1024 bit number. See Appendix C for further details.

Remark 3. The relaxed security conditions of trapdoor hash functions may lead to new types of signature schemes whose hash functions are based on multivariate polynomials. Most of the multivariate signature schemes proposed so far were broken by attacking their hidden inversion structure. In the new paradigm, there is no need to invert $h(m; r) = c$, and thus they may be more resistant to cryptanalytic attacks.

3 The Hash-Sign-Switch Paradigm

We now introduce our general method for combining any trapdoor hash family $(\mathcal{I}, \mathcal{H})$ and any signature scheme (G, S, V) to get an on-line/off-line signature scheme. For a security parameter k, we construct an on-line/off-line scheme (G', S', V'), as follows.

- **The Key Generation Algorithm G'.**
  1. Generate a pair (SK, VK) of signing key and verification key, by applying G to the input 1^k (where G is the key generation algorithm of the original scheme).
  2. Generate a pair (HK, TK) of hash key and trapdoor key, by applying $\mathcal{I}$ to the input 1^k (where $\mathcal{I}$ is the key generation algorithm of the trapdoor hash family).

  The signing key is (SK, HK, TK) and the verification key is (VK, HK).
- **The Signing Algorithm S'.** Given a signing key (SK, HK, TK), the signing algorithm operates as follows.
  1. *Off-line phase:*
     - Choose at random $(m', r') \in_R \mathcal{M} \times \mathcal{R}$, and compute $h_{HK}(m'; r')$ (using HK).
     - Run the signing algorithm S with the signing key SK to sign the message $h_{HK}(m'; r')$. Denote the output $S_{SK}(h_{HK}(m'; r'))$ by Σ.
     - Store the pair (m', r'), the hash value $h_{HK}(m'; r')$, and the signature Σ. (The hash value $h_{HK}(m'; r')$ is stored only to avoid its re-computation in the on-line phase).
  2. *On-line phase:* Given a message m, the on-line phase proceeds as follows.
     - Retrieve from memory the pair (m', r'), the hash value $h_{HK}(m'; r')$, and the signature Σ.
     - Find $r \in \mathcal{R}$ such that $h_{HK}(m; r) = h_{HK}(m'; r')$.
     - Send (r, Σ) [2] as a signature of m.
- **The Verification Algorithm V'.** To verify that the pair (r, Σ) is indeed a signature of the message m, with respect to the verification key (VK, HK), compute $h_{HK}(m; r)$ and use the verification algorithm V (of the original signature scheme) to check that Σ is indeed a signature of the hash value $h_{HK}(m; r)$ with the verification key VK.

We now analyze the security and the efficiency of the resultant on-line/off-line signature scheme.

3.1 Efficiency

The off-line phase of the signing algorithm consists of one evaluation of the trapdoor hash function and one invocation of the original signing algorithm. The verification algorithm of the on-line/off-line signature scheme consists of one evaluation of the trapdoor hash function and one invocation of the original verification algorithm. Hence, the additional overhead of the off-line signing phase and the verification algorithm is a single evaluation of the trapdoor hash function. The on-line phase consists of a single collision finding computation.

[2] Note that the signature (r, Σ) has the property that the distribution of r is computationally indistinguishable from uniform in $\mathcal{R}$, and that the distribution of Σ is identical to the distribution of $S_{SK}(h_{HK}(m; r))$.

Using the third type of trapdoor hash function presented in Section 2, evaluation requires one modular exponentiation, and collision finding requires about 0.1 modular multiplications. The length of the keys and the length of the signatures increase only by a factor of two, which is much better than in previous proposals.

3.2 Security

The general conversion technique proposed in this paper preserves the security of the original signature scheme, and even improves it in some respects since the opponent cannot control the random strings it is asked to sign during the off-line phase. We can thus prove that our on-line/off-line signature scheme is secure against adaptive chosen message attacks, even if the original signature scheme is secure only against generic chosen message attacks. Due to the practical emphasis of this work, we focus on exact security, rather than on asymptotic security.

Lemma 4. *Let (G, S, V) be a signature scheme and let $(\mathcal{I}, \mathcal{H})$ be a trapdoor hash family. Let (G', S', V') be the resultant on-line/off-line signature scheme. Suppose that (G', S', V') is existentially forgeable by a Q-adaptive chosen message attack in time T with success probability ϵ. Then one of the following cases holds:*

1. *There exists a probabilistic algorithm that given a hash key HK, finds collisions of h_{HK} in time $T + T_G + Q(T_\mathcal{H} + T_S)$ with success probability $\geq \frac{\epsilon}{2}$ (where T_G is the running time of G, $T_\mathcal{H}$ is the running time required to compute functions in $\mathcal{H}$, and T_S is the running time of S).*
2. *The original signature scheme (G, S, V) is existentially forgeable by a generic Q-chosen message attack in time $T + Q(T_\mathcal{H} + T_{COL}) + T_I$ with success probability $\geq \frac{\epsilon}{2}$ (where T_{COL} is the time required to find collisions of the trapdoor hash function given the hash key and the trapdoor key, and T_I is the running time of algorithm I).*

Proof. Suppose that $\mathcal{F}'$ is a probabilistic algorithm that given a verification key (HK, VK), forges a signature with respect to the signature scheme (G', S', V') by a Q-chosen message attack in time T with success probability ϵ. Let $\{m_i\}_{i=1}^{Q}$ denote the Q queries that the forger $\mathcal{F}'$ sends to the signing oracle, and let $\{(r_i, \Sigma_i)\}_{i=1}^{Q}$ denote the corresponding signatures produced by the oracle. Let $m, (r, \Sigma)$ denote the output of $\mathcal{F}'$. Since with probability $\geq \epsilon$, (r, Σ) is a valid signature of the message m (with respect to the on-line/off-line signature scheme (G', S', V')), it follows that

$$Pr[V_{VK}(h_{HK}(m; r), \Sigma) = 1] \geq \epsilon.$$

Hence, one of the following cases holds:

1. $Pr[V_{VK}(h_{HK}(m; r), \Sigma) = 1 \ \& \ \exists i \text{ s.t. } h_{HK}(m_i; r_i) = h_{HK}(m; r)] \geq \frac{\epsilon}{2}$.
2. $Pr[V_{VK}(h_{HK}(m; r), \Sigma) = 1 \ \& \ \forall i, h_{HK}(m_i; r_i) \neq h_{HK}(m; r)] \geq \frac{\epsilon}{2}$.

If case 1 holds, then we define a probabilistic algorithm $\mathcal{A}$ that given a hash key HK finds collisions of the hash function h_{HK}, as follows.

1. Generate a pair (SK, VK) of signing key and verification key, by applying G to the input 1^k (where G is the key generation algorithm of the original signature scheme).
2. Simulate the forger $\mathcal{F}'$ on the input (VK, HK), such that whenever $\mathcal{F}'$ queries the signing oracle S' with a query m_i, algorithm $\mathcal{A}$ operates as follows:
   - Choose at random $r_i \in_R \mathcal{R}$ and compute $h_{HK}(m_i; r_i)$.
   - Generate a valid signature of $h_{HK}(m_i; r_i)$ (with respect to the original signature scheme (G, S, V)), by using the known signing key SK. Denote the generated signature of $h_{HK}(m_i; r_i)$ by Σ_i.
   - Proceed in the simulation of $\mathcal{F}'$ as if the signature obtained by the signing oracle S' was (r_i, Σ_i).

Note that the distribution of the simulated oracle is identical to the distribution of the real oracle, and hence with probability $\geq \frac{\epsilon}{2}$, $\mathcal{A}$ succeeds in obtaining a message m and a pair (r, Σ), such that for every i, $m \neq m_i$, and there exists i such that $h_{HK}(m; r) = h_{HK}(m_i; r_i)$. Hence, $\mathcal{A}$ succeeds in finding collisions to the hash function h_{HK} with probability $\geq \frac{\epsilon}{2}$ in time $T + T_G + Q(T_{\mathcal{H}} + T_S)$.

If case 2 holds, we define a probabilistic algorithm $\mathcal{F}$ that forges a signature with respect to (G, S, V) by a generic Q-chosen message attack, as follows.

1. Generate a pair (HK, TK) of hash key and trapdoor key, by applying $\mathcal{I}$ to the input 1^k (where $\mathcal{I}$ is the key generation algorithm of the trapdoor hash family).
2. Choose at random Q pairs $(m_i', r_i') \in_R \mathcal{M} \times \mathcal{R}$ and compute $h_{HK}(m_i'; r_i')$. The set $\{h_{HK}(m_i'; r_i')\}_{i=1}^{Q}$ will be the set of queries to the signing oracle S.

Given a verification key VK and given a set of signatures $\{\Sigma_i\}_{i=1}^{Q}$ (where Σ_i is a signature of $h_{HK}(m_i; r_i)$ with respect to the verification key VK), $\mathcal{F}$ simulates the forger $\mathcal{F}'$ on input (VK, HK) as follows. When $\mathcal{F}'$ queries the oracle with a message m_i, $\mathcal{F}$ finds $r_i \in \mathcal{R}$ such that $h_{HK}(m_i; r_i) = h_{HK}(m_i'; r_i')$ and proceeds as if the signature obtained by the signing oracle S' was (r_i, Σ_i). Recall that r_i can be chosen such that if r_i' is uniformly distributed in $\mathcal{R}$ then r_i is computationally indistinguishable from uniform in $\mathcal{R}$. Hence, the distribution of the output of the simulated oracle is computationally indistinguishable from the distribution of the output of the real oracle. Thus, with probability $\geq \frac{\epsilon}{2}$, $\mathcal{F}$ obtains a message m and a pair (r, Σ) such that:

- $h_{HK}(m; r) \neq h_{HK}(m_i', r_i')$ for every $i = 1, \ldots, Q$.
- Σ is a valid signature of $h_{HK}(m; r)$ (with respect to the original signature scheme).

Hence $\mathcal{F}$ succeeds in forging a new signature with probability $\geq \frac{\epsilon}{2}$ in time $T + T_{\mathcal{I}} + Q(T_{\mathcal{H}} + T_{COL})$. □

Recalling the definitions of security, we get:

Theorem 1. *The resulting on-line/off-line signature scheme is secure against adaptive chosen message attacks, provided that the original scheme is secure against generic chosen message attacks.*

Our technique can be used to enhance the security of signature schemes even further. In particular, our conversion method can be used to convert any signature scheme which is secure only against *random message attacks* into a signature scheme which is secure against *adaptive chosen message attacks*. Recall that in the proof of Lemma 4, the signing oracle S' with a given query m_i was simulated as follows: Retrieve from memory the signature Σ_i of $h_{HK}(m'_i; r'_i)$ (obtained by the oracle), find an element r_i such that $h_{HK}(m_i; r_i) = h_{HK}(m'_i; r'_i)$, and output (r_i, Σ_i) as a signature of m_i. If the original scheme is only secure against random message attacks, then the forger $\mathcal{F}$ has access to an oracle that outputs pairs (c_i, Σ_i), where c_i is a random message (generated by the oracle) and Σ_i is a valid signature of c_i. Hence, using the same technique, to simulate the signing oracle S' with a given query m_i one needs to find r_i such that $h_{HK}(m_i; r_i) = c_i$. Thus, we need the trapdoor hash family to have the following inversion property: given a pair (HK, TK), a message $m \in \mathcal{M}$, and an element c in the image of h_{HK}, it is easy to find $r \in \mathcal{R}$ such that:

- $h_{HK}(m; r) = c$.
- The distribution of r is computationally indistinguishable from uniform in $\mathcal{R}$, provided that for every m the distribution of c is computationally indistinguishable from the distribution of $h_{HK}(m; r)$, where r is uniformly distributed in $\mathcal{R}$. [3]

By applying our on-line/off-line conversion method with such a trapdoor hash family, we can modify the proof of Lemma 4 to prove that the signature scheme obtained is secure against adaptive chosen message attacks, provided that the original scheme is secure against random message attacks.

References

1. Shimon Even, Oded Goldreich, and Silvio Micali, *On-line/off-line Digital Signatures*. In *Advances in Cryptology: Crypto '89*, pp 263-277. August 1990. Springer.
2. Shafi Goldwasser, Silvio Micali, and Ron Rivest, *A Digital Signature Scheme Secure Against Adaptive Chosen-Message Attacks, SIAM J. on Computing, 17*, pp 281-308, 1988.
3. Hugo Krawczyk and Tal Rabin, *Chameleon Signatures*. In *Symposium on Network and Distributed Systems Security (NDSS '00)*, pp 143-154, February 2000, Internet Society.
4. Gary Miller, *Riemann's Hypothesis and Tests for Primality, J. Comp. Sys. Sci.*, 13:300-317, 1976.

[3] Note that there is an implicit assumption here that for every two messages m_1, m_2 the distributions $h_{HK}(m_1; r_1)$ and $h_{HK}(m_2; r_2)$ are computationally indistinguishable, where r_1 and r_2 are uniformly distributed in $\mathcal{R}$.

5. Michael Rabin, *Digitized Signatures as Intractable as Factorization, Technical Report MIT/LCS TR-212*, January 1979.
6. Ron Rivest, Adi Shamir, and Len Adleman, *A Method of Obtaining Digital Signatures and Public-Key Cryptosystems, CACM, 21(2)*, pp 120-126, February 1978.

A Proof of Lemma 1

Proof. 1. *Efficiency:* Clearly, given a hash key n and a pair $(m;r) \in \mathcal{M} \times QR_n$, the function $h(m;r) = 4^m r^{2^{|m|}}$ $(mod\ n)$ can be computed in polynomial time.

2. *Collision resistance:* Assume to the contrary, that there exists a probabilistic polynomial time algorithm that given a hash key n outputs two pairs $(m_1, r_1), (m_2, r_2) \in \mathcal{M} \times QR_n$ such that $m_1 \neq m_2$ and $h_{HK}(m_1, r_1) = h_{HK}(m_2, r_2)$, with a probability which is not negligible. Let i be the smallest index of a bit where m_1 and m_2 differ (i.e., $m_1[i] \neq m_2[i]$ and $m_1[j] = m_2[j]$ for all $j < i$). Such a bit exists due to the suffix-free property of $\mathcal{M}$. Since we assume that the result of the hash function on (m_1, r_1) and (m_2, r_2) is the same and that $m_1[j] = m_2[j]$ for all $j < i$, and since f_0, f_1 are permutations, it follows that

$$f_{m_1[i]} \circ \cdots \circ f_{m_1[|m_1|]}(r_1) = f_{m_2[i]} \circ \cdots \circ f_{m_2[|m_2|]}(r_2).$$

Thus, we found a pair of values r'_1 and r'_2 for which $f_{m_1[i]}(r'_1) = f_{m_2[i]}(r'_2)$. As proven in [2], the existence of such claws for (f_0, f_1) contradicts the Factoring Assumption.

3. *Trapdoor collisions:* Given a pair $(m_1, r_1) \in \mathcal{M} \times QR_n$ and any additional message $m_2 \in \mathcal{M}$, a value $r_2 \in QR_n$ such that $h_{HK}(m_1; r_1) = h_{HK}(m_2; r_2)$ is given by

$$r_2 = (f^{-1}_{m_2[1]} \circ f^{-1}_{m_2[2]} \circ \cdots \circ f^{-1}_{m_2[|m_2|]}(h_{HK}(m_1; r_1))).$$

Given the trapdoor key $TK = (p, q)$, the functions f_0^{-1}, f_1^{-1} are computable in polynomial time, and therefore the value of r_2 is also computable in polynomial time. It remains to note that since f_0, f_1 are permutations on QR_n, it follows that if r_1 is uniformly distributed in QR_n then r_2 is also uniformly distributed in QR_n.

□

B Proof of Lemma 2

Proof. 1. *Efficiency:* Clearly, given a hash key $HK = (p, g, y)$ and a pair $(m, r) \in Z_q \times Z_q$, the function $h_{HK}(m, r) = g^m y^r$ $(mod\ p)$ is computable in polynomial time.

2. *Collision resistance:* Assume to the contrary, that there exists a probabilistic polynomial time algorithm that given a hash key $HK = (p, g, y)$, outputs two pairs $(m_1, r_1), (m_2, r_2) \in Z_q \times Z_q$ such that $m_1 \neq m_2$ and $h_{HK}(m_1, r_1) = h_{HK}(m_2, r_2)$, with a probability which is not negligible. The discrete log of y with respect to the basis g can be calculated in polynomial time from the output, as follows. Let α denote the discrete log of y. Then

$$m_1 + \alpha r_1 = m_2 + \alpha r_2 \ (mod\ q).$$

The fact that $m_1 \neq m_2 \ (mod\ q)$ implies that $r_1 \neq r_2 \ (mod\ q)$, and thus $r_1 - r_2$ is invertible modulo the prime q. Hence, α can be computed in polynomial time as follows.

$$\alpha = (r_2 - r_1)^{-1}(m_1 - m_2) \ (mod\ q).$$

This contradicts the Discrete Log Assumption.
3. *Trapdoor collisions:* Assume that we are given a hash key (p, g, y) and a corresponding trapdoor key α. Given any pair $(m_1, r_1) \in Z_q \times Z_q$ and any additional message $m_2 \in Z_q$, we want to find $r_2 \in Z_q$ such that

$$g^{m_1} y^{r_1} = g^{m_2} y^{r_2} \ (mod\ p).$$

The value of r_2 can be calculated in polynomial time as follows.

$$r_2 = \alpha^{-1}(m_1 - m_2) + r_1 \ (mod\ q).$$

It remains to note that if r_1 is uniformly distributed in Z_q then r_2 is also uniformly distributed in Z_q.

□

C Proof of Lemma 3

Proof. 1. *Efficiency:* Clearly, given a hash key $HK = (n, g)$ and a pair $(m, r) \in Z_n \times Z_{\lambda(n)}$, the function $h_{HK}(m; r) = g^{m \circ r} \ (mod\ n)$ is computable in polynomial time.
2. *Collision resistance:* Assume to the contrary, that there exists a probabilistic polynomial time algorithm that on input $HK = (n, g)$ outputs two pairs $(m_1, r_1), (m_2, r_2) \in Z_n \times Z_{\lambda(n)}$ such that $g^{m_1 \circ r_1} = g^{m_2 \circ r_2} \ (mod\ n)$, with a probability which is not negligible. Denote by $x \stackrel{\text{def}}{=} m_1 \circ r_1 - m_2 \circ r_2$ (this equality is over $\mathbf{Z}$). $x \neq 0$ since $m_1 \neq m_2$. The fact that $g^x = 1 \ (mod\ n)$ implies that $\lambda(n)$ divides x. Thus, $\phi(n)$ divides $2x$ (Since $\phi(n) = (p-1)(q-1) = 4p'q' = 2\lambda(n)$). Hence, there exists a probabilistic polynomial time algorithm, that on input (n, g) outputs a multiple of $\phi(n)$. It is known [4] that from any multiple of $\phi(n)$ the factorization of n can be efficiently computed. So we found a probabilistic polynomial time algorithm that solves the Factoring Problem with a probability which is not negligible. This contradicts the Factoring Assumption.

3. *Trapdoor collisions:* Given a hash key $HK = (n, g)$, a pair $(m_1, r_1) \in Z_n \times Z_{\lambda(n)}$, and an additional message $m_2 \in Z_n$, we want to find $r_2 \in Z_{\lambda(n)}$ such that $g^{m_1 o r_1} = g^{m_2 o r_2} \ (mod\ n)$. Namely, we want to find $r_2 \in Z_{\lambda(n)}$ such that $2^k m_1 + r_1 = 2^k m_2 + r_2 \ mod\ \lambda(n)$. Given the trapdoor key $TK = (p, q)$, $\lambda(n)$ can be computed in polynomial time, and hence r_2 can be computed in polynomial time as follows.

$$r_2 = 2^k(m_1 - m_2) + r_1 \ (mod\ \lambda(n)).$$

It remains to note that if r_1 is uniformly distributed in $Z_{\lambda(n)}$ then r_2 is also uniformly distributed in $Z_{\lambda(n)}$

□

Remark 4. Each r is uniformly distributed in $Z_{\lambda(n)}$, and thus a polynomial number of signatures reveal a logarithmic number of the most significant bits in the secret $\lambda(n)$. However, this is not dangerous since the known n and the secret $\phi(n) = 2\lambda(n)$ have the same bits in their top halves.

Remark 5. The equation used to find collisions in the second and third trapdoor hash families look similar, but are based on different security assumptions (discrete log vs. factoring). This difference makes it possible to replace the multiplication operation $\alpha^{-1}(m_1 - m_2)$ by the simpler left shift operation $2^k(m_1 - m_2)$, which saves about half the total time. In addition, when the size of the modulus is 1024 bits and the size of the (hashed) $(m_1 - m_2)$ is 160 bits, the reduction of the 1184 bit result modulo a 1024 bit modulus is about 6 times faster than a standard reduction of a 2048 bit product modulo a 1024 bit modulus. Consequently, we estimate that software implementations of the collision finding procedure will be about ten times faster than performing a single modular multiplication of two 1024 bit numbers.

An Efficient Scheme for Proving a Shuffle

Jun Furukawa and Kazue Sako

NEC Corporation, 4-1-1 Miyazaki, Miyamae, Kawasaki 216-8555, Japan
j-furukawa@ay.jp.nec.com, k-sako@ab.jp.nec.com

Abstract. In this paper, we propose a novel and efficient protocol for proving the correctness of a shuffle, without leaking how the shuffle was performed. Using this protocol, we can prove the correctness of a shuffle of n data with roughly $18n$ exponentiations, where as the protocol of Sako-Kilian[SK95] required $642n$ and that of Abe[Ab99] required $22n \log n$. The length of proof will be only $2^{11}n$ bits in our protocol, opposed to $2^{18}n$ bits and $2^{14}n \log n$ bits required by Sako-Kilian and Abe, respectively. The proposed protocol will be a building block of an efficient, universally verifiable mix-net, whose application to voting system is prominent.

Keywords: Mix-net, Permutation, Electronic Voting, Universal Verifiability

1 Introduction

A mix-net[Ch81] scheme is useful for applications which require anonymity, such as voting. The core technique in a mix-net scheme is to execute multiple rounds of shuffling and decryption by multiple, independent mixers, so that the output decryption can not be linked to any of the input encryptions.

To ensure the correctness of the output, it is desirable to achieve the property of universal verifiability. However, proving the correctness of a shuffle without sacrificing unlinkability required a large amount of computation in the prior art. For example, [SK95] adopted a cut-and-choose method to prove the correctness. Abe[Ab99] took an approach to represent a shuffle using multiple pairwise permutations [1]. In practical terms, however, neither scheme is efficient enough to handle a large number of ciphertexts, say on the order of 10,000.

This paper proposes a novel, efficient scheme for proving the correctness of a shuffle. We take a completely different approach than that of [SK95] and [Ab99]. We represent a permutation by a matrix, and introduce two conditions which suffice to achieve a permutation matrix. We then present zero-knowledge proofs to prove the satisfiability of each condition. Moreover, these two proofs can be merged into one proof, resulting in a very efficient proof of a correct shuffle.

We also present here an analysis of the efficiency of our proof. Our proof requires roughly $18n$ exponentiations to prove the correctness of a n-data shuffle,

[1] Another approach, based on a verifiable secret exponent multiplication is described in [Ne01].

J. Kilian (Ed.): CRYPTO 2001, LNCS 2139, pp. 368–387, 2001.

where as the protocol of Sako-Kilian[SK95] required $642n$ and that of Abe[Ab99] required $22n \log n$. Using the computation tools in [HAC], the total computation cost necessary in our proof can be reduced to an equivalent of $5n$ exponentiations. The length of a proof will be only $2^{11}n$ bits in our protocol, opposed to $2^{18}n$ bits and $2^{14}n \log n$ bits required by Sako-Kilian and Abe, respectively.

Our paper is organized in the following way. In Section 2, we present the two conditions on a permutation matrix. In Section 4 we give zero-knowledge proofs for each of the two conditions, and discuss how these proofs are combined to achieve to prove the whole shuffle. In Section 5 we describe our protocol and in Section 6, we compare the efficiency of our protocol to prior work.

2 Basic Idea

2.1 Shuffling

Informally speaking, a shuffling is a procedure which on input of n ciphertexts $(E_1, E_2, \ldots, E_n)$, outputs n ciphertexts $(E'_1, E'_2, \ldots, E'_n)$ where:

- there exists a permutation ϕ s.t $D(E'_i) = D(E_{\phi^{-1}(i)})$ for all i. Here, D is a decryption algorithm for ciphertexts.
- Without the knowledge of D or ϕ, $(E_1, E_2, \ldots, E_n)$, and $(E'_1, E'_2, \ldots, E'_n)$ reveal no information on the permutation ϕ.

We consider the use of ElGamal cryptosystems, with public keys (p, q, g, y) and secret key $X \in \mathbf{Z_q}$ s.t. $y = g^X \bmod p$. [2]

Given n ciphertexts $\{E_i\} = \{(g_i, m_i)\}$, where all $\{g_i\}$ and $\{m_i\}$ have the order q, shuffled ciphertexts $\{E'_i\} = \{(g'_i, m'_i)\}$ can be obtained by

$$\begin{aligned} g'_i &= g^{r_i} \cdot g_{\phi^{-1}(i)} \bmod p \\ m'_i &= y^{r_i} \cdot m_{\phi^{-1}(i)} \bmod p \end{aligned} \tag{1}$$

using randomly generated $\{r_i\}$.

2.2 Permutation Matrix

We define a matrix (A_{ij}) to be a permutation matrix if it can be written as follow using some permutation function ϕ.

$$A_{ij} = \begin{cases} 1 \bmod q & \text{if } \phi(i) = j \\ 0 \bmod q & \text{otherwise.} \end{cases}$$

Using this permutation matrix, the equation (1) is equivalent to

$$(g'_i, m'_i) = (g^{r_i} \prod_{j=1}^{n} g_j^{A_{ji}}, y^{r_i} \prod_{j=1}^{n} m_j^{A_{ji}}) \bmod p. \tag{2}$$

In order to prove the correctness of the shuffle, we need to show the following two things.

[2] We assume, as usual, p and q are two primes s.t. $p = kq + 1$, where k is an integer, and g is an element that generates a subgroup $\mathbf{G_q}$ of order q in $\mathbf{Z_p^*}$.

1. For each pair $\{(g'_i, m'_i)\}$, the same r_i and (A_{ij}) has been used.
2. (A_{ij}) used is a permutation matrix.

The first property can be efficiently shown using a standard technique[Br93]. The contribution of this paper is to present a novel technique to prove the second property.

At first, we concentrate on proving the existence of a permutation matrix (A_{ij}) and $\{r_i\}$ when given $\{g_i\}$ and $\{g'_i\}$, s.t.

$$g'_i = g^{r_i} \prod_{j=1}^{n} g_j{}^{A_{ji}} \bmod p. \tag{3}$$

We thus need to prove the existence of such a permutation matrix. We begin by looking at necessary conditions which suffices to achieve a permutation matrix. The following is the key observation used to construct the proposed protocol.

Theorem 1. *A matrix $(A_{ij})_{(i,j=1,\ldots,n)}$ is a permutation matrix if and only if, for all i, j, and k, both*

$$\sum_{h=1}^{n} A_{hi}A_{hj} = \begin{cases} 1 \bmod q, & \textit{if } i = j \\ 0 \bmod q, & \textit{if } i \neq j \end{cases} \tag{4}$$

$$\sum_{h=1}^{n} A_{hi}A_{hj}A_{hk} = \begin{cases} 1 \bmod q & \textit{if } i = j = k \\ 0 \bmod q & \textit{if otherwise} \end{cases} \tag{5}$$

hold.

Notation 1 *For convenience, we define δ_{ij} and $\delta_{ijk} (i, j, k = 1, \ldots, n)$ to be, respectively,*

$$\delta_{ij} = \begin{cases} 1 & \textit{if } i = j \\ 0 & \textit{if } i \neq j \end{cases} \quad \textit{and} \quad \delta_{ijk} = \begin{cases} 1 & \textit{if } i = j = k \\ 0 & \textit{if otherwise.} \end{cases}$$

Proof. We first show that there is exactly one non-zero element in each row vector of (A_{ij}) and then, the same for each column vector.

Let C_i be a i-th column vector of the matrix $(A_{ij})_{(i,j=1,\ldots,n)}$. Then, from Equation (4), we see $(C_i, C_j) = \delta_{ij}$ where (A, B) is inner product of vectors A and B. This implies that $rank(A_{ij}) = n$, that is, there is at least one non-zero element in each row and each column. Next we consider a vector $C_i \odot C_j (i \neq j)$ where the operator $\odot$ is defined as $(a_1 \ldots a_n) \odot (b_1 \ldots b_n) = (a_1 b_1 \ldots a_n b_n)$. Define a vector $\hat{C} = \sum_{l=0}^{n} \kappa_l C_l$ for an arbitrary κ_l. From the fact $(\hat{C}, C_i \odot C_j) = \sum_{l=1} \kappa_l \delta_{lij} = 0$ and linear combinations of $\{C_l\}$ generate the space $\mathbf{Z_q}^n$, we obtain $C_i \odot C_j = \mathbf{0}$. This means for any h, i and j s.t. $i \neq j$, either $A_{hi} = 0$ or $A_{hj} = 0$. Therefore, the number of non-zero elements in each row vector of (A_{ij}) is at most 1, and thus exactly 1.

From the above observations, the matrix (A_{ij}) contains exactly n non-zero elements. Since $C_i \neq \mathbf{0}$ for all i, the number of non-zero element in each column vector is also 1. Thus, there is exactly one non-zero element in each row vector and each column vector of the matrix (A_{ij}) if Equations (4) and (5) hold.

The unique non-zero element e_i in $i-th$ row must be $e_i^2 = 1 \bmod q$ from Equation (4) and $e_i^3 = 1 \bmod q$ from Equation (5). This leads to $e_i = 1$ and that matrix $(A_{ij})_{(i,j=1,\ldots,n)}$ is a permutation matrix over $\mathbf{Z_q}$.

2.3 Outline of Main Protocol

Using Theorem 1, the main protocol can be constructed by the following proofs

Proof-1 a proof that given $\{g_i\}$ and $\{g_i'\}$, $\{g_i'\}$ can be expressed as eq.(3) using integers $\{r_i\}$ and a matrix that satisfies the first condition.

Proof-2 a proof that given $\{g_i\}$ and $\{g_i'\}$, $\{g_i'\}$ can be expressed as eq.(3) using integers $\{r_i\}$ and a matrix that satisfies the second condition.

Proof-3 a proof that integers $\{r_i\}$ and the matrix used in the above two proofs are identical.

Proof-4 For each pair (g_i', m_i'), the same r_i and $\{A_{ij}\}$ has been used.

In the Section 4, we provide protocols for **Proof-1** and **Proof-2**.

3 Security of the Protocol

We will prove that the main protocol is sound and zero-knowledge under computational assumption. More specifically, for the property of soundness, we can claim that if a verifier accepts the protocol, then either prover knows the permutation or he knows integers $\{a_i\}$ and a satisfying $g^a \prod_{i=1}^n g_i{}^{a_i} = 1$ with overwhelming probability. For the zero-knowledge property, we can construct a simulator and claim that if there is a distinguisher who can distinguish between a real transcript from the protocol and an output from the simulator, then this distinguisher can be used to solve the decisional Diffie-Hellman problem. We note that to make a shuffle secret, we already assume the hardness of the decisional Diffie-Hellman problem.

In the course of reduction, we use the following arguments. First, we define the following set.

Definition 1. *Define R_n^m to be the set of tuples of $n \times m$ elements in $\mathbf{G_q}$:*

$$I = (x_1^{(1)}, \ldots, x_1^{(m)}, x_2^{(1)}, \ldots, x_2^{(m)}, \ldots, x_n^{(1)}, \ldots, x_n^{(m)}).$$

We then define the subset D_n^m of R_n^m to be the set of tuples I satisfying

$$\log_{x_1^{(1)}} x_1^{(i)} = log_{x_j^{(1)}} x_j^{(i)} \bmod p$$

for all $i (i = 2, 3, .., m)$ and $j (j = 2, ..., n)$.

Definition 2. *We define the problem of distinguishing instances uniformly chosen from R_n^m and those from D_n^m by DDH_n^m.*

Note that the decisional Diffie-Hellman problem can be denoted as DDH_2^2. We claim that for any n and m the difficulty of DDH_n^m equals to the decisional Diffie-Hellman problem, by proving the following.

Lemma 1. *For any $n(\geq 2)$ and $m(\geq 2)$, if DDH_n^m is easy, then DDH_n^2 is easy.*

Lemma 2. *If for any $n(\geq 2)$, DDH_n^2 is easy then the decisional Diffie-Hellman is easy.*

Proofs for Lemma 1 and 2 are sketched in Appendix A

4 Proof-1 and Proof-2

In this section, we give two proofs that will be the building blocks of the main protocol.

4.1 Proving the First Condition (Proof-1)

The following protocol proves that given $\{g_i\}$ and $\{g'_i\}$, the prover knows $\{r_i\}$ and $\{A_{ij}\}$ s.t.

$$g'_i = g^{r_i} \prod_{j=1}^{n} g_j^{A_{ji}} \bmod p$$

$$\sum_{h=1}^{n} A_{hi} A_{hj} = \delta_{ij} \bmod q.$$

The main idea is to issue $s = \sum_{j=1}^{n} r_j c_j$ and $s_i = \sum_{j=1}^{n} A_{ij} c_j$ as a response to a challenge $\{c_j\}$ and let the verifier check

$$\sum_{i=1}^{n} {s_i}^2 = \sum_{j=1}^{n} {c_j}^2 \bmod q$$

$$g^s \prod_{i=1}^{n} {g_i}^{s_i} = \prod_{j=1}^{n} {g'_j}^{c_j} \bmod p.$$

However, this apparently leaks information on A_{ij}, so we need to add randomizers and commitments. By making the response $s = \sum_{j=1}^{n} r_j c_j + \alpha$ and $s_i = \sum_{j=1}^{n} A_{ij} c_j + \alpha_i$ using randomizers $\{\alpha_i\}$ and α, a verifier needs to check the following equation:

$$\sum_{i=1}^{n} {s_i}^2 = \sum_{j=1}^{n} {c_j}^2 + \sum_{j=1}^{n} B_j c_j + D \bmod q$$

where B_j and D are quadratic polynomials of $\{A_{ij}\}$ and α_i. Therefore these B_j and D, together with $g^\alpha \prod_{i=1}^{n} g_i^{\alpha_i}$, will be also sent in advance to enable verification. We further add another randomizer σ and modify the verification equation to be

$$\sum_{i=1}^{n} s_i^2 + \sigma s = \sum_{j=1}^{n} c_j^2 + \sum_{j=1}^{n} (B_j + \sigma r_j) c_j + (D + \sigma\alpha) \bmod q.$$

In order to hide the actual value of σ, $\{B_j + \sigma r_j\}$ and $D + \sigma\alpha$, this verification is computed over exponents. The below gives a complete description of the **Proof-1**.

Proof-1

Input:$p, q, g, \{g_i\}, \{g'_i\}$.

1. Prover ($\mathcal{P}$) generates random numbers $\sigma, \alpha, \{\alpha_i\}_{(i=1,\dots,n)} \in_R \mathbf{Z_q}$ and computes

$$\begin{aligned} w &= g^{\sigma} \bmod p \\ g' &= g^{\alpha} \prod_{j=1}^{n} g_j{}^{\alpha_j} \bmod p \qquad (6) \\ \dot{w}_i &= g^{\sum_{j=1}^{n} 2\alpha_j A_{ji} + \sigma r_i} \ (= g^{B_i + \sigma r_i}) \bmod p \quad i = 1, \dots, n \\ \dot{w} &= g^{\sum_{j=1}^{n} \alpha_j{}^2 + \sigma\alpha} \ (= g^{D + \sigma\alpha}) \bmod p \end{aligned}$$

   and sends $w, g', \{\dot{w}_i\}, \dot{w}$ $(i = 1, \dots, n)$ to $\mathcal{V}$.
2. $\mathcal{V}$ sends back randomly chosen $\{c_i\}_{(i=1,\dots,n)} \in_R \mathbf{Z_q}$ as a challenge.
3. $\mathcal{P}$ computes $s = \sum_{j=1}^{n} r_j c_j + \alpha \bmod q$
   and $s_i = \sum_{j=1}^{n} A_{ij} c_j + \alpha_i \bmod q (i = 1, \dots, n)$ and sends to $\mathcal{V}$.
4. $\mathcal{V}$ verifies the following:

$$g^s \prod_{j=1}^{n} g_j{}^{s_j} = g' \prod_{j=1}^{n} g_j'^{c_j} \bmod p \qquad (7)$$

$$w^s g^{\sum_{j=1}^{n} (s_j^2 - c_j^2)} = \dot{w} \prod_{j=1}^{n} \dot{w}_j{}^{c_j} \bmod p \qquad (8)$$

.

Properties of Proof-1

Theorem 2. **Proof-1** *is complete. That is, if $\mathcal{P}$ knows $\{r_i\}$ and $\{A_{ij}\}$ satisfying the first condition, $\mathcal{V}$ always accepts.*

Theorem 3. *If $\mathcal{V}$ accepts* **Proof-1** *with a non-negligible probability, then $\mathcal{P}$ either knows both $\{r_i\}$ and $\{A_{ij}\}$ satisfying the first condition, or can generate integers $\{a_i\}$ and a satisfying $g^a \prod_{i=1}^{n} g_i{}^{a_i} = 1$ with overwhelming probability.*

A sketch of Proof: Theorem 3 can be proved from the following lemmas, proofs of which are sketched in Appendix B.

Lemma 3. *If $\mathcal{V}$ accepts* **Proof-1** *with non-negligible probability, then $\mathcal{P}$ knows $\{A_{ij}\}$, $\{r_i\}$, $\{\alpha_i\}$, and α satisfying Equations (3) and (6).*

Lemma 4. *Assume $\mathcal{P}$ knows $\{A_{ij}\}$, $\{r_i\}$, $\{\alpha_i\}$, and α satisfying Equations (3) and (6). If $\mathcal{P}$ knows $\{s_i\}$ and s which satisfy Equation (7), and either $s \neq \sum_{j=1}^{n} r_j c_j + \alpha$ or $s_i \neq \sum_{j=1}^{n} A_{ij} c_j + \alpha_i$ for some i hold, then $\mathcal{P}$ can generate non-trivial integers $\{a_i\}$ and a satisfying $g^a \prod_{i=1}^{n} {g_i}^{a_i} = 1$ with overwhelming probability.*

Lemma 5. *Assume $\mathcal{P}$ knows $\{A_{ij}\}$, $\{r_i\}$, $\{\alpha_i\}$, and α satisfying Equations (3) and (6). If Equations (7) and (8) hold with non-negligible probability, then either Equation (4) hold or $\mathcal{P}$ can generate non-trivial integers $\{a_i\}$ and a satisfying $g^a \prod_{i=1}^{n} {g_i}^{a_i} = 1$ with overwhelming probability.*

□

Theorem 4. *We can construct a simulator of* **Proof-1** *such that if there is a distinguisher who can distinguish between a real transcript from the protocol and an output from the simulator, then we can solve the decisional Diffie-Hellman problem.*

A sketch of Proof: Given in Appendix B. □

4.2 Proving the Second Condition(Proof-2)

Analogous to **Proof-1**, the proof for the fact the prover knows $\{r_i\}$ and $\{A_{ij}\}$ s.t.

$$g'_i = g^{r_i} \prod_j g_j^{A_{ji}} \bmod p$$

$$\sum_{h=1}^{n} A_{hi} A_{hj} A_{hk} = \delta_{ijk} \bmod q$$

for $\{g_i\}$ and $\{g'_i\}$, is given as **Proof-2**.

Proof-2

Input:$p, q, g, \{g_i\}, \{g'_i\}$.

1. Prover ($\mathcal{P}$) generates random numbers $\rho, \tau, \alpha, \{\alpha_i\}, \lambda, \{\lambda_i\} \in_R \mathbf{Z_q}$ $(i = 1, \ldots, n)$ and computes

$$t = g^{\tau}, v = g^{\rho}, u = g^{\lambda}, u_i = g^{\lambda_i} \bmod p \quad i = 1, \ldots, n$$

$$g' = g^{\alpha} \prod_{j=1}^{n} {g_j}^{\alpha_j} \bmod p$$

$$\dot{t}_i = g^{\sum_{j=1}^{n} 3\alpha_j A_{ji} + \tau\lambda_i} \bmod p \quad i = 1, \ldots, n$$

$$\dot{v}_i = g^{\sum_{j=1}^{n} 3{\alpha_j}^2 A_{ji} + \rho r_i} \bmod p \quad i = 1, \ldots, n$$

$$\dot{v} = g^{\sum_{j=1}^{n} {\alpha_j}^3 + \tau\lambda + \rho\alpha} \bmod p$$

and sends $t, v, u, \{u_i\}, g', \{\dot{t}_i\}, \{\dot{v}_i\}, \dot{v}$ $(i = 1, \ldots, n)$, to $\mathcal{V}$.

2. $\mathcal{V}$ sends back randomly chosen $\{c_i\}_{(i=1,\ldots,n)} \in_R \mathbf{Z_q}$ as challenge.
3. $\mathcal{P}$ computes $s = \sum_{j=1}^n r_j c_j + \alpha \bmod q$, $s_i = \sum_{j=1}^n A_{ij} c_j + \alpha_i \bmod q (i = 1, \ldots, n)$, and $\lambda' = \sum_{j=1}^n \lambda_j c_j^2 + \lambda \bmod q$ and sends to $\mathcal{V}$.
4. $\mathcal{V}$ verifies the following:

$$g^s \prod_{j=1}^n g_j{}^{s_j} = g' \prod_{j=1}^n g_j'^{c_j} \bmod p$$

$$g^{\lambda'} = u \prod_{j=1}^n u_j{}^{c_j{}^2} \bmod p$$

$$t^{\lambda'} v^s g^{\sum_{j=1}^n (s_j^3 - c_j^3)} = \dot{v} \prod_{j=1}^n \dot{v}_j{}^{c_j} t_j{}^{c_j^2} \bmod p$$

Properties of Proof-2

We claim the following properties of **Proof-2**, which can be proved analogously to that of **Proof-1**.

Theorem 5. **Proof-2** *is complete. That is, if $\mathcal{P}$ knows $\{r_i\}$ and $\{A_{ij}\}$ satisfying the second condition, $\mathcal{V}$ always accepts.*

Theorem 6. *If $\mathcal{V}$ accepts* **Proof-2** *with a non-negligible probability, then $\mathcal{P}$ either knows both $\{r_i\}$ and $\{A_{ij}\}$ satisfying the second condition, or can generate integers $\{a_i\}$ and a satisfying $g^a \prod_{i=1}^n g_i{}^{a_i} = 1$ with overwhelming probability.*

Theorem 7. *We can construct a simulator of* **Proof-2** *such that if there is a distinguisher who can distinguish between a real transcript from the protocol and an output from the simulator, then we can solve the decisional Diffie-Hellman problem.*

4.3 Constructing the Main Protocol

In this subsection, we explain how our main protocol is constructed using these **proof-1** and **Proof-2**. It should be noted, that these proofs did not have the ordinary soundness property. That is, a prover knowing integers satisfying $g^a \prod_{i=1}^n g_i^{a_i} = 1$ can deceive verifiers as if he had shuffled correctly. Since $\{g_i\}$ is originally chosen by those who encrypted the messages, there is no control to assure that the prover does not know the relations among them. Therefore, we fix a set of basis $\{\tilde{g}, \tilde{g}_1, \ldots \tilde{g}_n\}$ independent from the input ciphertexts, in a way we can assure the relations among the basis unknown. In fact, under Discrete Logarithm Assumption, we can make it computationally infeasible to obtain such $\{a_i\}$ and a if we generate $\{\tilde{g}, \tilde{g}_1, \ldots \tilde{g}_n\}$ randomly[Br93]. This way it also suffices the requirement that the verifier should not know $\log_g \tilde{g}$ for zero-knowledge property.

We require the prover to perform the same permutation on the set of fixed basis as he did on the input ciphertexts. The prover proves that the permutation on the fixed basis $\{\tilde{g}, \tilde{g}_1, \dots \tilde{g}_n\}$ is indeed a permutation, and that he indeed applied the same permutation to the input ciphertext.

Using the above methodology, we need not provide **Proof-3** described in the Subsection 2.3. If a prover knows two different representations of an element using $\{\tilde{g}, \tilde{g}_1, \dots \tilde{g}_n\}$, it means that he knows the relations among the base $\{\tilde{g}, \tilde{g}_1, \dots \tilde{g}_n\}$ which is against the assumption. **Proof-4** is achieved using the standard techniques described in [Br93]. Therefore we are now equipped with building blocks to prove the correctness of a shuffle.

5 The Main Protocol

In the previous subsection we illustrated our protocol as a combination of **proof-1** and **proof-2**, mainly for comprehensiveness. The proofs can be executed in parallel, resulting in a three-round protocol with reduced communication complexity.

<u>Main Protocol</u>

Input:$p, q, g, y, \tilde{g}, \{\tilde{g}_i\}, \{(g_i, m_i)\}, \{(g'_i, m'_i)\}$.

1. Prover ($\mathcal{P}$) generates the following random numbers:
   $\sigma, \rho, \tau, \alpha, \alpha_i, \lambda, \lambda_i \in_R \mathbf{Z}_q$ $(i = 1, \dots, n)$
2. $\mathcal{P}$ computes the following:

$$t = g^\tau, v = g^\rho, w = g^\sigma, u = g^\lambda, u_i = g^{\lambda_i} \bmod p \quad i = 1, \dots, n$$

$$\tilde{g}'_i = \tilde{g}^{r_i} \prod_{j=1}^{n} \tilde{g}_j{}^{A_{ji}} \bmod p \quad i = 1, \dots, n \tag{9}$$

$$\tilde{g}' = \tilde{g}^\alpha \prod_{j=1}^{n} \tilde{g}_j{}^{\alpha_j} \bmod p \tag{10}$$

$$g' = g^\alpha \prod_{j=1}^{n} g_j{}^{\alpha_j} \bmod p$$

$$m' = y^\alpha \prod_{j=1}^{n} m_j{}^{\alpha_j} \bmod p$$

$$\dot{t}_i = g^{\sum_{j=1}^{n} 3\alpha_j A_{ji} + \tau\lambda_i} \bmod p \quad i = 1, \dots, n$$

$$\dot{v}_i = g^{\sum_{j=1}^{n} 3\alpha_j{}^2 A_{ji} + \rho r_i} \bmod p \quad i = 1, \dots, n$$

$$\dot{v} = g^{\sum_{j=1}^{n} \alpha_j{}^3 + \tau\lambda + \rho\alpha} \bmod p$$

$$\dot{w}_i = g^{\sum_{j=1}^{n} 2\alpha_j A_{ji} + \sigma r_i} \bmod p \quad i = 1, \dots, n$$

$$\dot{w} = g^{\sum_{j=1}^{n} \alpha_j{}^2 + \sigma\alpha} \bmod p.$$

3. $\mathcal{P}$ sends the following to the verifier $\mathcal{V}$:
   $t, v, w, u, \{u_i\}, \{\tilde{g}'_i\}, \tilde{g}', g', m', \{\dot{t}_i\}, \{\dot{v}_i\}, \dot{v}, \{\dot{w}_i\}, \dot{w}$ $(i = 1, \dots, n)$.

4. $\mathcal{V}$ sends back randomly chosen $\{c_i\}_{(i=1,\ldots,n)} \in_R \mathbf{Z_q}$ as a challenge.
5. $\mathcal{P}$ computes the following and sends them to $\mathcal{V}$.

$$s = \sum_{j=1}^{n} r_j c_j + \alpha, \quad s_i = \sum_{j=1}^{n} A_{ij} c_j + \alpha_i \mod q \quad i = 1, \ldots, n$$
$$\lambda' = \sum_{j=1}^{n} \lambda_j c_j^2 + \lambda \mod q$$

6. $\mathcal{V}$ verifies the following:

$$\tilde{g}^s \prod_{j=1}^{n} \tilde{g}_j{}^{s_j} = \tilde{g}' \prod_{j=1}^{n} \tilde{g}_j'^{c_j} \mod p \tag{11}$$
$$g^s \prod_{j=1}^{n} g_j{}^{s_j} = g' \prod_{j=1}^{n} g_j'^{c_j} \mod p \tag{12}$$
$$y^s \prod_{j=1}^{n} m_j{}^{s_j} = m' \prod_{j=1}^{n} m_j'^{c_j} \mod p \tag{13}$$
$$g^{\lambda'} = u \prod_{j=1}^{n} u_j{}^{c_j{}^2} \mod p \tag{14}$$
$$t^{\lambda'} v^s g^{\sum_{j=1}^{n}(s_j^3 - c_j^3)} = \dot{v} \prod_{j=1}^{n} \dot{v}_j{}^{c_j} t_j{}^{c_j^2} \mod p \tag{15}$$
$$w^s g^{\sum_{j=1}^{n}(s_j^2 - c_j^2)} = \dot{w} \prod_{j=1}^{n} \dot{w}_j{}^{c_j} \mod p \tag{16}$$

Theorem 8. *Main Protocol is complete. That is, if $\mathcal{P}$ knows $\{r_i\}$ and $\{A_{ij}\}$ satisfying the both conditions of Theorem 1, $\mathcal{V}$ always accepts.*

Theorem 9. *If $\mathcal{V}$ accepts Main Protocol with a non-negligible probability, then $\mathcal{P}$ knows $\{r_i\}$ and permutation matrix (A_{ij}) satisfying Equations (2), or can generate non-trivial integers $\{a_i\}$ and a satisfying $\tilde{g}^a \prod_{i=1}^{n} \tilde{g}_i{}^{a_i} = 1$ with overwhelming probability.*

Theorem 10. *We can construct a simulator of Main Protocol such that if there is a distinguisher who can distinguish between a real transcript from the protocol and an output from the simulator, then we can solve the decisional Diffie-Hellman problem.*

Proofs for Theorem 9 and 10 are sketched in Appendix C.

6 Discussions

In this section, we compare the efficiency of the proposed protocol described in Section 5 to the SK95 protocol in [SK95] and MiP-2 protocol in [Ab99]. To enable a fair comparison, we assume the security parameter of [SK95] to be 160 and lengths of p and q to be 1024 and 160 respectively.

We first compare them by the number of exponentiations used in each protocol, in the case of shuffling n ciphertexts. These are $22(n \log n - n + 1)$ for Abe's protocol, $642n$ for the SK95 protocol, and $18n + 18$ for the proposed protocol. If we adopt computation tools described in [HAC], such as the simultaneous multiple exponentiation algorithm and the fixed-base comb method, the number of exponentiations can be heuristically reduced to $11.2(n \log n - n + 1)$,$64n$, and $4.84n + 4.5$, respectively. The total number of bits needing to be transfered during the protocols is $13,248(n \log n - n + 1)$,$353,280n$, and $5,280n + 13,792$. The rounded-up numbers are shown in Table 1.

Table 1. Comparison of three protocols

	Abe (MiP-2)	SK95	This Paper
No. exponentiations	$22n \log n$	$642n$	$18n$
(heuristically adjusted)	$11n \log n$	$64n$	$5n$
No. communication bits	$2^{14} n \log n$	$2^{18} n$	$2^{11} n$

7 Conclusion

In this paper, we presented a novel method to prove the correctness of a shuffle, and demonstrated its efficiency. The proposed method requires only $18n$ exponentiations for shuffling n ciphertexts, where as previous methods required 35 times more, or required a higher order, $O(n \log n)$.

The proposed protocol can be used to build an efficient, universally verifiable voting system where the number of voters can scale up to the order of 10,000.

Acknowledgments. The authors would like to thank Tatsuaki Okamoto, Masayuki Abe, and Satoshi Obana for many helpful discussions.

References

[Ab99] M. Abe, *Mix-Networks on Permutation Networks*, Asiacrypt '99, LNCS 1716, 258-273 (1999)

[Br93] S. Brands, *An Efficient Off-line Electronic Cash System Based On The Representation Problem*, CWI Technical Report CS-R9323, (1993)

[Ch81] D. Chaum, *Untraceable Electronic Mail, Return Addresses, and Digital Pseudonyms*, Communications of the ACM, Vol.24, No.2 84-88 (1981)

[CDS94] R. Cramer, I. Damgård and B. Schoenmakers, *Proofs of Partial Knowledge and Simplified Design of Witness Hiding Protocols*, Crypto '94, LNCS 839, 174-187 (1994)

[HAC] A. Menezes, C. van Oorschot and S. Vanstone, *Handbook of Applied Cryptography*, CRC Press, 617-619

[Ne01] C.A. Neff, *Verifiable, Secret Shuffles of ElGamal Encrypted Data*, Initial version circulated Mar. 2000, current version submitted to ACMCCS 01

[OKST97] W. Ogata, K. Kurosawa, K. Sako and K. Takatani, *Fault tolerant anonymous channel*, 1st International Conference on Information and Communications Security (ICICS), LNCS 1334, 440-444 (1997)

[SK95] K. Sako and J. Kilian, *Receipt-free mix-type voting scheme –A practical solution to the implementation of voting booth*, Eurocrypt 95, LNCS 921, 393-403 (1995)

A DDH_n^m and DDH

Lemma 1. *For any $m(\geq 2)$ and $n(\geq 2)$, if DDH_n^m is easy, then DDH_n^2 is easy.*

Proof. We claim that if DDH_n^m is easy, then either DDH_n^{m-1} is easy or DDH_n^2 is easy. By induction we can prove the correctness of the lemma.

In order to prove the claim, we define the subset M_n^m of R_n^m to be the set of tuples

$$I = (x_1^{(1)}, \dots, x_1^{(m)}, x_2^{(1)}, \dots, x_2^{(m)}, \dots, x_n^{(1)}, \dots, x_n^{(m)})$$

satisfying

$$\log_{x_1^{(1)}} x_1^{(i)} = log_{x_j^{(1)}} x_j^{(i)} \bmod p$$

for all $i(i = 2, 3, .., m-1)$ and $j(j = 2, ..., n)$, but whether or not

$$\log_{x_1^{(1)}} x_1^{(m)} = log_{x_j^{(1)}} x_j^{(m)} \bmod q$$

holds for all $j(j = 2, ..., n)$ is arbitrary. Therefore, the set D_n^m is a subset of M_n^m.

It is clear that if DDH_n^m is easy, then we can either distinguish between the instances chosen uniformly from R_n^m and M_n^m or the instances chosen uniformly from M_n^m and D_n^m. In the former case, it means DDH_n^{m-1} is easy. We claim in the following that in the latter case DDH_n^2 is easy.

Assume M_n^m and D_n^m are distinguishable. For any $I_n^2 \in R_n^2$ s.t.

$$I_n^2 = (x_1^{(1)}, x_1^{(2)}, x_2^{(1)}, x_2^{(2)}, \dots, x_n^{(1)}, x_n^{(2)})$$

we transform it to $I_n^m \in R_n^m$

$$I_n^m = (x'^{(1)}_1, \dots, x'^{(m)}_1, x'^{(1)}_2, \dots, x'^{(m)}_2, \dots, x'^{(1)}_n, \dots, x'^{(m)}_n)$$

where

$$x'^{(i)}_j = \begin{cases} x_j^{(1)} & j = 1, \dots, n \text{ (if } i = 1) \\ (x_j^{(1)})^{z_i} \bmod p & j = 1, \dots, n \text{ (if } 2 \leq i \leq m-1) \\ x_j^{(2)} & j = 1, \dots, n \text{ (if } i = m) \end{cases}$$

with randomly chosen $\{z_i\}_{(i=2,\dots,m-1)}$ in $\mathbf{Z_q}$.

If I_n^2 is chosen uniformly from D_n^2, then I_n^m is distributed uniformly in D_n^m, and if I_n^2 is chosen uniformly from R_n^2, then I_n^m is distributed uniformly in M_n^m. Therefore if D_n^m and M_n^m is distinguishable, then we can solve DDH_n^2.

Lemma 2. *If DDH_n^2 $(n \geq 2)$ is easy then the decisional Diffie-Hellman problem (DDH_2^2) is easy.*

Proof. For any $I_2^2 = (x_1^{(1)}, x_1^{(2)}, x_2^{(1)}, x_2^{(2)}) \in R_2^2$, we transform it to $I_n^2 \in R_n^2$

$$I_n^2 = ({x'}_1^{(1)}, {x'}_1^{(2)}, {x'}_2^{(1)}, {x'}_2^{(2)}, \ldots, {x'}_n^{(1)}, {x'}_n^{(2)})$$

where

$$\begin{aligned}
&{x'}_1^{(1)} = x_1^{(1)},\ {x'}_1^{(2)} = x_1^{(2)},\ {x'}_2^{(1)} = x_2^{(1)},\ {x'}_2^{(2)} = x_2^{(2)} \\
&{x'}_j^{(1)} = (x_1^{(1)})^{z_j} \cdot (x_2^{(1)})^{w_j},\ \ {x'}_j^{(2)} = (x_1^{(2)})^{z_j} \cdot (x_2^{(2)})^{w_j} \bmod p \quad j = 3, \ldots, n
\end{aligned}$$

with randomly chosen $\{z_j\}$ and $\{w_j\} (j = 3, \ldots, n)$ in $\mathbf{Z_q}$.

If I_2^2 is chosen uniformly from D_2^2, then I_n^2 is distributed uniformly in D_n^2, and if I_2^2 is chosen uniformly from R_2^2, then I_n^2 is distributed uniformly in R_n^2. Therefore if DDH_n^2 is easy, then so is DDH_2^2.

B Properties of Proof-1

In this section, we sketch the proofs of the following theorems.

Theorem 3 (soundness). *If $\mathcal{V}$ accepts* **Proof-1** *with a non-negligible probability, then $\mathcal{P}$ either knows both $\{r_i\}$ and $\{A_{ij}\}$ satisfying the first condition, or can generate integers $\{a_i\}$ and a satisfying $g^a \prod_{i=1}^n {g_i}^{a_i} = 1$ with overwhelming probability.*

Theorem 4 (zero-knowledge). *We can construct a simulator of* **Proof-1** *such that if there is a distinguisher who can distinguish between a real transcript from the protocol and an output from the simulator, then we can solve the decisional Diffie-Hellman problem.*

B.1 Soundness

It is clear that Theorem 3 holds if Lemmas 3, 4 and 5 hold. We therefore prove the lemmas.

Lemma 3. *If $\mathcal{V}$ accepts* **Proof-1** *with non-negligible probability, then $\mathcal{P}$ knows $\{A_{ij}\}$, $\{r_i\}$, $\{\alpha_i\}$, and α satisfying Equations (3) and (6).*

A sketch of Proof: Define $\mathcal{C}_p$ as the space which is spanned by the vector $(1, c_1, c_2, \ldots, c_n)$ made of the challenges to which $\mathcal{P}$ can compute responses $s, \{s_i\}_{(i=1,\ldots,n)}$ such that Equation (7) holds. If the $\dim(\mathcal{C}_p) = n+1$, $\mathcal{P}$ can choose $n+1$ challenges which are linearly independent and obtain $\{A_{ij}\}_{(i,j=1,\ldots,n)}$, $\{r_i\}_{(i=1,\ldots,n)}$, $\{\alpha_i\}_{(i=1,\ldots,n)}$, and α which satisfies the relation:

$$s = \sum_{j=1}^{n} r_j c_j + \alpha, \quad s_i = \sum_{j=1}^{n} A_{ij} c_j + \alpha_i \bmod q \quad i = 1, \ldots, n$$

Such $\{A_{ij}\}$, $\{r_i\}$,$\{\alpha_i\}$, and α satisfies Equations (3) and (6) . If, $\dim(\mathcal{C}_p) < n+1$. The probability that $\mathcal{V}$ generates a challenge in $\mathcal{C}_p$ is at most $q^{n-1}/q^n = 1/q$, which is negligible. □

Lemma 4. *Assume $\mathcal{P}$ knows $\{A_{ij}\}$,$\{r_i\}$, $\{\alpha_i\}$, and α satisfying Equations (3) and (6). If $\mathcal{P}$ knows $\{s_i\}$ and s which satisfy Equation (7), and either $s \neq \sum_{j=1}^n r_j c_j + \alpha$ or $s_i \neq \sum_{j=1}^n A_{ij} c_j + \alpha_i$ for some i hold, then $\mathcal{P}$ can generate non-trivial integers $\{a_i\}$ and a satisfying $g^a \prod_{i=1}^n {g_i}^{a_i} = 1$ with overwhelming probability.*

Proof. The following gives a non-trivial representation of 1 using $g, \{g_i\}$.

$$g^{\sum_{j=1}^n s_j c_j + \alpha - s} \prod_{i=1}^n {g_i}^{\sum_{j=1}^n A_{ij} c_j + \alpha_i - s_i} = 1 \bmod p.$$

Lemma 5. *Assume $\mathcal{P}$ knows $\{A_{ij}\}$, $\{r_i\}$, $\{\alpha_i\}$, and α satisfying Equations (3) and (6). If Equations (7) and (8) hold with non-negligible probability, then either Equation (4) hold or $\mathcal{P}$ can generate non-trivial integers $\{a_i\}$ and a satisfying $g^a \prod_{i=1}^n {g_i}^{a_i} = 1$ with overwhelming probability.*

A sketch of Proof: From Lemma 4, If Equation (7) holds, then either

$$\begin{cases} s = \sum_{j=1}^n r_j c_j + \alpha \bmod q \\ s_i = \sum_{j=1}^n A_{ij} c_j + \alpha_i \bmod q \quad i = 1, \ldots, n \end{cases}$$

holds or $\mathcal{P}$ can generate non-trivial integers $\{a_i\}$ and a satisfying $g^a \prod_{i=1}^n {g_i}^{a_i} = 1$ with overwhelming probability. We concentrate on the former case. If Equation (8) holds, then

$$\sum_{i=1}^n \sum_{j=1}^n (\sum_{h=1}^n A_{hi} A_{hj} - \delta_{ij}) c_i c_j + \sum_{i=1}^n \left\{ (\sum_{j=1}^n 2\alpha_j A_{ji} + \sigma r_i) - \psi_i \right\} c_i$$
$$+ \left\{ (\sum_{j=1}^n \alpha_j^2 + \sigma \alpha) - \psi \right\} = 0 \quad \bmod q$$

where $\psi_i = \sum_{j=1}^n 2\alpha_j A_{ji} + \sigma r_i, \psi = \sum_{j=1}^n {\alpha_j}^2 + \sigma\alpha \bmod q$. If Equation (4) does not hold for some i and j, then the probability that Equation (8) holds is negligible. □

B.2 Zero-Knowledge

A sketch of Proof: We first give a construction of the simulator. We then prove that if there exists such a distinguisher then we can solve DDH_{n+1}^2. From Lemma 2, it means it is equivalent to solving the decisional Diffie-Hellman problem.

The Construction of the Simulator

We will construct the simulator $\mathcal{S}$ of the **Proof-1** with the input $p, q, g, \{g_i\}$, $\{g'_i\}$ as follows.

The simulator $\mathcal{S}$ first generates $s, \{s_i\}, \{c_i\} \in_R \mathbf{Z_q}$, $w, \{\dot{w}_i\} \in_R \mathbf{G_q}$ randomly. Then it computes $g', \dot{w}$ as the following.

$$g' = g^s \prod_{j=1}^{n} g_j{}^{s_j} g_j'^{-c_j} \bmod p$$

$$\dot{w} = w^s g^{\sum_{j=1}^{n}(s_j^2 - c_j^2)} \prod_{j=1}^{n} \dot{w}_j^{-c_j} \bmod p$$

The output of $\mathcal{S}$ is $(w, g', \{\dot{w}_i\}, \dot{w}, \{c_i\}, s, \{s_i\})$

A Distinguisher of D^2_{n+1} and R^2_{n+1}

We will then construct a distinguisher $\mathcal{D}'$ who can distinguish between the uniform instances of D^2_{n+1} and R^2_{n+1} if $\mathcal{S}$ can not simulate the **Proof-1**.

Let's say the instance $I = (x_1^{(1)}, x_1^{(2)}, \ldots, x_{n+1}^{(1)}, x_{n+1}^{(2)})$ was chosen uniformly from either D^2_{n+1} or R^2_{n+1}. Then this distinguisher will first generate $g_1, g_2, ..g_n$ as the constants used in **Proof-1** and let $g = x_1^{(1)}$.

It will then generate a random permutation matrix (A_{ji}) and compute

$$g'_i = x_{i+1}^{(1)} \prod_{j=1}^{n} g_j^{A_{ji}} \bmod p. \quad (i = 1, \ldots, n)$$

We note that $\{g'_i\}$ gives a random permutation of $\{g_i\}$.

Based on $g, \{g_i\}, \{g'_i\}$, the distinguisher $\mathcal{D}'$ is going to act as a simulator $\mathcal{S}'$ which simulates the simulator $\mathcal{S}$. More specifically, the simulator $\mathcal{S}'$ randomly generates $s, \{s_i\}, \{c_i\} \in_R \mathbf{Z_q}$ and computes

$$w = x_1^{(2)}$$

$$g' = g^s \prod_{j=1}^{n} g_j{}^{s_j} g_j'^{-c_j} \bmod p$$

$$\alpha_j = s_j - \sum_{k=1}^{n} A_{jk} c_k \bmod q \quad j = 1, \ldots, n$$

$$\dot{w}_i = x_{i+1}^{(2)} \prod_{j=1}^{n} g^{2\alpha_j A_{ji}} \bmod p \quad i = 1, \ldots, n$$

$$\dot{w} = w^s g^{\sum_{j=1}^{n}(s_j^2 - c_j^2)} \prod_{j=1}^{n} \dot{w}_j^{-c_j} \bmod p.$$

The simulator $\mathcal{S}'$ outputs

$$w, g', \{\dot{w}_i\}, \dot{w}, \{c_i\}, s, \{s_i\}. (i = 1, \ldots, n)$$

Lemma 6. *Simulator $\mathcal{S}'$ perfectly simulates* **Proof-1** *when $I \in_R D^2_{n+1}$.*

Sketch: We let

$$\log_{x_1^{(1)}} x_{i+1}^{(1)} = r_i, \quad \log_{x_1^{(1)}} x_1^{(2)} = \sigma.$$

Then it is clear that by randomly choosing $\{s_i\}$ and s, it gives the same distribution of the output as when $\{\alpha_i\}$ and α were first chosen randomly, and verifier honestly chooses random challenge $\{c_i\}$.

Lemma 7. *Simulator $\mathcal{S}'$ perfectly simulates $\mathcal{S}$ when $I \in_R R^2_{n+1}$.*

Since $\{x_i^{(2)}\}_{(i=1,\ldots,n+1)}$ are randomly chosen, it gives the same distribution when $\dot{w}_i$ and w are randomly chosen.

□

Therefore, if there exists a distinguisher $\mathcal{D}$ that distinguishes the output of the simulator $\mathcal{S}$ and a real transcript of **Proof-1**, then this distinguisher can be used to solve DDH^2_{n+1}.

C Properties of the Main Protocol

In this section, we discuss the properties of the main protocol. The completeness property is clear. We provide proofs for the soundness and the zero-knowledge property.

C.1 Soundness

Theorem 9. *If $\mathcal{V}$ accepts Main Protocol with a non-negligible probability, then $\mathcal{P}$ knows $\{r_i\}$ and permutation matrix (A_{ij}) satisfying Equations (2), or can generate non-trivial integers $\{a_i\}$ and a satisfying $\tilde{g}^a \prod_{i=1}^n \tilde{g}_i{}^{a_i} = 1$ with overwhelming probability.*

A sketch of Proof:
We can show $\mathcal{P}$'s knowledge of $\{A_{ij}\}, \{r_i\}, \{\alpha_i\}$, and α satisfying Equations (9) and (10) from the satisfiability of Equation (11), similar to Lemma 3. From the satisfiability of Equations (11) and (16), and additionally that of Equations (14) and (15), we can prove that the $\{A_{ij}\}$ satisfies the both conditions of Theorem 1, in a similar manner as proving Lemma 5. Thus Theorem 1 ensures that (A_{ij}) is a permutation matrix. The following lemma ensures that the same permutation matrix was applied to both $\{g_i\}$ and $\{m_i\}$ to achieve $\{g_i'\}$ and $\{m_i'\}$, yielding the correctness of the shuffle. □

Lemma 8. *Assume $\mathcal{P}$ knows $\{A_{ij}\}, \{r_i\}, \{\alpha_i\}$, and α satisfying Equations (9) and (10), and $\{s_i\}$ and s satisfying Equation (11). If Equations (12) and (13) hold with non-negligible probability, then either the relationships*

$$\begin{cases} g' = g^{\alpha} \prod_{j=1}^{n} g_j{}^{\alpha_j} \bmod p \\ g'_i = g^{r_i} \prod_{j=1}^{n} g_j{}^{A_{ji}} \bmod p \quad i = 1, \ldots, n \\ m' = y^{\alpha} \prod_{j=1}^{n} m_j{}^{\alpha_j} \bmod p \\ m'_i = y^{r_i} \prod_{j=1}^{n} m_j{}^{A_{ji}} \bmod p \quad i = 1, \ldots, n \end{cases} \tag{17}$$

hold or $\mathcal{P}$ can generate nontrivial integers $\{a_i\}$ and a satisfying $\tilde{g}^a \prod_{i=1}^{n} \tilde{g}_i{}^{a_i} = 1$ with overwhelming probability.

A sketch of Proof: Similarly to Lemma 4, we can ensure that

$$\begin{cases} s = \sum_{j=1}^{n} r_j c_j + \alpha \bmod q \\ s_i = \sum_{j=1}^{n} A_{ij} c_j + \alpha_i \bmod q \quad i = 1, \ldots, n \end{cases}$$

hold from the satisfiability of Equation (11) unless $\mathcal{P}$ can generate non-trivial integers $\{a_i\}$ and a satisfying $\tilde{g}^a \prod_{i=1}^{n} \tilde{g}_i{}^{a_i} = 1$.

If Equation (12) holds, then

$$1 = \frac{g^{\alpha} \prod_{i=1}^{n} g_i{}^{\alpha_i}}{g'} \prod_{j=1}^{n} \left(\frac{g^{r_j} \prod_{i=1}^{n} g_i{}^{A_{ij}}}{g'_j} \right)^{c_j} \bmod p.$$

If first two equations on Equations (17) does not hold, then the probability that Equation (12) hold is negligible. The same thing can be said for $m', \{m'_i\}_{(i=1,\ldots,n)}$ from the satisfiability of (13). □

C.2 Zero-Knowledge

Theorem 10. *We can construct a simulator of Main Protocol such that if there is a distinguisher who can distinguish between a real transcript from the protocol and an output from the simulator, then we can solve the decisional Diffie-Hellman problem.*

A sketch of Proof: We first give a construction of the simulator. We then prove that if there exists such a distinguisher then we can solve DDH^5_{n+1}. From Lemma 1 and 2, it means it is equivalent to solving the decisional Diffie-Hellman problem.

The Construction of the Simulator

We will construct the simulator $\mathcal{S}$ of the main protocol with the input $p, q, g, y, \tilde{g}, \{\tilde{g}_i\}, \{(g_i, m_i)\}, \{(g'_i, m'_i)\}$ as follows.

The simulator $\mathcal{S}$ first generates $s, \{s_i\}, \{c_i\}, \lambda' \in_R \mathbf{Z_q}$, t, v, w, $\{u_i\}$, $\{\dot{t}_i\}$, $\{\dot{v}_i\}$, $\{\dot{w}_i\}$, $\{\tilde{g}'_i\} \in_R \mathbf{G_q}$ randomly. Then it computes $\tilde{g}', g', m', u, \dot{v}, \dot{w}$ as the following.

$$u = g^{\lambda'} \prod_{j=1}^{n} u_j{}^{-c_j^2} \bmod p$$

$$\tilde{g}' = \tilde{g}^s \prod_{j=1}^{n} \tilde{g}_j{}^{s_j} \tilde{g}'_j{}^{-c_j} \bmod p$$

$$g' = g^s \prod_{j=1}^{n} g_j{}^{s_j} g'_j{}^{-c_j} \bmod p$$

$$m' = y^s \prod_{j=1}^{n} m_j{}^{s_j} m'_j{}^{-c_j} \bmod p$$

$$\dot{v} = t^{\lambda'} v^s g^{\sum_{j=1}^{n}(s_j^3 - c_j^3)} \prod_{j=1}^{n} \dot{t}_j^{-c_j^2} \dot{v}_j{}^{-c_j} \bmod p$$

$$\dot{w} = w^s g^{\sum_{j=1}^{n}(s_j^2 - c_j^2)} \prod_{j=1}^{n} \dot{w}_j^{-c_j} \bmod p.$$

The output of $\mathcal{S}$ is $\big(t, v, w, u, \{u_i\}, \{\tilde{g}'_i\}, \tilde{g}', g', m', \{\dot{t}_i\}, \{\dot{v}_i\}, \dot{v}, \{\dot{w}_i\}, \dot{w}, \{c_i\}, s, \{s_i\}, \lambda'\big)$.

A Distinguisher of D^5_{n+1} and R^5_{n+1}

We will then construct a distinguisher $\mathcal{D}'$ who can distinguish between the uniform instances of D^5_{n+1} and R^5_{n+1} if $\mathcal{S}$ can not simulate the main protocol.

Let's say the instance I

$$I = (x_1^{(1)}, x_1^{(2)}, \ldots, x_1^{(5)}, \ldots, x_{n+1}^{(1)}, x_{n+1}^{(2)}, \ldots, x_{n+1}^{(5)})$$

was chosen uniformly from either D^5_{n+1} or R^5_{n+1}. Then this distinguisher will first generate $g_1, m_1, g_2, m_2, \ldots, g_n, m_n, \tilde{g}_1, \tilde{g}_2, \ldots, \tilde{g}_n$ as the constants used in Main Protocol and let $X \in_R \mathbf{Z_q}, g = x_1^{(1)}, \tilde{g} = x_1^{(2)}, y = g^X \bmod p$. It will then generate a random permutation A_{ji} and a secret key $X \in_R \mathbf{Z_q}$ and compute

$$(g'_i, m'_i) = (x_{i+1}^{(1)} \prod_{j=1}^{n} g_j^{A_{ji}},\ (x_{i+1}^{(1)})^X \prod_{j=1}^{n} m_j^{A_{ji}}) \bmod p. \quad (i = 1, \ldots, n)$$

We note that $\{(g'_i, m'_i)\}$ gives a random shuffle of $\{(g_i, m_i)\}$.

Based on $g, y, \tilde{g}, \{\tilde{g}_i\}, \{(g_i, m_i)\}$ and $\{(g'_i, m'_i)\}$ the distinguisher $\mathcal{D}'$ is going to act as a simulator $\mathcal{S}'$ which simulates the simulator $\mathcal{S}$. More specifically, the simulator $\mathcal{S}'$ randomly generates

$$s, \{s_i\}, \{c_i\}, \lambda', \{\beta_i\} \in_R \mathbf{Z_q} \quad i = 1, \ldots, n$$

and computes

$$
\begin{aligned}
t &= x_1^{(3)},\ v = x_1^{(4)},\ w = x_1^{(5)} \\
u_i &= (x_{i+1}^{(1)})^{\beta_i} \bmod p \quad i = 1, \dots, n \\
u &= g^{\lambda'} \prod_{j=1}^{n} u_j^{-c_j^2} \bmod p \\
\tilde{g}'_i &= x_{i+1}^{(2)} \prod_{j=1}^{n} \tilde{g}_j^{A_{ji}} \bmod p \quad i = 1, \dots, n \\
\tilde{g}' &= \tilde{g}^{s} \prod_{j=1}^{n} \tilde{g}_j{}^{s_j} \tilde{g}'_j{}^{-c_j} \bmod p \\
g' &= g^{s} \prod_{j=1}^{n} g_j{}^{s_j} g'_j{}^{-c_j} \bmod p \\
m' &= y^{s} \prod_{j=1}^{n} m_j{}^{s_j} m'_j{}^{-c_j} \bmod p \\
\alpha_j &= s_j - \sum_{k=1}^{n} A_{jk} c_k \bmod q \quad j = 1, \dots, n \\
\dot{t}_i &= (x_{i+1}^{(3)})^{\beta_i} \prod_{j=1}^{n} g^{3\alpha_j A_{ji}} \bmod p \quad i = 1, \dots, n \\
\dot{v}_i &= x_{i+1}^{(4)} \prod_{j=1}^{n} g^{3\alpha_j{}^2 A_{ji}} \bmod p \quad i = 1, \dots, n \\
\dot{w}_i &= x_{i+1}^{(5)} \prod_{j=1}^{n} g^{2\alpha_j A_{ji}} \bmod p \quad i = 1, \dots, n \\
\dot{v} &= t^{\lambda'} v^{s} g^{\sum_{j=1}^{n}(r_j^3 - c_j^3)} \prod_{j=1}^{n} (\dot{t}_j^{-c_j^2} \dot{v}_j^{-c_j}) \bmod p \\
\dot{w} &= w^{s} g^{\sum_{j=1}^{n}(s_j^2 - c_j^2)} \prod_{j=1}^{n} \dot{w}_j^{-c_j} \bmod p.
\end{aligned}
$$

The simulator $\mathcal{S}'$ outputs $(t, v, w, u, \{u_i\}, \{\tilde{g}'_i\}, \tilde{g}', g', m', \{\dot{t}_i\}, \{\dot{v}_i\}, \dot{v}, \{\dot{w}_i\}, \dot{w}, \{c_i\}, s, \{s_i\}, \lambda')$.

Lemma 9. *Simulator $\mathcal{S}'$ perfectly simulates Main Protocol when $I \in_R D_{n+1}^5$.*

Sketch: We let

$$
\begin{aligned}
&\log_{x_1^{(1)}} x_{i+1}^{(1)} = r_i,\ \log_{x_1^{(1)}} (x_{i+1}^{(1)})^{\beta_i} = \lambda_i \\
&\log_{x_1^{(1)}} x_1^{(3)} = \tau,\ \log_{x_1^{(1)}} x_1^{(4)} = \rho,\ \log_{x_1^{(1)}} x_1^{(5)} = \sigma.
\end{aligned}
$$

This gives for $i = 1, .., n$,

$$x_{i+1}^{(1)} = g^{r_i}, (x_{i+1}^{(1)})^{\beta_i} = g^{\lambda_i}, x_{i+1}^{(2)} = \tilde{g}^{r_i},$$
$$(x_{i+1}^{(3)})^{\beta_i} = g^{\tau\lambda_i}, \; x_{i+1}^{(4)} = g^{\rho r_i}, \; x_{i+1}^{(5)} = g^{\sigma r_i}.$$

Therefore, it is clear that by randomly choosing $s, \{s_i\}, \lambda'$ and $\{\beta_i\}$, it gives the same distribution of the output as when α, $\{\alpha_i\}$, $\{\lambda_i\}$ and λ were first chosen randomly, and verifier honestly chooses random challenge $\{c_i\}$.

Lemma 10. *Simulator $\mathcal{S}'$ perfectly simulates $\mathcal{S}$ when $I \in_R R_{n+1}^5$*

Sketch: Since $x_i^{(2)}, x_i^{(3)}, x_i^{(4)}, x_i^{(5)} (i = 1, 2, ..., n+1)$ and $\beta_i (i = 1, 2, ..., n)$ are randomly chosen, it gives the same distribution when $\tilde{g}, t, v, w, \{\tilde{g}_i'\}, \{\dot{t}_i\}, \{\dot{v}_i\}, \{\dot{w}_i\}$ and $\{u_i\}$ are randomly chosen for $i = 1, 2, ..., n$.

□

Therefore, if there exists a distinguisher $\mathcal{D}$ that distinguishes the output of the simulator $\mathcal{S}$ and a real transcript of Main Protocol, then this distinguisher can be used to solve DDH_{n+1}^5.

D Alternative Notation

We present here an alternative notation of the variables. Since we have discussed the basis $\{g, g_1, \ldots, g_n\}$ throughout the paper, we can think of representing g by g_0. Similarly y by m_0 and $\tilde{g}$ by $\tilde{g}_0$. We can include the value of randomizers $\{r_i\}$, α_i and α, in the matrix by defining $A_{0i} = r_i$, $A_{i0} = \alpha_i$, and $A_{00} = \alpha$.

Treating a public key in a similar manner with input variables may be awkward, but it gives a compact representation to some of the variables, e.g,

$$g_\mu' = \prod_{\nu=0}^{n} g_\nu{}^{A_{\nu\mu}}, m_\mu' = \prod_{\nu=0}^{n} m_\nu{}^{A_{\nu\mu}}, \tilde{g}_\mu' = \prod_{\nu=0}^{n} \tilde{g}_\nu{}^{A_{\nu\mu}} \quad \mu = 0, \ldots, n.$$

Further suggestions for the alternative notation follows:

$$g_0' = g', m_0' = m', \tilde{g}_0' = \tilde{g}', s_0 = s, c_0 = 1, \lambda_0 = \lambda, u_0 = u$$
$$s_\mu = \sum_{\nu=0}^{n} A_{\mu\nu} c_\nu, \lambda' = \sum_{\nu=0}^{n} \lambda_\nu c_\nu^2, \prod_{\nu=0}^{n} g_\nu^{s_\nu} = \prod_{\nu=0}^{n} g'{}_\nu^{c_\nu}, g^{\lambda'} = \prod_{\nu=0}^{n} u_\nu^{c_\nu^2} \quad \mu = 0, \ldots, n.$$

An Identity Escrow Scheme with Appointed Verifiers

Jan Camenisch[1] and Anna Lysyanskaya[2]

[1] IBM Research
Zurich Research Laboratory
CH–8803 Rüschlikon
jca@zurich.ibm.com
[2] MIT LCS
545 Technology Square
Cambridge, MA 02139 USA
anna@theory.lcs.mit.edu

Abstract. An identity escrow scheme allows a member of a group to prove membership in this group without revealing any extra information. At the same time, in case of abuse, his identity can still be discovered. Such a scheme allows anonymous access control. In this paper, we put forward the notion of an identity escrow scheme with appointed verifiers. Such a scheme allows the user to only convince an appointed verifier (or several appointed verifiers) of his membership; but no unauthorized verifier can verify a user's group membership even if the user fully cooperates, unless the user is completely under his control. We provide a formal definition of this new notion and give an efficient construction of an identity escrow scheme with appointed verifiers provably secure under common number-theoretic assumptions in the public-key model.

Keywords. Identity escrow, group signatures, privacy protection, formal model for group signatures.

1 Introduction

As digital communication becomes the preferred means of information exchange, it becomes ever easier for those of questionable motivation to mine the accumulated data. Under these circumstances, both the importance and the challenge of protecting the privacy of individuals grow considerably. A number of cryptographic protocols that limit the information dispersed from accumulated data have been proposed. These are, for instance, anonymous voting protocols [5,30], anonymous payment schemes [6,18], and credential systems [9,16]. All these systems follow the principle of data minimization, i.e., a participant in the system can only learn as much information about the other participants as is necessary for the system to function properly.

In this context, group signatures [2,12,19] are an important building block. They allow a member of some group to sign anonymously on the group's behalf.

J. Kilian (Ed.): CRYPTO 2001, LNCS 2139, pp. 388–407, 2001.

Thus, a party receiving a signature can be sure that its originator is a member of the group, but receives no other information. However, in exceptional cases such as when the anonymity is misused and a legal dispute arises, a designated revocation manager has the power to reveal the unambiguous identity of the originator of the signature. At the same time, no one can misattribute a valid group signature. A concept dual to group signature schemes is that of identity escrow [32] schemes. They can be seen as group-member identification schemes with revocable anonymity. In fact, any group signature scheme can be turned into an identity escrow scheme and vice versa.

Group signatures can, for instance, be used by the purchasing department of a company to hide the internal structure of this department. All members of the department form a group, and sign all purchasing orders using group signatures. In case one day a sports car gets delivered instead of pencils, the department manager will be able to identify the culprit. Recently, group signature schemes were used to realize an anonymous credential system [9]. Here, being a member of some particular group meant possessing a particular credential. Hence, ownership of a credential can be proved anonymously. Other applications include bidding [20], electronic cash [33], and anonymous fingerprinting [7].

Group signature/identity escrow schemes with appointed verifiers, as proposed in this paper, go a step further: here a group member can prove his membership only to an appointed verifier but not to anyone else. There can be several different appointed verifiers for each member. This property of not being able to convince non-appointed parties is similar to receipt-freeness in electronic voting schemes, where a voter must not be able to prove to anyone how she voted, which is required to hinder vote-buying. We stress that this is different from the situation with so-called confirmer signatures [17] or designated-verifier proofs [31], where although signatures (resp., proofs) can only be verified by a designated party, the signer (resp., prover) would have the power to issue a signature (resp., proof) that is universally verifiable.

Appointed verifiers are useful for many applications of group signature and identity escrow schemes. As an example, consider a bank that issues a credential stating that the customer is eligible for a small business loan. The bank might want to have a guarantee that the customer cannot use this credential in order to obtain a better loan from a competing bank; or to use the loan money for something other than the business for which it was granted. Or, consider the purchasing department scenario outlined above. Naturally, different members of the department are authorized to conduct different kinds of transactions. Using a group signature scheme with appointed verifiers allows the department manager to ensure that employees can only order from the companies they are authorized. Finally, consider their use in a credential scheme. It is natural that for some types of credentials the user should not be able to show them to anyone except the intended verifier. This can be useful in preventing abuse of credentials as well as in controlling who can get to know to whom the credentials were issued.

Let us loosely outline how our identity escrow (group signature) scheme with appointed verifiers is constructed. To this end we first explain how efficient and provably secure group signature schemes [2,12] are realized. The public key of the group can be viewed as a public key of a suitable signature scheme. The group manager holds the secret key corresponding to this public key. To become a group member, a user chooses, as membership secret key, an element of a certain (algebraic) group. The user's identifier is computed as a one-way function of this key, for example through exponentiation in a group where computing the discrete logarithm is conjectured to be hard. The group manager signs (certifies) this identifier and sends back the signature to the new group member. This signature is the user's group membership certificate. To convince a verifier of her group membership, a user proves in zero-knowledge that she knows a membership certificate and the corresponding membership secret key. In case of a group signature scheme, this proof is turned into a signature scheme using the so-called Fiat-Shamir heuristic [26]. An identity escrow scheme constructed in this way is provably secure as long as the underlying signature scheme is secure. The corresponding group signature scheme is provably secure in the random oracle model. The challenge in designing an efficient identity escrow or group signature scheme is finding a signature scheme for the group manager and a format for membership secret keys and corresponding identifiers such that the proof of membership is efficient.

To extend such a scheme to an identity escrow system with appointed verifiers, we will have the group manager split the group membership certificate into two pieces. The first piece will be handed over to the user. The second piece will be encrypted under the appointed verifier's public key. It will be easy to fake a tuple that looks like the first piece of the membership certificate and the encryption of the second piece. Only the appointed verifier, under whose public key the encryption is carried out, will be able to verify that a given ciphertext corresponds to the second piece of a user's certificate. Together, the two pieces constitute an unforgeable group membership certificate. To prove group membership to the appointed verifier, the user could prove possession of his piece of the membership certificate as before, and then give the verifier a blinded version of the encrypted piece.

An adversary in this system can try to induce some verifier to accept an invalid user; or he can try to make it look as though some honest user participated in a shady transaction; or he can conduct a shady transaction and then try to avoid anonymity revocation; or he can try to convince another adversary, who is not an authorized verifier, that some user is a group member. We provide a formal definition of security against such attacks. For the first time, a formal model for identity escrow schemes along the lines of an ideal world specification, is given. These new identity escrow/group signature specifications are more rigorous than the ones that exist to date. As they are similar to the definitions from the multi-party computation literature [13,14,37], they integrate with this literature better than previous specifications did, and so such properties as composability of protocols can be better understood in this framework (but we do not address them here). Finally, we formally define the appointed-verifier property, i.e., the property that no proof system (A, B) exists in which A is a group member, B is not the appointed verifier, and yet A acts as a prover and B as a verifier for the statement that A is a group member, and the gap between the completeness and soundness of the system is non-negligible. Cryptographic problems of this flavor have not been sufficiently explored. While receipt-free voting is a relatively well-studied example [5,30], no formal definition of receipt-freeness has been given, and it is not well understood what gap between completeness and soundness for the adversary-verifier in receipt-free voting is satisfactory. Thus, we are the first to explore this in a formal way and to obtain a scheme that satisfies our strong and relatively natural definition.

We prove that, under the strong RSA assumption, the decisional composite residuosity assumption, and the decisional Diffie-Hellman assumptions, our scheme is secure and has the appointed verifier property.

2 The Model

In this section, we define an ideal identity escrow scheme with appointed verifiers. Here, an ideal trusted third party takes care of the proper functionality of the system. Our model captures all the properties of previous ones (without appointed verifiers) in a natural way. We then define what it means for a real system to match this specification. We define the system with one group and one revocation manager; extending it to multiple ones is straightforward. Extending the model to group signatures can be done as well.

The Ideal System. The ideal system, the functionality of which is ensured by an ideal trusted party T, is as follows:
Ideal parties: The trusted party T, the group manager M, a set of users $\mathcal{U}$, a set of verifiers $\mathcal{V}$, and the anonymity revocation manager R.
Ideal communication: All communication is routed through T. If the sender of a message wishes to be anonymous, he requests that T not reveal his identity to the recipient. Finally, a sender of a message may request that a session, i.e., a block of messages, be established between him and the recipient. This session then gets a session id *sid*.
Ideal operations for a general identity escrow scheme:

- *Join.* This operation is a session between a user U and the group manager M. M tells T that it wants user U to become a member of the group. The user confirms that he wants to be a member. Upon receiving this messages from M and U, T sends a key K_U to U for further transactions related to his group membership; he also notifies M of the success of the transaction.
- *Authenticate.* This operation is a session between a user and a verifier V. The user must send a tuple (K, sid, V, con) to T, where K denotes a key, *sid* denotes a session id, V is the name of the verifier, and *con* is the condition under which the identity of the participating user can be established. T verifies that K is a key that corresponds to some group member (not necessarily the user from whom the request originates). If so, T tells the verifier V that the user

with whom the verifier has session *sid* running is a member of the group. V then either accepts or rejects, and forwards his reply to T. (If T receives no reply that is equivalent to rejecting.) T then notifies the user of the verifier's output.

Identify. This operation is a session between the revocation manager R and the verifier V. V submits a tuple (*sid*, *con*) to T and to R. R asks T to confirm that *sid* was an Authenticate operation with revocation condition *con*. Then R may ask T to reveal to R the identity of the user who participated in session *sid*. Finally, R may ask T to reveal the user's identity to V.

Ideal operations for an appointed-verifier identity escrow scheme:

Join with appointed verifier. This operation is a session between a user U and the group manager M. As a result, M tells T that user U's membership can be confirmed to verifier V. The user receives a key K_U from T for further transactions related to authenticating his group membership to V.

Authenticate to appointed verifier. This is the same as in the general scheme, except that T will only carry this out with the appointed verifier V.

Convert. This operation is between a user and the appointed verifier V. V tells T that the user is now authorized to demonstrate group membership to other verifiers. T notifies the user of that fact.

Authenticate. This is the same as in the general scheme, except that T will only carry this out if the user is authorized to demonstrate group membership to all verifiers.

Identify. This is the same as in the general scheme.

Inputs and outputs of the ideal players: The ideal players are interactive probabilistic Turing machines. Prior to initiating a transaction, a player receives an input that tells it to do so. These inputs are produced externally. At the end of the lifetime of the system, each player outputs a list of interactions in which this player has participated and their outcome (success/failure).

The Real System. We make the following assumptions on the communication in the real-system: We are in the public-key model, i.e., each user has carried out a proof of knowledge of his secret key at the beginning of the lifetime of the system. It is possible to establish a session between an anonymous user and a verifier (in practice, this can be achieved by a so-called mix-network [15] or by onion-routing protocols [29]. The information transmitted over a channel cannot later be retrieved by some physical means (i.e., it does not stick around in routers and caches). This is necessary to make sure that one cannot demonstrate that one sent or received a given message. This can also be achieved in conjunction with the methods to get anonymous communication, e.g., by requiring the hosts to delete all processed data. The real system is implemented by cryptographic protocols.

Security vs. Appointed-Verifier Property. The usual way of defining security of a real system is to restrict the power of the real-world adversary to the power of an adversary that controls the same set of players in the ideal system. Security in this sense is exhibited by providing a simulator that translates the real-world adversary into one in the ideal world. Here, in addition to providing security in this sense, we have to also allow for the case where there are two adversaries, such that one is trying to convince the other of his relationship with other players. Therefore, two security properties must be satisfied.

Protecting the Honest Players. First, we have to guarantee simulator-based security for the honest parties.

The ideal-world (resp., real-world) adversary is a probabilistic polynomial-time Turing machine that can control some subset of ideal (resp., real) parties and participate in transactions on their behalf. In addition, the adversary controls the environment, i.e., he either explicitly gives input to other players as to the transactions to be carried out, or he specifies the probability distribution on these inputs.

At the end of the lifetime of the system, each player outputs the entire list of interactions in which this player has participated and their outcome (success/failure).

Let the ideal system be called *IS*, and its cryptographic implementation be called *CS*. Let $p = \texttt{poly}(k)$ be the number of players in the system with security parameter k. Let Z_i denote the output of the i-th player. In the real world, a public-key infrastructure has been securely set up (i.e., each party has produced a public key and proved knowledge of the corresponding secret key). Let P denote its public information; let a denote the collection of dishonest players' secret keys. (In case we are working in the absence of the public-key model, these are empty strings.) An identity escrow scheme is secure if the adversary $\mathcal{A}$ cannot distinguish whether he is interacting with the real-world honest players, or if in fact the system is implemented in the ideal world (so all the honest players are shielded because T protects them) and he is just interacting with a simulator. More formally, with "$D_1(1^k) \stackrel{c}{\approx} D_2(1^k)$" denoting the computational indistinguishability of the distributions D_1 and D_2:

Definition 1 (Secure identity escrow scheme). *CS is* secure *if there exists a simulator $\mathcal{S}$ (ideal-world adversary) such that for all interactive probabilistic polynomial-time real-world adversaries $\mathcal{A}$, for all sufficiently large k, we have:*

- *In the IS, $\mathcal{S}$ controls the same set of players as $\mathcal{A}$ does in CS.*
- *The inputs given by $\mathcal{S}$ to the ideal-world players are identical to those given by $\mathcal{A}$ to the real-world players.*
- *For all P,*

$$(\{Z_i^{CS}(1^k, P, s_i)\}_{i=1}^{p}, \mathcal{A}(1^k, P, a)) \stackrel{c}{\approx} (\{Z_i^{IS}(1^k, P, s_i)\}_{i=1}^{p}, \mathcal{S}^{\mathcal{A}}(1^k, P, a)) ,$$

where $\mathcal{S}$ is given black-box *access to $\mathcal{A}$.*

Comparison with previous models. It is easy to see that this ideal model captures the requirements *correctness*, *anonymity*, *unlinkability*, *traceability*, *exculpability/framing*, and *coalition-resistance* of previous models (e.g., [2]), i.e., that the trusted party T ensures them.

No Benefits for Dishonest Players that Mistrust Each Other. Informally, an identity escrow scheme is *appointed-verifier* if only the appointed verifier can be persuaded that a user is a member of the group. A formal definition is more complex. Formally, we have two adversaries, $\mathcal{A}$ and $\mathcal{B}$, and $\mathcal{A}$ tries to convince $\mathcal{B}$ that some player A it controls is a group member, even though $\mathcal{B}$ does not control the appointed verifier V. The appointed verifier property of the scheme makes it impossible for any proof system $(\mathcal{A}, \mathcal{B})$, where $\mathcal{A}$ acts as prover and $\mathcal{B}$ as verifier, to have a non-negligible gap between the completeness and the soundness of the system. However, in defining this property, we have to take into account that (1) $\mathcal{B}$ can apply to V to tell him whether a given user is a group member; and (2) $\mathcal{B}$ can become convinced of the truth of the statement by means that are independent on the system's implementation: for example, if A is the only user in the system, and V flashes a green light every time it recognizes a group member. Thus, a formalization of the appointed verifier property is bound to be technically involved.

The approach we will take to defining it is as follows: we will require that for any $\mathcal{A}$, there exists an efficient $\mathcal{D}$ such that whenever $\mathcal{A}$ can convince $\mathcal{B}$ that A has group membership with appointed verifier V, $\mathcal{D}$ can convince $\mathcal{B}$ of the same statement without access to group manager's M's messages pertaining to the corresponding *Join* operation. We will call $\mathcal{D}$ *the deceiver*, because it can deceive any verifier $\mathcal{B}$. However, $\mathcal{D}$ is not responsible if $\mathcal{B}$ has other ways, implementation-independent, of getting convinced. That is why, in the definition, we need an additional efficient machine, $\mathcal{F}$, called *the filter*, which sets up the relevant group membership on behalf of A, but shields $\mathcal{D}$ from this information. $\mathcal{F}$ guarantees that group manager M and verifier V have the same view whether $\mathcal{A}$ has a valid membership certificate or one faked by $\mathcal{D}$. Intuitively, if $\mathcal{B}$ cannot distinguish whether he is talking to $\mathcal{A}$, or to the deceiver $\mathcal{D}$, but can still tell whether or not A is a group member with appointed verifier V, then $\mathcal{B}$'s way of telling is implementation-independent, and arises from the way other parties, such as M and V, behave. We now proceed to formalize this idea.

Let $\mathcal{A}$ and $\mathcal{B}$ be the two adversaries, modeled by probabilistic polynomial-time interactive Turing machines. Let *ES* denote an event sequence in the cryptographic identity escrow scheme. We write C^{ES} for a machine C to denote the fact that these events may be scheduled one-by-one, maybe even by an adversary. Let P denote the public information of the public-key infrastructure. Let a denote the set of secret keys of the players controlled by $\mathcal{A}$. Let a' be an additional input to $\mathcal{A}$. By $A \in \mathcal{A}$ we denote that A is a player controlled by adversary $\mathcal{A}$. Let $L \subseteq \{(A, V) : A \in \mathcal{A}, V \notin \mathcal{A}, V \notin \mathcal{B}\}$ be a list of user-verifier pairs that is given as a challenge to $\mathcal{B}$. We say that such a list L is *good* for *ES*, $\mathcal{A}$ and $\mathcal{B}$, if in the sequence of events specified by *ES*, for all $(A, V) \in L$, V never performs the *Convert* and *Identify* operations for A and $\mathcal{B}$ and V have not engaged in the *Authenticate with appointed verifier* protocol in which V accepted such that a subsequent *Identify* operation, if carried out, will point to A.

Let $\mathcal{F}^{ES}(1^k, P, L, a, mode)$, $\mathcal{D}^{\mathcal{A},\mathcal{F}}(1^k, P, L, a, mode)$ be interactive Turing machines. The *mode* part of their input specifies their behavior as follows: There are two modes of operation, the *real* mode and the *fake* mode. In the *real* mode, $\mathcal{F}$ passes the messages received from *ES* on to $\mathcal{D}$, which in turn passes them on to $\mathcal{A}$. If $\mathcal{A}$ sends any messages to $\mathcal{B}$, $\mathcal{D}$ faithfully passes them.

In the *fake* mode $\mathcal{F}$ behaves as follows: If a session *sid* is a *Join with appointed verifier* of user $A \in \mathcal{A}$ for verifier V, where $(A, V) \in L$, then $\mathcal{F}$ does not pass $\mathcal{A}$'s messages for *sid* to M, and does not forward M's replies to $\mathcal{A}$ for this *sid*. Instead, $\mathcal{F}$ carries out the *Join* operation himself, on behalf of $\mathcal{A}$, possibly guided by additional input from $\mathcal{D}$. It then notifies $\mathcal{D}$ whether this *Join* was successful. If a session *sid* is an *Authenticate to appointed verifier* between A and V such that $(A, V) \in L$ and the corresponding *Join* has taken place, then $\mathcal{F}$ does not pass $\mathcal{A}$'s messages for *sid* to V, and does not forward V's replies to $\mathcal{D}$ for this *sid*. Instead, $\mathcal{F}$ carries out the *Authenticate* operation himself, on behalf of $\mathcal{A}$, possibly guided by additional input from $\mathcal{D}$. It then notifies $\mathcal{D}$ whether this *Authenticate* was successful. For all other sessions, $\mathcal{F}$ just passes all the messages to and from $\mathcal{D}$.

In the *fake* mode, $\mathcal{D}$ behaves as follows: For a session *sid* of *Join with appointed verifier* for user A and verifier V where $(A, V) \in L$, $\mathcal{D}$ will create fake messages and send them to $\mathcal{A}$ in place of the group manager's messages. For a session *sid* of *Authenticate to appointed verifier* $\mathcal{D}$ will decide whether this session is between user A and verifier V, $(A, V) \in L$. In case it is, $\mathcal{D}$ notifies $\mathcal{F}$, and possibly sends it additional information. $\mathcal{D}$ will then create messages to $\mathcal{A}$ in place of V's responses. For all other sessions, $\mathcal{D}$ passes all the messages to and from $\mathcal{A}$.

We stress that $\mathcal{D}$ does not have the ability to reset $\mathcal{B}$.

Definition 2 (Appointed verifier property). *An identity escrow scheme has the appointed-verifier property if there exist polynomial-time algorithms $\mathcal{D}$, $\mathcal{F}$ as described above, such that for all probabilistic polynomial-time (in their first input) adversaries $\mathcal{A}$,$\mathcal{B}$, for all P, a', b, for all sequences of events in the system ES, and for all good lists L,*

$$\mathcal{D}^{\mathcal{A}(1^k,P,a,a'),\mathcal{F}^{ES,\mathcal{B}^{ES}(1^k,b)}(1^k,P,L,a,real)}(1^k, P, L, a, real) \stackrel{c}{\approx}$$
$$\mathcal{D}^{\mathcal{A}(1^k,P,a,a'),\mathcal{F}^{ES,\mathcal{B}^{ES}(1^k,b)}(1^k,P,L,a,fake)}(1^k, P, L, a, fake)$$

3 High-Level Presentation of Our Construction

First, a public-key infrastructure is set up in which each user has a secret key x and, based on this secret, an identifier $\tilde{h}^x$, where $\tilde{h}$ is a generator of some group G. Other players in the system have their public keys set up as follows: The group manager's public key is a modulus $n = pq$ such that $p = 2p' + 1$ and

$q = 2q' + 1$, and p,q, p' and q' are all prime numbers, and five quadratic residues modulo n, denoted $(a_0, a_1, a_2, a_3, a_4)$. (The length of n depends on the size of the group G.) Each verifier has a public key for the Paillier cryptosystem. A revocation manager R for this scheme will have a Cramer-Shoup public key in G. The specifics of how these keys are set up are described in Section 5.1.

For a user with secret key x, a group membership certificate for an appointed verifier V, will be a quin-tuple (s, Z, c, u, e) such that each of these values lies in the correct integer interval, $u^{2e} = (a_0 a_1^s a_2^x Z a_4^c)^2$ holds, and c is the encryption of the value $\log_{a_3} Z \bmod n$ under V's public key. We show that such a certificate is hard to forge under the strong RSA assumption [3,11,23,27,28] and the assumption that computing discrete logarithms modulo a modulus of this form is hard. On the other hand, if c is not an encryption of $\log_{a_3} Z \bmod n$, then this certificate is easy to forge (Lemma 3). As V is the only entity that can check this, under the assumption that the Paillier cryptosystem is semantically secure, this is the first key step towards obtaining the appointed verifier property (the other key step is discussed at the end of this section). The fact that c is included in the certificate implies security for the verifier against adaptive attacks even though the Paillier encryption scheme as such is not secure against these attacks[1]. This membership certificate is issued via a protocol (between the user and the group manager), that does not allow the group manager to learn x and s, but only $\tilde{h}^x$ and $a_1^s a_2^x \bmod n$. This protocol is described in detail in Section 5.2.

To prove group membership to V, the user blinds c to obtain c', and blinds Z to obtain Z' in such a way that, if c is the encryption of $\log_{a_3} Z$, then c' is the encryption of $\log_{a_3} Z'$. This is why we use the Paillier cryptosystem: the additive homomorphism property of the Paillier scheme is crucial for this step. c' and Z' are given to the verifier. Further, the user proves knowledge of a tuple (x, s, c, Z, u, e, r) such that (s, Z, c, u, e) is a group membership tuple for key x, and r is the randomizer used to blind (c, Z) to obtain (c', Z'). In addition, to enable anonymity revocation, the user provides an encryption E of his identifier $\tilde{h}^x$ under the anonymity revocation manager's public key and proves that E is a valid encryption of an identifier that is based on the same x as the group membership certificate. These proofs are done using efficient statistical zero-knowledge discrete-logarithm-based proofs of knowledge. The fact that these proofs are zero-knowledge and that the user blinds c and Z give us anonymity for the user. These proofs are described in detail in Section 5.3. Finally, the verifier checks that (1) c' is an encryption of $\log_{a_3} Z'$, and (2) the user carried out the proofs correctly. If so, the verifier accepts.

To convert an appointed-verifier membership certificate into a universally verifiable membership certificate, the appointed verifier reveals $\log_{a^3} Z'$ to the user. Under the strong RSA assumption and the hardness of discrete logarithms

[1] This step resolves the following paradox: On the one hand, we want the encryption scheme to be malleable, so that the user can successfully blind the ciphertext c. On the other hand, we want it to be secure against adaptive attacks by malicious users. Thus c is created by the group manager.

modulo n, the resulting tuple, (x, s, z, c, u, e) is hard to forge (cf. full version of this paper [10]).

Let us finally discuss the second key element to achieve the appointed verifier property: requiring a user to verifiably encrypt, under her own public key, some of the secrets she uses in the *Authenticate to appointed verifier* protocol. This is necessary as, in essence, the definition for this property requires that no matter how adversary $\mathcal{A}$ behaves, and no matter how often and when $\mathcal{A}$ and $\mathcal{B}$ exchange messages, there is nothing $\mathcal{A}$ can convince $\mathcal{B}$ of that $\mathcal{D}$ (in fake mode) would not be able to convince him of either. Running in fake mode requires $\mathcal{D}$ to know a great deal about the internal information of $\mathcal{A}$. Traditionally, this would be realized by allowing $\mathcal{D}$ black-box access to $\mathcal{A}$ and the ability to rewind it. However, as we allow message exchanges between $\mathcal{A}$ and $\mathcal{B}$ at arbitrary times, arbitrarily interleaved with other executions, this is not possible as it would require $\mathcal{D}$ to have black-box access to other players as well (in particular those controlled by $\mathcal{B}$). Thus, $\mathcal{D}$ must somehow contain a knowledge extractor that does not rewind $\mathcal{A}$. $\mathcal{D}$ will instead extract what it needs to know from the verifiably encrypted secrets. Thus, we need the public-key model: in this model, $\mathcal{A}$ and, as a consequence, $\mathcal{D}$, will receive as input the secret keys of all the players controlled by $\mathcal{A}$.

4 Preliminaries

4.1 Proof Protocols and Corresponding Notation

We use notation introduced by Camenisch and Stadler [12] for the various proofs of knowledge of discrete logarithms and proofs of the validity of statements about discrete logarithms. For instance,

$$PK\{(\alpha,\beta,\gamma) : y = g^\alpha h^\beta \ \wedge\ \tilde{y} = \tilde{g}^\alpha \tilde{h}^\gamma \ \wedge\ (u \le \alpha \le v)\}$$

denotes a "*zero-knowledge* Proof *of* Knowledge *of integers* α, β, *and* γ *such that* $y = g^\alpha h^\beta$ *and* $\tilde{y} = \tilde{g}^\alpha \tilde{h}^\gamma$ *holds, where* $v < \alpha < u$," where $y, g, h, \tilde{y}, \tilde{g}$, and $\tilde{h}$ are elements of some groups $G = \langle g \rangle = \langle h \rangle$ and $\tilde{G} = \langle \tilde{g} \rangle = \langle \tilde{h} \rangle$. By convention, the Greek letters denote quantities the knowledge of which is being proved, while all other parameters are known to the verifier. Using this notation, a proof-protocol can be described by just pointing out its aim while hiding all details.

It is important that we use protocols that are *concurrent* zero-knowledge. They are characterized by remaining zero-knowledge even if several instances of the same protocol are run arbitrarily interleaved [24,25]. Damgård [24] shows that so-called Σ-protocols (this includes all the *PK*'s discussed above) can easily be made concurrent zero-knowledge in many practical scenarios, including the public-key model. We assume throughout that the latter technique is used with all *PK*'s.

4.2 Proving That a Commitment Contains a Paillier Encryption

Our scheme requires a proof that some value e is a Paillier encryption [34,35] of a value x that the prover knows, under a given Paillier public key (g, n), and a similar proof where the ciphertext e is not given as input to the verifier; instead only a Pedersen commitment [36] to ciphertext e is given. Protocols for carrying out the former proof have been realized [21]. The latter proof is, to the best of our knowledge, not found in the literature and is constructed as follows:

Let (g, n) be the public key of Paillier's encryption scheme. Assume that we are given a group $\hat{G} = \langle \hat{g} \rangle = \langle \hat{h} \rangle$ of order n^2. Let E be the commitment to a ciphertext, i.e., $E = \hat{g}^e \hat{h}^z$ where $e = g^x r^n \mod n^2$. Using the protocol denoted $PK\{(\alpha, \beta, \gamma) : E = \hat{g}^{g^\alpha \beta^n} \hat{h}^\gamma\}$ the prover can convince the verifier that E is a commitment to a Paillier encryption of some value she knows. The protocol is as follows.

1. The prover chooses $r_1 \in_R \mathbb{Z}_n$ and $r_2, r_3 \in_R \mathbb{Z}_{n^2}$, computes $t = \hat{g}^{g^{r_1} r_2^n} \hat{h}^{r_3}$ and sends t to the verifier.
2. The verifier chooses a $c \in_R \{0, 1\}$ and sends c to the prover.
3. The prover computes $s = r_1 - cx \mod n$, $u = r_2/r^c \mod n^2$, and $v = r_3 - czg^s u^n \mod n^2$ and sends s and u to the verifier.
4. The verifier checks whether $t = \hat{g}^{g^s u^n} \hat{h}^v$ if $c = 0$ and whether $t = E^{g^s u^n} \hat{h}^v$ otherwise.

It is easy to see that the proof is correct and honest-verifier zero-knowledge proof of knowledge.

4.3 Verifiable Encryption

Verifiable encryption [1,8], is a protocol between a prover and a verifier such that as a result of the protocol, on input public key E, and value v, the verifier obtains an encryption e of some value s under E such that $(w, y) \in \mathcal{R}$. For instance, $\mathcal{R}$ could be the relation $(w, g^w) \subset \mathbb{Z}_q \times G$. Generalizing the protocol of Asokan et al. [1], Camenisch and Damgård [8] provide a verifiable encryption scheme for a class of relations that, in particular, includes all discrete-logarithm relations that are of relevance in this paper. We denote verifiable encryption similarly as the *PK*'s, e.g., $e := VE(\mathsf{ElGamal}, (u, v))\{\xi : y = g^\xi\}$ denotes the verifiable encryption protocol for the ElGamal scheme, whereby $\log_g y$ is encrypted in e under public key (u, v). Note that e is not a single encryption, but the verifier's entire transcript of the protocol and contains several encryptions, commitments and responses of the underlying *PK*.

5 An Identity Escrow Scheme with Appointed Verifiers

5.1 Key and System Setup

Our protocols are realized in the public-key model, thus the initial setup is the public-key infrastructure in which each user has a public key and has proved knowledge of the secret key to some entity, say the *CA*. Specifically, some group $\tilde{G} = \langle \tilde{g} \rangle = \langle \tilde{h} \rangle$ of prime order $\tilde{q}$, such that $\log_{\tilde{g}} \tilde{h}$ is unknown. Also, each user has a secret key an $x \in_R \mathbb{Z}_q$, and a corresponding public key $\tilde{S}_U = \tilde{h}^x$. The user has submitted this $\tilde{S}_U$ to the *CA* of this public-key infrastructure and and has executed $PK\{(\alpha) : \tilde{S}_U = \tilde{h}^\alpha\}$ with the *CA*. The *CA* sends the user a signature on $\tilde{S}_U$ and publishes $\tilde{S}_U$ and the user's name.

In addition, to get security in case the protocols are executed concurrently, we assume that all zero-knowledge proofs (*PK*) are carried out using the construction due to Damgård [24]. This requires to initially set up public keys for a trapdoor commitment scheme.

Other security-related system parameters are as follows: the length ℓ_n of the RSA modulus of the group manager, integer intervals $\Gamma =]-2^{\ell_\Gamma}, 2^{\ell_\Gamma}[$, $\Delta =]-2^{\ell_\Delta}, 2^{\ell_\Delta}[$, $\Lambda =]2^{\ell_\Lambda}, 2^{\ell_\Lambda+\ell_\Sigma}[$ such that $\tilde{q} < 2^{\ell_\Gamma}$, $\ell_\Delta = \epsilon(4\ell_n + 3)$ and $\ell_\Gamma = 2\ell_n$, where $\epsilon > 1$ is a security parameter, and $\ell_\Lambda > \ell_\Sigma + \ell_\Delta + 4$. Furthermore, let ℓ_v be the length of the RSA modulus of the verifier for Paillier's encryption scheme [35]. We require that $2\ell_v < \ell_\Gamma$ holds. There further are ℓ_z and ℓ_r with $\ell_z > \epsilon\ell_r + 1$ and $\ell_z + \epsilon\ell_r + 1 < \ell_v$. Define the integer intervals $\Omega =]2^{\ell_z} - 2^{\ell_r}, 2^{\ell_z} + 2^{\ell_r}$, $\Phi =]-2^{\epsilon\ell_r}, 2^{\epsilon\ell_r}[$, and $\Omega' =]2^{\ell_z} - 2^{\epsilon\ell_r+1}, 2^{\ell_z} + 2^{\epsilon\ell_r+1}[$ (ℓ_r must be large enough to make computing an ℓ_r-bit discrete logarithm modulo an ℓ_n-bit RSA modulus hard, where the modulus is the product of two safe primes.)

The public key of the group manager consists of an ℓ_n-bit RSA modulus $n = pq = (2p' + 1)(2q' + 1)$ that is the product of two safe primes, and random elements $a_4, a_3, a_2, a_1, a_0, g, h \in_R QR_n$ of maximal order. The factorization of n is the group manager's secret key. The revocation manager sets up his public and secret key for the Cramer-Shoup encryption scheme [22] over $\tilde{G}$ (i.e., the group that comes from the public-key infrastructure), i.e., $x_1, \ldots, x_5 \in_R \mathbb{Z}_{\tilde{q}}$ are the secret keys and $(y_1 := \tilde{g}^{x_1}\tilde{h}^{x_2}, y_2 := \tilde{g}^{x_3}\tilde{h}^{x_4}, y_3 := \tilde{g}^{x_5})$ constitutes the public key. The revocation manager also publishes a collision-resistant hash function $\mathcal{H}$.

Each user also publishes an ℓ_n-bit RSA modulus n_U that is the product of two safe primes and two generators g_U and h_U of QR_n.

Each appointed verifier chooses a public key (n_v, g_v) of the Paillier encryption scheme, where n_v is an ℓ_v bit RSA modulus and $g_v = 1 + n_v \pmod{n_v^2}$. The verifier also publishes $\hat{G} = \langle \hat{g} \rangle = \langle \hat{h} \rangle$ of order n_v^2.

5.2 Joining with Appointed Verifier

In this protocol, aside from the public information, the user's input will be a secret key $x \in \Gamma$ and her identifier $\tilde{S}_U$ and her output will be a membership certificate tuple (s, Z, c, e, u) w.r.t. an appointed verifier V such that $s \in_R \Delta$, c is the encryption of $z = \log_{a_3} Z \bmod n$ under V's Paillier public key, $z \in \Omega$, $e \in \Lambda$ a prime, and $u^e = a_4^c Z a_2^x a_1^s a_0 \bmod n$. The group manager's input will be his secret key and all the public information in the system. His output is the user's identifier $\tilde{S}_U = \tilde{h}^x$ and also the values $S = a_1^s a_2^x$, z, c, e, u.

A secure two-party protocol that has this functionality is as follows:

1. User chooses a value $s_1 \in_R \Delta$. The integer s_1 will be the user's contribution to s. $r_x, r_s \in_R \{0,1\}^{2\ell_n}$ are also chosen. User sets $C_1 := g^{s_1} h^{r_s} \mod n$ and $C_2 := g^x h^{r_x} \mod n$, sends C_1, C_2, $\tilde{S}_U$, and the *CA*'s signature on $\tilde{S}_U$ to the *GM*, and serves as the prover to verifier *GM* in

$$PK\{(\alpha,\beta,\gamma,\delta) : C_1^2 \equiv (g^2)^\alpha (h^2)^\beta \ \wedge \ C_2^2 \equiv (g^2)^\gamma (h^2)^\delta \ \wedge \\ \tilde{S}_U = \tilde{h}^\gamma \ \wedge \ \alpha \in \Delta \ \wedge \ \gamma \in \Gamma\} \ .$$

2. *GM* checks the *CA*'s signature on $\tilde{S}_U$, chooses a random $s_2 \in_R \Delta$ and sends s_2 to U.
3. The user computes $s = (s_1 + s_2 \mod (2^{\ell_\Delta+1} - 1)) - 2^{\ell_\Delta} + 1$, ($s$ is the sum of s_1 and s_2, adjusted appropriately so as to fall in the interval Δ) and $\tilde{s} = \left\lfloor \frac{s_1+s_2}{2^{\ell_\Delta+1}-1} \right\rfloor$ ($\tilde{s}$ is the value of the carry resulting from the computation of s above). The user then sets $S := a_1^s a_2^x$ and sends S to *GM*.
4. Now, the user must show that S was formed correctly. To that end, she chooses $r_{\tilde{s}} \in_R \{0,1\}^{\ell_n}$, sets $C_3 := g^{\tilde{s}} h^{r_{\tilde{s}}}$, sends C_3 to *GM*, and executes

$$PK\{(\alpha,\beta,\gamma,\delta,\varepsilon,\zeta,\vartheta,\xi) : C_1^2 = (g^2)^\alpha (h^2)^\beta \ \wedge \ C_3^2 = (g^2)^\varepsilon (h^2)^\zeta \ \wedge \\ S^2 = (a_1^2)^\vartheta (a_2^2)^\gamma \ \wedge \ (C_1^2 (g^2)^{(r-2^{\ell_\Delta}+1)})/(C_3^2)^{(2^{\ell_\Delta+1}+1)} = (g^2)^\vartheta (h^2)^\xi \ \wedge \\ \tilde{S}_U = \tilde{h}^\gamma \ \wedge \ \gamma \in \Gamma \ \wedge \ \vartheta \in \Delta\}$$

as prover with the *GM*.
5. *GM* chooses $z \in_R \Omega$, a prime $e \in_R \Lambda$, computes $Z := a_3^z$ and $u := (a_4^c a_3^z S a_0)^{1/e} \pmod n$, encrypts z under the public key of the appointed verifier, i.e., chooses a random $r \in_R \mathbb{Z}_{n_v}$ and computes $c := g_v^z r^{n_v} \pmod{n_v^2}$. *GM* sends Z, u, e, and c to U.
6. User checks whether $u^e \equiv a_4^c Z a_2^x a_1^s a_0 \pmod n$, $e \in \Lambda$, and $c \in \mathbb{Z}_{n_v^2}$.
7. *GM* proves to the user that c indeed encrypts $\log_{a_3} Z$ and that this value lies in Ω. To this end *GM* chooses $\hat{r} \in_R \{0,1\}^{\ell_n}$, computes $\hat{Z} := g_U^z h_U^{\hat{r}}$, sends $\hat{Z}$ to U and carries out the protocol

$$PK\{(\alpha,\beta,\gamma,\delta) : \tilde{S}_U = \tilde{h}^\gamma \ \vee \ (c \equiv g_v^\alpha \beta^n \pmod{n_v^2}) \ \wedge \\ Z^2 \equiv (a_3^2)^\alpha \pmod n \ \wedge \ \hat{Z}^2 \equiv (g_U^2)^\alpha (h_U^2)^\delta \pmod{n_U} \ \wedge \ \alpha \in \Omega)\}$$

as the prover with the user.
8. *GM* stores $S, \tilde{S}_U, u, e, c, z$ and the user's name in its database.
9. *GM* and the user go home, listen to music, and have coffee, tea, and cake.

Remark: In step 7 the *GM* proves that it knows either the user's secret $x = \log_{\tilde{h}} \tilde{S}_U$ or that c is an encryption of $\log_{a_3} Z$ so as to leave no evidence to the user that the protocol took place.

5.3 Authenticate to an Appointed Verifier

This is a protocol between a user and an appointed verifier. The user's input is the public information, the membership certificate issued as described above, and a revocation condition *con* which specifies under which conditions the user's identity may be discovered. The verifier's input, aside from the public information, is his Paillier secret key. The verifier's output is *con*, and an encryption of the user's identifier under the revocation manager's public key with condition *con*. The verifier accepts if the user succeeds in proving knowledge of a valid membership certificate, and in proving that this membership certificate was issued to the user whose encrypted identifier is provided. The protocol is as follows:

1. The user and the verifier agree on a revocation condition *con*.
2. The user first blinds the ciphertext c, i.e., chooses random $\tilde{r}_1 \in_R \varPhi$ and $\tilde{r}_2 \in_R \mathbb{Z}_{n_v}$ and computes $\tilde{c} := c g_v^{\tilde{r}_1} \tilde{r}_2^{n_v^2} \pmod{n_v}$ and $\tilde{Z} := Z a_3^{\tilde{r}_1} \pmod{n}$, then sends $\tilde{c}$ and $\tilde{Z}$ to the verifier.
3. The user computes a blinded public key for use with verifiable encryption, i.e., she chooses a random $w \in_R \mathbb{Z}_{\tilde{q}}$, computes $\tilde{u} := \tilde{h}^w$ and $\tilde{v} := \tilde{S}_U^w$ (hence $\tilde{v} = \tilde{u}^x$), and sends $\tilde{u}$, $\tilde{v}$ to the verifier.
4. The user chooses $r_1, r_2, r_3 \in_R \mathbb{Z}_{n^2}$ and $\hat{r} \in_R \mathbb{Z}_{n_v^2}$ and computes $T_1 = uh^{r_1} \bmod n$, $T_2 = g^{r_1} h^{r_2} \bmod n$, $T_3 = g^{\tilde{r}_1} h^{r_3} \bmod n$, and $\hat{T} = \hat{g}^{g_v^{\tilde{r}_1} \tilde{r}_2^{n_v}} \hat{h}^{\hat{r}}$. ($T_1$ serves as a blinded u, and T_2 is an additional commitment which will be used to prove that T_1 was formed correctly. $\hat{T}$ and T_3 are needed to show that the ciphertext c was blinded in the same way as Z.) Then the user computes the encryption E of his identifier under condition *con*, as follows: he chooses $r_4 \in_R \mathbb{Z}_q$ and sets $E := (E_1, E_2, E_3, E_4)$, where $E_1 := \tilde{g}^{r_4}$, $E_2 := \tilde{h}^{r_4}$, $E_3 := \tilde{h}^x y_3^{r_4}$, $E_4 := y_1^{r_4} y_2^{r_4 \mathcal{H}(E_1 \| E_2 \| E_3 \| con)}$. The user sends $(T_1, T_2, T_3, \hat{T}, E)$ to the verifier.
5. The user serves as prover to the verifier in

$$\mathit{VE}(\mathsf{ElGamal}, (\tilde{u}, \tilde{v}))\{(\varrho, \vartheta, \varsigma) : \ \hat{T} = \hat{g}^{g_v{}^{\vartheta} \varrho^n} \hat{h}^{\mu} \ \wedge \ T_3^2 = (g^2)^{\vartheta} (h^2)^{\varsigma} \ \wedge \ \vartheta \in \varPhi\}$$

and in

$$\begin{aligned} PK\{(\alpha, \beta, \gamma, \delta, \zeta, \varepsilon, \varphi, \xi, \nu, \mu, \psi, \vartheta, \varsigma) : \ 1 = (T_2^2)^{\alpha} \Big(\frac{1}{g^2}\Big)^{\varepsilon} \Big(\frac{1}{h^2}\Big)^{\psi} \ \wedge \\ a_0^2 \tilde{Z}^2 = (T_1^2)^{\alpha} \Big(\frac{1}{a_1^2}\Big)^{\beta} \Big(\frac{1}{a_2^2}\Big)^{\nu} (a_3^2)^{\vartheta} \Big(\frac{1}{a_4^2}\Big)^{\varphi} \Big(\frac{1}{h^2}\Big)^{\varepsilon} \ \wedge \ T_2^2 = (g^2)^{\delta} (h^2)^{\zeta} \ \wedge \\ T_3^2 = (g^2)^{\vartheta} (h^2)^{\varsigma} \ \wedge \ \hat{g}^{\tilde{c}} = \hat{T}^{\varphi} \hat{h}^{\kappa} \ \wedge \ \tilde{v} = \tilde{u}^{\nu} \ \wedge \ \tilde{u} = \tilde{h}^{\gamma} \ \wedge \\ E_1 = \tilde{g}^{\xi} \ \wedge \ E_2 = \tilde{h}^{\xi} \ \wedge \ E_3 = \tilde{h}^{\nu} y_3^{\xi} \ \wedge \ E_4 = (y_1 y_2^{\mathcal{H}(E_1 \| E_2 \| E_3 \| con)})^{\xi} \wedge \\ \alpha \in \varLambda \wedge \beta \in \varDelta \wedge \nu \in \varGamma \wedge \vartheta \in \varPhi \wedge \varphi \in [1, n_v^2 - 1]\} \ . \end{aligned}$$

6. The verifier decrypts $\tilde{c}$ to get $\tilde{z}$ and checks whether $\tilde{Z} = a_3^{\tilde{z}} \pmod{n}$ and whether $\tilde{z} \in \varOmega'$.

Let us consider the efficiency of the above verifiable encryption protocol. Recall that verifiable encryption works by repeating the underlying *PK* sufficiently many times, e.g., $k = 80$ times. Assuming that exponentiation with a $2\ell_n$-bit modulus corresponds to about 8 exponentiations with an ℓ_n-bit modulus, the total computational load of both the prover and the verifier for the verifiable encryption protocol amounts to $17k$ exponentiations with an ℓ_n-bit modulus and about 42 exponentiations with an ℓ_n-bit modulus for the *PK*. On the verifier's side, this load can be considerably reduced by applying so-called batch verification [4].

5.4 Convert and Authenticate

This paragraph briefly discusses how an appointed verifier can convert an appointed-verifier membership certificate into an ordinary membership certificate and how a group member can then convince anyone of her group membership.

To convert a certificate, the user and the verifier first carry out the *authenticate with appointed verifier* operation. If this operation is successful, the verifier can provide the user with the decryption of $\tilde{c}$. This will allow the user to compute the value z encrypted as c. Thus she holds values (x, s, z, c, u, e) such that $u^{2e} = (a_4^c a_3^z a_2^x a_1^s a_0)^2 \bmod n$, i.e., a valid group membership certificate. Proving possession of this certificate, i.e., authenticating as a group member to any verifier, can now be done similarly to the way it is done for an appointed verifier above. The only difference is that there is no encryption $\tilde{c}$ and no commitments T_3 and $\hat{T}$, and hence the corresponding parts in the proof-protocol are dropped: First, steps 2 and 5 are no longer needed; second, in step 4 the verifiable encryption protocol is not needed and in the *PK* the first term of the expression proved is replaced by $a_0^2 = (T_1^2)^\alpha (\frac{1}{a_1^2})^\beta (\frac{1}{a_2^2})^\nu (\frac{1}{a_3^2})^\vartheta (\frac{1}{a_4^2})^\varphi (\frac{1}{h^2})^\varepsilon$ while the terms $T_3^2 = (g^2)^\vartheta (h^2)^\varsigma$, $\hat{g}^{\tilde{c}} = \hat{T}^\varphi \hat{h}^\kappa$, $\tilde{v} = \tilde{u}^\nu$, and $\tilde{u} = \tilde{h}^\gamma$ are dropped. The fact that the verifiable encryption protocol is no longer needed makes the whole protocol much more efficient as it was the bulk of the computational load.

5.5 Anonymity Revocation

Upon a request $E = (E_1, E_2, E_3, E_4)$ and *con*, the revocation manager checks whether $E_4 = E_1^{x_1 + x_3 \mathcal{H}(E_1 \| E_2 \| E_3 \| con)} E_2^{x_2 + x_4 \mathcal{H}(E_1 \| E_2 \| E_3 \| con)}$ and whether the revocation condition *con* is fulfilled. If these checks succeed, he returns $\hat{S} := E_3 / E_1^{x_5}$. If E was produced in an *Authenticate to an Appointed Verifier* or an *Authenticate* protocol, $\hat{S}$ will match the identifier $\tilde{S}_U$ of the user who took part in the protocol.

5.6 Proof of Security and Appointed Verifier Property

We outline how security is proven and state the important theorems and lemmas. For details and all the proofs we refer to the full version of this paper [10].

Protecting the Honest Players. Security for the honest players is proven by providing a simulator that satisfies Definition 1. The simulator will create cryptographic instantiations for the honest parties. For every transaction between the adversary and an honest party, the simulator will execute its cryptographic part on behalf of these honest parties. If the cryptographic implementation of a protocol prescribes that a real-world honest player should behave in a way that is different from the underlying ideal-world player, then the simulator rejects. (This can happen if an adversary succeeds in proving group membership in such a way that the simulator is unable to extract a secret key to which a membership certificate was issued in a previous transaction. As a result, an ideal trusted party would tell the ideal verifier to reject the adversary's user, while the cryptographic implementation would dictate the real-world verifier to accept.)

This simulator is constructed [10] in the usual way, with the following subtle difference: in the *Authenticate* protocol, when an honest user interacts with a dishonest verifier, the simulator does not get to know which user it is and hence does not know which user to simulate towards the the verifier. There are two cases to consider here, one where the revocation manager is honest and one where he is not. For brevity we will address only the former case here: The simulator forms a ciphertext E that is an encryption of 0 the revocation manager's public key. He then creates a random public key $P = (\tilde{u}, \tilde{v})$ for the verifiable encryption and chooses $\tilde{r}_1 \in_R \Phi$, $\tilde{r}_2 \in_R \mathbb{Z}_{n_v}$, and T_1,T_2 and $\hat{T}$ at random from their corresponding domains. Then, the simulator sends $(Z', c', T_1, T_2, \hat{T}, E, P)$ to the adversary and carries out the verifiable encryption protocol:

$$VE(\mathsf{ElGamal}, (\tilde{u}, \tilde{v}))\{(\varrho, \vartheta, \varsigma) : \ \hat{T} = \hat{g}^{{g_v}^{\vartheta} \varrho^n} \hat{h}^{\mu} \ \wedge \ T_3^2 = (g^2)^{\vartheta} (h^2)^{\varsigma} \ \wedge \ \vartheta \in \Phi\}$$

with the adversary and finally runs the simulator for the view of the verifier in the group membership proof protocol described in Section 5.3.

The following lemma follows from the semantic security of the verifiable encryption scheme, as well as from adaptive chosen-ciphertext security of the encryption scheme under which the users' identifiers are encrypted [10].

Lemma 1. *Either the simulator produces a computationally indistinguishable view, or it rejects. The computational indistinguishability is under the decisional Diffie-Hellman assumption for the group over which the Cramer-Shoup encryption of the identifiers is done.*

The only thing left to prove security is to show that the simulator almost never rejects. We observe that the only case when the simulator rejects is when the adversary demonstrates group membership for an unauthorized user-verifier pair. We show [10] that if this simulator rejects non-negligibly often, then either there exists a polynomial-time algorithm for forging membership certificates (thus violating the strong RSA assumption or the discrete logarithm assumption), or there exists a polynomial-time algorithm for cracking the Paillier cryptosystem, or there exists a way to circumvent the knowledge extractor for one of the proofs of knowledge:

Lemma 2. *Under the strong RSA assumption, the hardness of discrete logarithms modulo a safe prime product, and the security of Paillier cryptosystem, the simulator rejects with only negligible probability.*

Putting everything together, we get:

Theorem 1. *Under standard number-theoretic assumptions, the construction presented in Section 5 is an identity escrow scheme with security guarantee for honest users, as required by Definition 1.*

Appointed Verifier Property. Given the public key $(n, a_0, a_1, a_2, a_3, a_4, g, h)$ of the group manager, and public key $(\tilde{n}, \tilde{g})$ of the appointed verifier V, for any given S, it is easy to create a tuple (Z, c, u, e) such that no one except V can distinguish it from a valid membership certificate. Create such a tuple as follows (call this procedure the *forger*): choose any $r \in_R \mathbb{Z}_{\tilde{n}^2}$, set $c := r^{\tilde{n}} \bmod \tilde{n}^2$ (c is simply the encryption of 0 under the verifier's public key), $u \in_R QR_n$, $e \in_R \Lambda$, and set $Z := u^e / a_4^c S a_0$.

Lemma 3. *Under the assumption that the Paillier cryptosystem is semantically secure, for all $x \in \Gamma$, the tuple (s, Z, c, e, u) such that $s \in_R \Delta$, and (Z, c, e, u) are created by the forger above on input $S = a_1^s a_2^x$, is indistinguishable from a valid membership certificate created by querying oracle O that, on input S, carries out step 5 of the* Join with appointed verifier *protocol.*

Proof. Let D_1 be the distribution of fake certificates as above, and D_2 be the distribution of valid certificates. Suppose that a distinguisher existed. Then we break the security of the Paillier cryptosystem as follows: we give the reduction access to the secret keys of the group manager. The reduction chooses a random $z \in \Omega$ and asks the encryption oracle to give it an encryption of either 0 or z. It is easy to see that if the oracle returns an encryption of 0, then the resulting tuple will be distributed according to D_1, while if the oracle returns an encryption of z, then the resulting tuple will be distributed according to D_2. Thus we can use the distinguisher for D_1 and D_2 to break the semantic security of the Paillier cryptosystem. □

Based on this way of forging a single membership certificate, we can now build a deceiver $\mathcal{D}$. In *fake* mode, on input a list L, $\mathcal{D}$ does not forward the messages pertaining to *Join with appointed verifier* for user A and verifier V if $(A, V) \in L$. Instead, he impersonates the group manager GM to $\mathcal{A}$. $\mathcal{D}$ proceeds as follows: it conducts steps 1 through 4 of the *Join with appointed verifier* protocol exactly the same way as GM would to get an input S. Then it creates a fake certificate (Z, c, e, u) using the forger described above. As the secret key $x = \log_{\tilde{h}} \tilde{S}_U$ of user A was given to $\mathcal{D}$ as input, $\mathcal{D}$ succeeds in carrying out the PK in step 7. It then stores this certificate.

For $(A, V) \notin L$, $\mathcal{D}$ forwards all the messages, and, in case of a successfully carried out *Join*, stores the certificate.

When $\mathcal{A}$ engages in *sid* that is an *Authenticate to appointed verifier* with some verifier V, $\mathcal{D}$ proceeds as follows (recall that verifiable encryption is by itself a three-move proof of knowledge): it first receives, from $\mathcal{A}$, all messages up to step 5 and buffers them. Then, it receives the first message of the *VE* protocol, and in particular the ciphertext $\tilde{c}$ and the value $\tilde{Z}$. By the properties of *VE*, this first message contains an ElGamal encryption under $(\tilde{v}, \tilde{u})$ of values $\tilde{r}_1$ and $\tilde{r}_2$. It checks whether $\tilde{v} = \tilde{u}^x$ for some secret key x of a player $\mathcal{A}$ controls. If this is not the case, then it knows that the verifier will reject anyway– so it forwards the message to V. If it finds the right x, then it decrypts the first message of the verifiable encryption and obtains $\tilde{r}_1$ and $\tilde{r}_2$. If the first message of the verifiable encryption is invalid, it detects that and then it knows that V will reject, so it forwards $\mathcal{A}$'s message to V. It then sets $c := \tilde{c}/(g_v^{\tilde{r}_1} \tilde{r}_1^{n_v})$. It then looks up a membership certificate that contains the ciphertext c. If it fails to find one, it knows that the verifier will reject – so it forwards the message to V. If it finds one, and it is a valid membership certificate, then it forwards all the messages between $\mathcal{A}$ and V for this *sid*.

If it is a fake membership certificate that includes ciphertext c, it checks whether this certificate also includes the value $Z := \tilde{Z}/(a_3^{\tilde{r}_1})$. If it does not, then $\mathcal{D}$ knows that the verifier will reject anyway – so it forwards the message to V.

Otherwise, this first message of $\mathcal{A}$ is valid. Since $\mathcal{D}$ has the valid membership certificate for (A, V), $\mathcal{D}$ tells $\mathcal{F}$ to send the first valid message of an *Authenticate to appointed verifier* to V. Then $\mathcal{D}$ simulates V for $\mathcal{A}$: it creates a challenge message and sends it to $\mathcal{A}$. If $\mathcal{A}$ responds to the message so as to correctly complete the corresponding proof of knowledge and verifiable encryption, then $\mathcal{D}$ tells $\mathcal{F}$ to send V a message that corresponds to a valid response to V's challenge. Otherwise, $\mathcal{D}$ tells $\mathcal{F}$ to send to V a message that does not constitute a valid response. After that, V either responds to $\mathcal{F}$ with an accept or reject. $\mathcal{F}$ forwards that response to $\mathcal{D}$, which in turn sends it to $\mathcal{A}$.

It is easy to see that the following lemma holds [10]:

Lemma 4. *Under the assumption that the Paillier cryptosystem is semantically secure, the strong RSA assumption, and the assumption that computing discrete logarithms modulo a safe prime product is hard, the following holds: Provided that V never performs the* Convert *and* Identify *operation for A, if the probability that $\mathcal{B}$ accepts when talking to $\mathcal{D}$ in* real *mode differs non-negligibly from the probability that $\mathcal{B}$ accepts when talking to $\mathcal{D}$ in* fake *mode, then: $\mathcal{B}$ and verifier V have engaged in the* Authenticate with appointed verifier *protocol in which V accepted such that a subsequent* Identify *operation, if carried out, will point to A.*

Using Lemma 4, the following is immediate by Definition 2:

Theorem 2. *Under standard number-theoretic assumptions, the construction presented in Section 5 is an identity escrow scheme with the appointed verifier property, as required by Definition 2.*

6 Concluding Remarks

We note that in order to implement several identity escrow schemes at the same time using our methods, the set-up, apart from the public-key infrastructure, has to be repeated for each instance. In particular, the public keys of the verifiers will have to be different for each instance. It is an interesting question whether it would be possible to avoid this and yet have a practical construction that is secure against adaptive attacks. It is also interesting whether the public-key model can be eliminated from the picture.

An appointed-verifier identity escrow scheme is only the first step towards a bigger goal of realizing protocols in which it is provably hard to convince an unauthorized party of the truth of some statement. It would be interesting to apply our methods in the context of electronic voting and consider existing voting schemes and how close they come to satisfying an appropriate modification of our definition, and, if a gap appears, whether the techniques developed in this paper could resolve it.

Acknowledgments. The second author acknowledges the support of an NSF graduate fellowship and of the Lucent Technologies GRPW program.

References

1. N. Asokan, V. Shoup, and M. Waidner. Optimistic fair exchange of digital signatures. *IEEE Journal on Selected Areas in Communications*, 18(4):591–610, 2000.
2. G. Ateniese, J. Camenisch, M. Joye, and G. Tsudik. A practical and provably secure coalition-resistant group signature scheme. In *CRYPTO 2000*, vol. 1880 of *LNCS*, pp. 255–270. Springer Verlag, 2000.
3. N. Barić and B. Pfitzmann. Collision-free accumulators and fail-stop signature schemes without trees. In *EUROCRYPT '97*, vol. 1233 of *LNCS*, pp. 480–494.
4. M. Bellare, J. A. Garay, and T. Rabin. Fast batch verification for modular exponentiation and digital signatures. In *EUROCRYPT '98*, vol. 1403 of *LNCS*, pp. 236–250. Springer Verlag, 1998.
5. J. C. Benaloh and D. Tuinstra. Receipt-free secret-ballot elections (extended abstract). In *Proc. 26th STOC*, pp. 544–553. ACM, 1994.
6. S. Brands. Untraceable off-line cash in wallets with observers. In *CRYPTO '93*, vol. 773 of *LNCS*, pp. 302–318, 1993.
7. J. Camenisch. Efficient anonymous fingerprinting with group signatures. In *ASIACRYPT 2000*, vol. 1976 of *LNCS*, pp. 415–428. Springer Verlag, 2000.
8. J. Camenisch and I. Damgård. Verifiable encryption, group encryption, and their applications to group signatures and signature sharing schemes. In *ASIACRYPT 2000*, vol. 1976 of *LNCS*, pp. 331–345, 2000.
9. J. Camenisch and A. Lysyanskaya. Efficient non-transferable anonymous multi-show credential system with optional anonymity revocation. In *EUROCRYPT 2001*, vol. 2045 of *LNCS*, pp. 93–118. Springer Verlag, 2001.
10. J. Camenisch and A. Lysyanskaya. An identity escrow scheme with appointed verifiers. `http://eprint.iacr.org/2001`, 2001.
11. J. Camenisch and M. Michels. A group signature scheme with improved efficiency. In *ASIACRYPT '98*, vol. 1514 of *LNCS*, pp. 160–174. Springer Verlag, 1998.
12. J. Camenisch and M. Stadler. Efficient group signature schemes for large groups. In *CRYPTO '97*, vol. 1296 of *LNCS*, pp. 410–424. Springer Verlag, 1997.

13. R. Canetti. *Studies in Secure Multiparty Computation and Applications.* PhD thesis, Weizmann Institute of Science, Rehovot 76100, Israel, June 1995.
14. R. Canetti. Security and composition of multi-party cryptographic protocols. *Journal of Cryptology*, 13(1):143–202, 2000.
15. D. Chaum. Untraceable electronic mail, return addresses, and digital pseudonyms. *Communications of the ACM*, 24(2):84–88, February 1981.
16. D. Chaum. Security without identification: Transaction systems to make big brother obsolete. *Communications of the ACM*, 28(10):1030–1044, Oct. 1985.
17. D. Chaum. Designated confirmer signatures. In *EUROCRYPT '94*, vol. 950 of *LNCS*, pp. 86–91. Springer Verlag Berlin, 1994.
18. D. Chaum, A. Fiat, and M. Naor. Untraceable electronic cash. In *CRYPTO '88*, vol. 403 of *LNCS*, pp. 319–327. Springer Verlag, 1990.
19. D. Chaum and E. van Heyst. Group signatures. In *EUROCRYPT '91*, vol. 547 of *LNCS*, pp. 257–265. Springer-Verlag, 1991.
20. L. Chen and T. P. Pedersen. New group signature schemes. In *EUROCRYPT '94*, vol. 950 of *LNCS*, pp. 171–181. Springer-Verlag, 1995.
21. R. Cramer, I. Damgård, and J. B. Nielsen. Multiparty computation from threshold homomorphic encryption. Manuscript. Available from `http://eprint.iacr.org`.
22. R. Cramer and V. Shoup. A practical public key cryptosystem provably secure against adaptive chosen ciphertext attack. In *CRYPTO '98*, vol. 1642 of *LNCS*, pp. 13–25, Berlin, 1998. Springer Verlag.
23. R. Cramer and V. Shoup. Signature schemes based on the strong RSA assumption. In *Proc. 6th ACM CCS*, pp. 46–52. ACM press, nov 1999.
24. I. Damgård. Efficient concurrent zero-knowledge in the auxiliary string model. In *EUROCRYPT 2000*, vol. 1807 of *LNCS*, pp. 431–444. Springer Verlag, 2000.
25. C. Dwork and A. Sahai. Concurrrent zero-knowledge: Reducing the need for timing constraints. In *CRYPTO '98*, vol. 1642 of *LNCS*, pp. 105–120, 1998.
26. A. Fiat and A. Shamir. How to prove yourself: Practical solution to identification and signature problems. In *CRYPTO '86*, vol. 263 of *LNCS*, pp. 186–194, 1987.
27. E. Fujisaki and T. Okamoto. Statistical zero knowledge protocols to prove modular polynomial relations. In *CRYPTO '97*, vol. 1294 of *LNCS*, pp. 16–30, 1997.
28. R. Gennaro, S. Halevi, and T. Rabin. Secure hash-and-sign signatures without the random oracle. In *EUROCRYPT '99*, vol. 1592 of *LNCS*, pp. 123–139, 1999.
29. D. M. Goldschlag, M. G. Reed, and P. F. Syverson. Onion routing for anonymous and private internet connections. *Communications of the ACM*, 42(2):84–88, 1999.
30. M. Hirt and K. Sako. Efficient receipt-free voting based on homomorphic encryption. In *EUROCRYPT 2000*, vol. 1807 of *LNCS*, pp. 539–556, 2000.
31. M. Jakobsson, K. Sako, and R. Impagliazzo. Designated verifier proofs and their applications. In *EUROCRYPT '96*, vol. 1233 of *LNCS*, 1996.
32. J. Kilian and E. Petrank. Identity escrow. In *CRYPTO '98*, vol. 1642 of *LNCS*, pp. 169–185, Berlin, 1998. Springer Verlag.
33. A. Lysyanskaya and Z. Ramzan. Group blind digital signatures: A scalable solution to electronic cash. In *Proc. Financial Cryptography*, 1998.
34. T. Okamoto and S. Uchiyama A new public-key cryptosystem as secure as factoring. In *EUROCRYPT '98*, vol. 1403 of *LNCS*, pp. 308–318, 1998.
35. P. Paillier. Public-key cryptosystems based on composite residuosity classes. In *EUROCRYPT '99*, vol. 1592 of *LNCS*, pp. 223–239. Springer Verlag, 1999.
36. T. P. Pedersen. Non-interactive and information-theoretic secure verifiable secret sharing. In *CRYPTO '91*, vol. 576 of *LNCS*, pp. 129–140. Springer Verlag, 1992.
37. B. Pfitzmann and M. Waidner. Composition and integrity preservation of secure reactive systems. In *Proc. 7th ACM CCS*, pp. 245–254. ACM press, nov 2000.

Session-Key Generation Using Human Passwords Only

Oded Goldreich* and Yehuda Lindell

Department of Computer Science and Applied Math,
Weizmann Institute of Science, Rehovot, ISRAEL.
{oded,lindell}@wisdom.weizmann.ac.il

Abstract. We present session-key generation protocols in a model where the legitimate parties share *only* a human-memorizable password. The security guarantee holds with respect to probabilistic polynomial-time adversaries that control the communication channel (between the parties), and may omit, insert and modify messages at their choice. Loosely speaking, the effect of such an adversary that attacks an execution of our protocol is comparable to an attack in which an adversary is only allowed to make a constant number of queries of the form "is w the password of Party A". We stress that the result holds also in case the passwords are selected at random from a small dictionary so that it is feasible (for the adversary) to scan the entire directory. We note that prior to our result, it was not clear whether or not such protocols were attainable without the use of random oracles or additional setup assumptions.

1 Introduction

This work deals with the oldest and probably most important problem of cryptography: enabling *private and reliable* communication among parties that use a public communication channel. Loosely speaking, *privacy* means that nobody besides the legitimate communicators may learn the data communicated, and *reliability* means that nobody may modify the contents of the data communicated (without the receiver detecting this fact). Needless to say, a vast amount of research has been invested in this problem. Our contribution refers to a difficult and yet natural setting of two parameters of the problem: the *adversaries* and the *initial set-up*.

We consider only probabilistic polynomial-time adversaries. Still even within this framework, an important distinction refers to the type of adversaries one wishes to protect against: *passive* adversaries only eavesdrop the channel, whereas *active* adversaries may also omit, insert and modify messages sent over the channel. Clearly, reliability is a problem only with respect to active adversaries (and holds by definition w.r.t passive adversaries). *We focus on active adversaries.*

* Supported by the MINERVA Foundation, Germany.

J. Kilian (Ed.): CRYPTO 2001, LNCS 2139, pp. 408–432, 2001.

The second parameter mentioned above is the initial set-up assumptions. Some assumption of this form must exist or else there is no difference between the legitimate communicators, called Alice and Bob, and the adversary (which may otherwise initiate a conversation with Alice pretending to be Bob). We list some popular initial set-up assumptions and briefly discuss what is known about them.

Public-key infrastructure: Here one assumes that each party has generated a secret-key and deposited a corresponding public-key with some trusted server(s). The latter server(s) may be accessed at any time by any user.

It is easy to establish private and reliable communication in this model (cf. [15,33]). (However, even in this case, one may want to establish "session keys" as discussed below.)

Shared (high-quality) secret keys: By *high-quality keys* we mean strings coming from distributions of high min-entropy (e.g., uniformly chosen 56-bit (or rather 192-bit) long strings, uniformly chosen 1024-bit primes, etc). Furthermore, these keys are selected by a suitable program, and cannot be memorized by humans.

In case a pair of parties shares such a key, they can conduct private and reliable communication (cf., [9,36,19,4]).

Shared (low-quality) secret passwords: In contrast to high-quality keys, *passwords* are strings that may be easily selected, memorized and typed-in by humans. An illustrating (and simplified) example is the case in which the password is selected uniformly from a relatively small dictionary; that is, the password is uniformly distributed in $\mathcal{D} \subset \{0,1\}^n$, where $|\mathcal{D}| = \text{poly}(n)$.

Note that using such a password in the role of a cryptographic key (in schemes as mentioned above) will yield a totally insecure scheme. A more significant observation is that the adversary may try to guess the password, and initiate a conversation with Alice pretending to be Bob and using the guessed password. So nothing can prevent the adversary from successfully impersonating Bob with probability $1/|\mathcal{D}|$. *But can we limit the adversary's success to about this much?*

The latter question is the focus of this paper.

Session-keys: The problem of establishing private and reliable communication is commonly reduced to the problem of generating a secure session-key (a.k.a "authenticated key exchange"). Loosely speaking, one seeks a protocol by which Alice and Bob may agree on a key (to be used throughout the rest of the current communication session) so that this key will remain unknown to the adversary.[1] Of course, the adversary may prevent such agreement (by simply blocking all communication), but this will be detected by either Alice or Bob.

[1] We stress that many famous key-exchange protocols, such as the one of Diffie and Hellman [15], refer to a passive adversary. In contrast, this paper refers to active adversaries.

1.1 What Security May Be Achieved Based on Passwords

Let us consider the related (although seemingly easier) task of *mutual authentication.* Here Alice and Bob merely want to establish that they are talking to one another. Repeating an observation made above, we note that if the adversary initiates $m \leq |\mathcal{D}|$ instances of the mutual authentication protocol, guessing a different password in each of them, then with probability $m/|\mathcal{D}|$ it will succeed in impersonating Alice to Bob (and furthermore find the password). The question posed above is rephrased here as follows:

> *Can one construct a password-based scheme in which the success probability of any probabilistic polynomial-time impersonation attack is bounded by $O(m/|\mathcal{D}|) + \mu(n)$, where m is the number of sessions initiated by the adversary, and $\mu(n)$ is a negligible function in the security parameter n?*

We resolve the above question in the affirmative. That is, assuming the existence of trapdoor one-way permutations, *we prove that schemes as above do exist* (for any $\mathcal{D}$ and specifically for $|\mathcal{D}| = \text{poly}(n)$). Our proof is constructive. We actually provide a protocol of comparable security for the more demanding goal of *authenticated session-key generation.*

Password-based authenticated session-key generation: Our definition for the task of authenticated session-key generation is based on the simulation paradigm. That is, we require that a secure protocol emulates an ideal execution of a session-key generation protocol (cf. [1,29,12]). In such an ideal execution, a trusted third party hands identical, uniformly distributed session-keys to the honest parties. The only power given to the adversary in this ideal model is to prevent the trusted party from handing keys to one of both parties. (We stress that, in this ideal model, the adversary learns *nothing* of the parties' joint password or output session-key).

Next, we consider a real execution of a protocol (where there is no trusted party and the adversary has full control over the communication channel between the honest parties). In general, a protocol is said to be secure if real-model adversaries can be emulated in the ideal-model such that the output distributions are computationally indistinguishable. Since in a password-only setting the adversary can always succeed with probability $1/|\mathcal{D}|$, it is impossible to achieve computational indistinguishability between the real model and above-described ideal model (where the adversary has zero probability of success). Therefore, in the context of a password-only setting, an authenticated session-key generation protocol is said to be secure if the above-mentioned ideal-model emulation results in an output distribution that can be distinguished from a real execution by (a gap of) at most $O(1/|\mathcal{D}|) + \mu(n)$.

Main result (informally stated): *Assuming the existence of trapdoor one-way permutations, there exists a secure authenticated session-key generation protocol in the password-only setting.*

The above (informal) definition implies the intuitive properties of authenticated session-key generation (e.g., security of the generated session-key and of the initial password). In particular, the output session-key can be distinguished from a random key by (a gap of) at most $O(1/|\mathcal{D}|)+\mu(n)$.[2] Similarly, the distinguishing gap between the parties' joint password and a uniformly distributed element in $\mathcal{D}$ is at most $O(1/|\mathcal{D}|)+\mu(n)$. (As we have mentioned, the fact that the adversary can distinguish with gap $O(1/|\mathcal{D}|)$ is an inherent limitation of password-based security.) The parties are also guaranteed that, except with probability $O(1/|\mathcal{D}|)+\mu(n)$, they either end-up with the same session-key or detect that their communication has been tampered with. Our definition also implies additional desirable properties of session-key protocols such as forward secrecy and security in the case of session-key loss (or known-key attacks). Furthermore, our protocol provides improved (i.e., negligible gap) security in case the adversary only eavesdrops the communication (during the protocol execution).

We mention that a suitable level of indistinguishability (of the real and ideal executions) holds when m sessions (referring to the same password) are conducted sequentially: in this case the distinguishing gap is $O(m/|\mathcal{D}|)+\mu(n)$ rather than $O(1/|\mathcal{D}|)+\mu(n)$ (which again is optimal). This holds also when any (polynomial) number of *other sessions w.r.t independently distributed passwords* are conducted concurrently to the above m sessions.

Caveat: Our protocol is proven secure only when assuming that the *same pair of parties* (using the same password) does not conduct several *concurrent* executions of the protocol. We stress that concurrent sessions of other pairs of parties (or of the same pair using a different password), are allowed. See further discussion in Sections 1.4 and 2.5.

1.2 Comparison to Prior Work

The design of secure mutual authentication and key-exchange protocols is a major effort of the applied cryptography community. In particular, much effort has been directed towards the design of *password-based* schemes that should withstand active attacks.[3] An important restricted case of the mutual authentication problem is the *asymmetric* case in which a human user authenticates himself to a server in order to *access* some service. The design of secure *access control* mechanisms based only on passwords is widely recognized as a central problem of computer practice and as such has received much attention.

[2] This implies that when using the session-key as a key to a MAC, the probability that the adversary can generate a valid MAC-tag to a message not sent by the legitimate party is small (i.e., $O(1/|\mathcal{D}|)$). Likewise, when using the session-key for private-key encryption, the adversary learns very little about the encrypted messages: for every partial-information function, the adversary can guess the value of the function applied to the messages with only small (i.e., $O(1/|\mathcal{D}|)$) advantage over the a-priori probability.

[3] A specific focus of this research has been on preventing *off-line dictionary attacks.* In such an off-line attack, the adversary records its view from past protocol executions and then scans the dictionary for a password consistent with this view. If checking consistency in this way is possible and the dictionary is small, then the adversary can derive the correct password. Clearly, a secure session-key generation protocol (as imformally defined above) withstands any off-line dictionary attack.

The first protocol suggested for password-based session-key generation was by Bellovin and Merritt [5]. This work was very influential and became the basis for much future work in this area [6,34,24,27,31,35]. However, these protocols have not been proven secure and their conjectured security is based on mere heuristic arguments. Despite the strong need for secure password-based protocols, the problem was not treated rigorously until quite recently. For a survey of works and techniques related to password authentication, see [28,26] (a brief survey can be found in [23]).

A first rigorous treatment of the *access control* problem was provided by Halevi and Krawczyk [23]. They actually considered an *asymmetric* hybrid model in which one party (the server) may hold a high-quality key and the other party (the human) may only hold a password. The human is also assumed to have secure access to a corresponding public-key of the server (either by reliable access to a reliable server or by keeping a "digest" of that public-key, which they call a *public-password*). The Halevi–Krawczyk model capitalizes on the asymmetry of the access control setting, and is inapplicable to settings in which communication has to be established between two humans (rather than a human and a server). Furthermore, requiring the human to keep the unmemorizable public-password (although not secretly) is undesirable even in the access control setting. Finally, we stress that the Halevi–Krawczyk model is a *hybrid* of the "shared-key model" and the "shared-password model" (and so their results don't apply to the "shared-password model"). Thus, it is of both theoretical and practical interest to answer the original question as posed above (i.e., without the public-password relaxation): *Is it possible to implement a secure access control mechanism (and authenticated key-exchange) based only on passwords*?

Positive answers to the original problem have been provided *in the random oracle model*. In this model, all parties are assumed to have oracle access to a totally random (universal) function [3]. Secure (password-based) access control schemes in the random oracle model were presented in [2,11]. The common interpretation of such results is that *security is $\mathcal{LIKELY}$ to hold* even if the random oracle is replaced by a ("reasonable") concrete function known explicitly to all parties. We warn that this interpretation is not supported by any sound reasoning. Furthermore, as pointed out in [14], there exist protocols that are secure in the random oracle model but become insecure if the random function is replaced by *any* specific function (or even a function uniformly selected from any family of functions).

To summarize, this paper is the first to present session-key generation (as well as mutual authentication) protocols *based only on passwords* (i.e., in the shared-password model), using only standard cryptographic assumptions (e.g., the existence of trapdoor one-way permutations, which in turn follows from the intractability assumption regarding integer factorization). We stress that prior to this work it was not clear whether such protocols exist at all (i.e., outside of the random oracle model).

Necessary conditions for mutual authentication: Halevi and Krawczyk [23] proved that mutual-authentication in the shared-password model implies (unauthenticated) secret-key exchange, which in turn implies one-way functions. Consequently, Boyarsky [10] pointed out that, in the shared-password model, mutual-authentication implies Oblivious Transfer.

1.3 Techniques

One central idea underlying our protocol is due to Naor and Pinkas [30]. They suggested the following protocol for the case of *passive* adversaries, using a secure protocol for polynomial evaluation.[4] In order to generate a session-key, party A first chooses a random linear polynomial $Q(\cdot)$ over a large field (which contains the dictionary of passwords). Next, A and B execute a secure polynomial evaluation in which B obtains $Q(w)$, where w is their joint password. The session-key is then set to equal $Q(w)$.

In [10] it was suggested to make the above protocol secure against *active* adversaries, by using non-malleable commitments. This suggestion was re-iterated to us by Moni Naor, and in fact our work grew out of his suggestion. In order to obtain a protocol secure against active adversaries, we augment the abovementioned protocol of [30] by several additional mechanisms. Indeed, we use non-malleable commitments [16], but in addition we also use a *specific* zero-knowledge proof [32], ordinary commitment schemes [7], a *specific* pseudorandom generator (of [9,36,8]), and message authentication schemes (MACs). The analysis of the resulting protocol is very complicated, *even when the adversary initiates a single session.* As explained below, we believe that these complications are unavoidable given the current state-of-art regarding *concurrent execution* of protocols.

Although not explicit in the problem statement, the problem we deal with actually concerns *concurrent* executions of a protocol. Even in case the adversary attacks a single session among two legitimate parties, its ability to modify messages means that it may actually conduct two *concurrent* executions of the protocol (one with each party).[5] Concurrent executions of some protocols were analyzed in the past, but these were relatively simple protocols. Although the high-level structure of our protocol can be simply stated in terms of a small number of modules, the currently known implementations of some of these modules are quite complex. Furthermore, these implementations are NOT known to be secure when two copies are executed *concurrently.* Thus, at the current state of affairs, the analysis cannot proceed by applying some composition theorems to (two-party) protocols satisfying some concurrent-security properties (because suitable concurrently-secure protocols and composition theorems are currently unknown). Instead, we have to analyze our protocol directly. We do so by reducing the analysis of (two concurrent executions of) our protocol to the analysis of non-concurrent executions of *related* protocols. Specifically, we show how a

[4] In the polynomial evaluation functionality, party A has a polynomial $Q(\cdot)$ over some finite field and Party B has an element x of the field. The evaluation is such that A learns nothing, and B learns $Q(x)$; i.e., the functionality is defined by $(Q, x) \mapsto (\lambda, Q(x))$.

[5] Specifically, the adversary may execute the protocol with Alice while claiming to be Bob, concurrently to executing the protocol with Bob while claiming to be Alice, where these two executions refer to the same joint Alice–Bob password.

successful adversary in the concurrent setting contradicts the security requirements in the non-concurrent setting. Such "reductions" are performed several times, each time establishing some property of the original protocol. Typically, the property refers to one of the two concurrent executions, and it is shown to hold even if the adversary is given some secrets of the legitimate party in the second execution. This is done by giving these secrets to the adversary, enabling him to effectively emulate the second execution internally. Thus, only the first execution remains and the relevant property is proven (in this standard non-concurrent setting). See Section 4 for an illustration of some of these proof techniques.

1.4 Discussion

We view our work as a theoretical study of the *very possibility* of achieving private and reliable communication among parties that share only a secret (low-quality) password and communicate over a channel that is controlled by an active adversary. Our main result is a demonstration of the *feasibility* of this task. That is, we demonstrate the *feasibility* of performing *session-key generation based only on* (low-quality) *passwords.* Doing so, this work is merely the first (rigorous) step in a research project directed towards providing a good solution to this practical problem. We discuss two aspects of this project that require further study.

Concurrent executions: Our protocol is proven secure only when the *same pair of parties* (using the same password) does not conduct several *concurrent* executions of the protocol. (We do allow concurrent executions that use different passwords.) Thus, actual use of our protocol requires a mechanism for ensuring that the same password is never used in concurrent executions. A simple mechanism enforcing the above is to disallow a party to enter an execution with a particular password if less than Δ units of time have passed since a previous execution with the same password. Furthermore, an execution must be completed within Δ units of time; that is, if Δ time units have elapsed then the execution is suspended. See Section 2.5 for further details. Indeed, it is desirable not to employ such a timing mechanism, and to prove that security holds also when many executions are conducted concurrently using the same password.

Efficiency: It is indeed desirable to have more efficient protocols than the one presented here. Some of our techniques may be useful towards this goal.

1.5 Independent Work

Independently of our work, Katz, Ostrovsky and Yung [25] presented a protocol for session-key generation based on passwords. Their protocol is incomparable to ours. On one hand, their protocol uses a stronger set-up assumption (i.e., public parameters selected by a trusted party), and a seemingly stronger intractability assumption (i.e., the Decisional Diffie-Hellman). On the other hand, their protocol seems practical and is secure in an unrestricted concurrent setting.

Recall that the thrust of our work is in demonstrating the feasibility of performing session-key generation based on passwords only (i.e., without any additional set-up assumptions).

2 Formal Setting

In this section we present notation and definitions that are specific to our setting, culminating in a definition of Authenticated Session-Key Generation. Given these, we state our main result.

2.1 Basic Notations

Typically, C denotes the *channel* (probabilistic polynomial-time adversary) via which parties A and B communicate. We adopt the notation of Bellare and Rogaway [4] and model the communication by giving C oracle access to A and B. We stress that, as in [4], these oracles have memory and model parties who participate in a session-key generation protocol. Unlike in [4], when A and B share a single password, C has oracle access to only a single copy of each party. We denote by $C^{A(x),B(y)}(\sigma)$, an execution of C (with auxiliary input σ) when it communicates with A and B, holding respective inputs x and y. Channel C's output from this execution is denoted by $\mathsf{output}(C^{A(x),B(y)}(\sigma))$.

The password dictionary is denoted by $\mathcal{D} \subseteq \{0,1\}^n$, and is fixed for the entire discussion. We let $\epsilon = \frac{1}{|\mathcal{D}|}$. We denote by U_n the uniform distribution over strings of length n. For a set S, we denote $x \in_R S$ when x is chosen *uniformly* from S. We use "ppt" as shorthand for probabilistic polynomial time. We denote an unspecified negligible function by $\mu(n)$. That is, for every polynomial $p(\cdot)$ and for all sufficiently large n's, $\mu(n) < \frac{1}{p(n)}$. For functions f and g (defined over the integers), we denote $f \approx g$ if $|f(n) - g(n)| < \mu(n)$. Finally, we denote computational indistinguishability by $\stackrel{c}{\equiv}$.

A security parameter n is often implicit in our notation and discussions. Thus, for example, by the notation $\mathcal{D}$ for the dictionary, our intention is actually $\mathcal{D}_n$ (where $\mathcal{D}_n \subseteq \{0,1\}^n$). Recall that we make no assumptions regarding the size of $\mathcal{D}_n$, and in particular it may by polynomial in n.

2.2 $(1-\epsilon)$-Indistinguishability and Pseudorandomness

Extending the standard definition of computational indistinguishability [22,36], we define the concept of $(1-\epsilon)$-indistinguishability. Two ensembles are $(1-\epsilon)$-indistinguishable if for every ppt machine, the probability of distinguishing between them (via a single sample) is at most negligibly greater than ϵ. (Note that $(1-\epsilon)$-indistinguishability is not preserved under multiple samples, but for efficiently constructible ensembles $(1-\epsilon)$-indistinguishability implies $(1-m\epsilon)$-indistinguishability of sequences of m samples.) Thus, computational indistinguishability coincides with 1-indistinguishability. The formal definition is as follows.

Definition 1 ($(1-\epsilon)$-indistinguishability): *Let $\epsilon : \mathsf{N} \to [0,1]$ be a function, and let $\{X_n\}_{n\in\mathsf{N}}$ and $\{Y_n\}_{n\in\mathsf{N}}$ be probability ensembles, so that for any n the distribution X_n (resp., Y_n) ranges over strings of length polynomial in n. We say that the ensembles are* $(1-\epsilon)$-indistinguishable, *denoted $\{X_n\}_{n\in\mathsf{N}} \stackrel{\epsilon}{\equiv} \{Y_n\}_{n\in\mathsf{N}}$, if for every probabilistic polynomial time distinguisher D, and all auxiliary information $z \in \{0,1\}^{\mathrm{poly}(n)}$*

$$|\Pr[D(X_n, 1^n, z) = 1] - \Pr[D(Y_n, 1^n, z) = 1]| < \epsilon + \mu(n)$$

We say that $\{X_n\}_{n\in\mathsf{N}}$ is $(1-\epsilon)$-pseudorandom if it is $(1-\epsilon)$-indistinguishable from $\{U_n\}_{n\in\mathsf{N}}$. The definition of pseudorandom functions [19] is similarly extended to $(1-\epsilon)$-pseudorandom functions.

2.3 Authenticated Session-Key Generation: Definition and Discussion

The problem of password-based authenticated session-key generation can be cast as a three-party functionality involving honest parties A and B, and an adversary C. Parties A and B should input their joint password and receive identical, uniformly distributed session-keys. On the other hand, the adversary C should have no output (and specifically should not obtain information on the password or output session-key). Furthermore, C should have no power to maliciously influence the outcome of the protocol (and thus, for example, cannot affect the choice of the key or cause the parties to receive different keys). However, recall that in a real execution, C controls the communication line between the (honest) parties. Thus, it can block all communication between A and B, and cause any protocol to fail. This (unavoidable) adversarial capability is modeled in the functionality by letting C input a single bit b indicating whether or not the execution is to be successful. Specifically, if $b = 1$ (i.e., success) then both A and B receive the above-described session-key. On the other hand, if $b = 0$ then A receives a session-key, whereas B receives a special abort symbol $\perp$ instead.[6] We stress that C is given no ability to influence the outcome beyond determining this single bit (i.e., b). In conclusion, the problem of password-based session-key generation is cast as the following three-party functionality:

$$(w_A, w_B, b) \mapsto \begin{cases} (U_n, U_n, \lambda) & \text{if } b = 1 \text{ and } w_A = w_B, \\ (U_n, \perp, \lambda) & \text{otherwise.} \end{cases}$$

where w_A and w_B are A and B's respective passwords.

Our definition for password-based authenticated session-key generation is based on the "simulation paradigm" (cf. [1,29,12]). That is, we require a secure protocol to emulate an *ideal* execution of the above session-key generation functionality. In such an ideal execution, communication is via a trusted third party

[6] This lack of symmetry in the definition is inherent as it is not possible to guarantee that A and B both terminate with the same "success/failure bit". For sake of simplicity, we (arbitrarily) choose to have A always receive a uniformly distributed session-key and to have B always output $\perp$ when $b = 0$.

who receives the parties inputs and (honestly) returns to each party its output, as designated by the functionality.

An important observation in the context of password-based security is that, in a real execution, an adversary can always attempt impersonation by simply guessing the secret password and participating in the protocol, claiming to be one of the parties. If the adversary's guess is correct, then impersonation always succeeds (and, for example, the adversary knows the generated session-key). Furthermore, by executing the protocol with one of the parties, the adversary can verify whether or not its guess is correct, and thus can learn information about the password (e.g., it can rule out an incorrect guess from the list of possible passwords). Since the dictionary may be small, this information learned by the adversary in a protocol execution may not be negligible at all. Thus, we cannot hope to obtain a protocol that emulates an ideal-model execution (in which C learns *nothing*) up to computational indistinguishability. Rather, the inherent limitation of password-based security is accounted for by (only) requiring that a real execution can be simulated in the ideal model such that the output distributions (in the ideal and real models) are $(1-O(\epsilon))$-*indistinguishable* (rather than 1-indistinguishable), where (as defined above) $\epsilon = 1/|\mathcal{D}|$.

We note that the above limitation applies only to active adversaries who control the communication channel. Therefore, in the case of a passive (eavesdropping) adversary, we demand that the ideal and real model distributions be computationally indistinguishable (and not just $(1 - O(\epsilon))$-indistinguishable). We now define the ideal and real models and present the formal definition of security.

The ideal model: Let $\hat{A}$ and $\hat{B}$ be honest parties and let $\hat{C}$ be any ppt ideal-model adversary (with arbitrary auxiliary input σ). An ideal-model execution proceeds in the following phases:

Initialization: A password $w \in_R \mathcal{D}$ is uniformly chosen from the dictionary and given to both $\hat{A}$ and $\hat{B}$.

Sending inputs to trusted party: $\hat{A}$ and $\hat{B}$ both send the trusted party the password they have received in the initialization stage. The adversary $\hat{C}$ sends either 1 (denoting a successful protocol execution) or 0 (denoting a failed protocol execution).

The trusted party answers all parties: In the case $\hat{C}$ sends 1, the trusted party chooses a uniformly distributed string $k \in_R \{0,1\}^n$ and sends k to both $\hat{A}$ and $\hat{B}$. In the case $\hat{C}$ sends 0, the trusted party sends $k \in_R \{0,1\}^n$ to $\hat{A}$ and $\perp$ to $\hat{B}$. In both cases, $\hat{C}$ receives no output.[7]

The ideal distribution is defined as follows:

$$\mathsf{ideal}_{\hat{C}}(\mathcal{D},\sigma) \stackrel{\text{def}}{=} (w, \text{output}(\hat{A}), \text{output}(\hat{B}), \text{output}(\hat{C}(\sigma)))$$

where $w \in_R \mathcal{D}$ is the input given to $\hat{A}$ and $\hat{B}$ in the initialization phase. Thus,

[7] Since $\hat{A}$ and $\hat{B}$ are always honest, we need not deal with the case that they hand the trusted third party different passwords.

$$\mathsf{ideal}_{\hat{C}}(\mathcal{D},\sigma) = \begin{cases} (w, U_n, U_n, \text{output}(\hat{C}(\sigma))) & \text{if send}(\hat{C}(\sigma)) = 1, \\ (w, U_n, \bot, \text{output}(\hat{C}(\sigma))) & \text{otherwise.} \end{cases}$$

where $\text{send}(\hat{C}(\sigma))$ denotes the value sent by $\hat{C}$ (to the trusted party), on auxiliary input σ.

The real model: Let A and B be honest parties and let C be any ppt real-model adversary with arbitrary auxiliary input σ. As in the ideal model, the real model begins with an initialization stage in which both A and B receive an identical, uniformly distributed password $w \in_R \mathcal{D}$. Then, the protocol is executed with A and B communicating via C.[8] The execution of this protocol is denoted $C^{A(w),B(w)}(\sigma)$ and we augment C's view with the accept/reject decision bits of A and B (this decision bit denotes whether a party's private output is a session-key or $\bot$). This formal requirement is necessary, since in practice this information can be implicitly understood from whether or not the parties continue communication after the session-key generation protocol has terminated. (We note that in our specific formulation, A always accepts and thus it is only necessary to provide C with the decision-bit output by B.) The real distribution is defined as follows:

$$\mathsf{real}_C(\mathcal{D},\sigma) \stackrel{\text{def}}{=} (w, \text{output}(A), \text{output}(B), \text{output}(C^{A(w),B(w)}(\sigma)))$$

where $w \in_R \mathcal{D}$ is the input given to A and B in the initialization phase.

The definition of security: Loosely speaking, the definition requires that a secure protocol (in the real model) emulates the ideal model (in which a trusted party participates). This is formulated by saying that adversaries in the ideal model are able to simulate the execution of a real protocol, so that the input/output distribution of the simulation is $(1 - O(\epsilon))$-indistinguishable from in a real execution. We further require that passive adversaries can be simulated in the ideal-model so that the output distributions are computationally indistinguishable (and not just $(1 - O(\epsilon))$-indistinguishable).[9]

Definition 2 (password-based authenticated session-key generation): *A protocol for* password-based authenticated session-key generation *is secure if the following two requirements hold:*

[8] We stress that there is a fundamental difference between the real model as defined here and as defined in standard multi-party computation. Here, the parties A and B do *not have the capability of communicating directly with each other.* Rather, A can only communicate with C and likewise for B. This is in contrast to standard multi-party computation where all parties have direct communication links or where a broadcast channel is used.

[9] A passive adversary is one that does not modify, omit or insert any messages sent between A or B. That is, it can only eavesdrop and thus is limited to analyzing the transcript of a protocol execution between two honest parties. Passive adversaries are also referred to as *semi-honest* in the literature (e.g., in [21]).

1. Passive adversaries: *For every ppt real-model* passive *adversary C there exists a ppt ideal-model adversary $\hat{C}$ such that for every dictionary $\mathcal{D} \subseteq \{0,1\}^n$ and every auxiliary input $\sigma \in \{0,1\}^{\text{poly}(n)}$*

$$\left\{\mathsf{ideal}_{\hat{C}}(\mathcal{D},\sigma)\right\}_{\mathcal{D},\sigma} \stackrel{\text{c}}{\equiv} \left\{\mathsf{real}_{C}(\mathcal{D},\sigma)\right\}_{\mathcal{D},\sigma}$$

2. Arbitrary (active) adversaries: *For every ppt real-model adversary C there exists a ppt ideal-model adversary $\hat{C}$ such that for every dictionary $\mathcal{D} \subseteq \{0,1\}^n$ and every auxiliary input $\sigma \in \{0,1\}^{\text{poly}(n)}$*

$$\left\{\mathsf{ideal}_{\hat{C}}(\mathcal{D},\sigma)\right\}_{\mathcal{D},\sigma} \stackrel{O(\epsilon)}{\equiv} \left\{\mathsf{real}_{C}(\mathcal{D},\sigma)\right\}_{\mathcal{D},\sigma}$$

where $\epsilon \stackrel{\text{def}}{=} \frac{1}{|\mathcal{D}|}$. We stress that the constant in $O(\epsilon)$ is a universal one.

Properties of Definition 2: Definition 2 asserts that the joint input/output distribution from a real execution is at most "$O(\epsilon)$-far" from an ideal execution in which the adversary learns nothing (and has no influence on the output except to cause B to reject). This immediately implies that the output session-key is $(1 - O(\epsilon))$-pseudorandom (which, as we have mentioned, is the best possible for password-based key generation). Thus, if such a key is used for encryption then for any (partial information) predicate P, the probability that an adversary learns $P(m)$ given the ciphertext $E(m)$ is at most $O(\epsilon) + \mu(n)$ greater than the a-priori probability (when the adversary is not given $E(m)$). Likewise, if the key is used for a message authentication code (MAC), then the probability that an adversary can generate a correct MAC-tag on a message not sent by A or B is at most negligibly greater than $O(\epsilon)$. We stress that the security of the output session-key does not deteriorate with its usage; that is, it can be used for polynomially-many encryptions or MACs and the security remains $O(\epsilon)$. Another important property of Definition 2 is that, except with probability $O(\epsilon)$, (either one party detects failure or) both parties terminate with the *same* session-key.

Definition 2 also implies that the password used remains $(1-O(\epsilon))$-indistinguishable from a randomly chosen (new) password $\tilde{w} \in_R \mathcal{D}$. (This can be seen from the fact that in the ideal model, the adversary learns nothing of the password w, which is part of the ideal distribution.) In particular, this implies that a secure protocol is resistant to offline dictionary attacks (whereby an adversary scans the dictionary in search of a password that is "consistent" with its view of a protocol execution).

Other desirable properties of session-key protocols are also guaranteed by Definition 2. Specifically, we mention *forward secrecy* and security in the face of *loss of session-keys* (also known as *known-key attacks*). Forward secrecy states that the session-key remains secure even if the password is revealed after the protocol execution. Analogously, security in the face of loss of session-keys means that the password and the current session-key maintain their security even if prior session-keys are revealed. These properties are immediately implied by the fact that, in the ideal-model, there is no dependence between the session-key and the password and between session-keys from different sessions. Thus, learning the password does not compromise the security of the session-key and visa versa.[10]

An additional property that is desirable is that of *intrusion detection*. That is, if the adversary modifies any message sent in a session, then with probability at least $(1 - O(\epsilon))$ this is detected and at least one party rejects. This property is not guaranteed by Definition 2 itself; however, it does hold for our protocol. Combining this with Item 1 of Definition 2 (i.e., the requirement regarding passive adversaries), we conclude that in order for C to take advantage of its ability to learn "$O(\epsilon)$-information" C must expose itself to the danger of being detected with probability $1 - O(\epsilon)$.

Finally, we observe that the above definition also enables mutual-authentication. This is because A's output session-key is always $(1 - O(\epsilon))$-pseudorandom to the adversary. As this key is secret, it can be used for explicit authentication via a (mutual) challenge/response protocol.[11] By adding such a step to any secure session-key protocol, we obtain explicit mutual-authentication.

Augmenting the definition: Although Definition 2 seems to capture all that is desired from authenticated session-key generation, there is a subtlety that it fails to address (as pointed out by Rackoff to the authors of [4]). The issue is that the two parties do not necesssarily terminate the session-key generation protocol simultaneously, and so one party may terminate the protocol and start using the session-key while the other party is still executing instructions of the session-key generation protocol (i.e., determining its last message). In this extended abstract, we note only that Definition 2 can be augmented to deal with this issue, and that our protocol is secure also with respect to the augmented definition. A full treatment of this issue is provided in the full version of the paper.

2.4 Our Main Result

Given Definition 2, we can now formally state our main result.

Theorem 3 *Assuming the existence of trapdoor permutations, there exist secure protocols for password-based authenticated session-key generation.*

2.5 Multi-session Security

The definition above relates to two parties executing a session-key generation protocol once. Clearly, we are interested in the more general case where many different parties run the protocol any number of times. It turns out that any

[10] The independence of session-keys from different sessions relates to the multi-session case, which is discussed in Section 2.5. For now, it is enough to note that the protocol behaves as expected in that after t executions of the real protocol, the password along with the outputs from all t sessions are $(1 - O(t\epsilon))$-indistinguishable from t ideal executions.

[11] It is easy to show that such a key can be used directly to obtain a $(1 - O(\epsilon))$-pseudorandom function, which can then be used in a standard challenge/response protocol.

protocol that is secure for a single invocation between two parties (i.e., as in Definition 2), is secure in the multi-party and sequential invocation case.

Many Invocations by Two Parties. Let A and B be parties who invoke t sequential executions of a session-key generation protocol. Given that we wish that an adversary gains no more than $O(1)$ password guesses upon each invocation, the security upon the t'th invocation should be $O(t\epsilon)$. That is, we consider ideal and real distributions consisting of the outputs from all t executions. Then, we require that these distributions be $(1 - O(t\epsilon))$-indistinguishable. It can be shown that any secure protocol for password-based authenticated session-key generation maintains $O(t\epsilon)$ security after t sequential invocations. Details are given in the full version of this work.

Sequential vs Concurrent Executions for Two Parties: Our solution is proven secure only if A and B do not invoke *concurrent* executions of the session-key generation protocol (with the same password). We stress that a scenario whereby the adversary invokes B twice or more (sequentially) during a single execution with A is not allowed. Therefore, in order to actually use our protocol, some mechanism must be used to ensure that such concurrent executions do not take place. This can be achieved by having A and B wait Δ units of time between protocol executions (where Δ is greater than the time taken to run a single execution). Note that parties do not usually need to initiate session-key generation protocols immediately one after the other. Therefore, this delay mechanism need only be employed when an attempted session-key generation execution fails. This means that parties not "under attack" by an adversary are not inconvenienced in any way.

We note that this limitation does *not* prevent the parties from opening a number of different (independently-keyed) communication lines. They may do this by running the session-key protocol *sequentially*, once for each desired communication line. However, in this case, they incur a delay of Δ units of time between each execution. Alternatively, they may run the protocol once and obtain a $(1 - O(\epsilon))$-pseudorandom session-key. This key may then be used as a shared, high-quality key for (concurrently) generating any polynomial number of $(1 - O(\epsilon))$-pseudorandom session-keys; one for each communication line (simple and efficient protocols exist for this task, see [4]).

Many Parties. In the case where many parties execute the session-key protocol simultaneously, we claim that for m invocations of the protocol (which must be sequential for the same pair of parties and may be *concurrent* otherwise), the security is $O(m\epsilon)$. We assume that different pairs of parties (executing concurrently) have independently distributed passwords. Then, the security is derived from the single-session case by noting that sessions with independently distributed passwords can be perfectly simulated by an adversary.

3 Our Session-Key Generation Protocol

All arithmetic below is over the finite field $\mathrm{GF}(2^n)$ which is identified with $\{0,1\}^n$. In our protocol, we use a secure protocol for evaluating *non-constant, linear polynomials* (actually, we could use any 1–1 Universal$_2$ family of hash functions). This protocol involves two parties A and B; party A has a non-constant, linear polynomial $Q(\cdot) \in \{0,1\}^{2n}$ and party B has a string $x \in \{0,1\}^n$. The functionality is $(Q,x) \mapsto (\lambda, Q(x))$; that is, A receives nothing and B receives the value $Q(x)$ (and nothing else). The fact that A is supposed to input a non-constant, linear polynomial can be enforced by simply mapping all possible input strings to the set of such polynomials (this convention is used for all references to polynomials from here on). We actually augment this functionality by having A also input a commitment to the polynomial Q (i.e., $c_A \in \mathrm{Commit}(Q)$) and its corresponding decommitment r (i.e., $c_A = C(Q,r)$). Furthermore, B also inputs a commitment value c_B. The augmentation is such that if $c_A \neq c_B$, then B receives a special failure symbol. This is needed in order to tie the polynomial evaluation to a value previously committed to in the main (higher level) protocol. The functionality is defined as follows:

Definition 4 (augmented polynomial evaluation):

- **Input:** *Party A inputs a commitment c_A and its corresponding decommitment r, and a linear, non-constant polynomial Q. Party B inputs a commitment c_B and a value x.*
- **Output:**
  1. Correct Input Case: *If $c_A = c_B$ and $c_A = C(Q,r)$, then B receives $Q(x)$ and A receives nothing.*
  2. Incorrect Input Case: *If $c_A \neq c_B$ or $c_A \neq C(Q,r)$, then B receives a special failure symbol, denoted $\perp$, and A receives nothing.*

We note that by [37,21], this functionality can be securely computed (observe that the input conditions can be checked in polynomial time because A also provides the decommitment r).

3.1 The Protocol

Let f be a one-way permutation and b a hard-core of f.

Protocol 5 (password-based authenticated session-key generation)

- **Input:** Parties A and B begin with a joint password w, which is supposed to be uniformly distributed in $\mathcal{D}$.
- **Output:** A and B each output an accept/reject bit as well as session-keys k_A and k_B respectively (where k_A "should" equal k_B).
- **The Protocol:**
  1. **Stage 1: (Non-Malleable) Commit**
     a) A chooses a random, linear, non-constant polynomial Q over $\mathrm{GF}(2^n)$.

b) A and B engage in a *non-malleable* (perfectly binding) commitment protocol in which A commits to the string $(Q, w) \in \{0,1\}^{3n}$. Denote the random coins used by B in the commitment protocol by r_B and denote B's view of the execution of the commitment protocol by $NMC(Q, w)$.[12]
Following the commitment protocol, B sends his random coins r_B to A. (This has no effect on the security, since the commitment scheme is perfectly binding and the commitment protocol has already terminated.)

2. **Stage 2: Pre-Key Exchange** – In this stage the parties "exchange" strings τ_A and τ_B, from which the output session-keys (as well as validation checks) are *derived.* Thus, τ_A and τ_B are called pre-keys.
   a) A sends B a commitment $c = C(Q, r)$, for a randomly chosen r.
   b) A and B engage in an augmented polynomial evaluation protocol. A inputs Q and (c, r); B inputs w and c.
   c) We denote B's output by τ_B. (Note that τ_B is supposed to equal $Q(w)$.)
   d) A internally computes $\tau_A = Q(w)$.
3. **Stage 3: Validation**
   a) A sends the string $y = f^{2n}(\tau_A)$ to B.
   b) A proves to B in zero-knowledge that she input the same polynomial in both the non-malleable commitment (performed in Stage 1) and the ordinary commitment (performed in Stage 2(a)), and that the value y is "consistent" with the non-malleable commitment. Formally, A proves the following statement:

   There exists a string $(X_1, x_2) \in \{0,1\}^{3n}$ and random coins $r_{A,1}, r_{A,2}$ (where $r_{A,1}$ and $r_{A,2}$ are A's random coins in the non-malleable and ordinary commitments, respectively) such that
      i. B's view of the non-malleable commitment, $NMC(Q, w)$, is identical to the receiver's view of a non-malleable commitment to (X_1, x_2), where the sender and receiver's respective random coins are $r_{A,1}$ and r_B. (Recall that r_B denotes B's random coins in the non-malleable commitment.)[13]
      ii. $c = C(X_1, r_{A,2})$, and
      iii. $y = f^{2n}(X_1(x_2))$.

[12] Recall that B's view consists of his random coins and all messages received during the commitment protocol execution.

[13] The view of a protocol execution is a function of the parties' respective inputs and random strings. Therefore, (X_1, x_2), $r_{A,1}$ and r_B define a single possible view. Furthermore, recall that B sent r_B to A following the commitment protocol. Thus A has $NMC(Q, w)$ (which includes r_B), the committed-to value (Q, w) and $r_{A,1}$, enabling her to efficiently prove the statement.

The zero-knowledge proof used here is the *specific* zero-knowledge proof of Richardson and Kilian [32], with a *specific* setting of parameters.[14]

c) Let t_A be the entire session transcript as seen by A (i.e., the sequence of all messages sent and received by A) and let MAC_k be a message authentication code, keyed by k. Then, A computes $k_1(\tau_A) \stackrel{\text{def}}{=} b(\tau_A) \cdots b(f^{n-1}(\tau_A))$, and sends $m = MAC_{k_1(\tau_A)}(t_A)$ to B.

4. **Decision Stage**

a) A always accepts and outputs $k_2(\tau_A) \stackrel{\text{def}}{=} b(f^n(\tau_A)) \cdots b(f^{2n-1}(\tau_A))$.

b) B accepts if and only if all the following conditions are fulfilled:
- $y = f^{2n}(\tau_B)$, where y is the string sent by A to B in Step 3(a) above and τ_B is B's output from the polynomial evaluation. (Note that if $\tau_B = \perp$ then no y fulfills this equality, and B always rejects.)
- B accepts the zero-knowledge proof in Step 3(b) above, and
- $\text{Verify}_{k_1(\tau_B)}(t_B, m) = 1$, where t_B is the session-transcript as seen by B, the string m is the alleged MAC-tag that B receives, and verification is with respect to the MAC-key defined by $k_1(\tau_B) = b(\tau_B) \cdots b(f^{n-1}(\tau_B))$.

If B accepts, then he outputs $k_2(\tau_B) = b(f^n(\tau_B)) \cdots b(f^{2n-1}(\tau_B))$, otherwise he outputs $\perp$. (Recall that the accept/reject decision bit is considered a public output.)

We stress that A and B always accept or reject based solely on these criteria, and that they do not halt (before this stage) even if they detect malicious behavior.

See Figure 1 below for a schematic diagram of Protocol 5.

In our description of the protocol, we have referred only to parties A and B. That is, we have ignored the existence (and possible impact) of the channel C. That is, when A sends a string z to B, we "pretend" that B actually received z and not something else. In a real execution, this may not be the case at all. In the actual analysis we will subscript every value by its owner, as we have done for τ_A and τ_B in the protocol. For example, we shall say that in Step 3(a), A sends a string y_A and the string received by B is y_B.

3.2 Motivation for the Security of the Protocol

The central module of Protocol 5 is the secure polynomial evaluation. This, in itself, is enough for achieving security against *passive* channels only. Specifically,

[14] The setting of parameters referred to relates to the number of iterations m in the first part of the Richardson-Kilian proof. We set m to equal the number of rounds in all other parts of our protocol plus any non-constant function of the security parameter.

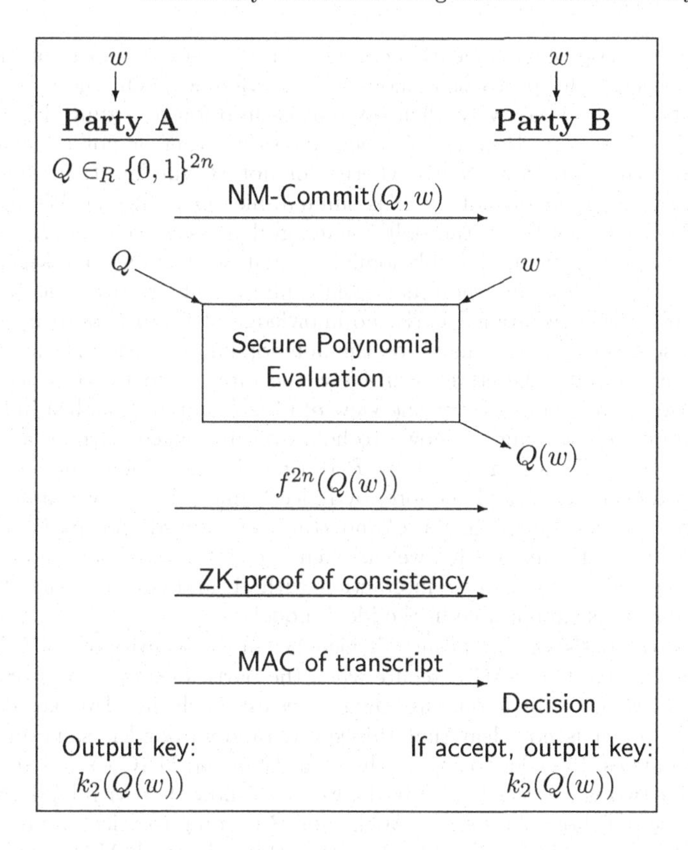

Fig. 1. Schematic Diagram of the Protocol.

consider the following protocol. Party A chooses a random, linear polynomial Q and inputs it into a secure polynomial evaluation with party B who inputs the joint password w. By the definition of the polynomial evaluation, B receives $Q(w)$ and A receives nothing. Next, A internally computes $Q(w)$ (she can do this as she knows both Q and w), and both parties use this value as the session-key. The key is uniformly distributed (since Q is random and linear) and due to the secrecy requirements of the polynomial evaluation, the protocol reveals nothing of w or $Q(w)$ to a passive eavesdropper C (since otherwise this would also be revealed to party A who should learn nothing from the evaluation).

One key problem in extending the above argument to our setting (where C may be active) is that the security definitions of two-party computation guarantee nothing about the simulatability of C's view in this concurrent setting. We now provide some intuition into how simulation of our protocol is nevertheless achieved. First, assume that the MAC-value sent by A at the conclusion of the protocol is such that unless C behaved passively (and relayed all message without modification), then B rejects (with some high probability). Now, if C

behaves passively, then B clearly accepts (as in the case of honest parties A and B that execute the protocol without any interference). On the other hand, if C does not behave passively, then (by our assumption regarding the security of the MAC) B rejects. However, C itself knows whether or not it behaved passively and therefore can predict whether or not B will reject. In other words, the accept/reject bit output by B is simulatable (by C itself). We proceed by observing that this bit is the only meaningful message sent by B during the protocol: apart from in the polynomial evaluation, the only messages sent by B are as the *receiver* in a non-malleable commitment protocol and the *verifier* in a zero-knowledge proof (clearly, no knowledge of the password w is used by B in these protocols). Furthermore, the polynomial evaluation is such that only B receives output. Therefore, intuitively, the input used by B is not revealed by the execution; equivalently, the view of C is (computationally) independent of B's input w (this can be shown to hold even in our concurrent setting). We conclude that all messages sent by B during the execution can be simulated without knowledge of w. Therefore, by indeed simulating B, we can reduce the *concurrent* scenario involving A, C and B to a (standard) two-party setting between A and C. In this setting, we can then apply standard tools and techniques for simulating C's view in its interaction with A, and conclude that the entire real execution is simulatable in the ideal model.

Thus, the basis for simulating C's view lies in the security of the MAC in our scenario. Indeed, the MAC is secure when the parties using it (a priori) share a random MAC-key; but in our case the parties establish the MAC-key during the protocol, and it is not clear that this key is random nor the same in the view of both parties. In order to justify the security of the MAC (in our setting), we show that two properties hold. Firstly, we must show that with high probability either A and B hold the same MAC key or B is going to reject anyhow (and C knows this). Secondly, we need to show that this (identical) MAC-key held by A and B has "sufficient pseudorandomness" to prevent C from successfully forging a MAC. The proof of these properties (especially the first one) is very involved and makes up a major part of the proof, which is presented in the full version of this work.

3.3 Properties of Protocol 5

The main properties of Protocol 5 are captured by the following theorem.

Theorem 6 *Protocol 5 constitutes a secure protocol for password-based authenticated session-key generation (as defined in Definition 2).*

All the cryptographic tools used in Protocol 5 can be securely implemented assuming the existence of trapdoor permutations. Thus, at the very least, Theorem 6 implies the feasibility result captured by Theorem 3.

Unfortunately, due to lack of space in this abstract, we do not provide a proof of Theorem 6. However, a demonstration of some of the proof techniques used to prove Theorem 6 is provided in Section 4.

4 An Illustration of Our Proof Techniques

In this section, we illustrate our proof techniques for a simplified scenario in which A and B execute a secure polynomial evaluation only, while communicating via an adversarial channel C. Recall that the polynomial evaluation functionality is defined (in the stand-alone setting) by $(Q, x) \mapsto (\lambda, Q(x))$. That is, A has a polynomial $Q(\cdot)$ over some finite field and B has an element x in that field. The evaluation is such that A learns nothing while B obtains $Q(x)$. In the scenario that we are considering, A's input is a random, linear polynomial and B's input is a random password $w \in_R \mathcal{D}$ (as is the case in Protocol 5).

Recall that in this setting C may omit, insert and modify any message sent between A and B. Thus, in a sense C conducts two *separate* executions of the polynomial evaluation: one with A in which C impersonates B (called the (A, C)-execution), and one with B in which C impersonates A (called the (C, B)-execution). These two executions are carried out *concurrently* (by C), and there is no explicit execution between A and B.

We remind the reader that the definition of (stand-alone) secure two-party computation *does not apply* to the concurrent setting that we consider here. Furthermore, there are currently no tools for dealing with (general) *concurrent* computation in the two-party case. Therefore, our analysis of these executions uses specific properties of the protocol to remove the concurrency and obtain a reduction to the stand-alone setting. That is, we show how an adversarial success in the concurrent setting can be translated into a related adversarial success in the stand-alone setting. This enables us to analyze the adversary's capability in the concurrent setting, based on the security of two-party stand-alone protocols.

In order to demonstrate our proof techniques, we show that C learns "little" of w and $Q(w)$ from the above concurrent execution. Our formal statement of this has an ideal-model/real-model flavor. Specifically, we show that for every ppt adversary C interacting with A and B, there exists a *non-interactive* ppt machine $\hat{C}$ (who receives no input or output), such that $\{w, Q(w), \mathsf{output}(C^{A(Q),B(w)}\}$ is $(1-\epsilon)$-indistinguishable from $\{w, U_n, \mathsf{output}(\hat{C})\}$.[15] (Recall that $C^{A(Q),B(w)}$ denotes an execution of C with A and B holding respective inputs Q and w.) One can think of C as being a real-model adversary and $\hat{C}$ an ideal-model adversary, where in this ideal model $\hat{C}$ sends no input to the trusted third party and likewise receives no output. We note that such a view is rather simplistic as we claim nothing here regarding the outputs of A and B from the execution (as is usually required in secure computation). In other words, here we prove a statement regarding privacy, but make no claims to correctness; for example, there is no guarantee that C does not maul or skew the parties' outputs in some undesired way. Formally, we prove the following:

Theorem 7 (illustration): *For every ppt adversarial channel C interacting with A and B, there exists a ppt machine $\hat{C}$ (interacting with nobody) such that for every dictionary $\mathcal{D} \subseteq \{0,1\}^n$,*

[15] As in Definition 2, this implies that following the execution, with respect to C's view, the password w is $(1-\epsilon)$-indistinguishable from a (new) randomly chosen password $\tilde{w}$. It also implies that the value $Q(w)$ (used in Protocol 5 to derive the MAC and session keys) is $(1-\epsilon)$-pseudorandom with respect to C's view.

$$\left\{w, Q(w), \mathsf{output}(C^{A(Q),B(w)})\right\} \stackrel{\epsilon}{\equiv} \left\{w, U_n, \mathsf{output}(\hat{C})\right\}$$

where $w \in_R \mathcal{D}$, Q is a random linear polynomial, and $\epsilon = \frac{1}{|\mathcal{D}|}$.

Proof: We prove the theorem by first showing how the (C, B) execution can be simulated so that C's view in the simulation is negligibly close to in a real interaction. Then, we remain with a stand-alone execution between A and C only. In this scenario, we apply the standard definition of secure two-party computation to conclude that C learns at most "ϵ-information" about w and $Q(w)$. The fact that the (C, B) execution can be simulated is formally stated as follows (in the statement of the lemma below, $C'^{A(Q)}$ denotes a stand-alone execution of C with A upon input Q):

Lemma 8 (simulating the (C, B) execution): *For every ppt adversary C interacting with both A and B, there exists a ppt adversary C' interacting with A only, such that for every dictionary $\mathcal{D} \subseteq \{0,1\}^n$,*

$$\left\{w, Q(w), \mathsf{output}(C^{A(Q),B(w)})\right\} \stackrel{c}{\equiv} \left\{w, Q(w), \mathsf{output}(C'^{A(Q)})\right\}$$

where $w \in_R \mathcal{D}$ and Q is a random linear polynomial.

Proof: Loosely speaking, we prove this lemma by showing that B's role in the (C, B) execution can be simulated without any knowledge of w. Thus, C' is able to simulate B's role for C and we obtain the lemma. We begin by showing that C learns nothing of B's input w from the (C, B) polynomial evaluation. This is trivial in a stand-alone setting by the definition of the functionality; here we claim that it also holds in our concurrent setting. Formally, we show that if B were to use some fixed $w' \in \mathcal{D}$ instead of the password w, then this is indistinguishable to C (when also interacting concurrently with A). That is,

$$\left\{w, Q(w), \mathsf{output}(C^{A(Q),B(w)})\right\} \stackrel{c}{\equiv} \left\{w, Q(w), \mathsf{output}(C^{A(Q),B(w')})\right\} \quad (1)$$

where $w \in_R \mathcal{D}$ is a random password and $w' \in \mathcal{D}$ is fixed. Later, we will use Eq. (1) in order to show how C' simulates the (C, B) execution for C. First, we prove Eq. (1) by reducing C's concurrent execution with A and B to a stand-alone two-party setting between C and B only. This reduction is obtained by giving the adversary C the polynomial Q. Now, C has A's entire input and can *perfectly emulate* the (A, C) execution by itself. Formally, there exists an adversary C'', given auxiliary input Q, and interacting with B only, such that the following two equations hold:

$$\left\{w, Q(w), \mathsf{output}(C^{A(Q),B(w)})\right\} \equiv \left\{w, Q(w), \mathsf{output}(C''^{B(w)}(Q))\right\} \quad (2)$$

$$\left\{w, Q(w), \mathsf{output}(C^{A(Q),B(w')})\right\} \equiv \left\{w, Q(w), \mathsf{output}(C''^{B(w')}(Q))\right\} \quad (3)$$

where $C''^{B(w)}(Q)$ denotes a stand-alone execution of C'' (given input Q) with B (who has input w). Machine C'' works by playing A's role in the (A, C)-execution and forwarding all messages belonging to the (C, B)-execution between C and B (notice that C'' can play A's role because it knows Q). We therefore remain with a stand-alone setting between C'' (given auxiliary input Q) and B, in which B inputs either w or w' into the polynomial evaluation. In this stand-alone setting, the security of the polynomial evaluation guarantees that C'' can distinguish the input cases with at most negligible probability. That is, for every ppt adversary C'', it holds that

$$\left\{w, Q(w), \mathsf{output}(C''^{B(w)}(Q))\right\} \stackrel{c}{\equiv} \left\{w, Q(w), \mathsf{output}(C''^{B(w')}(Q))\right\} \quad (4)$$

Eq. (1) then follows by combining Equations (2), (3) and (4). In summary, we have shown that even in our concurrent setting where C interacts with both A and B, the adversary C cannot distinguish the cases that B inputs w or w'.

We are now ready to show how C' works (recall that C' interacts with A only and its aim is to simulate a concurrent execution with A and B for C). Machine C' begins by selecting an arbitrary $w' \in \mathcal{D}$. Then, C' perfectly emulates an execution of $C^{A(Q),B(w')}$ by playing B's role in the (C, B) execution and forwarding all messages belonging to the (A, C) execution between A and C (C' can play B's role here because w' is known to it). By Eq. (1) we conclude that this emulation is computationally indistinguishable from a real execution of $C^{A(Q),B(w)}$. This completes the proof of the lemma. ■

(We remark that the proof of Lemma 8 is typical of many of our proofs. Our goal is to obtain a reduction from the concurrent setting to the stand-alone setting between A and C, and we obtain this reduction by simulating B. However, in order to show that this simulation is "good" we first reduce the concurrent setting to a stand-alone setting *between C and B by simulating A*.)

It remains to show that C''s view of its (stand-alone) interaction with A can be simulated and that in this interaction, C' learn at most "ϵ-information" about w and $Q(w)$. Formally,

Lemma 9 (simulating the (A, C') stand-alone execution): *For every ppt adversary C' interacting with A, there exists a ppt machine $\hat{C}$ (interacting with nobody) such that for every dictionary $\mathcal{D} \subseteq \{0,1\}^n$,*

$$\left\{w, Q(w), \mathsf{output}(C'^{A(Q)})\right\} \stackrel{\epsilon}{\equiv} \left\{w, U_n, \mathsf{output}(\hat{C})\right\}$$

where $w \in_R \mathcal{D}$, Q is a random linear polynomial and $\epsilon = \frac{1}{|\mathcal{D}|}$.

Proof: The setting of this lemma is already that of standard two-party computation. Therefore, the security definition of two-party computation can be applied directly in order to prove the lemma. We sketch this more standard proof for the sake of completeness. We begin by showing that

$$\left\{w, Q(w), \mathsf{output}(C'^{A(Q)})\right\} \stackrel{\epsilon}{\equiv} \left\{w, U_n, \mathsf{output}(C'^{A(Q)})\right\} \quad (5)$$

In order to prove Eq. (5), recall that the security of the polynomial evaluation implies that the receiver (here played by C') can learn nothing beyond the value of $Q(\cdot)$ at a single point selected by C'. We denote this point by w_C. Then, in the case that $w_C \neq w$, the values $Q(w)$ and U_n are identically distributed (by the pairwise independence of random linear polynomials). That is, unless $w_C = w$, machine C' learns nothing of the value $Q(w)$. However, since w is uniformly distributed in $\mathcal{D}$, the probability that $w_C = w$ is at most ϵ. This means that, given C''s view, $Q(w)$ can be distinguished from U_n with probability at most ϵ.

We are now ready to define the (non-interactive) machine $\hat{C}$. Machine $\hat{C}$ works by first choosing a random linear polynomial $\hat{Q}$. Next, $\hat{C}$ perfectly emulates $C'^{A(\hat{Q})}$ by playing A's role in the execution with C ($\hat{C}$ uses the polynomial $\hat{Q}$ as A's input). Finally $\hat{C}$ outputs whatever C' does. Since w and U_n are independent of the polynomials Q and $\hat{Q}$, it follows that

$$\left\{w, U_n, \mathsf{output}(C'^{A(Q)})\right\} \equiv \left\{w, U_n, \mathsf{output}(\hat{C})\right\} \tag{6}$$

The lemma follows by combining Equations (5) and (6). ■

Combining Lemmas 8 and 9, we obtain Theorem 7. ■

We reiterate that Theorem 7 relates only to the secrecy of the password w and value $Q(w)$. Unlike Definition 2, it does *not say anything* about the outputs of the parties A and B. Furthermore, the model is significantly simplified by the fact that there is no public accept/reject bit output by the parties (as discussed in Section 3.2, simulating this bit is the most involved part of our proof). Thus, unfortunately, the above proof is merely an illustration of some of our techniques used in proving Theorem 6.

Acknowledgements. We would like to thank Moni Naor for suggesting this problem to us and for his valuable input in the initial stages of our research. We are also grateful to Alon Rosen for much discussion and feedback throughout the development of this work. We also thank Jonathan Katz for helpful discussion. Finally, we would like to thank Ran Canetti, Shai Halevi and Tal Rabin for discussion that led to a significant simplification of the protocol.

References

1. D. Beaver. Secure Multi-party Protocols and Zero-Knowledge Proof Systems Tolerating a Fault Minority. *Journal of Cryptology*, Vol. 4, pages 75–122, 1991.
2. M. Bellare, D. Pointcheval and P. Rogaway. Authenticated Key Exchange Secure Against Dictionary Attacks. In *EuroCrypt 2000*, Springer-Verlag (LNCS 1807), pages 139–155, 2000.
3. M. Bellare and P. Rogaway. Random Oracles are Practical: A Paradigm for Designing Efficient Protocols. In *1st Conf. on Computer and Communications Security*, ACM, pages 62–73, 1993.
4. M. Bellare and P. Rogaway. Entity Authentication and Key Distribution. In *CRYPTO'93*, Springer-Verlag (LNCS 773), pages 232–249, 1994.
5. S. M. Bellovin and M. Merritt. Encrypted key exchange: Password-based protocols secure against dictionary attacks. In *Proceedings of the ACM/IEEE Symposium on Research in Security and Privacy*, pages 72–84, 1992.

6. S. M. Bellovin and M. Merritt. Augmented encrypted key exchange: A password-based protocol secure against dictionary attacks and password file compromise. In *Proceedings of the 1st ACM Conference on Computer and Communication Security*, pages 244–250, 1993.
7. M. Blum. Coin Flipping by Phone. *IEEE Spring COMPCOM*, pages 133–137, February 1982.
8. M. Blum and S. Goldwasser. An Efficient Probabilistic Public-Key Encryption Scheme which hides all partial information. In *CRYPTO'84*, Springer-Verlag (LNCS 196), pages 289–302.
9. M. Blum and S. Micali. How to Generate Cryptographically Strong Sequences of Pseudo-Random Bits. *SICOMP*, Vol. 13, pages 850–864, 1984. Preliminary version in *23rd FOCS*, 1982.
10. M. Boyarsky. Public-key Cryptography and Password Protocols: The Multi-User Case. In *Proceedings of the 6th ACM Conference on Computer and Communication Security*, 1999.
11. V. Boyko, P. MacKenzie and S. Patel. Provably Secure Password-Authenticated Key Exchange Using Diffie-Hellman. In *EuroCrypt 2000*, Springer-Verlag (LNCS 1807), pages 156–171, 2000.
12. R. Canetti. Security and Composition of Multi-party Cryptographic Protocols. *Journal of Cryptology*, Vol. 13, No. 1, pages 143–202, 2000.
13. R. Canetti. A unified framework for analyzing security of protocols. Cryptology ePrint Archive, Report No. 2000/067, 2000. Available from `http://eprint.iacr.org`.
14. R. Canetti, O. Goldreich, and S. Halevi. The Random Oracle Methodology, Revisited. In *Proc. of the 30th STOC*, pages 209–218, 1998.
15. W. Diffie, and M.E. Hellman. New Directions in Cryptography. *IEEE Trans. on Info. Theory*, IT-22 (Nov. 1976), pages 644–654.
16. D. Dolev, C. Dwork, and M. Naor. Non-Malleable Cryptography. *SIAM Journal on Computing*, January 2000.
17. U. Feige and A. Shamir. Witness Indistinguishability and Witness Hiding Protocols. In *22nd STOC*, pages 416–426, 1990.
18. O. Goldreich. *Secure Multi-Party Computation.* Manuscript. Preliminary version, 1998. Available from `http://www.wisdom.weizmann.ac.il/~oded/pp.html`.
19. O. Goldreich, S. Goldwasser, and S. Micali. How to Construct Random Functions. *JACM*, Vol. 33, No. 4, pages 792–807, 1986.
20. O. Goldreich and A. Kahan. How To Construct Constant-Round Zero-Knowledge Proof Systems for NP. Journal of Cryptology, Vol. 9, pages 167–189, 1996.
21. O. Goldreich, S. Micali and A. Wigderson. How to Play any Mental Game – A Completeness Theorem for Protocols with Honest Majority. In *19th STOC*, pages 218–229, 1987. For details see [18].
22. S. Goldwasser and S. Micali. Probabilistic Encryption. *JCSS*, Vol. 28, No. 2, pages 270–299, 1984.
23. S. Halevi and H. Krawczyk. Public-Key Cryptography and Password Protocols. In *ACM Conference on Computer and Communications Security*, 1998.
24. D. P. Jablon. Strong password-only authenticated key exchange. *SIGCOMM Comput. Commun. Rev.*, Vol 26, No. 5, pages 5–26, 1996.
25. J. Katz, R. Ostrovsky and M. Yung. Practical Password-Authenticated Key Exchange Provably Secure under Standard Assumptions. In *Eurocrypt 2001.*

26. C. Kaufman, R. Perlman and M. Speciner. *Network Security.* Prentice Hall, 1997.
27. S. Lucks. Open key exchange: How to defeat dictionary attacks without encrypting public keys. In *Proceedings of the Workshop on Security Protocols*, Ecole Normale Superieure, 1997.
28. A. Menezes, P. Van Oorschot and S. Vanstone. *Handbook of Applied Cryptography.* CRC Press, 1997.
29. S. Micali and P. Rogaway. Secure Computation. Unpublished manuscript, 1992. Preliminary version in *Crypto'91*, Springer-Verlag (LNCS 576), 1991.
30. M. Naor and B. Pinkas. Oblivious Transfer and Polynomial Evaluation. In *31st STOC*, pages 245-254, 1999.
31. S. Patel. Number theoretic attacks on secure password schemes. In *Proceedings of the 1997 IEEE Symposium on Security and Privacy*, pages 236–247, 1997.
32. R. Richardson and J. Kilian. On the Concurrent Composition of Zero-Knowledge Proofs. In *EuroCrypt99*, pages 415–431.
33. R. Rivest, A. Shamir and L. Adleman. A Method for Obtaining Digital Signatures and Public Key Cryptosystems. *CACM*, Vol. 21, Feb. 1978, pages 120–126.
34. M. Steiner, G. Tsudi and M. Waidner. Refinement and extension of encrypted key exchange. *ACM SIGOPS Oper. Syst. Rev.*, Vol. 29, 3, pages 22–30, 1995.
35. T. Wu. The secure remote password protocol. In *1998 Internet Society Symposium on Network and Distributed System Security*, pages 97–111, 1998.
36. A.C. Yao. Theory and Application of Trapdoor Functions. In *23rd FOCS*, pages 80–91, 1982.
37. A.C. Yao. How to Generate and Exchange Secrets. In *27th FOCS*, pages 162–167, 1986.

Cryptanalysis of RSA Signatures with Fixed-Pattern Padding

Eric Brier[1], Christophe Clavier[1], Jean-Sébastien Coron[2], and David Naccache[2]

[1] Gemplus Card International
Parc d'Activités de Gémenos, B.P. 100, 13881 Gémenos Cedex, France
{eric.brier, christophe.clavier}@gemplus.com
[2] Gemplus Card International
34 rue Guynemer, 92447 Issy-les-Moulineaux, France
{jean-sebastien.coron, david.naccache}@gemplus.com

Abstract. A fixed-pattern padding consists in concatenating to the message m a fixed pattern P. The RSA signature is then obtained by computing $(P|m)^d \bmod N$ where d is the private exponent and N the modulus. In Eurocrypt '97, Girault and Misarsky showed that the size of P must be at least half the size of N (in other words the parameter configurations $|P| < |N|/2$ are insecure) but the security of RSA fixed-pattern padding remained unknown for $|P| > |N|/2$. In this paper we show that the size of P must be at least two-thirds of the size of N, *i.e.* we show that $|P| < 2|N|/3$ is insecure.

Keywords: RSA signatures, fixed-pattern padding, affine redundancy.

1 Introduction

RSA was invented in 1977 by Rivest, Shamir and Adleman [8], and is now the most widely used public-key cryptosytem. RSA is commonly used for providing privacy and authenticity of digital data, and securing web traffic between servers and browsers.

A very common practice for signing with RSA is to first hash the message, add some padding, and then raise the result to the power of the decryption exponent. This paradigm is the basis of numerous standards such as PKCS #1 v2.0 [9].

In this paper, we consider RSA signatures with fixed-pattern padding, without using a hash function. To sign a message m, the signer concatenates a fixed padding P to the message, and the signature is obtained by computing:

$$s = (P|m)^d \bmod N$$

where d is the private exponent and N the modulus.

More generally, we consider RSA signatures in which a simple affine redundancy is used. To sign a message m, the signer first computes:

$$R(m) = \omega \cdot m + a \quad \text{where} \quad \begin{cases} w \text{ is the multiplicative redundancy} \\ a \text{ is the additive redundancy} \end{cases} \tag{1}$$

J. Kilian (Ed.): CRYPTO 2001, LNCS 2139, pp. 433–439, 2001.

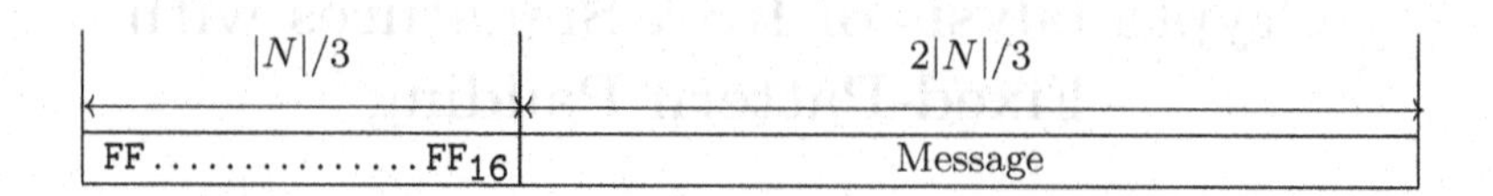

Fig. 1. Example of an RSA padding forgeable by De Jonge and Chaum's method where $\omega = 1$ and $a = \texttt{FF...FF 00...00}_{16}$

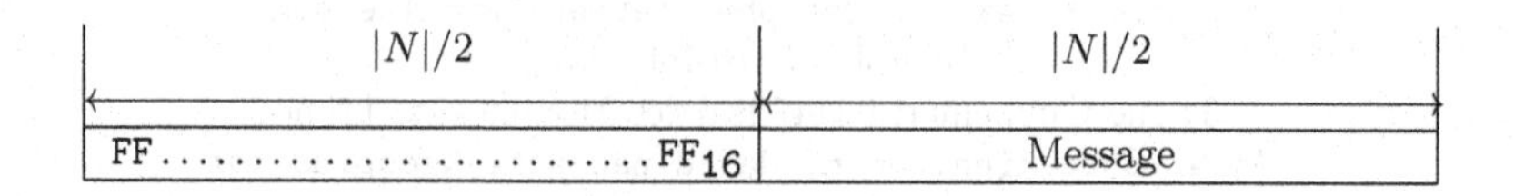

Fig. 2. Example of an RSA padding forgeable by Girault and Misarsky's method where $\omega = 1$ and $a = \texttt{FF...FF 00...00}_{16}$

The signature of m is then:

$$s = R(m)^d \bmod N$$

A left-padded redundancy scheme $P|m$ is obtained by taking $\omega = 1$ and $a = P{\cdot}2^\ell$, whereas a right-padding redundancy scheme $m|P$ is obtained by taking $\omega = 2^\ell$ and $a = P$.

No proof of security is known for RSA signatures with affine redundancy, and several attacks on such formats have appeared (see [6] for a thorough survey). At Crypto '85, De Jonge and Chaum [1] exhibited a multiplicative attack against RSA signatures with affine redundancy, based on the extended Euclidean algorithm. Their attack applies when the multiplicative redundancy ω is equal to one and the size of the message is at least two-thirds of the size of the RSA modulus N.

$$|\text{message}| \succ \frac{2}{3}|N|$$

For example, a signature can be forged if one uses the affine redundancy of figure 1.

De Jonge and Chaum's attack was extended by Girault and Misarsky [2] at Eurocrypt '97, using Okamoto-Shiraishi's algorithm [7], which is an extension of the extended Euclidean algorithm. They increased the field of application of multiplicative attacks on RSA signatures with affine redundancy as their attack applies to any value of ω and a, when the size of the message is at least half the size of the modulus (refer to figure 2 for an illustration):

$$|\text{message}| \succ \frac{1}{2}|N|$$

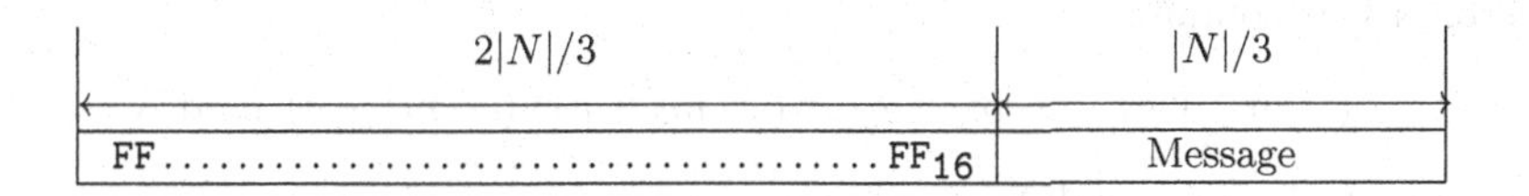

Fig. 3. Example of an RSA padding forgeable by our technique where the ω is equal to one and $a = \texttt{FF...FF 00...00}_{16}$

Girault and Misarsky also extended the multiplicative attacks to RSA signatures with modular redundancy:

$$R(m) = \omega_1 \cdot m + \omega_2 \cdot (m \bmod b) + a \tag{2}$$

where ω_1, ω_2 is the multiplicative redundancy, a is the additive redundancy and b is the modular redundancy. In this case, the size of the message must be at least half the size of the modulus plus the size of the modular redundancy.

Finally, Girault and Misarsky's attack was extended by Misarsky [5] at Crypto '97 to a redundancy function in which the message m and the modular redundancy $m \bmod b$ can be split into different parts, using the LLL algorithm [4]. The attack applies when the size of the message is at least half the size of the modulus plus the size of the modular redundancy.

In this paper, we extend Girault and Misarsky's attack against RSA signatures with affine redundancy to messages of size as small as one third of the size of the modulus, as illustrated in figure 3.

$$|\text{message}| \succ \frac{1}{3}|N|$$

As Girault and Misarsky's attack, our attack applies for any w and a and runs in polynomial time. However, our attack is existential only, as we cannot choose the message the signature of which we forge, whereas Girault and Misarsky's attack is selective: they can choose the message which signature is forged.

2 The New Attack

In this section we extend Girault and Misarsky's multiplicative attack on RSA signatures with affine redundancy, to messages of size as small as one third of the size of N. A multiplicative attack is an attack in which the redundancy function of a message can be expressed as a multiplicative combination of the redundancy functions of other messages. So we look for four distinct messages m_1, m_2, m_3 and m_4, each as small as one third of the size of the modulus, such that:

$$R(m_1) \cdot R(m_2) = R(m_3) \cdot R(m_4) \bmod N \tag{3}$$

Then, using the signatures of m_2, m_3 and m_4, one can forge the signature of m_1 by:

$$R(m_1)^d = \frac{R(m_3)^d \cdot R(m_4)^d}{R(m_2)^d} \bmod N$$

From (3) we obtain:

$$(\omega \cdot m_1 + a) \cdot (\omega \cdot m_2 + a) = (\omega \cdot m_3 + a) \cdot (\omega \cdot m_4 + a) \bmod N$$

Denoting $P = a/\omega \bmod N$, we obtain:

$$(P + m_1) \cdot (P + m_2) = (P + m_3) \cdot (P + m_4) \bmod N$$

and letting:

$$\begin{array}{ll} t = m_3 & y = m_2 - m_3 \\ x = m_1 - m_3 & z = m_4 - m_1 - m_2 + m_3 \end{array} \tag{4}$$

we obtain:

$$((P + t) + x) \cdot ((P + t) + y) = (P + t) \cdot ((P + t) + x + y + z) \bmod N$$

which simplifies into:

$$x \cdot y = (P + t) \cdot z \bmod N \tag{5}$$

Our goal is consequently to find four integers x, y, z and t, each as small as one third of the size of N, satisfying equation (5).

First, we obtain two integers z and u such that

$$P \cdot z = u \bmod N \text{ with } \begin{cases} -N^{\frac{1}{3}} < z < N^{\frac{1}{3}} \\ 0 < u < 2 \cdot N^{\frac{2}{3}} \end{cases}$$

As noted in [3], this is equivalent to finding a good approximation of the fraction P/N, and can be done efficiently by developing it in continued fractions, *i.e.* applying the extended Euclidean algorithm to P and N. A solution is found such that $|z| < Z$ and $0 < u < U$ if $Z \cdot U > N$, which is the case here with $Z = N^{\frac{1}{3}}$ and $U = 2 \cdot N^{\frac{2}{3}}$.

We then select an integer y such that $N^{\frac{1}{3}} \leq y \leq 2 \cdot N^{\frac{1}{3}}$ and $\gcd(y, z) = 1$. We find the non-negative integer $t < y$ such that:

$$t \cdot z = -u \bmod y$$

which is possible since $\gcd(y, z) = 1$. Then we take

$$x = \frac{u + t \cdot z}{y} \leq 4N^{\frac{1}{3}}$$

and obtain:

$$P \cdot z = u = x \cdot y - t \cdot z \bmod N$$

which gives equation (5), with x, y, z and t being all smaller than $4 \cdot N^{\frac{1}{3}}$. From x, y, z, t we derive using (4) four messages m_1, m_2, m_3 and m_4, each of size one third the size of N:

$$\begin{array}{ll} m_1 = x + t & m_2 = y + t \\ m_3 = t & m_4 = x + y + z + t \end{array} \tag{6}$$

Since $-N^{1/3} < z < N^{1/3}$ and $y \geq N^{1/3}$, we have $y + z > 0$, which gives using $u \geq 0$:

$$x + t = \frac{u + t \cdot (y + z)}{y} \geq 0$$

which shows that the four integers m_1, m_2, m_3 and m_3 are non-negative, and we have

$$R(m_1) \cdot R(m_2) = R(m_3) \cdot R(m_4) \bmod N$$

The complexity of our attack is polynomial in the size of N. In the appendix we give an example of such a forgery computed using RSA Laboratories' official 1024-bits challenge-modulus RSA-309.

3 Extension to Selective Forgery

The attack of the previous section is only existential: we can not choose the message to be forged. In this section we show how we can make the forgery selective, but in this case the attack is no longer polynomial. Let m_3 be the message which signature must be forged. Letting x, y, z and t as in (4), we compute two integers z and u such that

$$(P + t) \cdot z = u \bmod N \quad \text{with} \quad \begin{cases} -N^{\frac{1}{3}} < z < N^{\frac{1}{3}} \\ 0 < u < 2 \cdot N^{\frac{2}{3}} \end{cases}$$

We then factor u, and try to write u as the product $x \cdot y$ of two integers of roughly the same size, so that eventually we have four integers x, y, z, t of size roughly one third of the size of the modulus, with:

$$x \cdot y = (P + t) \cdot z \bmod N$$

which gives again

$$R(m_1) \cdot R(m_2) = R(m_3) \cdot R(m_4) \bmod N$$

The signature of m_3 can now be forged using the signatures of m_1, m_2 and m_4. For a 512-bit modulus the selective forgery attack is truly practical. For a 1024-bit modulus the attack is more demanding but was still implemented with success.

4 Conclusion

We have extended Girault and Misarsky's attack on RSA signatures with affine redundancy: we described a chosen message attack against RSA signatures with affine redundancy for messages as small as one third of the size of the modulus. Consequently, when using a fixed padding $P|m$ or $m|P$, the size of P must be at least two-thirds of the size of N. Our attack is polynomial in the length of the modulus. It remains an open problem to extend this attack to even smaller messages (or, equivalently, to bigger fixed-pattern constants): we do not know if there exists a polynomial time attack against RSA signatures with affine redundancy for messages shorter than one third of the size of the modulus. However, we think that exploring to what extent affine padding is malleable increases our understanding of RSA's properties and limitations.

Acknowledgements. We would like to thank Christophe Tymen, Pascal Paillier, Helena Handschuh and Alexey Kirichenko for helpful discussions and the anonymous referees for their constructive comments.

References

1. W. De Jonge and D. Chaum, *Attacks on some RSA signatures.* Proceedings of Crypto '85, LNCS vol. 218, Springer-Verlag, 1986, pp. 18-27.
2. M. Girault and J.-F. Misarksy, *Selective forgery of RSA signatures using redundancy*, Proceedings of Eurocrypt '97, LNCS vol. 1233, Springer-Verlag, 1997, pp. 495–507.
3. M. Girault, P. Toffin and B. Vallée, *Computation of approximation L-th roots modulo n and application to cryptography*, Proceedings of Crypto '88, LNCS vol. 403, Springer-Verlag, 1988, pp. 100–117.
4. A. K. Lenstra, H.W. Lenstra and L. Lovász, *Factoring polynomials with rational coefficients*, Mathematische Annalen, vol. 261, n. 4, 1982, pp. 515–534.
5. J.-F. Misarsky, *A multiplicative attack using LLL algorithm on RSA signatures with redundancy*, Proceedings of Crypto '97, LNCS vol. 1294, Springer-Verlag, pp. 221–234.
6. J.-F. Misarsky, *How (not) to design RSA signature schemes*, Public-key cryptography (PKC), Springer-Verlag, Lectures notes in computer science 1431, pp. 14–28, 1998.
7. T. Okamoto and A. Shiraishi, *A fast signature scheme based on quadratic inequalities*, Proc. of the 1985 Symposium on Security and Privacy, April 1985, Oakland, CA.
8. R. Rivest, A. Shamir and L. Adleman, *A method for obtaining digital signatures and public key cryptosystems*, CACM 21, 1978.
9. RSA Laboratories, PKCS #1 : *RSA cryptography specifications*, version 2.0, September 1998.

A A Practical Forgery

We describe a practical forgery with $\omega = 1$ and $a = 2^{1023} - 2^{352}$, the modulus N being RSA Laboratories official challenge RSA-309, which factorisation is still unknown.

$N = \text{RSA-309}$

= bdd14965 645e9e42 e7f658c6 fc3e4c73 c69dc246 451c714e b182305b 0fd6ed47
d84bc9a6 10172fb5 6dae2f89 fa40e7c9 521ec3f9 7ea12ff7 c3248181 ceba33b5
5212378b 579ae662 7bcc0821 30955234 e5b26a3e 425bc125 4326173d 5f4e25a6
d2e172fe 62d81ced 2c9f362b 982f3065 0881ce46 b7d52f14 885eecf9 03076ca5

$R(m_1)$ = 7fffffff ffffffff ffffffff ffffffff ffffffff ffffffff ffffffff ffffffff
ffffffff ffffffff ffffffff ffffffff ffffffff ffffffff ffffffff ffffffff
ffffffff ffffffff ffffffff ffffffff ffffffff 00415df4 ca4219b6 ea5fa8e4
e2eabcfc 61348b80 e7ccbac7 3d1f5cc7 249e1519 9412886a f76220c6 d1409cd6

$R(m_2)$ = 7fffffff ffffffff ffffffff ffffffff ffffffff ffffffff ffffffff ffffffff
ffffffff ffffffff ffffffff ffffffff ffffffff ffffffff ffffffff ffffffff
ffffffff ffffffff ffffffff ffffffff ffffffff 00127f44 f753253a a0348be7
826e893f 693032db c2194dbb 3b81e1c2 630b66d3 1448a3f4 7fd2d34f b28aefd6

$R(m_3)$ = 7fffffff ffffffff ffffffff ffffffff ffffffff ffffffff ffffffff ffffffff
ffffffff ffffffff ffffffff ffffffff ffffffff ffffffff ffffffff ffffffff
ffffffff ffffffff ffffffff ffffffff ffffffff 00781bd4 e0c918a7 308fcff7
8f64044c a35b4937 36cd37d7 93f281b5 fdd0a951 52a0479b 57dd73b2 25b6df85

$R(m_4)$ = 7fffffff ffffffff ffffffff ffffffff ffffffff ffffffff ffffffff ffffffff
ffffffff ffffffff ffffffff ffffffff ffffffff ffffffff ffffffff ffffffff
ffffffff ffffffff ffffffff ffffffff ffffffff 000919fd 86e5afce 7fc11c94
0e0827c8 03be05bb 71f8de48 c61d6d5f 0feb036d a1ff2f8b 5f596108 3d142538

We obtain:

$$R(m_1) \cdot R(m_2) = R(m_3) \cdot R(m_4) \bmod N$$

where messages m_1, m_2, m_3 and m_4 are as small as one third of the size of the modulus.

Correlation Analysis of the Shrinking Generator

Jovan D. Golić

GEMPLUS
Rome CryptoDesign Center, Technology R&D
Via Pio Emanuelli 1, 00143 Rome, Italy
jovan.golic@gemplus.com

Abstract. The shrinking generator is a well-known keystream generator composed of two linear feedback shift registers, LFSR_1 and LFSR_2, where LFSR_1 is clock-controlled according to regularly clocked LFSR_2. A probabilistic analysis of the shrinking generator which shows that this generator can be vulnerable to a specific fast correlation attack is conducted. The first stage of the attack is based on a recursive computation of the posterior probabilites of individual bits of the regularly clocked LFSR_1 sequence when conditioned on a given segment of the keystream sequence. Theoretical analysis shows that these probabilities are significantly different from one half and can hence be used for reconstructing the initial state of LFSR_1 by iterative probabilistic decoding algorithms for fast correlation attacks on regularly clocked LFSR's. In the second stage of the attack, the initial state of LFSR_2 is reconstructed in a similar way, which is based on a recursive computation of the posterior probabilites of individual bits of the LFSR_2 sequence when conditioned on the keystream sequence and on the reconstructed LFSR_1 sequence.

Keywords. Stream ciphers, unconstrained irregular clocking, posterior probabilities, fast correlation attacks.

1 Introduction

The shrinking generator [1] is a well-known keystream generator for stream cipher applications. It consists of only two linear feedback shift registers (LFSR's). The clock-controlled LFSR, LFSR_1, is irregularly clocked according to the clock-control LFSR, LFSR_2, which is regularly clocked. More precisely, at each time, both LFSR's are clocked once and the bit produced by LFSR_1 is taken as the output bit if the clock-control bit produced by LFSR_2 is equal to 1. Otherwise, the output bit is not produced. The output sequence is thus a nonuniformly decimated LFSR_1 sequence. It is recommended in [1] that the LFSR initial states and the feedback polynomials be defined by the secret key. Under certain conditions, the output sequences possess a long period, a high linear complexity, and good statistical properties.

As pointed out in [1], a basic divide-and-conquer attack on the shrinking generator is the linear consistency attack [17] on LFSR_2 which requires the exhaustive search through all possible initial states and feedback polynomials of

J. Kilian (Ed.): CRYPTO 2001, LNCS 2139, pp. 440–457, 2001.

$LFSR_2$. On the other hand, a probabilistic correlation attack targeting $LFSR_1$ which requires the exhaustive search through all possible initial states and feedback polynomials of $LFSR_1$ is proposed in [4] and analyzed by computer simulations in [15]. A reduced complexity method based on searching for specific subsequences of the output sequence is suggested in [9], but both the complexity and the required keystream segment length are exponential in the length of $LFSR_1$.

It is shown in [3] that the output sequence may have a detectable linear statistical weakness if the feedback polynomial of $LFSR_1$ has low-weight polynomial multiples of moderately large degrees. It is suggested in [5] that this weakness may even be used for recovering the $LFSR_1$ feedback polynomial. A theoretical framework for a fast correlation attack targeting the initial state of $LFSR_1$ is also proposed in [5], but the attack is not implemented as it requires a search for specific polynomial multiples of the $LFSR_1$ feedback polynomial.

The objective of this paper is to investigate if the initial states of $LFSR_1$ and $LFSR_2$ can be reconstructed by an algorithm that would not require the exhaustive search through all possible initial states and whose complexity can be sufficiently small even for large LFSR lengths. The LFSR feedback polynomials are assumed to be known. The basic point of our approach is to consider the posterior probabilites of individual bits of the regularly clocked $LFSR_1$ sequence when conditioned on a given segment of the keystream sequence. In the probabilistic model where the LFSR sequences are assumed to be independent and purely random,[1] a recursion and an explicit expression for computing these probabilities with complexity quadratic in the keystream segment length are both derived. A theoretical analysis shows that the computed posterior probabilities can be significantly different from one half for a purely random output sequence. In a more general probabilistic model, in which the $LFSR_1$ sequence is assumed to be a sequence of independent, not necessarily uniformly distributed, binary random variables, it is proved that the posterior probabilities can be recursively computed with complexity cubic in the keystream segment length.

Accordingly, as these probabilities represent soft-valued estimates of the corresponding bits of the regularly clocked $LFSR_1$ sequence, they can be used in an iterative probabilistic decoding algorithm for fast correlation attacks on regularly clocked LFSR's (e.g., see [11], [12], and [8]). It is known that the complexity of such an algorithm primarily depends on the degrees and numbers of low-weight polynomial multiples of the feedback polynomial of $LFSR_1$ which, according to [10], [7], and [14], may also contain an additional number of concentrated nonzero terms. The initial state of $LFSR_1$ can thus be recovered. A more sophisticated method in which the posterior probabilities are iteratively updated by intertwining the probabilistic decoding with the recursive computation is also introduced.

In addition, a composite method that effectively enhances the posterior probabilities for longer keysteam segments is proposed. Essentially, it consists in

[1] A sequence of independent uniformly distributed random variables over a finite set is called purely random.

dividing a longer keystream segment into subsegments of equal length, in computing the posterior probabilities for the subsegments, and then in combining these posterior probabilities appropriately.

If the posterior probabilities corresponding to a given keystream sequence are not sufficiently different from one half, they can be computed for subsequences of the keystream sequence obtained by discarding the initial segment of variable length until the significant posterior probabilities are obtained. This will improve the performance of the fast correlation attacks explained above, but the length of the initial LFSR_1 segment has to be guessed. For the initial output segment of length $j-1$, one has to make $O(\sqrt{2j})$ guesses around the expected value $2j-1$. Moreover, one can thus also search for the outstanding posterior probabilities and then apply an information set decoding algorithm to recover the LFSR_1 initial state. The success of such an algorithm is independent of the LFSR_1 feedback polynomial, but the achievable complexity is still exponential in the length of LFSR_1. This improves the reduced complexity method [9].

The second point of our approach is to consider the posterior probabilites of individual bits of the regularly clocked LFSR_2 sequence when conditioned on a given segment of the keystream sequence and on the reconstructed LFSR_1 sequence, as suggested in [9]. It is proved that these probabilities can be recursively computed with complexity cubic in the keystream segment length, thus showing that the expression given in [9] is incorrect. As the LFSR_1 sequence is assumed to be known, the computed posterior probabilities are more distinguished from one half than in the case of LFSR_1. This makes the reconstruction much easier. Consequently, the initial state of LFSR_2 can be recovered either by an iterative probabilistic decoding algorithm or by a simple information set decoding algorithm using a subset of the probabilities close to zero or one.

Section 2 contains an overview of known results concerning the posterior probabilities of blocks of LFSR_1 bits. The results regarding the posterior probabilities of individual LFSR_1 and LFSR_2 bits are presented in Sections 3 and 4, respectively. These posterior probabilities are theoretically analyzed in Section 5. The combined fast correlation attacks are proposed in Section 6, and conclusions are given in Section 7. Proofs of two underlying theorems are presented in Appendices A and B.

2 Posterior Probabilities of Blocks of LFSR_1 Bits

We use the notation $A = a_1, a_2, \ldots$ for a binary sequence, A_k for its subsequence $a_k, a_{k+1}, \ldots$, A^n for its prefix $(a_i)_{i=1}^n = a_1, a_2, \ldots a_n$, and A_k^n for its subsequence $(a_i)_{i=k}^n = a_k, a_{k+1}, \ldots, a_n$. If its length is finite, then A is called a string. Let $w(A)$ and $d(A)$ denote the numbers of 1's and 0's in A, respectively. For simplicity, we keep the same notation for random variables and their values.

Let X, C, and Y denote the output sequences of LFSR_1, LFSR_2, and the shrinking generator itself, respectively. In a general model, let X and C be arbitrary binary sequences. Then Y is obtained from X by the nonuniform decimation according to C, that is, a bit x_i is deleted from X iff $c_i = 0$. Accordingly,

Y is a function of X and C, $Y = F(X, C)$, where the length of Y may be finite and is equal to $w(C)$. Thus Y^n is a function of X and C, $Y^n = F^n(X, C)$, for any $1 \leq n \leq w(C)$. If $w(C) = 0$, then Y is not produced. If $w(C^n) = l \geq 1$ and $c_n = 1$, then $y_l = x_n$. It follows that y_n is a function of X_n and C, $f_n(X_n, C)$.

We assume a probabilistic model where X and C are independent and purely random binary sequences. It then follows that the output sequence Y is also purely random. We are first interested in deriving the posterior probability $\Pr\{X^n \mid Y\}$ which is in this model equal to $\Pr\{X^n \mid Y^n\}$. To this end, according to [4], define the following conditional probability for prefixes of X and Y

$$Q(e, s) \stackrel{def}{=} \Pr\{Y^s, d(C^{e+s}) = e \mid X^{e+s}\}. \tag{1}$$

It is in fact the probability that Y^s is obtained by deleting e bits from a given string X^{e+s}. The permissible values of s and e are $0 \leq s \leq n$ and $0 \leq e \leq n - s$, where Y^0 denotes an empty set and, formally, $Q(0, 0) = 1$. This probability can be computed recursively by

$$Q(e, s) = \frac{1}{2} Q(e - 1, s) + \frac{1}{2} \delta(x_{e+s}, y_s) Q(e, s - 1) \tag{2}$$

where the terms on the right-hand side corresponding to unpermissible values of e or s (i.e., for $e = 0$ or $s = 0$) are assumed to be equal to zero (see [4] and Appendix B). Here, $\delta(i, j)$ or $\delta_{i,j}$ is the Kronecker symbol, i.e., $\delta(i, j) = 1$ if $i = j$ and $\delta(i, j) = 0$ if $i \neq j$.

Consequently, we have

$$\begin{aligned} \Pr\{Y^n \mid X^n\} &= \sum_{e=0}^{n} \Pr\{Y^n, d(C^n) = e \mid X^n\} \\ &= \sum_{e=0}^{n} \Pr\{Y_{n-e+1}^n \mid Y^{n-e}, d(C^n) = e, X^n\} Q(e, n - e) \\ &= \sum_{e=0}^{n} 2^{-e} Q(e, n - e) \end{aligned} \tag{3}$$

in view of the fact that, on the condition that $d(C^n) = e$, the string Y_{n-e+1}^n is obtained by decimating X_{n+1} according to C_{n+1}, where X_{n+1} and C_{n+1} remain to be mutually independent and purely random (even when conditioned on X^n and Y^{n-e}). Therefore, under the given conditions, Y_{n-e+1}^n remains to be uniformly distributed. Further, as X^n and Y^n are both uniformly distributed, we have

$$\Pr\{X^n \mid Y^n\} = \Pr\{Y^n \mid X^n\} = \sum_{e=0}^{n} 2^{-e} Q(e, n - e) \tag{4}$$

which is computed in $O(n^2)$ time and $O(n)$ space. The probability (4) can be found in [9], and also corresponds to the probability derived in [6] for the alternating step generator, because the nonuniform decimation of a purely random sequence can be regarded as the inverse operation to the nonuniform interleaving of two purely random sequences which is inherent to this generator.

For ease of computation, one can introduce $N(e,s) = 2^{e+s}\,Q(e,s)$ which represents the number of clock-control strings C^{e+s} that result in Y^s from a given X^{e+s}. These integers can be computed by the recursion

$$N(e,s) = N(e-1,s) + \delta(x_{e+s}, y_s)\,N(e,s-1). \tag{5}$$

Then

$$\Pr\{X^n \mid Y^n\} = 2^{-n}\sum_{e=0}^{n} 2^{-e}\,N(e,n-e). \tag{6}$$

It is proposed in [4] to use the probability $Q(m-n,n)$, where $m \approx 2n$, in order to reconstruct the LFSR_1 initial state from a given keystream segment Y^n. This probability is computed in $O(n(m-n)) = O(n^2)$ time. Statistical experiments from [15] show that $n \approx 20r_1$ is sufficient for a successful reconstruction.[2] Here, $Q(m-n,n)$ is used as a measure of correlation between Y^n and X^m, where X^m is produced from an assumed LFSR_1 initial state. It would be interesting to compare $Q(m-n,n)$ with the posterior probability (4) with respect to the minimum keystream segment length and the complexity required. However, the exhaustive search over all possible LFSR_1 initial states is required for both measures. It is worth mentioning that a conclusion from [9] that the required n is independent of r_1 is incorrect, because, according to the deletion channel capacity argument, n must be linear in r_1 (see [4] and [15]).

3 Posterior Probabilities of Individual LFSR_1 Bits

In this section, the posterior probabilities of individual bits of the regularly clocked LFSR_1 sequence when conditioned on a given segment of the keystream sequence are introduced. In Section 3.1, it is shown that these probabilities can be computed recursively in a probabilistic model in which the LFSR_2 sequence is assumed to be purely random, the LFSR_1 sequence is assumed to be a sequence of independent binary random variables, and both sequences are assumed to be mutually independent. This general model is relevant for a fast correlation attack on LFSR_1 in which the posterior probabilities are iteratively updated by intertwining the recursive computation with a probabilistic decoding algorithm used in fast correlation attacks on regularly clocked LFSR's. In Section 3.2, a special case of this model in which the LFSR_1 sequence is assumed to be purely random is considered. This case is especially relevant for a fast correlation attack on LFSR_1 in which the posterior probabilities recursively computed in the first stage are then processed by an iterative probabilistic decoding algorithm in the second stage.

[2] The length of LFSR_i is denoted as r_i, $i = 1, 2$.

3.1 General Probabilistic Model

Generalize the probabilistic model from Section 2 in such a way that a prefix of X need not be purely random. More precisely, let X be a sequence of independent binary random variables (bits) such that $\Pr\{x_i = 1\} = p_i$ for $1 \le i \le n$ and $\Pr\{x_i = 1\} = 0.5$ for $i > n$, where n is a given positive integer. Our objective here is to determine the posterior probabilities $\hat{p}_i = \Pr\{x_i = 1 \mid Y^n\}$ for $1 \le i \le n$. It follows that

$$\hat{p}_i \;=\; p_i\,\frac{\Pr\{Y^n \mid x_i = 1\}}{\Pr\{Y^n\}}. \tag{7}$$

The problem is how to compute the probabilities $\Pr\{Y^n \mid x_i = 1\}$ and $\Pr\{Y^n\}$ efficiently. To this end, introduce the following partial probabilities, for prefixes of Y,

$$P_i(e,s) \stackrel{def}{=} \Pr\{Y^s, d(C^{e+s}) = e \mid x_i = 1\} \tag{8}$$

$$P(e,s) \stackrel{def}{=} \Pr\{Y^s, d(C^{e+s}) = e\} \tag{9}$$

for $0 \le s \le n$ and $0 \le e \le n-s$, where formally $P(0,0) = 1$ and $P_i(0,0) = 1$.

The following theorem, proved in Appendix A, shows that the partial probabilities can be computed recursively and then used to obtain the desired posterior probabilities by (7).

Theorem 1. *For any given Y^n and each $1 \le i \le n$, we have*

$$\hat{p}_i \;=\; p_i\,\frac{\sum_{e=0}^{n} 2^{-e}\, P_i(e, n-e)}{\sum_{e=0}^{n} 2^{-e}\, P(e, n-e)} \tag{10}$$

where the partial probabilities are determined recursively by

$$\begin{aligned} P_i(e,s) &= \frac{1}{2}\, P_i(e-1,s) \\ &+ \frac{1}{2}\left(\delta_{i,e+s} y_s + (1-\delta_{i,e+s})(y_s p_{e+s} + (1-y_s)(1-p_{e+s}))\right) P_i(e,s-1) \end{aligned} \tag{11}$$

$$P(e,s) = \frac{1}{2}\, P(e-1,s) \;+\; \frac{1}{2}\left(y_s p_{e+s} + (1-y_s)(1-p_{e+s})\right) P(e,s-1) \tag{12}$$

for $0 \le s \le n$, $0 \le e \le n-s$, and $(e,s) \ne (0,0)$, from the initial values $P_i(0,0) = P(0,0) = 1$. (The terms on the right-hand sides of these equations corresponding to unpermissible values of e or s, i.e., for $e = 0$ or $s = 0$, are assumed to be equal to zero.)

The time and space complexities of the corresponding algorithm are clearly $O(n^3)$ and $O(n)$, respectively. The algorithm may thus be feasible even if n is large. For computational convenience, the multiplicative factor 0.5 can be removed from the recursions without affecting the values of the posterior probabilities. The time complexity can be reduced to $O(n^2\sqrt{n})$ if $P_i(e,s)$ and $P(e,s)$ are computed approximately, only for $O(\sqrt{2s})$ values of e around s.

3.2 Purely Random String Probabilistic Model

Consider now the model in which X is a purely random sequence. It is a particular instance of the general model from Section 3.1 in which $p_i = 0.5$, $1 \leq i \leq n$. In this model, the recursion (12) can be explicitly solved as $P(e,s) = \binom{e+s}{e} 2^{-(e+2s)}$, so that $\Pr\{Y^n\} = 2^{-n}$, as to be expected. Accordingly, the posterior probabilities can be computed by the following corollary to Theorem 1.

Corollary 1. *If X is purely random, then for any given Y^n and each $1 \leq i \leq n$, we have*

$$\hat{p}_i = 2^{n-1} \sum_{e=0}^{n} 2^{-e} P_i(e, n-e) \tag{13}$$

where the partial probability is determined recursively by

$$P_i(e,s) = \frac{1}{2} P_i(e-1,s) + \frac{1}{4}(1 + \delta_{i,e+s}(2y_s - 1)) P_i(e,s-1) \tag{14}$$

for $0 \leq s \leq n$, $0 \leq e \leq n-s$, and $(e,s) \neq (0,0)$, from the initial value $P_i(0,0) = 1$.

Further simplification and an explicit expression can be obtained by using the fact that X is purely random. Namely, in a similar way as (34) in Appendix A, we obtain

$$\begin{aligned}
\Pr\{Y^n \mid x_i = 1\} &= \sum_{e=0}^{i} \Pr\{Y^n, d(C^i) = e \mid x_i = 1\} \\
&= \sum_{e=0}^{i} \Pr\{Y^n_{i-e+1}, Y^{i-e}, d(C^i) = e \mid x_i = 1\} \\
&= \sum_{e=0}^{i} \Pr\{Y^n_{i-e+1} \mid Y^{i-e}, d(C^i) = e, x_i = 1\} \, P_i(e, i-e) \\
&= 2^{-(n-i)} \sum_{e=0}^{i} 2^{-e} P_i(e, i-e) \; = \; 2^{-(n-i)} \Pr\{Y^i \mid x_i = 1\}.
\end{aligned} \tag{15}$$

As a consequence, we have

$$\Pr\{x_i = 1 \mid Y^n\} = \Pr\{x_i = 1 \mid Y^i\}. \tag{16}$$

Also, it follows that

$$P_i(e, i-e) = \frac{1}{2} P(e-1, i-e) + \frac{1}{2} y_{i-e} P(e, i-e-1) \tag{17}$$

where $P(e,s) = 2^{-(e+2s)} M(e,s)$, $M(e,s) = \binom{e+s}{e}$, and the binomial coefficients can be computed recursively by

$$M(e,s) = M(e-1,s) + M(e,s-1) \tag{18}$$

for $0 \le s \le n-1$, $0 \le e \le n-1-s$, and $(e,s) \ne (0,0)$, from the initial value $M(0,0) = 1$. Then (13) and (17) imply that

$$\hat{p}_i = \frac{1}{2} 2^{-i} \sum_{e=0}^{i} \left(\binom{i-1}{e-1} + 2 \binom{i-1}{e} y_{i-e} \right). \tag{19}$$

Finally, we obtain the following theorem.

Theorem 2. *If X is purely random, then for any given Y^n and each $1 \le i \le n$, we have*

$$\hat{p}_i = \frac{1}{2} \left(\frac{1}{2} + 2^{-(i-1)} \sum_{e=0}^{i-1} \binom{i-1}{e} y_{i-e} \right). \tag{20}$$

The time and space complexities of the algorithm corresponding to Theorem 2 are $O(n^2)$ and $O(n)$, respectively, where the binomial coefficients can be recursively precomputed in $O(n^2)$ time by using (18). However, (20) shows that $\hat{p}_i$ can be numerically approximated with an arbitrarily small error by using only $O(\sqrt{i-1}/2)$ values of e around $(i-1)/2$. This reduces the time complexity to $O(n\sqrt{n})$.

The following immediate corollary to Theorem 2 shows that the posterior probabilities cannot approach 0 or 1.

Corollary 2. *If X is purely random, then for any given Y^n and each $1 \le i \le n$, we have*

$$\frac{1}{4} \le \hat{p}_i \le \frac{3}{4} \tag{21}$$

where the lower and upper bounds are achieved if and only if Y^i consists of all 0's and of all 1's, respectively.

4 Posterior Probabilities of Individual $LFSR_2$ Bits

In this section, it is shown that the posterior probabilities of individual bits of the regularly clocked $LFSR_2$ sequence when conditioned on a given segment of the keystream sequence and on a segment of the reconstructed $LFSR_1$ sequence can be computed recursively with complexity cubic in the segment length.

Assuming that X and C are independent and purely random, our objective is to determine the posterior probabilities $\hat{q}_i = \Pr\{c_i = 1 \mid Y^n, X^n\}$ for $1 \le i \le n$. It follows that

$$\hat{q}_i = \frac{1}{2} \frac{\Pr\{Y^n \mid c_i = 1, X^n\}}{\Pr\{Y^n \mid X^n\}}. \tag{22}$$

In Section 2, it is shown that $\Pr\{Y^n \mid X^n\}$ can be computed recursively. The problem is how to compute $\Pr\{Y^n \mid c_i = 1, X^n\}$ efficiently. Similarly to (1), define the following conditional probability for prefixes of X and Y

$$Q_i(e,s) \stackrel{def}{=} \Pr\{Y^s, d(C^{e+s}) = e \mid c_i = 1, X^{e+s}\} \tag{23}$$

for $0 \le s \le n$ and $0 \le e \le n-s$, with $Q_i(0,0) = 1$.

The following theorem, proved in Appendix B, shows that this probability can be computed recursively and then used to obtain the desired posterior probabilities by (22). This theorem shows that the expression for the posterior probabilities given in [9] is incorrect, not only in general, but also in a special case of the probabilities $\Pr\{c_i = 1 \mid Y^i, X^i\}$.

Theorem 3. *For any given Y^n and X^n and each $1 \leq i \leq n$, we have*

$$\hat{q}_i = \frac{1}{2} \frac{\sum_{e=0}^{n} 2^{-e} Q_i(e, n-e)}{\sum_{e=0}^{n} 2^{-e} Q(e, n-e)} \tag{24}$$

where $Q(e,s)$ and $Q_i(e,s)$, respectively, are determined recursively by (2) and by

$$Q_i(e,s) = \frac{1}{2}(1-\delta_{i,e+s})\, Q_i(e-1,s) + \frac{1}{2}(1+\delta_{i,e+s})\, \delta(x_{e+s}, y_s)\, Q_i(e,s-1) \tag{25}$$

for $0 \leq s \leq n$, $0 \leq e \leq n-s$, and $(e,s) \neq (0,0)$, from the initial value $Q_i(0,0) = 1$.

The time and space complexities of the corresponding algorithm are clearly $O(n^3)$ and $O(n)$, respectively. For ease of computation, one can introduce the integers $N_i(e,s) = 2^{e+s} Q_i(e,s)$ which can be computed by the recursion

$$N_i(e,s) = (1-\delta_{i,e+s})\, N_i(e-1,s) + (1+\delta_{e,i+s})\, \delta(x_{e+s}, y_s)\, N_i(e,s-1). \tag{26}$$

Then

$$\hat{q}_i = \frac{1}{2} \frac{\sum_{e=0}^{n} 2^{-e} N_i(e, n-e)}{\sum_{e=0}^{n} 2^{-e} N(e, n-e)} \tag{27}$$

where the integers $N(e,s)$ satisfy the recursion (5). The time complexity can be reduced to $O(n^2\sqrt{n})$ if $N_i(e,s)$ and $N(e,s)$ are computed approximately, only for $O(\sqrt{2s})$ values of e around s.

5 Analysis of Posterior Probabilities

The posterior probabilities of individual LFSR_1 bits computed according to Theorem 2 may be useful for reconstructing the unknown LFSR_1 sequence from a known segment of the output sequence if they are sufficiently different from one half. According to Theorem 2 and Corollary 2, the posterior probability $\hat{p}_i$ will be close to 1/4 (3/4) if there is an output segment of length relatively close to $\sqrt{i-1}/2$ around the position $(i-1)/2$ in the output string such that the relative number of 0's (1's) on this segment is considerably different from one half. More generally, if Y^j is relatively unbalanced, that is, if the relative number of 0's in Y^j is considerably different from one half, then most of the posterior probabilities of bits in X^{2j} will be significant.

As $\hat{p}_i$ depends on the output string Y^i, it is interesting to analyze the average value of the absolute difference $|\Delta\hat{p}_i| = |\hat{p}_i - 0.5|$ over purely random Y^i. In view of (20), we get

$$|\Delta\hat{p}_i| = \frac{1}{2} 2^{-(i-1)} |\sum_{e=0}^{i-1} \binom{i-1}{e} (y_{i-e} - 0.5)|. \tag{28}$$

Exact analysis of (28) appears to be difficult. However, the following approximate analysis establishes that $|\Delta\hat{p}_i|$ is significantly different from zero for a uniformly distributed Y^i.

The analysis is based on approximating a binomial distribution $B(n, 0.5)$ by a uniform distribution, with the same expected value and standard deviation, over a segment of length $\sqrt{3n}$ centered around $0.5n$. Consequently, let $I(i)$ denote a segment of length $m(i) \approx \sqrt{3(i-1)}$ centered around $0.5(i+1)$. Then (28) reduces to

$$\begin{aligned} |\Delta\hat{p}_i| &\approx \frac{1}{2} \frac{1}{m(i)} |\sum_{j \in I(i)} (y_j - 0.5)| \\ &\approx \frac{1}{2} \frac{1}{m(i)} |m_1(i) - 0.5m(i)| \end{aligned} \tag{29}$$

where $m_1(i)$ is the number of 1's in Y^i on the segment $I(i)$. Now, as $m_1(i)$ is binomially distributed, we further get the following average values over Y^i

$$|m_1(i) - 0.5m(i)|_{\text{av}} \approx \frac{1}{\sqrt{2\pi}} \sqrt{m(i)} \tag{30}$$

$$\begin{aligned} |\Delta\hat{p}_i|_{\text{av}} &\approx \frac{1}{2\sqrt{2\pi}} \frac{1}{\sqrt{m(i)}} \\ &\approx \frac{1}{2\sqrt{2\pi\sqrt{3}}} \frac{1}{\sqrt[4]{i-1}} \approx 0.1515 \frac{1}{\sqrt[4]{i}}. \end{aligned} \tag{31}$$

Except maybe for the multiplicative constant, the approximation is very good for $i \geq 100$. Thus, as i increases, it turns out that $|\Delta\hat{p}_i|_{\text{av}}$ decreases approximately like $0.1515/\sqrt[4]{i}$. The decrease is to be expected, because of a loss of synchronization between the original and the decimated sequence. However, it may be surprising that the decrease is very slow, so that the posterior probabilities remain significant even for relatively large values of i. For example, $|\Delta\hat{p}_i|_{\text{av}}$ is approximately 0.01515 for $i = 10000$ and 0.01 for $i = 50000$.

The posterior probabilities of individual LFSR_2 bits computed according to Theorem 3 depend on both the output sequence and on the reconstructed LFSR_1 sequence. They are harder to analyze theoretically, but should be much more different from one half than the posterior probabilities of individual LFSR_1 bits, because the LFSR_1 sequence is assumed to be known. They can be used for reconstructing the unknown LFSR_2 sequence from a known segment of the output sequence and a segment of the reconstructed LFSR_1 sequence.

6 Combined Fast Correlation Attacks

It is assumed that the LFSR feedback polynomials and a sufficiently long segment of the keystream sequence, in the known-plaintext scenario, are known. The objective of cryptanalysis is to reconstruct the secret-key-dependent initial states of LFSR_1 and LFSR_2 by an algorithm whose complexity can be relatively small even for large LFSR lengths.

6.1 Basic Attack on LFSR_1

Let Y^n be a given segment of the keystream sequence and let X^n be the corresponding segment of the regularly clocked output sequence of LFSR_1 whose initial state is to be recovered. The basic attack on LFSR_1 consists of two stages.

In the first stage, compute the posterior probabilities of individual bits of X^n by using the probabilistic model in which the input strings are assumed to be purely random. This is achieved in $O(n\sqrt{n})$ time by applying Theorem 2 from Section 3.2. The obtained sequence of posterior probabilities, $(\hat{p}_i)_{i=1}^n$, is a soft-valued estimate of X^n. A hard estimate, $\bar{X}^n = (\bar{x}_i)_{i=1}^n$, of X^n can be obtained by applying the maximum posterior probability decision rule for individual bits, i.e., $\bar{x}_i = 1$ if $\hat{p}_i > 0.5$ and $\bar{x}_i = 0$ otherwise. Therefore

$$\Pr\{\bar{x}_i \neq x_i \mid Y^i\} = \min(\hat{p}_i, 1 - \hat{p}_i). \tag{32}$$

The correlation coefficient between $\bar{x}_i$ and x_i, conditioned on Y^i, is then

$$c_i = 1 - 2\Pr\{\bar{x}_i \neq x_i \mid Y^i\} = |1 - 2\hat{p}_i|. \tag{33}$$

The analysis conducted in Section 5 shows that the expected value of c_i over Y^i slowly decreases approximately like $0.303/\sqrt[4]{i}$ as i increases. So, it remains to be significantly large even for relatively large i such as $i = 10000$.

In the second stage, X^n is reconstructed from $(\hat{p}_i)_{i=1}^n$ by using the LFSR_1 linear recursion. Equation (32) means that $\bar{X}^n$ can be modeled as a noisy output of a time-varying binary symmetric channel when X^n is applied to its input, where the errors are approximately independent. As X^n is a codeword of the corresponding (truncated cyclic) linear block code, the problem of reconstructing X^n is thus essentially a decoding problem. It can be solved by using parity-check based iterative probabilistic decoding algorithms for fast correlation attacks on regularly clocked LFSR's (e.g., see [11], [12], and [8]). The time-variant correlation coefficient should improve the performance of these attacks.

It is known that the complexity of fast correlation attacks on a regularly clocked LFSR and the required output string length n mainly depend on the magnitude of the correlation coefficient and on the degrees and numbers of low-weight polynomial multiples of the LFSR feedback polynomial (e.g., see [11], [13], [7], and [8]). Successful fast correlation attacks are reported in [8], for random feedback polynomials, and in [16], for low-weight feedback polynomials, for the correlation coefficients as small as 2/15 and 1/16, respectively. For the shrinking generator, according to Section 5, the expected value of the correlation coefficient

c_i is considerably different from zero even if i is relatively large. For example, this expected value is approximately equal to 1/10, 1/20, 1/35, and 1/50 for $i = 100$, 1000, 10000, and 50000, respectively.

Since the expected value of c_i slowly decreases as i increases, it is of interest to keep n reasonably small. To this end, the so-called parity checks with memory [10] (also see [7]) or the parity checks sharing a given number of bits in common [14] may be utilized. In conclusion, the second stage of the basic fast correlation attack on the shrinking generator may be successful for a large class of LFSR$_1$ feedback polynomials.

If an information set decoding (e.g., error-free sliding window) technique is applied at the end, then the reconstructed string $\hat{X}^n$ will satisfy the LFSR$_1$ recursion, but should be tested for correlation with $\bar{X}^n$. Alternatively, one may use the posterior probability (4) of blocks of LFSR$_1$ bits as a measure of correlation.

6.2 Iterative Attack on LFSR$_1$

The iterative probabilistic decoding algorithms in the second stage of the basic attack from Section 6.1 iteratively update the posterior probabilities of individual bits of X^n. Therefore, the basic attack can be (considerably) improved if the first stage of the attack is incorporated in iterations of the iterative probabilistic decoding algorithm chosen. For example, we propose an iterative attack whose first iteration coincides with the basic attack and every subsequent iteration consists of two stages. First, update the posterior probabilities of individual bits of X^n by Theorem 1 from Section 3.1 where the posterior probabilities from the preceding iteration are used as the prior probabilities. Second, update the posterior probabilities of individual bits of X^n by applying the iterative probabilistic decoding algorithm.

6.3 Composite Attack on LFSR$_1$

As the posterior probabability $\hat{p}_i$ slowly approaches one half as i increases, it makes sense to divide a longer keystream segment into subsegments of equal length, to compute the posterior probabilities for the subsegments, and then to combine these posterior probabilities appropriately.

To this end, consider m overlapping output subsegments $Y_{jn+1}^{jn+2n+\tau_j}$, $0 \leq j \leq m-1$, where $\tau_j \approx \sqrt{2(j+1)n}$, $0 \leq j \leq m-2$, and $\tau_{m-1} = 0$. Compute $2n+\tau_j$ posterior probabilities for the corresponding LFSR$_1$ segment $X_{i_j+1}^{i_j+2n+\tau_j}$, for each $0 \leq j \leq m-1$. Here, $i_0 = 0$ and for $j > 0$, i_j is unknown, but is expected to be around $2jn+1$ within an interval of length proportional to $\sqrt{2jn}$. So, a segment of $2mn$ posterior probabilities can be composed by guessing i_j, $1 \leq j \leq m-1$, and by taking the posterior probabilities more different from one half for the overlapping parts of the LFSR$_1$ subsegments. Additional τ_j bits for the j-th subsegment serve to fill in a possible gap between the j-th and $(j+1)$-th subsegments. As $\hat{p}_i$ slowly changes with i, the method is not sensitive to $m-1$ guesses of unknown positions i_j.

Finally, a fast correlation attack is run by using the composed segment of $2mn$ consecutive posterior probabilities. It has to be run for each of about $\sqrt{(m-1)!}(2n)^{(m-1)/2}$ guesses. For example, $n \leq 20000$ and $m \leq 5$ are realistic choices of the parameters.

6.4 Subsequence Attack on LFSR$_1$

Suppose that the posterior probabilities corresponding to a given keystream segment Y^n are not sufficiently different from one half, because the length n required for the success of fast correlation attacks explained above is too large. One can then compute the posterior probabilities for a number of subsequences of the keystream sequence obtained by discarding the initial segment of variable length until more significant posterior probabilities are obtained. This will improve the performance of the fast correlation attacks, but the length of the initial LFSR$_1$ segment has to be guessed. More precisely, if a segment $X_{j'}^{j'+n-1}$ of the LFSR$_1$ sequence is reconstructed from the output segment Y_j^{j+n-1}, one has to make $O(\sqrt{2j})$ guesses around the expected value $2j$ in order to find the unknown initial position j'. The number of tested subsequences is j/δ if one skips $\delta - 1$ output bits at a time. Testing can be simplified by searching for relatively unbalanced output subsequences instead of the significant posterior probabilities.

In particular, one can also search for about r_1, not necessarily consecutive, outstanding posterior probabilities (close to 1/4 or 3/4) and then apply an information set decoding algorithm to recover the LFSR$_1$ initial state, where the posterior probability (4) of blocks of LFSR$_1$ bits is used as a measure of correlation. The success of such an algorithm is independent of the LFSR$_1$ feedback polynomial, but, according to the information set decoding arguments, the achievable complexity cannot be smaller than about $2^{0.5r_1}$ corresponding steps. This improves the reduced complexity method [9] based on specific subsequences of the output sequence. Namely, as the class of usable subsequences is effectively enlarged, the required keystream segment length, around $2^{0.5r_1}$, can be considerably reduced. The expression given in [9] is approximative, whereas the accurate expression for the posterior probabilities is provided by Theorem 2. Moreover, the need for guessing the length of the initial LFSR$_1$ segment is overlooked in [9].

6.5 Reinitialization Attack on LFSR$_1$

Suppose that for resynchronization purposes the shrinking generator is reinitialized by bitwise addition of a reinitialization vector to the secret-key-controlled LFSR initial states, in view of the fact that the nonlinear next-state function prevents the resynchronization attack [2]. The posterior probabilities of individual LFSR$_1$ bits produced from the secret-key-controlled initial state can then be computed for different initialization vectors and all combined into values more different from one half, so that the corresponding fast correlation attack is easier.

6.6 Attack on $LFSR_2$

After reconstructing a candidate initial state of $LFSR_1$, the initial state of $LFSR_2$ can be recovered by computing the posterior probabilities of individual $LFSR_2$ bits by Theorem 3 from Section 4. More precisely, the posterior probabilities of individual bits of C^m are computed in $O(m^2\sqrt{m})$ time from given Y^m and reconstructed $\hat{X}^m$, $m \leq n$. Here, C^m is the corresponding segment of the regularly clocked output sequence of $LFSR_2$ whose initial state is to be recovered. As $\hat{X}^m$ is assumed to be known, the obtained posterior probabilities are much more distinguished from one half than in the case of $LFSR_1$. The reconstruction problem is then much easier and m can be much smaller than n. The posterior probabilities can be further enhanced by the reinitialization method described in Section 6.5. Accordingly, the initial state of $LFSR_2$ can be reconstructed by iterative probabilistic decoding algorithms in the same way as in the basic attack on $LFSR_1$ explained in Section 6.1. Moreover, as the posterior probabilities can be close to 0 or 1, simple information set decoding algorithms may also be applicable.

One should repeat the attack on $LFSR_2$ for several small phase shifts, positive or negative, of the reconstructed $LFSR_1$ sequence until the correct initial states of both LFSR's are reconstructed. Note that the number of solutions for the LFSR initial states is the number of 0's in a cycle of the $LFSR_2$ sequence preceding the first clock-control bit equal to 1 (see [15]).

7 Conclusions

The introduced probabilistic analysis of the shrinking generator shows that the irregularly clocked LFSR's, unlike a common belief in the open literature, may be vulnerable to fast correlation attacks. The analysis can be generalized to deal with arbitrary keystream generators based on clock-controlled LFSR's.

In order to reconstruct the initial state of the clock-controlled LFSR, $LFSR_1$, in the shrinking generator, the new idea is to compute the posterior probabilities of individual bits of the regularly clocked $LFSR_1$ sequence when conditioned on a given segment of the output sequence. Perhaps surprisingly, a theoretical analysis indicates that these probabilities can be significantly different from one half even for relatively long segments of the $LFSR_1$ sequence. Accordingly, the initial state of $LFSR_1$ may be recovered by a fast correlation attack, applicable to a regularly clocked LFSR, based on the computed posterior probabilities. It is known that such an attack can be successful for certain LFSR feedback polynomials. More sophisticated fast correlation attacks including the iterative attack, the composite attack, the subsequence attack, and the reinitialization attack are also proposed.

The initial state of the clock-control LFSR, $LFSR_2$, can be reconstructed in a similar way, but based on the computed posterior probabilities of individual bits of the regularly clocked $LFSR_2$ sequence when conditioned on a given segment of the output sequence and on a segment of the reconstructed $LFSR_1$ sequence. As these probabilities are more distinguished from one half, the corresponding fast correlation attack is easier.

Appendix

A Proof of Theorem 1

To prove (10), we start from (7). First, in view of (8), we get

$$\begin{aligned}\Pr\{Y^n \mid x_i = 1\} &= \sum_{e=0}^{n} \Pr\{Y^n, d(C^n) = e \mid x_i = 1\} \\ &= \sum_{e=0}^{n} \Pr\{Y_{n-e+1}^n, Y^{n-e}, d(C^n) = e \mid x_i = 1\} \\ &= \sum_{e=0}^{n} \Pr\{Y_{n-e+1}^n \mid Y^{n-e}, d(C^n) = e, x_i = 1\}\, P_i(e, n-e) \\ &= \sum_{e=0}^{n} 2^{-e}\, P_i(e, n-e). \end{aligned} \tag{34}$$

Namely, on the condition that $d(C^n) = e$, the string Y_{n-e+1}^n is obtained by decimating X_{n+1} according to C_{n+1}, where X_{n+1} and C_{n+1} are mutually independent and purely random even when conditioned on x_i and Y^{n-e}. Therefore, under the given conditions, Y_{n-e+1}^n is uniformly distributed. Similarly, in view of (9), we have

$$\Pr\{Y^n\} = \sum_{e=0}^{n} 2^{-e}\, P(e, n-e). \tag{35}$$

Consequently, (7) together with (34) and (35) result in (10).

As for the recursions, we only prove (11), whereas (12) is proved analogously. For $(e, s) \neq (0, 0)$, (8) results in

$$\begin{aligned} P_i(e, s) &= \Pr\{Y^s, d(C^{e+s}) = e \mid x_i = 1, c_{e+s} = 0\} \cdot \Pr\{c_{e+s} = 0 \mid x_i = 1\} \\ &\quad + \Pr\{Y^s, d(C^{e+s}) = e \mid x_i = 1, c_{e+s} = 1\} \cdot \Pr\{c_{e+s} = 1 \mid x_i = 1\} \\ &= \Pr\{Y^s, d(C^{e+s-1}) = e - 1 \mid x_i = 1, c_{e+s} = 0\} \cdot \frac{1}{2} \\ &\quad + \Pr\{Y^s, d(C^{e+s-1}) = e \mid x_i = 1, c_{e+s} = 1\} \cdot \frac{1}{2}. \end{aligned} \tag{36}$$

Now, as $d(C^{e+s-1})$ is independent of c_{e+s}, and Y^s is independent of c_{e+s} on the condition that $d(C^{e+s-1}) = e - 1$, we get

$$\begin{aligned}\Pr\{Y^s, d(C^{e+s-1}) = e - 1 \mid x_i = 1, c_{e+s} = 0\} &= \\ \Pr\{Y^s, d(C^{e+s-1}) = e - 1 \mid x_i = 1\} &= P_i(e - 1, s). \end{aligned} \tag{37}$$

On the other hand, if $c_{e+s} = 1$ and $d(C^{e+s-1}) = e - 1$, then $y_s = x_{e+s}$. Thus, we get

$$\begin{aligned}
\Pr\{&Y^s, d(C^{e+s-1}) = e \mid x_i = 1, c_{e+s} = 1\} \\
&= \Pr\{x_{e+s} = y_s, Y^{s-1}, d(C^{e+s-1}) = e \mid x_i = 1, c_{e+s} = 1\} \\
&= \Pr\{x_{e+s} = y_s \mid Y^{s-1}, d(C^{e+s-1}) = e, x_i = 1, c_{e+s} = 1\} \\
&\quad \cdot \Pr\{Y^{s-1}, d(C^{e+s-1}) = e \mid x_i = 1, c_{e+s} = 1\} \qquad (38) \\
&= \Pr\{x_{e+s} = y_s \mid x_i = 1\} \\
&\quad \cdot \Pr\{Y^{s-1}, d(C^{e+s-1}) = e \mid x_i = 1\} \qquad (39) \\
&= (\delta_{i,e+s} y_s + (1 - \delta_{i,e+s})(y_s p_{e+s} + (1 - y_s)(1 - p_{e+s}))) \cdot P_i(e, s-1). \qquad (40)
\end{aligned}$$

The first line of (39) follows from the first line of (38) because x_{e+s} is independent of C^{e+s} and, on the condition that $d(C^{e+s-1}) = e$, it is also independent of Y^{s-1}. In addition, as $d(C^{e+s-1})$ is independent of c_{e+s} and Y^{s-1} is independent of c_{e+s} on the condition that $d(C^{e+s-1}) = e$, the second line of (39) follows from the second line of (38).

Equation (11) directly follows from (36), (37), and (40). If $e = 0$, then the first term on the right-hand side of (11) is omitted, and if $s = 0$, then the second term on the right-hand side of (11) is omitted. The correct values of $P_i(1, 0)$ and $P_i(0, 1)$ are both obtained from the initial value $P_i(0, 0) = 1$.

B Proof of Theorem 3

The proof is essentially similar to the proof of Theorem 1, but should be conducted carefully. To prove (24), we start from (22). First, in view of (23), we get

$$\begin{aligned}
\Pr\{&Y^n \mid c_i = 1, X^n\} \\
&= \sum_{e=0}^{n} \Pr\{Y^n, d(C^n) = e \mid c_i = 1, X^n\} \\
&= \sum_{e=0}^{n} \Pr\{Y^n_{n-e+1}, Y^{n-e}, d(C^n) = e \mid c_i = 1, X^n\} \\
&= \sum_{e=0}^{n} \Pr\{Y^n_{n-e+1} \mid Y^{n-e}, d(C^n) = e, c_i = 1, X^n\}\, Q_i(e, n-e) \\
&= \sum_{e=0}^{n} 2^{-e}\, Q_i(e, n-e). \qquad (41)
\end{aligned}$$

Namely, on the condition that $d(C^n) = e$, the string Y^n_{n-e+1} is obtained by decimating X_{n+1} according to C_{n+1}, where X_{n+1} and C_{n+1} are mutually independent and purely random even when conditioned on c_i and Y^{n-e}. Therefore, under the given conditions, Y^n_{n-e+1} is uniformly distributed. Note that (3) is similarly derived from (1). Consequently, (22) together with (41) and (3) result in (24).

As for the recursions, we note that the proof of (2) is similar to the proof of (25) given below. For $(e,s) \neq (0,0)$, (23) results in

$$\begin{aligned} Q_i(e,s) &= \Pr\{Y^s, d(C^{e+s}) = e \mid c_i = 1, X^n, c_{e+s} = 0\} \cdot \Pr\{c_{e+s} = 0 \mid c_i = 1, X^n\} \\ &\quad + \Pr\{Y^s, d(C^{e+s}) = e \mid c_i = 1, X^n, c_{e+s} = 1\} \cdot \Pr\{c_{e+s} = 1 \mid c_i = 1, X^n\} \\ &= \Pr\{Y^s, d(C^{e+s-1}) = e-1 \mid c_i = 1, X^n, c_{e+s} = 0\} \cdot \frac{1}{2}(1 - \delta_{i,e+s}) \\ &\quad + \Pr\{Y^s, d(C^{e+s-1}) = e \mid c_i = 1, X^n, c_{e+s} = 1\} \cdot \frac{1}{2}(1 + \delta_{i,e+s}) \qquad (42) \end{aligned}$$

where the conditional probability in the first term is computed only for $i \neq e+s$.

Now, as $d(C^{e+s-1})$ is independent of c_{e+s}, and Y^s is independent of c_{e+s} on the condition that $d(C^{e+s-1}) = e-1$, we get that for $i \neq e+s$

$$\begin{aligned} &\Pr\{Y^s, d(C^{e+s-1}) = e-1 \mid c_i = 1, X^n, c_{e+s} = 0\} \\ &\quad = \Pr\{Y^s, d(C^{e+s-1}) = e-1 \mid c_i = 1, X^n\} = Q_i(e-1,s). \qquad (43) \end{aligned}$$

On the other hand, if $c_{e+s} = 1$ and $d(C^{e+s-1}) = e-1$, then $y_s = x_{e+s}$. Thus, we get

$$\begin{aligned} &\Pr\{Y^s, d(C^{e+s-1}) = e \mid c_i = 1, X^n, c_{e+s} = 1\} \\ &\quad = \Pr\{x_{e+s} = y_s, Y^{s-1}, d(C^{e+s-1}) = e \mid c_i = 1, X^n, c_{e+s} = 1\} \\ &\quad = \Pr\{x_{e+s} = y_s \mid Y^{s-1}, d(C^{e+s-1}) = e, c_i = 1, X^n, c_{e+s} = 1\} \\ &\qquad \cdot \Pr\{Y^{s-1}, d(C^{e+s-1}) = e \mid c_i = 1, X^n, c_{e+s} = 1\} \qquad (44) \\ &\quad = \Pr\{x_{e+s} = y_s \mid x_{e+s}\} \\ &\qquad \cdot \Pr\{Y^{s-1}, d(C^{e+s-1}) = e \mid c_i = 1, X^n\} \qquad (45) \\ &\quad = \delta(x_{e+s}, y_s) \cdot Q_i(e, s-1). \qquad (46) \end{aligned}$$

The first line of (45) follows from the first line of (44) as x_{e+s} is contained in X^n. In addition, as $d(C^{e+s-1})$ is independent of c_{e+s} and Y^{s-1} is independent of c_{e+s} on the condition that $d(C^{e+s-1}) = e$, the second line of (45) follows from the second line of (44).

Equation (25) directly follows from (42), (43), and (46). If $e = 0$, then the first term on the right-hand side of (25) is omitted, and if $s = 0$, then the second term on the right-hand side of (25) is omitted. The correct values of $Q_i(1,0)$ and $Q_i(0,1)$ are both obtained from the initial value $Q_i(0,0) = 1$.

References

1. D. Coppersmith, H. Krawczyk, and Y. Mansour, "The shrinking generator," Advances in Cryptology - CRYPTO '93, *Lecture Notes in Computer Science*, vol. 773, pp. 22-39, 1993.
2. J. Daemen, R. Govaerts, and J. Vandewalle, "Resynchronization weakness in synchronous stream ciphers," Advances in Cryptology - EUROCRYPT '93, *Lecture Notes in Computer Science*, vol. 765, pp. 159-167, 1994.

3. J. Dj. Golić, "Intrinsic statistical weakness of keystream generators," Advances in Cryptology - ASIACRYPT '94, *Lecture Notes in Computer Science*, vol. 917, pp. 91-103, 1995.
4. J. Dj. Golić and L. O'Connor, "Embedding and probabilistic correlation attacks on clock-controlled shift registers," Advances in Cryptology - EUROCRYPT '94, *Lecture Notes in Computer Science*, vol. 950, pp. 230-243, 1995.
5. J. Dj. Golić, "Towards fast correlation attacks on irregularly clocked shift registers," Advances in Cryptology - EUROCRYPT '95, *Lecture Notes in Computer Science*, vol. 921, pp. 248-262, 1995.
6. J. Dj. Golić and R. Menicocci, "Edit probability correlation attack on the alternating step generator," Sequences and their Applications - SETA '98, *Discrete Mathematics and Theoretical Computer Science*, C. Ding, T. Helleseth, and H. Niederreiter eds., Springer-Verlag, pp. 213-227, 1999.
7. J. Dj. Golić, "Iterative probabilistic decoding and parity checks with memory," *Electronics Letters*, vol. 35(20), pp. 1721-1723, Sept. 1999.
8. J. Dj. Golić, M. Salmasizadeh, and E. Dawson, "Fast correlation attacks on the summation generator," *Journal of Cryptology*, vol. 13, pp. 245-262, 2000.
9. T. Johansson, "Reduced complexity correlation attacks on two clock-controlled generators," Advances in Cryptology - ASIACRYPT '98, *Lecture Notes in Computer Science*, vol. 1514, pp. 342-357, 1998.
10. T. Johansson and F. Jonnson, "Improved fast correlation attacks on stream ciphers via convolutional codes," Advances in Cryptology - EUROCRYPT '99, *Lecture Notes in Computer Science*, vol. 1592, pp. 347-362, 1999.
11. W. Meier and O. Staffelbach, "Fast correlation attacks on certain stream ciphers," *Journal of Cryptology*, vol. 1, pp. 159-176, 1989.
12. M. J. Mihaljević and J. Dj. Golić, "A comparison of cryptanalytic principles based on iterative error-correction," Advances in Cryptology - EUROCRYPT '91, *Lecture Notes in Computer Science*, vol. 547, pp. 527-531, 1991.
13. M. J. Mihaljević and J. Dj. Golić, "Convergence of a Bayesian iterative error-correction procedure on a noisy shift register sequence," Advances in Cryptology - EUROCRYPT '92, *Lecture Notes in Computer Science*, vol. 658, pp. 124-137, 1993.
14. M. J. Mihaljević, M. P. C. Fossorier, and H. Imai, "A low-complexity and high-performance algorithm for the fast correlation attack," Fast Software Encryption - New York 2000, *Lecture Notes in Computer Science*, vol. 1978, pp. 196-212, 2001.
15. L. Simpson, J. Dj. Golić, and E. Dawson, "A probabilistic correlation attack on the shrinking generator," Information Security and Privacy - Brisbane '98, *Lecture Notes in Computer Science*, vol. 1438, pp. 147-158, 1998.
16. L. Simpson, J. Dj. Golić, M. Salmasizadeh, and E. Dawson, "A fast correlation attack on multiplexer generators," *Information Processing Letters*, vol. 70, pp. 89-93, 1999.
17. K. Zeng, C. H. Yang, and T. R. N. Rao, "On the linear consistency test (LCT) in cryptanalysis with applications," Advances in Cryptology - CRYPTO '89, *Lecture Notes in Computer Science*, vol. 435, pp. 164-174, 1990.

Nonlinear Vector Resilient Functions

Jung Hee Cheon

International Research center for Information Security (IRIS)
Information and Communications University (ICU), Taejon, Republic of Korea
jhcheon@icu.ac.kr
http://vega.icu.ac.kr/~jhcheon

Abstract. An (n, m, k)-resilient function is a function $f : \mathbb{F}_2^n \to \mathbb{F}_2^m$ such that every possible output m-tuple is equally likely to occur when the values of k arbitrary inputs are fixed by an adversary and the remaining $n-k$ input bits are chosen independently at random. In this paper we propose a new method to generate a $(n + D + 1, m, d - 1)$-resilient function for any non-negative integer D whenever a $[n, m, d]$ linear code exists. This function has algebraic degree D and nonlinearity at least $2^{n+D} - 2^n \lfloor\sqrt{2^{n+D+1}}\rfloor + 2^{n-1}$. If we apply this method to the simplex code, we can get a $(t(2^m - 1) + D + 1, m, t2^{m-1} - 1)$-resilient function with algebraic degree D for any positive integers m, t and D. Note that if we increase the input size by D in the proposed construction, we can get a resilient function with the same parameter except algebraic degree increased by D.

Keywords: Resilient functions, nonlinearity, correlation immunity, linearized polynomials

1 Introduction

An (n, m, k)-resilient function is a function $f : \mathbb{F}_2^n \to \mathbb{F}_2^m$ such that every possible output m-tuple is equally likely to occur when the values of k arbitrary inputs are fixed by an adversary and the remaining $n-k$ input bits are chosen independently at random. The concept was introduced by Chor *et al.* in [8] and independently by Bennett *et al.* in [1]. It was called just *a resilient function* in those references. We call it a vector resilient function when we need to distinguish it from a resilient function with $m = 1$ since the term 'a resilient function' was regarded as a balanced correlation immune function, i.e. a resilient function with $m = 1$ in recent references [17,21]. The application area of this function includes fault-tolerant distributed computing [8], privacy amplification [1,2] and a combining generator for stream ciphers. A resilient function is also closely related to the coloring problem [9] to find the smallest k such that $(2^m; n, k)$-coloring exists. *$(2^m; n, k)$-coloring is a coloring of the n-dimensional Boolean cube with 2^m colors such that in every k-dimensional subcube each color appears $2^k/2^m$ times.*

Almost all of works on resilient functions with few exceptions [5,20,23] deals with linear resilient functions or resilient functions with a single bit output. In

J. Kilian (Ed.): CRYPTO 2001, LNCS 2139, pp. 458–469, 2001.

[10,3], they focused on finding a bound on a resiliency of a vector Boolean function with algebraic degree one. In [6,7,16,17,18,21], they focused on constructing a resilient function with a single bit output having as high as nonlinearity as possible. In [23], Zhang and Zheng proposed a method to construct a nonlinear vector resilient function from a linear vector resilient function by permuting nonlinearly its output bits. This method gives an easy transformation from a linear resilient function to a nonlinear resilient function, but has a disadvantage that a resilient function with m bit output constructed by the method has algebraic degree at most m. In [20], Stinson and Massey proposed nonlinear resilient functions, which are the counterexamples of the conjecture: *If there exist a resilient function with certain parameters, then there exists a linear resilient function with the same parameters.* They proposed infinitely many functions, but it covers only special parameters.

In this paper, we propose a new method to construct nonlinear vector resilient functions using linearized polynomial. A linearized polynomial $R(x)$ is a polynomial over $\mathbb{F}_{2^n}$ such that every term of $R(x)$ has degree of a power of 2. An equivalent definition is that the set of roots of $R(x)$ in its splitting field forms a vector space over $\mathbb{F}_2$. Given positive integers n, m and D, let d to be the minimal distance of certain m-dimensional linear code with length n. If we take a linearized polynomial $R(x)$ whose roots forms a n-dimensional subspace of $\mathbb{F}_{2^{n+D+1}}$, then some projection of $R(x)^{-1}+x$ to $\mathbb{F}_{2^m}$ is a $(n+D+1, m, d-1)$-resilient function under the basis whose dual contains a subset generating the set of roots of $R(x)$. We can easily find such a projection using a $[n, m, d]$ linear code. Such a function has algebraic degree D and nonlinearity at least $2^{n+D}-2^n\lfloor\sqrt{2^{n+D+1}}\rfloor+2^{n-1}$. To sum up, we can construct a $(n+D+1, m, d-1)$-resilient function with algebraic degree D whenever a $[n, m, d]$ linear code exists. Observe that by increasing the input size by D we can construct a resilient function with the same parameter except algebraic degree increased by D.

A simplex code is a $[2^m-1, m, 2^{m-1}]$ linear code, whose minimal distance is maximal. By concatenating each codeword t times, we get a $[t(2^m-1), m, t2^{m-1}]$ linear code. Using this code, we can construct a $(t(2^m-1)+D+1, m, t2^{m-1}-1)$-resilient function with algebraic degree D for any positive integers m, t and D. It has nonlinearity greater than or equal to

$$2^{t(2^m-1)+D} - 2^{t(2^m-1)}\lfloor\sqrt{2^{t(2^m-1)+D+1}}\rfloor + 2^{t(2^m-1)-1}.$$

In Section 2, we introduce some notation and definitions of cryptographic properties. In Section 3, we propose a new method to construct a resilient function from a linearized polynomial. In Section 4, we prove the algebraic degree of the proposed resilient function. In Section 5, we deal with nonlinearity. In Section 6, we generalize the method in Section 3 into a vector resilient function. In Section 7, we apply the proposed vector resilient function for a combining generator with multi-bit output, a kind of stream cipher. We conclude in Section 8.

2 Boolean Functions and Nonlinearity

Let E be a vector space of finite dimension n over the finite field $\mathbb{F}_2$. A function f from E into $\mathbb{F}_2$ is called *a Boolean function*. The cardinality of the set $\{x \in E | f(x) = 1\}$ is called the *weight* of f and denoted by $\mathrm{wt}(f)$. The degree of f, denoted by $\deg(f)$, is the maximal value of the degrees of the terms of f when expressed in the reduced form, called the *algebraic normal form*. A function with degree 1 is called an affine function. The *Hamming distance* between two function f and g is the weight of $f + g$. The minimal distance between f and any affine function from E into $\mathbb{F}_2$ is the *nonlinearity* of f, that is:

$$\mathcal{N}(f) = \min_{\phi \in \Gamma} \mathrm{wt}(f + \phi) \tag{1}$$

where Γ is the set of all affine functions over E.

A function $F : E \to \mathbb{F}_{2^m}$ is called *a vector Boolean function*. Note that if a basis of $\mathbb{F}_{2^m}$ over $\mathbb{F}_2$ is specified, there are the unique Boolean function f_i's such that $F = (f_1, f_2, \cdots, f_m)$. We denotes by $b \cdot F$ the Boolean function $b_1 f_1 + b_2 f_2 + \cdots + b_n f_n$ for $b = (b_1, b_2, \cdots, b_m) \in \mathbb{F}_{2^m}$. Using this notation, we can write Γ as follows:

$$\Gamma = \{a \cdot x + \delta | a \in E, \delta \in \mathbb{F}_2\}. \tag{2}$$

Definition 1. *The nonlinearity $\mathcal{N}(F)$ of a Boolean function $F : E \to \mathbb{F}_{2^m}$ is defined as*

$$\mathcal{N}(F) = \min_{b \in \mathbb{F}_{2^m}^*} N(b \cdot F) = \min_{b \in \mathbb{F}_{2^m}^*, \phi \in \Gamma} wt(b \cdot F + \phi) \tag{3}$$

where Γ is the set of all affine functions over E. Or equivalently,

$$\mathcal{N}(F) = \min_{a \in E, b \in \mathbb{F}_{2^m}^*, \delta \in \mathbb{F}_2} wt(b \cdot F + a \cdot x + \delta). \tag{4}$$

The Walsh-Hadamard transformation of a Boolean function f is defined as

$$W_f(a) = \sum_{x \in E} (-1)^{f(x) + a \cdot x}, \quad a \in E. \tag{5}$$

Since $W_f(a) = wt(f(x) + a \cdot x) - wt(f(x) + a \cdot x + 1)$, we have

$$\mathcal{N}(f) = 2^{n-1} - \frac{1}{2} \max_{a \in E} |W_f(a)|. \tag{6}$$

Definition 2. *A Boolean function $f : E \to \mathbb{F}_2$ is called a k-th order correlation immune function if $W_f(a) = 0$ for all $a \in E$ with $0 < wt(a) \leq k$. A k-th order correlation immune function is called a k-resilient function if it is balanced(i.e. $W_f(0) = 0$).*

Definition 3. *A vector Boolean function $F : E \to \mathbb{F}_{2^m}$ is called a k-resilient function or a (n, m, k)-resilient function for the dimension n of E if $b \cdot F$ is a k-resilient function for any $b \in \mathbb{F}_{2^m}^*$.*

3 Resiliency

Throughout this paper, let $q = 2^n$ for a positive integer n. A polynomial in $\mathbb{F}_q[x]$ is called a linearized polynomial if each of its terms has degree of a power of 2 [14]. Let $R(x) = \sum_{i=0}^{h} A_i x^{2^i}$ $(A_i \in \mathbb{F}_{2^n})$ be a linearized polynomial over $\mathbb{F}_{2^n}$ and $N_R(\mathbb{F}_q) = \{x \in \mathbb{F}_q | R(x) = 0\}$ be the set of zeros of $R(x)$ which forms a subspace of $\mathbb{F}_q$. From now on, we define the inversion function $R(x)^{-1}$ to be $R(x)^{2^n-2}$. Note that if we represent $a, b \in \mathbb{F}_q$ by a basis and its dual basis, respectively, we have $a \cdot b = Tr[ab]$ where $Tr[\cdot]$ is the trace function from $\mathbb{F}_q$ to $\mathbb{F}_2$.

Lemma 1. *[11] Let $a, b \in \mathbb{F}_q$, $R(x)$ a linearized polynomial and $F(x) = 1/R(x)$. If $Tr[ax]$ does not vanish identically on $N_R(\mathbb{F}_q)$, then*

$$W_{Tr[bF(x)]}(a) = 0.$$

Proof. Suppose $x_0 \in \mathbb{F}_q \setminus N_R(\mathbb{F}_q)$. For $x = x_0 + x'$ with $x' \in N_R(\mathbb{F}_q)$, we have $Tr[ax + \frac{b}{R(x)}] = Tr[ax_0 + \frac{b}{R(x_0)}] + Tr[ax']$ and this is zero for $\#N_R(\mathbb{F}_q)/2$ elements x'. Since a half of elements of each coset of $N_R(\mathbb{F}_q)$ satisfies $Tr[ax + \frac{b}{R(x)}] = 0$, we have $W_{Tr[bF]}(a) = 0$.

Using this, we can derive the following.

Theorem 1. *Let $R(x)$ be a linearized polynomial such that $N_R(\mathbb{F}_q)$ is generated by $\{\xi_1, \xi_2, \cdots, \xi_w\}$ for some $w > 0$, and let $F(x) = 1/R(x) + cx$ for $c \in \mathbb{F}_q$. Suppose $\mathcal{B} = \{\xi_1, \xi_2, \cdots, \xi_n\}$ is a basis of $\mathbb{F}_q$ and $\hat{\mathcal{B}} = \{\widehat{\xi_1}, \widehat{\xi_2}, \cdots, \widehat{\xi_n}\}$ its dual basis. Then $Tr[bF]$ is a $(t-1)$-resilient function under the basis $\mathcal{B}$ if the projection of bc on $\langle \widehat{\xi_1}, \widehat{\xi_2}, \cdots, \widehat{\xi_w} \rangle$ has weight t.*

Observe that the maximum of t is w.

Proof. Let $a = \sum_{i=0}^{n} a_i \hat{\xi_i}$ and $bc = \sum_{i=1}^{n} b_i \hat{\xi_i}$. If we write $f(x) = Tr[b(1/R(x) + cx)]$, we have

$$\begin{aligned} W_f(a) \neq 0 &\Leftrightarrow Tr[(a + bc)x] = 0 \quad \text{on } N_R(\mathbb{F}_q) \\ &\Leftrightarrow Tr[(a + bc)\xi_i] = 0 \quad \text{for } 1 \leq i \leq w \\ &\Leftrightarrow a_i = b_i \quad \text{for } 1 \leq i \leq w \end{aligned}$$

Since t elements of b_i for $1 \leq i \leq w$ is equal to one, we have $W_f(a) = 0$ for all a with $0 \leq wt(a) < t$, which proves the $(t-1)$-resiliency of $Tr[bF]$.

Example 1. Let $q = 2^8$ and $V = \{\xi_1, \xi_2, \xi_3, \xi_4\}$ a set of linearly independent elements of $\mathbb{F}_q$, and let $R(x) = \prod(x - \xi)$ where ξ ranges over all linear combinations of elements of V. Suppose $\mathcal{B} = \{\xi_1, \xi_2, \cdots, \xi_8\}$ is a basis of $\mathbb{F}_q$ and $\hat{\mathcal{B}} = \{\widehat{\xi_1}, \widehat{\xi_2}, \cdots, \widehat{\xi_8}\}$ its dual basis. Then $f(x) = Tr[(\widehat{\xi_1} + \widehat{\xi_2} + \widehat{\xi_3} + \widehat{\xi_4})(\frac{1}{R(x)} + x)]$ is a 3-resilient function under the basis $\hat{\mathcal{B}}$.

4 Algebraic Degree

Theorem 2. *Let $w \geq 0$. Consider a linearized polynomial $R(x) = \prod(x - \xi)$ where ξ ranges over all elements of a w-dimensional subspace V of $\mathbb{F}_q$. Then $F(x) = \frac{1}{R(x)}$ has the algebraic degree $n - 1 - w$.*

Proof. First, we claim that $F(x)$ has the algebraic degree $\leq n - 1 - w$. We use the induction on w. For $w = 0$, it is trivial since $F(x) = 1/x$ has the algebraic degree $n - 1$. Assume that the claim holds for all dimension less than w. Let W be a $(w-1)$-dimensional subspace of V, $\alpha \in V \setminus W$ and $S(x) = \prod_{\zeta \in W}(x - \zeta)$. Then we have

$$\frac{1}{R(x)} = \frac{1}{S(x)S(x+\alpha)} = \frac{1}{S(x) + S(x+\alpha)} \left(\frac{1}{S(x)} + \frac{1}{S(x+\alpha)} \right). \tag{7}$$

Note that $f(x) + f(x + a)$ has algebraic degree less than that of f for any Boolean function f and $a \in \mathbb{F}_q$. Since $S(x)$ is a linearized polynomial and so has the algebraic degree 1, $S(x) + S(x + \alpha)$ is a nonzero constant for $\alpha \in W$. By the induction hypothesis, $\frac{1}{S(x)}$ has algebraic degree $\leq n - 1 - (w - 1) = n - w$. Hence $F(x)$ has algebraic degree less than $n - w$ which proves the claim.

Now we prove the equality. Suppose that there is a w-dimensional subspace V such that $\frac{1}{R(x)}$ has algebraic degree less than $n - w - 1$. Take a basis $B = \langle \xi_1, \xi_2, \cdots, \xi_n \rangle$ of $\mathbb{F}_q$ where $\xi_1, \xi_2, \cdots, \xi_w$ generates V. Take $R_w(x) = R(x)$ and $R_{i+1}(x) = R_i(x)R_i(x + \xi_i)$ for $w \leq i < n - 1$. By the same deduction with (7), $1/R_{i+1}(x)$ has algebraic degree less than $1/R_i(x)$ for $w \leq i < n - 1$. Thus, $1/R_{n-1}(x)$ has algebraic degree less than $(n - 1) - (n - 1) = 0$. That is, $1/R_{n-1}(x) = 0$ should be zero for all $x \in \mathbb{F}_q$ which implies $R_{n-1}(x) = 0$ for all $x \in \mathbb{F}_q$. This is a contradiction because R_{n-1} has only 2^{n-1} roots. Therefore we have the theorem.

Observe that if V has the dimension w, we can derive a $(w - 1)$-resilient function with the algebraic degree $n - w - 1$ from $F(x) = 1/R(x)$. From the Siegenthaler's inequality [19], we have $\deg f \leq n - 1 - (w - 1) = n - w$ for every component function f of $1/R(x)$. Thus, our resilient function has one less algebraic degree than the maximal degree achieved by $(w-1)$-resilient functions in $\mathbb{F}_q$.

5 Nonlinearity

Consider a non-singular complete curve given by $y^2 + y = ax + \frac{b}{R(x)}$ for $a, b \in \mathbb{F}_q$. By Hurwitz-Zeuthen formula, it has the genus $g = 2^h - \delta_{a,0}$ where h is the degree of $R(x)$ and the Kronecker delta $\delta_{a,0}$ is one if and only if $a = 0$. Using the Hasse-Weil bound on the number of points of an algebraic curve, we can get the following lemma.

Lemma 2. *Let $R(x)$ be a linearized polynomial such that $N_R(\mathbb{F}_q)$ is generated by $\{\xi_1, \xi_2, \cdots, \xi_w\}$ for some $0 < w < n$. Let $a, b \in \mathbb{F}_q$ and $b \neq 0$. Let C be a complete non-singular curve over $\mathbb{F}_q$ given by $y^2 + y = ax + \frac{b}{R(x)}$. Then we have*

$$|\#C(\mathbb{F}_q) - q - 1| \leq 2g\sqrt{q},$$

where $g = 2^w - \delta_{a,0}$ is the genus of the curve C.

Theorem 3. *Let $R(x)$ be a linearized polynomial such that $N_R(\mathbb{F}_q)$ is generated by $\{\xi_1, \xi_2, \cdots, \xi_w\}$ for some $0 < w < n$. Then we have*

$$\mathcal{N}(\frac{1}{R(x)}) \geq 2^{n-1} - 2^w \lfloor \sqrt{2^n} \rfloor + 2^{w-1}.$$

Proof. Let $F(x) = 1/R(x)$ and $b \in \mathbb{F}_q^*$.

Assume $a \neq 0$. The complete non-singular curve C given by $y^2 + y = ax + b/R(x)$ has a point at the infinity and a point on each of roots of $R(x)$. Otherwise, it has 2 points whenever $Tr[ax + b/R(x)] = 0$. Hence we have

$$\#C(\mathbb{F}_q) = 2\#\{x \in \mathbb{F}_q | Tr[ax + \frac{b}{R(x)}] = 0\} + 2^w + 1. \tag{8}$$

Assume $a = 0$. The complete non-singular curve C given by $y^2 + y = b/R(x)$ has two points at the infinity and a point on each of roots of $R(x)$. Otherwise, it has 2 points whenever $Tr[ax + b/R(x)] = 0$. Hence we have

$$\#C(\mathbb{F}_q) = 2\#\{x \in \mathbb{F}_q | Tr[ax + \frac{b}{R(x)}] = 0\} + 2^w + 2. \tag{9}$$

Observe that $\#C(\mathbb{F}_q) - 1 - \delta_{a,0}$ is divisible by 2^{w+1} from Corollary (1.5) in [11]. Since $W_{b \cdot F}(a) = 2\#\{x \in \mathbb{F}_q | Tr[ax + \frac{b}{R(x)}] = 0\} - q = \#C(\mathbb{F}_q) - 1 - \delta_{a,0} - 2^w - q$, we can write $W_{Tr[bF]}(a) = s \cdot 2^{w+1} - 2^w$ for some integer s.

On the other hand, by Lemma 2, for all a we have

$$|\#C(\mathbb{F}_q) - q - 1| = |s \cdot 2^{w+1} + \delta_{a,0}| \leq 2(2^w - \delta_{a,0})\sqrt{q}.$$

That is, we have $|s| \leq \lfloor \sqrt{q} \rfloor$.

Combining them, we find that the maximum of $|W_{Tr[bF]}(a)|$ is bounded by $2^{w+1}\lfloor \sqrt{q} \rfloor - 2^w$. From 6, we get the theorem.

Observe that this bound of nonlinearity is very tight for small w, but not so good for large w.

6 Vector Resilient Functions

We begin with some basic terminology of coding theory [13]. A *linear code* C is a linear subspace of $\mathbb{F}_2^n$. An element of C is called a *codeword*. The *minimum distance* of C is defined as the minimum of weights of all nonzero codewords in C. A $[n, m, d]$ *code* is a m-dimensional linear code of length n with minimum distance d.

Suppose W is a vector space generated by $\{e_1, e_2, e_3\}$. Then $V = \langle e_1+e_2+e_3 \rangle$ is a $[3, 1, 3]$ linear code since V has one nonzero element $e_1 + e_2 + e_3$ with weight 3. If we define $V = \langle e_1 + e_2, e_2 + e_3 \rangle$, it is a $(3, 2, 2)$ linear code since every nonzero element of V has weight 2.

Theorem 4. *Let $R(x)$ be a linearized polynomial such that $N_R(\mathbb{F}_q)$ is generated by $\{\xi_1, \xi_2, \cdots, \xi_w\}$ for some $w > 0$, and $F(x) = 1/R(x) + x$. Suppose $\mathcal{B} = \{\xi_1, \xi_2, \cdots, \xi_n\}$ is a basis of $\mathbb{F}_q$ and $\hat{\mathcal{B}} = \{\widehat{\xi_1}, \widehat{\xi_2}, \cdots, \widehat{\xi_n}\}$ its dual basis. For $1 \le m \le w$, let $B_1, B_2, \cdots, B_m$ be elements of the vector space $\mathbb{F}_q$ with the basis $\hat{\mathcal{B}}$ whose projection on $\langle \widehat{\xi_1}, \widehat{\xi_2}, \cdots, \widehat{\xi_w} \rangle$ forms a $[w, m, d]$ linear code. Then $(Tr[B_1 F], Tr[B_2 F], \cdots, Tr[B_m F])$ is a $(d-1)$-resilient function under the basis $\mathcal{B}$.*

Proof. Any component function of $(Tr[B_1 F], Tr[B_2 F], \cdots, Tr[B_m F])$ is written as $Tr[BF]$ for $B = \sum_{i=0}^{m} b_i B_i$ with $b_i \in \mathbb{F}_2$. Observe that the projection of such B on $\langle \widehat{\xi_1}, \widehat{\xi_2}, \cdots, \widehat{\xi_w} \}\rangle$ has weight greater than or equal to d. Hence $B \cdot F$ is a $(d-1)$-resilient function by Theorem 1. Since every component function is a $(d-1)$-resilient function, so does $(Tr[B_1 F], Tr[B_2 F], \cdots, Tr[B_m F])$.

Using Theorem 4, we can construct a (n, m, k)-resilient function from $\mathbb{F}_2^n$ to $\mathbb{F}_2^m$ when $k = d(w, m) - 1$ for some w with $0 < m \le w < n$ as Algorithm 1.

Algorithm 1 (Construct a vector resilient function)

1. **Input n, m and k such that $k = d(w, m) - 1$ for some w with $0 < m \le w < n$.**
2. **Take a set $V = \{\xi_1, \xi_2, \cdots, \xi_w\}$ of w linearly independent elements of $\mathbb{F}_{2^n}$. Let $\mathcal{B} = \{\xi_1, \xi_2, \cdots, \xi_n\}$ is a basis of $\mathbb{F}_{2^n}$ and $\hat{\mathcal{B}} = \{\widehat{\xi_1}, \widehat{\xi_2}, \cdots, \widehat{\xi_n}\}$ its dual basis.**
3. **Assume a $[w, m, d]$ linear code is generated by $\{c_i, c_2, \cdots, c_m\}$ where $c_i = [c_{i1}, c_{i2}, \cdots, c_{iw}]$ and $c_{iw} \in \mathbb{F}_2$. Compute $B_i = \sum_{j=1}^{m} c_{ij} \widehat{\xi_i}$.**
4. **Let $F(x) = 1/R(x) + x$ for $R(x) = \prod_\zeta (x - \zeta)$ where ζ ranges over all elements of the subspace generated by V. Compute $Tr[B_i F(x)]$ for $1 \le i \le m$.**
5. **Output a k-resilient function**

$$S(x) = (Tr[B_1 F(x)], Tr[B_2 F(x)], \cdots, Tr[B_m F])$$

from $\mathbb{F}_2^n$ to $\mathbb{F}_2^m$ by taking the basis $\mathcal{B}$ for $\mathbb{F}_{2^n}$.

The following is an example of Algorithm 1.

Example 2. Let $q = 2^8$ and $V = \{\xi_1, \xi_2, \xi_3\}$ a set of linearly independent elements of $\mathbb{F}_q$, and let $R(x) = \prod(x - \xi)$ where ξ ranges over all linear combinations of elements of V. Let $\mathcal{B} = \{\xi_1, \xi_2, \cdots, \xi_n\}$ is a basis of $\mathbb{F}_q$ and $\hat{\mathcal{B}} = \{\widehat{\xi_1}, \widehat{\xi_2}, \cdots, \widehat{\xi_n}\}$

its dual basis. Then (f_1, f_2) is a (8,2,1)-resilient function under the basis $\mathcal{B}$ where $f_1 = Tr[(\widehat{\xi_1} + \widehat{\xi_2})(\frac{1}{R(x)} + x)]$ and $f_2 = Tr[(\widehat{\xi_2} + \widehat{\xi_3})(\frac{1}{R(x)} + x)]$.

If we combine Theorem 2, 3 and 4, we can get the following Theorem.

Theorem 5. *Assume $0 < m \leq n$ and a $[n, m, d]$ linear code exists. For any nonnegative integer D, there exists a $(n + D + 1, m, d - 1)$-resilient function with algebraic degree D, whose nonlinearity is greater than or equal to $2^{n+D} - 2^n\lfloor\sqrt{2^{n+D+1}}\rfloor + 2^{n-1}$.*

Note that for any positive integer there exists a $[2^m - 1, m, 2^{m-1}]$ code, so called a simplex code, which has the maximal value of minimal distances for m-dimensional linear codes with length $2^m - 1$. Concatenating each codeword t times gives a $[t(2^m - 1), m, t2^{m-1}]$ linear code. If we apply this code to Theorem 5, we get the following result.

Corollary 1. *For any positive integers m, t and D, there is a $(t(2^m - 1) + D + 1, m, t2^{m-1} - 1)$-resilient function with algebraic degree D and nonlinearity greater than or equal to*

$$2^{t(2^m-1)+D} - 2^{t(2^m-1)}\lfloor\sqrt{2^{t(2^m-1)+D+1}}\rfloor + 2^{t(2^m-1)-1}.$$

Given positive integers n and m, we define the maximal resiliency $\kappa(n, m)$ to be the maximal value of resiliency k such that a (n, m, k)-resilient function exists. Chor *et al.* [8] showed that $\kappa(n, 2) = \lfloor\frac{2n}{3}\rfloor - 1$. For general m, Friedman [10] showed that given positive integers n and m the maximal resiliency $\kappa(n, m)$ satisfies

$$\kappa(n, m) \leq n - 1 - \frac{n(2^m - 2)}{2(2^m - 1)}. \quad (10)$$

Bierbrauer *et al.* [3] showed that a $[n, m, d]$ linear code can be used to construct a $(n, m, d - 1)$-resilient function. Combining this with (10), we find that $\kappa(t(2^m - 1), m) = t2^{m-1} - 1$. On the other hand, if we consider linear resilient functions, i.e. $D = 1$, in Corollary 1, the proposed construction gives $(t(2^m - 1) + 2, m, t2^{m-1} - 1)$-resilient function which has 2 bit larger input length with the same output size and resiliency. By this construction, however, for any positive integer D we can construct a resilient function of algebraic degree D with the same parameter by increasing the input size by D bits.

In [23], authors proposed a method to construct a nonlinear vector resilient function from a linear vector resilient function by permuting nonlinearly its output bits. That is,

Let F be a linear (n, m, k)-resilient function and G a permutation on $\mathbb{F}_2^m$ whose nonlinearity is $\mathcal{N}_G$. Then $P = G \cdot F$ is a (n, m, k)-resilient function such that

1. *the nonlinearity $\mathcal{N}_P$ of P satisfies $\mathcal{N}_P = 2^{n-m}\mathcal{N}_G$ and*
2. *the algebraic degree of P is the same as that of G.*

A vector Boolean function with m bit output generated by this method has an algebraic degree less than m while our method can generate a resilient function with algebraic degree up to $n-2-m$. The largest nonlinearity achieved by a permutation on $\mathbb{F}_2^m$ is $2^{m-1}-2^{(m-1)/2}$ [15]. Thus, such (n,m,k)-resilient function has nonlinearity $\leq 2^{n-1}-2^{n-(m+1)/2}$. Hence resilient functions constructed by the proposed method have larger bound of nonlinearity for small m than the previous method. Another obstacle of the previous method is to find a nonlinear permutation, which is not easy for even m except x^{-1}.

Generally, it is not easy to obtain the maximum value of m given n and d or the maximal value of d given n and m. For small n, m, however, there is a table [4] for the maximum value $d(n,m)$ of d such that a $[n,m,d]$ linear code exists. Refer to the appendix for $1 \leq n \leq 15$ and $1 \leq m \leq 6$. These maximum values of the minimum distances gives the maximal resiliency k of (n,m,k)-resilient functions with the algebraic degree D constructed by Algorithm 1. In Table 1, 0-resiliency means balancedness.

Table 1. The maximum resiliency k of proposed (n,m,k)-resilient functions with the algebraic degree D.

$m \setminus n$	$2+D$	$3+D$	$4+D$	$5+D$	$6+D$	$7+D$	$8+D$	$9+D$	$10+D$	$11+D$	$12+D$	$13+D$	$14+D$
1	0	1	2	3	4	5	6	7	8	9	10	11	12
2		0	1	1	2	3	3	4	5	5	6	7	7
3			0	1	1	2	3	3	3	4	5	5	6
4				0	1	1	2	3	3	3	4	5	5
5					0	1	1	1	2	3	3	3	4
6						0	1	1	1	2	3	3	3

7 Stream Ciphers

One of the most widely used design for stream cipher is a combination generator. A combination generator consists of several linear feedback shift registers(LFSRs) whose output sequences are combined by a nonlinear Boolean function, called a combining function. To resist against the well-known correlation attack, a combining function should be resilient. Fig. 1 is an example of a stream cipher with multi-bit output where KGSs are key stream generators and F is a combining function.

To get a high linear complexity, we use feedback shift registers with carry operation (FCSRs) [12] as KSGs instead of LFSRs in a combining generator. Let n be the number of FCSRs with k registers and m the number of output bits. By Theorem 5, we can construct a $(w+D+1, m, d-1)$-resilient function for any non-negative integer D whenever a $[w,m,d]$ linear code exists. The function has algebraic degree D and nonlinearity at least $2^{w+D} - 2^w \lfloor \sqrt{2^{w+D+1}} \rfloor + 2^{w-1}$. We use this vector resilient function as a combining function.

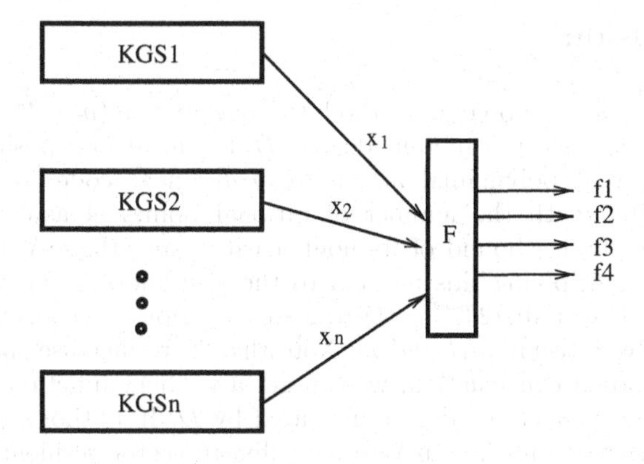

Fig. 1. A stream cipher with multi-bit output

Note that correlation attack has complexity $O(2^{kd})$ when the combining function is $(d-1)$-resilient. On the other hand, linear complexity attack has complexity $O(M^3)$ for a cipher with linear complexity M. Since every FCSR has linear complexity 2^k and the combining function has algebraic degree $n-w-1$, we have $M = 2^{k(n-w-1)}$. Hence when $d(w,m) \approx 3(n-w-1)$, two complexities are similar.

For example, if we let $n-w-1=2$ and $d=5$, the complexity becomes $O(2^{3k})$. In that case, we have $w=9$ for $m=2$ and $w=m+8$ for $m \geq 3$. That is, if $k=20$, we can design ciphers with the following feature. Here the complexity is against the linear complexity attack and the correlation attack for a linear combination of output bits.

However, if we consider a correlation attack using a nonlinear combination of output bits, the complexity might be different. In that case, the maximum correlation coefficient [22] should be considered. Currently, we don't know the maximum correlation of the proposed vector resilient functions. It would be interesting problem to compute them.

Table 2. Input v.s. Output with the fixed Resiliency

Input(n)	Output(m)	Dim(w)	Alg. Deg.(D)	Resiliency(k)	Complexity
12	2	9	2	5	2^{120}
14	3	11	2	5	2^{120}
15	4	12	2	5	2^{120}
17	5	14	2	5	2^{120}
18	6	15	2	5	2^{120}
19	7	16	2	5	2^{120}
21	9	18	2	5	2^{120}

8 Conclusion

In this paper we proposed a method to construct a $(n + D + 1, m, d - 1)$-resilient function with algebraic degree D for arbitrary positive integer D using a linearized polynomial and a $[n, m, d]$ linear code. Since its nonlinearity is related with the number of rational points of associated algebraic curves, we can find a bound of its nonlinearity using Hasse-Weil bound of algebraic curves. Applying this method to the well-known simplex code gives a $(t(2^m - 1) + D + 1, m, t2^{m-1} - 1)$-resilient function with algebraic degree D for any positive integers m, t and D. Note that if we increase the input size by D in the proposed construction, we can get a resilient function with the same parameter except algebraic degree increased by D. In author's knowledge, this method is the first one to generate a nonlinear vector resilient function with larger algebraic degree than the output size.

Acknowledgements. The author would like to thank Dr. Seongtaek Chee, Prof. Joseph Silverman, and Prof. Kyeongcheol Yang for helpful discussion and comments. This work was supported by postdoctoral fellowship program from Korea Science and Engineering Foundation (KOSEF).

References

1. C. Bennett, G. Brassard, and J. Robert, "Privacy Amplification by Public Discassion," SIAM J. Computing, Vol. 17, pp.210-229, 1988.
2. C. Bennett, G. Brassard, C. Crépeau, and U. Maurer, "Generalized Privacy Amplification," IEEE Trans. on Information Theory, Vol. 41, No. 6, pp. 1915-1923, 1995.
3. J. Bierbrauer, K. Gopalakrishnan, and D. Stinson, "Bounds on Resilient Functions and Orthogonal Arrays, " Proc. of Crypto'94, pp.247-256, Springer-Verlag, 1994.
4. A. Brouwer and T. Verhoeff, "An Updated Table of Mimimum-Distance Bounds for Binary Linear Codes," IEEE Trans. on Infomation Theory, Vol. 39, No. 2, pp.662-677, 1993.
5. J. Cheon and S. Chee, "Elliptic Curves and Resilient Functions," Proc. of ICISC'00, pp.64-72, 2000.
6. P. Camion, C. Carlet, P. Charpin, and N. Sendrier, "On Correlation Immune Functions," Proc. of Crypto'91, pp.86-100, Springer-Verlag, 1992.
7. S. Chee, S. Lee, D. Lee, and S. Sung, "On the Correlation Immune Functions and their Nonlinearity," Proc. of Asiacrypt'96, pp.232-243, Springer-Verlag, 1996.
8. B. Chor, O. Goldreich, J. Hastad, J. Friedman, S. Rudich, and R. Smolensky, "The Bit Extraction Problem or t-Resilient Functions," IEEE Symposium on Foundations of Computer Science, Vol. 26, pp. 396-407, 1985.
9. K. Friedl and S.C. Tsai, "Two Results on the Bit Extraction Problem", Discrete Applied Mathematics, Vol 99, pp. 443–454, 2000
10. J. Friedman, "On the Bit Extraction Problem," Proc. of 33rd IEEE Symposium on Foundations of Computer Science, pp.314-319, 1992.
11. G. van der Geer and M. van der Vlugt, "Trace Codes and Families of Algebraic Curves," Math. Z., Vol. 209, pp.307-315, Springer-Verlag, 1992.

12. A. Klapper and M. Goresky, "Feedback Shift Registers, Combiners with Memory, and 2-adic Span," Journal of Cryptology, Vol. 10, Springer-Verlag, pp. 111-147, 1997.
13. J.H. van Lint, *Intoroduction to Coding Theory*, Springer-Verlag, 1992.
14. R. Lidl and H. Niederreiter, *Finite Fields*, Cambridge University Press, 1997.
15. K. Nyberg, "S-Boxes and Round Functions with Controllable Linearity and Differential Uniformity," Proc. of the Second Fast Software Encryption, pp. 111 - 130, Springer-Verlag, 1994.
16. E. Pasalic and T. Johansson, "Further Results on the Relation Between Nonlinearity and Resiliency for Boolean Functions," Proc. of IMA Conference on Cryptography and Coding, pp. 35-44, LNCS 1746, Springer-Verlag, 1999.
17. P. Sarkar and S. Maitra, "Nonlinearity Bounds and Constructions of Resilient Boolean Functions," Proc. of Crypto'00, pp. 515-532, Springer-Verlag, 2000.
18. J. Seberry, X. Zhang and Y. Zheng, "On Constructions and Nonlinearity of Correlation Immune Boolean Functions," Eurocrypt'93, pp. 181-199, Springer-Verlag, 1993.
19. T. Siegenthaler, "Correlation-Immunity of Nonlinear Combining Functions for Cryptographic Applications," IEEE Transactions on Information Theory, IT-30(5), pp.776-780, 1984.
20. D. Stinson and J. Massey, "An Infinite Class of Counterexamples to a Conjecture Concerning Nonlinear Resilient Functions," Journal of Cryptology, Vol 8, No. 3, pp.167-173, Springer-Verlag, 1995.
21. Y. Tarannikov, "On Resilient Boolean Functions with Maximum Possible Nonlinearity," Proc. of Indocrypt'00, pp.19-30, Springer-Verlag, 2000.
22. M. Zhang and A. Chan, "Maximum Correlation Analysis of Nonlinear S-boxes in Stream Ciphers," Proc. of Crypto2000, pp. 501-514, Springer-Verlag, 2000.
23. X. Zhang and Y. Zheng, "Cryptographically Resilient Functions," IEEE Trans. Inform. Theory, Vol 43, No 5, pp. 1740-1747, 1997.

Appendix: Minimum Distance of Linear Codes

For given $n, m \leq 127$, there is a table [4] for the maximum value of d such that a $[n, m, d]$ linear code exists. Some of them are as below:

Table 3. The maximum d such that a $[n, m, d]$ linear code exists.

$m \setminus n$	1	2	3	4	5	6	7	8	9	10	11	12	13	14	15
1	1	2	3	4	5	6	7	8	9	10	11	12	13	14	15
2		1	2	2	3	4	4	5	6	6	7	8	8	9	10
3			1	2	2	3	4	4	4	5	6	6	7	8	8
4				1	2	2	3	4	4	4	5	6	6	7	8
5					1	2	2	2	3	4	4	4	5	6	7
6						1	2	2	2	3	4	4	4	5	6

New Public Key Cryptosystem Using Finite Non Abelian Groups

Seong-Hun Paeng, Kil-Chan Ha,
Jae Heon Kim, Seongtaek Chee, and Choonsik Park

National Security Research Institute
161 Kajong-dong, Yusong-gu, Taejon, 305-350, KOREA
{shpaeng,kcha,jaeheon,chee,csp}@etri.re.kr

Abstract. Most public key cryptosystems have been constructed based on abelian groups up to now. We propose a new public key cryptosystem built on finite non abelian groups in this paper. It is convertible to a scheme in which the encryption and decryption are much faster than other well-known public key cryptosystems, even without no message expansion.
Furthermore a signature scheme can be easily derived from it, while it is difficult to find a signature scheme using a non abelian group.

1 Introduction

Most frequently used problems in the public key cryptosystems are the factorization problem [19] and the discrete logarithm problem (DLP). Cryptosystems based on these problems have been built on abelian groups [5,3,8,12,13]. In Crypto 2000, Ko et al. proposed a new public cryptosystem based on Braid groups, which are non abelian groups. To authors' best knowledge, it was the first practical public key cryptosystem based on non abelian groups.

When we use a non abelian group G for a public key cryptosystem, we need to consider the following problems related to the word problem:

- How do we express a message as an element of G?
- Can every element of G be represented in a unique way for a given expression?

If an element of G is not represented in a unique way, then a plaintext and a deciphertext may not be the same. Therefore the second problem is very important when we use a non abelian group for a public key system. Matrix groups and semi-direct product of abelian groups are examples of non abelian groups which have such expressions.

In this paper, we suggest a new cryptosystem based on such a finite non abelian group G. Our PKC is based on DLP in the inner automorphism group

$$Inn(G) = \{Inn(g) \mid g \in G\},$$

where $Inn(g)(x) = gxg^{-1}$. The advantages of our PKC are as follows:

J. Kilian (Ed.): CRYPTO 2001, LNCS 2139, pp. 470–485, 2001.

- We can apply our encryption scheme to G even if DLP and the (special) conjugacy problem in G are not hard problems.
- Parameter selections are much easier than those in ECC [12,13] and XTR [8].
- We can increase the speed of the encryption and decryption. More precisely, when m is a message and g^a is the public key in ElGamal-type encryption [5,12,13], $(g^{ab}m, g^b)$ should be sent to a receiver and it is crucial that different random integers b should be used to encrypt different messages. In our scheme, we can fix b without loss of security so that we can increase the speed of the encryption and decryption. Moreover no message expansion is required.
- It is easy to make a signature scheme with our PKC: In general, it is not easy to find a signature scheme using an infinite non abelian group such as a braid group [11].

If we fix b, our PKC is about 30 times faster than RSA for a 32-bit public exponent in RSA encryption scheme and is about 200 times faster in decryption.

2 Preliminaries

2.1 Semi-direct Product

From some given groups, we can easily make new non abelian groups using semi direct products. Recall the definition of the semi-direct product:

Definition 1. *(Semi-direct product) Let G and H be given groups and $\theta : H \to Aut(G)$ be a homomorphism, where $Aut(G)$ is the automorphism group of G. Then semi-direct product $G \times_\theta H$ is the set*

$$G \times H = \{(g, h) |\ g \in G,\ \ h \in H\}$$

together with the multiplication map

$$(g_1, h_1) \cdot (g_2, h_2) = (g_1\theta(h_1)(g_2), h_1h_2).$$

Since

$$(g, h)^{-1} = (\theta(h^{-1})(g^{-1}), h^{-1}),$$

we have

$$(e_G, h_1)(g_2, e_H)(e_G, h_1)^{-1} = (\theta(h_1)(g_2), e_H), \tag{2.1}$$

where e_G, e_H are the identity elements of G and H, respectively. So G can be considered as a normal subgroup of $G \times_\theta H$. If $\theta(H) \neq Id$, then $G \times_\theta H$ is a non abelian group even if G and H are abelian.

Example 1. (1) Most familiar example of the semi-direct product is the isometry group on Euclidean space $\mathbb{R}^n$. This group is the semi-direct product of the translational isometry group $\mathbb{R}^n$ and the orthogonal group $O(n, \mathbb{R})$.

(2) It is a well known fact that $\mathrm{Aut}(\mathbb{Z}_n) = \mathbb{Z}_n^*$, where $\mathbb{Z}_n^*$ is the multiplicative group of $\mathbb{Z}_n$. Since $\mathbb{Z}_4^* \simeq \mathbb{Z}_2$, there exists a non constant homomorphism, in fact, an isomorphism of $\mathbb{Z}_2$ into $\mathbb{Z}_4^*$. Thus $\mathbb{Z}_4 \times_\theta \mathbb{Z}_2$ is a non abelian group.
(3) If G is a non abelian group, then there exists a natural homomorphism from G to $\mathrm{Aut}(G)$. Precisely,

$$\begin{aligned} Inn : G &\to \mathrm{Aut}(G) \\ g &\mapsto Inn(g),\ Inn(g)(h) = ghg^{-1}. \end{aligned} \tag{2.2}$$

We call $Inn(g)$ an inner automorphism. It is easy to check that $\ker(Inn)$ is the center of G. Recall that the center of G is the set $\{z | \ [z, g] = zgz^{-1}g^{-1} = e_G$ for all $g \in G\}$. If G is an abelian group, Inn is a constant map and so $G \times_{Inn} G = G \times G$. But if G is a non abelian group, then $G \times_{Inn} G$ is an interesting extension of G.

If we apply a semi-direct product to p-groups inductively, we can make a nilpotent group [7]. It is a well known fact that nilpotent groups can be expressed in a unique way as a direct product of abelian groups. The above $\mathbb{Z}_4 \times_\theta \mathbb{Z}_2$ in Example 1. (1) is a nilpotent group with order 8.

2.2 Conjugacy Problem

One of the most important characteristics of non abelian groups distinguished from abelian groups is that Inn is not constant, i.e. there exist two distinct elements which are congruent to each other.

Definition 2. (1) *For arbitrary $x, y \in G$, the conjugacy problem is to find $w \in G$ such that $wxw^{-1} = y$.*
(2) *For a given $Inn(g)$, the special conjugacy problem is to find g' satisfying $Inn(g') = Inn(g)$.*

There are many groups where the word problem is known to be solvable in polynomial time while there is no known polynomial time algorithm to solve the conjugacy problem (the braid group is an example) [1]. If G is a non abelian group and its conjugacy problem is hard in G, we can consider the following cryptosystem. Let $\{\delta_i\}$ be a set of generators of G. Let g be an element of G. The public key is $\{\epsilon_i = g\delta_i g^{-1}\}$ and the secret key is g. Mathematically, the public key can be expressed as $Inn(g)$. Then the ciphertext is $E = Inn(g)(m)$ and the deciphertext is $g^{-1}Eg$ ([1] or [20]). In order to use such an encryption scheme, every element of G should be easily expressible as a product of δ_i's. If an element of G is also easily expressible as a product of ϵ_i's, then we also obtain $Inn(g^{-1})$ immediately. Since $g^{-1}Eg = Inn(g^{-1})(E)$, we can decrypt without knowing g. Thus it is essential that elements of G should not be easily expressible as products of ϵ_i's.

This system depends on the difficulty of finding g' satisfying $Inn(g') = Inn(g)$ for a given $Inn(g)$, i.e. the above system is based on the special conjugacy problem. Unfortunately, we know few finite non abelian groups to which we can apply the above system. For example, the general linear group $\mathrm{GL}(2, \mathbb{Z}_p)$ and the special linear group $\mathrm{SL}(2, \mathbb{Z}_p)$ are non abelian groups on which the (special) conjugacy problem is not difficult (see Appendix A).

Remark 1. If we use DLP in $\mathrm{SL}(2, \mathbb{Z}_p)$, we choose $g \in \mathrm{SL}(2, \mathbb{Z}_p)$ whose order is divided by p. The order of $\mathrm{SL}(2, \mathbb{Z}_p)$ is $|\mathrm{SL}(2, \mathbb{Z}_p)| = p(p-1)(p+1)$. Such elements which we are aware of are the conjugates of $I + c\delta_{12}$ and $I + c\delta_{21}$, where $c \in \mathbb{Z}_p$ and δ_{ij} is a matrix whose entries are all zero except the (i, j)-entry which is 1. Let $g = A(I + \delta_{12})A^{-1}$ for

$$A = \begin{pmatrix} a & b \\ c & d \end{pmatrix} \in \mathrm{SL}(2, \mathbb{Z}_p).$$

Then we have

$$\begin{aligned} g^m &= A(I + m\delta_{12})A^{-1} \\ &= \begin{pmatrix} ad - bc - mac & ma^2 \\ -mc^2 & ad - bc + mac \end{pmatrix} = \begin{pmatrix} 1 - mac & ma^2 \\ -mc^2 & 1 + mac \end{pmatrix}. \end{aligned} \tag{2.3}$$

Consider DLP in the cyclic group $\langle g^m \rangle$. Since $(1, 2)$-component of g^m and g^{ml} are ma^2 and mla^2, respectively, we can obtain $l \mod p$. Hence DLP in $\langle g^m \rangle$ is not a hard problem. The (special) conjugacy problem and DLP are not hard problems in $\mathrm{SL}(2, \mathbb{Z}_p)$.

3 New Cryptosystem

In this section, we suggest a new encryption scheme which are based on DLP in the inner automorphism group.

Let G be a non abelian group with non trivial center $Z(G)$. We assume that $Z(G)$ is not small. Let g be an element of G.

Proposed encryption scheme. Let $\{\gamma_i\}$ be a set of generators of G. Since $Inn(g)$ is a homomorphism, $Inn(g)$ is obtained if we know $Inn(g)(\gamma_i)$, i.e. if we express m as $\gamma_{j_1} \cdots \gamma_{j_n}$, then $Inn(g)(m) = Inn(g)(\gamma_{j_1}) \cdots Inn(g)(\gamma_{j_n})$. Therefore we can represent $Inn(g)$ by $\{Inn(g)(\gamma_i)\}$. The basic scheme is the following:

- public key : $Inn(g)$, $Inn(g^a)$
- secret key : a

Encryption

1. Alice expresses a plaintext $m \in G$ as a product of γ_i's.
2. Alice chooses an arbitrary b and computes $(Inn(g^a))^b$, i.e. $\{(Inn(g^a))^b(\gamma_i)\}$.
3. Alice computes $E = Inn(g^{ab})(m) = (Inn(g^a))^b(m)$.
4. Alice computes $\phi = Inn(g)^b$, i.e. $\{Inn(g^b)(\gamma_i)\}$.
5. Alice sends (E, ϕ).

Decryption

1. Bob expresses E as a product of γ_i's.
2. Bob computes ϕ^{-a}, i.e. $\{\phi^{-a}(\gamma_i)\}$.
3. Bob computes $\phi^{-a}(E)$.

To implement our scheme, we should express $Inn(g^a)$ with small bits. Since G is a finitely generated group, $Inn(g^a)$ is expressed by $\{Inn(g^a)(\gamma_i)\}$ for a generator set $\{\gamma_i\}$. If we do not have a fast algorithm to express $\gamma \in G$ by a product of generators, we cannot express $Inn(g^a)$ actually. In the next section, we will introduce a non abelian group to which our scheme can be applied. (Precisely, see 4.3.)

Although our scheme looks like an ElGamal-type, we may not change b for each encryption. In ElGamal-type encryption based on abelian groups (e.g. ECC), we must change b for each encryption. (If a fixed b is used, we can obtain $m_1^{-1}m_2 = (m_1g^{ab})^{-1}m_2g^{ab}$).) But in our scheme, it is impossible to obtain $m_1^{-1}m_2$ from $Inn(g^b)(m_1)$ and $Inn(g^b)(m_2)$. Thus we may fix b. As we see in section 4.3, this fact will be very useful for fast encryption and decryption scheme.

Due to the non commutativity of braid groups, the cryptosystem using braid groups has a difficulty in making a signature scheme. However, our scheme enables us to make a signature scheme easily (e.g. Nyberg-Rueppel type signature) even if G is non abelian.

Now we consider the method to find a from the given $Inn(g)$ and $Inn(g^a)$. First, we solve DLP in $\langle Inn(g)\rangle$ directly. The index calculus is the most efficient known method to solve DLP [4]. But its application is too restrictive to be applied to general cyclic groups. It seems that the index calculus cannot be applied to the group $\langle Inn(g)\rangle$. In general cases, expected run times for solving DLP are $O(\sqrt{p})$, where p is the order of a cyclic subgroup.

Secondly, we solve DLP in $\langle g\rangle$. If we assume that the special conjugacy problem is not a hard problem, we can find g_0 satisfying $Inn(g_0) = Inn(g^a)$. We can easily verify that $g_0 = g^a z$ for some $z \in Z(G)$. If $|Z(G)|$ is large enough, then it is almost impossible to determine whether $g^a z$ is an element of $\langle g\rangle$. Then even if DLP in G may be easy, we cannot apply any algorithm to solve DLP in G.

We should be careful in the choice of a plaintext and g. If $[m, g] = e_G$, then $E = g^{ab}mg^{-ab} = m$. In particular, if m is a central element, then $E = m$ so m should not be chosen in the center. Also if g is a central element, then $Inn(g)$ is the identity map and so $E = m$. We should select a non central element g.

We should note that there may be other attacks depending on G as we see in section 5.

Remark 2. In the above scheme, $E = Inn(g^{ab})(m)$, E and m are contained in the same conjugacy class. Assume that E is a ciphertext of either m_0 or m_1, which are not contained in the same conjugacy class. Then an adversary can find the right plaintext by examining the conjugacy class of E. To avoid this attack, we can use a padding method in the encryption (see Remark 4 and [16]). It also makes fast encryption and decryption scheme (which fixes b) non deterministic.

4 Construction of a Non Abelian Group

4.1 An Example of Non Abelian Group $\mathbf{SL(2,\mathbb{Z}_p)\times_\theta \mathbb{Z}_p}$

If we use the semi-direct product, we can construct many non abelian groups with non trivial center as in section 5. But it is not easy to construct a non abelian group on which our system is secure. We modify $\mathrm{SL}(2,\mathbb{Z}_p)$ by a semi-direct product as follows. There exists a cyclic subgroup $\langle\alpha\rangle$ with order p of $\mathrm{SL}(2,\mathbb{Z}_p)$, e.g. $I+\delta_{12}$. Let

$$G=\mathrm{SL}(2,\mathbb{Z}_p)\times_\theta \mathbb{Z}_p,$$

where

$$\theta = Inn\circ\theta_1 : \mathbb{Z}_p \to \mathrm{Aut}(\mathrm{SL}(2,\mathbb{Z}_p))$$

and θ_1 is an isomorphism from $\mathbb{Z}_p$ to $\langle\alpha\rangle$. Then $\theta(y)(x)=\theta_1(y)x\theta_1(y)^{-1}$. Now we solve the conjugacy equations in G. Let $g=(x,y)$. Computing the conjugate of (a,b), we obtain that

$$(x,y)(a,b)(x,y)^{-1}=(x\theta(y)(a)\theta(b)(x^{-1}),b). \tag{4.4}$$

If $b=0$, we have

$$(x,y)(a,0)(x,y)^{-1}=(x\theta(y)(a)x^{-1},0)=((x\theta_1(y))a(x\theta_1(y))^{-1},0). \tag{4.5}$$

If we solve the special conjugacy problem in $\mathrm{SL}(2,\mathbb{Z}_p)$ as we see in Appendix A, we can obtain $x\theta_1(y)$. Let $(x_1,y_1)\in G$ such that $x_1\theta_1(y_1)=x\theta_1(y)$. For $b\neq 0$, if we use the fact that $\mathbb{Z}_p$ is an abelian group and θ_1 is a homomorphism, we can easily verify that

$$\begin{aligned}x_1\theta(y_1)(a)\theta(b)(x_1^{-1}) &= x_1\theta_1(y_1)a\theta_1(y_1)^{-1}\theta_1(b)x_1^{-1}\theta_1(b)^{-1}\\ &= (x_1\theta_1(y_1))a\theta_1(y_1)^{-1}\theta_1(b)\theta_1(y_1)\theta_1(y)^{-1}x^{-1}\theta_1(b)^{-1}\\ &= (x\theta_1(y))a\theta_1(-y_1+b+y_1)\theta_1(y)^{-1}x^{-1}\theta_1(b)^{-1}\\ &= x\theta_1(y)a\theta_1(b)\theta_1(y)^{-1}x^{-1}\theta_1(b)^{-1}\\ &= x\theta_1(y)a\theta_1(y)^{-1}\theta_1(b)x^{-1}\theta_1(b)^{-1}\\ &= x\theta(y)(a)\theta(b)(x^{-1}).\end{aligned} \tag{4.6}$$

It can be easily verified that if $x_1\theta_1(y_1)=-x\theta_1(y)$, then the above equation also holds. Also note that the center of $Z(\mathrm{SL}(2,\mathbb{Z}_p))=\pm I$. Hence the set of solutions for the special conjugacy problem is

$$S=Inn^{-1}(Inn(g))=\{(x_1,y_1)|\ y_1\in\mathbb{Z}_p,\ x_1=\pm x\theta_1(y-y_1)\}. \tag{4.7}$$

The cardinality of S, $|S|$ is $2p$. Note that if $Inn(g)=Inn(g_1)$, then $Inn(g^{-1}g_1)=Id$. It means that $g^{-1}g_1$ is an element of the center of G. Also for any central element z, $Inn(gz)=Inn(g)$. So we know that $S=Inn^{-1}(Inn(g))=gZ(G)$ and

$$Z(G)=\{(x_1,y_1)|\ y_1\in\mathbb{Z}_p,\ x_1=\pm\theta_1(-y_1)\}. \tag{4.8}$$

The cardinality of the center of G is $2p$. Note that the probability to choose m and g in the center is smaller than $2p/p^3 = 2/p^2 \approx 0$ and $2p/p^4 = 2/p^3 \approx 0$, respectively.

For a given $Inn(g)$, m satisfying $[g,m] = e_G$ is a fixed point, i.e. $Inn(g)^{ab}(m) = m$. The cardinality of $Z[g] = \{m \mid [g,m] = e_G\}$ is $2p^2$ if we choose g of order p [16] and thus the probability to choose m in $Z[g]$ is smaller than $2p^2/p^3 = 2/p \approx 0$.

Remark 3. From Theorem 2 in section 4.3, DLP in G is reduced to a linear equation $ny = Y$ for given $y \neq 0, Y$, and so it is an easy problem.

4.2 Parameter Selections

We will apply the above scheme to $G = \mathrm{SL}(2, \mathbb{Z}_p) \times_\theta \mathbb{Z}_p$. Since the last component is invariant under the conjugation, we must take the message in $\mathrm{SL}(2, \mathbb{Z}_p)$ (see (4.4)).

In [20], we see

$$\{T = \begin{pmatrix} 1 & 1 \\ 0 & 1 \end{pmatrix},\ S = \begin{pmatrix} 0 & -1 \\ 1 & 0 \end{pmatrix}\}$$

is a generator set of $\mathrm{SL}(2, \mathbb{Z})$ and hence it is also a generator set of $\mathrm{SL}(2, \mathbb{Z}_p)$. Moreover there exists an algorithm which finds a decomposition of each $g \in \mathrm{SL}(2, \mathbb{Z}_p)$ as a product of T, S [2],[20], i.e.

$$g = S^{i_0} T^{j_1} S T^{j_2} \cdots S T^{j_n} S^{i_{n+1}},$$

where i_0, i_{n+1} is either 0 or 1 and $j_k = \pm 1, \pm 2 \cdots$.

Theorem 1. *If $g \in SL(2, \mathbb{Z}_p)$ with non zero $(2,1)$-entry,*

$$g = T^{j_1} S T^{j_2} S T^{j_3}.$$

Proof. By computing $T^{j_1} S T^{j_2} S T^{j_3}$, we obtain

$$\begin{pmatrix} 1 & j_1 \\ 0 & 1 \end{pmatrix} \begin{pmatrix} 0 & -1 \\ 1 & 0 \end{pmatrix} \begin{pmatrix} 1 & j_2 \\ 0 & 1 \end{pmatrix} \begin{pmatrix} 0 & -1 \\ 1 & 0 \end{pmatrix} \begin{pmatrix} 1 & j_3 \\ 0 & 1 \end{pmatrix} = \begin{pmatrix} j_1 j_2 - 1 & j_1 j_2 j_3 - j_3 - j_1 \\ j_2 & j_2 j_3 - 1 \end{pmatrix}.$$

From this equation and the fact that $\mathbb{Z}_p$ is a field, we can find j_1, j_2, j_3 such that $g = T^{j_1} S T^{j_2} S T^{j_3}$ for any $g \in \mathrm{SL}(2, \mathbb{Z}_p)$. (Since every element of $\mathrm{SL}(2, \mathbb{Z}_p)$ is determined when three entries are determined, we only need to consider 3 entries.)

Note that since $\mathbb{Z}$ is not a field, the above theorem does not hold in $\mathrm{SL}(2, \mathbb{Z})$.

Since $\{(T,0), (S,0), (I,1)\}$ is a set of generators of G, we can obtain $Inn(g)$ if we know gTg^{-1}, gSg^{-1} and $g(I,1)g^{-1}$. Since $m \in \mathrm{SL}(2, \mathbb{Z}_p)$ and $\mathrm{SL}(2, \mathbb{Z}_p)$ is a normal subgroup of G, the restriction of $Inn(g)$ to $\mathrm{SL}(2, \mathbb{Z}_p)$, $Inn(g)|_{\mathrm{SL}(2,\mathbb{Z}_p)}$ can be considered as an automorphism of $\mathrm{SL}(2, \mathbb{Z}_p)$. Hence the public key is $Inn(g)|_{\mathrm{SL}(2,\mathbb{Z}_p)}$ and $Inn(g^a)|_{\mathrm{SL}(2,\mathbb{Z}_p)}$, precisely. In order to express $Inn(g)|_{\mathrm{SL}(2,\mathbb{Z}_p)}$, we only need to know $\{gTg^{-1}, gSg^{-1}\}$.

We choose $\theta_1(1)$ among elements of $\mathrm{SL}(2,\mathbb{Z}_p)$ whose order is p, e.g. $I+\delta_{12}$.

We compute the order of $g=(x,y)\in G$. If $y\neq 0$, then the order of g is a multiple of p.

Theorem 2. *For* $(x,y)\in G$,

$$(x,y)^n = ((x\theta_1(y))^n\theta_1(y)^{-n}, ny).$$

Proof. We prove this using induction. For $n=1$, it is clear. We assume that Theorem 2 holds for $n=k$. Then we obtain that

$$\begin{aligned}(x,y)^{k+1} &= (x,y)^k(x,y) = ((x\theta_1(y))^k\theta_1(b)^{-k}, ky)(x,y)\\ &= ((x\theta_1(y))^k\theta_1(y)^{-k}\theta(y)^k(x),(k+1)y)\\ &= ((x\theta_1(y))^k\theta_1(y)^{-k}\theta_1(y)^k x\theta_1(y)^{-k},(k+1)y) \qquad (4.9)\\ &= ((x\theta_1(y))^k(x\theta_1(y))\theta_1(y)^{-(k+1)},(k+1)y)\\ &= ((x\theta_1(y))^{k+1}\theta_1(y)^{-(k+1)},(k+1)y),\end{aligned}$$

which completes the proof.

We may choose $g=(x,y)$ satisfying $x\theta_1(y)\in A(I+c\delta_{12})A^{-1}$ for some fixed $c\in\mathbb{Z}_p$ and $A\in\mathrm{SL}(2,\mathbb{Z}_p)$. Then we obtain that the order of $Inn(g)$ is p by Theorem 2. If we choose g arbitrarily and the order of g is not fixed, then the security may be increased since we should know the order of a given cyclic group to apply a known algorithm for DLP, i.e. we should solve DLP under the assumption that the order of g is pd for each $d|(p+1)(p-1)$.

4.3 Security and Efficiency

Security of the system. We check the security of our system against solving DLP in $\langle Inn(g)\rangle$ directly. From the public data, $Inn(g)$ and $Inn(g^a)$, we solve DLP to obtain the secret key a. In this case, it seems that the fastest algorithm (index calculus) to solve DLP cannot be applied since $\langle Inn(g)\rangle$ is contained in $\mathrm{Aut}(G)\subset\mathrm{End}(G)\subset G^G$, where $\mathrm{End}(G)$ is the endomorphism group of G and G^G is the set of all function from G to G. We cannot apply the index calculus to any of them since they are not even expressed as matrix groups.

So an expected run time for solving DLP is $O(\sqrt{p})$-group operations if the order of g is p. (In order to increase the security of the system, we can choose g with an order which is a multiple of p. If $p(p+1)=p_1^{e_1}\cdots p_n^{e_n}$, then the total number of divisors of $p+1$ is $(e_1+1)\cdots(e_n+1)$. To find the order of g, we need $(e_1+1)\cdots(e_n+1)$-trials, and it takes $(e_1+1)\cdots(e_n+1)O(\sqrt{p})$-group operations [17].)

Now we check the security of our system against the second method in section 3. As we see in Appendix A, the special conjugacy problem in G is not a hard problem. Let $S = \{g_1 | \, Inn(g_1) = Inn(g^a)\}$. We can immediately obtain a from $g = (x, y)$ and $g^a = (X, Y)$ since if

$$(x, y)^a = ((x\theta_1(y))^n \theta_1(y)^{-n}, ay) = (X, Y), \tag{4.10}$$

we only need to solve $ay = Y$ for solving DLP for g and g^a. But since $|S| = 2p$, we need $O(p)$-trials to find g^a in S. So it is less efficient than finding a from $Inn(g)$ and $Inn(g^a)$ directly.

For DLP to be a hard problem in $\langle Inn(g) \rangle$, we choose 160 bit prime p. Then the security of our system is comparable to 1024-bit RSA. (An expected run time for solving DLP in $\langle Inn(g) \rangle$ and for factorization in 1024-bit RSA is about 2^{87} and 2^{80}, respectively.)

If we compare our system with RSA and XTR, our system has the following advantage. In RSA and XTR, an expected run time to find the private key from the public key is subexponential, $L[n, 1/3, 1.923]$. In our system, it takes an exponential run time $O(\sqrt{p})$ as ECC.

Number of multiplications in $\mathbb{Z}_p$. Now we consider the number of multiplications in $\mathbb{Z}_p$ required for computing $Inn(g^b)$ from $Inn(g)$. We can express $Inn(g)(S)$ and $Inn(g)(T)$ as $T^{j_1}ST^{j_2}ST^{j_3}$ and $T^{l'_1}ST^{l'_2}ST^{l'_3}$, respectively. Each of them takes 2-multiplications by Theorem 1. Then

$$\begin{aligned} Inn(g^2)(S) &= Inn(g)(T^{j_1}ST^{j_2}ST^{j_3}) \\ &= (Inn(g)(T))^{j_1}(Inn(g)(S))(Inn(g)(T))^{j_2}(Inn(g)(S))(Inn(g)(T))^{j_3} \end{aligned}$$

and

$$\begin{aligned} Inn(g^2)(T) &= Inn(g)(T^{l_1}ST^{l_2}ST^{l_3}) \\ &= (Inn(g)(T))^{l_1}(Inn(g)(S))(Inn(g)(T))^{l_2}(Inn(g)(S))(Inn(g)(T))^{l_3}. \end{aligned}$$

From (2.3) in Remark 1, we can obtain $(Inn(g)(T))^j$ from $Inn(g)(T)$ with 4-multiplications. More precisely, if

$$Inn(g)(T) = \begin{pmatrix} x & y \\ z & w \end{pmatrix},$$

then

$$(Inn(g)(T))^j = \begin{pmatrix} 1 - j(1-x) & jy \\ jz & 1 + j(w-1) \end{pmatrix}.$$

It takes 92 multiplications for computing $Inn(g^2)(S)$ and $Inn(g^2)(T)$. So it takes about $92 \log_2 p$ multiplications for computing $Inn(g^b)$ from $Inn(g)$. Also $92 \log_2 p$ multiplications are needed to compute $Inn(g^{ab})$ from $Inn(g^a)$. So number of multiplications for encryption is $184 \log_2 p$. Since one multiplication needs $O((\log_2 p)^2)$-bit operations [9], the encryption needs about $184(\log_2 p)^3 C \approx 8 \times 10^8 C$-bit operations for some constant C. In 1024-bit RSA, it takes $(\log_2 n)^3 C \approx (1024)^3 C \approx 10^9 C$-bit operations. If the public exponent in RSA encryption scheme is 32-bit number, then it takes $3.2 \times 10^7 C$-bit operations.

Fast encryption and decryption. We can reduce the number of bit operations as follows. Assume that Bob wants to send an encrypted message to Alice. Then Bob computes $Inn(g^a)^b$ and $Inn(g^b)$ for a fixed b and send $Inn(g^b)$ to Alice. As we see in section 3, we may fix b, i.e. contrary to ElGamal encryption, we cannot obtain $m_1^{-1}m_2$ from $Inn(g^b)(m_1)$ and $Inn(g^b)(m_2)$ in our scheme . Alice computes $Inn(g^b)^{-a}$. Bob will encrypt a message m as $E = Inn(g^a)^b(m)$ and send E to Alice. Alice will decrypt E by computing $Inn(g^b)^{-a}(E)$.

In order to compute $Inn(g^a)^b(m)$ from given $Inn(g^a)^b$ and m, it takes 46 multiplications, and so it takes about $1.2 \times 10^6 C$-bit operations in encryption. Even if 32-bit public exponent is used in RSA, $3.2 \times 10^7 C$-bit operations are needed in encryption. Encryption of our system is about 30 times faster than 1024-bit RSA.

In decryption of our system, we need the same number of multiplications as the encryption. In decryption of RSA, it takes about $2.5 \times 10^8 C$-bit operations even if we use the Chinese Remainder Theorem. Thus decryption of our system is 200 times faster than that of RSA.

If we compare our system with ECC, our system has an advantage in the decryption too. In ECC, since b is not fixed, precomputations of g^b is impossible. Then the number of multiplications for decryption in 170-bit ECC are 1900, respectively. Then it is about 40 times faster than ECC.

In ECC, it needs $O(\log_2 p)$ multiplications in decryption, and thus the number of multiplications will increase linearly with respect to the number of bits $\log_2 p$. The decryption of our system always needs 46 multiplications which are independent of the size of p. In Table 1, we roughly compare the number of multiplications for decryption in our system with ECC. Note that the cryptosystems in the same row have the same securities roughly.

This fast scheme can be useful in many applications.

Table 1. Comparison of run times for decryption with ECC(multiplications)

	our PKC(r-bit)	r-bit ECC
$r = 170$	46	1900
$r = 240$	46	2700
$r = 310$	46	3500

Remark 4. We can encrypt a message with a padding as follows (see also [16]). Let $M \neq 0$ be a message and r_1, r_2 be random numbers in $\mathbb{Z}_p$. We encrypt $m = \begin{pmatrix} M & r_1 \\ r_2 & \frac{1+r_1r_2}{M} \end{pmatrix} \in \mathrm{SL}(2, \mathbb{Z}_p)$. Then m can be an element of any conjugacy class by varying r_1, r_2. It prevents an adversary from determining the right plaintext among two given plaintexts by examining the conjugacy class of E. Furthermore, since b is fixed, the encryption and the decryption is also fast but the encryption scheme is not deterministic.

Expression and key size. Since $Inn(g^a)(T)$ and $Inn(g^a)(S)$ can be considered as elements of $\mathrm{SL}(2,\mathbb{Z}_p)$, we can express them by three entries. Since $Inn(g^a)(T)$ can be expressed by $3\log_2 p$-bit, $6\log_2 p$-bit are needed to express $Inn(g^a)$. If p is a 160-bit prime number, then it takes 960-bit to express $Inn(g^a)$. So we can express the public key with smaller size than RSA.

The secret key size is $\log_2 p \approx 160$-bit, and so it is much smaller than 1024-bit RSA.

5 Other Examples

5.1 The General Linear Group $\mathrm{GL}(k,\mathbb{Z}_p)$

One of the most familiar non abelian group is the general linear group $\mathrm{GL}(k,\mathbb{Z}_p)$. Since cI is a central element for any $c \neq 0$, the center of $\mathrm{GL}(k,\mathbb{Z}_p)$ is sufficiently large, i.e. $|Z(\mathrm{GL}(k,\mathbb{Z}_p)| \geq p/2$. We know that $Inn(g)$ can be represented by a linear map on the $k \times k$-matrix ring [15]. So we can represent $Inn(g)$ by a $k^2 \times k^2$-matrix, $R(g)$. So the DLP on $\langle Inn(g)\rangle$ is convertible to the DLP on the $k^2 \times k^2$-matrix ring.

We must be careful in the choice of g. Considering an attack using the determinant [15], we choose g whose order is much larger than p (e.g. $p(p-1)$). It would be better to choose g satisfying that $\det(R(g)) = 1$. Also the characteristic polynomial of $R(g)$ should be irreducible.

5.2 Other Constructions

We introduce some methods to obtain non abelian groups. For a given non abelian group G, we can obtain a new non abelian group $Inn(G)$ as we see in previous sections. Also $Inn(Inn(G))$ can be obtained from $Inn(G)$. Inductively we can make many non abelian groups from a given non abelian group. Since $Inn(G) = G/Z(G)$, this method reduces the size of a given group.

Extensions of non abelian groups is obtained as follows. First, Let θ_1 be a homomorphism on G. (It may be the identity map.) We define θ as follows:

$$\begin{aligned} \theta = Inn \circ \theta_1 : G &\to Aut(G) \\ g &\mapsto Inn(\theta_1(g)) \end{aligned}$$

Then we construct an extension of G, $\bar{G} = G \times_\theta G$. We can easily obtain $Z(\bar{G}) = \{(X,Y) \in G \times_\theta G \mid x, y \in Z(G)\}$. If we use the group G in section 4, $|Z(\bar{G})| = 4p^2$.

Secondly, Let G be a non abelian group and H be a subgroup of automorphism group $Aut(G)$. We construct a non abelian group naturally. Let $\theta = \mathrm{Id}$. Then we can easily obtain $G \times_\theta H$. For example, we can always obtain $Inn(G) = G/Z(G)$ and $G \times_\theta Inn(G)$. If we know other subgroups of $Aut(G)$, we can construct many useful non abelian groups.

Nilpotent group $G = (\mathbb{Z}_p \times \mathbb{Z}_p) \times_\theta \mathbb{Z}_p$. Since $\mathrm{Aut}(\mathbb{Z}_p \times \mathbb{Z}_p) = \mathrm{GL}(2, \mathbb{Z}_p)$, we can make the following non abelian group. There exists an injective homomorphism

$$\theta : \mathbb{Z}_p \to \mathrm{SL}(2, \mathbb{Z}_p).$$

Then we can construct $G = (\mathbb{Z}_p \times \mathbb{Z}_p) \times_\theta \mathbb{Z}_p$. Since G is a p-group, it is a nilpotent group. Hence G has a non trivial center and its cardinality is at least p.

In this case, a generator set of G is $\{e_1 = (1,0,0), e_2 = (0,1,0), e_3 = (0,0,1)\}$ and we can easily express any elements of G as a product of e_i's. Then

$$Inn((X,y))(e_1) = (\theta(y)((1,0)), 0)$$

and

$$Inn((X,y))(e_2) = (\theta(y)((0,1)), 0).$$

So $\theta(y) \in \mathrm{SL}(2, \mathbb{Z}_p)$ can be easily obtained. If $g^a = (X', y^a)$ and $g = (X, y)$, then we can obtain $\theta(y)^a$ and $\theta(y)$. We can solve DLP in $\mathrm{SL}(2, \mathbb{Z}_p)$ as in the Remark 1. Hence the cryptosystem in section 3 is not secure in $G = (\mathbb{Z}_p \times \mathbb{Z}_p) \times_\theta \mathbb{Z}_p$.

Since the variables X, y are separated, this phenomenon occurs. We note here that X, y are separated since the subgroup $\mathbb{Z}_p \times \mathbb{Z}_p$ is abelian. To prevent the separation of variables, we suggest the following non abelian group.

Semi-direct product $G = (\mathbb{Z}_p \times_{\theta_1} \mathbb{Z}_q) \times_{\theta_2} \mathbb{Z}_q$. We replace the abelian group $\mathbb{Z}_p \times \mathbb{Z}_p$ by $\mathbb{Z}_p \times_{\theta_1} \mathbb{Z}_q$, where q is a prime satisfying $q|(p-1)$. Then we can prevent the separation of variables. Since $\mathrm{Aut}(\mathbb{Z}_p) \cong \mathbb{Z}_p^* \cong \mathbb{Z}_{p-1}$, we can make $\mathbb{Z}_p \times_{\theta_1} \mathbb{Z}_q$, where θ_1 is an injective homomorphism from $\mathbb{Z}_q$ to $\mathbb{Z}_p^*$. We denote $\mathbb{Z}_p \times_{\theta_1} \mathbb{Z}_q$ by H. Then H is not abelian.

We will apply the same method as in section 4.2. We can consider $\mathbb{Z}_q$ as a subgroup of H. Its conjugate is also a cyclic subgroup of order q. Let K be one of the conjugates of $\mathbb{Z}_q$ in H. Then there exists an isomorphism θ' from $\mathbb{Z}_q$ to K, and $\theta_2 = Inn \circ \theta'$.

Equations (4.5),(4.6), (4.7) also hold in $G = (\mathbb{Z}_p \times_{\theta_1} \mathbb{Z}_q) \times_{\theta_2} \mathbb{Z}_q$. Then we can find the center of G of order q as in 4.2.

In this case, we denote a generator set of $\mathbb{Z}_p \times_{\theta_1} \mathbb{Z}_q$ by $e_1 = (1,0,0)$ and $e_2 = (0,1,0)$. Since $\mathbb{Z}_p$ is a normal subgroup of G, we assume that $Inn(g)(e_1) = (a_1, 0, 0)$ and $Inn(g)(e_2) = (b_1, b_2, 0)$. We can prove that

$$(z,w)^k = (\frac{\theta_1(w)^k - 1}{\theta_1(w) - 1} z, kw) = (\frac{\theta_1(1)^{wk} - 1}{\theta_1(1)^w - 1} z, kw)$$

for $(z,w) \in \mathbb{Z}_p \times_{\theta_1} \mathbb{Z}_q$ by induction. Then we have for $g = (x_1, x_2, y)$,

$$Inn(g^2)(e_1, 0) = Inn(g)(a_1, 0, 0) = (Inn(g)(e_1))^{a_1} = (a_1, 0, 0)^{a_1} = (a_1^2, 0, 0)$$

and

$$\begin{aligned} Inn(g^2)(e_2,0) &= Inn(g)(b_1,b_2,0) = (Inn(g)(e_1))^{b_1}(Inn(g)(e_2))^{b_2} \\ &= (a_1,0,0)^{b_1}(b_1,b_2,0)^{b_2} = (a_1b_1,0,0)(\frac{\theta_1(1)^{b_2^2}-1}{\theta_1(1)^{b_2}-1}b_1,b_2^2,0) \\ &= (a_1b_1 + \frac{\theta_1(1)^{b_2^2}-1}{\theta_1(1)^{b_2}-1}b_1,b_2^2,0). \end{aligned}$$

From this, we obtain that $Inn(g^k)(e_1) = a_1^k \in \mathbb{Z}_p$. Since $H = \mathbb{Z}_p \times_{\theta_1} \mathbb{Z}_q$ is not an abelian group, the order of $\theta_1(1)$ is q. Thus DLP in $\langle Inn(g) \rangle$ can be reduced to DLP in $\mathbb{Z}_p$, and so the cryptosystem in section 4 is not secure in $G = (\mathbb{Z}_p \times_{\theta_1} \mathbb{Z}_q) \times_{\theta_2} \mathbb{Z}_q$.

The reason of this phenomenon is $\mathbb{Z}_p$ is an abelian normal subgroup. If α is a generator of an abelian normal subgroup, then $Inn(g)(\alpha) = \alpha^s$ for some s and $Inn(g^k)(\alpha) = \alpha^{s^k}$. So we can reduce DLP in $\langle Inn(g) \rangle$ to DLP in $\langle \alpha \rangle \subset \mathbb{Z}_p$. If we use $Inn(Inn(G))$ instead of $Inn(G)$, we can avoid such an attack.

6 Concluding Remarks

We have presented a novel public key cryptosystem (based on a finite non abelian groups) and suggested some examples of finite non abelian groups. There may be other non abelian groups to be used in our system. However we must be careful in applying a non abelian group to our cryptosystem in order that the cryptosystem is secure. As we see in section 5, we should check the following:

- The existence of abelian normal subgroup reduces the security of the cryptosystems. So any abelian normal subgroup must be of small order.
- The algorithm to express an element of G as a product of generators must be efficient.
- Since $Inn(g)$ is expressed as $\{Inn(g)(\gamma_i) \in G \mid \gamma_i \text{ is a generator}\}$, both the number of generators and bits needed to express an element of G must be of small order.

We may use other homomorphisms from G to $\mathrm{Aut}(G)$ instead of the inner automorphism (if exists). Also we can consider the DLP in the endomorphism group $End(G)$.

If we know any representation of G, G can be considered as a subset of a large matrix group up to the kernel (we call a homomorphism from G to a matrix group a representation of G). Hence the representation of G is very useful for cryptosystem as in section 3. If we use DLP in a subgroup $\langle g \rangle$ of a non abelian group and a representation R of G, it would be better to choose $\det(R(g)) = 1$ [15].

Acknowledgment. We would like to thank to our colleagues in NSRI and Dr. Bae Eun Jung for their useful comments. Also we would like to express our gratitude to professor Hong-Jong Kim and Professor Ki-Suk Lee for their kind advice.

References

1. I. Anshel, M. Anshel, D. Goldfeld *An algebraic method for public-key cryptography*, Mathematical Research Letters 6 (1999), 1–5
2. S. Blackburn, S. Galbraith *Cryptanalysis of two cryptsystems based on group actions*, Proc. ASIACRYPT' 99 (2000), 52–61
3. A. E. Brower, R. Pellikaan, E. R. Verheul *Doing more with fewer bits*, Proc. ASIACRYPT' 99 (2000), 321–332
4. D. Coopersmith, A. M. Odlzyko, R. Schroeppel *Discrete logarithms in GF(p)* , Algorithmica, 1 (1986), 1–15
5. T. ElGamal *A public key cryptosystem and a signature scheme based on discrete logarithms* , IEEE Transactions andInformation Theory, 31 (1985), 469–472
6. S. Flannery *Cryptography:An investigation of a new algorithm vs. the RSA*, http://cryptome.org/flannery-cp.pdf, 1999
7. T. W. Hungerford *Algebra*, Springer-Verlag
8. A. K. Lenstra, E. R. Verheul. *The XTR public key system*, Proc. Crypto 2000 (2000), 1–20
9. A. J. Menezes, P. C. Van Oorshot, S. A. Vanstone *Handbook of applied cryptography*, CRC press, 1997
10. R. Lidl, H. Niederreiter *Introduction to finite fields and their application*, Cambridge University press, 1986
11. K. H. Ko, S. J. Lee, J. H. Cheon, J. W. Han, J. -S. Kang, C. Park *New public-key cryptosystem using braid groups*, Proc. Crypto 2000 (2000), 166–184
12. N. Koblitz *Elliptic curve cryptosystems*, Mathematics of Computation, 48 (1987), 203–209
13. V. Miller *Use of elliptic curves in cryptography*, Proc. Crypto 85 (1986), 417–426
14. K. Nyberg, R. Rueppel *A new signature scheme based on DSA giving message recovery*, 1st ACM Conference on Computer and Communications Security, (1993), 58–61
15. S.-H. Paeng, J.-W. Han, B. E. Jung *"The security of XTR in view of the determinant"*, preprint, 2001
16. S.-H. Paeng *" A provably secure public key cryptosystem using finite non abelian groups"*, preprint, 2001
17. S. C. Pohlig, M. E. Hellman *An improved algorithm for computing logarithms over GF(p) and its cryptographic significance*, IEEE Transactions on Information Theory, 23 (1978), 106–110
18. J. M. Pollard *Monte Carlo methods for index computation (mod p)*, Mathematics of computation, 32 (1978), 918–924
19. R. L. Rivest, A. Shamir, L. M. Adleman *A method for digital signature and public-key cryptosystems*, Communications of the ACM, 21 (1978), 120–126
20. A. Yamamura *Public key cryptsystems using the modular group*, PKC'98, 203–216

Appendix A : Special Conjugacy Problem in Matrix Groups

Let G be a matrix group, for example $\mathrm{GL}(2, R)$ or $\mathrm{SL}(2, R)$, where $R = \mathbb{Z}$ or $\mathbb{Z}_p$ for a prime number p. We will solve the special conjugacy problem in G. Let

$$A = \begin{pmatrix} a & b \\ c & d \end{pmatrix}, \; X = \begin{pmatrix} x & y \\ z & w \end{pmatrix}.$$

We will find X from XAX^{-1} for $A \in G$. Let

$$A = \begin{pmatrix} a & b \\ 0 & d \end{pmatrix} \text{ and } XAX^{-1} = \begin{pmatrix} \alpha & \beta \\ \gamma & \delta \end{pmatrix}.$$

From the above equation, we obtain the following linear equations,

$$\begin{aligned} ax &= \alpha x + \beta z \\ az &= \gamma x + \delta z \\ bx + dy &= \alpha y + \beta w \\ bz + dw &= \gamma y + \delta w. \end{aligned}$$

From the first equation, we can easily obtain the ratio of x to z, i.e. $(a-\alpha)x = \beta z$. (Note that we cannot obtain other ratios as we see in Example 2.)

Similarly, if we solve the conjugacy equation for $XA'X^{-1}$ and

$$A' = \begin{pmatrix} a' & b' \\ c' & d' \end{pmatrix}, c' \neq 0$$

we can also get another linear system. If we replace βz by $(a-\alpha)x$, we can obtain remaining ratios between x, y, z and w. So we can solve the special conjugacy problem in G easily. By Example 2, we can easily understand the procedure.

Note that the conjugacy problems in $\mathrm{SL}(2, R)$ or $\mathrm{GL}(2, R)$ are not difficult since we can obtain at most two linear equations by the conjugacy equation.

Example 2. In [20], the author suggested a public key system using $\mathrm{SL}(2, \mathbb{Z})$. It was shown that this system is not secure in [2]. For the point based scheme in [2], we can find the secret key if we solve the conjugacy equations directly as above. Let

$$A = \begin{pmatrix} 1 & -1 \\ 1 & 0 \end{pmatrix} \text{ and } B = \begin{pmatrix} 0 & -1 \\ 1 & 0 \end{pmatrix}.$$

Then $\{A, B\}$ is a generator set of $\mathrm{SL}(2, \mathbb{Z})$. Furthermore, $A^3 = B^2 = -I$, and so every element in $\mathrm{SL}(2, \mathbb{Z})$ can be expressed as the normal form $\pm A^{i_1} B \cdots A^{i_{n-1}} B A^{i_n}$, where $i_j = 0, 1$ or 2. In the public key system suggested in [2], they use a semi-group generated by $\{V_1 = (BA)^i,\ V_2 = (BA^2)^j\}$ for given $i, j \geq 2$. The public key is $\{MV_1M^{-1}, MV_2M^{-1}\}$ and the secret key is M. In order to find the secret key from the public key, we must solve the conjugacy equations. For example, let

$$V_1 = (BA)^2 = \begin{pmatrix} 1 & 0 \\ -2 & 1 \end{pmatrix}, V_2 = (BA^2)^2 = \begin{pmatrix} 1 & -2 \\ 0 & 1 \end{pmatrix} \text{ and } M = \begin{pmatrix} 3 & 1 \\ 5 & 2 \end{pmatrix}$$

Then the public key is

$$MV_1M^{-1} = \begin{pmatrix} -3 & 2 \\ -8 & 5 \end{pmatrix}, MV_1M^{-1} = \begin{pmatrix} 31 & -18 \\ 50 & -29 \end{pmatrix}.$$

Put

$$M = \begin{pmatrix} x & y \\ z & w \end{pmatrix}$$

and find M by solving the conjugacy equation for V_1 and MV_1M^{-1}. We obtain the following linear equations:

$$4x - 2y - 2z = 0$$

$$4y - 2w = 0$$

$$8x - 4z - 2w = 0$$

$$8y - 4w = 0.$$

Then we have $2y = w$. (Check that we cannot obtain other ratios from these equations.)

If we solve the conjugacy equation for V_2 and MV_2M^{-1}, we obtain that $5x = 3z$ and $x + 15y - 9w = 0$. Replacing w by $2y$, we have $x = 3y$ so $2x = 3w$. Hence we obtain the secret key

$$M = C \begin{pmatrix} 3 & 1 \\ 5 & 2 \end{pmatrix}$$

for some C. Since $\det(M) = 1$, we obtain that $C = 1$.

We should note that the dimension of solutions in $\mathrm{GL}(2, R)$ is always larger than 1. From one conjugacy equation, we can obtain at most two linearly independent equations. Combining two conjugacy equations, we obtain three linearly independent equations and one dimensional solutions. In $\mathrm{SL}(2, \mathbb{Z}_p)$, we obtain only one solution. We can apply the same method to any other V_1, V_2 which are suggested in [2].

Pseudorandomness from Braid Groups

Eonkyung Lee, Sang Jin Lee, and Sang Geun Hahn

Department of Mathematics,
Korea Advanced Institute of Science and Technology,
Taejon 305-701, Republic of Korea
{eklee,sjlee,sghahn}@mathx.kaist.ac.kr

Abstract. Recently the braid groups were introduced as a new source for cryptography. The group operations are performed efficiently and the features are quite different from those of other cryptographically popular groups. As the first step to put the braid groups into the area of pseudorandomness, this article presents some cryptographic primitives under two related assumptions in braid groups. First, assuming that the conjugacy problem is a one-way function, say f, we show which particular bit of the argument x is pseudorandom given $f(x)$. Next, under the decision Ko-Lee assumption, we construct two provably secure pseudorandom schemes: a pseudorandom generator and a pseudorandom synthesizer.

1 Introduction

The notions of pseudorandomness and onewayness which are closely related are quite important in modern cryptography [8,1,17,12]. These concepts are informally stated as: (i) A distribution is *pseudorandom* if no efficient algorithm can distinguish it from the uniform distribution [26]. (ii) A function is *one-way* if it is easy to evaluate but hard to invert [9].

Recently, some mathematically hard problems in braid groups have been proposed as new candidates for cryptographic one-way functions [2,19]. A braid group B_n is an infinite non-commutative group naturally arising from geometric braids composed of n strands. One of the famous problems in braid groups is the *conjugacy problem*: Given $(\alpha, \beta) \in B_n \times B_n$, find (*or* determine whether there exists) $\chi \in B_n$ such that $\beta = \chi^{-1}\alpha\chi$. This problem was first introduced in the 1920s, and no polynomial-time algorithm is known for $n \geq 5$. A variant of this problem was first applied to cryptography to build a key agreement scheme by Anshel *et al.* [2].

Ko *et al.* [19] introduced another variant of this problem: Given $\alpha, \chi_1^{-1}\alpha\chi_1$, $\chi_2^{-1}\alpha\chi_2 \in B_n$, where χ_1 and χ_2 are contained in some known subgroups of B_n so that $\chi_1\chi_2 = \chi_2\chi_1$, find $\chi_2^{-1}\chi_1^{-1}\alpha\chi_1\chi_2 \in B_n$. For convenience, we call this problem the *Ko-Lee problem*. The Ko-Lee problem looks like the Diffie-Hellman problem in their structures, but it does not in their internal properties because of the different characteristics of the braid groups from finite commutative groups. For instance, a braid group is non-commutative and it has no finite subgroup except for the trivial subgroup. As the basis of the Ko-Lee problem,

J. Kilian (Ed.): CRYPTO 2001, LNCS 2139, pp. 486–502, 2001.

they introduced, by restricting the conjugacy problem to a smaller braid group, the (n,m)-*generalized conjugacy problem* (GCP): Given $(\alpha,\beta) \in B_n \times B_n$ and $m(\leq n)$, find $\chi \in B_m$ such that $\beta = \chi^{-1}\alpha\chi$. Like the conjugacy problem, the GCP and the Ko-Lee problem have no polynomial-time solving algorithm yet.

The motivation for this article is that the braid groups have potential for a good source to enrich cryptography from the point of view of their features and efficient operations. In the sequel to key agreement schemes [2,19] and a public-key cryptosystem [19], we discuss how to construct cryptographic primitives in the area of pseudorandomness from the two related assumptions in braid groups: the intractability assumptions of the conjugacy and the Ko-Lee problems. We call the latter the *Ko-Lee assumption* (KL-Assumption).

1.1 The Ko-Lee Problem

As a basic pseudorandom primitive, a pseudorandom generator is informally defined to be an efficient algorithm expanding short random bit sequences into long pseudorandom bit sequences [26,8].

Naor *et al.* [23] first introduced the notion of pseudorandom synthesizer as a stronger one than pseudorandom generator in the following sense: While a pseudorandom generator, G, guarantees the pseudorandomness of $\{G(z_i)\}_{1\leq i\leq n}$ only when $z_1,\ldots,z_n$ are chosen uniformly and *independently*, a pseudorandom synthesizer, S, guarantees the pseudorandomness of $\{S(z_i)\}_{1\leq i\leq n}$ even when the z_i's are not *completely independent*. Loosely speaking, a pseudorandom synthesizer is a two variable function $S(\cdot,\cdot)$, so that if polynomially many random assignments are chosen to both variables, $(x_1,\ldots,x_m)$ and $(y_1,\ldots,y_m)$, then the output of S on all the combinations of these assignments, $(S(x_i,y_j))_{1\leq i,j\leq m}$, is pseudorandom.

Our Result: From the KL-Assumption, we formally derive a decisional version mentioned to refer to the security of the braid public-key cryptosystem [19]. Under the decision Ko-Lee assumption (DKL-Assumption), we construct a pseudorandom generator and a pseudorandom synthesizer and show that they are provably secure.

1.2 The Conjugacy Problem

The Ko-Lee problem was originally proposed as a variant of the conjugacy problem to induce a trapdoor one-way function (for a public-key cryptosystem). However, it looks easier to solve than the conjugacy problem. Since pseudorandomness needs no trapdoor, the conjugacy problem itself can be considered.

If f is a one-way function, every bit of the argument x cannot be easily computed from $f(x)$. A natural question is whether there is a specific bit of x which is not distinguished from a random bit by any efficient algorithm given $f(x)$. This question was first addressed by Blum *et al.* [8]. Demonstrating such a pseudorandom bit for the discrete exponentiation function, they introduced the notion of hard-core predicate as a cryptographically useful tool. Loosely speaking, a *hard-core predicate* b of a function f is a polynomial-time computable boolean predicate such that $b(x)$ is hard to predict from $f(x)$. So far, two kinds of hard-core predicates have been proposed. On the one hand, for a few one-way function f's, there has been discovered a particular bit of x, the so-called *hard-core bit*, which is the source of $b(x)$ by the unique characteristic of f [8,1]. For instance, Alexi *et al.* [1] showed that $b(x)$ points to the least significant bit of x for the RSA and the Rabin functions. On the other hand, for any one-way function, one can make a hard-core predicate by Goldreich-Levin's construction [14]. More precisely, for any one-way function f, the inner-product mod 2 of x and r is a hard-core of $g(x,r) \stackrel{\text{def}}{=} (f(x), r)$. To distinguish these two kinds of hard-core predicates, we call the former kind the *peculiar* one and the latter kind the *generic* one.

Considering that among a number of known one-way functions only the RSA, the Rabin, and the discrete exponentiation functions have their peculiar hard-core predicates, it is interesting to find it for the conjugacy problem. It indicates which bit of the solution is equally difficult to compute as the entire solution.

The conjugacy problem in braid groups is quite different from those above one-way functions in the sense that it is not a group homomorphism. Since such a property is the basis for the construction of the previous peculiar hard-core predicates, we should take a completely different way to construct a peculiar hard-core for the conjugacy problem.

Our Result: We first present a collection of one-way functions, CNJ, under the intractability assumption of the $(n, n-1)$-GCP, which is almost the conjugacy problem from a computational complexity point of view. And we present *two* hard-core bits of CNJ. Using one of them, we construct a peculiar hard-core predicate, INF, and prove that predicting $\text{INF}(x)$ from $\text{CNJ}(x)$ is as hard as inverting $\text{CNJ}(x)$. Likewise the other hard-core bit.

1.3 Outline

In §2, we introduce some notations and briefly describe the braid groups. In §3, we examine the bit security in the conjugacy problem (§3.1), present a collection of one-way functions based on that problem (§3.2), and construct a hard-core predicate of the one-way function (§3.3). In §4, we construct a pseudorandom generator (§4.1) and a pseudorandom synthesizer (§4.2).

2 Preliminaries

2.1 Notations

Basic notation: Let **N** and **Z** denote the set of all natural numbers and the set of all integers, respectively. For any bit-string x, $||x||$ denotes its length (i.e. the number of bits in x). For a finite set S, $|S|$ denotes the cardinality of S and $||S||$ denotes the maximum among the bit-lengths of elements of S. The notation $(a_{i,j})_{1\le i\le n, 1\le j\le m}$ denotes an $(n \times m)$-matrix whose (i,j)-entry is $a_{i,j}$.

Probability notation: The following notations are based on [16,15,3].

A probability distribution $\mathcal{D}$ on a finite set S assigns a probability $\mathcal{D}(s) \geq 0$ to each $s \in S$, and thus $\sum_{s \in S} \mathcal{D}(s) = 1$. For a distribution $\mathcal{D}$, $[\mathcal{D}]$ denotes the support of $\mathcal{D}$ (the set of elements of positive probability). If a random variable x is distributed according to $\mathcal{D}$ on S, we write $x \overset{\mathcal{D}}{\leftarrow} S$, or simply $x \leftarrow \mathcal{D}$ if the set S is obvious from the context. The notation $x_1, \ldots, x_n \leftarrow \mathcal{D}$ indicates that n random variables $x_1, \ldots, x_n$ are independently distributed according to $\mathcal{D}$ on S.

If f is a function mapping S to a set T, then $\langle f(x) : x \leftarrow \mathcal{D} \rangle$ is a random variable that defines a distribution $\mathcal{E}$, where for all $t \in T$, $\mathcal{E}(t) = \sum_{s \in S, f(s)=t} \mathcal{D}(s)$.

If $\mathcal{A}$ is a probabilistic algorithm, then for any input $x, y, \ldots$ the notation $\mathcal{A}(x, y, \ldots)$ refers to the probability distribution induced by its internal random coin tosses. So if $x \leftarrow \mathcal{D}, y \leftarrow \mathcal{E}, \ldots$ are random variables, then $\langle \mathcal{A}(x, y, \ldots) : x \leftarrow \mathcal{D}; y \leftarrow \mathcal{E}; \ldots \rangle$ represents the random variable distributed according to $\mathcal{D}, \mathcal{E}, \ldots$ and its internal random coin tosses.

We let $x \overset{u}{\leftarrow} S$ indicate that x is uniformly distributed on S; i.e., for all $s \in S$, $\Pr[x = s : x \overset{u}{\leftarrow} S] = 1/|S|$.

For probability distributions $\mathcal{D}, \mathcal{E}, \ldots$, the notation $\Pr[p(x, y, \ldots) : x \leftarrow \mathcal{D}; y \leftarrow \mathcal{E}; \cdots]$ denotes the probability that the predicate $p(x, y, \ldots)$ is true after the (ordered) execution of the algorithms $x \leftarrow \mathcal{D}, y \leftarrow \mathcal{E}$, etc..

PPTA is short for "probabilistic polynomial time algorithm in its input length(s)".

2.2 The Braid Groups

In this section, we briefly review some basic material for braid groups. See [6, 10,7] for details. For each integer $n \geq 2$, the n-braid group B_n is defined by the following group presentation

$$B_n = \left\langle \sigma_1, \ldots, \sigma_{n-1} \;\middle|\; \begin{array}{ll} \sigma_i\sigma_j\sigma_i = \sigma_j\sigma_i\sigma_j & \text{if } |i-j| = 1 \\ \sigma_i\sigma_j = \sigma_j\sigma_i & \text{if } |i-j| \geq 2 \end{array} \right\rangle.$$

The integer n is called the *braid index* and each element of B_n is called an *n-braid.* An n-braid has the following geometric interpretation: it is a set of disjoint n strands which run essentially to the same direction (our convention is vertical direction). The multiplication $\alpha\beta$ of two braids α and β is the braid obtained by positioning α on the top of β, the identity e_n is the braid consisting of n straight vertical strands, and the inverse of α is the reflection of α with respect to a horizontal plane. Examples are given in Figure 1 (a,b,c). Henceforth, let σ_i denote only a generator of the corresponding braid group.

B_n^+ denotes the monoid defined by the generators and relations in the above presentation, and its elements are called *positive n-braids.* To each permutation $\pi = b_1 b_2 \cdots b_n$, we associate a positive n-braid obtained by connecting the upper i-th point to the lower b_i-th point by a straight line. Such braids as this are called *permutation braids* or *canonical factors.* The permutation n-braid corresponding to the permutation $(n)(n-1)\cdots(2)(1)$ is called the *fundamental braid* and denoted by Δ_n. See Figure 1 (d) for example. For $\alpha \in B_n^+$, define two sets $S(\alpha) = \{i \mid \alpha = \sigma_i\beta \text{ for some } \beta \in B_n^+\}$ and $F(\alpha) = \{i \mid \alpha = \beta\sigma_i \text{ for some } \beta \in B_n^+\}$.

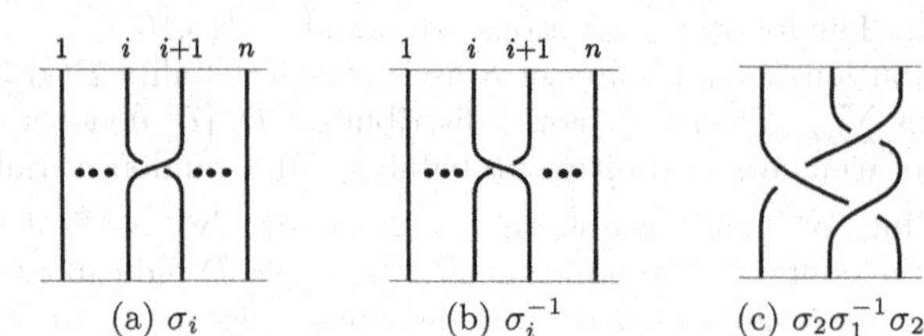

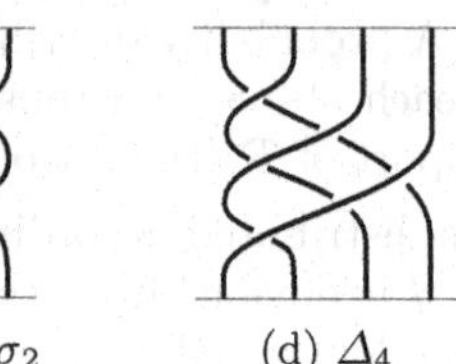

(a) σ_i (b) σ_i^{-1} (c) $\sigma_2\sigma_1^{-1}\sigma_2$ (d) Δ_4

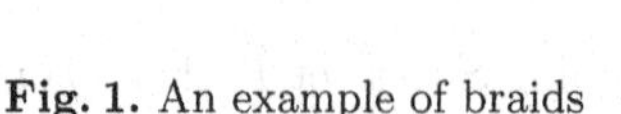

Fig. 1. An example of braids

Every braid $\chi \in B_n$ has a unique decomposition called the *left-canonical form*, $\chi = \Delta_n^u \chi_1 \cdots \chi_k$, where $u \in \mathbf{Z}$ and χ_i's are permutation braids except for e_n and Δ_n such that $F(\chi_i) \supset S(\chi_{i+1})$. In this article, all the braids are supposed to be in their left-canonical forms. Hence, for $\alpha, \beta \in B_n$, $\alpha\beta$ means the left-canonical form of $\alpha\beta$ and so it is hard to guess its original factor α or β from $\alpha\beta$.

For $m < n$, B_m is regarded as the subgroup of B_n generated only by $\sigma_1, \ldots, \sigma_{m-1}$ of B_n, and so $\Delta_m (\in B_n)$ is a permutation n-braid corresponding to a permutation $(m)(m-1)\cdots(2)(1)(m+1)(m+2)\cdots(n)$.

Due to [10,7], braid groups with all their operations—multiplication, inversion, converting into left-canonical forms—are efficiently handled by computers.

3 Hard-Core Predicate

From the intractability assumption of the conjugacy problem, one can naturally derive a one-way function, $\text{CNJ}_\alpha : B_n \longrightarrow B_n$, defined by $\text{CNJ}_\alpha(\chi) = \chi^{-1}\alpha\chi$, where $\alpha \in B_n$.

Our goal in this section is to construct a *peculiar* hard-core predicate of CNJ_α. Therefore, we should discover for CNJ_α the hard-core bit of a braid into which the one-wayness of CNJ_α is transformed.

Notice that we are in different situation from previous ones for the following reasons: (i) A braid is not naturally expressed as a digit. (ii) CNJ_α is not a group homomorphism. By (i), we should find a different type of bit from the least significant bit (for RSA, Rabin) [1] or the most significant bit (for discrete exponentiation function) [8]. Since such a bit must be an invariant of a braid, let us consider the left-canonical form. Recall that any braid $\chi \in B_n$ is *uniquely* expressed in its left-canonical form $\chi = \Delta_n^u \chi_1 \cdots \chi_p$. Here, each of the integers u, p, and $u+p$ is called the *infimum*, the *canonical-length*, and the *supremum* of χ and denoted by $\inf(\chi)$, $\text{len}(\chi)$, and $\sup(\chi)$, respectively. Because they are invariants of a braid, the hard-core bit may be derived from some of them. In contrast to (ii), the homomorphic property of the other one-way functions was essential to find their hard-core bits [8,1]. Therefore, we should approach our problem in a new way.

3.1 Candidates for the Hard-Core Bit

The following two propositions show the key properties of the infimum and the supremum to be the hard-core bits.

Proposition 1. *Let* $\chi = \Delta_n^u \varphi \in B_n$*, where* $\varphi \in B_n^+ - \Delta_n B_n^+$*. Then for any generator* σ_i *of* B_n*,*

$$\inf(\chi\sigma_i^{-1}) = \begin{cases} \inf(\chi) & \text{if } i \in F(\varphi) \\ \inf(\chi) - 1 & \text{otherwise.} \end{cases}$$

Proof. Note that for any $\chi_1, \chi_2 \in B_n$, $\inf(\chi_1\chi_2) \geq \inf(\chi_1) + \inf(\chi_2)$. Using this, we get $\inf(\chi) - 1 \leq \inf(\chi\sigma_i^{-1}) \leq \inf(\chi)$. Thus it suffices to show that $\inf(\chi\sigma_i^{-1}) = \inf(\chi)$ if and only if $i \in F(\varphi)$. If $i \in F(\varphi)$, then $\varphi = \varphi_1\sigma_i$ for some $\varphi_1 \in B_n^+ - \Delta_n B_n^+$ and $\inf(\chi\sigma_i^{-1}) = \inf(\Delta_n^u\varphi_1) = u = \inf(\chi)$. Conversely, if $\inf(\chi\sigma_i^{-1}) = \inf(\chi)$, then $\chi\sigma_i^{-1} = \Delta_n^u\varphi_2$ for some $\varphi_2 \in B_n^+ - \Delta_n B_n^+$. This implies that $\varphi = \varphi_2\sigma_i$ and so $i \in F(\varphi)$. □

Proposition 2. *Let* $\Delta_n^u\chi_1 \cdots \chi_k$ *be the left-canonical form of* $\chi \in B_n$*. Then for any generator* σ_i *of* B_n*,*

$$\sup(\chi\sigma_i) = \begin{cases} \sup(\chi) + 1 & \text{if } i \in F(\chi_k) \\ \sup(\chi) & \text{otherwise.} \end{cases}$$

Proof. If $i \in F(\chi_k)$, then it is clear that $\sup(\chi\sigma_i) = \sup(\chi) + 1$. Otherwise, $\chi_k\sigma_i$ is a permutation braid, so that $\sup(\chi\sigma_i) \leq u + k = \sup(\chi)$. Since $\sup(\chi\sigma_i) \geq \sup(\chi)$, we have $\sup(\chi\sigma_i) = \sup(\chi)$. □

From now on, we consider only the infimum. By Proposition 2, the supremum can be dealt with similarly to the infimum.

Proposition 1 shows a clue to finding a hard-core bit for the conjugacy problem in the following way: Loosely speaking, given $(\alpha, \chi^{-1}\alpha\chi)$, if an adversary is allowed to access to an oracle $\mathcal{INF}$ which on input $(\alpha, \zeta^{-1}\alpha\zeta)$ outputs $\inf(\zeta) \bmod 2$ for all $\zeta \in B_n$, then (s)he can detect the last generator of χ by comparing $\mathcal{INF}(\alpha, \chi^{-1}\alpha\chi)$ with $\mathcal{INF}(\alpha, \sigma_i\chi^{-1}\alpha\chi\sigma_i^{-1})$. In the recursive way, (s)he finally obtains the entirety of χ.

The existence of $\mathcal{INF}$ assumes that $\zeta_1^{-1}\alpha\zeta_1 = \zeta_2^{-1}\alpha\zeta_2$ implies $\inf(\zeta_1) = \inf(\zeta_2) \bmod 2$. However, it does not always happen. For example, if $\alpha = \Delta_n$ and $\zeta_2 = \Delta_n\zeta_1$, then $\zeta_1^{-1}\alpha\zeta_1 = \zeta_2^{-1}\alpha\zeta_2$ but $\inf(\zeta_2) = \inf(\zeta_1) + 1$. Since α has a major influence on the complexity of the conjugacy problem, α cannot be arbitrarily chosen but must satisfy some property.

Definition 1. *We say that* $\alpha \in B_n$ *is* centralizer-free in B_m *if for any* $\chi \in B_m$ $(m < n)$*,* $\chi\alpha = \alpha\chi$ *implies* $\chi = e_m$*.*

Note that if α is centralizer-free in B_m, then $\zeta_1^{-1}\alpha\zeta_1 = \zeta_2^{-1}\alpha\zeta_2$ $(\zeta_1, \zeta_2 \in B_m)$ implies $\zeta_1 = \zeta_2$, and hence $\inf(\zeta_1) = \inf(\zeta_2)$.

We claim that if we choose $\alpha \in B_n$ at random, then it is centralizer-free in B_{n-1} with negligible exceptions. Because the argument needs dynamics of disc homeomorphims, which seems beyond the scope of this article, we briefly list some known facts.

Fact 1. Braids are classified into three dynamical types [20,4]—periodic, reducible, pseudo-Anosov—by the Nielsen-Thurston classification of surface automorphisms [25,24,11,5]. The periodic and the reducible types are of extremely special forms and the pseudo-Anosov one is of typical form [25].

Fact 2. The pseudo-Anosov n-braids are centralizer-free in B_{n-1} (See [21]).

It seems that if we choose at random an n-braid α with p canonical factors, then it is pseudo-Anosov with probability almost $1 - \frac{1}{n^p}$.

The following proposition shows that the least significant bit of the infimum has potential for the hard-core bit for CNJ_α.

Proposition 3. *Let $\alpha \in B_n$ be centralizer-free in B_{n-1} and $\mathcal{INF}$ be as above. Then CNJ_α is inverted for all $\chi \in B^+_{n-1} - \Delta_{n-1}B^+_{n-1}$ by invoking $\mathcal{INF}$ polynomial in $(n, \text{len}(\chi))$ times.*

Proof. We exhibit a basic algorithm that inverts CNJ_α by making calls to $\mathcal{INF}$. Using Proposition 1, the algorithm on input $(\alpha, \chi^{-1}\alpha\chi)$ finds χ generator-by-generator from right to left of χ. In the middle of the execution, the variable χ' will contain the right half of the generators of χ and the variable β' is such that $\text{CNJ}_\alpha^{-1}(\beta')$ = the left half of the χ. The algorithm, abstractly, transfers the last generator of $\text{CNJ}_\alpha^{-1}(\beta')$ in front of χ' until $\text{CNJ}_\alpha^{-1}(\beta') = e_{n-1}$, and thus all of χ is reconstructed in χ'.

1. $\beta' \leftarrow \chi^{-1}\alpha\chi$; $\chi' \leftarrow e_{n-1}$.
2. for $i = 1$ to $n-2$ do

   2.1. if $\mathcal{INF}(\alpha, \sigma_i\beta'\sigma_i^{-1}) = \mathcal{INF}(\alpha, \beta')$, then
   $\chi' \leftarrow \sigma_i\chi'$; $\beta' \leftarrow \sigma_i\beta'\sigma_i^{-1}$.

   2.2. if $\beta' = \alpha$, then go to step 3,
   else, go to step 2.
3. output χ'.

Note that every n-permutation braid is composed of at most $\frac{n(n-1)}{2}$ generators of B_n. So, the running time of the above algorithm is $\mathcal{O}(n^3 \text{len}(\chi) T)$, where T is the running time of $\mathcal{INF}$. □

3.2 Construction of a Collection of One-Way Functions, CNJ

The original definition of one-way function refers to a single function operating on an infinite domain like $f : \{0,1\}^* \longrightarrow \{0,1\}^*$. This formulation is suitable for an abstract discussion. However, for practical purposes, an infinite collection of functions each operating on a finite domain is more adequate. In this context, this section describes a collection of one-way functions under the intractability assumption of the conjugacy problem. Recall the formal definition of a collection of one-way functions.

Definition 2 ([13]). *Let I be an index set and for each $i \in I$ let D_i be a finite domain.* A collection of one-way functions *is a set $F = \{f_i : D_i \longrightarrow \{0,1\}^*\}_{i \in I}$ satisfying the following conditions:*

Cond 1. *There exists a PPTA $\mathcal{I}$ which on input 1^n outputs $i \in I \cap \{0,1\}^n$.*
Cond 2. *There exists a PPTA $\mathcal{D}$ which on input $i \in I$ outputs $x \in D_i$.*
Cond 3. *There exists a polynomial-time algorithm that on input $(i,x) \in I \times D_i$ outputs $f_i(x)$.*
Cond 4. *For every PPTA $\mathcal{A}$, every polynomial P, and all sufficiently large n's,*

$$\Pr[f_i(z) = f_i(x) : i \leftarrow \mathcal{I}(1^n); x \leftarrow \mathcal{D}(i); z \leftarrow \mathcal{A}(i, f_i(x))] < \tfrac{1}{P(n)}.$$

Intuitively, the (n,m)-GCP becomes harder as m increases because B_m is a subgroup of B_n. As mentioned in §1, the (n,m)-GCP is a by-product of the KL-Assumption which is based on the $(n,\frac{n}{2})$-GCP [19]. However, one-way functions have no problem to be constructed from the conjugacy problem itself. To construct a hard-core predicate, from the discussion in §3.1 we consider the $(n,n-1)$-GCP which is almost the conjugacy problem in terms of computational complexity.

The hardness of the $(n,n-1)$-GCP depends on the braid index n, and the actual bound of the canonical-lengths of braids it takes. So it is natural and practical to take both the braid index and the canonical-length as its security parameter.

Notation. For $n \in \mathbf{N}$ and $i \le j \in \mathbf{Z}$, let $[i,j]_n \stackrel{\text{def}}{=} \{\chi \in B_n \mid \inf(\chi) \ge i, \sup(\chi) \le j\}$.

Construction 1. *Let $I \stackrel{\text{def}}{=} \{(n,p) \mid n,p \in \mathbf{N}\}$ be an index set.*

- *$\forall k = (n,p) \in I$, let $I_k \stackrel{\text{def}}{=} \{\alpha \in B_n^+ - \Delta_n B_n^+ \mid \mathrm{len}(\alpha) = p\}$ be an instance set. Let $\mathcal{IG}$ be a probabilistic algorithm that on input $(1^n, 1^p)$, where $k = (n,p) \in I$, outputs an element of I_k.*
- *$\forall k = (n,p) \in I$, let $D_k \stackrel{\text{def}}{=} [-p,p]_{n-1}$. Let $\mathcal{DG}$ be a probabilistic algorithm that on input $(1^n, 1^p)$, where $k = (n,p) \in I$, outputs an element of D_k.*
- *$\forall k = (n,p) \in I, \forall \alpha \in I_k$, define an instance function $\mathrm{CNJ}_\alpha : D_k \longrightarrow B_n$ by $\mathrm{CNJ}_\alpha(\chi) = \chi^{-1}\alpha\chi$.*
- *$\forall k = (n,p) \in I$, let F_k be the random variable defined on $\{\mathrm{CNJ}_\alpha\}_{\alpha \in I_k}$ distributed according to $\mathcal{IG}(1^n, 1^p)$.*
- *Let $\mathrm{CNJ} \stackrel{\text{def}}{=} \{F_k\}_{k \in I}$.*

CNJ clearly satisfies Cond 3 because given $(\alpha,\chi) \in I_{n,p} \times D_{n,p}$, one can compute the left-canonical form of $\chi^{-1}\alpha\chi$ in time $\mathcal{O}(p^2 n \log n)$ [10,19]. Now we check Cond 1,2. Notice that to satisfy Cond 4, $\mathcal{DG}(1^n, 1^p)$ cannot be mainly concentrated on polynomially many (in k) elements [13].

The proof of Theorem 3 in [19] is followed by the next corollary.

Corollary 1. *There exists a PPTA whose outputs, on input $(1^n, 1^p)$, are distributed uniformly over a subset, S, of $\{\chi \in B_n^+ - \Delta_n B_n^+ \mid \mathrm{len}(\chi) = p\}$, where $|S| \ge (\lfloor \frac{n-1}{2} \rfloor !)^p$.*

Therefore, we can have $\mathcal{IG}$ and $\mathcal{DG}$ satisfy Cond 1,2,4 under the intractability assumption of the $(n,n-1)$-GCP. Furthermore, from this corollary and from the discussion of α in §3.1, CNJ_α can be regarded as $1-1$ for all sufficiently large $k = (n,p)$'s in I and a randomly chosen α by $\mathcal{IG}(1^n, 1^p)$. Hereafter, saying *large* k means large n and large p.

3.3 Construction of a Hard-Core Predicate, INF

This section constructs a hard-core predicate of CNJ. Recall the original definition of a hard-core predicate.

Definition 3 ([13]). *A polynomial-time computable predicate* $b : \{0,1\}^* \longrightarrow \{0,1\}$ *is called a* hard-core *of* $f : \{0,1\}^* \longrightarrow \{0,1\}^*$ *if for every PPTA* $\mathcal{A}$*, every positive polynomial* P*, and all sufficiently large* n*'s in* $\mathbf{N}$

$$\Pr[\mathcal{A}(f(x)) = b(x) : x \overset{u}{\leftarrow} \{0,1\}^n] < \tfrac{1}{2} + \tfrac{1}{P(n)}.$$

Notice that, given $(\alpha, \chi^{-1}\alpha\chi)$, to retrieve $\chi \in D_{n,p}$ we must know $\inf(\zeta)$ mod 2 from $(\alpha, \zeta^{-1}\alpha\zeta)$ for many ζ's in B_{n-1} which are closely related to χ. However, any finite subset of B_{n-1} except for $\{e_{n-1}\}$ is not a group. So it happens that for some χ's in $D_{n,p}$, some ζ's are not in $D_{n,p}$. For this reason, the domain of hard-core predicate is defined slightly different from the corresponding one of CNJ.

For every $k = (n,p) \in I$, consider a slightly enlarged set of D_k,

$$\bar{D}_k \overset{\text{def}}{=} D_k \cup \{\chi\sigma_i^{-1} \mid \chi \in D_k, i \in \{1, \ldots, n-2\}\}.$$

Thus, $D_k = [-p,p]_{n-1} \subset \bar{D}_k \subset [-(p+1),p]_{n-1} \subset D_{n,p+1}$.

Notation. $\sigma_0 \overset{\text{def}}{=} e_n$.

For every $k = (n,p) \in I$, define a PPTA $\overline{\mathcal{DG}}(1^n, 1^p)$ in the following order:

$$\chi \leftarrow \mathcal{DG}(1^n, 1^p);\ i \overset{u}{\leftarrow} \{0,1,\ldots,n-2\};\ \text{output } \chi\sigma_i^{-1}.$$

Using the infimum and $\bar{D}_k$, we now define a collection of boolean predicates

$$\text{INF} = \{\text{INF}_k : \bar{D}_k \longrightarrow \{0,1\}\}_{k\in I} \ \text{ by } \ \text{INF}_k(\chi) = \inf(\chi) \bmod 2.$$

The following lemma is crucial to our main result. It shows, for a random choice $\chi \in \bar{D}_k$, how to turn a PPTA that predicts correctly $\text{INF}_k(\chi)$ from $\text{CNJ}_\alpha(\chi)$ with probability non-negligibly higher than 1/2 into a PPTA predicting almost correctly.

Lemma 1. *For an infinite subset* F *of* I*, let* $\mathcal{A}$ *be a PPTA and* P *be a positive polynomial such that for all* $k = (n,p) \in F$

$$\Pr[\mathcal{A}(1^n, 1^p, \alpha, \chi^{-1}\alpha\chi) = \text{INF}_k(\chi) : \alpha \leftarrow \mathcal{IG}(1^n, 1^p); \chi \leftarrow \overline{\mathcal{DG}}(1^n, 1^p)] \geq \tfrac{1}{2} + \tfrac{1}{P(k)}.$$

Then for any positive polynomial Q*, there exists a PPTA* $\mathcal{C}$ *such that for all* $k = (n,p) \in F$

$$\Pr[\mathcal{C}(1^n, 1^p, \alpha, \chi^{-1}\alpha\chi) = \text{INF}_k(\chi) : \alpha \leftarrow \mathcal{IG}(1^n, 1^p); \chi \leftarrow \overline{\mathcal{DG}}(1^n, 1^p)] \geq 1 - \tfrac{1}{Q(k)}.$$

Proof. For every $k \in F$, let $N = N(k) \overset{\text{def}}{=} \frac{1}{4}P(k)^2Q(k)$. On every input $(1^n, 1^p, \alpha, \chi^{-1}\alpha\chi)$, where $k = (n,p) \in F, \alpha \in [\mathcal{IG}(1^n, 1^p)]$, and $\chi \in [\overline{\mathcal{DG}}(1^n, 1^p)]$, $\mathcal{C}$ executes the following algorithm:

1. Invoke $\mathcal{A}$ on input $(1^n, 1^p, \alpha, \chi^{-1}\alpha\chi)$ independently N-times. And let $\mathcal{A}^{(i)}$ be the i-th invoking of $\mathcal{A}$ for each $i \in \{1, \ldots, N\}$.
2. If $\sum_{i=1}^{N} \mathcal{A}^{(i)}(1^n, 1^p, \alpha, \chi^{-1}\alpha\chi) \geq \frac{N}{2}$, output 1. Otherwise, output 0.

For every $k = (n,p) \in I$ and every $i \in \{1, \ldots, N\}$, define a PPTA $\zeta_i^{\mathcal{A}}(1^n, 1^p, \cdot, \cdot)$ induced by $\mathcal{A}$ as

$$\zeta_i^{\mathcal{A}}(1^n, 1^p, \alpha, \chi^{-1}\alpha\chi) = \begin{cases} 1 & \text{if } \mathcal{A}^{(i)}(1^n, 1^p, \alpha, \chi^{-1}\alpha\chi) \neq \text{INF}_k(\chi) \\ 0 & \text{otherwise,} \end{cases}$$

where $\alpha \leftarrow \mathcal{IG}(1^n, 1^p)$; $\chi \leftarrow \overline{\mathcal{DG}}(1^n, 1^p)$.

The independence of $\langle\{\mathcal{A}^{(i)}(1^n, 1^p, \alpha, \chi^{-1}\alpha\chi)\}_{1 \leq i \leq N} : \alpha \leftarrow \mathcal{IG}(1^n, 1^p); \chi \leftarrow \overline{\mathcal{DG}}(1^n, 1^p)\rangle$ yields the independence of $\langle\{\zeta_i^{\mathcal{A}}(1^n, 1^p, \alpha, \chi^{-1}\alpha\chi)\}_{1 \leq i \leq N} : \alpha \leftarrow \mathcal{IG}(1^n, 1^p); \chi \leftarrow \overline{\mathcal{DG}}(1^n, 1^p)\rangle$. And for every $i \in \{1, \ldots, N\}$

$$\begin{aligned} &\Pr\left[\zeta_i^{\mathcal{A}}(1^n, 1^p, \alpha, \chi^{-1}\alpha\chi) = 1 : \alpha \leftarrow \mathcal{IG}(1^n, 1^p); \chi \leftarrow \overline{\mathcal{DG}}(1^n, 1^p)\right] \\ &= \Pr\left[\mathcal{A}(1^n, 1^p, \alpha, \chi^{-1}\alpha\chi) \neq \text{INF}_k(\chi) : \alpha \leftarrow \mathcal{IG}(1^n, 1^p); \chi \leftarrow \overline{\mathcal{DG}}(1^n, 1^p)\right] \\ &\leq \tfrac{1}{2} - \tfrac{1}{P(k)}. \end{aligned}$$

So $\langle\{\zeta_i^{\mathcal{A}}(1^n, 1^p, \alpha, \chi^{-1}\alpha\chi)\}_{1 \leq i \leq N} : \alpha \leftarrow \mathcal{IG}(1^n, 1^p); \chi \leftarrow \overline{\mathcal{DG}}(1^n, 1^p)\rangle$ are independent and identically distributed random variables with common binomial distribution $B(1,p)$, where $p \leq \frac{1}{2} - \frac{1}{P(k)}$.

From $\mathrm{E}[\zeta_i^{\mathcal{A}}(1^n, 1^p, \alpha, \chi^{-1}\alpha\chi) : \alpha \leftarrow \mathcal{IG}(1^n, 1^p); \chi \leftarrow \overline{\mathcal{DG}}(1^n, 1^p)] \leq \frac{1}{2} - \frac{1}{P(k)}$ and by applying Chebyshev's inequality, we get

$$\begin{aligned} &\Pr\left[\frac{1}{N}\sum_{i=1}^{N} \zeta_i^{\mathcal{A}}(1^n, 1^p, \alpha, \chi^{-1}\alpha\chi) \geq \frac{1}{2} : \alpha \leftarrow \mathcal{IG}(1^n, 1^p); \chi \leftarrow \overline{\mathcal{DG}}(1^n, 1^p)\right] \\ &\leq P(k)^2 \mathrm{Var}\left[\frac{1}{N}\sum_{i=1}^{N} \zeta_i^{\mathcal{A}}(1^n, 1^p, \alpha, \chi^{-1}\alpha\chi) : \alpha \leftarrow \mathcal{IG}(1^n, 1^p); \chi \leftarrow \overline{\mathcal{DG}}(1^n, 1^p)\right]. \end{aligned}$$

Because $\langle\{\zeta_i^{\mathcal{A}}(1^n, 1^p, \alpha, \chi^{-1}\alpha\chi)\}_{1 \leq i \leq N} : \alpha \leftarrow \mathcal{IG}(1^n, 1^p); \chi \leftarrow \overline{\mathcal{DG}}(1^n, 1^p)\rangle$ are pairwise independent and because

$$\mathrm{Var}[\zeta_i^{\mathcal{A}}(1^n, 1^p, \alpha, \chi^{-1}\alpha\chi) : \alpha \leftarrow \mathcal{IG}(1^n, 1^p); \chi \leftarrow \overline{\mathcal{DG}}(1^n, 1^p)] < \tfrac{1}{4},$$

it follows that

$$\mathrm{Var}\left[\frac{1}{N}\sum_{i=1}^{N} \zeta_i^{\mathcal{A}}(1^n, 1^p, \alpha, \chi^{-1}\alpha\chi) : \alpha \leftarrow \mathcal{IG}(1^n, 1^p); \chi \leftarrow \overline{\mathcal{DG}}(1^n, 1^p)\right] < \tfrac{1}{4N}.$$

Thus,

$$\Pr\left[\tfrac{1}{N}\textstyle\sum_{i=1}^{N} \zeta_i^{\mathcal{A}}(1^n, 1^p, \alpha, \chi^{-1}\alpha\chi) \geq \tfrac{1}{2} : \alpha \leftarrow \mathcal{IG}(1^n, 1^p); \chi \leftarrow \overline{\mathcal{DG}}(1^n, 1^p)\right] < \tfrac{1}{Q(k)}.$$

That is to say,

$$\Pr[\mathcal{C}(1^n, 1^p, \alpha, \chi^{-1}\alpha\chi) = \text{INF}_k(\chi) : \alpha \leftarrow \mathcal{IG}(1^n, 1^p); \chi \leftarrow \overline{\mathcal{DG}}(1^n, 1^p)] \geq 1 - \tfrac{1}{Q(k)}.$$

□

By this lemma and by the basic algorithm in Proposition 3, we get the following result:

Theorem 1. INF *is a hard-core predicate of* CNJ.

Proof. Assume that there exist a PPTA $\mathcal{A}$, an infinite subset F of I, and a positive polynomial P such that for all $k = (n, p) \in F$

$$\Pr[\mathcal{A}(1^n, 1^p, \alpha, \chi^{-1}\alpha\chi) = \text{INF}_k(\chi) : \alpha \leftarrow \mathcal{IG}(1^n, 1^p); \chi \leftarrow \overline{\mathcal{DG}}(1^n, 1^p)] \geq \tfrac{1}{2} + \tfrac{1}{P(k)}.$$

From Lemma 1, there is a PPTA $\mathcal{C}$ such that for all $k = (n, p) \in F$

$$\Pr[\mathcal{C}(1^n, 1^p, \alpha, \chi^{-1}\alpha\chi) = \text{INF}_k(\chi) : \alpha \leftarrow \mathcal{IG}(1^n, 1^p); \chi \leftarrow \overline{\mathcal{DG}}(1^n, 1^p)] \geq 1 - \tfrac{1}{2pn^3}.$$

Fix $k = (n, p) \in F$. Using the basic algorithm in Proposition 3, on input $(1^n, 1^p, \alpha, \chi^{-1}\alpha\chi)$, where $\alpha \leftarrow \mathcal{IG}(1^n, 1^p)$; $\chi \leftarrow \mathcal{DG}(1^n, 1^p)$, $\mathcal{M}$ executes the following algorithm:

1. $\beta' \leftarrow \chi^{-1}\alpha\chi$; $\chi' \leftarrow e_{n-1}$.
2. for $u = -p$ to p do
2.1. if $\beta' = \Delta_{n-1}^{-u}\alpha\Delta_{n-1}^{u}$, then go to step 4.
3. for $j = 1$ to $n-1$ do
3.1. $i \overset{u}{\leftarrow} \{0, 1, \ldots, n-2\}$.
3.2. if $\mathcal{C}(1^n, 1^p, \alpha, \sigma_i\beta'\sigma_i^{-1}) = \mathcal{C}(1^n, 1^p, \alpha, \beta')$, then
$\chi' \leftarrow \sigma_i\chi'$; $\beta' \leftarrow \sigma_i\beta'\sigma_i^{-1}$,
else go to step 3.
3.3. for $u = -p$ to p do
3.3.1. if $\beta' = \Delta_{n-1}^{-u}\alpha\Delta_{n-1}^{u}$, then go to step 4.
3.4. go to step 3.
4. output $\Delta_{n-1}^{u}\chi'$.

Each repetition of the above algorithm makes two calls to $\mathcal{C}$ independently and the number of repetitions of the algorithm is at most $p(n-1)^3$. By the definition of $\overline{\mathcal{DG}}$ and $\mathcal{M}$, for all $k = (n, p) \in F$

$$\Pr[\varsigma^{-1}\alpha\varsigma = \chi^{-1}\alpha\chi : \alpha \leftarrow \mathcal{IG}(1^n, 1^p); \chi \leftarrow \mathcal{DG}(1^n, 1^p); \varsigma \leftarrow \mathcal{M}(1^n, 1^p, \alpha, \chi^{-1}\alpha\chi)] > \tfrac{1}{2pn^3}.$$

□

Notice that hard-core predicates are used to construct pseudorandom generators in some cases by Blum-Micali's general method [8]. Loosely speaking, if $l : \mathbf{N} \longrightarrow \mathbf{N}$ is a stretching function and $f : \{0,1\}^n \longrightarrow \{0,1\}^n$ is a $1-1$ one-way function with a hard-core b, then $G(s) \overset{\text{def}}{=} b(x_1)b(x_2)\cdots b(x_{l(n)})$ is a pseudorandom generator, where $x_0 = s$ and $x_i = f(x_{i-1})$ for $i = 1, \ldots, l(n)$. This method does not apply to INF because $\text{CNJ}_\alpha(D_k)$ is much larger than D_k. From the fact that most known one-way functions in braid groups (see [19]) do not preserve their finite domains, hard-core predicates in braid groups seem to have no relation to this method.

4 Pseudorandom Schemes

The original KL-Assumption is as follows:

> Given a triplet $(\alpha, \chi^{-1}\alpha\chi, \psi^{-1}\alpha\psi)$ of elements in B_n, where $\chi \in \langle\sigma_1, \ldots, \sigma_{\lfloor\frac{n}{2}\rfloor-1}\rangle$ and $\psi \in \langle\sigma_{\lfloor\frac{n}{2}\rfloor+1}, \ldots, \sigma_{n-1}\rangle$, it is computationally infeasible to find $\psi^{-1}\chi^{-1}\alpha\chi\psi$.

Let $m(n) \stackrel{\text{def}}{=} \lfloor\frac{n}{2}\rfloor$. For every $k = (n,p) \in I$, let m mean $m(n)$ and let $LD_k \stackrel{\text{def}}{=} [-p,p]_m$. Consider a group monomorphism $\tau : B_{n-m} \longrightarrow B_n$ defined by $\tau(\sigma_i) = \sigma_{m+i}$ for $i = 1, \ldots, n-m-1$. Then $\tau(B_{n-m}) = \langle\sigma_{m+1}, \ldots, \sigma_{n-1}\rangle$ is a subgroup of B_n isomorphic to B_{n-m}. Let $RD_k \stackrel{\text{def}}{=} \tau([-p,p]_{n-m})$. Here, we defined $m(n)$ as $\lfloor\frac{n}{2}\rfloor$ for notational convenience. Instead, it can take any number around this. From the definition of LD_k and RD_k, for every $k = (n,p) \in I$ and every $(\chi,\psi) \in LD_k \times RD_k$, it follows that: (i) $\chi\psi = \psi\chi$, (ii) $\chi\psi \in [-p,p]_n$. (i) is trivial. (ii) uses the fact that there exists $\zeta \in B_n^+$ such that $\Delta_n = \Delta_m\tau(\Delta_{n-m})\zeta$.

For every $k = (n,p) \in I$ and every $\alpha \in I_k$, let $R_{k,\alpha} \stackrel{\text{def}}{=} \{\zeta^{-1}\alpha\zeta \mid \zeta \in [-p,p]_n\}$. Using these notations, the DKL-Assumption is stated as follows:

[The DKL-Assumption]

For every PPTA $\mathcal{A}$, every positive polynomial P, and all sufficiently large $k = (n,p)$'s in I,

$$\Big|\Pr\big[\mathcal{A}(\alpha,\chi^{-1}\alpha\chi,\psi^{-1}\alpha\psi,\psi^{-1}\chi^{-1}\alpha\chi\psi)=1 : \alpha\leftarrow\mathcal{IG}(1^n,1^p); \chi\stackrel{u}{\leftarrow}LD_k; \psi\stackrel{u}{\leftarrow}RD_k\big]$$
$$-\Pr\big[\mathcal{A}(\alpha,\chi^{-1}\alpha\chi,\psi^{-1}\alpha\psi,\beta)=1 : \alpha\leftarrow\mathcal{IG}(1^n,1^p); \chi\stackrel{u}{\leftarrow}LD_k; \psi\stackrel{u}{\leftarrow}RD_k; \beta\stackrel{u}{\leftarrow}R_{k,\alpha}\big]\Big|$$
$$< \frac{1}{P(k)}.$$

Actually, there is no known PPTA sampling χ from LD_k uniformly at random. However, from Corollary 1, one can construct a PPTA $\mathcal{LDG}$ such that for every $k = (n,p) \in I$, $\mathcal{LDG}(1^n, 1^p)$ is uniformly distributed on $[\mathcal{LDG}(1^n, 1^p)] \subset LD_k$. Moreover, for every polynomial Q, $\|[\mathcal{LDG}(1^n, 1^p)]\| > Q(k)$ for all sufficiently large $k = (n,p)$'s in I. So, in this section saying that $\chi \stackrel{u}{\leftarrow} LD_k$ implicitly means two folds. On the one hand, we have such a $\mathcal{LDG}$ as this. On the other hand, $\chi \leftarrow \mathcal{LDG}(1^n, 1^p)$. In other words, LD_k means $[\mathcal{LDG}(1^n, 1^p)]$ in a probabilistic sense. Likewise, let us view $\chi \stackrel{u}{\leftarrow} RD_k$ and $\chi \stackrel{u}{\leftarrow} R_{k,\alpha}$ in this way.

Under this DKL-Assumption, this section constructs a pseudorandom generator and a pseudorandom synthesizer which are similar to those based on the decision Diffie-Hellman assumption [22]. Since the securities are proved typically by the standard hybrid techniques [13,16,22,23], we only sketch them.

4.1 Pseudorandom Generator

Recall the formal definition of pseudorandom generator.

Definition 4 ([26,8]). *A deterministic polynomial-time algorithm,* $G : \{0,1\}^* \longrightarrow \{0,1\}^*$, *is called a* pseudorandom generator *if there exists a stretching function,* $l : \mathbf{N} \longrightarrow \mathbf{N}$, *so that for all* $x \in \{0,1\}^*$, $||G(x)|| = l(||x||)$ *and if for every PPTA* $\mathcal{A}$, *every positive polynomial* P, *and all sufficiently large* n*'s in* $\mathbf{N}$

$$\left|\Pr[\mathcal{A}(G(x)) = 1 : x \stackrel{u}{\leftarrow} \{0,1\}^n] - \Pr[\mathcal{A}(r) = 1 : r \stackrel{u}{\leftarrow} \{0,1\}^{l(n)}]\right| < \tfrac{1}{P(n)}.$$

The idea of this section is as follows: Given $(\alpha, \chi^{-1}\alpha\chi)$ for $\alpha \in B_n, \chi \in B_m$, it looks hard to find χ even if we know $(\psi_i^{-1}\alpha\psi_i, \chi^{-1}\psi_i^{-1}\alpha\psi_i\chi)$'s for polynomially many ψ_i's randomly chosen in $\tau(B_{n-m})$.

Notation. For every $k \in I$ and every $\alpha \in I_k$, let $LR_{k,\alpha} \stackrel{\text{def}}{=} \{\chi^{-1}\alpha\chi \mid \chi \in LD_k\}$.

Definition 5 ($\mathcal{PGIG}_{KL}$). *An* instance generator $\mathcal{PGIG}_{KL}$ *is a probabilistic algorithm that on input* $(1^n, 1^p, 1^l)$, *where* $k = (n,p) \in I$ *and* $l \in \mathbf{N}$, *executes the following:*

$$\alpha \leftarrow \mathcal{IG}(1^n, 1^p);\ \alpha_1, \ldots, \alpha_l \stackrel{u}{\leftarrow} LR_{k,\alpha};\ output\ (\alpha, \alpha_1, \ldots, \alpha_l).$$

By the definition of $\mathcal{IG}$ in §3.2, $\mathcal{PGIG}_{KL}$ clearly runs in polynomial in (k,l) time.

Construction 2. *Let* $l : I \longrightarrow \mathbf{N}$ *be a polynomial. For every* $k = (n,p) \in I$, $\alpha \in I_k$, $\boldsymbol{\alpha} = (\alpha_1, \ldots, \alpha_l) \in (LR_{k,\alpha})^l$, *define* $g_{\alpha,\boldsymbol{\alpha}} : RD_k \longrightarrow (R_{k,\alpha})^l$ *by* $g_{\alpha,\boldsymbol{\alpha}}(\psi) = (\psi^{-1}\alpha_1\psi, \ldots, \psi^{-1}\alpha_l\psi)$, *where* $l = l(k)$. *Let* G_k *be the random variable that assumes as values the function* $g_{\alpha,\boldsymbol{\alpha}}$, *where the distribution of* $(\alpha, \boldsymbol{\alpha})$ *is* $\mathcal{PGIG}_{KL}(1^n, 1^p, 1^l)$. *Let* $G_{KL} \stackrel{\text{def}}{=} \{G_k\}_{k \in I}$.

The following result shows that G_{KL} is pseudorandom at least as secure as the DKL-Assumption.

Theorem 2. *If the DKL-Assumption holds, then for every PPTA* $\mathcal{A}$, *every positive polynomial* P, *and all sufficiently large* $k = (n,p)$*'s in* I,

$$\begin{aligned} &\Big|\Pr\big[\mathcal{A}(g_{\alpha,\boldsymbol{\alpha}}(\psi)) = 1 : (\alpha, \boldsymbol{\alpha}) \leftarrow \mathcal{PGIG}_{KL}(1^n, 1^p, 1^l); \psi \stackrel{u}{\leftarrow} RD_k\big] \\ &\quad - \Pr\big[\mathcal{A}(\beta_1, \ldots, \beta_l) = 1 : \alpha \leftarrow \mathcal{IG}(k); \beta_1, \ldots, \beta_l \stackrel{u}{\leftarrow} R_{k,\alpha}\big]\Big| \\ &< \tfrac{1}{P(k)}, \end{aligned}$$

where $l = l(k)$.

Sketch of Proof. Fix $k = (n,p) \in I$ and let $l = l(k)$. First, define a PPTA $\mathcal{M}$, on input $\langle \alpha, \chi^{-1}\alpha\chi, \psi^{-1}\alpha\psi, \tilde{\beta}\rangle$ where $\alpha \in I_k$, $\chi \in LD_k$, $\psi \in RD_k$, and $\tilde{\beta} \in R_{k,\alpha}$, from $\mathcal{A}$ as:

1. $J \stackrel{u}{\leftarrow} \{1, \ldots, l\}$.
2. $\chi_1, \ldots, \chi_{J-1} \stackrel{u}{\leftarrow} LD_k$; $\beta_{J+1}, \ldots, \beta_l \stackrel{u}{\leftarrow} R_{k,\alpha}$.
3. $H \stackrel{\text{def}}{=} \langle \chi_1^{-1}\psi^{-1}\alpha\psi\chi_1, \ldots, \chi_{J-1}^{-1}\psi^{-1}\alpha\psi\chi_{J-1}, \tilde{\beta}, \beta_{J+1}, \ldots, \beta_l\rangle$.

4. Output $\mathcal{A}(H)$.

Next, for each $i \in \{1, \dots, l\}$, define the i-th hybrid distribution

$$H_i^{k,l} = \langle \psi^{-1}\alpha_1\psi, \dots, \psi^{-1}\alpha_i\psi, \beta_{i+1}, \dots, \beta_l \rangle,$$

where $(\alpha, \alpha_1, \dots, \alpha_i) \leftarrow \mathcal{PGIG}_{KL}(1^n, 1^p, 1^i)$; $\psi \overset{u}{\leftarrow} RD_k$; $\beta_{i+1}, \dots, \beta_l \overset{u}{\leftarrow} R_{k,\alpha}$.

Then we get that

$$\Big|\Pr\big[\mathcal{M}(\alpha, \chi^{-1}\alpha\chi, \psi^{-1}\alpha\psi, \psi^{-1}\chi^{-1}\alpha\chi\psi) = 1 : \alpha \leftarrow \mathcal{IG}(1^n, 1^p); \chi \overset{u}{\leftarrow} LD_k; \psi \overset{u}{\leftarrow} RD_k\big]$$

$$- \Pr\big[\mathcal{M}(\alpha, \chi^{-1}\alpha\chi, \psi^{-1}\alpha\psi, \beta) = 1 : \alpha \leftarrow \mathcal{IG}(1^n, 1^p); \chi \overset{u}{\leftarrow} LD_k; \psi \overset{u}{\leftarrow} RD_k; \beta \overset{u}{\leftarrow} R_{k,\alpha}\big]\Big|$$

$$= \frac{1}{l}\Big|\Pr\big[\mathcal{A}(H_l^{k,l}) = 1 : (\alpha, \alpha_1, \dots, \alpha_l) \leftarrow \mathcal{PGIG}_{KL}(1^n, 1^p, 1^l); \psi \overset{u}{\leftarrow} RD_k\big]$$

$$- \Pr\big[\mathcal{A}(H_0^{k,l}) = 1 : \alpha \leftarrow \mathcal{IG}(1^n, 1^p); \beta_1, \dots, \beta_l \overset{u}{\leftarrow} R_{k,\alpha}\big]\Big|.$$

Using these, the theorem can be proved by contradiction. □

So, G_{KL} generates pseudorandom sequences of braids in $R_{k,\alpha}$. A pseudorandom generator can be constructed from G_{KL} by making use of the leftover hash lemma and pairwise independent hash functions [18,16,22].

The expansion property of the pseudorandom generator depends on the choice of $l(\cdot)$. Namely, $l(\cdot)$ should satisfy: $l(k)\log_2|R_{k,\alpha}| > 2||RD_k||$. Using the fact that $|R_{k,\alpha}| \geq |LD_k| \cdot |RD_k|$, $l(n,p) = 2pn$ suffices.

4.2 Pseudorandom Synthesizer

Although the notion of pseudorandom synthesizer was first introduced by Naor *et al.* [23] as a useful tool to get a parallel construction of a pseudorandom function, it is important itself as another type of pseudorandom generator. More precisely, pseudorandom synthesizers may be useful for software implementations of pseudorandom generators because from a pseudorandom synthesizer a pseudorandom generator with long output length can be easily defined and subsequences of its output can be computed directly.

Recall the formal definition of a pseudorandom synthesizer:

Notation ([23]). Let $f : \{0,1\}^{2n} \longrightarrow \{0,1\}^l$ be any function, and let $x = (x_1, \dots, x_k)$ and $y = (y_1, \dots, y_m)$ be two sequences of n-bit strings. We define $\mathbf{C}_f(x, y)$ to be the $(k \times m)$-matrix $(f(x_i, y_j))_{i,j}$.

Definition 6 ([23]). *Let $l : \mathbf{N} \longrightarrow \mathbf{N}$ be any function, and let $S : \{0,1\}^* \times \{0,1\}^* \longrightarrow \{0,1\}^*$ be a polynomial-time computable function such that for every $x, y \in \{0,1\}^n$, $||S(x,y)|| = l(n)$. Then S is a* pseudorandom synthesizer *if for every PPTA $\mathcal{A}$, every two positive polynomials P and m, and all sufficiently large n's*

$$|\Pr[\mathcal{A}(\mathbf{C}_S(x,y)) = 1] - \Pr[\mathcal{A}((r_{i,j})_{1 \leq i,j \leq m}) = 1]| < \frac{1}{P(n)},$$

where $m = m(n)$ and $x_1, \dots, x_m, y_1, \dots, y_m \overset{u}{\leftarrow} \{0,1\}^n$; $x = (x_1, \dots, x_m)$, $y = (y_1, \dots, y_m)$; $r_{1,1}, \dots, r_{m,m} \overset{u}{\leftarrow} \{0,1\}^{l(n)}$.

As mentioned in §1.1, the notion of pseudorandom synthesizer is stronger because pseudorandom synthesizers require that $\{S(z_i)\}_{1\le i\le m^2}$ remains pseudorandom even when the z_i's are of the form $\{x_i \circ y_j\}_{1\le i,j\le m}$, where $\circ$ stands for x concatenated with y. If $l(n) > 2n$ for all $n \in \mathbf{N}$, a pseudorandom synthesizer directly becomes a pseudorandom generator with $m(n) = 1$. However, every pseudorandom generator is not a pseudorandom synthesizer (See [23] for example).

Now we construct a pseudorandom synthesizer based on the DKL-Assumption.

Construction 3. *For every $k = (n,p) \in I$ and every $\alpha \in I_k$, define $s_\alpha : LD_k \times RD_k \longrightarrow R_{k,\alpha}$ by $s_\alpha(\chi,\psi) = \psi^{-1}\chi^{-1}\alpha\chi\psi$. Let S_k be the random variable that assumes as values the function s_α according to the distribution, $\mathcal{IG}(1^n,1^p)$. Let $S_{KL} \stackrel{\text{def}}{=} \{S_k\}_{k\in I}$.*

Then we get the following result:

Theorem 3. *If the DKL-Assumption holds, then for every PPTA $\mathcal{A}$, every positive polynomials l, P, and all sufficiently large $k = (n,p)$'s in I,*

$$|\Pr[\mathcal{A}(\mathbf{C}_{s_\alpha}(\chi,\psi)) = 1] - \Pr[\mathcal{A}((\gamma_{i,j})_{1\le i,j\le l}) = 1]| < \tfrac{1}{P(k)},$$

where $l = l(k)$ and $\alpha \leftarrow \mathcal{IG}(1^n,1^p)$; $\chi_1,\ldots,\chi_l \stackrel{u}{\leftarrow} LD_k$; $\chi = (\chi_1,\ldots,\chi_l)$; $\psi_1,\ldots,\psi_l \stackrel{u}{\leftarrow} RD_k$; $\psi = (\psi_1,\ldots,\psi_l)$; $\gamma_{1,1},\ldots,\gamma_{l,l} \stackrel{u}{\leftarrow} R_{k,\alpha}$.

Sketch of Proof. Fix $k = (n,p) \in I$ and let $l = l(k)$. First, define a PPTA $\mathcal{M}$, on input $\langle \alpha, \chi^{-1}\alpha\chi, \psi^{-1}\alpha\psi, \tilde{\beta}\rangle$ where $\alpha \in I_k$, $\chi \in LD_k$, $\psi \in RD_k$, and $\tilde{\beta} \in R_{k,\alpha}$, from $\mathcal{A}$ as:

1. $J \stackrel{u}{\leftarrow} \{1,\ldots,l^2\}$.
2. Compute J_1, J_2 such that $1 \le J_1, J_2 \le l$ and $J = l(J_1-1)+J_2$.
3. Let $\chi_{J_1} \stackrel{\text{def}}{=} \chi$ and $\psi_{J_2} \stackrel{\text{def}}{=} \psi$.
4. $\chi_1,\ldots,\chi_{J_1-1} \stackrel{u}{\leftarrow} LD_k$; $\psi_1,\ldots,\psi_{J_2-1},\psi_{J_2+1},\ldots,\psi_l \stackrel{u}{\leftarrow} RD_k$; $\beta_{J+1},\ldots,\beta_{l^2} \stackrel{u}{\leftarrow} R_{k,\alpha}$.
5. Define the $(l\times l)$-matrix $H = (h_{i,j})_{1\le i,j\le l}$ to be

$$h_{i,j} = \begin{cases} \psi_j^{-1}\chi_i^{-1}\alpha\chi_i\psi_j & \text{if } l(i-1)+j < J, \\ \tilde{\beta} & \text{if } l(i-1)+j = J, \\ \beta_w & \text{if } w \stackrel{\text{def}}{=} l(i-1)+j > J. \end{cases}$$

6. Output $\mathcal{A}(H)$.

Next, for each $0 \le r \le l^2$, define the r-th hybrid distribution $H_r^{k,l} = (h_{i,j})_{1\le i,j\le l}$ to be

$$h_{i,j} = \begin{cases} \psi_j^{-1}\chi_i^{-1}\alpha\chi_i\psi_j & \text{if } l(i-1)+j \le r, \\ \beta_w & \text{if } w \stackrel{\text{def}}{=} l(i-1)+j > r, \end{cases}$$

where $\alpha \leftarrow \mathcal{IG}(1^n, 1^p)$; $\chi_1, \ldots, \chi_l \overset{u}{\leftarrow} LD_k$; $\psi_1, \ldots, \psi_l \overset{u}{\leftarrow} RD_k$; $\beta_{r+1}, \ldots, \beta_{l^2} \overset{u}{\leftarrow} R_{k,\alpha}$.

Then we get that

$$\Big|\Pr\big[\mathcal{M}(\alpha, \chi^{-1}\alpha\chi, \psi^{-1}\alpha\psi, \psi^{-1}\chi^{-1}\alpha\chi\psi) = 1 : \alpha \leftarrow \mathcal{IG}(1^n, 1^p); \chi \overset{u}{\leftarrow} LD_k; \psi \overset{u}{\leftarrow} RD_k\big]$$
$$- \Pr\big[\mathcal{M}(\alpha, \chi^{-1}\alpha\chi, \psi^{-1}\alpha\psi, \beta) = 1 : \alpha \leftarrow \mathcal{IG}(1^n, 1^p); \chi \overset{u}{\leftarrow} LD_k; \psi \overset{u}{\leftarrow} RD_k; \beta \overset{u}{\leftarrow} R_{k,\alpha}\big]\Big|$$
$$= \frac{1}{l^2}\Big|\Pr\big[\mathcal{A}(H_{l^2}^{k,l}) = 1 : \alpha \leftarrow \mathcal{IG}(1^n, 1^p); \chi_1, \ldots, \chi_l \overset{u}{\leftarrow} LD_k; \psi_1, \ldots \psi_l \overset{u}{\leftarrow} RD_k\big]$$
$$- \Pr\big[\mathcal{A}(H_0^{k,l}) = 1 : \alpha \leftarrow \mathcal{IG}(1^n, 1^p); \beta_1, \ldots, \beta_{l^2} \overset{u}{\leftarrow} R_{k,\alpha}\big]\Big|.$$

Using these, the theorem can be proved by contradiction. □

5 Concluding Remarks

This article has considered two related hard problems in braid groups: the conjugacy and the Ko-Lee problems, which are believed to be computationally infeasible in our current state of knowledge.

Assuming that the conjugacy problem is one-way, we have presented two peculiar hard-core predicates that are provably secure using the infimum and the supremum of a braid. This means that, given $(\alpha, \chi^{-1}\alpha\chi)$, predicting the least significant bit of $\inf(\chi)$ (*or* $\sup(\chi)$) is as hard as the entirety of χ.

Under the decision Ko-Lee assumption, we have proposed two practical pseudorandom schemes, a pseudorandom generator and a pseudorandom synthesizer, that are provably secure.

Braid groups are quite different from the other groups which have been dealt with so far. So the known methods to turn hard-core predicates into pseudorandom generators and to turn pseudorandom generators or pseudorandom synthesizers into pseudorandom function generators cannot be applied naively. Therefore, a natural line for further research is to study how to get these next cryptographic primitives from our results.

Acknowledgement. We wish to thank the members of the CRYPTO committee. The first two authors were supported in part by the Ministry of Science and Technology under the National Research Laboratory Grant 1999 program. And the third author was supported in part by the Korea IT Industry Promotion Agency under the Information Technologies Research Center program.

References

1. W. Alexi, B. Chor, O. Goldreich, and C.P. Schnorr, *RSA and Rabin functions: certain parts are as hard as the whole*, SIAM J. Comput. **17** (1988) 194–209.
2. I. Anshel, M. Anshel, and D. Goldfeld, *An algebraic method for public-key cryptography*, Math. Res. Lett. **6** (1999) 287–291.
3. M. Bellare and P. Rogaway, *Random oracles are Practical: a Paradigm for Designing Efficient Protocols*, In 1st Annual Conference on Computer and Communications Security, ACM (1993) 62–73.

4. D. Bernardete, Z. Nitecki and M. Gutierrez, *Braids and the Nielsen-Thurston classification*, J. Knot theory and its ramifications **4** (1995) 549–618.
5. M. Bestvina and M. Handel, *Train Tracks for surface automorphisms*, Topology **34** (1995) 109–140.
6. J. Birman, *Braids, links and the mapping class group*, Ann. Math. Studies 82, Princeton Univ. Press (1974).
7. J.S. Birman, K.H. Ko, and S.J. Lee, *New approaches to the world and conjugacy problem in the braid groups*, Advances in Math. **139** (1998) 322–353.
8. M. Blum and S. Micali, *How to generate cryptographically strong sequences of pseudorandom bits*, SIAM J. Comput. **13** (1984) 850–864.
9. W. Diffie and M.E. Hellman, *New Directions in Cryptography*, IEEE Trans. on Info. Theory, IT-22 (1976) 644-654.
10. D.B.A. Epstein, J.W. Cannon, D.F. Holt, S.V.F. Levy, M.S. Patterson, and W. Thurston, *Word processing in groups*, Jones and Barlett, Boston and London (1992).
11. A. Fathi, F. Laudenbach, and V. Poénaru, *Travaux de Thurston sur les surfaces*, Astérisque (1979) 66–67.
12. J.B. Fischer and J. Stern, *An Efficient Pseudo-Random Generator Provably as Secure as Syndrome Decoding*, Proc. Eurocrypt '96, LNCS 1070, Springer-Verlag (1996) 245–255.
13. O. Goldreich, *Foundation of Cryptography—Fragments of a Book*, Available at `http://www.theory.lcs.mit.edu/~oded/frag.html` (1995).
14. O. Goldreich and L.A. Levin, *Hard-core Predicates for any One-Way Function*, 21st STOC (1989) 25–32.
15. S. Goldwasser, S. Micali, and R. Rivest, *A Digital signature scheme secure against adaptive chosen-message attacks*, SIAM J. Comput. **17** (1988) 281–308.
16. J. Hastad, R. Impaglizo, L.A. Levin, and M. Luby, *A Pseudorandom Generator from any One-way Function*, SIAM J. Comput. **28** (1999) 1364–1396.
17. R. Impagliazzo and M. Naor, *Efficient cryptographic schemes provably as secure as subset sum*, Proc. IEEE 30th Symp. on Found. of Comput. Sci. (1989) 231–241.
18. R. Impagliazzo and D. Zuckerman, *Reclying random bits*, Proc. 30th IEEE Symposium on Foundations of Computer Science (1989) 248–253.
19. K.H. Ko, S.J. Lee, J.H. Cheon, J.W. Han, J.S. Kang, and C. Park, *New Public-key Cryptosystem Using Braid Groups*, Proc. Crypto 2000, LNCS 1880, Springer-Verlag (2000) 166–183.
20. J. Los, *Pseudo-Anosov maps and invariant train track in the disc: a finite algorithm*, Proc. Lond. Math. Soc. **66** (1993) 400-430.
21. J. McCarthy, *Normalizers and centralizers of psuedo-Anosov mapping clases*, Available at `http://www.mth.msu.edu/~mccarthy/research/`.
22. M. Naor and O. Reingold, *Number-Theoretic constructions of efficient pseudo-random functions*, Proc. 38th IEEE Symp. on Foundations of Computer Science (1997) 458–467.
23. M. Naor and O. Reingold, *Synthesizers and their application to the parallel construction of pseudo-random functions*, J. of Computer and Systems Sciences **58** (1999) 336–375.
24. J. Nielsen, In *Collected papers of J. Nielsen*, Birkhauser (1986).
25. W. P. Thurston, *On the geometry and dynamics of diffeomorphisms of surfaces*, Bull. AMS **19** (1988) 417-431.
26. A. Yao, *Theory and applications of trapdoor functions*, Proc. 23rd IEEE Symp. of Found. of Comput. Sci. (1982) 80–91.

On the Cost of Reconstructing a Secret, or VSS with Optimal Reconstruction Phase

Ronald Cramer[1], Ivan Damgård[1], and Serge Fehr[2*]

[1] Aarhus University, BRICS
{cramer,ivan}@brics.dk
[2] ETH Zürich, Switzerland
fehr@inf.ethz.ch

Abstract. Consider a scenario where an l-bit secret has been distributed among n players by an honest dealer using some secret sharing scheme. Then, if all players behave honestly, the secret can be reconstructed in one round with zero error probability, and by broadcasting nl bits.
We ask the following question: how close to this ideal can we get if up to t players (but not the dealer) are corrupted by an adaptive, active adversary with unbounded computing power? - and where in addition we of course require that the adversary does not learn the secret ahead of reconstruction time. It is easy to see that $t = \lfloor (n-1)/2 \rfloor$ is the maximal value of t that can be tolerated, and furthermore, we show that the best we can hope for is a one-round reconstruction protocol where every honest player outputs the correct secret or "failure". For any such protocol with failure probability at most $2^{-\Omega(k)}$, we show a lower bound of $\Omega(nl + kn^2)$ bits on the information communicated. We further show that this is tight up to a constant factor.
The lower bound trivially applies as well to VSS schemes, where also the dealer may be corrupt. Using generic methods, the scheme establishing the upper bound can be turned into a VSS with efficient reconstruction. However, the distribution phase becomes very inefficient. Closing this gap, we present a new VSS protocol where the distribution complexity matches that of the previously best known VSS, but where the reconstruction phase meets our lower bound up to a constant factor. The reconstruction is a factor of n better than previous VSS protocols. We show an application of this to multi-party computation with pre-processing, improving the complexity of earlier similar protocols by a factor of n.

1 Introduction

The concept of *secret-sharing* (introduced by Shamir [13]) is of fundamental importance: in practical data security, as a way to protect a secret simultaneously from exposure and from being lost; and theoretically, as the basis for building general multi-party secure protocols.

In the original setting of Shamir, a *dealer* distributes a secret, say an l-bit string, to n players, by privately sending a *share* to each player. The computation

* Supported by the Swiss SNF, project no. SPP 2000-055466.98.

J. Kilian (Ed.): CRYPTO 2001, LNCS 2139, pp. 503–523, 2001.

of the shares is done w.r.t. a threshold value t, where $1 \leq t \leq n$. Later, some subset of the players can attempt to reconstruct the secret by pooling their shares. A secret sharing scheme must ensure *privacy*, i.e., an adversary who sees up to t of the shares learns no information about the secret, and *correctness*, i.e., the secret can always be reconstructed from a set of at least $t+1$ shares.

Here, we will first consider a more adversarial setting where up to t of the players (but not the dealer) may be corrupted by an active, adaptive and unbounded adversary, in particular, corrupted players may contribute incorrect shares (or nothing) in the reconstruction phase. We still require privacy, and also correctness in the sense that the honest players can reconstruct the correct secret. Consider the following question. How much information must be sent in order for such a scheme to work? This question is interesting only if $n/3 \leq t < n/2$, since otherwise the problem is either "too hard" or "too easy": if $t \geq n/2$ the problem clearly cannot be solved, and if $t < n/3$, standard methods (see [2]) immediately give an optimal solution with zero error probability.

Somewhat surprisingly, little work seems to have been done on the case of $n/3 \leq t < n/2$ (although upper bounds follow from known protocols [12,4]). It is easy to see that for t in this range, one cannot construct a scheme where the correct secret is *always* reconstructed. At best one can make a scheme where every honest player outputs the correct secret or "failure", where the latter happens with probability only $2^{-\Omega(k)}$, where k is a security parameter. For schemes that achieve this for the maximal value of t, i.e. $t = \lfloor (n-1)/2 \rfloor$, and where the reconstruction is completed in a single round, we show a lower bound of $\Omega(nl + kn^2)$ bits on the amount of information sent in the reconstruction. This may be seen as an answer to the question "what does it cost to get the best possible security in a minimal number of rounds?". No such bound was known previously, and it holds even for schemes that are not efficient.

We refer to the type of scheme we just described as *Honest-Dealer VSS*. This is because the well-known concept of Verifiable Secret Sharing (VSS), introduced in [6], is essentially what we just described, except that also the dealer can be corrupt. In VSS, distributing the secret may then take the form of an interactive, several rounds protocol. One usually assumes that a private channel connects every pair of players and that a broadcast channel is available[1]. A secure VSS must, in addition to what we required above, also ensure that immediately after the distribution phase, some value of the secret is uniquely defined (even if the dealer is corrupt) and that this value will be reconstructed (with overwhelming probability). Note that the standard definition of VSS is slightly weaker than ours in that it allows honest players to reconstruct (with small probability) an incorrect value of the secret, even if the adversary was passive in the distribution phase. However, all known VSS protocols for our communication model (see e.g. [12,4]) satisfy or can trivially be modified to satisfy our stronger definition.

Our lower bound for Honest-Dealer VSS trivially applies also to VSS (we cannot expect to do better in a more adversarial situation).

[1] The latter can be simulated by the private ones if $t < n/3$, but must be assumed as a separate primitive otherwise.

For an honest dealer, we use known results on authentication codes to show that the lower bound is tight up to a constant factor (even if we count the total information sent). This scheme establishing the upper bound is computationally efficient and can - at least in principle - be turned into a VSS, since the honest dealer could always be replaced by a secure multi-party computation using generic methods (e.g. [12,4]). This, however, is not a satisfactory solution: while reconstruction would be the same complexity as before, the distribution would become extremely inefficient in comparison. To close this gap, we present a new VSS protocol where the complexity of the distribution matches that of the previously best known VSS for our scenario [4], but where the reconstruction meets our lower bound. This beats previous VSS protocols by a factor of n.

We show an application of this to multi-party computation with pre-processing, introduced in [1], where the n players ultimately want to compute a function f on private inputs $x_1, \dots, x_n$. In order to do this more efficiently than starting from scratch, the players are allowed to a pre-processing and store some information obtained in this phase *before* the function and the inputs become known. The computation phase of our protocol has communication complexity $O(n^2k|C|)$, where $|C|$ is the size of the circuit to be computed. This improves the computation phase of earlier similar protocols by a factor of n without increasing the complexity of the pre-processing.

In the appendix, we sketch how our results for a dishonest minority generalize for almost all t in the range $n/3 \leq t < n/2$ and observe that already an arbitrarily small linear gap between t and $n/2$ allows to reduce the communication complexity of the reconstruction by a factor of n. Using methods from [5], we also show how to generalize our schemes to provide security against any (non-threshold) Q^2 adversary (see [9]), improving known results by a factor of at least n. Finally, we look at the case where the reconstruction is allowed to use more than one round of interaction and observe, using results from [7], that the amount of information sent by the honest dealer can be brought down to $n(n+k)$ bits, at the expense of a significantly more inefficient reconstruction phase.

2 Communication Model

Throughout the paper, we consider the *secure-channels model with broadcast* [12], i.e. there is a set $\mathcal{P} = \{P_1, \dots, P_n\}$ of n players plus a so called *dealer* D, every two entities being connected by a secure, untappable channel, and there is a broadcast channel available. We assume an *active* adversary with unbounded computing power that can corrupt up to a certain number t out of the n players in $\mathcal{P}$ plus the dealer D. An adversary is *rushing*, if he can learn the messages sent by the honest players in each round before deciding on the messages for corrupted players in this round. Finally, the adversary can either be *static* or *adaptive*, the former meaning that he has to corrupt the players before the protocol execution and the latter that he can corrupt players at his will during the protocol execution, depending on what he has seen so far. Throughout the paper, we consider a *security parameter* k.

3 Single-Round Honest-Dealer VSS

We first model the general communication pattern for VSS schemes where the dealer is guaranteed to be honest and whose reconstruction phase consists of a single round of communication. We will call such a scheme *Single-Round Honest-Dealer VSS*. Our main point of interest is the communication complexity of the reconstruction phase of such a scheme. Consider schemes of the following general form, and assume an active adversary who corrupts up to t of the n players P_i, *but not the dealer* (this is also known as *robust secret sharing*).

Distribution Phase: The honest dealer generates shares $s_i = (k_i, y_i)$, $i = 1 \dots n$, according to a fixed and publicly known conditional probability distribution $P_{S_1 \cdots S_n | S}(\cdots | s)$, where s is the secret. Privately he sends s_i to player P_i.

Reconstruction Phase: Each player P_i is required to broadcast $\tilde{y}_i$, which is supposedly equal to y_i. Locally and by some fixed (possibly probabilistic) method, each player P_i decides on the secret s based on his private k_i and on the broadcast $\tilde{y}_1, \dots, \tilde{y}_n$, i.e., either outputs a value $\tilde{s}$, hopefully equal to s, or outputs "failure".

It is not difficult to see that in fact we may always and without loss of generality assume our schemes of interest to be of this form (please refer to Appendix A).

For each of the at most t corrupted players P_j, the adversary can broadcast a manipulated $\tilde{y}_j$, which may depend arbitrarily on the private information $s_j = (k_j, y_j)$ of those corrupted players, or broadcast nothing at all in some cases ("crash faults"). Note though that for at least $n - t$ $\tilde{y}_i$'s it holds that $\tilde{y}_i = y_i$. If additionally the adversary is rushing, he can choose to "speak last" in the reconstruction phase. This means that in principle any corrupted shares may additionally depend on the information broadcast by the honest players, in particular they may depend on the secret s. By contrast, a non-rushing adversary is one who selects the corrupted shares before the start of the reconstruction phase. Note that security against non-rushing adversaries makes sense in a communication model enhanced with a "simultaneous broadcast channel", i.e., one by means of which all players broadcast their information at the same time.

We define our notion of security. Assume an active adversary that corrupts at most t of the n players but not the dealer. Additionally, the adversary can be static or adaptive, and rushing or non-rushing. A Single-Round Honest-Dealer VSS scheme is $(t, n, 1-\delta)$*-secure* if the following holds.

Privacy: As a result of the distribution phase, the adversary gains no information about the secret s distributed by the honest dealer.

$(1-\delta)$*-Correctness*: In the reconstruction phase, each uncorrupted player outputs either the correct secret s or "failure", where for every player the latter happens with probability at most $\delta < 1$, independent of s.
In the special case that the adversary introduces only crash-faults or remains passive, all honest players recover the correct secret s with probability 1.

As mentioned in the Introduction, we focus on the case of a *dishonest minority*, i.e., $t = \lfloor (n-1)/2 \rfloor$, the maximal value of t for which $(t, n, 1-\delta)$-security is achievable. For the corresponding results for a (nearly) arbitrary t in the range $n/3 \leq t < n/2$, we refer to Appendix C. Note that the case $t < n/3$ is completely understood: zero failure probability and optimally efficient communication can be achieved by a combination of Shamir's secret sharing scheme and standard efficient error correction techniques [2].

We stress that our definition of security captures the best one can achieve in this setting. Negligible error δ^m is achieved by m parallel repetitions. More importantly, it only differs from perfect security in the sense that there is a (small) probability that some player does not reconstruct the secret and outputs "failure" instead. This is unavoidable in the presence of an arbitrary (not necessarily rushing) active adversary, as is easy to see (please refer to Appendix B). Furthermore, existing Honest-Dealer VSS schemes like [12] ("secret sharing when the dealer is a knight") fulfill our security definition without any changes in the required communication.

A seemingly stronger security definition would require agreement among the honest players in all cases, i.e., they all recover the correct secret or they all output "failure", where the latter would happen with probability at most δ. However, this is impossible to achieve in a single round reconstruction phase with a *rushing adversary*, as we show in Appendix B. [2]

Note also that the reconstruction procedure in our definition is completely general in that it does not dictate how the correct secret is recovered by the honest players. The definition merely states that from all broadcast and from his private information, an honest player can reconstruct the secret. In particular, in our definition it need not be the case that an honest player, using his private information, "filters out" false shares and reconstructs the secret from the "good" ones, as it is the case for known schemes [12,4] and the one we present later.

4 Lower Bound on Reconstruction Complexity

We prove the following lower bound. Note that the standard definitions of entropy, conditional entropy, mutual information and conditional mutual information are used throughout this section. We refer to [3] for an excellent introduction to information theory.

Theorem 1. *For any family of Single-Round Honest-Dealer VSS schemes, $(t, n, 1-\delta)$-secure against an active, rushing adversary, the following holds. If $t = \lfloor (n-1)/2 \rfloor$ and $\delta \in 2^{-\Omega(k)}$ for a security parameter k, then the total information broadcast in the reconstruction phase is lower bounded by $\Omega(nH(S) + kn^2)$.*

Note that it is immaterial whether the adversary is adaptive or not.

In the following, we will call K_i the *key* and Y_i the *public share* of player P_i. Theorem 1 follows immediately from

[2] In Appendix E, we argue that agreement is possible in the presence of a non-rushing adversary. Agreement can be achieved in all cases by adding one extra round of communication.

Proposition 1. *Let $S_1 = (K_1, Y_1), \ldots, S_n = (K_n, Y_n)$ be distributed according to the Single-Round Honest-Dealer VSS scheme. Then, in case of an odd n, the size of any public share Y_i is lower bounded by*

$$H(Y_i) \in \Omega(H(S) + kn)\,,$$

while for an even n, it is the size $H(Y_iY_j)$ of every pair $Y_i \neq Y_j$ that is lower bounded by $\Omega(H(S) + kn)$.

We will only prove the case of an odd n, i.e., $n = 2t + 1$; the proof for an even n, i.e. $n = 2t+2$, goes accordingly. But before going into the proof, consider the following Lemma, which states a well known result from Authentication Theory, which can be found in various literature starting with [14] (for a very general treatment of Authentication Theory consult [11]).

Lemma 1. *Let K, M, Y and Z be random variables (typically key, message, tag and public information of an authentication scheme) with joint distribution P_{KMYZ} such that M is independent of K and Z but uniquely defined by Y and Z. Then, knowing Z, one can compute $\tilde{Y}$, consistent with K and Z with probability*

$$p_I \geq 2^{-I(K;Y|Z)}\,.$$

Also, knowing Z and Y, one can compute $\tilde{\tilde{Y}}$, consistent with K and Z and a $\tilde{\tilde{M}} \neq M$ with probability

$$p_S \geq 2^{-H(K|Z)}\,.$$

In the context of Authentication Theory, $\tilde{Y}$ describes an *impersonation* and $\tilde{\tilde{Y}}$ a *substitution* attack, and p_I and p_S are the corresponding success probabilities.

In the proof of Proposition 1, we apply the following Corollary, which follows from the fact that a successful impersonation attack is also a successful substitution attack with probability at least $1/2$, assumed that M is uniformly distributed among a set of cardinality at least two.

Corollary 1. *Let K, M, Y and Z be as above, except that M is required to be uniformly distributed among a non-trivial set. Then, knowing Z, one can compute $\tilde{Y}$, consistent with K and Z and a $\tilde{M} \neq M$ with probability*

$$p_S \geq 2^{-I(K;Y|Z)-1}\,.$$

Proof of Proposition 1: Since by the privacy of the scheme the public share Y_i is independent of S and hence $H(Y_i)$ does not depend on the distribution of S, we can assume P_S to be the uniform distribution. Furthermore, for symmetry reasons, we can focus on the public share of the player P_{t+1}.
Let $i \in \{1, \ldots, t\}$ be arbitrary but fixed, and consider an adversary corrupting the first $i-1$ players $P_1, \ldots, P_{i-1}$ as well the player P_{t+1}. One of the goals of the adversary could be to substitute P_{t+1}'s public share Y_{t+1} by a false share $\tilde{Y}_{t+1}$ that is consistent with the public shares $Y_1, \ldots, Y_t$ of the first t

players and player P_i's key K_i (and maybe even the keys $K_1, \ldots, K_{i-1}$), but that leads to an incorrect secret $\tilde{S} \neq S$. Indeed, if the adversary succeeds in this attack, from player P_i's point of view, the $t+1$ public shares $Y_1, \ldots, Y_t, \tilde{Y}_{t+1}$ *could* come from honest and the t shares $Y_{t+2}, \ldots, Y_n$ from corrupted players. Hence, P_i clearly cannot compute the correct secret with certainty, and so outputs "failure". Therefore, the success probability of this attack is at most $\delta \in 2^{-\Omega(k)}$. On the other hand however, according to the above Corollary, applied to $K = K_i$, $M = S$, $Y = Y_{t+1}$ and $Z = (K_1, \ldots, K_{i-1}, Y_1, \ldots, Y_t)$, the success probability is at least $p_S \geq 2^{-I(K_i;Y_{t+1}|K_1 \cdots K_{i-1} Y_1 \cdots Y_t)-1}$. Therefore, we have $I(K_i; Y_{t+1}|K_1 \cdots K_{i-1} Y_1 \cdots Y_t) \in \Omega(k)$. This holds for every $i \in \{1, \ldots, t\}$, and hence, using the chain rule for mutual information, we get

$$I(K_1 \cdots K_t; Y_{t+1}|Y_1 \cdots Y_t) = \sum_{i=1}^{t} I(K_i; Y_{t+1}|Y_1 \cdots Y_t K_1 \cdots K_{i-1}) \in \Omega(kt)$$

and therefore $H(Y_{t+1}) \geq I(K_1 \cdots K_t; Y_{t+1}|Y_1 \cdots Y_t) \in \Omega(kt) = \Omega(kn)$.

As $S_1, \ldots, S_t$ gives no information about S, but $S_1, \ldots, S_t, Y_{t+1}$ determines S, we also have $H(Y_{t+1}) \geq H(S)$, and hence $H(Y_{t+1}) \in \Omega(H(S) + kn)$. □

In Appendix E we illustrate the power of rushing by giving an example of a concrete scheme secure against a non-rushing adversary, that beats the lower bound, and sketch a tight lower bound result. We also briefly discuss the minimal complexity of the distribution phase of schemes secure against a rushing adversary.

5 Tightness of the Lower Bound

We first describe a very natural, generic construction of a Single-Round Honest-Dealer VSS and then present a particular instantiation that meets the lower bound from the previous section. Rabin and Ben-Or [12] first considered a solution of this type. The scheme below differs from theirs only in the choice of the authentication code (which, however, will be relevant later on).

Let a $(t+1, n)$-threshold secret-sharing scheme be given as well as an authentication scheme, e.g. based on a family of strongly universal hash functions $\{h_\kappa\}_{\kappa \in \mathcal{K}}$ (see e.g. [15]). To share a secret s, the dealer D generates shares $s_1, \ldots, s_n$ according to the secret sharing scheme, and, for each pair of players P_i, P_j, he selects a random authentication key $\kappa_{ij} \in \mathcal{K}$ which will be sent to P_j who will later use it to verify a share contributed by P_i. Then D computes for each share s_i and for each P_j the authentication tag $y_{ij} = h_{\kappa_{ij}}(s_i)$ that should be revealed by P_i at reconstruction time to convince P_j that P_i's share s_i is valid. D then simply sends shares, tags and keys privately to the players who own them. To reconstruct, every player broadcasts his share together with the tags (or, alternatively, sends to every player his share and the corresponding tag), and verifies the authenticity of the received shares using his keys.

We use Shamir's secret sharing scheme [13] over a field F with $|F| > n$, and the well-known family of hash functions $h_{(\alpha,\beta)}(X) = \alpha X + \beta$ defined over F. The success probability of a substitution attack of the corresponding authentication scheme is $1/|F|$. It follows that the probability of player P_i accepting a false share from another player is $1/|F|$, and hence the probability of player P_i not reconstructing the correct secret is at most $t/|F|$. By comparing all the accepted shares with the reconstructed sharing polynomial and outputting "failure" in case of inconsistencies, he makes sure not to output an incorrect secret. Hence, choosing F such that $|F|$ is in $2^{\Theta(k)}$ (assuming n to be at most polynomial in k), we have the following upper bound, already achieved in [12].

Theorem 2. *For $t = \lfloor (n-1)/2 \rfloor$, there exists a Single-Round Honest-Dealer VSS scheme, $(t, n, 1-2^{-\Omega(k)})$-secure against an adaptive and rushing adversary, with a total communication complexity of $O(kn^2)$ bits.*

A remark concerning the authentication code. The choice of the code is not completely arbitrary, since it is important for our later purposes that computation of tags has low arithmetic complexity (here one multiplication and one addition over F) and that the tags are *linear* if α is fixed, as shown in Section 7.1.

6 Upper Bound in the Presence of a Corrupted Dealer

In this section, we present a VSS scheme with a one-round reconstruction, where the complexity of the distribution phase matches that of the previous best known VSS for our scenario [4], but where the reconstruction phase meets our lower bound up to a constant factor. This is at least a factor of n better than previous VSS protocols.

6.1 Definition

Since now the dealer might be corrupt as well and so the distribution of the secret takes the form of an interactive protocol, the adversary can not only intrude faults in the reconstruction, but also in the distribution. Therefore, our definition operates with two error probabilities, which for a concrete scheme do not have to be equal: first the probability that the distribution fails to work as supposed, and second the probability that the reconstruction fails, even though the distribution succeeded.

Assume an active adversary that corrupts at most t of the n players plus the dealer (respectively, including the dealer, in case he is one of the players). Additionally, the adversary can be static or adaptive, and rushing or non-rushing. Consider a scheme with an arbitrary distribution phase resulting in every player P_i holding a key k_i and a public share y_i and with a one-round reconstruction phase as in the honest dealer case. We call such a scheme $(t, n, 1-\beta, 1-\delta)$-*secure* if, except with probability β (taken over the coin flips during the distribution), the following holds.

Privacy: As long as the dealer remains honest, the adversary gains no information about the shared secret s as a result of the distribution phase.

$(1-\delta)$-*Correctness:* Once all currently uncorrupted players complete the distribution phase, there exists a fixed value s' such that in the reconstruction phase each uncorrupted player outputs either s' or "failure", where for every player the latter happens with probability at most $\delta < 1$, independent of s'. If the dealer remains uncorrupted during the distribution, then $s' = s$.
In the special case that the adversary introduces only crash-faults or remains passive, all honest players recover s' with probability 1.

Again, existing VSS schemes essentially fulfill our stronger definition, in particular the most efficient solution known, [4], fulfills it without any changes in the required communication, while the [12] protocol requires some straightforward modifications.

6.2 Towards VSS with Optimized Reconstruction

The security of the scheme from the last section evidently completely breaks down in case the dealer is corrupted. In the distribution phase, he could hand out inconsistent shares and inconsistent authentication tags, and, in the reconstruction phase, since he knows all the keys, he could compute correct tags for false shares. This would allow him to disrupt the reconstruction and even to actually cause different secrets to be reconstructed (see the analysis in [4] of WSS from [12]). To remedy this, we have to ensure that the players that remain honest receive consistent shares, and that they accept each others shares at reconstruction, while rejecting false shares. Of course, as mentioned in the introduction, this could in principal be achieved by replacing the dealer of the Honest-Dealer VSS by a general MPC. This, however, would result in a rather inefficient distribution phase. Also the following approach seems to be no satisfactory solution because of the same reason. We force the dealer to distribute consistent shares $s_1, \ldots, s_n$ by doing a "two-dimensional sharing" as in [2] or [4] and then every tag y_{ij} for a share s_i is computed in a multi-party fashion, such that it is guaranteed to be correct and the corresponding key is only known to the verifier P_j. Again, doing general MPC would result in a rather inefficient distribution phase; however, the following points provide some intuition as to why the full generality of MPC protocols is not needed, and instead we can do a *specialized* MPC.

1. A "two-dimensional sharing" from [2] or [4] not only ensures that the uncorrupted players hold consistent shares, but also that every share s_i is again correctly shared. Hence, one input to the MPC, s_i, is already correctly shared.
2. We only have to guarantee that a tag is computed correctly, if the player who will later verify it is honest at distribution time. At reconstruction, a corrupted player can always claim a tag to be invalid, even if it were good. For this reason, full VSS of the authentication key will not be necessary.
3. The function to be computed uses only one multiplication and one addition. This will allow us to do the distributed multiplication locally, i.e. no resharing as in [8] will be needed.

6.3 The CDDHR VSS Sharing Protocol

To describe the sharing protocol from [4], we start by reviewing the concept of Information Checking (IC), introduced in [12]. In essence, an IC scheme provides unconditionally secure "signatures" with limited transferability. More concretely, it allows a *sender* S to provide a *transmitter* T (also called *intermediary*) with a message m and a "signature" σ, such that T can later pass (m, σ) on to a *recipient* R, claiming that m originates with S. The signature σ enables R to verify this. We use the notation $\sigma_m(S, T; R)$ to refer to such a signature. Although in reality the "signing" procedure is an interactive protocol involving all three players and using a broadcast channel, we abuse language slightly and simply say that S "sends the signature $\sigma_m(S, T; R)$ to T". IC must fulfill the following requirements, except with some small error probability. If T and R are uncorrupted, then R indeed accepts T's message m (*consistency*). If, on the other hand, S and R are uncorrupted, then R rejects any message $m' \neq m$ (*correctness*). Finally, if S and T are uncorrupted, then R gets no information on m before T passes (m, σ) on to him (*secrecy*). It is easy to extend this concept and the corresponding protocols to multiple recipients, say $R_1, \ldots, R_n$, by simply executing the single recipient protocol for each possible recipient. We then use the notation $\sigma_m(S, T) = (\sigma_m(S, T; R_1), \ldots, \sigma_m(S, T; R_n))$. For a formal definition and technical details, please refer to [12,4].

Please recall that the IC-signatures from [4] over a field F have the following *linearity* properties. If T holds two signatures $\sigma_m(S, T; R)$ and $\sigma_{m'}(S, T; R)$ and if λ is known to R and T, then T can compute a signature $\sigma_{m+m'}(S, T; R)$ for $m + m'$ and a signature $\sigma_{\lambda m}(S, T; R)$ for λm. This holds analogously in the multi-recipient case. As to efficiency, generating a signature $\sigma_m(S, T; R)$ costs $O(\log |F|)$ bits of communication, generating a signature $\sigma_m(S, T)$ with n recipients costs $O(n \log |F|)$ bits of communication. Furthermore, the secrecy condition holds perfectly while correctness and consistency hold with probability $1 - 2^{-\log |F|}$ for a single-recipient and $1 - 2^{-\log |F| + \log(n)}$ for a multi-recipient signature.

We present the VSS sharing protocol from [4], which we will call PreShare, in a slightly modified version. Namely, for ease of exposition, we use a *symmetrical* polynomial and we omit the signatures made by the dealer (since these are needed only to catch a corrupted dealer early on).

Protocol PreShare

1. To share a secret $s \in F$, the dealer chooses a random symmetrical bivariate polynomial f of degree at most t in both variables with s as constant coefficient, i.e. $f(0, 0) = s$.
2. To every player P_i, the dealer privately sends the *actual share* $s_i = f(i, 0)$ and the sharing $s_{i1} = f(i, 1), \ldots, s_{in} = f(i, n)$ of s_i. [3]

[3] In the descriptions of all the protocols, whenever a player expects to receive a message from another player, but no message arrives or it is not in the right format, he takes some fixed default value as received message.

3. For every two players P_i and P_j, the following is done. P_i sends s_{ij} together with a signature $\sigma_{s_{ij}}(P_i, P_j) = (\sigma_{s_{ij}}(P_i, P_j; P_1), \dots, \sigma_{s_{ij}}(P_i, P_j; P_n))$ to P_j. If $s_{ij} \neq s_{ji}$, then P_j broadcasts a complaint, to which the dealer has to answer by broadcasting s_{ji}. If this value does not coincide with P_j's s_{ji}, then P_j accuses the dealer publicly who then has to broadcast P_j's share s_j and subshares $s_{j1}, \dots, s_{jn}$.[4]
4. If at some point, the broadcast information is inconsistent, the players take some publicly known default sharing.

This protocol stands as a VSS sharing protocol on its own (but with "expensive" reconstruction, as argued earlier). The proof of this fact is based on the following observations. Please refer to [4] or the appendix.

Proposition 2. *After the execution of* Pre Share, *every honest P_i holds a share s_i and signed sub-shares $s_{i1} \dots s_{in}$ such that*

1. *If the dealer remains honest, then the adversary has no information about the secret s.*
2. *The sub-shares $s_{i1} \dots s_{in}$ of any honest player P_i are a correct sharing of s_i, and $s_{ij} = s_{ji}$ holds for all P_i and P_j who remain honest.*
3. *The shares s_i of the honest players are correct shares of a unique value s', which is the secret s if the dealer remains honest.*
4. *For any (honest or dishonest) player P_j, the sub-shares s_{ij} of the honest players P_i are correct shares of P_j's share s_j, which is well defined by the shares s_i of the honest players.*

The communication complexity of this Pre Share protocol is $O(n^3 \log |F|)$ bits, the dealer essentially distributes n^2 sub-shares and each of these sub-shares is signed, where signing costs $O(n \log |F|)$ bits of communication per signature.

6.4 Computing Tags by a Specialized MPC

Consider now a fixed player P_i after the execution of Pre Share, holding his share s_i and the corresponding sub-shares $s_{i1}, \dots, s_{in}$ with signatures $\sigma_{s_{i1}}(P_1, P_i), \dots, \sigma_{s_{in}}(P_n, P_i)$. We now want to compute authentication tags $y_{ij} = \alpha_{ij} \cdot s_i + \beta_{ij}$ for s_i as they are computed by the dealer in the Honest-Dealer VSS protocol, but without letting the dealer know the keys, $(\alpha_{ij}, \beta_{ij})$ should only be known to P_j.

At the heart, there is the following problem. A player P wants to compute the tag $y = \alpha \cdot m + \beta$ for his secret message m with respect to a player V's secret key α, β. As already mentioned earlier, this will be done by a *specialised* MPC.

We assume that P's message m is already correctly shared by shares $m_1, \dots, m_n$ and that P holds signatures $\sigma_{m_1}(P_1, P; V), \dots, \sigma_{m_n}(P_n, P; V)$, verifiable by V. If the protocol Pre Share from the previous section has been executed, and if P's message m stands for P_i's share s_i, then this is fulfilled with $m_k = s_{ik}$ and $\sigma_{m_k}(P_k, P; V) = \sigma_{s_{ik}}(P_k, P_i; P_j)$.

[4] Of course, broadcast values do not have to be signed anymore; however, for simpler notation, we assume that also broadcast sub-shares s_{ij} are signed by $\sigma_{s_{ij}}(P_i, P_j)$.

Protocol MP Auth

1. V chooses a random polynomial f_α of degree at most t with $f_\alpha(0) = \alpha$ and a random polynomial f_β of degree at most $2t$ with $f_\beta(0) = \beta$. For every player P_k, V sends the shares $\alpha_k = f_\alpha(k)$ and $\beta_k = f_\beta(k)$ to P_k together with signatures $\sigma_{\alpha_k}(V, P_k; P)$ and $\sigma_{\beta_k}(V, P_k; P)$, verifiable by P.
2. Every player P_k, having received the shares α_k and β_k with the corresponding signatures and holding the share m_k of m, computes $y_k = \alpha_k \cdot m_k + \beta_k$ and, using the linearity property of the signatures, the corresponding signature $\sigma_{y_k}(V, P_k; P)$ [5] and passes y_k and $\sigma_{y_k}(V, P_k; P)$ on to P, who verifies the signature (see point 3. in Section 6.2).
3. If P receives all the y_k and all the signatures are good, then he can reconstruct y by interpolation, i.e. by computing a polynomial f_y of degree at most $2t$ with $f_y(k) = y_k$ for all P_k and computing $y = f_y(0)$.
   If some signature $\sigma_{y_k}(V, P_k; P)$ is not correct, then before computing y as above, P passes m_k and $\sigma_{m_k}(P_k, P; V)$ on to V, who verifies the signature and in case of a good signature returns $y_k = \alpha_k \cdot m_k + \beta_k$ to P (see point 2. in Section 6.2 for the case V refuses).

Proposition 3. *Under the assumptions stated before the protocol, the following holds except with probability* $2^{-\log|F|+O(\log n)}$.

1. *If P and V remain honest during the execution, then* $y = \alpha \cdot m + \beta$.
2. *If P remains honest, then the adversary learns nothing about m.*
3. *If V remains honest, then the adversary learns nothing about α.*

Hence, the tag y can be thought of being computed by some honest player.

Proof. We will prove 1., 2. and 3. under the assumption that the security properties of the signatures hold without error probability; this proves the claim.

1. Let f_m be the polynomial of degree at most t with $f_m(k) = m_k$ and hence $f_m(0) = m$. The n shares $y_k = \alpha_k \cdot m_k + \beta_k$ define a unique polynomial f_y of degree at most $2t$ with $f_y(k) = y_k$ and $f_y(0) = y = \alpha \cdot m + \beta$, namely $f_y = f_\alpha \cdot f_m + f_\beta$. So, if all n players P_k behave and send y_k with the correct signature to P, then P can compute f_y and hence y. If on the other hand some corrupted player P_k misbehaves and sends an incorrect y_k to P (or an incorrect signature or nothing at all), then P recognizes this and gets the correct y_k from V. Hence, even in this case P gets all the correct y_k and can therefore reconstruct y.
2. We assume wlog that V is corrupted. If all the corrupted players P_k follow the protocol, then the adversary definitely gets no information at all. If some corrupted player P_k misbehaves (e.g. by sending a bad y_k), then the adversary only learns m_k, which he already knows.
3. We assume that P is corrupted. Note that the adversary does not learn anything new by asking V for a y_k in step 3., since the correct value m_k must be sent to V (otherwise V would not accept the signature and return nothing).

[5] Note that m_k is known to both P_k and P.

We have to show that the adversary's view of this protocol gives no information about α. The adversary's view, excluding the signatures, consists of $m, m_1, \ldots, m_n, y_1, \ldots, y_n$ and α_k and β_k for $P_k \in A$, where A is the set of corrupted players, with $y_k = \alpha_k \cdot m_k + \beta_k$. Consider the polynomial $d_\alpha(X) = \prod_{P_k \in A}(k-X)/k$ of degree t and the polynomial $d_\beta = -d_\alpha \cdot f_m$ of degree at most $2t$. Note that $d_\alpha(0) = 1$ and $d_\beta(0) = -m$ and $d_\alpha(k) = 0 = d_\beta(k)$ for all P_k in A. This implies that if f_α and f_β are the sharing polynomials for α and β, then for any α', β' with $\alpha' \cdot m + \beta' = y$, the polynomials $f_{\alpha'} = f_\alpha + (\alpha' - \alpha)d_\alpha$ and $f_{\beta'} = f_\beta + (\alpha' - \alpha)d_\beta$ are sharing polynomials for α' and β', consistent with the adversary's view. Note that $f_{\beta'}(0) = \beta - (\alpha' - \alpha)m = y - \alpha' \cdot m = \beta'$. Since f_α and f_β are randomly chosen with $f_\alpha(0) = \alpha$ and $f_\beta(0) = \beta$, the adversary's view of the protocol, excluded the signatures, is independent of α. This together with the secrecy property of the signatures proves the claim. □

The communication complexity of one execution of MP Auth is $O(n \log |F|)$ bits. Namely, V essentially shares α and β. Note that the signatures involved are signatures verifiable by one player, hence they only cost $O(\log |F|)$ bits of communication.

6.5 The VSS Protocol

The VSS sharing protocol that meets the lower bound of Theorem 1 now works as follows. First, Pre Share is applied to the secret and then, by applying MP Auth to the shares, the sub-shares and signatures are stripped off and replaced by tags for the actual shares:

Protocol Share

1. The above protocol Pre Share is executed on the secret s. As a result, every player P_i holds a share s_i, sub-shares $s_{i1}, \ldots, s_{in}$ and signatures $\sigma_{s_{i1}}(P_1, P_i)$, $\ldots, \sigma_{s_{in}}(P_n, P_i)$.
2. For every player P_i, tags $y_{i1}, \ldots, y_{in}$ for s_i are computed by executing MP Auth with every player P_j on the message s_i and P_j's randomly chosen key $(\alpha_{ij}, \beta_{ij})$.

Note that all the sub-shares s_{ij} and signatures $\sigma_{s_{ij}}(P_j, P_i)$ are only temporarily used and can be deleted at the end of the protocol. For the reconstruction, as in the honest-dealer case, only the shares, the tags and the keys are needed.

Theorem 3. *For $t = \lfloor (n-1)/2 \rfloor$, there exists a Verifiable Secret Sharing scheme, $(t, n, 1-2^{-\Omega(k)}, 1-2^{-\Omega(k)})$-secure against an adaptive and rushing adversary, with a sharing complexity of $O(kn^3)$ and a single-round reconstruction of complexity $O(kn^2)$.*

Proof sketch: We can take the above scheme over a field F with $|F| \in 2^{\Theta(k)}$. Secrecy and correctness follow from Propositions 2 and 3. The communication complexity of the Pre Share protocol is $O(kn^3)$, of the MP Auth protocol it is

$O(kn)$. Therefore, the communication complexity of the sharing protocol, which calls Pre Share once and MP Auth n^2-times, is $O(kn^3)$. The communication complexity of the reconstruction is as in the Honest-Dealer VSS $O(kn^2)$ bits. □

7 Applications to MPC with Pre-processing

As an application of the above described VSS scheme, we will now present a general MPC protocol in the pre-processing model [1]. Our protocol is secure against an active, adaptive adversary who can corrupt up to $t = \lfloor (n-1)/2 \rfloor$, a minority, of the players. The idea behind MPC with pre-processing, introduced by Beaver [1], is to do as much work as possible in a *pre-processing phase*, before the inputs and even the circuit [6] are known. This is to reduce the work and the assumptions on the communication network required in the *computation phase* when the inputs and circuit have actually become available.

This is based on circuit randomization and a generic construction that can be applied to any general MPC protocol based on a VSS with certain linearity properties explained below. The computation phase doesn't require secure channels, it only consists of broadcasting information and performing the local computations necessary for VSS reconstructions. It should therefore be clear that MPC in the pre-processing model benefits from VSS with optimized reconstruction.

The required linearity properties are as follows. If s and s' are two VSS'ed secrets and λ a public constant, then the players should be able to locally compute VSS shares of $s + s'$ and $\lambda \cdot s$ (if this is the case then the scheme is called *homomorphic*) and of $s + \lambda$. Before showing that our VSS has these properties, we sketch the protocol for general MPC with pre-processing. Assume that adequate upperbounds on the number of inputs and multiplication gates in the future circuit are known. In the pre-processing phase, each player chooses a sufficient number of independent random values a and VSS'es them. Next, the players jointly prepare a sufficient number of random triples r, r' and r'' such that $r'' = rr'$ and such that each of these values is VSS'ed. Note that mutual randomness is easily achieved by having players VSS random values, and taking the sum of those as a mutually random value. By the linearity property, this random value is effectively VSS'ed. By invoking the general MPC protocol, products can be securely computed with the result VSS'ed.
In the computation phase, inputs and circuit are known. Assume for simplicity that each player has a single private input value. Each player then takes his actual private input s, and simply broadcasts the difference $\epsilon = a - s$ between this input s and the random value a he VSS'ed in the pre-processing phase. Subsequently, all players locally compute their shares in s from the shares in a they hold and the now public value ϵ. In the computation phase, the addition gates are handled locally while to multiply two shared values s and s', a fresh precomputed random triple (r, r', r'') is taken, the differences $\delta = s - r$ and $\delta' = s' - r'$ are revealed by invoking the reconstruction of VSS. Since

[6] Usually, the function that is to be securely computed is given as an arithmetic circuit

$ss' = (r+\delta)(r'+\delta') = rr' + \delta' r + \delta r' + \delta\delta' = r'' + \delta' r + \delta r' + \delta\delta'$, every player P_i can locally compute a share of ss' from the shares of r, r' and r'' and the values δ and δ'. Note that linearity of the VSS facilitates all of these steps.

7.1 Applying Our VSS to MPC with Pre-processing

We first argue that our VSS can be made to have the required linearity properties. Note that Shamir shares trivially possess these properties, so it suffices to focus on the authentication code. As mentioned in Section 5, the only thing we need to do is to fix throughout the computation the values α that are part of the verification keys (α, β). Indeed, if y and y' are authentication tags for m and m' with keys (α, β) and (α, β'), respectively, then for every $\lambda \in F$, $\lambda \cdot y + y'$ is an authentication tag for the message $\lambda \cdot m + m'$ with key $(\alpha, \lambda \cdot \beta + \beta')$. Namely, $\alpha \cdot (\lambda \cdot m + m') + (\lambda \cdot \beta + \beta') = \lambda \cdot (\alpha \cdot m + \beta) + (\alpha \cdot m' + \beta') = \lambda \cdot y + y'$. Analogue, it can be shown that y is an authentication tag for the message $m + \lambda$ with key $(\alpha, \beta - \alpha \cdot \lambda)$. Furthermore, it is not difficult to see by induction that after l authentications and verifications with the same α, the substitution probability still is $l/(|F| - l + 1)$ (see e.g. [4]).

For a field F with $|F| \in 2^{\Theta(k)}$, the protocol now works as follows. In the pre-processing phase, the random input values a are treated just as above, based on our VSS. In order to prepare the random triples, we use the general MPC techniques of [4] to prepare triples r, r' and r'' with $r'' = rr'$ as described earlier. This results in a VSS of these values according to [4] (i.e., according to the protocol PreShare from Section 6.3). We can convert these to sharings as they would have been produced by our VSS, we simply apply the protocol MPAuth (see Section 6) to get shares according to Share. Hence, all necessary pre-processing information will be shared according to our VSS. The computation phase can now proceed based on the reconstruction phase of our VSS.

As to efficiency, generating the sharings of r and r' consists essentially of $O(n)$ executions of PreShare, and thus this has complexity $O(kn^3)$ bits. The computation of the sharing of r'' costs according to [4] $O(kn^4)$ bits of communication, assuming everyone coorperates. Multi-party computing the tags is negligible compared to the rest, namely $O(kn^3)$. Hence, we have a best case complexity of $O(kn^4)$. If a corrupted player refuses to coorperate, then the easiest thing to do is to exclude the player and restart the computation. This will allow the adversary to slow down the computation by at most a factor linear in n. [7] Hence we have

Theorem 4. *Let C be an arithmetic circuit over a field F with M multiplication gates, where $|F| \in 2^{\Theta(k)}$. Communicating $O(Mkn^5)$ bits in a pre-processing phase, there exists a MPC protocol, secure, except with probability $2^{-\Omega(k)+M}$, against a rushing adversary who can adaptively corrupt up to $t = \lfloor (n-1)/2 \rfloor$ of the players, computing the circuit C with $O(Mkn^2)$ bits of comunication.*

[7] Instead of restarting, one could also reconstruct the share(s) of the caught cheater, if needed. This way, the adversary cannot slow down the computation substantially, resulting in a pre-processing complexity of $O(Mkn^4)$ instead of $O(Mkn^5)$.

The most efficient previously known protocol for MPC with pre-processing in our model is based on [4]. Note that this would result in a pre-processing phase with complexity of the same order as in our case. However, due to VSS with optimized reconstruction, we gain an efficiency improvement of a multiplicative factor n in the computation phase of our protocol.

References

1. D. Beaver. Efficient multiparty protocols using circuit randomization. In *CRYPTO '91*, LNCS 576, pages 420–432. Springer-Verlag, 1992.
2. M. Ben-Or, S. Goldwasser, and A. Widgerson. Completeness theorems for non-cryptographic fault-tolerant distributed computation. In *20th Annual ACM Symposium on the Theory of Computing*, pages 1–10, 1988.
3. R.E. Blahut. *Priciples and Practice of Information Theory.* Addison-Wesley, 1987.
4. R. Cramer, I. Damgard, S. Dziembowski, M. Hirt, and T. Rabin. Efficient multiparty computations secure against an adaptive adversary. In *EUROCRYPT '99*, LNCS 1592. Springer-Verlag, 1999.
5. R. Cramer, I. Damgaard, and U. Maurer. General secure multi-party computation from any linear secret-sharing scheme. In *EUROCRYPT 2000*, LNCS 1807. Springer-Verlag, 2000.
6. B. Chor, S. Goldwasser, S. Micali, and B. Awerbuch. Verifiable secret sharing and achieving simultaneity in the presence of faults (extended abstract). In *26th Annual Symposium on Foundations of Computer Science*, pages 383–395, 1985.
7. S. Cabello, C. Padró, and G. Sáez. Secret sharing schemes with detection of cheaters for a general access structure. In *Proceedings of the 12th International Symposium on Fundamentals of Computation Theory, FCT '99*, LNCS 1233, pages 185–193, 1999.
8. R. Gennaro, M.O. Rabin, and T. Rabin. Simplified VSS and fast-track multiparty computations with applications to threshold cryptography. In *17th ACM Symposium on Principles of Distributed Computing*, 1998.
9. M. Hirt and U. Maurer. Complete characterization of adversaries tolerable in secure multi-party computation (extended abstract). In *16th ACM Symposium on Principles of Distributed Computing*, pages 25–34, 1997.
10. M. Karchmer and A. Wigderson. On span programs. In *8th Annual Conference on Structure in Complexity Theory (SCTC '93)*, pages 102–111, 1993.
11. U. Maurer. Authentication theory and hypothesis testing. *IEEE Transaction on Information Theory*, 2000.
12. T. Rabin and M. Ben-Or. Verifiable secret sharing and multiparty protocols with honest majority (extended abstract). In *21th Annual ACM Symposium on the Theory of Computing*, pages 73–85, 1989.
13. A. Shamir. How to share a secret. *Communications of the Association for Computing Machinery*, 22(11):612–613, 1979.
14. G.J. Simmons. Authentication theory/coding theory. In *CRYPTO '84*, LNCS 196, pages 411–431. Springer-Verlag, 1985.
15. D.R. Stinson. *Cryptography — Theory and Practice.* Number ISBN 0-8493-8521-0. CRC Press, 1995.

A Communication Pattern from Section 3: Justification

When justifying the claim that the proposed communication pattern is most general, it should be kept in mind that we are interested in the complexity of the reconstruction phase, and that all "re-modeling" operations are allowed as long as they do not affect the complexity of reconstruction (apart from constant factors).

By the assumption that the dealer is honest, we may assume without loss of generality that the distribution phase only consists of the dealer sending private information s_i to each of the players P_i, i.e., any secure distributed computation carried out by the players in the distribution phase could as well be carried out by the honest dealer, without consequences to the complexity of the reconstruction phase. Similarly, we may assume that in the reconstruction phase each player P_i merely broadcasts a piece of information, y_i, that only depends on the private information s_i received from the dealer. Namely, at the cost of at most a constant factor of increased communication, private channels can be simulated by one-time pads, the keys of which are distributed by the honest dealer. In fact, it can be assumed that in general $s_i = (k_i, y_i)$, where y_i is required to be broadcast in the reconstruction phase, and each player P_i makes a local (possibly probabilistic) decision on the secret s based on the broadcast information and his private k_i.

B Impossibility Lemmas from Section 3

Lemma 2. *There exists a static, non-rushing adversary such that with non-zero probability some honest players output "failure" in the reconstruction phase.*

Proof. Given that $t \geq n/3$, let B, A_0, A_1 be an arbitrary disjoint partition of $\{1, \ldots, n\}$ such that $|B| = t$ and $1 \leq |A_0|, |A_1| \leq t$. We show a strategy for the adversary that forces all players in B to output "failure" with non-zero probability. The adversary corrupts the players in A_0, selects a random secret $\tilde{s}$ and randomly guesses the shares $s_i = (k_i, y_i)$ held by the players in B. By the privacy of the scheme and assuming that he guessed the shares correctly and that $s \neq \tilde{s}$ (which both happens with non-zero probability), he can sample random shares $\tilde{s}_j$ for the corrupted players, so that these, together with the shares of the players in B, are consistent with the secret $\tilde{s}$, and have the same distribution as when sent by the honest dealer. It is now clear that in the reconstruction phase (assumed that the adversary guessed the shares correctly and that $s \neq \tilde{s}$), every player in B has to output "failure". Indeed, the players in B must definitely not output the incorrect secret $\tilde{s}$. On the other hand, if some player in B outputs the correct secret s (with positive probability), then by corrupting the players in A_1 instead of A_0, but otherwise playing the corresponding game, the adversary creates the same view for the players in B, however with the correct and the incorrect secrets exchanged, and hence this player would now output the incorrect secret (with positive probability), which is a contradiction. □

Lemma 3. *There exists a static, rushing adversary such that such that with non-zero probability some honest player recovers the secret in the reconstruction phase, while some other honest player outputs "failure".*

Proof. Consider the case $t > n/3$. Let B, A_0, A_1 be an arbitrary disjoint partition of $\{1, \ldots, n\}$ such that $1 \leq |A_0| \leq t-1$ and $1 \leq |A_1| \leq t$. Note that $2 \leq |B| \leq t$. Let p, q be distinct members of B. We consider the same adversary as before, except that in the reconstruction phase, the adversary "rushes", and waits until the players in B have broadcast their y_i's. He then makes a guess for player p's private k_p, and broadcasts random $\tilde{y}_j$'s for the players, consistent with k_p and with the y_i's of the players in B and a random secret different from the correct one (which he knows by now). For similar reasons as before we conclude that player p does not reconstruct the secret if the guess for k_p was correct. However, in that case player q must reconstruct the secret with positive probability: for if not, corrupting A_0 *and* player p (note that this amounts to at most t corruptions), the adversary would not have to guess k_p anymore, and hence there would be a strategy that makes at least one honest player output "failure" in the reconstruction with probability equal to 1. This contradicts correctness. □

C Non-maximal t

In the main body of this paper, we have only considered a maximal t in the interesting range $n/3 \leq t < n/2$. We will now state the generalizations of the Theorems 1 to 4 for a (nearly) arbitrary t in this range. The corresponding proofs are similar but technically more involved.

Theorem 1'. *For any family of Single-Round Honest-Dealer VSS schemes, $(t, n, 1-\delta)$-secure against an arbitrary active, rushing adversary, the following holds. Let k be a security parameter and let $\epsilon > 0$ be an arbitrary constant. If $\delta = 2^{-\Omega(k)}$ and $n/3 \cdot (1+\epsilon) \leq t < n/2$ then the total information broadcast in the reconstruction phase is lower bounded by $\Omega((nH(S) + kn^2)/(n-2t))$.*

Note that already an arbitrarily small linear gap between t and $n/2$ reduces the lower bound by a factor of n. The following Theorem shows that the reconstruction complexity indeed reduces by a factor of n for such a t (at least in case of a security parameter k slightly larger than linear in n).

Theorem 2'. *For $n/3 \leq t < n/2$ and $k = \Omega((n-2t)\log(t))$, there exists a Single-Round Honest-Dealer VSS scheme, $(t, n, 1-2^{-\Omega(k)})$-secure against an adaptive and rushing adversary, with a total communication complexity of $O(kn^2/(n-2t))$ bits.*

The according holds for VSS.

Theorem 3'. *For $n/3 \leq t < n/2$ and $k = \Omega((n-2t)\log(t))$, there exists a Single-Round VSS scheme, $(t, n, 1-2^{-\Omega(k)}, 1-2^{-\Omega(k)})$-secure against an adaptive and rushing adversary, with a sharing complexity of $O(kn^3)$ and a reconstruction complexity of $O(kn^2/(n-2t))$.*

Applied to MPC with preprocessing, we achieve

Theorem 4'. *Let C be an arithmetic circuit over a field F with M multiplication gates, where $|F| \in 2^{\Theta(k)/(n-2t)}$, $n/3 \leq t < n/2$ and $k = \Omega((n-2t)\log(t))$. Communicating $O(Mkn^5)$ bits in a pre-processing phase, there exists a MPC protocol, secure, except with probability $2^{-\Omega(k)+M}$, against an adversary who can adaptively corrupt up to t of the players, computing the circuit C with $O(Mkn^2/(n-2t))$ bits of communication.*

D General Adversaries

We now go beyond security against a dishonest *minority* by sketching how to adjust our VSS and MPC protocols to be secure against a *general Q^2-adversary* [9], i.e. against an adversary who can corrupt any subset of players in a given family of subsets, where no two subsets in the family cover the full player set.

By replacing the bivariate polynomial sharing in Pre Share by the information-theoretic commitment/WSS protocol from [5] based on monotone span programs [10], we are in the same position as described by Proposition 2, except that 4. is not guaranteed, i.e. the share s_i of a corrupted player P_i is not necessarily correctly shared by the sub-shares s_{ji} of the honest players P_j. But this can easily be achieved by doing another level of sharing: every player P_i shares his share s_i with the WSS protocol from [5] where every player P_j insists that the share they get of s_i is the sub-share s_{ji}.
In the MP Auth protocol, replacing the threshold sharings of the values α and β by sharings based on monotone span program sharings [10] with multiplication, and using the fact that these can be constructed from ordinary monotone span programs with only constant overhead [5], Proposition 3 remains intact.
This results in a VSS scheme secure against a general Q^2-adversary. Furthermore, the sharing and reconstruction complexities are $O(knm^2)$ and $O(knm)$ bits, respectively, where $m \geq n$ is the size of the monotone span program, while the respective complexities of the general adversary VSS scheme suggested in [4] are both $O(knm^3)$ bits (even though one could achieve $O(knm^2)$ using their techniques in a more elaborate way).

Based on this general adversary VSS scheme, similar to the previous section, one can achieve a general MPC protocol, secure against a general Q^2 adversary, which in the pre-processing model has a communication complexity of $O(Mknm)$ bits, compared to $O(Mknm^3)$ (respectively $O(Mknm^2)$), which would be achieved by the general adversary MPC protocol from [4].

E The Power of Rushing (Honest Dealer Case)

We show that our tight lower bound from Section 4 does not hold if the adversary does not rush, and instead selects the corrupted shares he will broadcast in the reconstruction phase before it has started. We also sketch a lower bound and outline some applications, namely to a scenario in which the amount of information sent in the *distribution phase* is to be minimized.

Let F be a finite field with $|F| > n$, and take Shamir's $(t+1, n)$-threshold scheme defined over F. Cabello, Padró and Sáez [7] have proposed the following so-called robust secret sharing scheme. To share a secret s in this scheme, the honest dealer selects a random field element ρ, independently generates full sets of Shamir-shares for the secrets s, ρ and $\rho \cdot s$, and privately distributes the shares to the players.

Given a set A of at least $t+1$ shares (which possibly contains corrupted shares), consider the three values s', ρ' and τ' that are computed by applying the reconstruction procedure of Shamir's scheme to the shares in A. The crucial observation is that if $s \neq s'$ and if the corrupted shares are *independently distributed from* ρ, the probability that $s' \cdot \rho' = \tau'$ is at most $1/|F|$. Hence, given for instance a trusted party available for reconstruction, connected with each player by an independent private channel, the independence requirement is satisfied and although the secret may not always be reconstructed from a qualified set, a corrupted secret is detected with high probability.

We note the following application of this scheme in our scenario of a non-rushing adversary. By assumption, this is an adversary who chooses the corrupted shares before the reconstruction phase. This ensures that the independence requirement stated before is satisfied. Let k be a security parameter and $t < n/2$ and assume additionally that $|F| \geq 2^{n+k}$. Let the distribution phase be according to the scheme of [7]. Consider an arbitrary set A of $t+1$ shares revealed in the reconstruction phase. If A consists exclusively of shares of honest players, then the secret s_A reconstructed by the procedure above would certainly be the correct secret s. Else, either a failure would be detected, or with probability at most 2^{-n-k}, a secret $s_A \neq s$ is accepted based on the shares in A. Let V denote the set of all distinct *accepted* "secrets" s_A, by quantifying over all sets A. Note that $s \in A$. Now each honest player simply computes V, and outputs "failure" if V has more than one element, and s otherwise. This way all honest players are in agreement, and the probability with which they output "failure" is clearly at most $2^n \cdot 2^{-n-k} = 2^{-k}$.

For the case that $k > n$ and $n = 2t+1$, we now sketch an argument showing that the distribution phase of this scheme is optimal, up to constant factors. A basic result in secret sharing says, informally speaking, that the size of individual shares is at least the size of the secret, and hence the question that remains is whether the error probability ϵ of the above scheme is optimal. We define an adversary who flips a random coin and either corrupts the first t players, or the last t players. In either case, he makes a random guess $\tilde{S}_{t+1}$ for the share S_{t+1} that player P_{t+1} received from the honest dealer in the distribution phase, deletes the correct shares received from the dealer by the corrupted players, and instead chooses random corrupted shares, consistent with his guess $\tilde{S}_{t+1}$ and with a random secret $\tilde{s}$. Assuming that the correct secret s was chosen at random by the honest dealer, if the adversaries' guess for player P_{t+1}'s share is correct, then there is no way for any reconstruction procedure to distinguish between s and

$\tilde{s}$. Hence, in order for $\log 1/\epsilon$ to be $O(k)$, the size of each individual share must be $\Omega(k)$.

Although it is generally not very realistic to assume that the adversary is not rushing, it is possible to construct a "simultaneous broadcast" channel on top of the "secure channels with broadcast model". Namely, simply have each player first VSS their values, e.g. by using the schemes of [12,4], after which all VSS's are opened. Using the concrete scheme above, this procedure would ensure that shares are "broadcast simultaneously", and hence that the required independence is achieved, at the cost of increased complexity of the reconstruction phase and use of private channels in that phase. The advantage, however, is that the efficiency of the *distribution phase* has been substantially improved.

F Proof of Proposition 2

1. First note that the adversary does not gain any new information by making players complain. Let A be the set of players who have been corrupted during the execution of PreShare. The existence the symmetrical polynomial

$$d(X,Y) = \prod_{P_i \in A} \frac{(X-i)(Y-i)}{i^2}$$

   of degree t, with $d(0,0) = 1$ and $d(i,\cdot) = d(\cdot,i) = 0$ for all $P_i \in A$, implies that for every $s' \in F$, the number of bivariate symmetrical polynomials of degree at most t with s' as constant coefficient and consistent with the adversary's view is the same. Therefore, as f is chosen at random, the shares and sub-shares of the corrupted players give no information about the secret s. The claim now follows from the secrecy property of the signatures.
2. If this was not the case, then there would have been complaining.
3. Let the set A consist of $t+1$ honest players. Their shares define a unique secret s'. Let now A' consist of the players in A and a further honest player (if there are only $t+1$ honest players, then we are finished anyway). Let λ_i, $i \in A$, be the reconstruction coefficients for the players in A and λ'_i, $i \in A'$, for the players in A'. So we have $s' = \sum_{i \in A} \lambda_i s_i$ and (according to 2.) $s_k = \sum_{i \in A} \lambda_i s_{ki} = \sum_{j \in A'} \lambda'_j s_{kj}$ for all $k \in A'$. It follows that $\sum_{k \in A'} \lambda'_k s_k = \sum_{k \in A'} \lambda'_k \sum_{i \in A} \lambda_i s_{ki} = \sum_{i \in A} \lambda_i \sum_{k \in A'} \lambda'_k s_{ki} = \sum_{i \in A} \lambda_i \sum_{k \in A'} \lambda'_k s_{ik} = \sum_{i \in A} \lambda_i s_i = s'$, hence the shares of the players in A' are still consistent. Inductively, it follows that the shares of all honest players are consistent and define a unique secret s'.
4. Can be shown with a similar argumentation as above using the fact that every share s_j can be written as a fix linear combination $\sum_k \mu_k s_k$ of the shares of the honest players P_k. □

Secure and Efficient Asynchronous Broadcast Protocols*
(Extended Abstract)

Christian Cachin, Klaus Kursawe, Frank Petzold**, and Victor Shoup

IBM Research, Zurich Research Laboratory
CH-8803 Rüschlikon, Switzerland
{cca,kku,sho}@zurich.ibm.com

Abstract. Broadcast protocols are a fundamental building block for implementing replication in fault-tolerant distributed systems. This paper addresses secure service replication in an asynchronous environment with a static set of servers, where a malicious adversary may corrupt up to a threshold of servers and controls the network. We develop a formal model using concepts from modern cryptography, give modular definitions for several broadcast problems, including reliable, atomic, and secure causal broadcast, and present protocols implementing them. Reliable broadcast is a basic primitive, also known as the Byzantine generals problem, providing agreement on a delivered message. Atomic broadcast imposes additionally a total order on all delivered messages. We present a randomized atomic broadcast protocol based on a new, efficient multi-valued asynchronous Byzantine agreement primitive with an external validity condition. Apparently, no such efficient asynchronous atomic broadcast protocol maintaining liveness and safety in the Byzantine model has appeared previously in the literature. Secure causal broadcast extends atomic broadcast by encryption to guarantee a causal order among the delivered messages. Our protocols use threshold cryptography for signatures, encryption, and coin-tossing.

1 Introduction

Broadcast protocols are a fundamental building block for fault-tolerant distributed systems. A group of servers can offer some service in a fault-tolerant way by using the state machine replication technique, which will mask the failure of any individual server or a fraction of them. In the model with Byzantine faults considered here, faulty servers may exhibit arbitrary behavior or even be controlled by an adversary.

* This work was supported by the European IST Project MAFTIA (IST-1999-11583). However, it represents the view of the authors. The MAFTIA project is partially funded by the European Commission and the Swiss Department for Education and Science.

** Frank Petzold has since left IBM and can be reached at petzold@hepe.com.

J. Kilian (Ed.): CRYPTO 2001, LNCS 2139, pp. 524–541, 2001.

In this paper, we present a modular approach for building robust broadcast protocols that provide reliability (all servers deliver the same messages), atomicity (a total order on the delivered messages), and secure causality (a notion that ensures no dishonest server sees a message before it is scheduled by the system). An important building block is a new protocol for multi-valued Byzantine agreement with "external validation." Our focus is on methods for distributing secure, trusted services on the Internet with the goal of increasing their availability and security. Cryptographic operations are exploited to a greater extent than previously for such protocols because we consider them to be relatively cheap, in particular compared to the message latency on the Internet.

We do not make any timing assumptions and work in a purely asynchronous model with a static set of servers and no probabilistic assumptions about message delays. Our protocols rely on a trusted dealer that is used once to set up the system, but they do not use any additional external constructs later (such as failure detectors or stability mechanisms). We view this as the standard cryptographic model for a distributed system with Byzantine faults. These choices maintain the safety of the service even if the network is temporarily disrupted. This model also avoids the problem of having to assume synchrony properties and to fix timeout values for a network that is controlled by an adversary; such choices are difficult to justify if safety and also security depend on them.

Despite the practical appeal of the asynchronous model, not much research has concentrated on developing efficient asynchronous protocols or implementing practical systems that need consensus or Byzantine agreement. Often, developers of distributed systems avoid the approach because of the result of Fischer, Lynch, and Paterson [9], which shows that consensus is not reachable by protocols that use an a priori bounded number of steps, even with crash failures only. But the implications of this result should not be overemphasized. In particular, there are randomized solutions that use only a *constant expected* number of asynchronous "rounds" to reach agreement [15,7,3]. Moreover, by employing modern, efficient cryptographic techniques and by resorting to the random oracle model, this approach has recently been extended to a practical yet provably secure protocol for cryptographic Byzantine agreement that withstands the maximal possible corruption [6].

Two basic broadcast protocols are *reliable broadcast* (following Bracha and Toueg [4]), which ensures that all servers deliver the same messages, and a variation of it that we call *consistent broadcast*, which only provides agreement among the actually delivered messages. The consistent broadcast primitive used here is particularly useful in connection with a *verifiability* property for the delivered messages, which ensures that a party can transfer a "proof of delivery" to another party in a single flow.

The efficient randomized agreement protocols mentioned before work only for binary decisions (or for decisions on values from small sets). In order to build distributed secure applications, this is not sufficient. One also needs agreement on values from large sets, in particular for ordering multiple messages. We propose a new *multi-valued Byzantine agreement* protocol with an *external* validity condition and show how it can be used for implementing *atomic broadcast.* External validity ensures that the decision value is acceptable to the particular application that requests agreement; this corrects a drawback of earlier agreement protocols for multi-valued agreement, which could decide on illegal values. Both protocols use digital signatures and additional cryptographic techniques.

The multi-valued Byzantine agreement protocol invokes only a constant expected number of binary Byzantine agreement sub-protocols on average and achieves this by using a cryptographic common coin protocol in a novel way. It withstands the maximal possible corruption of up to one third of the parties and has expected quadratic message complexity (in the number of parties), which is essentially optimal.

Our atomic broadcast protocol guarantees that a message from an honest party cannot be delayed arbitrarily by an adversary as soon as a minimum number of honest parties are aware of that message. The protocol invokes one multi-valued Byzantine agreement per batch of payload messages that is delivered. An analogous reduction of atomic broadcast to consensus in the crash-fault model has been described by Chandra and Toueg [8], but it cannot be directly transferred to the Byzantine setting.

We also define and implement a variation of atomic broadcast called *secure causal atomic broadcast.* This is a robust atomic broadcast protocol that tolerates a Byzantine adversary and also provides secrecy for messages up to the moment at which they are guaranteed to be delivered. Thus, client requests to a trusted service using this broadcast remain confidential until they are answered by the service and the service processes the requests in a causal order. This is crucial in our asynchronous environment for applying the state machine replication method to services that involve confidential data.

Secure causal atomic broadcast works by combining an atomic broadcast protocol with robust threshold decryption. The notion and a heuristic protocol were proposed by Reiter and Birman [17], who called it "secure atomic broadcast" and also introduced the term "input causality" for its main property. Recent progress in threshold cryptography allows us to present an efficient robust protocol together with a security proof in the random oracle model.

In accordance with the comprehensive survey of fault-tolerant broadcasts by Hadzilacos and Toueg [10], we define and implement our protocols in a modular way, with reliable and consistent broadcasts and Byzantine agreement as primitives. This leads to the following layered architecture:

<table>
<tr><td colspan="2">Secure Causal Atomic Broadcast</td></tr>
<tr><td colspan="2">Atomic Broadcast</td></tr>
<tr><td colspan="2">Multi-valued Byzantine Agreement</td></tr>
<tr><td>Broadcast Primitives</td><td>Byzantine Agreement</td></tr>
</table>

Important for the presentation of our broadcast protocols is our formal model of a modular protocol architecture, where a number of potentially corrupted parties communicate over an insecure, asynchronous network; it uses complexity-theoretic concepts from modern cryptography. This makes it possible to easily integrate the formal notions for encryption, signatures, and other cryptographic tools with distributed protocols. The model allows for quantitative statements about the running time and the complexity of protocols; the essence of our definition is to bound the number of steps taken by participants on behalf of a protocol *independently* from network behavior. In view of the growing importance of cryptography for secure distributed protocols, a unified formal model for both is a contribution that may be of independent interest.

Organization of the Paper. For lack of space, only the most important results are described in this extended abstract. It begins with a brief account of the formal model and definitions for binary Byzantine agreement and consistent broadcast. Then it presents validated Byzantine agreement and an implementation for the multi-valued case, which is extended to atomic broadcast. More details, in particular the formal model, detailed definitions and proofs, the discussion of related work, and the descriptions of reliable broadcast and secure causal atomic broadcast, can be found in the full version [5].

2 Model

2.1 Overview of the Formal Model

Our system consists of a collection of n interactive Turing machines, of which t are (statically) corrupted by an adversary, modeled by an arbitrary Turing machine. There is a trusted dealer that has distributed some cryptographic keys initially, but it is not used later. Our model differs in two respects from other models traditionally used in distributed systems with Byzantine faults: (1) In order to use the proof techniques of complexity-based cryptography, our model is *computational*: all parties and the adversary are constrained to perform only feasible, i.e., polynomial-time, computations. This is necessary for using formal notions from cryptography in a meaningful way. (2) We make no assumptions about the network at all and leave it under complete control of the adversary. Our protocols work only to the extent that the adversary delivers messages faithfully. In short, *the network is the adversary.* The differences become most apparent in the treatment of termination, for which we use more concrete conditions that together imply the traditional notion of "eventual" termination.

We define termination by bounding a *statistic* measuring the amount of work that honest, uncorrupted parties do on behalf of a protocol. In particular, we use the communication complexity of a protocol for this purpose, which is defined as the length of all protocol messages that are "associated" to the protocol instance. We use the term *protocol message* for messages that the parties send to each other to implement a protocol, in contrast to the *payload messages* that are the subject of the (reliable, consistent, atomic ...) broadcasts among all parties. The specification of a protocol requires certain things to happen under the condition that all protocol messages have been delivered; thus, bounding the length (and also the number) of protocol messages generated by uncorrupted parties ensures that the protocol has actually terminated under this condition.

As usual in cryptography, we prove security with respect to all polynomial-time adversaries. Our notion of an *efficient* (deterministic) protocol requires that the statistic is *uniformly bounded* by a fixed polynomial independent of the adversary. We also define the corresponding notion of a *probabilistically uniformly bounded statistic* for randomized protocols; the expected running time of such a protocol can be derived from this. Both notions are closed under modular composition of protocols, which is not trivial for randomized protocols.

Our model uses the adversary in two roles: to invoke new instances of a protocol through input actions (as an application might do) and to deliver protocol messages (modeling the network).

For simplicity, all protocol messages delivered by the adversary are assumed to be authenticated (implementing this is straightforward in our model).

2.2 Byzantine Agreement

We give the definition of *(binary) Byzantine agreement* (or *consensus* in the crash-fault model) here as it is needed for building atomic broadcast protocols. It can be used to provide agreement on independent *transactions.*

The Byzantine agreement protocol is activated when the adversary delivers an input action to P_i of the form $(ID, \texttt{in}, \texttt{propose}, v)$, where $v \in \{0,1\}$. When this occurs, we say P_i *proposes* v for transaction ID. A party terminates the Byzantine agreement protocol (for transaction ID) by generating an output action of the form $(ID, \texttt{out}, \texttt{decide}, v)$. In this case, we say P_i *decides* v for transaction ID. Let any protocol message with tag ID or $ID|\ldots$ that is generated by an honest party be *associated* to the agreement protocol for ID.

Definition 1 (Byzantine agreement). *A protocol solves* Byzantine agreement *if it satisfies the following conditions except with negligible probability:*

Validity: *If all honest parties that are activated on a given ID propose v, then any honest party that terminates for ID decides v.*[1]

Agreement: *If an honest party decides v for ID, then any honest party that terminates decides v for ID.*

Liveness: *If all honest parties have been activated on ID and all associated messages have been delivered, then all honest parties have decided for ID.*

Efficiency: *For every ID, the communication complexity for ID is probabilistically uniformly bounded.*

2.3 Cryptographic Primitives

Apart from ordinary digital signature schemes, we use collision-free hashing, pseudo-random generators, robust non-interactive dual-threshold signatures [18], threshold public-key encryption schemes [19], and a threshold pseudo-random function [13,6]. Definitions can be found in the full version [5].

[1] We use the term "validity" for this condition in accordance with most of the literature on Byzantine agreement. Alternatively, one might also adopt the terminology of fault-tolerant broadcasts [10] and instead call it "integrity," to emphasize that it is a general safety condition (in contrast to a liveness condition).

3 Broadcast Primitives: Verifiable Consistent Broadcast

Our multi-valued agreement protocol builds on top of a *consistent broadcast* protocol, which is a relaxation of Byzantine *reliable broadcast* [12]. Consistent broadcast provides a way for a distinguished party to send a message to all other parties such that two parties never deliver two conflicting messages for the same sender and sequence number. In other words, it maintains consistency among the actually delivered payloads with the same senders and sequence numbers, but makes no provisions that two parties do deliver the payloads. Such a primitive has also been used by Reiter [16].

Broadcasts are parameterized by a tag ID, which can also be thought of as identifying a broadcast "channel," augmented by the identity of the sender, j, and by a sequence number s. We restrict the adversary to submit a request for consistent broadcast tagged with $ID.j.s$ to P_i only if $i = j$ and at most once for every sequence number.

A consistent broadcast protocol is activated when the adversary delivers an input action to P_j of the form $(ID.j.s, \texttt{in}, \texttt{c-broadcast}, m)$, with $m \in \{0,1\}^*$ and $s \in \mathbb{N}$. When this occurs, we say P_j *consistently broadcasts* m *tagged with* $ID.j.s$. Only the sender P_j is activated like this. The other parties are activated when they perform an explicit *open* action for instance $ID.j.s$ in their role as receivers (this occurs implicitly in our system model when they **wait for** an output tagged with $ID.j.s$).

A party terminates a consistent broadcast of m tagged with $ID.j.s$ by generating an output action of the form $(ID.j.s, \texttt{out}, \texttt{c-deliver}, m)$. In this case, we say P_i *consistently delivers* m *tagged with* $ID.j.s$. For brevity, we also the terms *c-broadcast* and *c-deliver*.

Definition 2 (Authenticated Consistent Broadcast). *A protocol for* authenticated consistent broadcast *satisfies the following conditions except with negligible probability:*

Validity: *If an honest party has* c-broadcast m *tagged with* $ID.j.s$*, then all honest parties* c-deliver m *tagged with* $ID.j.s$*, provided all honest parties have been activated on* $ID.j.s$ *and the adversary delivers all associated messages.*

Consistency: *If some honest party* c-delivers m *tagged with* $ID.j.s$ *and another honest party* c-delivers m' *tagged with* $ID.j.s$*, then* $m = m'$*.*

Authenticity: *For all* ID*, senders* j*, and sequence numbers* s*, every honest party* c-delivers *at most one message* m *tagged with* $ID.j.s$*. Moreover, if* P_j *is honest, then* m *was previously* c-broadcast *by* P_j *with sequence number* s*.*

Efficiency: *For any* ID*, sender* j*, and sequence number* s*, the communication complexity of instance* $ID.j.s$ *is uniformly bounded.*

The provision that the "adversary delivers all associated messages" is our quantitative counterpart to the traditional "eventual" delivery assumption.

A party P_i that has delivered a payload message using consistent broadcast may want to inform another party P_j about this. Such information might be useful to P_j if it has not yet delivered the message, but can exploit this knowledge to deliver the payload message itself, maintaining consistency. We call this property the *verifiability* of a consistent broadcast.

Informally, we use *verifiability* like this: when P_j claims that it is not yet in a state to *c-deliver* a particular payload message m, then P_i can send a single protocol message to P_i and when P_j processes this, it will *c-deliver* m immediately.

Definition 3 (Verifiability). *A consistent broadcast protocol is called* verifiable *if the following holds, except with negligible probability: When an honest party has* c-delivered *m tagged with $ID.j.s$, then it can produce a single protocol message M that it may send to other parties such that any other honest party will* c-deliver *m tagged with $ID.j.s$ upon receiving M (provided the other party has not already done so before).*

We call M the message that *completes* the verifiable broadcast. This notion implies that there is a polynomial-time computable predicate $V_{ID.j.s}$ that the receiving party can apply to an arbitrary bit string for checking if it constitutes a message that completes a verifiable broadcast tagged with $ID.j.s$.

A protocol for verifiable authenticated consistent broadcast (denoted VCBC) is given in the full version [5]. It is inspired by the "echo broadcast" of Reiter [16] and based on a threshold signature scheme. Its message complexity is $O(n)$ and its bit complexity is $O(n(|m|+K))$, assuming the length of a threshold signature and a signature share is at most K bits.

4 Validated Byzantine Agreement

The standard notion of validity for Byzantine agreement implements a binary decision and requires that only if *all* honest parties propose the same value, this is also the agreement value. No particular outcome is guaranteed otherwise. Obviously, this still ensures that the agreement value was proposed by *some* honest party for the binary case. But it does not generalize to multi-valued Byzantine agreement, and indeed, all previous protocols for multi-valued agreement [15,20, 14] may fall back to a default value in this case, and decide for a value that *no* honest party proposed.

We solve this problem by introducing an *external validity* condition, which requires that the agreement value is legal according to a global, polynomial-time computable predicate, known to all parties and determined by the particular higher-level application.

Validated Byzantine agreement generalizes the primitive of *agreement on a core set* [2], which is used in the information-theoretic model for a similar purpose (a related protocol was also developed by Ben-Or and El-Yaniv [1]).

4.1 Definition

Suppose there is a global polynomial-time computable predicate Q_{ID} known to all parties, which is determined by an external application. Each party may propose a value v that should satisfy Q_{ID} and perhaps contains validation information. The agreement domain is not restricted to binary values.

A validated Byzantine agreement protocol is activated by a message of the form $(ID, \texttt{in}, \texttt{v-propose}, v)$, where $v \in \{0,1\}^*$. When this occurs, we say P_i *proposes* v for transaction ID. We assume the adversary activates all honest parties on a given ID at most once. W.l.o.g., honest parties propose values that satisfy Q_{ID}.

A party terminates a validated Byzantine agreement protocol by generating a message of the form $(ID, \texttt{out}, \texttt{v-decide}, v)$. In this case, we say P_i *decides* v for transaction ID.

We say that any protocol message with tag ID that was generated by an honest party is *associated* to the validated Byzantine agreement protocol for ID. An agreement protocol may also invoke sub-protocols for low-level broadcasts or for Byzantine agreement; in this case, all messages associated to those protocols are associated to ID as well (such messages have tags with prefix $ID|\ldots$).

Definition 4 (Validated Byzantine Agreement). *A protocol solves* validated Byzantine agreement *with predicate Q_{ID} if it satisfies the following conditions except with negligible probability:*

External Validity: *Any honest party that terminates for ID decides v such that $Q_{ID}(v)$ holds.*

Agreement: *If some honest party decides v for ID, then any honest party that terminates decides v for ID.*

Liveness: *If all honest parties have been activated on ID and all associated messages have been delivered, then all honest parties have decided for ID.*

Integrity: *If all parties follow the protocol, and if some party decides v for ID, then some party proposed v for ID.*

Efficiency: *For every ID, the communication complexity for ID is probabilistically uniformly bounded.*

A variation of the validity condition is that an application may prefer one class of decision values over others. Such an agreement protocol may be *biased* and *always* choose the preferred class in cases where other values would have been valid as well.

Validated Byzantine agreement is often used with arguments that consist of a "value" part v and a separate "proof" π that establishes the validity of v. If v is a single bit, we call this the problem of *binary validated agreement*; a protocol for this task is used below.

In fact, we will need a binary validated agreement protocol that is "biased" towards 1. Its purpose is to detect whether there is some validation for 1, so it suffices to guarantee termination with output 1 if $t+1$ honest parties know the corresponding information at the outset. Formally, a *binary validated Byzantine agreement protocol biased towards 1* is a protocol for validated Byzantine agreement on values in $\{0,1\}$ such that the following condition holds:

Biased Validity: If at least $t+1$ honest parties propose $v = 1$ then any honest party that terminates for ID decides $v = 1$.

We describe two related protocols for multi-valued validated Byzantine agreement below: Protocol VBA, described in Section 4.3, needs $O(n)$ rounds and invokes $O(n)$ binary agreement sub-protocols; this can be improved to a constant expected number of rounds, resulting in Protocol VBAconst, which is described in Section 4.4. But first we discuss the binary case.

4.2 Protocols for the Binary Case

It is easy to see that any binary asynchronous Byzantine agreement protocol can be adapted to external validity and can also be biased.

For example, in the protocol of Cachin, Kursawe, and Shoup [6] one has to "justify" the pre-votes of round 1 with a valid "proof" π. The logic of the protocol guarantees that either a decision is reached immediately or the validations for 0 and for 1 are seen by all parties in the first two rounds. Furthermore, the protocol can be biased towards 1 by modifying the coin such that it always outputs 1 in the first round.

4.3 A Protocol for the Multi-valued Case

We now describe Protocol VBA that implements multi-valued validated Byzantine agreement.

The basic idea is that every party proposes its value as a candidate value for the final result. One party whose proposal satisfies the validation predicate is then selected in a sequence of binary Byzantine agreement protocols and this value becomes the final decision value. More precisely, the protocol consists of the following steps.

Echoing the proposal (lines 1–4): Each party P_i *c-broadcasts* the value that it proposes to all other parties using verifiable authenticated consistent broadcast. This ensures that all honest parties obtain the same proposal value for any particular party, even if the sender is corrupted. Then P_i waits until it has received $n-t$ proposals satisfying Q_{ID} before entering the agreement loop.

Agreement loop (lines 5–20): One party is chosen after another, according to a fixed permutation Π of $\{1,\ldots,n\}$. Let a denote the index of the party selected in the current round (P_a is called the "candidate"). Each party P_i carries out the following steps for P_a:

1. Send a `v-vote` message to all parties containing 1 if P_i has received P_a's proposal (including the proposal in the vote) and 0 otherwise (lines 6–11).
2. Wait for $n-t$ `v-vote` messages, but do not count votes indicating 1 unless a valid proposal from P_a has been received—either directly or included in the `v-vote` message (lines 12–13).
3. Run a *binary* validated Byzantine agreement biased towards 1 to determine whether P_a has properly broadcast a valid proposal. Vote 1 if P_i has received a valid proposal from P_a and add the protocol message

Protocol VBA for party P_i, tag ID, and validation predicate Q_{ID}

LET $V_{ID|a}((v,\rho))$ BE THE FOLLOWING PREDICATE:

$$V_{ID|a}((v,\rho)) \equiv (v=0) \textbf{ or } (v=1 \textbf{ and } \rho \text{ completes the verifiable authenticated c-broadcast of a message } (\texttt{v-echo}, w_a) \text{ with tag } ID.a.0 \text{ such that } Q_{ID}(w_a) \text{ holds})$$

UPON RECEIVING MESSAGE $(ID, \texttt{in}, \texttt{v-propose}, w)$:

1: *verifiably authenticatedly c-broadcast* message $(\texttt{v-echo}, w)$ tagged with $ID|\texttt{vcbc}.i.0$
2: $w_j \leftarrow \perp \quad (1 \le j \le n)$
3: **wait for** $n-t$ messages $(\texttt{v-echo}, w_j)$ to be *c-delivered* with tag $ID|\texttt{vcbc}.j.0$ from distinct P_j such that $Q_{ID}(w_j)$ holds
4: $l \leftarrow 0$
5: **repeat**
6: $l \leftarrow l+1; a \leftarrow \Pi(l)$
7: **if** $w_a = \perp$ **then**
8: send the message $(ID, \texttt{v-vote}, a, 0, \perp)$ to all parties
9: **else**
10: let ρ be the message that completes the c-broadcast with tag $ID|\texttt{vcbc}.a.0$
11: send the message $(ID, \texttt{v-vote}, a, 1, \rho)$ to all parties
12: $u_j \leftarrow \perp; \rho_j \leftarrow \perp \quad (1 \le j \le n)$
13: **wait for** $n-t$ messages $(ID, \texttt{v-vote}, a, u_j, \rho_j)$ from distinct P_j such that $V_{ID|a}((u_j, \rho_j))$ holds
14: **if** there is some $u_j = 1$ **then**
15: $v \leftarrow (1, \rho_j)$
16: **else**
17: $v \leftarrow (0, \perp)$
18: propose v for $ID|a$ in binary validated Byzantine agreement biased towards 1, with predicate $V_{ID|a}$
19: **wait for** the agreement protocol to decide some (b, σ) for $ID|a$
20: **until** $b = 1$
21: **if** $w_a = \perp$ **then**
22: use σ to complete the verifiable authenticated c-broadcast with tag $ID|\texttt{vcbc}.a.0$ and *c-deliver* $(ID, \texttt{v-echo}, w_a)$
23: output $(ID, \texttt{out}, \texttt{v-decide}, w_a)$
24: **halt**

Fig. 1. Protocol VBA for multi-valued validated Byzantine agreement.

that completes the verifiable broadcast of P_a's proposal to validate this vote. Otherwise, if P_i has received $n - t$ v-vote messages containing 0, vote 0; no additional information is needed. If the agreement decides 1, exit from the loop (lines 14–20).

Delivering the chosen proposal (lines 21–24): If P_i has not yet *c-delivered* the broadcast by the selected candidate, obtain the proposal from the value returned by the Byzantine agreement.

The full protocol is shown in Figure 1.

Theorem 1. *Given a protocol for biased binary validated Byzantine agreement and a protocol for verifiable authenticated consistent broadcast, Protocol* VBA *provides multi-valued validated Byzantine agreement for $n > 3t$.*

The message complexity of Protocol VBA is $O(tn^2)$ if Protocol VCBC [5] is used for verifiable consistent broadcast and the binary validated Byzantine agreement is implemented according to Section 4.2.

If all parties propose v and π that are together no longer than L bits, the communication complexity in the above case is $O(n^2(tK + L))$, assuming the length of a threshold signature and a signature share is at most K bits. For a constant fraction of corrupted parties, however, both values are cubic in n.

4.4 A Constant-Round Protocol for Multi-valued Agreement

In this section we present Protocol VBAconst, which is an improvement of the protocol in the previous section that guarantees termination within a constant expected number of rounds. The drawback of Protocol VBA above is that the adversary knows the order Π in which the parties search for an acceptable candidate, i.e., one that has broadcast a valid proposal. Although at least one third of all parties are guaranteed to be accepted, the adversary can choose the corruptions and schedule messages such that none of them is examined early in the agreement loop.

The remedy for this problem is to choose Π randomly during the protocol *after* making sure that enough parties are already committed to their votes on the candidates. This is achieved in two steps. First, one round of commitment exchanges is added before the agreement loop. Each party must commit to the votes that it will cast by broadcasting the identities of the $n - t$ parties from which it has received valid v-echo messages (using at least authenticated consistent broadcast). Honest parties will later only accept v-vote messages that are consistent with these commitments. The second step is to determine the permutation Π using a threshold coin-tossing scheme that outputs a pseudo-random value, after enough votes are committed. Taken together, these steps ensure that the fraction of parties which are guaranteed to be accepted are distributed randomly in Π, causing termination in a constant expected number of rounds.

The details of Protocol VBAconst are described in Figure 2 as modifications to Protocol VBA.

Protocol VBAconst **for party** P_i**, tag** ID**, and validation predicate** Q_{ID}

Modify Protocol VBA for party P_i, tag ID, and validation predicate Q_{ID} as follows:

1. Initialize and distribute the shares for an $(n, t+1)$-threshold coin-tossing scheme C_1 with k''-bit outputs during system setup. Recall that this defines a pseudorandom function F. Let G be a pseudorandom generator according to Section 2.3.
2. Include the following instructions between lines 3 and 4 of Protocol VBA, before entering the agreement loop:
   1: $c_j \leftarrow \begin{cases} 1 & \text{if } w_j \neq \bot \\ 0 & \text{otherwise} \end{cases} \quad (1 \leq j \leq n)$
   2: $C \leftarrow [c_1, \ldots, c_n]$
   3: *authenticatedly c-broadcast* the message $(\texttt{v-commit}, C)$ tagged with $ID|\texttt{cbc}.i.0$
   4: $C_j \leftarrow \bot \quad (1 \leq j \leq n)$
   5: **wait for** $n - t$ messages $(\texttt{v-commit}, C_j)$ to be *c-delivered* with tag $ID|\texttt{cbc}.j.0$ such that at least $n - t$ entries in C_j are 1
   6: generate a *coin share* γ of the coin $ID|\texttt{vba}$ and send the message $(ID, \texttt{v-coin}, \gamma)$ to all parties
   7: **wait for** $t + 1$ v-coin messages containing shares of the coin $ID|\texttt{vba}$ and combine these to get the value $S = F(ID|\texttt{vba}) \in \{0,1\}^{k''}$
   8: choose a random permutation Π, using the pseudorandom generator G with seed S.
3. Modify the condition for accepting v-vote messages (line 13) inside the agreement loop such that $(\texttt{v-vote}, a, 0, \bot)$ from P_j is accepted only if C_j is known and $C_j[a] = 0$. (This involves also waiting for additional messages $(\texttt{v-commit}, C_j)$ to be *c-delivered* as above.)

Fig. 2. Protocol VBAconst for multi-valued validated Byzantine agreement.

Theorem 2. *Given a protocol for biased binary validated Byzantine agreement and a protocol for verifiable consistent broadcast, Protocol* VBAconst *provides multi-valued validated Byzantine agreement for* $n > 3t$ *and invokes a constant expected number of binary Byzantine agreement sub-protocols.*

The expected message complexity of Protocol VBAconst is $O(n^2)$ if Protocol VCBC [5] is used for consistent verifiable broadcast and the binary validated Byzantine agreement is implemented according to Section 4.2.

If all parties propose v and π that are together no longer than L bits, the expected communication complexity in the above case is $O(n^3 + n^2(K + L))$, assuming a digital signature is K bits. The n^3-term, which results from broadcasting the commitments, has actually a very small hidden constant because the commitments can be represented as bit vectors.

5 Atomic Broadcast

Atomic broadcast guarantees a total order on messages such that honest parties deliver all messages with a common tag in the same order. It is well known that protocols for atomic broadcast are considerably more expensive than those for reliable broadcast because even in the crash-fault model, atomic broadcast is equivalent to consensus [8] and cannot be solved by deterministic protocols. The atomic broadcast protocol given here builds directly on multi-valued validated Byzantine agreement from the last section.

5.1 Definition

Atomic broadcast ensures that all messages broadcast with the same tag ID are delivered in the same order by honest parties; in this way, ID can be interpreted as the name of a broadcast "channel." The total order of atomic broadcast yields an implicit labeling of all messages.

An atomic broadcast is activated when the adversary delivers an input message to P_i of the form $(ID, \texttt{in}, \texttt{a-broadcast}, m)$, where $m \in \{0,1\}^*$. When this occurs, we say P_i *atomically broadcasts m with tag ID*. "Activation" here refers only to the broadcast of a particular payload message; the broadcast channel ID must be opened before the first such request.

A party terminates an atomic broadcast of a particular payload by generating an output message of the form $(ID, \texttt{out}, \texttt{a-deliver}, m)$. In this case, we say P_i *atomically delivers m with tag ID*. To distinguish atomic broadcast from other forms of broadcast, we will also use the terms *a-broadcast* and *a-deliver*.

The acknowledgement mechanism needed for composition of atomic broadcast with other protocols is omitted from this extended abstract.

Again, the adversary must not request an *a-broadcast* of the same payload message from any particular party more than once for each ID (however, several parties may *a-broadcast* the same message).

Atomic broadcast protocols should be *fair* so that a payload message m is scheduled and delivered within a reasonable (polynomial) number of steps after it is *a-broadcast* by an honest party. But since the adversary may delay the sender arbitrarily and *a-deliver* an a priori unbounded number of messages among the remaining honest parties, we can only provide such a guarantee when at least $t+1$ honest parties become "aware" of m. Our definitions of validity and of fairness require actually that only after $t+1$ honest parties have *a-broadcast* some payload, it will be delivered within a reasonable number of steps. This is also the reason for allowing multiple parties to *a-broadcast* the same payload message—a client application might be able to satisfy this precondition through external means and achieve guaranteed fair delivery in this way. Fairness can be interpreted as a termination condition for the broadcast of a particular payload m.

The *efficiency* condition (which ensures fast termination) for atomic broadcast differs from the protocols discussed so far because the protocol for a particular tag cannot terminate on its own. It merely stalls if no more undelivered payload messages are in the system and must be terminated externally. Thus, we cannot define efficiency using the absolute number of protocol messages generated. Instead we measure the progress of the protocol with respect to the number of messages that are *a-delivered* by honest parties. In particular, we require that the number of associated protocol messages does not exceed the number of *a-delivered* payload messages times a polynomial factor, independent of the adversary.

We say that a protocol message is *associated* to the atomic broadcast protocol with tag ID if and only if the message is generated by an honest party and tagged with ID or with a tag $ID|\ldots$ starting with ID. In particular, this encompasses all messages of the atomic broadcast protocol with tag ID generated by honest parties and all messages associated to basic broadcast and Byzantine agreement sub-protocols invoked by atomic broadcast.

Fairness and efficiency are defined using the number of payload messages in the "implicit queues" of honest parties. We say that a payload message m is *in the implicit queue of a party P_i (for channel ID)* if P_i has *a-broadcast* m with tag ID, but *no honest party* has *a-delivered* m tagged with ID. The *system queue* contains any message that is in the implicit queue of *some* honest party. We say that one payload message in the implicit queue of an honest party P_i is *older* than another if P_i *a-broadcast* the first message before it *a-broadcast* the second one.

When discussing implicit queues at particular points in time, we consider a sequence of events $E_1, \ldots, E_{k'''}$ during the operation of the system, where each event but the last one is either an *a-broadcast* or *a-delivery* by an honest party. The phrase "at time τ" for $1 \leq \tau \leq k'''$ refers to the point in time just *before* event E_τ occurs.

Definition 5 (Atomic Broadcast). *A protocol for* atomic broadcast *satisfies the following conditions except with negligible probability:*

Validity: *There are at most t honest parties with non-empty implicit queues for some channel ID, provided the adversary opens channel ID for all honest parties and delivers all associated messages.*

Agreement: *If some honest party has* a-delivered *m tagged with ID, then all honest parties* a-deliver *m tagged with ID, provided the adversary opens channel ID for all honest parties and delivers all associated messages.*

Total Order: *Suppose an honest party P_i has* a-delivered *$m_1, \ldots, m_s$ with tag ID, a distinct honest party P_j has* a-delivered *$m'_1, \ldots, m'_{s'}$ with tag ID, and $s \leq s'$. Then $m_l = m'_l$ for $1 \leq l \leq s$.*

Integrity: *For all ID, every honest party* a-delivers *a payload message m at most once tagged with ID. Moreover, if all parties follow the protocol, then m was previously* a-broadcast *by some party with tag ID.*

Fairness: *Fix a particular protocol instance with tag ID. Consider the system at any point in time τ_0 where there is a set $\mathcal{T}$ of $t+1$ honest parties with non-empty implicit queues, let $\mathcal{M}$ be the set consisting of the oldest payload message for each party in $\mathcal{T}$, and let S_0 denote the total number of distinct payload messages* a-delivered *by any honest party so far. Define a random variable U as follows: let U be the total number of distinct payload messages* a-delivered *by honest parties at the point in time when the first message in $\mathcal{M}$ is* a-delivered *by any honest party, or let $U = S_0$ if this never occurs. Then $U - S_0$ is uniformly bounded.*

Efficiency: *For a particular protocol instance with tag ID, let X denote its communication complexity, and let Y be the total number of distinct payload messages that have been* a-delivered *by any honest party with tag ID. Then, at any point in time, the random variable $X/(Y+1)$ is probabilistically uniformly bounded.*

5.2 A Protocol for Atomic Broadcast

Our Protocol ABC for atomic broadcast uses a secure digital signature scheme $\mathcal{S}$ and proceeds as follows. Each party maintains a FIFO queue of not yet *a-delivered* payload messages. Messages received to *a-broadcast* are appended to this queue whenever they are received. The protocol proceeds in asynchronous global rounds, where each round r consists of the following steps:

1. Send the first payload message w in the current queue to all parties, accompanied by a digital signature σ in an `a-queue` message.
2. Collect the `a-queue` messages of $n-t$ distinct parties and store them in a vector W, and propose W for validated Byzantine agreement.
3. Perform multi-valued Byzantine agreement with validation of a vector of tuples $W = [(w_1, \sigma_1), \ldots, (w_n, \sigma_n)]$ through the predicate $Q_{ID|\mathsf{abc}.r}(W)$ which is true if and only if for at least $n-t$ distinct tuples j, the string σ_j is a valid $\mathcal{S}$-signature on $(ID, \texttt{a-queue}, r, j, w_j)$ by P_j.
4. After deciding on a vector V of messages, deliver the union of all payload messages in V according to a deterministic order; proceed to the next round.

In order to ensure liveness of the protocol, there are actually two ways in which the parties move forward to the next round: when a party receives an *a-broadcast* input message (as stated above) and when a party with an empty queue receives an `a-queue` message of another party pertaining to the current round. If either of these two messages arrive and contain a yet undelivered payload message, and if the party has not yet sent its own `a-queue` message for the current round, then it enters the round by appending the payload to its queue and sending an `a-queue` message to all parties.

The detailed description of Protocol ABC is given in Figure 3. The FIFO queue q is an ordered list of values (initially empty). It is accessed using the operations *append*, *remove*, and *first*, where *append*(q, m) inserts m into q at the end, *remove*(q, m) removes m from q (if present), and *first*(q) returns the first element in q. The operation $m \in q$ tests if an element m is contained in q.

A party waiting at the beginning of a round simultaneously **waits for** an `a-broadcast` and an `a-queue` message containing some $w \notin d$ in line 2. If it receives an *a-broadcast* request, the payload m is appended to q. If only a suitable `a-queue` protocol message is received, the party makes w its own message for the round, but does not append it to q.

Theorem 3. *Given a protocol for multi-valued validated Byzantine agreement and assuming $\mathcal{S}$ is a secure signature scheme, Protocol* ABC *provides atomic broadcast for $n > 3t$.*

Protocol ABC for party P_i and tag ID

LET $Q_{ID|\mathtt{abc}.r}$ BE THE FOLLOWING PREDICATE:

$$Q_{ID|\mathtt{abc}.r}([(w_1,\sigma_1),\ldots,(w_n,\sigma_n)]) \equiv \text{for at least } n-t \text{ distinct } j,\ \sigma_j \text{ is a valid } \mathcal{S}\text{-signature by } P_j \text{ on } (ID, \mathtt{a\text{-}queue}, r, j, w_j)$$

INITIALIZATION:

$q \leftarrow []$ {FIFO queue of messages to *a-broadcast*}
$d \leftarrow \emptyset$ {set of *a-delivered* messages}
$r \leftarrow 0$ {current round}

UPON RECEIVING MESSAGE $(ID, \mathtt{in}, \mathtt{a\text{-}broadcast}, m)$:

if $m \notin d$ **and** $m \notin q$ **then**
  append(q, m)

FOREVER:

1: $w_j \leftarrow \bot; \sigma_j \leftarrow \bot \quad (1 \le j \le n)$
2: **wait for** $q \neq []$ **or** a message $(ID, \mathtt{a\text{-}queue}, r, l, w_l, \sigma_l)$ received from P_l such that $w_l \notin d$ and σ_l is a valid signature from P_l
3: **if** $q \neq []$ **then**
4:   $w \leftarrow$ *first*(q)
5: **else**
6:   $w \leftarrow w_l$
7: compute a digital signature σ on $(ID, \mathtt{a\text{-}queue}, r, i, w)$
8: send the message $(ID, \mathtt{a\text{-}queue}, r, i, w, \sigma)$ to all parties
9: **wait for** $n-t$ messages $(ID, \mathtt{a\text{-}queue}, r, j, w_j, \sigma_j)$ such that σ_j is a valid signature from P_j (including the message from P_l above)
10: $W \leftarrow [(w_1,\sigma_1),\ldots,(w_n,\sigma_n)]$
11: propose W for multi-valued validated Byzantine agreement for $ID|\mathtt{abc}.r$ with predicate $Q_{ID|\mathtt{abc}.r}$
12: **wait for** the validated Byzantine agreement protocol to decide some $V = [(v_1,\tau_1),\ldots,(v_n,\tau_n)]$ for $ID|\mathtt{abc}.r$
13: $b \leftarrow \bigcup_{j=1}^{n} v_j$
14: **for** $m \in (b \setminus d)$, in some deterministic order **do**
15:   output $(ID, \mathtt{out}, \mathtt{a\text{-}deliver}, m)$
16:   $d \leftarrow d \cup \{m\}$
17:   *remove*(q, m)
18: $r \leftarrow r+1$

Fig. 3. Protocol ABC for atomic broadcast using multi-valued validated Byzantine agreement.

The message complexity of Protocol ABC to broadcast one payload message m is dominated by the number of messages in the multi-valued validated Byzantine agreement; the extra overhead for atomic broadcast is only $O(n^2)$ messages. The same holds for the communication complexity, but the proposed values have length $O(n(|m| + K))$, assuming digital signatures of length K bits.

With Protocol VBAconst from Section 4.4, the total expected message complexity is $O(n^2)$ and the expected communication complexity is $O(n^3|m|)$ for an atomic broadcast of a single payload message.

6 Secure Causal Atomic Broadcast

Secure causal atomic broadcast is a useful protocol for building secure applications that use state machine replication in a Byzantine setting. It provides atomic broadcast, which ensures that all recipients receive the same sequence of messages, and also guarantees that the payload messages arrive in an order that maintains "input causality," a notion introduced by Reiter and Birman [17]. Informally, input causality ensures that a Byzantine adversary may not ask the system to deliver any payload message that depends in a meaningful way on a yet undelivered payload sent by an honest client. This is very useful for delivering client requests to a distributed service in applications that require the contents of a request to remain secret until the system processes it. Input causality is related to the standard causal order, which goes back to Lamport [11]; causality is a useful safety property for distributed systems with crash failures, but is actually not well defined in the Byzantine model [10].

Input causality can be achieved if the sender encrypts a message to broadcast with the public key of a threshold cryptosystem for which all parties share the decryption key [17]. The ciphertext is then broadcast using an atomic broadcast protocol; after delivering it, all parties engage in an additional round to recover the message from the ciphertext.

The definition and an implementation of secure causal atomic broadcast on top of atomic broadcast can be found in the full version [5].

References

1. M. Ben-Or and R. El-Yaniv, "Interactive consistency in constant time." Manuscript, 1991.
2. M. Ben-Or, R. Canetti, and O. Goldreich, "Asynchronous secure computation," in *Proc. 25th Annual ACM Symposium on Theory of Computing (STOC)*, 1993.
3. P. Berman and J. A. Garay, "Randomized distributed agreement revisited," in *Proc. 23th International Symposium on Fault-Tolerant Computing (FTCS-23)*, pp. 412–419, 1993.
4. G. Bracha and S. Toueg, "Asynchronous consensus and broadcast protocols," *Journal of the ACM*, vol. 32, pp. 824–840, Oct. 1985.
5. C. Cachin, K. Kursawe, F. Petzold, and V. Shoup, "Secure and efficient asynchronous broadcast protocols." Cryptology ePrint Archive, Report 2001/006, Mar. 2001. `http://eprint.iacr.org/`.

6. C. Cachin, K. Kursawe, and V. Shoup, "Random oracles in Constantinople: Practical asynchronous Byzantine agreement using cryptography," in *Proc. 19th ACM Symposium on Principles of Distributed Computing (PODC)*, pp. 123–132, 2000. Full version available from Cryptology ePrint Archive, Report 2000/034, `http://eprint.iacr.org/`.
7. R. Canetti and T. Rabin, "Fast asynchronous Byzantine agreement with optimal resilience," in *Proc. 25th Annual ACM Symposium on Theory of Computing (STOC)*, pp. 42–51, 1993. Updated version available from `http://www.research.ibm.com/security/`.
8. T. D. Chandra and S. Toueg, "Unreliable failure detectors for reliable distributed systems," *Journal of the ACM*, vol. 43, no. 2, pp. 225–267, 1996.
9. M. J. Fischer, N. A. Lynch, and M. S. Paterson, "Impossibility of distributed consensus with one faulty process," *Journal of the ACM*, vol. 32, pp. 374–382, Apr. 1985.
10. V. Hadzilacos and S. Toueg, "Fault-tolerant broadcasts and related problems," in *Distributed Systems* (S. J. Mullender, ed.), New York: ACM Press & Addison-Wesley, 1993. An expanded version appears as Technical Report TR94-1425, Department of Computer Science, Cornell University, Ithaca NY, 1994.
11. L. Lamport, "Time, clocks, and the ordering of events in a distributed system," *Communications of the ACM*, vol. 21, pp. 558–565, July 1978.
12. L. Lamport, R. Shostak, and M. Pease, "The Byzantine generals problem," *ACM Transactions on Programming Languages and Systems*, vol. 4, pp. 382–401, July 1982.
13. M. Naor, B. Pinkas, and O. Reingold, "Distributed pseudo-random functions and KDCs," in *Advances in Cryptology: EUROCRYPT '99* (J. Stern, ed.), vol. 1592 of *Lecture Notes in Computer Science*, Springer, 1999.
14. K. J. Perry, "Randomized Byzantine agreement," *IEEE Transactions on Software Engineering*, vol. 11, pp. 539–546, June 1985.
15. M. O. Rabin, "Randomized Byzantine generals," in *Proc. 24th IEEE Symposium on Foundations of Computer Science (FOCS)*, pp. 403–409, 1983.
16. M. Reiter, "Secure agreement protocols: Reliable and atomic group multicast in Rampart," in *Proc. 2nd ACM Conference on Computer and Communications Security*, 1994.
17. M. K. Reiter and K. P. Birman, "How to securely replicate services," *ACM Transactions on Programming Languages and Systems*, vol. 16, pp. 986–1009, May 1994.
18. V. Shoup, "Practical threshold signatures," in *Advances in Cryptology: EUROCRYPT 2000* (B. Preneel, ed.), vol. 1087 of *Lecture Notes in Computer Science*, pp. 207–220, Springer, 2000.
19. V. Shoup and R. Gennaro, "Securing threshold cryptosystems against chosen ciphertext attack," in *Advances in Cryptology: EUROCRYPT '98* (K. Nyberg, ed.), vol. 1403 of *Lecture Notes in Computer Science*, Springer, 1998.
20. R. Turpin and B. A. Coan, "Extending binary Byzantine agreement to multivalued Byzantine agreement," *Information Processing Letters*, vol. 18, pp. 73–76, 1984.

Soundness in the Public-Key Model

Silvio Micali and Leonid Reyzin

Laboratory for Computer Science
Massachusetts Institute of Technology
Cambridge, MA 02139
reyzin@theory.lcs.mit.edu
http://theory.lcs.mit.edu/~reyzin

Abstract. The public-key model for interactive proofs has proved to be quite effective in improving protocol efficiency [CGGM00]. We argue, however, that its soundness notion is more subtle and complex than in the classical model, and that it should be better understood to avoid designing erroneous protocols. Specifically, for the public-key model, we

- identify four *meaningful* notions of soundness;
- prove that, under *minimal* complexity assumptions, these four notions are *distinct;*
- identify *the exact soundness notions* satisfied by prior interactive protocols; and
- identify the *round complexity* of some of the new notions.

1 Introduction

THE BARE PUBLIC-KEY MODEL FOR INTERACTIVE PROOFS. A novel protocol model, which we call the *bare public-key (BPK) model,* was introduced by Canetti, Goldreich, Goldwasser and Micali in the context of resettable zero-knowledge [CGGM00]. Although introduced with a specific application in mind, the BPK model applies to interactive proofs in general, regardless of their knowledge complexity. The model simply assumes that the verifier has a public key, PK, that is registered before any interaction with the prover begins. No special protocol needs to be run to publish PK, and no authority needs to check any property of PK. It suffices for PK to be a string known to the prover, and chosen by the verifier prior to any interaction with him.

The BPK model is very simple. In fact, it is a *weaker* version of the frequently used public-key infrastructure (PKI) model, which underlies any public-key cryptosystem or digital signature scheme. In the PKI case, a secure association between a key and its owner is crucial, while in the BPK case no such association is required. The single security requirement of the BPK model is that a bounded number of keys (chosen beforehand) are "attributable" to a given user. Indeed, having a prover $\mathcal{P}$ work with an incorrect public key for a verifier $\mathcal{V}$ does not affect soundness nor resettable zero-knowledgeness; at most, it may affect completeness. (Working with an incorrect key may only occur when

J. Kilian (Ed.): CRYPTO 2001, LNCS 2139, pp. 542–565, 2001.

an active adversary is present— in which case, strictly speaking, completeness does not even apply: this fragile property only holds when all are honest.)

Despite its apparent simplicity, the BPK model is quite powerful. While resettable zero-knowledge (RZK) protocols exist both in the standard and in the BPK models [CGGM00], only in the latter case can they be constant-round, at least in a black box sense (even the weaker notion of concurrent zero knowledge [DNS98] is not black-box implementable in a constant number of rounds [CKPR01]). Indeed, the BPK model was introduced precisely to improve the round efficiency of RZK protocols.

THE PROBLEM OF SOUNDNESS IN THE BARE PUBLIC-KEY MODEL. Despite its simple mechanics, we argue that the soundness property of the bare public-key model has not been understood, and indeed is more complex than in the classical case.

In the classical model for interactive proofs, soundness can be defined quite easily: essentially, there should be no efficient malicious prover $\mathcal{P}^*$ that can convince $\mathcal{V}$ of the verity of a false statement with non-negligible probability. This simple definition suffices regardless of whether $\mathcal{P}^*$ interacts with the verifier only once, or several times in a sequential manner, or several times in a concurrent manner. The reason for this sufficiency is that, in the standard model, $\mathcal{V}$ is polynomial-time and has no "secrets" (i.e., all of its inputs are known to $\mathcal{P}^*$). Thus, if there were a $\mathcal{P}^*$ successful "against a multiplicity of verifiers," then there would also be a malicious prover successful against a single verifier $\mathcal{V}$: it would simply let $\mathcal{P}^*$ interact with $\mathcal{V}$ while "simulating all other verifiers."

In the BPK model, however, $\mathcal{V}$ has a secret key SK, corresponding to its public key PK. Thus, $\mathcal{P}^*$ could potentially gain some knowledge about SK from an interaction with $\mathcal{V}$, and this gained knowledge might help $\mathcal{P}^*$ to convince $\mathcal{V}$ of a false theorem in a subsequent interaction. Therefore,

> *in the BPK model, the soundness property may be affected by the type of interaction a malicious prover is entitled to have with the verifier, as well as the sheer number of these interactions.*

In addition, other totally new issues arise in the BPK model. For example, should $\mathcal{P}^*$ be allowed to determine the exact false statement of which it tries to convince $\mathcal{V}$ before or after it sees PK? Should $\mathcal{P}^*$ be allowed to change that statement after a few interactions with $\mathcal{V}$?

In sum, an increased use of the BPK model needs to be coupled with a better understanding of its soundness properties in order designing protocols that are unsound (and thus insecure) or "too sound" (and thus, potentially, less efficient than otherwise possible). This is indeed the process we start in this paper.

FOUR NOTIONS OF SOUNDNESS IN THE BARE PUBLIC-KEY MODEL. Having identified the above issues, we formalize four *meaningful* notions of soundness in the BPK model. (These notions correspond in spirit to the commonly used notions of zero knowledge in the standard model. That is, the ways in which a malicious prover is allowed to interact with the honest verifier correspond to those in which a malicious verifier is allowed to interact with the honest prover in various notions of zero knowledgeness.) Roughly speaking, here are the four notions, each of which implies the previous one:

1. **one-time soundness,** when $\mathcal{P}^*$ is allowed a single interaction with $\mathcal{V}$ per theorem statement;
2. **sequential soundness**, when $\mathcal{P}^*$ is allowed multiple but sequential interactions with $\mathcal{V}$;
3. **concurrent soundness,** when $\mathcal{P}^*$ is allowed multiple interleaved interactions with the same $\mathcal{V}$; and
4. **resettable soundness,** when $\mathcal{P}^*$ is allowed to reset $\mathcal{V}$ with the same random tape *and* interact with it concurrently.

All four notions are meaningful. Sequential soundness (the notion implicitly used in [CGGM00]) is certainly a very natural notion, and concurrent and resettable soundness are natural extensions of it. As for one-time soundness, it is also quite meaningful when it is possible to enforce that a prover who fails to convince the verifier of the verity of a given statement S does not get a second chance at proving S. (E.g., the verifier may memorize the theorem statements for which the prover failed; or make suitable use of timestamps.)

These four notions of soundness apply both to interactive proofs (where a malicious prover may have unlimited power [GMR89]) and argument systems (where a malicious prover is restricted to polynomial time [BCC88]).

Separating the Four Notions. We prove that the above four notions are not only meaningful, but also *distinct.* Though conceptually important, these separations are technically simple. They entail exhibiting three protocols, each satisfying one notion but not the next one; informally, we prove the following theorems.

Theorem 1. *If one-way functions exist, there is a compiler-type algorithm that, for any language L, and any interactive argument system for L satisfying one-time soundness, produces another interactive argument system for the same language L that satisfies one-time soundness but not sequential soundness.*

Theorem 2. *If one way functions exist, there is a compiler-type algorithm that, for any language L, and any argument system for L satisfying sequential soundness, produces another argument system for the same language L that satisfies sequential soundness but not concurrent soundness.*

Theorem 3. *There exists a compiler-type algorithm that, for any language L, and any interactive proof (or argument) system for L satisfying concurrent soundness, produces another interactive proof (respectively, argument) system for the same language L that satisfies concurrent soundness but not resettable soundness.*

Note that our separation theorems hold with complexity assumptions that are indeed minimal: the third theorem holds *unconditionally;* while the first and second rely only on the existence of one-way functions. (This is why Theorems 1 and 2 only hold for bounded provers).

Realizing that there exist separate notions of soundness in the BPK model is crucial to avoid errors. By relying on a single, undifferentiated, and intuitive notion of soundness, one might design a BPK protocol sound in settings where malicious provers are limited in their interactions, while someone else might erroneously use it in settings where malicious provers have greater powers.

THE EXACT SOUNDNESS OF PRIOR PROTOCOLS IN THE BPK MODEL. Having realized that there are various notions of soundness and that it is important to specify which one is satisfied by any given protocol, a natural question arises: *what type of soundness is actually enjoyed by the already existing protocols in the BPK model?*

There are right now two such protocols: the original RZK argument proposed in [CGGM00] and the 3-round RZK argument of [MR01] (the latter holding in a BPK model with a counter). Thus we provide the following answers:

1. *The CGGM protocol is sequentially sound, and probably no more than that.* That is, while it is sequentially sound, we provide evidence that it is NOT concurrently sound.
2. *The MR protocol is exactly concurrently sound.* That is, while it is concurrently sound, we prove that it is NOT resettably sound.
   (As we said, the MR protocol works in a stronger public-key model, but all our notions of soundness easily extend to this other model.)

THE ROUND COMPLEXITY OF SOUNDNESS IN THE BPK MODEL. Since we present four notions of soundness, each implying the previous one, one may conclude that only the last one should be used. However, we shall argue that achieving a stronger notion of soundness requires using more rounds. Since rounds perhaps are the most expensive resource in a protocol, our lowerbounds justify using weaker notions of soundness whenever possible.

To begin with, we adapt an older lowerbound of [GK96] to prove the following theorem.

Theorem 4. *Any (resettable or not) black-box ZK protocol satisfying concurrent soundness in the BPK model for a language L outside of BPP requires at least four rounds.*

However, whether such an RZK protocol exists remains an open problem. A consequence of the above lowerbound is that, in any application in which four rounds are deemed to be too expensive, one needs either to adopt a stronger model (e.g., the public-key model with counter of [MR01]), or to settle for 3-round protocols satisfying a weaker soundness property[1]. We thus provide such a protocol; namely, we prove the following theorem.

[1] It is easy to prove that one cannot obtain fewer rounds than three, using the theorem from [GO94] stating that, in the standard model, 2-round auxiliary-input ZK is impossible for non-trivial languages.

Theorem 5. *Assuming the security of RSA with large prime exponents against subexponentially-strong adversaries, for any $L \in$ NP, there exists a 3-round black-box RZK protocol in the BPK model that possesses one-time, but not sequential, soundness.*

Whether the BPK model allows for 3-round, sequentially sound, ZK protocols remains an open problem. It is known that four rounds suffice in the standard model for ZK protocols [FS89], and therefore also in the BPK model. However, in the following theorem we show that in the BPK model four rounds suffice even for resettable ZK.

Theorem 6. *Assuming there exist certified trapdoor permutation families*[2] *secure against subexponentially-strong adversaries, for any $L \in$ NP, there exists a 4-round black-box RZK protocol in the BPK model that possesses sequential soundness.*

2 Four Notions of Soundness

Note: For the sake of brevity, in this section we focus exclusively on arguments, rather than proofs (i.e., the malicious prover is limited to polynomial time, and soundness is computational). All the currently known examples of protocols in the BPK model are arguments anyway, because they enable a malicious prover to cheat if it can recover the secret key *SK* from the public key *PK*. Our definitions, however, can be straightforwardly modified for proofs. (Note that the BPK model does not rule out interactive proofs: in principle, one can make clever use of a verifier public key that has no secrets associated with it.)

In this section, we formally define soundness in the BPK model, namely that a malicious prover should be unable to get the verifier to accept a false statement.[3] For the sake of brevity, we focus only on soundness. The notions of completeness (which is quite intuitive) and resettable zero-knowledgeness (previously defined in [CGGM00]) are provided in Appendix A

The Players

Before providing the definitions, we need to define the parties to the game: the honest $\mathcal{P}$ and $\mathcal{V}$ and the various malicious impostors. Let

[2] A trapdoor permutation family is *certified* if it is easy to verify that a given function belongs to the family.

[3] It is possible to formalize the four notions of soundness by insisting that the verifier give zero knowledge to the (one-time, sequential, concurrent or resetting) malicious prover. This would highlight the correspondence of our soundness notions to the notions of zero-knowledge, and would be simpler to define, because the definitions of zero-knowledge are already well established. However, such an approach is an overkill, and would result in unnecessarily restrictive notions of soundness in the BPK model: we do not care if the prover gains knowledge so long as the knowledge does not allow the prover to cheat.

- A *public file* F be a polynomial-size collection of records (id, PK_{id}), where id is a string identifying a verifier, and PK_{id} is its (alleged) public key.
- An *(honest) prover* $\mathcal{P}$ *(for a language* L*)* be an interactive deterministic polynomial-time TM that is given as inputs (1) a security parameter 1^n, (2) a n-bit string $x \in L$, (3) an auxiliary input y, (4) a public file F, (5) a verifier identity id, and (6) a random tape ω.
- An *(honest) verifier* $\mathcal{V}$ be an interactive deterministic polynomial-time TM that works in two stages. In stage one (the *key-generation* stage), on input a security parameter 1^n and random tape r, $\mathcal{V}$ outputs a public key PK and the corresponding secret key SK. In stage two (the *verification* stage), on input SK, and n-bit string x and a random string ρ, $\mathcal{V}$ performs an interactive protocol with a prover, and outputs "accept x" or "reject x." For simplicity of exposition, fixing SK and ρ, one can view the verification stage of $\mathcal{V}$ as a *non-interactive* TM that is given x and the entire history of the messages already received in the interaction, and outputs the next message to be sent, or "accept x"/"reject x." This view allows one to think of $\mathcal{V}(SK, \rho)$ as a simple deterministic oracle, which is helpful in defining the notion of resettable soundness below (however, we will use the interactive view of $\mathcal{V}$ in defining one-time, sequential and concurrent soundness).
- A *s-sequential malicious prover* $\mathcal{P}^*$ for a positive polynomial s be a probabilistic polynomial-time TM that, on first input 1^n, runs in at most $s(n)$ stages, so that
  1. In stage 1, $\mathcal{P}^*$ receives a public key PK and outputs a string x_1 of length n.
  2. In every even stage, $\mathcal{P}^*$ starts in the final configuration of the previous stage and performs a single interactive protocol: it outputs outgoing messages and receives incoming messages (the machine with which it performs the interactive protocol will be specified below, in the definition of sequential soundness). It can choose to abort an even stage at any point and move on to the next stage by outputting a special message.
  3. In every odd stage $i > 1$, P^* starts in the final configuration of the previous stage and outputs a string x_i of length n.
- An *s-concurrent malicious prover* $\mathcal{P}^*$, for a positive polynomial s, be a probabilistic polynomial-time TM that, on inputs 1^n and PK, performs at most $s(n)$ interactive protocols as follows:
  1. If $\mathcal{P}^*$ is already running $i-1$ interactive protocols $1 \le i-1 < s(n)$, it can output a special message "Start x_i," where x_i is a string of length n.
  2. At any point it can output a message for any of its (at most $s(n)$) interactive protocols (the protocol is unambiguously identified in the outgoing message). It then immediately receives the party's response and continues.
- An *s-resetting malicious prover* $\mathcal{P}^*$, for a positive polynomial s, be a probabilistic polynomial-time TM that, on inputs 1^n and PK, gets access to $s(n)$ oracles for the verifier (to be precisely specified below, in the definition of resettable soundness).

The Definitions

A pair $(\mathcal{P}, \mathcal{V})$ can satisfy one or more of the four different notions of soundness defined below. We note that each subsequent notion trivially implies the previous one.

For the purposes of defining one-time and sequential soundness, we consider the following procedure for a given s-sequential malicious prover $\mathcal{P}^*$, a verifier $\mathcal{V}$ and a security parameter n.

Procedure Sequential-Attack

1. Run the key-generation stage of $\mathcal{V}$ on input 1^n and a random string r to obtain PK, SK.
2. Run first stage of $\mathcal{P}^*$ on inputs 1^n and PK to obtain an n-bit string x_1.
3. For i ranging from 1 to $s(n)/2$:
   3.1 Select a random string ρ_i.
   3.2 Run the $2i$-th stage of $\mathcal{P}^*$, letting it interact with the verification stage of $\mathcal{V}$ with input SK, x_i, ρ_i.
   3.3 Run the $(2i+1)$-th stage of $\mathcal{P}^*$ to obtain an n-bit string x_i.

Definition 1. $(\mathcal{P}, \mathcal{V})$ *satisfies* one-time soundness for a language L *if for all positive polynomials s, for all s-sequential malicious provers $\mathcal{P}^*$, the probability that in an execution of Sequential-Attack, there exists i such that* $1 \le i \le s(n)$, *$x_i \notin L$, $x_j \neq x_i$ for all $j < i$ and $\mathcal{V}$ outputs "accept x_i" is negligible in n.*

Sequential soundness differs from one-time soundness only in that the malicious prover is allowed to have $x_i = x_j$ for $i < j$.

Definition 2. $(\mathcal{P}, \mathcal{V})$ *satisfies* sequential soundness for a language L *if for all positive polynomials s, for all s-sequential malicious provers $\mathcal{P}^*$, the probability that in an execution of Sequential-Attack, there exists i such that* $1 \le i \le s(n)$, *$x_i \notin L$, and $\mathcal{V}$ outputs "accept x_i" is negligible in n.*

For the purposes of defining concurrent soundness, we consider the following procedure for a given s-concurrent malicious prover $\mathcal{P}^*$, a verifier $\mathcal{V}$ and a security parameter n.

Procedure Concurrent-Attack

1. Run the key-generation stage of $\mathcal{V}$ on input 1^n and a random string r to obtain PK, SK.
2. Run $\mathcal{P}^*$ on inputs 1^n and PK.
3. Whenever $\mathcal{P}^*$ outputs "Start x_i," select a fresh random string ρ_i and let the i-th machine with which $\mathcal{P}^*$ interacts be the verification stage of $\mathcal{V}$ on inputs SK, x_i, ρ_i.

Of course, the multiple instances of $\mathcal{V}$ are "unaware" and independent of each other, because they are started with fresh random strings.

Definition 3. $(\mathcal{P}, \mathcal{V})$ *satisfies* concurrent soundness for a language L *if for all positive polynomials s, for all s-concurrent malicious provers $\mathcal{P}^*$, the probability that in an execution of Concurrent-Attack, $\mathcal{V}$ ever outputs "accept x" for $x \notin L$ is negligible in n.*

Finally, for the purposes of defining resettable soundness, we consider the following procedure for a given s-resetting malicious prover $\mathcal{P}^*$, a verifier $\mathcal{V}$ and a security parameter n.

Procedure Resetting-Attack

1. Run the key-generation stage of $\mathcal{V}$ on input 1^n and a random string r to obtain PK, SK.
2. Run $\mathcal{P}^*$ on inputs 1^n and PK.
3. Generate $s(n)$ random strings ρ_i for $1 \leq i \leq s(n)$.
4. Let $\mathcal{P}^*$ interact with oracles for the second stage of the verifier, the i-th oracle having input SK, ρ_i.

Note that concurrent soundness and resettable soundness differ in one crucial aspect: for the former, every instance of $\mathcal{V}$ is an interactive TM that keeps state between rounds of communication, and thus cannot be rewound; whereas for the latter, every instance of $\mathcal{V}$ is just an oracle, and thus can effectively be rewound.

Definition 4. $(\mathcal{P}, \mathcal{V})$ *satisfies* resettable soundness for a language L *if for all positive polynomials s, for all s-resetting malicious provers $\mathcal{P}^*$, the probability that in an execution of Resetting-Attack, $\mathcal{P}^*$ ever receives "accept x" for $x \notin L$ from any of the oracles is negligible in n.*

3 Separating the Four Notions

The Common Idea

Given a protocol $(\mathcal{P}, \mathcal{V})$ that satisfies the i-th soundness notion (for $i = 1, 2$, or 3), we deliberately weaken the verifier to come up with a protocol $(\mathcal{P}', \mathcal{V}')$ that does not satisfy the $(i+1)$-th soundness notion, but still satisfies the i-th. In each case, we add rounds at the beginning of the $(\mathcal{P}, \mathcal{V})$ (and sometimes information to the keys) that have nothing to do with the language or the theorem being proven. At the end of these rounds, either $\mathcal{V}'$ accepts, or $(\mathcal{P}', \mathcal{V}')$ proceed with the protocol $(\mathcal{P}, \mathcal{V})$. In each case, it will be easy for a malicious prover for the $(i+1)$-th notion of soundness to get $\mathcal{V}'$ to accept at the end of these additional rounds.

To prove that the resulting protocol $(\mathcal{P}', \mathcal{V}')$ still satisfies the i-th notion of soundness, it will suffice to show that if a malicious prover $\mathcal{P}'^*$ for $(\mathcal{P}', \mathcal{V}')$ exists, then it can be used to construct a malicious prover $\mathcal{P}^*$ for $(\mathcal{P}, \mathcal{V})$. In each case, this is easily done: $\mathcal{P}^*$ simply simulates the additional rounds to $\mathcal{P}'^*$ (one also has to argue that $\mathcal{V}'$ interacting with $\mathcal{P}'^*$ is unlikely to accept during these additional rounds).

Finally, to ensure that zero-knowledgeness of $(\mathcal{P}, \mathcal{V})$ is not affected, during the additional rounds the honest $\mathcal{P}'$ will simply send some fixed values to $\mathcal{V}'$ and disregard the values sent by $\mathcal{V}'$.

Each of the subsections below described the specific additional information in the keys and the additional rounds. We do not provide the details of proofs, as they can be easily derived from the discussion above.

Proof of Theorem 1

Let F be a pseudorandom function [GGM86]; we denote by $F_s(x)$ the output of F with seed s on input x. Note that such functions exist assuming one-way functions exist [HILL99]. Let x denote the theorem that the prover is trying to prove to the verifier.

Add to Key Gen: Generate random n-bit seed s; add s to the secret key SK.

Add $\mathcal{P}$ Step: Set $\beta = 0$; send β to the verifier.
Add $\mathcal{V}$ Step: If $\beta = F_s(x)$, accept and stop. Else send $F_s(x)$ to prover.

Note that a sequential malicious prover can easily get $\mathcal{V}'$ to accept: it finds out the value of $F_s(x)$ in the first interaction, and sets $\beta = F_s(x)$ for the second. If, on the other hand, the malicious prover is not allowed to use the same x twice, then it cannot predict $F_s(x)$ before sending β, and thus cannot get $\mathcal{V}'$ to accept.

Proof of Theorem 2

Let (SigKeyGen, Sign, Ver) be a signature scheme secure against adaptive chosen message attacks [GMR88]. Note that such a scheme exists assuming one-way functions exist [Rom90].

Add to Key Gen: Generate a key pair $(SigPK, SigSK)$ for the signature scheme; add $SigPK$ to the public key PK and $SigSK$ to the secret key SK.

Add 1st $\mathcal{P}$ Step: Set $M = 0$, and send M to the verifier.
Add 1st $\mathcal{V}$ Step: 1. Send a signature s of M to the prover.
2. Let M' be random n-bit string; send M' to prover.

Add 2nd $\mathcal{P}$ Step: Set $s' = 0$. Send s' to the verifier.
Add 2nd $\mathcal{V}$ Step: If s' is a valid signature of M', then accept and stop.

Note that a concurrent malicious prover can easily get $\mathcal{V}'$ to accept. It starts a protocol with $\mathcal{V}'$, sends $M = 0$, receives M' from $\mathcal{V}$, and then pauses the protocol. During the pause, it starts a second protocol, and sends $M = M'$ to $\mathcal{V}'$ to obtain a signature s of M' in first message from $\mathcal{V}'$. It then resumes the first protocol, and sends $s' = s$ to $\mathcal{V}'$ as its second message, which $\mathcal{V}'$ accepts.

Also note that a sequential malicious prover will most likely not be able to come up with a valid signature of M', because of the signature scheme's security against adaptive chosen message attacks.

Proof of Theorem 3

Add $\mathcal{P}$ Step: Set β be the string of n zeroes; send β to the verifier.
Add $\mathcal{V}$ Step: Set α be a random string.
If $\beta = \alpha$, accept and stop. Else send α to the prover.

Note that a resetting malicious prover can easily get $\mathcal{V}'$ to accept: it finds out the value of α in the first interaction, then resets $\mathcal{V}'$ with the same random tape (and hence the same α, because α comes from $\mathcal{V}$'s random tape) and sets $\beta = \alpha$ for the second interaction. A concurrent malicious prover, on the other hand, knows nothing about α when it determines β, and thus cannot get $\mathcal{V}'$ to accept.

Note that this separation holds in the standard model as well—we never used the BPK model in this proof.

4 The "Exact" Soundness of Existing BPK Protocols

There are only two known protocols in the BPK model, the original one of [CGGM00] and the one of [MR01] (the latter actually working in a slightly stronger model). Thus we need to understand which notions of soundness they satisfy.

The CGGM Protocol Is Sequentially but Probably Not Concurrently Sound

Although [CGGM00] did not provide formal definitions of soundness in the BPK model, their soundness proof essentially shows that their protocol is sequentially sound. However, let us (sketchily) explain why it will probably not be possible to prove their protocol concurrently sound.

The CGGM protocol begins with $\mathcal{V}$ proving to $\mathcal{P}$ knowledge of the secret key by means of parallel repetitions of a three-round proof of knowledge subprotocol. The subprotocol is as follows: in the first round, $\mathcal{V}$ sends to $\mathcal{P}$ a *commitment*; in the second round, $\mathcal{P}$ sends to $\mathcal{V}$ a one-bit *challenge*; in the third round, $\mathcal{V}$ sends to $\mathcal{P}$ a *response*. This is repeated k times in parallel in order to reduce the probability of $\mathcal{V}$ cheating to roughly 2^{-k}.

In order to prove soundness against a malicious prover $\mathcal{P}^*$, these parallel repetitions of the subprotocol need to be simulated to $\mathcal{P}^*$ (by a simulator that does not know the secret key). The best known simulation techniques for this general type of proof of knowledge run in time roughly 2^k. This exponential in k simulation time is not a concern, because of their use of "complexity leveraging" in the proof of soundness. Essentially, the soundness of their protocol relies on an underlying much harder problem: for instance, one that is assumed to take more than 2^{3k} time to solve. Thus, the soundness of the CGGM protocol is proved by contradiction: by constructing a machine from $\mathcal{P}^*$ that runs in time $2^k < 2^{3k}$ and yet solves the underlying harder problem.

A concurrent malicious prover $\mathcal{P}^*$, however, may choose to run L parallel copies of $\mathcal{V}$. Thus, to prove soundness against such a $\mathcal{P}^*$, the proof-of-knowledge subprotocol would have to be simulated Lk times in parallel, and this simulation would take roughly 2^{Lk} time. If $L > 3$, then we will not be able to solve the underlying hard problem in time less than 2^{3k}, and thus will not be able to derive any contradiction.

Thus, barring the emergence of a polynomial-time simulation for parallel repetitions of 3-round proofs of knowledge (or a dramatically new proof technique for soundness), the CGGM protocol is not provably concurrently sound.

The MR Protocol Is Concurrently but Not Resettably Sound

The protocol in [MR01] extends the BPK model with a *counter*. Namely, there is an a-priori polynomial bound B that limits the total number of times the verifier executes the protocol, and the verifier maintains state information from one interaction to the next via a counter (that can be tested and incremented in a single atomic operation).

Our soundness notions easily extend to the MR model as well, and their soundness proof can be easily modified to yield that their protocol is concurrently sound in the new model. However, let us (sketchily) prove here that the MR protocol is not resettably sound.

In the MR protocol, verifier $\mathcal{V}$ publishes a public key for a trapdoor commitment scheme, and then proves knowledge of the trapdoor using non-interactive zero-knowledge proof of knowledge (NIZKPK), relative to a jointly generated string σ. It is easy to see that in the MR protocol, if $\mathcal{P}^*$ could learn $\mathcal{V}$'s trapdoor, then he could force $\mathcal{V}$ to accept a false theorem. The knowledge-extraction requirement of the NIZKPK system guarantees that, by properly selecting σ, one could extract the trapdoor from the proof. Now, a malicious resetting prover $\mathcal{P}^*$ has total control over σ. Indeed, in the MR protocol σ is the exclusive-or of two strings: $\sigma_{\mathcal{P}}$ provided by the prover in the first round, and $\sigma_{\mathcal{V}}$ provided by the verifier in the second round. Thus, $\mathcal{P}^*$ simply finds out $\sigma_{\mathcal{V}}$ by running the protocol once, then resets $\mathcal{V}$ and provides $\sigma_{\mathcal{P}}$ such that the resulting $\sigma = \sigma_{\mathcal{V}} \oplus \sigma_{\mathcal{P}}$ will equal the string that allows $\mathcal{P}^*$ to extract the trapdoor.

5 The Cost of Soundness in Zero-Knowledge Proofs

The BPK model was introduced to save rounds in RZK protocols, but has itself introduced four notions of soundness. We have already shown that these notions are formally separated. Now, we show that they also have quite different algorithmic requirements: namely, stronger notions of soundness for ZK protocols require more rounds to be implemented. More precisely, we show a lowerbound, namely that concurrently sound black-box ZK requires four or more rounds, and two upperbounds, namely that one-time-sound RZK can be achieved in three rounds (which can be shown optimal using the standard-model lowerbound of [GO94]), and that sequential RZK can be achieved in four rounds.

Note that our lowerbound in the BPK model is not contradicted by the existence of the 3-round concurrently-sound protocol of [MR01], which is in a stronger model, where the verifier has a counter.

We derive our lowerbound in the BPK model, where there are different notions of soundness, from the older one of Goldreich and Krawczyk [GK96] for black-box ZK in the *standard* model, where one-time, sequential and concurrent soundness coincide. Thus, somehow, their proof can be extended to verifiers that have public and secret keys, though (as clear from our upperbound) *this extension fails to apply to some types of soundness.* This point is important to understanding soundness in the BPK model, and we'll try to highlight it when sketching the lowerbound proof below.

Our bounds are not tight: we do not know whether 4-round concurrently sound RZK protocols exist, nor whether 3-round sequentially sound ZK protocols exist. Before our work, however, the gap was even wider: the CGGM — sequentially sound— RZK protocol had 8 rounds without preprocessing, though it could be easily reduced to 5 rounds.

5.1 No Concurrent Soundness for Black-Box ZK in Three Rounds

Theorem 4 *Any (resettable or not) black-box ZK protocol satisfying concurrent soundness in the BPK model for a language L outside of BPP requires at least four rounds.*

Proof Sketch. The Goldreich and Krawczyk's proof that, for languages outside of BPP, there are no three-round protocols that are black-box zero-knowledge in the standard model, proceeds by contradiction. Assuming the existence of a black-box zero-knowledge simulator M, it constructs a BPP machine $\bar{M}$ for L. Recall that M interacts with a verifier in order to output the verifier's view. On input x, $\bar{M}$ works essentially as follows: it simply runs M on input x, simulating a verifier to it. For this simulation, $\bar{M}$ uses the algorithm of the honest verifier $\mathcal{V}$ and the messages supplied by M, but ignores the random strings supplied by M and uses its own random strings (if the same message is given twice by M, then $\bar{M}$ uses the same random string—thus making the verifier appear deterministic to M). If the view that M outputs at the end is accepting, then $\bar{M}$ concludes that $x \in L$. Otherwise, it concludes that $x \notin L$.

To show that $\bar{M}$ is a BPP machine for L, Goldreich and Krawczyk demonstrate two statements: that if $x \in L$, M is likely to output an accepting conversation, and that if $x \notin L$, M is unlikely to output an accepting conversation. The first statement follows because, by zero-knowledgeness, M's output is indistinguishable from the view generated by the true prover and the true verifier on input x, and, by completeness, this view is accepting. The second statement follows from soundness: if M can output an accepting conversation for $x \notin L$, then one can construct a malicious prover $\mathcal{P}^*$ that can convince $\mathcal{V}$ of the false statement "$x \in L$." Such a $\mathcal{P}^*$ needs in essence to "execute M" and simply let it interact with $\mathcal{V}$.

Having $\mathcal{P}^*$ execute M requires some care. At first glance, because simulator M is capable of resetting the verifier, it would seem that, in order to execute M, also $\mathcal{P}^*$ should have this capability. However, for 3-round protocols only, [GK96] show that

(∗) $\mathcal{P}^*$ can execute M without resetting $\mathcal{V}$, so long as it has one-time access to $\mathcal{V}$.

Notice that by the term "one-time access" we make retroactive use of our modern terminology: [GK96] make no mention of one-time provers, because they work in the standard model. However, this terminology allows us to separate their proof of $(*)$ into two distinct steps:

$(*')$ $\mathcal{P}^*$ can execute M so long as it has concurrent access to $\mathcal{V}$; and
$(*'')$ losing only a polynomial amount of efficiency, concurrent access to $\mathcal{V}$ is equivalent to one-time access.

Tedious but straightforward analysis shows that $(*')$ and the rest of their proof — except for $(*'')$— carries through in the BPK model (where the 3-round protocol is modified to include verifier key generation, and public and secret verifier keys are then involved). Step $(*'')$, however, only holds in the standard model (where, as we pointed out, one-time, sequential and concurrent soundness coincide).

In sum, therefore, once verifier keys are introduced, one is left with a concurrent prover. □

5.2 One-Time Sound RZK in Three Rounds

Theorem 5 *Assuming the security of RSA with large prime exponents against subexponentially-strong adversaries, for any $L \in$ NP, there exists a 3-round black-box RZK protocol in the BPK model that possesses one-time, but not sequential, soundness.*

Proof Sketch. The proof of the theorem is constructive: we demonstrate such a protocol $(\mathcal{P}, \mathcal{V})$.

BASIC TOOLS. The protocol $(\mathcal{P}, \mathcal{V})$ relies on three techniques: a pseudorandom function PRF [GGM86], a verifiable random functions VRF [MRV99], and a non-interactive zero-knowledge (NIZK) proof system (NIP, NIV) [BFM88,BDMP91]. Note that both PRFs and NIZKs can be constructed using general assumptions [HILL99,FLS99], and it is only for VRFs that we need the specific RSA assumption (which is formally stated in Appendix B.3).

The definitions of NIZKs and VRFs are recalled recalled in Appendix B. Here we briefly introduce the notation:

- The keys *VRFPK*, *VRFSK* for VRF are produced by VRFGen. The evaluation is performed by VRFEval, and the proof is computed by VRFProve. The verification is performed by VRFVer.
- The NIZK proof with security parameter n requires a shared random string σ of length NIσLen(n). The proof is computed by NIP and verified by NIV. The shared string and the proof together can by simulated by NIS.

The construction works for any language L for which an NIZK proof system exists, and, therefore, for all of NP.

This construction also uses "complexity leveraging" [CGGM00], although in a somewhat unusual way. Namely, let α be the pseudorandomness constant for VRF (that is, the output of the VRFEval is indistinguishable from random for circuits of size 2^{k^α}, where k is VRF the security parameter). Let γ_1 be the following constant: for all sufficiently large n, the length of the NIZK proof Π for $x \in L$ of length n is upper bounded by n^{γ_1}. Let γ_2 be the following constant: for all sufficiently large n, the length of the NP-witness y for $x \in L$ of length n is upper bounded by n^{γ_2}. We then set $\gamma = \max(\gamma_1, \gamma_2)$, and $\epsilon > \gamma/\alpha$. We use NIZK with security parameter n and VRF with a (larger) security parameter $k = n^\epsilon$. This ensures that one can enumerate all potential NIZK proofs Π, or all potential NP-witnesses y, in time 2^{n^γ}, which is less than the time it would take to break the residual pseudorandomness of VRF (because $2^{n^\gamma} < 2^{k^\alpha}$).

THE PROTOCOL. For a security parameter n, $\mathcal{V}$ generates a key pair for the VRF with output length NIσLen(n) and security parameter $k = n^\epsilon$. *VRFSK* is $\mathcal{V}$'s secret key, and *VRFPK* is $\mathcal{V}$'s public key.

Public File: A collection F of records $(id, VRFPK_{id})$, where $VRFPK_{id}$ is allegedly the output of VRFGen(1^k)

Common Input: An element $x \in L$

$\mathcal{P}$ Private Input: The NP-witness y for $x \in L$; $\mathcal{V}$'s id and the file F; a random string ω

$\mathcal{V}$ Private Input: A secret key *SK*

$\mathcal{P}$ Step One:
1. Using the string ω as a seed for PRF, generate a string σ_P of length NIσLen(n) from the inputs x, y and id.
2. Send $\sigma_\mathcal{P}$ to $\mathcal{V}$.

$\mathcal{V}$ Step One:
1. Compute a string $\sigma_\mathcal{V}$ of length NIσLen(n) as $\sigma_\mathcal{V} = \text{VRFEval}(\mathit{VRFSK}, x)$, and the VRF proof $\mathit{pf} = \text{VRFProve}(\mathit{VRFSK}, x)$.
2. Send $\sigma_\mathcal{P}$ and *pf* to $\mathcal{P}$.

$\mathcal{P}$ Step Two:
1. Verify that $\sigma_\mathcal{V}$ is correct by invoking $\text{VRFVer}(\mathit{VRFPK}, x, \tau, \mathit{pf})$. If not, abort.
2. Let $\sigma = \sigma_\mathcal{V} \oplus \sigma_\mathcal{P}$. Using NIP$(\sigma, x, y)$, compute and send to $\mathcal{V}$ the proof Π of the statement "$x \in L$."

$\mathcal{V}$ Step Two:
1. Let $\sigma = \sigma_\mathcal{V} \oplus \sigma_\mathcal{P}$. Using NIV$(\sigma, x, \Pi)$, verify if Π is valid If so, accept. Else reject.

As far as we know, the above protocol is the first application of VRFs. The very strong properties of this new tool yield surprisingly simple proofs of one-time soundness and resettable zero-knowledgeness.

COMPLETENESS AND RZK. As usual, completeness of our protocol is easily verified. The RZK property can be shown in a way similar to (and simpler than) the way is shown in [CGGM00] and [MR01]. One simply builds an RZK simulator who finds out VRFEval(*VRFSK*, x) for every pair (*VRFPK*, x) that $\mathcal{V}^*$ is likely to input to $\mathcal{P}$, and then rewinds and uses the NIZK simulator NIS(x) just like the sequential malicious prover described above.

SOUNDNESS. First of all, note that soundness of our protocol is provably not sequential, because $\sigma_{\mathcal{V}}$ depends only on the input x, and hence will repeat if $\mathcal{V}$ is run with the same x twice. Thus, once a sequential malicious prover $\mathcal{P}^*$ knows $\sigma_{\mathcal{V}}$, it can run the NIZK simulator NIS(x) to obtain (σ', Π'), restart with the same x, and use $\sigma'_{\mathcal{P}} = \sigma' \oplus \sigma_{\mathcal{V}}$ as its first message and Π' as its second message.

To show one-time soundness, first assume (for simplicity) that $\mathcal{P}^*$ interacts with $\mathcal{V}$ only once (we will deal with the general case later). Then we will construct a machine $T = (T_J, T_E)$ to break the residual pseudorandomness of the VRF (see the definition of VRF in Appendix B). Namely, given the public key *VRFPK* of a VRF with security parameter k, T_J runs the first stage of $\mathcal{P}^*$ on input *VRFPK* to receive a string x. It then checks if $x \in L$ by simply enumerating all potential NP witnesses y in time $2^{n^{\gamma_2}}$. If it is, then T_J outputs (x, state), where $\mathit{state} = 0$. Otherwise, it runs the second stage of $\mathcal{P}^*$ to receive $\sigma_{\mathcal{P}}$, and outputs (x, state), where $\mathit{state} = (x, \sigma_{\mathcal{P}})$.

Now, T_E receives v, and T_E's job is to find out whether v is a random string or VRFEval(*VRFSK*, x). If $\mathit{state} = 0$, then T_E simply guesses at random. Otherwise, $\mathit{state} = (x, \sigma_{\mathcal{P}})$. Let $\sigma = \sigma_{\mathcal{P}} \oplus v$. If v is a random string, then σ is also random, so most likely there is no NIZK proof Π of the statement "$x \in L$" with respect to σ (by soundness of the NIZK proof system). Otherwise, $v = \sigma_{\mathcal{V}}$, so, if $\mathcal{P}^*$ has a better than negligible probability of success, then there is a better than negligible probability that Π exists with respect to σ. Thus, T_E simply searches whether a proof Π exists (in time $2^{n^{\gamma_1}}$) to determine whether v is random or the output of VRFEval.

Complexity leveraging is crucial here: we are using the fact that the VRF is "stronger" than the non-interactive proof system. Otherwise, the output of VRFProve (which the prover gets, but T does not) could help a malicious prover find Π. By using a stronger VRF, we are ensuring that such Π will most likely not even exist.

Now we address the general case, when $\mathcal{P}^*$ is allowed $s(n)$ sequential interactions with $\mathcal{V}$, and wins if $\mathcal{V}$ accepts at least one of them (say, the i-th one) for $x_i \notin L$. Then T_J simply guesses, at random, the conversation number i for which $\mathcal{P}^*$ will succeed, and simulates conversations before the i-th one by querying VRFEval and VRFProve on x_j for $j < i$ (it is allowed to do so, because, in one-time soundness, $x_j \neq x_i$). □

5.3 Sequentially Sound RZK in Four Rounds

Theorem 6 *Assuming there exist certified trapdoor permutation families*[4] *secure against subexponentially-strong adversaries, for any* $L \in$ NP, *there exists*

[4] A trapdoor permutation family is *certified* if it is easy to verify that a given function belongs to the family.

a 4-round black-box RZK protocol in the BPK model that possesses sequential soundness.

Proof Sketch. The proof is, again, constructive. The construction is a modification of the CGGM protocol (which has 8 rounds, and can easily be modified to have 5 by combining the first three rounds with later rounds).

MAIN IDEAS. The CGGM protocol starts with a three-round proof of knowledge subprotocol in which $\mathcal{V}$ proves to $\mathcal{P}$ knowledge of the secret key. After that, $\mathcal{P}$ proves to $\mathcal{V}$ that a graph is three-colorable using a five-round protocol.

Our main idea is to replace the five-round protocol with a single round using non-interactive zero-knowledge. The first three rounds are then used both for the proof of knowledge and for agreeing on a shared random auxiliary string σ needed for the NIZK proof. To agree on σ, $\mathcal{V}$ sends to $\mathcal{P}$ an encryption of a random string $\sigma_{\mathcal{V}}$, $\mathcal{P}$ sends to $\mathcal{V}$ its own random string $\sigma_{\mathcal{P}}$, and then $\mathcal{V}$ reveals $\sigma_{\mathcal{V}}$ (and the coins used to encrypt it). The string σ is computed as $\sigma_{\mathcal{P}} \oplus \sigma_{\mathcal{V}}$.

Thus, $\mathcal{V}$'s key pair is simply a key pair for an encryption scheme. The protocol is zero-knowledge essentially for the same reasons that the CGGM protocol is zero-knowledge: because the simulator can extract the decryption key from the proof of knowledge and thus find out $\sigma_{\mathcal{V}}$ before needing to submit $\sigma_{\mathcal{P}}$. This will allow it to select σ as it wishes and thus use the NIZK simulator.

The protocol is sequentially sound because if the theorem is false, then with respect to only a negligible portion of the possible strings σ does a NIZK proof of the theorem exist. Thus, if a malicious prover $\mathcal{P}^*$, after seeing only an encryption of $\sigma_{\mathcal{V}}$, is able to come up with $\sigma_{\mathcal{P}}$ such that the NIZK proof exists with respect to the resulting string $\sigma = \sigma_{\mathcal{P}} \oplus \sigma_{\mathcal{V}}$, then one can use $\mathcal{P}^*$ to break the security of the encryption scheme.

The computational assumption for our protocol follows from the fact that trapdoor permutations are sufficient for encryption [GM84,Yao82,GL89], certified trapdoor permutations are sufficient for NIZKs [FLS99], one-way permutations are sufficient for the proof of knowledge [Blu86] (which is the same as in the CGGM protocol) and one-way functions are sufficient for PRFs [HILL99].

DETAILS OF THE CONSTRUCTION. This construction, like the previous one, works for any languages L for which an NIZK proof system exists. Hence it works for all $L \in NP$.

The protocol below relies on parallel executions of three-round proofs of knowledge, which are performed in exactly the same way as in [CGGM00]. We also use "complexity leveraging," in a way similar to our three-round one-time-sound construction. Namely, let α be the indistinguishability constant for the encryption scheme (that is, the encryptions of two different strings are indistinguishable from each other for circuits of size 2^{k^α}, where k is the security parameter). Let γ_1 be the following constant: for all sufficiently large n, the length of the NIZK proof Π for x of length n is upper bounded by n^{γ_1}. Let γ_2 be following constant: n parallel repetitions of the proof-of-knowledge subprotocol can be simulated in time less that $2^{n^{\gamma_2}}$. We then set $\gamma = \max(\gamma_1, \gamma_2)$, and $\epsilon > \gamma/\alpha$.

We use NIZK with security parameter n and perform n parallel repetitions of the proof-of-knowledge subprotocol, while the encryption scheme has a (larger) security parameter $k = n^\epsilon$. This ensures that one can enumerate all potential NIZK proofs Π and simulate the proof of knowledge subprotocol in time 2^{n^γ}, which is less than the time it would take to break the indistinguishability of the encryption scheme (because $2^{n^\gamma} < 2^{k^\alpha}$).

THE PROTOCOL. For a security parameter n, the verifier $\mathcal{V}$ generates a pair $(EncPK, EncSK)$ of keys for the encryption scheme with security parameter $k = n^\epsilon$. $EncSK$ is $\mathcal{V}$'s secret key, and $EncPK$ is $\mathcal{V}$'s public key.

Public File: A collection F of records $(id, EncPK_{id})$, where $EncPK_{id}$ is allegedly the output of $\mathcal{V}$'s key generation
Common Inputs: An element $x \in L$
$\mathcal{P}$ Private Input: The NP-witness y for $x \in L$; $\mathcal{V}$'s id and the file F; a random string ω
$\mathcal{V}$ Private Input: A secret key $EncSK$; a random string ρ

$\mathcal{V}$ Step One:
1. Generate a random string $\sigma_\mathcal{V}$ of length $\mathrm{NI}\sigma\mathrm{Len}(n)$.
2. Encrypt $\sigma_\mathcal{V}$, using a portion ρ_E of the input random string ρ, to get a ciphertext c. Send c to $\mathcal{P}$.
3. Generate and send to $\mathcal{P}$ the first message of the n parallel repetitions of the proof of knowledge of $EncSK$.

$\mathcal{P}$ Step One:
1. Using the input random string ω as a seed for PRF, generate a sufficiently long "random" string from the input to be used in the remaining computation by $\mathcal{P}$.
2. Generate and send to $\mathcal{V}$ random string $\sigma_\mathcal{P}$ of length $\mathrm{NI}\sigma\mathrm{Len}(n)$.
3. Generate and send to $\mathcal{V}$ the second message of the n parallel repetitions of the proof of knowledge of $EncSK$.

$\mathcal{V}$ Step Two:
1. Send $\sigma_\mathcal{V}$ and the coins ρ_E used to encrypt it to $\mathcal{P}$.
2. Generate and send the third message of the n parallel repetitions of the proof of knowledge of $EncSK$.

$\mathcal{P}$ Step Two:
1. Verify that $\sigma_\mathcal{V}$ encrypted with coins ρ_E produces ciphertext c.
2. Verify the n parallel repetitions proof of knowledge of $EncSK$.
3. If both verifications hold, let $\sigma = \sigma_\mathcal{V} \oplus \sigma_\mathcal{P}$. Using the NIZK prover $\mathrm{NIP}(\sigma, x, y)$, compute and send to $\mathcal{V}$ the proof Π of the statement "$x \in L$."

$\mathcal{V}$ Step Three: Let $\sigma = \sigma_\mathcal{V} \oplus \sigma_\mathcal{P}$. Using the NIZK verifier $\mathrm{NIV}(\sigma, x, \Pi)$, verify if Π is valid. If so, accept. Else reject.

COMPLETENESS AND RZK. Completeness of this protocol is, as usual, easily verified. The proof of resettable zero-knowledgeness is very similar to that of

[CGGM00]: once the simulator recovers SK from the proof of knowledge, it can find out $\sigma_{\mathcal{V}}$ before having to send $\sigma_{\mathcal{P}}$, and thus can run the NIZK simulator to get (σ, Π) and set $\sigma_{\mathcal{P}} = \sigma \oplus \sigma_{\mathcal{V}}$.

SOUNDNESS. Sequential soundness can be shown as follows. Suppose $\mathcal{P}^*$ is a malicious prover that can make $\mathcal{V}$ accept a false theorem with probability $p(n)$ (where the probability is taken over the coin tosses of the $\mathcal{V}$ and $\mathcal{P}^*$). First, assume (for simplicity) that $\mathcal{P}^*$ interacts with $\mathcal{V}$ only once (we will deal with the general case of a sequential malicious prover later).

We will use $\mathcal{P}^*$ to construct an algorithm A that breaks the encryption scheme. A is given, as input, the public key PK for the encryption scheme. Its job is to pick two strings τ_0 and τ_1, receive an encryption of τ_b for a random bit b and tell whether $b = 0$ or $b = 1$. It picks τ_0 and τ_1 simply as random strings of length $\mathrm{NI}\sigma\mathrm{Len}(n)$. Let c be the encryption of τ_b. Then A publishes PK as its public key, runs the first stage of $\mathcal{P}^*$ to receive x, and initiates a protocol with the second stage of $\mathcal{P}^*$.

In the first message, A sends c for the encryption of $\sigma_{\mathcal{V}}$ (for the proof-of-knowledge subprotocol, A uses the simulator, which runs in time $2^{n^{\gamma_2}}$). It then receives $\sigma_{\mathcal{P}}$ from $\mathcal{P}^*$, computes $\sigma_i = \sigma_{\mathcal{P}} \oplus \tau_i$ and determines (by exhaustive search, which takes time $2^{n^{\gamma_1}}$) if there exists an NIZK proof Π_i for the statement $x \in L$ with respect to σ_i (for $i = 0, 1$). If Π_i exists and Π_{1-i} does not, then A outputs $b = i$. If neither Π_0 nor Π_1 exists, or if both exist, then A outputs a random guess for b.

We now need to compute the probability that A correctly guessed b. Of course, by construction,

$$\begin{aligned}\Pr[A \text{ outputs } b] = \Pr[\exists \Pi_b \text{ and } \nexists \Pi_{1-b}] + \Pr[\exists \Pi_b \text{ and } \exists \Pi_{1-b}]/2 + \\ \Pr[\nexists \Pi_b \text{ and } \nexists \Pi_{1-b}]/2 .\end{aligned}$$

Note that $\Pr[\exists \Pi_b \text{ and } \exists \Pi_{1-b}] + \Pr[\nexists \Pi_b \text{ and } \nexists \Pi_{1-b}] = 1 - (\Pr[\exists \Pi_b \text{ and } \nexists \Pi_{1-b}] + \Pr[\nexists \Pi_b \text{ and } \exists \Pi_{1-b}])$. Therefore,

$$\Pr[A \text{ outputs } b] = 1/2 - \Pr[\nexists \Pi_b \text{ and } \exists \Pi_{1-b}]/2 + \Pr[\exists \Pi_b \text{ and } \nexists \Pi_{1-b}]/2 .$$

Note that the either of the events $\nexists \Pi_b$ and $\nexists \Pi_{1-b}$ can occur only if $x \notin L$, by completeness of the NIZK system. Therefore,

$$\begin{aligned}\Pr[A \text{ outputs } b] &= 1/2 - \Pr[\nexists \Pi_b \text{ and } \exists \Pi_{1-b} \text{ and } x \notin L]/2 + \\ &\quad \Pr[\exists \Pi_b \text{ and } \nexists \Pi_{1-b} \text{ and } x \notin L]/2 \\ &= 1/2 - \Pr[\nexists \Pi_b \text{ and } \exists \Pi_{1-b} \text{ and } x \notin L]/2 \\ &\quad + \Pr[\exists \Pi_b \text{ and } x \notin L]/2 - \Pr[\exists \Pi_b \text{ and } \exists \Pi_{1-b} \text{ and } x \notin L]/2 \\ &\geq 1/2 + p(n)/2 - \Pr[\exists \Pi_{1-b} \text{ and } x \notin L] .\end{aligned}$$

However, τ_{1-b} is picked uniformly at random and $\mathcal{P}^*$ receives no information about it, so the string $\sigma_{1-b} = \sigma_{\mathcal{P}} \oplus \tau_{1-b}$ is distributed uniformly at random, so, by soundness of NIZK, $\Pr[\exists \Pi_{1-b} \text{ and } x \notin L]$ is negligible in n. Thus, A's advantage is only negligibly less than $p(n)/2$.

Now we address the case of sequential malicious provers. Suppose $\mathcal{P}^*$ is an s-sequential malicious prover. Then $\mathcal{P}^*$ initiates at most $s(n)$ *sequential* conversations and wins if $\mathcal{V}$ accepts at least one of them for $x \notin L$. Then A simply guesses, at random, the conversation for which $\mathcal{P}^*$ will succeed, and simply simulates the other conversations by using the simulator for the proof of knowledge and honestly encrypting random strings. Only for that conversation does it use the procedure described above. This reduces A's advantage by a polynomial factor of at most $s(n)$. □

References

[BCC88] Gilles Brassard, David Chaum, and Claude Crépeau. Minimum disclosure proofs of knowledge. *Journal of Computer and System Sciences*, 37(2):156–189, October 1988.

[BDMP91] Manuel Blum, Alfredo De Santis, Silvio Micali, and Giuseppe Persiano. Noninteractive zero-knowledge. *SIAM Journal on Computing*, 20(6):1084–1118, December 1991.

[BFM88] Manuel Blum, Paul Feldman, and Silvio Micali. Non-interactive zero-knowledge and its applications (extended abstract). In *Proceedings of the Twentieth Annual ACM Symposium on Theory of Computing*, pages 103–112, Chicago, Illinois, 2–4 May 1988.

[Blu86] Manuel Blum. How to prove a theorem so no one else can claim it. In *Proc. of the International Congress of Mathematicians, Berkeley, CA*, pages 1444–1451, 1986.

[CGGM00] Ran Canetti, Oded Goldreich, Shafi Goldwasser, and Silvio Micali. Resettable zero-knowledge. In *Proceedings of the Thirty-Second Annual ACM Symposium on Theory of Computing*, Portland, Oregon, 21–23 May 2000. Updated version available at the Cryptology ePrint Archive, record 1999/022, `http://eprint.iacr.org/`.

[CKPR01] Ran Canetti, Joe Kilian, Erez Petrank, and Alon Rosen. Black-box concurrent zero-knowledge requires $\tilde{\Omega}(\log n)$ rounds. In *Proceedings of the Thirty-Second Annual ACM Symposium on Theory of Computing*, Crete, Greece, 6–8 July 2001.

[DNS98] Cynthia Dwork, Moni Naor, and Amit Sahai. Concurrent zero knowledge. In *Proceedings of the Thirtieth Annual ACM Symposium on Theory of Computing*, pages 409–418, Dallas, Texas, 23–26 May 1998.

[FLS99] Uriel Feige, Dror Lapidot, and Adi Shamir. Multiple non-interactive zero knowledge proofs under general assumptions. *SIAM Journal on Computing*, 29(1):1–28, 1999.

[FS89] Uriel Feige and Adi Shamir. Zero knowledge proofs of knowledge in two rounds. In G. Brassard, editor, *Advances in Cryptology—CRYPTO '89*, volume 435 of *Lecture Notes in Computer Science*, pages 526–545. Springer-Verlag, 1990, 20–24 August 1989.

[GGM86] Oded Goldreich, Shafi Goldwasser, and Silvio Micali. How to construct random functions. *Journal of the ACM*, 33(4):792–807, October 1986.

[GK96] Oded Goldreich and Hugo Krawczyk. On the composition of zero-knowledge proof systems. *SIAM Journal on Computing*, 25(1):169–192, February 1996.

[GL89] O. Goldreich and L. Levin. A hard-core predicate for all one-way functions. In *Proceedings of the Twenty First Annual ACM Symposium on Theory of Computing*, pages 25–32, Seattle, Washington, 15–17 May 1989.

[GM84] S. Goldwasser and S. Micali. Probabilistic encryption. *Journal of Computer and System Sciences*, 28(2):270–299, April 1984.

[GMR88] Shafi Goldwasser, Silvio Micali, and Ronald L. Rivest. A digital signature scheme secure against adaptive chosen-message attacks. *SIAM Journal on Computing*, 17(2):281–308, April 1988.

[GMR89] Shafi Goldwasser, Silvio Micali, and Charles Rackoff. The knowledge complexity of interactive proof systems. *SIAM Journal on Computing*, 18:186–208, 1989.

[GO94] Oded Goldreich and Yair Oren. Definitions and properties of zero-knowledge proof systems. *Journal of Cryptology*, 7(1):1–32, 1994.

[HILL99] J. Håstad, R. Impagliazzo, L.A. Levin, and M. Luby. Construction of pseudorandom generator from any one-way function. *SIAM Journal on Computing*, 28(4):1364–1396, 1999.

[MR01] Silvio Micali and Leonid Reyzin. Min-round resettable zero knowledge in the public-key model. In Birgit Pfitzmann, editor, *Advances in Cryptology—EUROCRYPT 2001*, volume 2045 of *Lecture Notes in Computer Science*, pages 373–393. Springer-Verlag, 6–10 May 2001.

[MRV99] Silvio Micali, Michael Rabin, and Salil Vadhan. Verifiable random functions. In *40th Annual Symposium on Foundations of Computer Science*, pages 120–130, New York, October 1999. IEEE.

[Rom90] John Rompel. One-way functions are necessary and sufficient for secure signatures. In *Proceedings of the Twenty Second Annual ACM Symposium on Theory of Computing*, pages 387–394, Baltimore, Maryland, 14–16 May 1990.

[Yao82] A. C. Yao. Theory and application of trapdoor functions. In *23rd Annual Symposium on Foundations of Computer Science*, pages 80–91, Chicago, Illinois, 3–5 November 1982. IEEE.

A Definitions of Completeness and RZK

Completeness for a pair $(\mathcal{P}, \mathcal{V})$ is defined the usual way. Consider the following procedure for $(\mathcal{P}, \mathcal{V})$, a string $x \in L$ of length n and a string y.

Procedure Normal-Interaction

1. Run the key-generation stage of $\mathcal{V}$ on input 1^n and a random string r to obtain PK, SK.
2. Pick any id, and let F be a public file that contains the record (id, PK).
3. Pick strings ω and ρ at random and run $\mathcal{P}$ on inputs $1^n, x, y, id, \omega$, and the verification stage of $\mathcal{V}$ on inputs SK, x, ρ, letting them interact.

Definition 5. *A pair* $(\mathcal{P}, \mathcal{V})$ *is* complete *for an NP-language* L *if for all* n*-bit strings* $x \in L$ *and their NP-witnesses* y*, the probability that in an execution of Normal-Interaction* $\mathcal{V}$ *outputs "accept" differs from 1 by a quantity negligible in* n.

The notion of resettable zero-knowledgeness is a bit harder to define. We do not describe the motivation and intuition behind RZK and instead refer the reader to the original exposition of [CGGM00]. Also, note that here we define only black-box RZK (because it is the notion most relevant to this paper). That is, we demand that there exist a single simulator that works for all malicious verifiers $\mathcal{V}^*$ (given oracle access to $\mathcal{V}^*$).

We introduce a few more players before formally stating the definition. Let

— An *honest prover* $\mathcal{P}$, for the purposes of defining RZK, be viewed as a *non-interactive* TM that is given, in addition to the inputs given in Section 2, the entire history of the messages already received in the interaction, and outputs the next message to be sent. Fixing all inputs, this view allows one to think of $\mathcal{P}(1^n, x, y, F, id, \omega)$ as a simple deterministic oracle that outputs the next message given the history of the interaction.

— An *(s,t)-resetting malicious verifier* $\mathcal{V}^*$, for any two positive polynomials s and t, be a TM that runs in two stages so that, on first input 1^n,
  1. In stage 1, $\mathcal{V}^*$ receives $s(n)$ values $x_1, \ldots, x_{s(n)} \in L$ of length n each, and outputs an arbitrary public file F and a list of $s(n)$ identities $id_1, \ldots, id_{s(n)}$.
  2. In stage 2, $\mathcal{V}^*$ starts in the final configuration of stage 1, is given oracle access to $s(n)^3$ provers, and then outputs its "view" of the interactions: its random string and the messages received from the provers.
  3. The total number of steps of $\mathcal{V}^*$ in both stages is at most $t(n)$.

— A *black-box simulator* M be a polynomial-time machine that is given oracle access to $\mathcal{V}^*$. By this we mean that it can run $\mathcal{V}^*$ multiple times, each time picking $\mathcal{V}^*$'s inputs, random tape and (because $\mathcal{V}^*$ makes oracle queries itself) the answers to all of $\mathcal{V}^*$'s queries. M is also given $s(n)$ values $x_1, \ldots, x_{s(n)} \in L$ as input.

Now we can formally define the resettable-zero-knowledgeness property.

Definition 6. $(\mathcal{P}, \mathcal{V})$ *is* black-box resettable zero-knowledge for an NP-language L *if there exists a simulator M such that for every pair of positive polynomials (s,t), for every (s,t)-resetting verifier $\mathcal{V}^*$, for every $x_1, \ldots, x_{s(n)} \in L$ and their corresponding NP-witnesses $y_1, \ldots, y_{s(n)}$, the following probability distributions are indistinguishable (in time polynomial in n):*

1. *The output of $\mathcal{V}^*$ obtained from the experiment of choosing $\omega_1, \ldots, \omega_{s(n)}$ uniformly at random, running the first stage of $\mathcal{V}^*$ to obtain F, and then letting $\mathcal{V}^*$ interact in its second stage with the following $s(n)^3$ instances of $\mathcal{P}$: $\mathcal{P}(x_i, y_i, F, id_k, \omega_j)$ for $1 \le i, j, k \le s(n)$.*
2. *The output of M with input $x_1, \ldots, x_{s(n)}$ interacting with $\mathcal{V}^*$.*

B Tools

B.1 Probabilistic Notation

(The following is taken verbatim from [BDMP91] and [GMR88].) If $A(\cdot)$ is an algorithm, then for any input x, the notation "$A(x)$" refers to the probability space that assigns to the string σ the probability that A, on input x, outputs σ. The set of strings having a positive probability in $A(x)$ will be denoted by "$\{A(x)\}$". If S is a probability space, then "$x \stackrel{R}{\leftarrow} S$" denotes the algorithm which assigns to x an element randomly selected according to S. If F is a finite set, then the notation "$x \stackrel{R}{\leftarrow} F$" denotes the algorithm that chooses x uniformly from F.

If p is a predicate, the notation $\mathrm{PROB}[x \stackrel{R}{\leftarrow} S; y \stackrel{R}{\leftarrow} T; \cdots : p(x, y, \cdots)]$ denotes the probability that $p(x, y, \cdots)$ will be true after the ordered execution of the algorithms $x \stackrel{R}{\leftarrow} S$; $y \stackrel{R}{\leftarrow} T; \cdots$. The notation $[x \stackrel{R}{\leftarrow} S; y \stackrel{R}{\leftarrow} T; \cdots : (x, y, \cdots)]$ denotes the probability space over $\{(x, y, \cdots)\}$ generated by the ordered execution of the algorithms $x \stackrel{R}{\leftarrow} S$, $y \stackrel{R}{\leftarrow} T, \cdots$.

B.2 Non-interactive Zero-Knowledge Proofs

Non-interactive zero-knowledge (NIZK) proofs for any language $L \in \mathrm{NP}$ were put forward and exemplified in [BFM88,BDMP91]. Ordinary ZK proofs rely on interaction. NIZK proofs replace interaction with a random *shared string*, σ, that enters the view of the verifier that a simulator must reproduce. Whenever the security parameter is 1^n, σ's length is $\mathrm{NI}\sigma\mathrm{Len}(n)$, where $\mathrm{NI}\sigma\mathrm{Len}$ is a fixed, positive polynomial.

Let us quickly recall their definition, adapted for polynomial-time provers.

Definition 7. *Let* NIP *(non-interactive prover) and* NIV *(non-interactive verifier) be two probabilistic polynomial-time algorithms, and let* $\mathrm{NI}\sigma\mathrm{Len}$ *be a positive polynomial. We say that* (NIP, NIV) *is a NIZK argument system for an NP-language L if*

1. Completeness. $\forall\ x \in L$ *of length n, σ of length* $\mathrm{NI}\sigma\mathrm{Len}(n)$, *and NP-witness y for x,*

$$\mathrm{PROB}[\Pi \stackrel{R}{\leftarrow} \mathrm{NIP}(\sigma, x, y) : \mathrm{NIV}(\sigma, x, \Pi) = \mathrm{YES}] = 1.$$

2. Soundness. $\forall\ x \in L$ *of length n,*

$$\mathrm{PROB}[\sigma \stackrel{R}{\leftarrow} \{0,1\}^{\mathrm{NI}\sigma\mathrm{Len}(n)} : \exists\ \Pi \text{ s. t. } \mathrm{NIV}(\sigma, x, \Pi) = \mathrm{YES}]$$

is negligible in n.

3. Zero-Knowledgeness. *There exists a probabilistic polynomial-time simulator* NIS *such that, $\forall$ sufficiently large n, $\forall\ x$ of length n and NP-witness y for x, the following two distributions are indistinguishable by any polynomial-time adversary:*

$$[(\sigma', \Pi') \stackrel{R}{\leftarrow} \mathrm{NIS}(x) : (\sigma', \Pi')] \text{ and}$$
$$[\sigma \stackrel{R}{\leftarrow} \{0,1\}^{\mathrm{NI}\sigma\mathrm{Len}(n)} ; \Pi \stackrel{R}{\leftarrow} \mathrm{NIP}(\sigma, x, y) : (\sigma, \Pi)]$$

The authors of [BDMP91] show that non-interactive zero-knowledge proofs exist for all NP languages under the quadratic residuosity assumption. The authors of [FLS99] show the same under a general assumptions: namely, that certified trapdoor permutations exist (a family of trapdoor permutations is *certified* if it is easy to tell that a given function belongs to the family). We refer the reader to these papers for details.

B.3 Verifiable Random Functions

A family of verifiable random functions (VRFs), as proposed in [MRV99], is essentially a pseudorandom function family with the additional property that the correct value of a function on an input can not only be computed by the owner of the seed, but also *proven* to be the unique correct value. The proof can be verified by anyone who knows the public key corresponding to the seed.

More precisely, a VRF is a quadruple of functions. The function VRFGen generates a key pair $(\mathit{VRFSK}, \mathit{VRFPK})$. The function $\mathrm{VRFEval}(\mathit{VRFSK}, x)$ computes the pseudorandom output v; the function $\mathrm{VRFProve}(\mathit{VRFSK}, x)$ computes pf_x, the proof that v is correct. This proof can be verified by anyone who knows the *VRFPK* by using $\mathrm{VRFVer}(\mathit{VRFPK}, x, v, pf_x)$; moreover, no matter how maliciously *VRFPK* is constructed, for each x, there exists at most one v for which a valid proof pf_x exists. The pseudorandomness requirement states that, for all the points for which no proof has been provided, the function $\mathrm{VRFEval}(\mathit{VRFSK}, \cdot)$ remains indistinguishable from random. The following formal definition is almost verbatim from [MRV99].

Definition 8. *Let* VRFGen, VRFEval, VRFProve, *and* VRFVer *be polynomial-time algorithms (the first and last are probabilistic, and the middle two are deterministic). Let $a: \mathbb{N} \to \mathbb{N} \cup \{*\}$ and $b: \mathbb{N} \to \mathbb{N}$ be any two functions that are computable in time* poly(n) *and bounded by a polynomial in n (except when a takes on the value $*$).*

We say that (VRFGen, VRFEval, VRFProve, VRFVer) *is a* verifiable pseudorandom function (VRF) *with* input length $a(n)$,[5] *and* output length $b(n)$ *if the following properties hold:*

1. *The following two conditions hold with probability $1 - 2^{-\Omega(n)}$ over the choice of $(\mathit{VRFPK}, \mathit{VRFSK}) \stackrel{R}{\leftarrow} \mathrm{VRFGen}(1^n)$:*
   a) *(Domain-Range Correctness): $\forall x \in \{0,1\}^{a(n)}$, $\mathrm{VRFEval}(\mathit{VRFSK}, x) \in \{0,1\}^{b(n)}$.*
   b) *(Complete Provability): $\forall x \in \{0,1\}^{a(k)}$, if $v = \mathrm{VRFEval}(\mathit{VRFSK}, x)$ and $pf = \mathrm{VRFProve}(\mathit{VRFSK}, x)$, then*

$$\mathrm{PROB}[\mathrm{VRFVer}(\mathit{VRFPK}, x, v, pf) = \mathrm{YES}] > 1 - 2^{-\Omega(k)}$$

   (this probability is over the coin tosses of VRFVer*).*

[5] When $a(n)$ takes the value $*$, it means that the VRF is defined for inputs of all lengths. Specifically, if $a(n) = *$, then $\{0,1\}^{a(n)}$ is to be interpreted as the set of all binary strings, as usual.

2. *(Unique Provability): For every $VRFPK$, x, v_1, v_2, pf_1, and pf_2 such that $v_1 \neq v_2$, the following holds for either $i = 1$ or $i = 2$:*

$$\text{PROB}[\text{VRFVer}(VRFPK, x, v_i, pf_i) = \text{YES}] < 2^{-\Omega(k)}$$

(this probability is also over the coin tosses of VRFVer*).*

3. *(Residual Pseudorandomness): Let $\alpha > 0$ be a constant. Let $T = (T_E, T_J)$ be any pair of algorithms such that $T_E(\cdot, \cdot)$ and $T_J(\cdot, \cdot, \cdot)$ run for a total of at most 2^{n^α} steps when their first input is 1^n. Then the probability that T succeeds in the following experiment is at most $1/2 + 1/2^{n^\alpha}$:*
   a) *Run* $\text{VRFGen}(1^n)$ *to obtain* $(VRFPK, VRFSK)$.
   b) *Run* $T_E^{\text{VRFEval}(VRFSK, \cdot), \text{VRFProve}(VRFSK, \cdot)}(1^n, VRFPK)$ *to obtain the pair* $(x, state)$.
   c) *Choose* $r \overset{R}{\leftarrow} \{0, 1\}$.
      i. *if* $r = 0$, *let* $v = \text{VRFEval}(VRFSK, x)$.
      ii. *if* $r = 1$, *choose* $v \overset{R}{\leftarrow} \{0, 1\}^{b(n)}$.
   d) *Run* $T_J^{\text{VRFEval}(VRFSK, \cdot), \text{VRFProve}(VRFSK, \cdot)}(1^n, v, state)$ *to obtain guess.*
   e) $T = (T_E, T_J)$ succeeds *if $x \in \{0, 1\}^{a(n)}$, guess $= r$, and x was not asked by either T_E or T_J as a query to* $\text{VRFEval}(VRFSK, \cdot)$ *or* $\text{VRFProve}(VRFSK, \cdot)$.

   We call α the pseudorandomness constant.

The authors of [MRV99] show how to construct VRFs based on the following variant of the RSA assumption. (We refer the reader to that paper for details of the construction.) Let PRIMES_n be the set of the n-bit primes, and RSA_n be the set of composite integers that are the product of two primes of length $\lfloor (n-1)/2 \rfloor$.

The RSA' Subexponential Hardness Assumption: There exists a constant α such that, if A is any probabilistic algorithm which runs in time 2^{n^α} when its first input is 1^n, then,

$$\text{PROB}[m \overset{R}{\leftarrow} \text{RSA}_n \, ; \, x \overset{R}{\leftarrow} \mathbb{Z}_m^* \, ; \, p \overset{R}{\leftarrow} \text{PRIMES}_{n+1} \, ; \, y \overset{R}{\leftarrow} A(1^n, m, x, p) : \\ y^p = x \pmod{m}] < 1/2^{n^\alpha}.$$

Robust Non-interactive Zero Knowledge

Alfredo De Santis[1], Giovanni Di Crescenzo[2], Rafail Ostrovsky[2], Giuseppe Persiano[1], and Amit Sahai[3]

[1] Dipartimento di Informatica ed Applicazioni, Università di Salerno, Baronissi (SA), Italy.
ads@dia.unisa.it, giuper@dia.unisa.it

[2] Telcordia Technologies, Inc., Morristown, NJ, USA.
giovanni@research.telcordia.com, rafail@research.telcordia.com

[3] Department of Computer Science, Princeton University. Princeton, NJ 08544.
sahai@cs.princeton.edu

Abstract. Non-Interactive Zero Knowledge (NIZK), introduced by Blum, Feldman, and Micali in 1988, is a fundamental cryptographic primitive which has attracted considerable attention in the last decade and has been used throughout modern cryptography in several essential ways. For example, NIZK plays a central role in building provably secure public-key cryptosystems based on general complexity-theoretic assumptions that achieve security against chosen ciphertext attacks. In essence, in a multi-party setting, given a fixed common random string of polynomial size which is visible to all parties, NIZK allows an arbitrary polynomial number of Provers to send messages to polynomially many Verifiers, where each message constitutes an NIZK proof for an arbitrary polynomial-size NP statement.

In this paper, we take a closer look at NIZK in the multi-party setting. First, we consider *non-malleable* NIZK, and generalizing and substantially strengthening the results of Sahai, we give the first construction of NIZK which remains non-malleable after polynomially-many NIZK proofs. Second, we turn to the definition of standard NIZK itself, and propose a strengthening of it. In particular, one of the concerns in the technical definition of NIZK (as well as non-malleable NIZK) is that the so-called "simulator" of the Zero-Knowledge property is allowed to pick a *different* "common random string" from the one that Provers must actually use to prove NIZK statements in real executions. In this paper, we propose a new definition for NIZK that eliminates this shortcoming, and where Provers and the simulator use the *same* common random string. Furthermore, we show that both standard and *non-malleable* NIZK (as well as NIZK Proofs of Knowledge) can be constructed achieving this stronger definition. We call such NIZK **Robust NIZK** and show how to achieve it. Our results also yields the simplest known public-key encryption scheme based on general assumptions secure against adaptive chosen-ciphertext attack (CCA2).

J. Kilian (Ed.): CRYPTO 2001, LNCS 2139, pp. 566–598, 2001.

1 Introduction

INTERACTIVE ZERO-KNOWLEDGE. Over the last two decades, Zero-Knowledge (ZK) as defined by Goldwasser, Micali, and Rackoff [21] has become a fundamental cryptographic tool. In particular, Goldreich, Micali and Wigderson [20] showed that any NP statement can be proven in *computational* [1] ZK (see also [16]). Though ZK was originally defined for use in two-party interactions (i.e., between a single Prover and a single Verifier), ZK was shown to be useful in a host of situations where multiple parties could be involved, especially in the multi-party secure function evaluation, first considered by Goldreich, Micali and Wigderson [19]. Informally, one reason the notion of interactive ZK has been so pervasive is that in the single Prover/Verifier case, ZK essentially guarantees that any poly-time Verifier after interacting with the Prover in a ZK protocol learns absolutely nothing. Thus, informally speaking, whatever a poly-time Verifier can do after verifying a ZK protocol, it could also have done before such a ZK interaction. However, in a multiparty setting, perhaps not surprisingly, the standard two-party definition of ZK does not guarantee what we would intuitively expect from "zero knowledge': that the polynomial-time Verifier after observing such proofs can not (computationally) do anything that he was not able to do before such a proofs. Essentially, two important problems were pointed out in the literature:

One problem, formally defined by Dolev, Dwork and Naor [13] is that of *malleability*, which informally means that an adversary who takes part in some ZK interaction can also interact with other parties and can exploit fragments of ZK interactions to prove something that he was not able to prove before. Indeed, this is a real problem to which [13] propose a solution that requires polylogarithmic overhead in the number of rounds of communication. It is not known how to reduce the number of rounds further in their solution.

Another problem of ZK in the multi-party setting, pointed out by Dwork, Naor and Sahai [14], is that verifiers can "collaborate" when talking to provers, and the ZK property must be guaranteed even in concurrent executions. Indeed, unless one introduce changes in the model such as timing assumptions, in terms of the number of rounds, it was shown that a polylogarithmic number of rounds is both necessary [6] and sufficient [25] to guarantee concurrent ZK.

NON-INTERACTIVE ZERO-KNOWLEDGE (NIZK): A way to reduce the number of rounds in a ZK proof (to just a single message from Prover to Verifier) was

[1] Recall that several variants of ZK have been considered in the literature, in terms of the strength of the *soundness* condition and the strength of the *simulation*. In terms of the quality of the simulation, *perfect*; *statistical*; and *computational* ZK are defined [21]. In terms of *soundness* two variants were considered: ZK *proofs*, where the proof remains valid even if an infinitely-powerful Prover is involved [21,20] and ZK *arguments*, where it is required that only polynomially-bounded Provers cannot cheat (except with negligible probability), given some complexity assumption [3,26]. For ZK proofs for languages outside BPP were shown to imply the existence of one-way functions for perfect, statistical [30] (see also [34]) as well as computational [31] variants of ZK.

proposed by Blum, Feldman and Micali [2] by changing the model as follows: we assume that a *common random reference string* is available to all players. The Prover sends a single message to Verifier, which constitutes "non-interactive zero-knowledge" (NIZK) proof. In [2] it was shown that any NP statement has a NIZK proof. Extending [2], Blum, De Santis, Micali and Persiano [1] showed how a Prover can prove polynomially many proofs based on algebraic assumptions. Feige, Lapidot and Shamir further refined the definition of NIZK and constructed[2] multiple-proof NIZK based on general assumptions [15]. De Santis and Persiano extended NIZK notion to NIZK Proofs of Knowledge (NIZK-PK)[3] [8].

Again, although the notion of NIZK was defined in a two-party setting, it quickly found applications in settings with many parties, in particular where the same reference string may be used by multiple parties (see e.g. [13,28,4,22]). Because of the non-interactive nature of NIZK proofs, many multi-party issues that appear in ZK, do not arise in NIZK; for example the problem of concurrent zero-knowledge is completely gone[4]!

The definition of NIZK proposed by [2,1,15], essentially provides the following guarantee: What one can output after seeing NIZK proofs is indistinguishable from what one can output without seeing any proofs, *if you consider the reference string as part of the output.* Thus, the standard notion of NIZK says that as long as one can simulate proofs *together* with random-looking reference strings, this satisfies the notion of NIZK. This definition, however, leaves open the question of what to do about output as it relates to the *particular* reference string that is being used by a collection of parties. Since the NIZK simulator produces its own different random string, its output would make sense only relative to the reference string that it chose, different from the one used by the provers. [5] One of the contributions of this paper is to strengthen the notion of NIZK to insist that the simulator works with the *same* totally random string that all the Provers work with.

NIZK proofs are broadcastable and transferable – that is, a single proof string can be broadcasted or transferred from verifier to verifier to convince multiples parties of the validity of a statement. However, transferability causes a new problem: a user who have seen an NIZK proof (of a hard problem) can now "prove" (by simply copying) what he was not able to prove before. Indeed,

[2] Efficiency improvements to these constructions were presented in [24,9,10].

[3] In the same paper [8] defined *dense cryptosystems* and showed that dense cryptosystems and NIZK proofs of membership for NP are sufficient in order to construct NIZK-PK for all of NP. This assumption was shown to be *necessary* for NIZK-PK in [11]. (Dense cryptosystemes were also shown to be equivalent to *extractable commitment* [11].)

[4] In fact, non-malleable commitment also becomes much easier to deal with in the non-interactive setting [12]. Also, though it is not always thought of as a multi-party issue, the problem of resettable zero-knowledge [5] is also easily dealt with for NIZK as well.

[5] Indeed, it seems quite unfair to let the simulator get away with ignoring the actual reference string!

more generally the problem of *malleability* does remain for NIZK proofs: With respect to a particular (fixed) reference string, after seeing some NIZK proofs, the adversary may be able to construct new proofs that it could not have been able to otherwise. Sahai introduced *non-malleable* NIZK in [33] where he shows how to construct NIZK which remains non-malleable only as long as the number of proofs seen by any adversary is bounded. In this paper (among other contributions) we continue and extend his work, strengthening the notion and the constructions of non-malleability and removing the limitation on the number of proofs. (For further discussion on malleability issues in multi-party situations, see Appendix A.)

OUR RESULTS: First, we consider the following notion of NIZK. The *sampling* algorithm produces a common random string together with auxiliary information. (We insist that the common random string comes from a uniform (or nearly uniform) distribution). Polynomially-bounded provers use this common random string to produce polynomially-many NIZK messages for some NP language. We insist that the simulator, given the *same* common random string, together with auxiliary information, can produce the proofs of theorems which are computationally indistinguishable from the proofs produced by honest provers *for the same reference string.* We call this notion *same-string* NIZK.

We show two facts regarding same-string NIZK: (1) same-string NIZK Proofs (i.e. where the prover is infinitely powerful) are impossible for any hard-on-average NP-complete languages (2) same-string NIZK Arguments (i.e. where the prover is computationally bounded) are possible given any one-way trapdoor permutation.

Next, we turn to non-malleability for NIZK, and a notion related to non-malleability called *simulation-soundness* first defined by Sahai [33]. The simulation-soundness requirement is that a polynomially-bounded prover can not prove false theorems even after seeing simulated proofs of any statements (including false statements) of its choosing. Sahai achieves non-malleability and simulation-soundness only with respect to a bounded number of proofs. In this paper, we show that assuming the existence of one-way trapdoor permutations, we can construct NIZK proof systems which remain simulation-sound even after the prover sees any polynomial number of simulated proofs[6]. Combined with [33] this also gives the simplest known construction of CCA2-secure public-key cryptosystem based on one-way trapdoor permutations.

In dealing with non-malleability, we next turn to NIZK Proofs of Knowledge (NIZK-PK), introduced by De Santis and Persiano[8]. We use NIZK-PK to propose a strengthening of the definition of non-malleability for NIZK, based

[6] We note that we can also achieve a form of non-malleability (as opposed to simulation soundness) for NIZK proofs of membership based only on trapdoor permutations. This non-malleability would also hold against any polynomial number of proofs, however the non-malleability achieved satisfies a weaker definition than the one we propose based on NIZK-PK (and in particular, the resulting NIZK proof would only be a proof of membership and not a proof of knowledge). We omit the details of this in these proceedings.

on NP-witnesses (which, in particular, implies the earlier definition [33]). We provide constructions which show that for any polynomial-time adversary, even after the adversary has seen any polynomial number of NIZK proofs for statements of its choosing, the adversary does not gain the ability to prove any new theorems it could not have produced an NP witness for prior to seeing any proofs, except for the ability to duplicate proofs it has already seen. This construction requires the assumption that trapdoor permutations exist and that public-key encryption schemes exist with an inverse polynomial density of valid public keys (called *dense cryptosystems*). Such dense cryptosystems exist under most common intractability assumptions which give rise to public-key encryption, such as the RSA assumption, Quadratic Residuosity, Diffie-Hellman [8] and factoring [11]. (In fact, in the context of NIZK-PK, we cannot avoid using such dense cryptosystems since they were shown to be *necessary* for any NIZK-PK [11].)

Finally, we call NIZK arguments that are both non-malleable and same-string NIZK **Robust NIZK**.

We highlight the contributions of our results:

- For NIZK arguments, we give the first construction where the simulator uses the same common random string as used by all the provers.
- Our Robust-NIZK proof systems are non-malleable with regard to *any* polynomial number of proofs seen by the adversary and with respect to the same proof-system. (We contrast this with the previous result of [33] which proves non-malleability against only a bounded number of proofs, and in fact the length of the reference string grew quadratically in the bound on the the number of proofs the adversary could see.) In our result, in contrast, the length of the reference string depends only on the security parameter.
- Our non-malleable NIZK definition and construction based on NIZK-PK achieves a very strong guarantee: We require that one can obtain an explicit NP witness for any statement that the adversary can prove after seeing some NIZK proofs. Thus, it intuitively matches our notion of what NIZK should mean: that the adversary cannot prove anything "new" that he was not able to prove before (except for copying proofs in their entirety).
- Finally, our construction yields the simplest known public-key encryption scheme based on general assumptions which is secure against adaptive chosen-cyphertext attacks (CCA2).

We point out some new techniques used to establish our results. All previous work on non-malleability in a non-interactive setting under general assumptions [13,12,33] used a technique called "unduplicatable set selection". Our first construction provides the first non-malleability construction based on general assumptions which *does not* use "unduplicatable set selection" at all, and rather relies on a novel use of pseudo-random functions of [18]. In our second construction, we show how to generalize the unduplicatable set selection technique to a technique we call "hidden unduplicatable set selection," and use this to build our proofs. Both techniques are novel, and may have further applications.

ORGANIZATION. In Section 2, we both recall old definitions as well as give the new definitions of this paper. In Section 3, we present our first construction of Robust NIZK and non-malleable NIZK (and NIZK-PK) proofs. In Section 4, we present our second construction which uses different techniques and a yields non-malleable NIZK and NIZKPK.

2 Preliminaries and Definitions

We use standard notations and conventions for writing probabilistic algorithms and experiments. If A is a probabilistic algorithm, then $A(x_1, x_2, \ldots; r)$ is the result of running A on inputs $x_1, x_2, \ldots$ and coins r. We let $y \leftarrow A(x_1, x_2, \ldots)$ denote the experiment of picking r at random and letting y be $A(x_1, x_2, \ldots; r)$. If S is a finite set then $x \leftarrow S$ is the operation of picking an element uniformly from S. $x := \alpha$ is a simple assignment statement. By a "non-uniform probabilistic polynomial-time adversary," we always mean a circuit whose size is polynomial in the security parameter. All adversaries we consider are non-uniform. (Thus, we assume our assumptions, such as the existence of one-way functions, also hold against non-uniform adversaries.)

In this section, we will formalize the notions of non-malleable, same-string and robust NIZK proofs. We will also define an extension of simulation soundness.

2.1 Basic Notions

We first recall the definition of an (efficient, adaptive) single-theorem NIZK proof systems [1,2,15,8]. Note that since we will always use the now-standard adaptive notion of NIZK, we will suppress writing "adaptive" in the future. We will also only concentrate on efficiently realizable NIZK proofs, and so we will suppress writing "efficient" as well. This first definition only guarantees that a single proof can be simulated based on the reference string. Note that our definition uses "Strong Soundness," based on Strong NIZK Proofs of Knowledge defined in [8] and a similar notion defined in [28], where soundness is required to hold even if the adversary may chose its proof after seeing the randomly selected reference string. Note that the constructions given in [15], for instance, meet this requirement. We simultaneously define the notion of an NIZK argument, in a manner completely analogous to the definition of an interactive ZK argument.

Definition 1 (NIZK [15]). $\Pi = (\ell, \mathrm{P}, \mathcal{V}, \mathcal{S} = (\mathcal{S}_1, \mathcal{S}_2))$ *is a* single-theorem NIZK proof system (resp., argument) *for the language* $L \in \mathrm{NP}$ *with witness relation* R *if:* ℓ *is a polynomial, and* $\mathrm{P}, \mathcal{V}, \mathcal{S}_1, \mathcal{S}_2$ *are all probabilistic polynomial-time machines such that there exists a negligible function* α *such that for all* k*:*

(Completeness): *For all* $x \in L$ *of length* k *and all* w *such that* $R(x, w) =$ `true` *, for all strings* σ *of length* $\ell(k)$*, we have that* $\mathcal{V}(x, \mathrm{P}(x, w, \sigma), \sigma) =$ `true` *.*

(Soundness): *For all unbounded (resp., polynomial-time) adversaries A, if $\sigma \in \{0,1\}^{\ell(k)}$ is chosen randomly, then the probability that $A(\sigma)$ will output (x,p) such that $x \notin L$ but $\mathcal{V}(x,p,\sigma) = \texttt{true}$ is less than $\alpha(k)$.*

(Single-Theorem Zero Knowledge): *For all non-uniform probabilistic polynomial-time adversaries $A = (A_1, A_2)$, we have that*

$$|\Pr[\mathsf{Expt}_A(k) = 1] - \Pr\left[\mathsf{Expt}_A^S(k) = 1\right]| \leq \alpha(k),$$

where the experiments $\mathsf{Expt}_A(k)$ and $\mathsf{Expt}_A^S(k)$ are defined as follows:

$\mathsf{Expt}_A(k)$:	$\mathsf{Expt}_A^S(k)$:
$\Sigma \leftarrow \{0,1\}^{\ell(k)}$	$(\Sigma, \tau) \leftarrow \mathcal{S}_1(1^k)$
$(x, w, s) \leftarrow A_1(\Sigma)$	$(x, w, s) \leftarrow A_1(\Sigma)$
$p \leftarrow \mathrm{P}(x, w, \Sigma)$	$p \leftarrow \mathcal{S}_2(x, \Sigma, \tau)$
$\texttt{return}\ A_2(p, s)$	$\texttt{return}\ A_2(p, s)$

To define a notion of NIZK where any polynomial number of proofs can be simulated, we change the Zero-knowledge condition as follows:

Definition 2 (unbounded NIZK [15]). *$\Pi = (\ell, \mathrm{P}, \mathcal{V}, \mathcal{S} = (\mathcal{S}_1, \mathcal{S}_2))$ is an* unbounded NIZK proof system *for the language $L \in$ NP if Π is a single-theorem NIZK proof system for L and furthermore: there exists a negligible function α such that for all k:*

(Unbounded Zero Knowledge): *For all non-uniform probabilistic polynomial-time adversaries A, we have that $|\Pr[\mathsf{Expt}_A(k) = 1] - [\mathsf{Expt}_A^S(k) = 1]| \leq \alpha(k)$, where the experiments $\mathsf{Expt}_A(k)$ and $\mathsf{Expt}_A^S(k)$ are defined as follows:*

$\mathsf{Expt}_A(k)$:	$\mathsf{Expt}_A^S(k)$:
$\Sigma \leftarrow \{0,1\}^{\ell(k)}$	$(\Sigma, \tau) \leftarrow \mathcal{S}_1(1^k)$
$\texttt{return}\ A^{P(\cdot,\cdot,\Sigma)}(\Sigma)$	$\texttt{return}\ A^{S'(\cdot,\cdot,\Sigma,\tau)}(\Sigma)$

where $S'(x, w, \Sigma, \tau) \stackrel{\text{def}}{=} \mathcal{S}_2(x, \Sigma, \tau)$.

Definition 3. *We say that an NIZK argument system is* same-string *NIZK if the (unbounded) zero knowledge requirement above is replaced with the following requirement: there exists a negligible function α such that for all k:*

(Same-String Zero Knowledge): *For all non-uniform probabilistic polynomial-time adversaries A, we have that $|\Pr[X = 1] - \Pr[Y = 1]| \leq \alpha(k)$, where X and Y are as defined in (and all probabilities are taken over) the experiment $\mathsf{Expt}(k)$ below:*

$\mathsf{Expt}(k)$:
$(\Sigma, \tau) \leftarrow \mathcal{S}_1(1^k)$
$X \leftarrow A^{P(\cdot,\cdot,\Sigma)}(\Sigma)$
$Y \leftarrow A^{S'(\cdot,\cdot,\Sigma,\tau)}(\Sigma)$

where $S'(x, w, \Sigma, \tau) \stackrel{\text{def}}{=} \mathcal{S}_2(x, \Sigma, \tau)$.

(Same-String Zero Knowledge, cont.): *The distribution on Σ produced by $\mathcal{S}_1(1^k)$ is the uniform distribution over $\{0,1\}^{\ell(k)}$.*

Remark 1. We make two observations regarding the definition of same-string NIZK:

- As done in [15], the definition could equivalently be one that states that with all but negligible probability over the choices of common random reference strings, the simulation is computationally indistinguishable from real proofs supplied by the prover. We omit the details for lack of space.
- On the other hand, the definition above differs from the standard definition on unbounded zero knowledge only in the new requirement that the simulator produce truly uniform reference strings. It is easy to verify that all other changes are cosmetic.
- In the next theorem, we show why we must speak only of same-string NIZK *arguments*, and not NIZK Proofs.

Theorem 1. *If one-way functions exist, then there cannot exist same-string (adaptive) NIZK Proof systems for any NP-complete language L, even for single-theorem NIZK. In fact, this result extends to any language that is hard-on-average with respect to an efficiently samplable distribution.*

Proof. **(Sketch)** We only sketch the proof of this impossibility result. Assume that one-way functions exist, and that a same-string (adaptive) single-theorem NIZK Proof system exists for an NP-complete language L. We will show a contradiction to the soundness of the NIZK Proof System. First we note that the existence of one-way functions and Cook's theorem implies that there is a probabilistic polynomial-time algorithm M such that for all non-uniform polynomial-time machines A, if $x \leftarrow M(1^k)$, the probability that A correctly decides whether $x \in L$ is only negligibly more than 1/2. It is implicit in the previous statement that with probability close to 1/2, if $x \leftarrow M(1^k)$, then $x \notin L$.

This hardness condition also implies that, in particular, the simulator must output proofs that are accepted with all but negligible probability when given as input $x \leftarrow M(1^k)$. At the same time, because the NIZK system is both same-string (adaptive) NIZK, it must be that the reference strings output by $S_1(1^k)$ come from a uniform distribution.

Now, consider a cheating (unbounded) prover which, for any given random string, guesses the auxiliary information τ which maximizes the probability that the simulator outputs an accepting proof on inputs chosen according to $x \leftarrow M(1^k)$. Since the reference string that the prover encounters is also uniform, it follows that the cheating prover will have at least as high a probability of convincing a verifier to accept on input $x \leftarrow M(1^k)$. But we know that the simulator causes the verifier to accept with probability negligibly close to 1. This contradicts the (unconditional) soundness of the NIZK proof system, completing the proof.

We also define the notion of an NIZK proof of knowledge [8] for an NP language L with witness relation R. Informally, the idea is that in an NIZK proof of knowledge, one should be able to extract the NP witness directly from the proof if given some special information about the reference string. We capture this notion by defining an *extractor* which produces a reference string together with some auxiliary information. The distribution on reference strings is statistically close to the uniform distribution. Given the auxiliary information and an NIZK proof, one can efficiently extract the witness. [8] show how to turn any NIZK proof system into a proof of knowledge under the assumption that public-key encryption schemes exist with sufficiently high density of valid public keys (called dense cryptosystems). We now recall the formal definition:

Definition 4 (NIZK proof of knowledge [8]). $\Pi = (\ell, \mathrm{P}, \mathcal{V}, \mathcal{S} = (\mathcal{S}_1, \mathcal{S}_2), E = (E_1, E_2))$ *is a* NIZK proof (or argument) of knowledge *for the language* $L \in \mathrm{NP}$ *with witness relation* R *if:* Π *is an NIZK proof (or argument) system (of any type) for* L *and furthermore* E_1 *and* E_2 *are probabilistic polynomial-time machines such that there exists a negligible function* α *such that for all* k*:*

(Reference-String Uniformity): *The distribution on reference strings produced by*
$E_1(1^k)$ has statistical distance at most $\alpha(k)$ from the uniform distribution on $\{0,1\}^{\ell(k)}$.

(Witness Extractability): *For all adversaries A, we have that* $\left[\mathsf{Expt}^E_A(k)\right]| \geq \Pr\left[\mathsf{Expt}_A(k)\right] - \alpha(k)$, *where the experiments* $\mathsf{Expt}_A(k)$ *and* $\mathsf{Expt}^S_A(k)$ *are defined as follows:*

$\mathsf{Expt}_A(k)$:	$\mathsf{Expt}^E_A(k)$:
$\Sigma \leftarrow \{0,1\}^{\ell(k)}$	$(\Sigma, \tau) \leftarrow E_1(1^k)$
$(x, p) \leftarrow A(\Sigma)$	$(x, p) \leftarrow A(\Sigma)$
$\texttt{return}\ V(x, p, \Sigma)$	$w \leftarrow E_2(\Sigma, \tau, x, p)$
	$\texttt{return}\ \texttt{true}\ \textit{if}\ (x, w) \in R$

2.2 Non-malleable NIZK

We now proceed to define non-malleable NIZK. The intuition that our definition will seek to capture is to achive the strongest possible notion of non-malleability: "whatever an adversary can prove after seeing polynomially many NIZK proofs for statements of its choosing, it could have proven without seeing them, except for the ability to duplicate proofs."[7] Extending the notion of NIZK-PK of De Santis and Persiano [8] we define non-malleable NIZK-PK. We will make the definition with regard to simulated proofs, but note that one can make a similar definition with regard to actual proofs; we omit it due to lack of space.

[7] When interpreting the line "it could have proven without seeing them," we insist that an actual NP witness for the statement should be extractable from the adversary, which is a very strong NIZK-PK property.

Definition 5. [Non-Malleable NIZK] Let $\Pi = (\ell, \mathcal{P}, \mathcal{V}, \mathcal{S})$ be an unbounded NIZK proof system for the NP language L with witness relation W. We say that Π is a *non-malleable NIZK proof system (or argument)*[8] for L if there exists a probabilistic polynomial-time oracle machine M such that:

For all non-uniform probabilistic polynomial-time adversaries A and for all non-uniform polynomial-time relations R, there exists a negligible function $\alpha(k)$ such that

$$\Pr\left[\mathsf{Expt}^{S}_{A,R}(k)\right] \leq \Pr\left[\mathsf{Expt}'_{A}(k)\right] + \alpha(k)$$

where $\mathsf{Expt}^{S}_{A,R}(k)$ and $\mathsf{Expt}'_{A,R}(k)$ are the following experiments:

$\mathsf{Expt}^{S}_{A,R}(k)$:	$\mathsf{Expt}'_{A,R}(k)$:
$(\Sigma, \tau) \leftarrow \mathcal{S}_1(1^k)$	
$(x, p, \text{aux}) \leftarrow A^{\mathcal{S}_2(\cdot, \Sigma, \tau)}(\Sigma)$	
Let Q be list of proofs given by S_2 above	$(x, w, \text{aux}) \leftarrow M^A(1^k)$
`return true` iff	`return true` iff
$(p \notin Q)$ `and`	
$(\mathcal{V}(x, p, \Sigma) = \texttt{true})$ `and`	$((x, w) \in W)$ `and`
$(R(x, \text{aux}) = \texttt{true})$	$(R(x, \text{aux}) = \texttt{true})$

We also consider (and strengthen) another notion for NIZK called simulation soundness [33] which is related to non-malleability, but also can be useful in applications – in particular, it suffices for building public-key encryption schemes secure against the strongest form of chosen-ciphertext attack (CCA2). The ordinary soundness property of proof systems states that with overwhelming probability, the prover should be incapable of convincing the verifier of a false statement. In this definition, we will ask that this remains the case even after a polynomially bounded party has seen any number of simulated proofs of his choosing. Note that simulation soundness is implied by our definition of non-malleability above.

Definition 6. [Unbounded Simulation-Sound NIZK] Let $\Pi = (\ell, \mathcal{P}, \mathcal{V}, \mathcal{S} = (\mathcal{S}_1, \mathcal{S}_2))$ be an unbounded NIZK proof system (or argument) for the language L. We say that Π is *simulation-sound* if for all non-uniform probabilistic polynomial-time adversaries A, we have that

$$\Pr\left[\mathsf{Expt}_{A,\Pi}(k)\right] \text{ is negligible in } k,$$

where $\mathsf{Expt}_{A,\Pi}(k)$ is the following experiment:

[8] To stress the main novelty of this definition, we will sometimes write "non-malleable in the explicit witness sense," to indicate that an explicit NP-witness can be extracted from any prover. We remark that our definition clearly implies the definition of [33].

$\mathsf{Expt}_{A,\Pi}(k)$:
$(\Sigma, \tau) \leftarrow \mathcal{S}_1(1^k)$
$(x, p) \leftarrow A^{\mathcal{S}_2(\cdot,\Sigma,\tau)}(\Sigma)$
Let Q be list of proofs given by $\mathcal{S}_2$ above
return true iff ($p \notin Q$ and $x \notin L$ and $\mathcal{V}(x, p, \Sigma) =$ **true**)

Definition 7. *We will call an NIZK argument that is non-malleable and has unbiased simulations a* robust NIZK *argument.*

3 First Construction

In this section, we exhibit our construction of NIZK proof systems that enjoy unbounded simulation-soundness. This construction is then readily modified using NIZK Proofs of Knowledge to construct proof systems with unbounded non-malleability (in the explicit witness sense), and robust NIZK arguments.

Assumptions needed. In order to construct our simulation-sound proof systems for some NP language L, we will require the existence of efficient single-theorem (adaptive) NIZK proof systems for a related language L', described in detail below. Such proof systems exist under the assumption that trapdoor permutations exist [15]. Further, we will require the existence of one-way functions. To construct the proof systems with full non-malleability, we will require efficient single-theorem (adaptive) NIZK proofs of knowledge for the language L'. Such proof systems exist under the assumption that dense cryptosystems exist and trapdoor permutations exist [8].

3.1 Ingredients

Let k be the security parameter. We first specify the ingredients used in our construction:

Commitment. We recall two elegant methods for constructing commitments. One, based on one-way permutations, will allow us to construct non-malleable NIZK arguments with unbiased simulations (*i.e.* robust NIZK). The other, which can be based merely on one-way functions, suffices to construct non-malleable NIZK proof systems.

The theorem of Goldreich and Levin [17] immediately yields the following bit commitment scheme from any one-way permutation f on k bits:

$$C(b, s) = (r, f(s)) \text{ where } r \in_R \{0,1\}^k \text{ such that } r \cdot s = b$$

Here, it should be that $s \in_R \{0,1\}^k$. Note that if $s = 0^k$ and $b = 1$, then no choice of r will allow for $r \cdot s = b$. In this case, r is chosen at random, but the commitment is invalid. Since invalid commitments can only occur with probability at most 2^{-k}, we can safely ignore this. To reveal the bit, the sender simply reveals s. Observe that the distribution $C(b, s)$ where both b and s are chosen uniformly has is precisely the uniform distribution over $\{0,1\}^{2k}$. We will sometimes write just $C(b)$ to mean $C(b, s)$ where $s \in_R \{0,1\}^k$. Note that in this commitment scheme, every string of length $2k$ corresponds to a commitment to *some unique* string.

On the other hand, we recall the bit commitment protocol of Naor [27] based on pseudorandom generators (which can be built from any one-way function [23]). Let G be a pseudorandom generator stretching k bits to $3k$ bits. The Naor commitment procedure commits to a bit b as follows:

$$C(b, s) = \begin{cases} (r, G(s)) & \text{if } b = 0 \\ (r, G(s) \oplus r) & \text{if } b = 1 \end{cases}$$

Here, $r \in_R \{0,1\}^{3k}$, and as above the string s should be selected uniformly at random among strings of length k. Again, we will sometimes write just $C(b)$ to mean $C(b, s)$ where $s \in_R \{0,1\}^k$. It is shown in [27] that if U and U' are both independent uniform distributions among strings of length $3k$, then the distributions (U, U'), $C(0)$, and $C(1)$ are all computationally indistinguishable (taken as ensembles of distributions indexed by k). Furthermore, it is clear that unless r is of the form $G(s_1) \oplus G(s_2)$ for some s_1 and s_2, there are no commitment strings that can arise as both commitments to 0 and commitments to 1. The probability of this being possible is thus less than 2^{-k} over the choices of r. Furthermore, the probability that a random sample from (U, U') could be interpreted as a commitment to any bit is at most 2^{-k} – in contrast to the one-way permutation based scheme above.

Pseudo-Random Functions. We also let $\{f_s\}_{s \in \{0,1\}^k}$ be a family of pseudorandom functions [18] mapping $\{0,1\}^*$ to $\{0,1\}^k$.

One-Time Signatures. Finally, let $(Gen, Sign, Ver)$ be a strong one-time signature scheme (see [29,33]), which can be constructed easily from universal one-way hash functions. Note that these objects can be constructed from one-way functions.

3.2 The Construction

Intuition. The NIZK system intuitively works as follows: First, a verification-key/signing-key pair (VK, SK) is chosen for the one-time signature scheme. Then the prover provides a NIZK proof that *either* x is in the language, *or* that the reference string actually specifies a hidden pseudo-random function and that some specified value is the output of this pseudo-random function applied to the verification key VK. Finally, this proof is itself signed using the signing key SK.

We now describe the proof system Π for L precisely. Note that a third possibility for the NIZK proof is added below; this is a technical addition which simplifies our proof of correctness.

- **Common random reference string.** The reference string consists of three parts Σ_1, Σ_2, and Σ_3.

1. Σ_1 is a string that we break up into k pairs $(r_1, c_1), \ldots, (r_k, c_k)$. If we use the one-way permutation-based commitments, each r_i and c_i are of length k; if we use the Naor commitment scheme, r_i and c_i are of length $3k$.
2. Σ_2 is a string of length $3k$.
3. Σ_3 is a string of length polynomial in k. The exact length of Σ_3 depends on an NIZK proof system described below.

- **Prover Algorithm.** We define the language L' to be the set of tuples $(x, u, v, \Sigma_1, \Sigma_2)$ such that at least one of the following three conditions hold:
  - $x \in L$
  - Σ_1 consists of commitments to the bits of the k bit string s, and $u = f_s(v)$: Formally, there exists $s = s_1 \ldots s_k$ with $s_i \in \{0,1\}$ for each i, and there exist $a_1, a_2, \ldots, a_k \in \{0,1\}^k$ such that $u = f_s(v)$ and such that for each i, (r_i, c_i) is a commitment under C to the bit s_i.
  - There exists $s \in \{0,1\}^k$ such that $\Sigma_2 = G(s)$

  We assume we have a single-theorem NIZK proof system for L' (which we denote Π'). Note that the length of the reference string Σ_3 should be $\ell_{\Pi'}(k)$. We now define the prover for L. On input x, a witness w, and the reference string $\Sigma = (\Sigma_1, \Sigma_2, \Sigma_3)$, the prover does the following:
  1. Use $Gen(1^k)$ to obtain a verification key / signing key pair (VK, SK) for the one-time signature scheme.
  2. Let u be uniformly selected from $\{0,1\}^k$.
  3. Using Σ_3 as the reference string and w as the witness, generate a single-theorem NIZK proof under Π' that $(x, u, VK, \Sigma_1, \Sigma_2) \in L'$. Denote this proof by π'.
  4. Output $(VK, x, u, \pi', Sign_{SK}(x, u, \pi'))$.

  As a sanity check, we observe that if $\Sigma = (\Sigma_1, \Sigma_2, \Sigma_3)$ is chosen uniformly, then the probability that Σ_1 can be interpreted as the commitment to any bits and the probability that Σ_2 is in the range of G are both exponentially small in k. Thus, with all but exponentially small probability over the choice of Σ_1 and Σ_2, a proof that $(x, u, VK, \Sigma_1, \Sigma_2) \in L'$ really does imply that $x \in L$.
- **Verifier Algorithm.** The verification procedure, on input the instance x and proof $(VK, x', u, \pi', \sigma)$, with respect to reference string $\Sigma = (\Sigma_1, \Sigma_2, \Sigma_3)$, proceeds as follows:
  1. Confirm that $x = x'$, and confirm the validity of the one-time signature — *i.e.* that $Ver_{VK}((x, u, \pi'), \sigma) = 1$.
  2. Verify that π' is a valid proof that $(x, u, VK, \Sigma_1, \Sigma_2) \in L'$.
- **Simulator Algorithm.** We now describe the two phases of the simulator algorithm. S_1 is the initial phase, which outputs a reference string Σ along with some auxiliary information τ. S_2 takes as input this auxiliary information, the reference string, and an instance x, and outputs a simulated proof for x. The intuition for the simulator is that it sets up the reference string to be such that a hidden pseudo-random function really is specified, and instead of proving that x is in the language, the simulator simply proves that it can evaluate this hidden pseudo-random function on the verification key of the signature scheme.

$\mathcal{S}_1(1^k)$:
  $s, \Sigma_2 \leftarrow \{0,1\}^{3k}$; $\Sigma_3 \leftarrow \{0,1\}^{\ell_{\Pi'}(k)}$
  $a_i \leftarrow \{0,1\}^k$ for $i = 1, \ldots, k$
  $g_i \leftarrow C(s_i, a_i)$ for $i = 1. \ldots, k$
  $\Sigma_1 = (g_1, g_2, \ldots, g_k)$
  `return` $\Sigma = (\Sigma_1, \Sigma_2, \Sigma_3)$ and
    $\tau = (s, a_1, \ldots, a_k)$

$\mathcal{S}_2(\tau = (s, a_1, \ldots, a_k), \Sigma = (\Sigma_1, \Sigma_2, \Sigma_3), x)$:
  $(VK, SK) \leftarrow Gen(1^k)$
  $u = f_s(VK)$
  Use Σ_3 as ref string and τ as witness to construct
    proof π' that $(x, u, VK, \Sigma_1, \Sigma_2) \in L'$
  $\sigma \leftarrow Sign_{SK}(x, u, \pi')$
  `return` (VK, x, u, π', σ)

Theorem 2. *If Π' is a single-theorem NIZK proof system for L', the proof system Π described above is either:*

- *an unbounded simulation-sound NIZK proof system for L if C is the Naor commitment scheme and one-way functions exist.*
- *an unbounded simulation-sound same-string NIZK argument for L with if C is the commitment scheme based on one-way permutations and one-way permutations exist.*

Proof. As they are standard, we only sketch the proofs for completeness, soundness, and zero-knowledge. We provide the proof of unbounded simulation soundness in full.

Completeness follows by inspection. For the case of NIZK proofs, soundness follows by the fact that if Σ is chosen uniformly at random, then the probability that Σ_1 can be interpreted as a commitment to any string is exponentially small, and likewise the probability that Σ_2 is in the image of the pseudorandom generator G is exponentially small. For the case of NIZK arguments, we will in fact establish not only soundness but the stronger simulation soundness property below.

In the case where C is based on a one-way permutation, we note that the simulator's distribution on Σ is exactly uniform, thus satisfying this property required by same-string NIZK.

The proof of unbounded zero-knowledge follows almost exactly techniques of [15]. First we note that if we modify the real prover experiment by replacing the uniform Σ_1 with the distribution from the simulation (which in the case where C is based on one-way permutations is no change at all), but keep the prover as is, then by the security of the commitment scheme, the views of the adversary are computationally indistinguishable. Now, [15] show that single-theorem NIZK implies unbounded witness-indistinguishability. Thus, since the simulator for Π uses only a different witness to prove the same statement, the view of the adversary in the simulator experiment is computationally indistinguishable from the view of the adversary in the modified prover experiment. Thus, unbounded zero-knowledge follows.

Unbounded simulation soundness – Overview. The proof of simulation soundness uses novel techniques based in part on a new application of pseudorandom functions to non-malleability. We also use a combination of techniques from [13, 33], [15], and [4]. As we do not use set selection at all, the proof is quite different from that techniques from [12,33]. The intuition is as follows: Through the use of the signature scheme, we know that any proof of a false theorem that the adversary might output which is different from the proofs provided by the simulator must use a verification key VK that is new. Otherwise, providing a valid signature would contradict the security of the signature scheme. Once we know that the verification key VK must be different, we observe that the only way to prove a false theorem with regard to the simulated reference string is to provide a value $u = f_s(VK)$. By considering several hybrid distributions, we show that this is impossible by the security of pseudorandom functions and the witness-indistinguishability of the NIZK proof system Π' for L'.

Unbounded simulation soundness – Full Proof. We recall from the definition of unbounded simulation soundness the adversary experiment, and substitute from our construction, to build experiment Expt_0.

$\mathsf{Expt}_0(1^k)$ **(Actual Adversary Experiment)**:
- Make Reference String $\Sigma = (\Sigma_1, \Sigma_2, \Sigma_3)$:
  - $\Sigma_2 \leftarrow \{0,1\}^{3k}$; $\Sigma_3 \leftarrow \{0,1\}^{\ell_{\Pi'}(k)}$
  - $s \leftarrow \{0,1\}^k$
  - $\Sigma_1 \leftarrow$ commitments to bits of s using randomness $a_1, \ldots, a_k$.
- Run adversary A. When asked for proof for x, do:
  - $(VK, SK) \leftarrow Gen(1^k)$
  - $u = f_s(VK)$
  - Use Σ_3 as ref string and $(s, a_1, \ldots, a_k)$ as witness to construct proof π' that $(x, u, VK, \Sigma_1, \Sigma_2) \in L'$
  - $\sigma \leftarrow Sign_{SK}(x, u, \pi')$
  - **return** (VK, x, u, π', σ)
- Let (x, π) be output of adversary.
- Let Q be list of proofs provided by simulator above.
- **return true** iff ($\pi \notin Q$ **and** $x \notin L$ **and** $\mathcal{V}(x, \pi, \Sigma) =$ **true**)

Let $\Pr[\mathsf{Expt}_0(1^k)] = p(k)$. We must show that $p(k)$ is negligible.

We denote the components of the proof π output by the adversary as (VK, x, u, π', σ). Let T be the list of verification keys output by the simulator. (Note that with all but exponentially small probability, these verification keys will all be distinct.) We first consider the probability $\Pr[\mathsf{Expt}_0(1^k)$ and $VK \in T]$.

In the case where this is true, we know that $\pi \notin Q$, and therefore this implies that the adversary was able to produce a message/signature pair for VK different than the one given by the simulator. Thus, if $\Pr[\mathsf{Expt}_0(1^k)$ and $VK \in T]$ is non-negligible, we can use it to forge signatures and break the (strong) one-time signature scheme. Thus, $\Pr[\mathsf{Expt}_0(1^k)$ and $VK \in T]$ is negligible. Since $p(k) = \Pr[\mathsf{Expt}_0(1^k)$ and $VK \in T] + \Pr[\mathsf{Expt}_0(1^k)$ and $VK \notin T]$, we now need only focus on the second probability. Let $p_0(k) = \Pr[\mathsf{Expt}_0(1^k)$ and $VK \notin T]$.

We now consider a second experiment, where we change the acceptance condition of the experiment:

$\mathsf{Expt}_1(1^k)$ **(Accept only if** $u = f_s(VK)$**):**
  Make Reference String $\Sigma = (\Sigma_1, \Sigma_2, \Sigma_3)$:
    $\Sigma_2 \leftarrow \{0,1\}^{3k}$; $\Sigma_3 \leftarrow \{0,1\}^{\ell_{\Pi'}(k)}$
    $s \leftarrow \{0,1\}^k$
    $\Sigma_1 \leftarrow$ commitments to bits of s using randomness $a_1, \ldots, a_k$.
  Run adversary A. When asked for proof for x, do:
    $(VK, SK) \leftarrow Gen(1^k)$
    $u = f_s(VK)$
    Use Σ_3 as ref string and $(s, a_1, \ldots, a_k)$ as witness
      to construct proof π' that $(x, u, VK, \Sigma_1, \Sigma_2) \in L'$
    $\sigma \leftarrow Sign_{SK}(x, u, \pi')$
    `return` (VK, x, u, π', σ)
  Let $(x, \pi = (VK, x, u, \pi', \sigma))$ be output of adversary.
  Let Q be list of proofs output by simulator above.
  Let T be list of verification keys output by simulator above.
  `return true` iff
    $(\pi \notin Q$ and $\mathcal{V}(x, \pi, \Sigma) =$ `true` and $VK \notin T$ and $u = f_s(VK))$

Now, let $p_1(k) = \Pr[\mathsf{Expt}_1(1^k)]$. In Expt_1, we insist that $VK \notin T$ and replace the condition that $x \notin L$ with $f_s(VK) = u$. Note that with these changes, the experiment can be implemented in polynomial-time. Now, by the fact that Π' is a proof system for L', we know that if $x \notin L$, then with overwhelming probability the only way the adversary's proof can be accepted is if $f_s(VK) = u$. (Recall that in all cases, Π' is an NIZK proof system, not an argument.) Thus, we have that $p_0(k) \leq p_1(k) + \alpha(k)$, where α is some negligible function.

We now consider a third experiment, where we change part of the reference string Σ_2 to make it pseudorandom:

Let $p_2(k) = \Pr[\mathsf{Expt}_2(1^k)]$. In Expt_2, the only change we made was to make Σ_2 be pseudorandom rather than truly random. Thus, we must have that $|p_2(k) - p_1(k)| \leq \alpha(k)$, where α is some negligible function. Otherwise, this would yield a distinguisher for the generator G.

We now consider a fourth experiment, where instead of providing proofs based on proving $u = f_s(VK)$, we provide proofs based on the pseudorandom seed for Σ_2:

Let $p_3(k) = \Pr[\mathsf{Expt}_3(1^k)]$. In Expt_3, the only change we made was to have the simulator use the seed for Σ_2 as the witness to generate its NIZK

$\mathsf{Expt}_2(1^k)$ **(Change Σ_2 to be pseudorandom)**:
Make Reference String $\Sigma = (\Sigma_1, \Sigma_2, \Sigma_3)$:
$d \leftarrow \{0,1\}^k$; Let $\Sigma_2 = G(d)$.
$\Sigma_3 \leftarrow \{0,1\}^{\ell_{\Pi'}(k)}$
$s \leftarrow \{0,1\}^k$
$\Sigma_1 \leftarrow$ commitments to bits of s using randomness $a_1, \ldots, a_k$.
Run adversary A. When asked for proof for x, do:
$(VK, SK) \leftarrow Gen(1^k)$
$u = f_s(VK)$
Use Σ_3 as ref string and $(s, a_1, \ldots, a_k)$ as witness
to construct proof π' that $(x, u, VK, \Sigma_1, \Sigma_2) \in L'$
$\sigma \leftarrow Sign_{SK}(x, u, \pi')$
`return` (VK, x, u, π', σ)
Let $(x, \pi = (VK, x, u, \pi', \sigma))$ be output of adversary.
Let Q be list of proofs output by simulator above.
Let T be list of verification keys output by simulator above.
`return true` iff
$(\pi \notin Q$ and $\mathcal{V}(x, \pi, \Sigma) =$ `true` and $VK \notin T$ and $u = f_s(VK))$

$\mathsf{Expt}_3(1^k)$ **(Use seed for Σ_2 to generate NIZK proofs)**:
Make Reference String $\Sigma = (\Sigma_1, \Sigma_2, \Sigma_3)$:
$d \leftarrow \{0,1\}^k$; Let $\Sigma_2 = G(d)$.
$\Sigma_3 \leftarrow \{0,1\}^{\ell_{\Pi'}(k)}$
$s \leftarrow \{0,1\}^k$
$\Sigma_1 \leftarrow$ commitments to bits of s using randomness $a_1, \ldots, a_k$.
Run adversary A. When asked for proof for x, do:
$(VK, SK) \leftarrow Gen(1^k)$
$u = f_s(VK)$
Use Σ_3 as ref string and d as witness
to construct proof π' that $(x, u, VK, \Sigma_1, \Sigma_2) \in L'$
$\sigma \leftarrow Sign_{SK}(x, u, \pi')$
`return` (VK, x, u, π', σ)
Let $(x, \pi = (VK, x, u, \pi', \sigma))$ be output of adversary.
Let Q be list of proofs output by simulator above.
Let T be list of verification keys output by simulator above.
`return true` iff
$(\pi \notin Q$ and $\mathcal{V}(x, \pi, \Sigma) =$ `true` and $VK \notin T$ and $u = f_s(VK))$

proof that $(x, u, VK, \Sigma_1, \Sigma_2) \in L'$. Note that this means that s and the randomness $a_1, \ldots, a_k$ are not used anywhere except to generate Σ_1. Now, [15] prove that any adaptive single-theorem NIZK proof system is also adaptive unbounded witness-indistinguishable (see [15] for the definition of witness-indistinguishable non-interactive proofs). The definition of adaptive unbounded witness-indistinguishability directly implies that $|p_3(k) - p_2(k)| \leq \alpha(k)$, where α is some negligible function.

We now consider a fifth experiment, where finally we eliminate all dependence on s by chosing Σ_1 independently of s:

$\mathsf{Expt}_4(1^k)$ **(Make Σ_1 independent of s):**
Make Reference String $\Sigma = (\Sigma_1, \Sigma_2, \Sigma_3)$:
$d \leftarrow \{0,1\}^k$; Let $\Sigma_2 = G(d)$.
$\Sigma_3 \leftarrow \{0,1\}^{\ell_{\Pi'}(k)}$
$s, s' \leftarrow \{0,1\}^k$
$\Sigma_1 \leftarrow$ commitments to bits of s' using randomness $a_1, \ldots, a_k$.
Run adversary A. When asked for proof for x, do:
$(VK, SK) \leftarrow Gen(1^k)$
$u = f_s(VK)$
Use Σ_3 as ref string and d as witness
to construct proof π' that $(x, u, VK, \Sigma_1, \Sigma_2) \in L'$
$\sigma \leftarrow Sign_{SK}(x, u, \pi')$
`return` (VK, x, u, π', σ)
Let $(x, \pi = (VK, x, u, \pi', \sigma))$ be output of adversary.
Let Q be list of proofs output by simulator above.
Let T be list of verification keys output by simulator above.
`return true` iff
$(\pi \notin Q$ `and` $\mathcal{V}(x, \pi, \Sigma) =$ `true` `and` $VK \notin T$ `and` $u = f_s(VK))$

Let $p_4(k) = \Pr[\mathsf{Expt}_4(1^k)]$. In Expt_4, we choose two independent uniformly random strings s, s' and make Σ_1 into a commitment to s' rather than s. This has the effect of making Σ_1 completely independent of the string s.

Suppose $s^0, s^1 \leftarrow \{0,1\}^k$; $b \leftarrow \{0,1\}$, and $\Sigma_1 \leftarrow$ commitments to bits of s^b. By the security of the commitment scheme (either by Naor[27] or Goldreich-Levin [17], depending on which scheme we use), we know that for every polynomial-time algorithm B, we have that $\Pr[B(s^0, s^1, \Sigma_1) = b] \leq \frac{1}{2} + \alpha(k)$, where α is some negligible function.

Consider the following algorithm B: On input s^0, s^1, Σ_1, execute Expt_4 (or equivalently Expt_3), except with $s = s^0$ and $s' = s^1$, and using the value of Σ_1 specified as input to B. Return 1 if the experiment succeeds.

Then:

$$\begin{aligned}\Pr[B = b] &= \frac{1}{2}\Pr[B = 1|b = 1] + \frac{1}{2}\Pr[B = 0|b = 0] \\ &= \frac{1}{2}(1 - p_4(k)) + \frac{1}{2}p_3(k) \\ &= \frac{1}{2} + \frac{1}{2}(p_3(k) - p_4(k))\end{aligned}$$

Thus, we have that $p_3(k) - p_4(k) \leq \alpha(k)$ for some negligible function α.

Finally, we consider the last experiment, where we replace the pseudorandom function f with a truly random function:

$\mathsf{Expt}_5(1^k)$ **(Replace f with truly random function):**
  Make Reference String $\Sigma = (\Sigma_1, \Sigma_2, \Sigma_3)$:
    $d \leftarrow \{0,1\}^k$; Let $\Sigma_2 = G(d)$.
    $\Sigma_3 \leftarrow \{0,1\}^{\ell_{\Pi'}(k)}$
    $s, s' \leftarrow \{0,1\}^k$
    $\Sigma_1 \leftarrow$ commitments to bits of s' using randomness $a_1, \ldots, a_k$.
  Run A with oracle to simulator. When asked for proof of x, do:
    $(VK, SK) \leftarrow Gen(1^k)$
    $u \leftarrow \{0,1\}^k$
    Use Σ_3 as ref string and d as witness
      to construct proof π' that $(x, u, VK, \Sigma_1, \Sigma_2) \in L'$
    $\sigma \leftarrow Sign_{SK}(x, u, \pi')$
    `return` (VK, x, u, π', σ)
  Let $(x, \pi = (VK, x, u, \pi', \sigma))$ be output of adversary.
  Let Q be list of proofs output by simulator above.
  Let T be list of verification keys output by simulator above.
  Let $u' \leftarrow \{0,1\}^k$
  `return true` iff
    ($\pi \notin Q$ `and` $\mathcal{V}(x, \pi, \Sigma) =$ `true` `and` $VK \notin T$ `and` $u = u'$)

Let $p_5(k) = \Pr[\mathsf{Expt}_5(1^k)]$. In Expt_5, we replace the pseudorandom function f_s with a truly random function F, which simply returns a truly random value at each query point. Note that since we only consider the case where $VK \notin T$, this means that $F(VK)$ will be a uniformly selected value (which we denote u') that is totally independent of everything the adversary sees. Thus, it follows that $p_5(k) \leq 2^{-k}$ since the probability that any value output by the adversary equals u' is at most 2^{-k}.

On the other hand, we will argue that $p_4(k)$ and $p_5(k)$ can only be negligibly apart by the pseudorandomness of $\{f_s\}$. Consider the following machine M which is given an oracle O to a function from $\{0,1\}^k$ to $\{0,1\}^k$: Execute experiment $\mathsf{Expt}_4(k)$ except replace any call to f_s with a call to the oracle. Note that s is not used in any other way in $\mathsf{Expt}_4(k)$. Return 1 iff the experiment succeeds.

Now, if the oracle provided to M is an oracle for f_s with $s \leftarrow \{0,1\}^k$, then $\Pr[M^O = 1] = p_4(k)$. If M is provided with an oracle for a truly random function F, then $\Pr[M^O = 1] = p_5(k)$. By the pseudorandomness of $\{f_s\}$, it follows that $|p_4(k) - p_5(k)| \leq \alpha(k)$ for some negligible function α.

In conclusion, we have that $p_5(k) \leq 2^{-k}$, and that $p_i(k) \leq p_{i+1}(k) + \alpha(k)$ for some negligible function α for each $i = 0, 1, 2, 3, 4$. Thus, $p_0(k) \leq \beta(k)$ for some negligible function β, which finally implies that $p(k)$ is negligible, completing the proof.

Theorem 3. *If the NIZK proof system Π' in the construction above is replaced by a single-theorem NIZK proof of knowledge for L', and assuming one-way* *functions exist, then Π is an unbounded non-malleable (in the explicit witness sense) NIZK proof system (or argument) for L. In particular if Π was also same-string NIZK, then Π is a Robust NIZK argument.*

Proof. (**Sketch**) This follows from essentially the same argument as was used above to prove that Π is unbounded simulation-sound. We sketch the details here.

To prove unbounded non-malleability in the explicit witness sense, we must exhibit a machine M that with oracle access to the adversary A produces an instance x, together with a witness w for membership of $x \in L$, satisfying some relation. Recall that since Π' is a proof of knowledge, there are extractor machines E_1 and E_2. We describe our machine M explicitly below:

> $M^A(1^k)$ **(Non-Malleability Machine)**:
> Make Reference String $\Sigma = (\Sigma_1, \Sigma_2, \Sigma_3)$:
> $(\Sigma_3, \tau) \leftarrow E_1(1^k)$
> $\Sigma_2 \leftarrow \{0,1\}^{3k}$
> $s \leftarrow \{0,1\}^k$
> $\Sigma_1 \leftarrow$ commitments to bits of s using randomness $a_1, \ldots, a_k$.
> Interact with $A(\Sigma)$. When asked for proof of x, do:
> $(VK, SK) \leftarrow Gen(1^k)$
> $u = f_s(VK)$
> Use Σ_3 as ref string and $(s, a_1, \ldots, a_k)$ as witness
> to construct proof π' that $(x, u, VK, \Sigma_1, \Sigma_2) \in L'$
> $\sigma \leftarrow Sign_{SK}(x, u, \pi')$
> `return` (VK, x, u, π', σ)
> Let $(x, \pi = (VK, x, u, \pi', \sigma), \text{aux})$ be output of adversary.
> Let $w' \leftarrow E_2(\Sigma, \tau, (x, u, VK, \Sigma_1, \Sigma_2), \pi')$
> If w' is a witness for $x \in L$, `return` (x, w', aux), else abort

M essentially executes $\mathsf{Expt}^S_{A,R}(k)$ from the definition of non-malleability, except using E_1 to generate Σ_3, (recall that this output of E_1 is distributed negligibly close to uniformly) and using E_2 to extract a witness from the NIZK proof for L'. We immediately see therefore that M will fail to meet the conditions of non-malleability only if there is a non-negligible probability that the witness w' returned by E_2 is not a witness for $x \in L$ and yet the proof π' is valid. By construction, with all but negligible probability over Σ_2 and Σ_3, this can only happen if w' is a witness for $u = f_s(VK)$. But the proof of simulation-soundness of Π implies that the adversary can output such a u with a valid proof π with only negligible probability. This shows that the probability of M's success is only negligibly different than the probability of success in the experiment $\mathsf{Expt}^S_{A,R}(k)$.

4 Second Construction

In this section, we exhibit our second construction of NIZK proof systems with unbounded adaptive non-malleability (in the explicit NP-witness sense). Our construction uses several tools, that can all be based on any NIZK proof of knowledge. In particular, this construction is based on a novel generalization of unduplicatable set selection [13,12,33] which we call *hidden undiplicatable set selection* which can be used to achieve unbounded non-malleability, and might be useful elsewhere. interest.

An informal description. As a starting point, we still would like to use the paradigm of [15] in order to be able to simulate arbitrarily many proofs, when requested by the adversary. In other words, we want to create a proof system where the simulator can use some "fake" witness to prove arbitrarily many theorems adaptively requested by an adversary but the adversary must use a "real" witness when giving a new proof.

One important step toward this goal is to use a new variation on the "unduplicatable set selection" technique (previously used in [13,12,33]). While in previous uses of unduplicatable set selection, the selected set was sent in the clear (for instance, being determined by the binary expansion of a commitment key or a signature public key), in our construction such a set is hidden.

Specifically, on input x, the prover picks a subset S of bits of the random string and proves that $x \in L$ or the subset S enjoys property P (to ensure soundness P is such that with overwhelming probability a subset of random bits does not enjoy P). The subset S is specified by a string s that is kept hidden from the verifier through a secure commitment. The same string s is used to specify a pseudo-random function f_s and the value of f_s on a random u is then used as source of randomness for the key generation of a signature scheme. To prevent that the adversary does not follow these instructions in generating the public key, our protocol requires that a non-interactive zero-knowledge proof for the correctness of this computation is provided. Thus, the prover actually produces two zero-knowledge proofs: the "real one" (in which he proves that $x \in L$ or the set S enjoys property P) and the "auxiliary proof" (in which he proves correctness of the construction). Finally, the two proofs are signed with the public key generated.

This way, the generation of the public key for the signature scheme is tied to the selected set S in the following sense: if an adversary tries to select the same set and the same input for the pseudo-random function as in some other proof he will be forced to use the same public key for the signature scheme (for which, however, she does not have a secret key).

Let us intuitively see why this protocol should satisfy unbounded non-malleable zero-knowledge. A crucial point to notice is that the simulator, when computing the multiple proofs requested by the adversary, will select a set of strings, set them to be pseudo-random and the remaining ones to be random, and always use this single selected set of strings, rather than a possibly different set for each proof, as done by a real prover; note however that the difference between these two cases is indistinguishable. As a consequence, the adversary, even after seeing many proofs, will not be able to generate a new proof without knowing its witness as we observe in the following three possible cases.

First, if the adversary tries to select a different set S' (from the one used in the simulation), then she is forced to use a random string. Therefore S' does not enjoy P and therefore she can produce a convincing real proof only if she has a witness for $x \in L$.

Second, if the adversary tries to select the same set of strings as the one used in the simulation and the same input for the pseudo-random function as at least in one of the proofs she has seen, then she is forced to use the same signature public key and therefore will have to forge a signaturewhich violates the security of the signature scheme used.

Third, if the adversary tries to select the same set of strings as the one used in the simulation and an input for the pseudo-random function different from all the proofs she has seen, she will either break the secrecy of the commitment scheme or the pseudorandomness of the pseudo-random function used.

Tools. We use the following tools:

1. A pseudo-random generator $g = \{g_n\}_{n \in \mathsf{N}}$ where $g_n : \{0,1\}^n \to \{0,1\}^{2n}$;
2. A pseudo-random family of functions $f = \{f_s\}_{s \in \mathsf{N}}$, where $f_s : \{0,1\}^{|s|} \to \{0,1\}^{|s|}$.
3. A commitment scheme (Commit,VerCommit).
   On input a n-bit string s and a n^a-bit random *reference* string σ, for a constant a, algorithm Commit returns a commitment key *com* and a decommitment key *dec* of length n^a. On input σ, s, com, dec, algorithm VerCommit returns 1 if *dec* is a valid decommitment key of *com* as s and $\perp$ otherwise.
4. A one-time strong signature scheme (KG,SM,VS).
   On input a random string r of length n^a for a constant a, algorithm KG returns a public key pk and a secret key sk of length n. On input pk, sk, a message m, algorithm SM returns a signature sig. On input pk, m, sig, algorithm VS returns 1 if sig is a valid signature of m or 0 otherwise.

In the description of our proof system we will use the following polynomial-time relations.

1. Let g be a pseudorandom generator that stretches random strings of length n into pseudorandom string of length $2n$. The domain of relation R_1 consists of a reference string σ, n pairs of $2n$-bit strings $(\tau_{i,0}, \tau_{i,1})_{i=1}^{n}$, and a commitment com such that com is the commitment of an n-bit string $s = s_1 \circ \cdots \circ s_n$ computed with reference string σ and for each $i = 1, \cdots, n$ there exists $\mathsf{seed}_i \in \{0,1\}^n$ such that $\tau_{i,s_i} = g_n(\mathsf{seed}_i)$. A witness for membership in the domain of R_1 consists of the decommitment key dec, the string s and the seeds $\mathsf{seed}_1, \cdots, \mathsf{seed}_n$.
2. Let KG be the key-generator algorithm of a secure signature scheme, $\{f_s\}$ a pseudorandom family of functions and g a pseudorandom generator that stretches random strings of length n into pseudorandom strings of length $2n$. The domain of relation R_2 consists of a public key pk, two reference strings σ_0 and σ_1, a commitment com, and an n-bit string u such that at least one of the following holds:

a) String com is the commitment of an n-bit string s computed using σ_1 as reference string and pk is the output of KG on input $f_s(u)$.
b) There exists an n-bit string r_0 such that $\sigma_0 = g(r_0)$.

Witnesses of membership into R_2 are of two forms: either consist of decommitment dec and string s or of string r_0 such that $\sigma_0 = g(r_0)$. We denote by (A_2, B_2) a NIZK proof system of knowledge for relation R_2. We denote by E_{02}, E_{12}, S_2 the simulator and extractor associated with (A_3, B_3).

3. Relation R_3 is the or of relation R_1 and relation R. We denote by (A_3, B_3) a NIZK proof system of knowledge for relation R_3. We denote by E_{03}, E_{13}, S_3 the simulator and extractor associated with (A_3, B_3).

The Construction. Let R be a polynomial-time relation.

- **Common input.** $x \in \{0,1\}^n$.
- **Common random reference string.** The reference string consists of five parts:
  $\Sigma_0, \Sigma_1, \Sigma_2, \Sigma_3$,and Σ_4,where $\Sigma_4 = (\Sigma_{4,1,0} \circ \Sigma_{4,1,1}) \circ \cdots \circ (\Sigma_{4,n,0} \circ \Sigma_{4,n,1})$.
- **Prover Algorithm.** On input a witness w such that $R(x, w) = 1$, do the following:
  1. Uniformly choose $s \in \{0,1\}^n$ and $u \in \{0,1\}^n$;
  2. let $(\mathsf{com}, \mathsf{dec}) = Commit(\Sigma_1, s)$;
  3. let $r = f_s(u)$ and $(pk, sk) = Gen(1^k, r)$;
  4. using reference string Σ_2, input $I_2 = (pk, \Sigma_0, \Sigma_1, \mathsf{com}, u)$ and and witness $W_2 = (\mathsf{dec}, s)$, generate an NIZK proof of knowledge π_2 of W_2 such that $R_2(I_2, W_2) = 1$;
  5. using reference string Σ_3, input $I_3 = (\Sigma_4, \mathsf{com}, x)$ and $W_3 = w$ as witness generate an NIZK proof of knowledge π_3 of W_3 that $R_3(I_3, W_3) = 1$;
  6. let $mes = (\mathsf{com}, u, \pi_2, \pi_3)$;
  7. compute signature $sig = Sign(pk, sk, mes)$ and output (mes, pk, sig).
- **Verifier Algorithm.** On input $(\mathsf{com}, u, \pi_2, \pi_3, sig)$ do the following:
  1. verify that sig is a valid signature of $(\mathsf{com}, u, \pi_2, \pi_3)$;
  2. verify that π_2 and π_3 are correct;
  3. if all these verification are satisfied then output: ACCEPT and halt, else output: REJECT and halt.

The above protocol, as written, can be used to show the following

Theorem 4. *If there exists an efficient NIZK proof of knowledge for an NP-complete language, then there exists (constructively) an unbounded non-malleable (in the explicit witness sense) NIZK proof system for any language in NP.*

Consider the above protocol, where NIZK proofs of knowledge are replaced by NIZK proofs of membership. The resulting protocol can be used to show the following

Theorem 5. *If there exists an efficient NIZK proof of membership for an NP-complete language, and there exist one-way functions, then there exists (constructively) an simulation-sound NIZK proof system for any language in NP.*

In Appendix B we present a proof of Theorem 4 (note that, as done for our first construction, we can use part of this proof to prove Theorem 5).

Acknowledgments. Part of this work done while the third author was visiting Universitá di Salerno and part was done while the fourth author was visiting Telcordia Technologies and DIMACS. We thank Shafi Goldwasser and Oded Goldreich for valuable discussions.

References

1. M. Blum, A. De Santis, S. Micali and G. Persiano, Non-Interactive Zero-Knowledge Proofs. *SIAM Journal on Computing*, vol. 6, December 1991, pp. 1084–1118.
2. M. Blum, P. Feldman and S. Micali, Non-interactive zero-knowledge and its applications. *Proceedings of the* 20th *Annual Symposium on Theory of Computing*, ACM, 1988.
3. G. Brassard, D. Chaum and C. Crépeau, *Minimum Disclosure Proofs of Knowledge.* JCSS, v. 37, pp 156-189.
4. M. Bellare, S.Goldwasser, New paradigms for digital signatures and message authentication based on non-interactive zero knowledge proofs. *Advances in Cryptology – Crypto* 89 *Proceedings*, Lecture Notes in Computer Science Vol. 435, G. Brassard ed., Springer-Verlag, 1989.
5. R. Canetti, O. Goldreich, S. Goldwasser, and S. Micali. Resettable Zero-Knowledge. ECCC Report TR99-042, revised June 2000. Available from `http://www.eccc.uni-trier.de/eccc/`. Preliminary version appeared in ACM STOC 2000.
6. R. Canetti, J. Kilian, E. Petrank, and A. Rosen Black-Box Concurrent Zero-Knowledge Requires $\tilde{\Omega}(\log n)$ Rounds. *Proceedings of the* -67th *Annual Symposium on Theory of Computing*, ACM, 1901.
7. R. Cramer and V. Shoup, A practical public key cryptosystem provably secure against adaptive chosen ciphertext attack. *Advances in Cryptology – Crypto* 98 *Proceedings*, Lecture Notes in Computer Science Vol. 1462, H. Krawczyk ed., Springer-Verlag, 1998.
8. A. De Santis and G. Persiano, Zero-knowledge proofs of knowledge without interaction. *Proceedings of the* 33rd *Symposium on Foundations of Computer Science*, IEEE, 1992.
9. A. De Santis, G. Di Crescenzo and G. Persiano, Randomness-efficient Non-Interactive Zero-Knowledge. Proceedings of 1997 *International Colloquium on Automata, Languagues and Applications* (ICALP 1997).
10. A. De Santis, G. Di Crescenzo and G. Persiano, Non-Interactive Zero-Knowledge: A Low-Randomness Characterization of NP. Proceedings of 1999 *International Colloquium on Automata, Languagues and Applications* (ICALP 1999).
11. A. De Santis, G. Di Crescenzo and G. Persiano, Necessary and Sufficient Assumptions for Non-Interactive Zero-Knowledge Proofs of Knowledge for all NP Relations. Proceedings of 2000 *International Colloquium on Automata, Languagues and Applications* (ICALP 2000).
12. G. Di Crescenzo, Y. Ishai, and R. Ostrovsky, Non-Interactive and Non-Malleable Commitment. *Proceedings of the* 30th *Annual Symposium on Theory of Computing*, ACM, 1998.
13. D. Dolev, C. Dwork, and M. Naor, Non-Malleable Cryptography. *Proceedings of the* -45th *Annual Symposium on Theory of Computing*, ACM, 1923 and SIAM Journal on Computing, 2000.

14. C. Dwork, M. Naor, and A. Sahai, Concurrent Zero-Knowledge. *Proceedings of the* 30th *Annual Symposium on Theory of Computing*, ACM, 1998.
15. U. Feige, D. Lapidot, and A. Shamir, Multiple non-interactive zero knowledge proofs based on a single random string. In *31st Annual Symposium on Foundations of Computer Science*, volume I, pages 308–317, St. Louis, Missouri, 22–24 October 1990. IEEE.
16. O. Goldreich, *Secure Multi-Party Computation*, 1998. First draft available at `http://theory.lcs.mit.edu/~oded`
17. O. Goldreich and L. Levin, *A Hard Predicate for All One-way Functions* . *Proceedings of the* 21st *Annual Symposium on Theory of Computing*, ACM, 1989.
18. O. Goldreich, S. Goldwasser and S. Micali, How to construct random functions. *Journal of the ACM*, Vol. 33, No. 4, 1986, pp. 210–217.
19. O. Goldreich, S. Micali, and A. Wigderson. How to play any mental game or a completeness theorem for protocols with honest majority. *Proceedings of the* 19th *Annual Symposium on Theory of Computing*, ACM, 1987.
20. O. Goldreich, S. Micali, and A. Wigderson. Proofs that Yield Nothing but their Validity or All Languages in NP have Zero-Knowledge Proof Systems. Journal of ACM 38(3): 691–729 (1991).
21. S. Goldwasser, S. Micali, and C. Rackoff, The knowledge complexity of interactive proof systems. *SIAM Journal on Computing*, 18(1):186–208, February 1989.
22. S. Goldwasser, R. Ostrovsky Invariant Signatures and Non-Interactive Zero-Knowledge Proofs are Equivalent. *Advances in Cryptology – Crypto* 92 *Proceedings*, Lecture Notes in Computer Science Vol. 740, E. Brickell ed., Springer-Verlag, 1992.
23. J. Håstad, R. Impagliazzo, L. Levin, and M. Luby, Construction of pseudorandom generator from any one-way function. SIAM Journal on Computing. Preliminary versions by Impagliazzo et. al. in *21st STOC* (1989) and Håstad in *22nd STOC* (1990).
24. J. Kilian, E. Petrank An Efficient Non-Interactive Zero-Knowledge Proof System for NP with General Assumptions, Journal of Cryptology, vol. 11, n. 1, 1998.
25. J. Kilian, E. Petrank Concurrent and Resettable Zero-Knowledge in Polylogarithmic Rounds. *Proceedings of the* -67th *Annual Symposium on Theory of Computing*, ACM, 1901
26. M. Naor, R. Ostrovsky, R. Venkatesan, and M. Yung. Perfect zero-knowledge arguments for NP can be based on general complexity assumptions. *Advances in Cryptology – Crypto* 92 *Proceedings*, Lecture Notes in Computer Science Vol. 740, E. Brickell ed., Springer-Verlag, 1992 and *J. Cryptology*, 11(2):87–108, 1998.
27. M. Naor, Bit Commitment Using Pseudo-Randomness, *Journal of Cryptology*, vol 4, 1991, pp. 151–158.
28. M. Naor and M. Yung, Public-key cryptosystems provably secure against chosen ciphertext attacks. *Proceedings of the* 22nd *Annual Symposium on Theory of Computing*, ACM, 1990.
29. M. Naor and M. Yung, "Universal One-Way Hash Functions and their Cryptographic Applications", *Proceedings of the* 21st *Annual Symposium on Theory of Computing*, ACM, 1989.
30. R. Ostrovsky One-way Functions, Hard on Average Problems and Statistical Zero-knowledge Proofs. In Proceedings of 6th Annual Structure in Complexity Theory Conference (STRUCTURES-91) June 30 – July 3, 1991, Chicago. pp. 51-59

31. R. OSTROVSKY, AND A. WIGDERSON One-Way Functions are Essential for Non-Trivial Zero-Knowledge. Appeared In Proceedings of the second Israel Symposium on Theory of Computing and Systems (ISTCS-93) Netanya, Israel, June 7th-9th, 1993.
32. C. RACKOFF AND D. SIMON, Non-interactive zero-knowledge proof of knowledge and chosen ciphertext attack. *Advances in Cryptology – Crypto* 91 *Proceedings*, Lecture Notes in Computer Science Vol. 576, J. Feigenbaum ed., Springer-Verlag, 1991.
33. A. SAHAI Non-malleable non-interactive zero knowledge and adaptive chosen-ciphertext security. *Proceedings of the* 40th *Symposium on Foundations of Computer Science*, IEEE, 1999
34. A. SAHAI AND S. VADHAN A Complete Problem for Statistical Zero Knowledge. Preliminary version appeared in *Proceedings of the* 38th *Symposium on Foundations of Computer Science*, IEEE, 1997. Newer version may be obtained from authors' homepages.

A Discussion of Usefulness of ZK in Multiparty Settings

Goldreich, Micali, and Wigderson [19] introduced a powerful paradigm for using zero-knowledge proofs in multiparty protocols. The idea is to use zero-knowledge proofs to force parties to behave according to a specified protocol in a manner that protects the secrets of each party. In a general sense, the idea is to include with each step in a protocol a zero-knowledge proof that the party has acted correctly. Intuitively, because each participant is providing a *proof*, they can only successfully give such a proof if they have, in truth, acted correctly. On the other hand, because their proof is *zero knowledge*, honest participants need not fear losing any secrets in the process of proving that they have acted correctly.

To turn this intuition into a proof that no secrets are lost, the general technique is to simulate the actions of certain parties without access to their secrets. The definition of zero knowledge (in both interactive and non-interactive settings) is based on the existence of a simulator which can produce simulated proofs of arbitrary statements. This often makes it easy to simulate the actions of parties (which we call the high-level simulation) as needed to prove that no secrets are lost.

The problem of malleability, however, can arise here in a subtle way. One feature of simulators for zero-knowledge proofs is that they can simulate proofs of false statements. In fact, this is often crucial in the high-level simulation of parties, because without knowing their secrets it is often not possible to actually follow the protocol they way they are supposed to. However, on the other hand, it may also be crucial in the high-level simulation that the proofs received by a simulated party be correct! As an example which arises in the context of chosen-ciphertext security for public-key encryption [28], consider the following: Suppose in a protocol, one party is supposed to send encryptions of a single message m under two different public keys K_1 and K_2. According to our paradigm, this party should also provide a zero-knowledge proof that indeed these two encryptions are encryptions of the same message. Now, suppose the receiver is supposed to know both decryption keys k_1 and k_2. But suppose that because we are simulating the receiver, we only know one key k_1. Suppose further that the simulator needs to decypher the message m in order to be able to continue the protocol. Now, if we could always trust proofs to be correct, knowing just one key would be enough, since we would know for sure that the two encryptions are encrypting the same message, and therefore the decryption of any one of them would provide us with m.

Here is where the malleability problem arises: Perhaps a simulated party occasionally provides simulated proofs of false statements. If the proof system is malleable, another party could turn around and provide the receiver above with two inconsistent encryptions and a false proof that they are consistent. Now, in this case, the behavior of the simulated party would be different from the behavior of the real party, because the simulator would not notice this inconsistency. Indeed, this very problem arises in the context of chosen-ciphertext security, and illustrates how malleable proofs can make it difficult to construct simulators. If we look more closely, we see that more specifically, the problem is the possibility that an adversary can use simulated proofs to construct proofs for false statements. Sahai [33] considered this problem by introducing the notion of a *simulation-sound* proof system, although he is not able to construct simulation-sound NIZK proof systems immune to any polynomial number of false proofs. (Note that our notion of non-malleability implies simulation soundness.) In this work, we show how to achieve simulation-sound NIZK proof systems immune to any polynomial number of false proofs. Our construction of such NIZK systems requires the assumption of one-way trapdoor permutations – a possibly weaker computational assumption then dense cryptosystems.

B Proof for Our Second Construction

First of all we need to show that the proposed protocol is an efficient NIZK proof system for the language equal to the domain of relation R; namely, that it satisfies the completeness and soundness requirements, and that the prover runs in polynomial-time, when given the appropriate witness. It is immediate to check that the properties of completeness and soundness are verified by the described protocol. In particular, for the completeness and the efficiency of the prover, note that since the honest prover has a witness for relation R, she can compute the proof π_3 in step 5 and make the verifier accept; for the soundness, note that if the input x is not in the domain of relation R then since the reference string is uniformly distributed, input I_3 is not in the domain of relation R_3 and therefore, from the soundness of (A_3, B_3), the verifier can be convinced with probability at most exponentially small.

In the rest of the proof, we prove the non-malleability property of our proof system. We start by presenting a construction for the adaptive simulator algorithm and the non-malleability machine, and then prove that, together with the above proof system, they satisfy the non-malleability property of Definition 5

The adaptive simulator algorithm. We now describe the simulator S algorithm for the proof system presented. S consists of two distinct machines: S_1, which constructs a reference string Σ along with some auxiliary information aux, and S_2 which takes as input Σ, aux and an instance x ad outputs a simulated proof π for x.

ALGORITHM $S_1(1^n)$.

1. Randomly choose $\Sigma_0 \in \{0,1\}^{2n}$, $\Sigma_1 \in \{0,1\}^{n^a}$ and Σ_2 and Σ_3;
2. randomly choose $s \in \{0,1\}^n$;
3. **for** $i = 1$ to n **do**
   randomly pick $\mathsf{seed}_i$from $\{0,1\}^n$;
   set $\Sigma_{4,i,s_i} = g(\mathsf{seed}_i)$;
   randomly pick $\Sigma_{4,i,1-s_i}$from $\{0,1\}^{2n}$;
4. set $\Sigma = \Sigma_0 \circ \Sigma_1 \circ \Sigma_2 \circ \Sigma_3 \circ \Sigma_4$;
5. set aux $= (s, \mathsf{seed}_1, \cdots \mathsf{seed}_n)$;
6. output (Σ, aux).

ALGORITHM $S_2(\Sigma, \text{aux}, x)$.

1. Write aux as aux $= (s, \mathsf{seed}_1, \cdots \mathsf{seed}_n)$;
2. compute $(\mathsf{com}, \mathsf{dec})$from $Commit(\Sigma_1, s)$;
3. randomly pick $u$from $\{0,1\}^n$ and compute $r = f_s(u)$;
4. compute $(pk, sk) = KG(r)$;
5. using reference string Σ_2, input $I_2 = (pk, \Sigma_0, \Sigma_1, \mathsf{com}, u)$ and witness $W_2 = (\mathsf{dec}, s)$, generate an NIZK proof of knowledge π_2 of W_2 such that $R_2(I_2, W_2) = 1$;
6. using reference string Σ_3, input $I_3 = (\Sigma_4, \mathsf{com}, x)$ and witness
   $W_3 = (\mathsf{dec}, s, \mathsf{seed}_1, \cdots, \mathsf{seed}_n)$ generate an NIZK proof of knowledge π_3 of W_3 such that $R_3(I_3, W_3) = 1$;
7. set $mes = (\mathsf{com}, u, \pi_2, \pi_3)$;
8. compute signature $sig = Sign(pk, sk, mes)$ and output (mes, pk, sig).

Note that the from the point of view of the adversary, the transcript output by the simulator S is indistinguishable from a real conversation with a prover, or otherwise either the secrecy of the commitment scheme or the security of the pseudorandom generator or the witness indstinguishability of the proof system used are violated. The proof of this is standard and is based on arguments from [15].

The non malleability machine M. The computation of the non-malleability machine M can be divided into three phases. During the first phase, M creates a reference string along with some auxiliary information to be used later; in the second phase M receives strings $x^1, \ldots, x^l$ from Adv and produces proofs $\pi^1, \ldots, \pi^l$; finally, in the third phase it receives a proof π^* for input x^* and extracts a witness w^* from π^*.

Input to M: security parameters 1^n.

Phase 1: Preprocessing.

0. Randomly choose $\Sigma_0 \in \{0,1\}^{2n}$;
1. randomly choose $\Sigma_1 \in \{0,1\}^{n^a}$;
2. run E_{20} on input 1^n to obtain Σ_2 along with auxiliary information aux_2;
3. run E_{30} on input 1^n to obtain Σ_3 along with auxiliary information aux_3;
4. randomly choose $s \in \{0,1\}^n$;
5. compute $(\mathsf{com}, \mathsf{dec}) = Commit(\Sigma_1, s)$;
6. for $i = 1$ to n do
   randomly pick seed_i from $\{0,1\}^n$;
   set $\Sigma_{4,i,s_i} = g(\mathsf{seed}_i)$;
   randomly pick $\Sigma_{4,i,1-s_i}$ from $\{0,1\}^{2n}$.

Phase 2: Interact with adversary Adv. When asked for proof of x^i, do:

1. compute $(\mathsf{com}^i, \mathsf{dec}^i)$from $Commit(\Sigma_1, s)$;
2. randomly pick $u^i$from $\{0,1\}^n$ and compute $r^i = f_s(u^i)$;
3. compute $(pk^i, sk^i) = KG(r^i)$;
4. using reference string Σ_2, input $I_2^i = (pk^i, \Sigma_0, \Sigma_1, \mathsf{com}^i, u^i)$ and witness $W_2^i = (\mathsf{dec}^i, s)$, generate an NIZK proof of knowledge π_2^i of W_2^i such that $R_2(I_2^i, W_2^i) = 1$;
5. using reference string Σ_3, input $I_3^i = (\Sigma_4, \mathsf{com}^i, x^i)$ and witness $W_3 = (\mathsf{dec}^i, s, \mathsf{seed}_1, \cdots, \mathsf{seed}_n)$ generate an NIZK proof of knowledge π_3^i of W_3^i such that $R_3(I_3^i, W_3^i) = 1$;
6. compute $mes^i = (\mathsf{com}^i, u^i, \pi_2^i, \pi_3^i)$;
7. compute signature $sig^i = Sign(pk^i, sk^i, mes^i)$ and output (mes^i, pk^i, sig^i).

Phase 3: Output. Receive (x^*, π^*) from the adversary and do:

1. let $W_3^* = E_{31}(\Sigma_3, \mathrm{aux}_3, x^*, \pi^*)$;
2. if W_3^* is a witness for $x \in L$ then return W_3^* else return $\perp$.

Next we prove the non-malleability property. Note that if the adversary is successful in producing a convincing new proof π^* then she is also producing a convincing proof of knowledge π_3^* that some input I_3 belongs to the domain of relation R_3. Using this proof, M can extract a witness W_3 such that $R_3(I_3, W_3) = 1$. By the construction of R_3, this witness is either a witness for R (in which case M is successful) or a witness for R_1. Therefore the non-malleability property of our proof system is proved by the following

Lemma 1. *The probability that, at Phase 3, M extracts from proof π^* a witness for relation R_1 is negligible.*

Proof. First of all we assume that the proof returned by the adversary is accepting (namely, both proofs π_2^*, π_3^* in π^* for relations R_2, R_3, respectively, are accepting), otherwise there is nothing to prove. We then consider the following cases and for each of them we show that the probability is negligible for otherwise we would reach a contradiction by showing that Adv can be used to contradict one of our original assumptions about the cryptographic tools used.
Case (a): The adversary has used a string s^* different from s.
Case (b): The adversary has used the same string s and a value u^* equal to u^j for some j.
Case (c): The adversary has used the same string s and a value u^* different from all u^i's.

Proof for Case (a). Suppose $s^* \neq s$ and let i be such that $s_i^* \neq s_i$. Then with very high probability there exists no seed_i^* such that $g(\mathsf{seed}_i^*) = \Sigma_{4,i,s_i^*}$. Therefore, there exists no witness W_3^* for I_3^* and relation R_1 and thus by the soundness of the proof system used the verifier will reject with very high probability.

Proof for Case (b). We denote by l the number of queries performed by Adv and by $u^1, \cdots, u^l$ the values used by M in answering the l queries of Adv and by u^* the value used by Adv in its proof π.

Assume that there exists $j \in \{1, \ldots, l\}$ such that $u^* = u^j$. Then, given that Adv has used the same pseudorandom functions, and that we are assuming that the proof π_2^* returned by Adv is accepting, it must be the case that Adv has used the same public key pk^j as M.

Therefore, if the proof π^* generated by Adv is different from the proofs produced by M during Phase 2, it can be for one of the following two reasons (a) π contains a tuple $(\mathsf{com}^*, u^*, \pi_2^*, \pi_3^*)$ different from the corresponding tuple $(\mathsf{com}^j, u^j, \pi_2^j, \pi_3^j)$ used by M to answer the j-th query or (b) exhibit a different signature.

In case (a), Adv can be used to violate the unforgeability of the signature scheme used as it manages to produce a message and to sign it without having access to the secret key for the signature scheme.

Case (b) is ruled out by the property of the signature scheme employed saying that, given message m and its signature sig, it is hard to provide a new signature of m that is different from sig.

Proof for Case (c). In this section we show that the probability that M obtains in Phase 3 a witness W for relation R_1 and that the proof produced by the adversary has used the same values s as M and a different u is negligible.

We consider a series of 4 polynomial-time experiments $\mathsf{Expt}_0, \ldots, \mathsf{Expt}_3$ with the event that $\mathsf{Expt}_0(1^n)$ gives 1 in output being exactly the experiment of M interacting with Adv we are interested in.

Thus, denoting by $p_i(n)$ the probability $\Pr[\mathsf{Expt}_i(1^n)] = 1$, we need to show that $p_0(n)$ is negligible. We do so, 1) by showing that the output of the experiments $\mathsf{Expt}_i(1^n)$ and $\mathsf{Expt}_{i+1}(1^n)$ are indistinguishable and thus $|p_i(n) - p_{i+1}(n)|$ is negligible for $i = 0, 1, 2$; 2) by showing that $p_3(n)$ is negligible.

1. $\mathsf{Expt}_0(1^n)$.
   $\mathsf{Expt}_0(1^n)$ is exactly experiment $\mathsf{Expt}'_{A,R}$, the experiment of the adversary interacting with algorithm M. We only modify Phase 3.

> *Phase 3: Output.* Receive (x^*, π^*) from Adv.
>
> 1. Write π^* as $\pi^* = (\mathsf{com}^*, u^*, \pi_2^*, \pi_3^*, pk^*, sig^*)$.
> 2. Let $W_2^* = E_{12}(\Sigma_2, aux_2, x, \pi_2)$.
> 3. Write W_2^* as $W_2^* = (\mathsf{dec}, s)$.
> 4. Let $W_3^* = E_{13}(\Sigma_3, aux_3, x, \pi_3)$.
> 5. If W_3^* is a witness for $x \in L$ then **output** 0.
> 6. Write W_3^* as $W_3^* = (\mathsf{dec}^*, s^*, \mathsf{seed}_1^*, \cdots, \mathsf{seed}_n^*)$.
> 7. Output 1 iff $s^* = s$ and $u^* \neq u^j$, for $j = 1, \cdots, l$.

2. $\mathsf{Expt}_1(1^n)$.
   In $\mathsf{Expt}_1(1^n)$ random string Σ_0 is the output of generator g_n on input a random n-bit string r_0 and the proofs at steps 4 and 5 of Phase 2 of M are produced using r_0 as witness.

> *Phase 1: Pre-Processing.* Similar to Phase 1 of M with step 0 replaced with the following.
> 0. Randomly choose $r_0 \in \{0,1\}^n$ and set $\Sigma_0 = g_n(r_0)$.
>
> *Phase 2: Interacting with adversary.* Receive x^i from Adv.
> Receive x^i from Adv.
> Modify steps 4 and 5 of Phase 2 of M in the following way:
>
> 4. using reference string Σ_2, input $I_2^i = (pk^i, \Sigma_0, \Sigma_1, \mathsf{com}^i, u^i)$ and witness $W_2^i = (r_0)$, generate an NIZK proof of knowledge π_2^i of W_2^i such that $R_2(I_2^i, W_2^i) = 1$;
> 5. using reference string Σ_3, input $I_3^i = (\Sigma_4, \mathsf{com}^i, x^i)$ and witness $W_3^i = (s, \mathsf{seed}_1, \cdots \mathsf{seed}_n)$ generate an NIZK proof of knowledge π_3^i of W_3^i such that $R_3(I_3^i, W_3^i) = 1$;
>
> *Phase 3: Output.* Same as Expt_0.

The output of Expt_0 and Expt_1 are indistinguishable for otherwise we would violate either the pseudorandomness of the generator g or the witness indistinguishability of the proof system. This can be viewed by consider an intermediate experiment in which Σ_0 is output of g but the proof do not use it as witness.

3. $\mathsf{Expt}_2(1^n)$.
Expt_2 differs from Expt_1 in the fact that pk is computed by KG on input a random value.

> *Phase 1: Pre-Processing.* Same as Expt_1.
> *Phase 2: Interact with the adversary.* Receive x^i from Adv.
> Modify step 3. of Phase 2 of M in the following way.
> 2. Randomly select r^i from $\{0,1\}^n$ and compute $(pk^i, sk^i) = KG(r^i)$.
>
> *Phase 3: Output.* Same as Expt_1.

To prove that the distribution of the output of Expt_1 and Expt_2 are indistinguishable we define experiments $\mathsf{Expt}_{2.j}$, for $j = 0, \cdots, l$. In the first j executions of Phase 2 of $\mathsf{Expt}_{2.j}$, the public file is computed as in Expt_1 and in the subsequent executions as in Expt_2. Thus distinguishing between the output of Expt_2 and Expt_1 implies the ability to distinguish between $\mathsf{Expt}_{2.\hat{j}}$ and $\mathsf{Expt}_{2.(\hat{j}+1)}$, for some $0 \leq \hat{j} \leq l-1$, which contradicts either the security of the commitment scheme or the pseudorandomness of f.
To substantiate this last claim, we consider the following three experiments. For sake of compactness, we look only at the relevant components of the proof, that is, the commitment com, the value u and the public key pk; we do not consider the remaining components since they stay the same in each experiment and their construction can be efficiently simulated.

$\mathsf{Expt}_a(1^n)$	$\mathsf{Expt}_b(1^n)$
1. Pick s, r at random from $\{0,1\}^n$.	1. Pick s, r at random from $\{0,1\}^n$.
2. Compute commitment com of s.	2. Compute commitment com of s.
3. Pick u at random from $\{0,1\}^n$.	3. Pick u at random from $\{0,1\}^n$.
4. Compute $pk = KG(f_s(u))$.	4. Compute $pk = KG(f_r(u))$.
5. Output (com, u, pk).	5. Output (com, u, pk).

$\mathsf{Expt}_c(1^n)$
a) Pick s, r at random from $\{0,1\}^n$.
b) Compute commitment com of s.
c) Pick u at random from $\{0,1\}^n$.
d) Compute $pk = KG(r)$.
e) Output (com, u, pk).

Now we have the following two observations:

Obs. 1 Expt_a and Expt_b are indistinguishable.
Suppose they are not and consider the following adversary A that contradicts the security of the commitment scheme. A receives two random n-bit strings s and r and a commitment *com* of either s or r and performs the following two steps. First A picks u at random from $\{0,1\}^n$ and then computes pk as $pk = KG(f_s(u))$.
Now notice that if *com* is a commitment of s then the triplet (com, u, pk) is distributed as in the output of $\mathsf{Expt}_a(1^n)$. On the other hand if *com* is a commitment of r, then (com, u, pk) is distributed as in the output of $\mathsf{Expt}_b(1^n)$.

Obs. 2 Expt_b and Expt_c are indistinguishable.
Suppose they are not and consider the following adversary A that contradicts the pseudorandomness of f. A has access to a black box that computes a function F that is either a completely random function f or a pseudorandom function f_r for some random n-bit string r. A performes the following steps to construct a triplet (com, u, pk). A picks s at random, computes a commitment *com* of s, picks u at random, feeds the black box u obtaining $t = F(u)$ and computes pk as $pk = KG(t)$.
Now notice that if F is a random function then then (com, u, pk) is distributed as in the output of $\mathsf{Expt}_c(1^n)$. On the other hand if F is a pseudorandom function f_r for some random r then (com, u, pk) is distributed as in the output of $\mathsf{Expt}_b(1^n)$.

By the above observations Expt_a (the simplified version of $\mathsf{Expt}_{2.\hat{j}}$) and Expt_c (the simplified version of $\mathsf{Expt}_{2.\hat{j}+1}$) are indistinguishable.

4. $\mathsf{Expt}_3(1^n)$.
Expt_3 differs from Expt_2 in the fact that a random string s' is committed to instead of string s.

> *Phase 1: Pre-Processing.* Same as Expt_2 with the following exception: step 4 is modified as follows:
> 4. randomly pick $s, s,' \in \{0,1\}^n$;
>
> *Phase 2: Interact with the adversary.* Receive x^i from Adv.
> Modify step 1 of M in the following way:
>
> 1. Compute $(com^i, dec^i) = Commit(\Sigma_1, s')$ uniformly choose $u^i \in \{0,1\}^n$.
>
> *Output.* Same as Expt_0.

The distributions of the output of Expt_3 and Expt_2 are indistinguishable for otherwise we could distinguish commitment.
Finally, observe that in $\mathsf{Expt}_3(1^n)$, what is seen by Adv is independent from s. Thus the probability that Adv guesses s is negligible. Therefore, $p_3(n)$ is negligible.

Author Index

Lecture Notes in Computer Science

For information about Vols. 1–2034
please contact your bookseller or Springer-Verlag

Vol. 2035: D. Cheung, G.J. Williams, Q. Li (Eds.), Advances in Knowledge Discovery and Data Mining – PAKDD 2001. Proceedings, 2001. XVIII, 596 pages. 2001. (Subseries LNAI).

Vol. 2036: S. Wermter, J. Austin, D. Willshaw (Eds.), Emergent Neural Computational Architectures Based on Neuroscience. X, 577 pages. 2001. (Subseries LNAI).

Vol. 2037: E.J.W. Boers et al. (Eds.), Applications of Evolutionary Computing. Proceedings, 2001. XIII, 516 pages. 2001.

Vol. 2038: J. Miller, M. Tomassini, P.L. Lanzi, C. Ryan, A.G.B. Tettamanzi, W.B. Langdon (Eds.), Genetic Programming. Proceedings, 2001. XI, 384 pages. 2001.

Vol. 2039: M. Schumacher, Objective Coordination in Multi-Agent System Engineering. XIV, 149 pages. 2001. (Subseries LNAI).

Vol. 2040: W. Kou, Y. Yesha, C.J. Tan (Eds.), Electronic Commerce Technologies. Proceedings, 2001. X, 187 pages. 2001.

Vol. 2041: I. Attali, T. Jensen (Eds.), Java on Smart Cards: Programming and Security. Proceedings, 2000. X, 163 pages. 2001.

Vol. 2042: K.-K. Lau (Ed.), Logic Based Program Synthesis and Transformation. Proceedings, 2000. VIII, 183 pages. 2001.

Vol. 2043: D. Craeynest, A. Strohmeier (Eds.), Reliable Software Technologies – Ada-Europe 2001. Proceedings, 2001. XV, 405 pages. 2001.

Vol. 2044: S. Abramsky (Ed.), Typed Lambda Calculi and Applications. Proceedings, 2001. XI, 431 pages. 2001.

Vol. 2045: B. Pfitzmann (Ed.), Advances in Cryptology – EUROCRYPT 2001. Proceedings, 2001. XII, 545 pages. 2001.

Vol. 2046: H. Daduna, Queueing Networks with Discrete Time Scale. X, 138 pages. 2001.

Vol. 2047: R. Dumke, C. Rautenstrauch, A. Schmietendorf, A. Scholz (Eds.), Performance Engineering. XIV, 349 pages. 2001.

Vol. 2048: J. Pauli, Learning Based Robot Vision. IX, 288 pages. 2001.

Vol. 2050: J.-Y. Le Boudec, P. Thiran, Network Calculus. XIX, 274 pages. 2001.

Vol. 2051: A. Middeldorp (Ed.), Rewriting Techniques and Applications. Proceedings, 2001. XII, 363 pages. 2001.

Vol. 2052: V.I. Gorodetski, V.A. Skormin, L.J. Popyack (Eds.), Information Assurance in Computer Networks. Proceedings, 2001. XIII, 313 pages. 2001.

Vol. 2053: O. Danvy, A. Filinski (Eds.), Programs as Data Objects. Proceedings, 2001. VIII, 279 pages. 2001.

Vol. 2054: A. Condon, G. Rozenberg (Eds.), DNA Computing. Proceedings, 2000. X, 271 pages. 2001.

Vol. 2055: M. Margenstern, Y. Rogozhin (Eds.), Machines, Computations, and Universality. Proceedings, 2001. VIII, 321 pages. 2001.

Vol. 2056: E. Stroulia, S. Matwin (Eds.), Advances in Artificial Intelligence. Proceedings, 2001. XII, 366 pages. 2001. (Subseries LNAI).

Vol. 2057: M. Dwyer (Ed.), Model Checking Software. Proceedings, 2001. X, 313 pages. 2001.

Vol. 2058: S. Brewster, R. Murray-Smith (Eds.), Haptic Human-Computer Interaction. Proceedings, 2000. XIII, 219 pages. 2001.

Vol. 2059: C. Arcelli, L.P. Cordella, G. Sanniti di Baja (Eds.), Visual Form 2001. Proceedings, 2001. XIV, 799 pages. 2001.

Vol. 2060: T. Böhme, H. Unger (Eds.), Innovative Internet Computing Systems. Proceedings, 2001. VIII, 183 pages. 2001.

Vol. 2062: A. Nareyek, Constraint-Based Agents. XIV, 178 pages. 2001. (Subseries LNAI).

Vol. 2064: J. Blanck, V. Brattka, P. Hertling (Eds.), Computability and Complexity in Analysis. Proceedings, 2000. VIII, 395 pages. 2001.

Vol. 2065: H. Balster, B. de Brock, S. Conrad (Eds.), Database Schema Evolution and Meta-Modeling. Proceedings, 2000. X, 245 pages. 2001.

Vol. 2066: O. Gascuel, M.-F. Sagot (Eds.), Computational Biology. Proceedings, 2000. X, 165 pages. 2001.

Vol. 2068: K.R. Dittrich, A. Geppert, M.C. Norrie (Eds.), Advanced Information Systems Engineering. Proceedings, 2001. XII, 484 pages. 2001.

Vol. 2070: L. Monostori, J. Váncza, M. Ali (Eds.), Engineering of Intelligent Systems. Proceedings, 2001. XVIII, 951 pages. 2001. (Subseries LNAI).

Vol. 2071: R. Harper (Ed.), Types in Compilation. Proceedings, 2000. IX, 207 pages. 2001.

Vol. 2072: J. Lindskov Knudsen (Ed.), ECOOP 2001 – Object-Oriented Programming. Proceedings, 2001. XIII, 429 pages. 2001.

Vol. 2073: V.N. Alexandrov, J.J. Dongarra, B.A. Juliano, R.S. Renner, C.J.K. Tan (Eds.), Computational Science – ICCS 2001. Part I. Proceedings, 2001. XXVIII, 1306 pages. 2001.

Vol. 2074: V.N. Alexandrov, J.J. Dongarra, B.A. Juliano, R.S. Renner, C.J.K. Tan (Eds.), Computational Science – ICCS 2001. Part II. Proceedings, 2001. XXVIII, 1076 pages. 2001.

Vol. 2075: J.-M. Colom, M. Koutny (Eds.), Applications and Theory of Petri Nets 2001. Proceedings, 2001. XII, 403 pages. 2001.